Digital Agriculture

P. M. Priyadarshan • Shri Mohan Jain •
Suprasanna Penna • Jameel M. Al-Khayri
Editors

Digital Agriculture

A Solution for Sustainable Food and Nutritional Security

Springer

Editors
P. M. Priyadarshan
Rubber Research Institute of India (formerly)
Kottayam, Kerala, India

Shri Mohan Jain
Department Agricultural Sciences
University of Helsinki
Helsinki, Finland

Suprasanna Penna
Amity Centre for Nuclear Biotechnology
Amity University Maharashtra
Mumbai, India

Jameel M. Al-Khayri
Department of Agricultural Biotechnology
King Faisal University
Al-Ahsa, Saudi Arabia

ISBN 978-3-031-43547-8 ISBN 978-3-031-43548-5 (eBook)
https://doi.org/10.1007/978-3-031-43548-5

This Springer imprint is published by the registered company Springer Nature Switzerland AG
The registered company address is: Gewerbestrasse 11, 6330 Cham, Switzerland

Paper in this product is recyclable.

Preface

The world population is increasing at dramatic propulsion and the arable land is decreasing at a faster pace. The looming climate change is expected to reduce total yield of crops by 15–20% and poses a formidable task to increase food production. Plant breeding has contributed significantly to sustainable food production by recombining the desired genes in new cultivars from the available gene pool. However, the yield potential of crops has plateaued threatening globally sustainable food production and feeding ever-growing population worldwide facing climatic changes. In addition, global warming may become disastrous to agriculture production and food supply chain, especially with the appearance of new insects, pests, and diseases, and some existing ones may disappear. New plant breeding technologies like transgenics, molecular-marker-assisted breeding, mutagenesis, and genome editing could contribute to sustaining crop production.

Agriculture that developed 12,000 years ago changed the way humans lived, switching from nomadic hunter-gatherer life styles to permanent settlements and farming. Agriculture went through three stages: traditional agriculture, technologically dynamic agriculture with low capital technology, and technologically dynamic agriculture with high capital technology. Currently, emerging digital technologies have the potential to be game-changers for traditional agricultural practices. These changes are popularly known as "Agriculture 4.0," indicating its role as the fourth major agricultural revolution. The World Economic Forum announced that the "Fourth Industrial Revolution" that includes agriculture will unfurl throughout the twenty-first century. Hence, the year 2000 marks the beginning of Agriculture 4.0. Digital agriculture (DA) is coming of age and has the motto to make farming profitable and sustainable through using information-cum-communication technologies and data science, ensuring safe and nutritious food affordable to all. The world's first entirely machine-operated crop was harvested in 2017, at an experimental farm run by researchers from Harper Adams University, in Edgmond village, UK. About 5 tons of spring barley was harvested from the world's first robotically tended farm. Everything including sowing, fertilizing, collecting samples, and harvesting was done by autonomous vehicles on the farm. This was a milestone in digital agriculture, many times described as "smart farming," or "e-agriculture."

DA originated from vertical farming and controlled environment agriculture (CEA). CEA uses advanced computer-based technologies of physically collecting information that is converted into a computer-readable language. This leads to the development of tools and sensors integrated into the Internet of Things (IoT) environment. Such innovations can enhance real-time analysis, machine learning, and artificial intelligence to enable the management of massive amounts of data, also known as big data. Artificial intelligence (AI) has greater potential in automated irrigation, soil sensing, weed management, and biocontrol or biostimulant applications spraying to enhance the productivity in digital sustainable farming for better economic benefits. Linear AI programming and yield mapping through machine learning help to uncover patterns hidden within large-scale data sets that can be used for crop planning and monitoring, production, and resource allocation. Light quality and intensity, and CO_2 levels directly affect photosynthesis, transpiration, water uptake, flowering, germination, internodal growth, etc. within the plant. These attributes along with fertigation are crucial to have an effective and economically controlled environment. Intelligent sensors, combined with visual data streams from drones, use AI to detect areas most infected with pests. e-Agriculture is emerging as a global community practice where people from all over the world exchange information, ideas, and resources on sustainable agriculture and rural development.

Some of the technologies predominantly used in DA are robotics, IoT and sensors, artificial intelligence (AI), drones, data analytics, remote sensing, and cloud connectivity. Robots can milk cows, pick strawberries, cut papayas and represent a global market share of over $5 billion. IoT and sensors have the ability to evaluate the environment inside the farm or the uptake of moisture from the soil in real time. AI already has a market value of $11.4 billion. AI competes with extension agents, farming experts, consultants, and professional expertise. More likely, AI will alter how those professions should function. Drones have the ability to go where humans can't and see things not readily observed from the ground which creates real insights into pest protection, fertilizer and herbicide application, irrigation, and harvest timing. Through data analytics, the world will store 175 zettabytes of data by 2025. Every step in agriculture like crop selection, cultivation method, harvesting, and supply chain management can be optimized by data analytics. Remote sensing is being used for mapping soil properties, classification of crop species, detection of crop water stress, monitoring of weeds and crop diseases, and mapping of crop yield, in addition to sensing climate change. DA can assist governments to improve their policy making and decisions to improve socio-economic, environmental, sustainable, and climate research applications to enhance the productivity and efficiency of a given system.

In this book, apart from introductory chapters, there are four sections dealing with vertical farming and nurseries, IoT (Internet of Things) in agriculture, digital agriculture roles in speed breeding/fast-forward breeding, precision agriculture technologies, and predictive agriculture. Soilless smart agriculture systems, various aspects of vertical farming, intelligent nutrient controlling systems, remote sensing in precision agriculture and climate change, satellite imagery and crop modelling applications of UAVs/drones, image-based plant phenotyping, smart IoT sensors and data science, digital yield predictions, crop phenomics and high-throughput phenotyping, speed breeding for crop improvement, digital agriculture and for protection against pests and diseases, sensors of plant health data analytics in agriculture, data science and artificial intelligence, sensing systems for precision agriculture, AI and machine learning models, predictive analytics and crop modelling for future climate change adaptation are some of the chapters. All chapters are thoroughly reviewed and revised before publication. We strongly believe this book will be beneficial to researchers, students, policy makers, agriculturists, and professionals working in high tech agro-industries.

We wish to profusely acknowledge Springer Nature for publishing this needy and timely book.

Kottayam, Kerala, India P. M. Priyadarshan
Helsinki, Finland Shri Mohan Jain
Mumbai, India Suprasanna Penna
Al-Ahsa, Saudi Arabia Jameel M. Al-Khayri

Acknowledgment

The editors extend their appreciation to the support of the Deanship of Scientific Research, Vice Presidency for Graduate Studies and Scientific Research, King Faisal University, Saudi Arabia through Project No. GRANT4662.

Contents

Digital Agriculture for the Years to Come

P. M. Priyadarshan, Suprasanna Penna, Shri Mohan Jain, and Jameel M. Al-Khayri

Abstract The agriculture industry has evolved significantly over the last 50 years. Technology developments have led to larger, quicker, and more productive farm equipment, enabling the more efficient cultivation of larger areas. Additionally, improved irrigation, fertilizers, and seed have helped farmers to increase crops. New technologies such as artificial intelligence, analytics, networked sensors, and others may increase yields even further, improve the efficiency of water and other inputs, and promote sustainability and resilience in cattle rearing and agricultural output. Implementing such cutting-edge technologies is known as agriculture 4.0. But, without a solid infrastructure for connectivity, none of this is practical. If connection is successfully implemented in the industry, agriculture may add $500 billion in value to the global GDP by 2030. This would lead to an increase of 7–9% over the anticipated total and greatly relieve the pressure currently imposed on farmers. It is one of just seven industries that will raise global GDP by $2 to $3 trillion over the next 10 years because of better connectivity. World population is expected to grow to 9.6 billion by 2050 that lead to significant increase in the demand for food. On the other hand, the availability of natural resources like freshwater and productive arable land is getting constrained year after year. Nearly 821 million people still suffer from hunger. Digital agriculture, also known as smart farming or e-agriculture, is the use of tools to collect, store, analyze, and disseminate electronic data and/or information in agriculture.

P. M. Priyadarshan (✉)
Rubber Research Institute of India (formerly), Kottayam, Kerala, India

S. Penna
Amity Centre for Nuclear Biotechnology, Amity University Maharashtra, Mumbai, India

S. M. Jain
University of Helsinki (formerly), Helsinki, Finland

J. M. Al-Khayri
Department of Agricultural Biotechnology, College of Agriculture and Food Sciences, King Faisal University, Al-Ahsa, Saudi Arabia

P. M. Priyadarshan et al. (eds.), *Digital Agriculture*,
https://doi.org/10.1007/978-3-031-43548-5_1

The present emphasis is on reducing water, energy, and material use in agriculture as access to water and material resources becomes more challenging due to climate change and population expansion. Controlled environment agriculture (CEA) can be used to grow vegetables and high-value commodities in any environment with outstanding water, soil, and fertilizer efficiency, since local production reduces transportation costs. Contrary to conventional field agriculture, CEA offers more effective nutrient usage while using up to 80% less land and nearly 90% less water. Keeping in view of the population progression, declining land resources, and climate vagaries, there is a need to develop selection methods with more accuracy and precision. The advancement of artificial intelligence (AI) in the past decade has offered great potential to augment the climate smart agriculture. Protected agriculture, as against open-field farming, offers a more conducive and manageable environment for crop growth through greenhouse technology, which is somewhat unrestricted by the natural environment and encourages the intensive and effective use of agricultural resources. Remote sensing (RS) is a diagnostic tool that can act as an early warning system. Due to recent developments in sensor technologies, data management, and data analytics, the agricultural community now has access to a number of RS choices. All digital technologies that can be used in agriculture to improve yield, plant protection and enhance nutritional quality are summarized here.

Keywords Digital agriculture · Vertical farming · Controlled environment agriculture · Sensors · IoT · Big data · Block chain · Supply chain · Robotics · Remote sensing

1 Introduction

The agriculture industry has changed significantly over the last 50 years. Technology developments have led to larger, quicker, and more productive farm equipment, enabling the more efficient cultivation of larger areas. Additionally, improved irrigation, fertilizers, and seed have helped farmers to increase crops. A new revolution in agriculture is currently taking place, one that is being fueled by connectivity and data (Mehrabi et al. 2021; Himesh et al. 2018). New technologies, such as artificial intelligence, analytics, networked sensors, and others, may increase yields even further, improve the efficiency of water and other inputs, and promote sustainability and resilience in cattle rearing and agricultural output (Javaid et al. 2022). Implementing such cutting-edge technologies is known as agriculture 4.0. (da Silveira and Amaral 2022). But, without a solid infrastructure for connectivity, none of this is practical. If connection is successfully implemented in the industry, agriculture may add $500 billion in value to the global GDP by 2030. This would lead to an increase of 7–9% over the anticipated total and greatly relieve the pressure currently imposed on farmers. It is one of just seven industries that will raise global GDP by $2 to $3 trillion over the next 10 years because of better connectivity (Goedde et al. 2020).

The world's agri-food system is increasingly subject to constraints, especially since it relies on a number of nonrenewable resources that are becoming scarcer (fresh water, phosphorus, oil, cultivable soil, etc.). This system will soon exert its impact over climate change, both directly (extreme weather events, drought, etc.) and indirectly (melting glaciers, proliferation and spread of harmful species of organisms and diseases, rising sea levels) (UNESCO 2019). The collapse of biodiversity in seeds, pollinators, crop auxiliaries, etc. are looming large that endangers many ecosystems (FAO 2019a, b). Conflicts over the use of land and water will also increase with the use of biomass for energy and the implementation of afforestation/reforestation programs to capture CO_2. This is also known as "negative emissions" technique that now substantiates all IPCC scenarios limiting the temperature increase to 2 °C. In addition, the yield of cereals deemed critical for food security as their yields seem to have reached a plateau (Maurel et al. 2022) (see Iddio et al. 2019 for a comprehensive review).

In many parts of the world, climate change has caused many irregular and extreme weather events (Li et al. 2021). Different parts of the world have begun experiencing intense drought, hurricanes and storms, and floods as a result of global warming (FAO 2021). Additionally, agricultural production success varies based on the complex environmental effects of global warming and climate change, both in the short and long term (Hatfield et al. 2011). Extreme heat, extreme cold, wetness, and dryness all have a deleterious impact on plants (Hatfield and Prueger 2015; FAO 2019a, b). Trade conflicts, epidemic and vegetative diseases, rising seed and fertilizer prices and wages, flash floods, heatwaves, and other weather variations all have a negative impact on agriculture. However, as evidenced by agriculture's contributions to greenhouse gas emissions, water pollution, and biodiversity loss (Springmann et al. 2018), major agricultural systems are on largely unsustainable trajectories. Countries must create policies and programs in a sustainable manner if they are to overcome these obstacles and satisfy future demands. The best ways to achieve sustainable agricultural development are to continue the process of innovation using contemporary genetic and information technologies to increase agricultural productivity while balancing economic, environmental, and social outcomes related to food and agricultural systems (Basso and Antle 2020).

Industrial agriculture has a substantial negative impact on the ecosystem as a result of human activity, which today controls almost all biogeochemical cycles on Earth (Park et al. 2016). Production from industrial agriculture is expected to rise by 100–110% between 2005 and 2030, which calls for substantial inputs of finite resources like fresh water, soil with sufficient sunlight, and nonrenewable nutrients (e.g., phosphorus) (Cordell et al. 2012). The degradation of soils, aquifer depletion, saltwater-intrusion, runoff and eutrophication, and emissions (e.g., CO_2, N_2O, etc.), contributing to global warming and resource scarcity are all effects of modern intensive agriculture (Cohen et al. 2022). In particular, this is because farmers frequently lack sufficient measuring, modeling, and dynamic control mechanisms to optimize inputs for plant growth (Dawson and Hilton 2011). Moreover, losses from farm to fork in the form of food waste can reach as high as 40% due to extended supply chains (Cohen et al. 2022).

The present emphasis is on reducing water, energy, and material use in agriculture as access to water and material resources becomes more challenging due to climate change and population expansion (Cohen et al. 2022). Regional decentralized controlled environment agriculture is one suggestion for enhancing the sustainability of vegetable production. The benefit of this is that practitioners can precisely control environmental effects, including nutrient application, water use efficiency, and lighting. Hydroponics or soilless culture feeds nutrients and water directly to the plant by employing recirculation (where the substrate is reused in the system until the nutrients are exhausted) or flow-through substrates (Silberbush et al. 2005). Controlled environment agriculture (CEA) can be used to grow vegetables and high-value commodities in any environment with outstanding water, soil, and fertilizer efficiency, since local production reduces transportation costs (Van Ginkel et al. 2017). Contrary to conventional field agriculture, CEA offers more effective nutrient usage while using up to 80% less land and nearly 90% less water (Carmassi et al. 2007).

In order to manage soil, climate, and genetic resources at the field and landscape scales, digital agriculture uses a collection of geospatial and digital information technologies that integrate sensors, analytics, and automation (Basso and Antle 2020). Big data, the Internet of Things (IoT), augmented reality, robotics, sensors, 3D printing, system integration, ubiquitous connectivity, artificial intelligence, machine learning, digital twins, and blockchain are just a few of the technologies that make up digitalization (Alm et al. 2016) (Table 1), which is anticipated to fundamentally alter daily life (Klerkx et al. 2019, food, fiber, and bioenergy supply chains and systems) and agricultural productivity processes. According to Rotz et al. (2019), the early indications of transition are already apparent.

Several concepts have emerged with digitalization in agricultural production systems, value chains and more broadly food systems. These include *smart farming, precision agriculture* or *precision farming, decision agriculture, digital agriculture, agriculture 4.0,* or what is referred to in French as *Agriculture Numérique* (i.e., *numerical agriculture*) (Rose and Chilvers 2018; Klerkx et al. 2019). On-farm management duties that take into account location, weather, behavior, phytosanitary status, consumption, energy use, prices, and economic data are all included in digitalization. This is done with the aid of sensors, equipment, drones, and satellites. Through ongoing monitoring or targeted big data science inquiries, the data so acquired is then utilized to understand the past, anticipate the future, and make more timely or correct judgments (Ingram and Maye 2020). These developments have mostly concentrated on deploying technologies for enhancing post-farmgate operations, postharvest quality monitoring, and real-time traceability (Rutten et al. 2013; Wolfert et al. 2017). This claim is supported by a variety of reviews on subjects such precision farming, big data analysis, drones, artificial intelligence, robots, 3D printing, and the Internet of Things (IoT), as well as their potential to enhance agricultural production systems, value chains, and food systems (Bertoglio et al. 2021). For instance, yield stability maps show regions that have consistently high production throughout time, regions with poor productivity, and other regions

Table 1 Technologies used in digital agriculture

Robotics	Agricultural robots are for increasing production yields. From drones to autonomous tractors to robotic arms, the technology is being deployed in creative and innovative applications. Some of the most common robots are the following: harvesting and picking, weed control, autonomous mowing, pruning, seeding, spraying and thinning, phenotyping, sorting and packing, and utility platforms. Robots can achieve to improve the size of yields and reduce waste from crops being left in the field
IoT and sensors	The Internet of Things is utility of Internet for various operational purposes. IoT becomes operational through sensors. Sensors are devices that detect and respond to changes in an environment. Inputs can come from a variety of sources such as light, temperature, motion, and pressure. Sensors provide valuable information through a network, and they can share data with other managerial information systems. The sensor attains a physical parameter and converts it into a signal suitable for processing (e.g., electrical, mechanical, optical). The output of the sensor is a signal which is converted to a human-readable form, like changes in characteristics, changes in resistance, capacitance, impedance, etc.
Artificial intelligence (AI)	In contrast to the intelligence exhibited by humans or other animals, artificial intelligence (AI) refers to the perception, synthesis, and inference of information made by computers. The term "intelligence" refers to the capacity for knowledge, reasoning, abstraction, and inference of meaning
Deep learning (DL)	It is simply a neural network with three or more layers and is a subset of machine learning. These neural networks make an effort to mimic how the human brain functions; however, they fall far short of being able to match it, enabling it to "learn" from vast volumes of data. Additional hidden layers can help to optimize and refine for accuracy even if a neural network with only one layer can still make approximation predictions
Drones and satellites	While drones record data in real time, they lack the hard drives necessary to store the vast amounts of digital data that satellites are designed to hold until they can be recovered and used. In order to provide more accurate measurements on a particular location, drones can also use GPS
Extended reality and the metaverse	Extended reality (XR) enables users to constantly access internet content thanks to the metaverse, which also makes considerable use of 3D visuals. From augmented reality (AR) to mixed reality (MR) to virtual reality (VR), XR technologies cover a broad range of immersive technologies
Virtual reality (VR)	With images and things that seem real, a virtual reality (VR) environment gives the user the impression that they are completely engrossed in their surroundings. A virtual reality headset, helmet, or other equipment is used to view this environment
Block chain	A blockchain is a type of distributed database or ledger – one of today's top tech trends – which means the power to update a blockchain is distributed between the nodes, or participants, of a public or private computer network. This is known as distributed ledger technology, or DLT
Data analytics	Analyzing data collections to identify trends and make judgments about the information they contain is known as data analytics (DA). Data analytics is increasingly carried out with the use of specialized hardware and software
Cloud connectivity	The capacity to connect two resources within a cloud, across clouds, and with on-premises data centers is referred to as cloud networking. A cloud service provider must offer the following three main forms of connectivity: site-to-cloud, the connection between cloud resources, and on-premises hardware

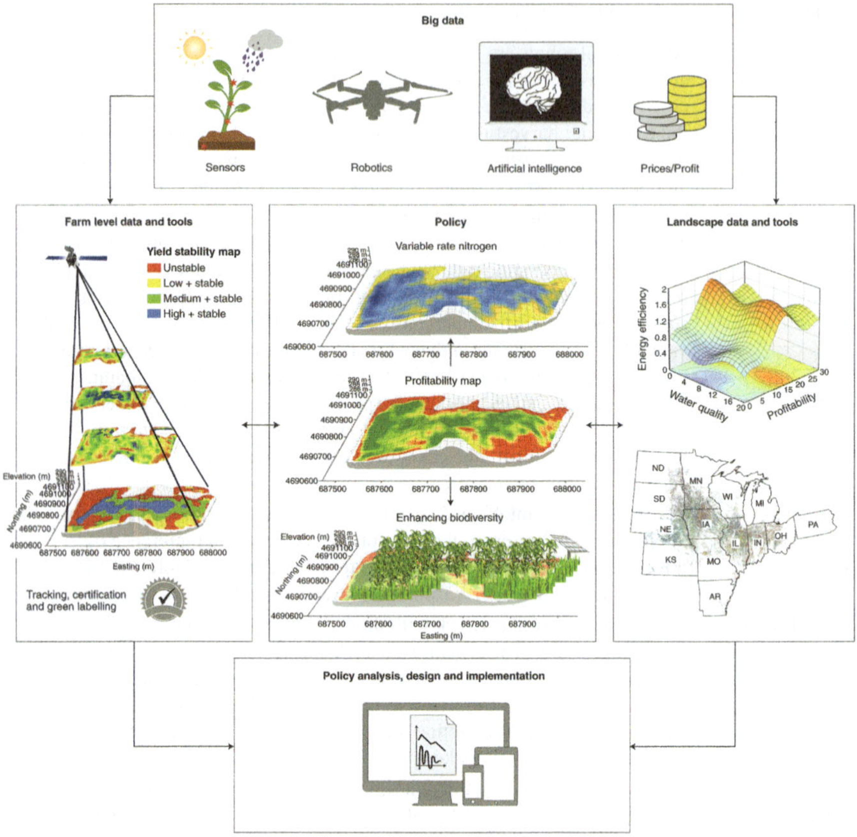

Fig. 2 DA in agricultural systems. DA can be used to design and implement sustainable agricultural systems at farm and landscape scales. With the use of stability maps, DA can help redesign fields or subareas within fields that are unprofitable or environmentally unsustainable and sustainably intensify high-yield areas of the field knowing that these can respond to more inputs. (Courtesy: Bruno Basso, Michigan State University; Nature Sustainability, doi: https://doi.org/10.1038/s41893-020-0510-0)

that have yields that fluctuate over time. Stability maps can be used by DA to redesign unprofitable or environmentally unsustainable fields or portions of fields, as well as sustainably intensify culture in high-yield regions that respond to more inputs (Fig. 2).

1.2 Controlled Environment Agriculture (CEA)

Controlled-environment agriculture (CEA), which deals with sophisticated horticultural practices and technological advancements, first gained popularity in the 1960s

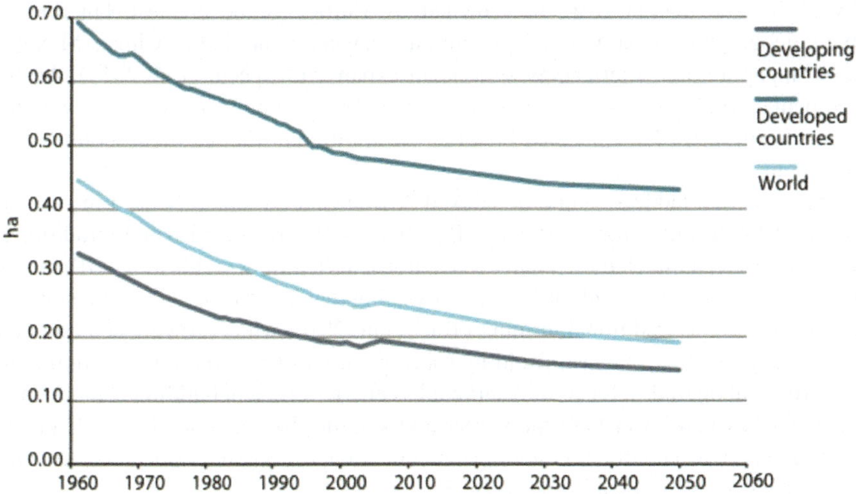

Fig. 3 Arable land per capita from 1960 to 2050. (FAO 2011)

(Hodges et al. 1968). Controlled environments (CEs) promote production efficiency, optimize plant yield, and enhance product quality by providing predictions on how plants will appear in their surroundings. The market has recently seen a rise in demand for locally sourced food. According to Eaves and Eaves (2018), this is accomplished through CEA, which covers small- (in-home production or indoor gardens), medium- (community gardens), or large-scale commercial operations. The ability to alter production environments to increase plant quality and output, lengthen growing seasons, and allow crop production under unfavorable climatic conditions (e.g., wind, rain, extremely high temperatures, and inadequate light) is a fundamental advantage of CEA. The two most prevalent types of CEs used in urban agriculture (UA) are greenhouses and plant factories (PFs). Due to the decreasing amount of arable land, such systems are unavoidable (Fig. 3). Controlling greenhouses presents a particular difficulty, because it calls for systems that can adapt to the microclimatic factors and constantly shifting environmental circumstances. The expense of heating and cooling greenhouses can account for 70–85% of the overall operating cost in northern latitudes and harsh climates (Engler and Krarti 2021).

1.2.1 CEA Facilities

All CEA facilities are included under the general term "urban agriculture" (UA). CEA, on the other hand, is merely a portion of UA as a whole that has conditioned spaces. There are many different kinds of CEA facilities, including greenhouses, plant factories, rooftop gardens, and vertical farms. The market for indoor farming was estimated to be worth 38.7 billion USD in 2022 and may rise to 96.6 billion by 2032. The market share now owned by Europe is the greatest, but due to their geographic constraints and economic development, India, China, Mexico, and Singapore are experiencing significant expansion (Specht et al. 2013). At the moment, greenhouses hold 70% of the market. The common CEAs include vertical farms, rooftop gardens, planned factories with artificial lighting, hydroponics, and aquaponics.

Scissor lifts, ladders, stairs, or stacked A-frames are frequently used in vertical farms (VFs) to raise crops vertically (Beacham 2019). In comparison to traditional farming, VFs can stack these plant beds to boost agricultural yields by 10–100 times (Fig. 4). The annual growth of VFs is so exacting that it reported using almost twice as much water as traditional farming (Tong et al. 2016) while growing at a rate that was nearly two times as fast. Building rooftop gardens (RTGs) requires minimal to no structural upgrades. RTGs are marketed as energy-efficient building components, because they can lower both the winter and summer heating and cooling loads. If implemented on a number of buildings, greening the roofs could also aid in reducing the impacts of urban heat islands (Fig. 5).

The advantages of RTGs are used in many building integrated agriculture (BIA) applications to produce energy-efficient CEAs (Benis et al. 2017; van Delden et al. 2021). Artificial lighting plant factories (PFALs) are often referred to as closed plant production systems (CPPS), which are totally sealed off from the outside

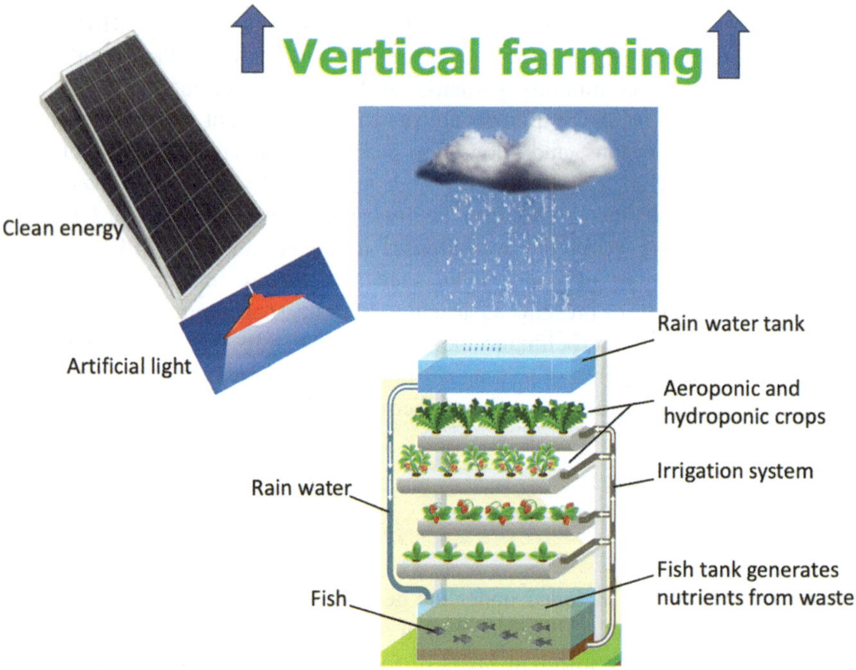

Fig. 4 Vertical farming

Fig. 5 RTGs (rooftop garden)

Fig. 6 Hydroponics

(Kozai 2019). They are often built in a building that resembles an airtight warehouse, with rows of tall, stacked plant beds that are illuminated artificially. Comparing the profitability of growing leafy vegetables in a greenhouse with a PFAL, the latter has an internal rate of return that can reach 35% (Eaves and Eaves 2018; Avgoustaki and Xydis 2020). There is some crossover between different CEA facilities, as a completely insulated VF or RTG could also be considered to be a PFAL (Zhuang et al. 2022).

Hydroponic crop cultivation has a number of potential advantages, including as separation from soil- or water-borne problems (such as nematodes, salinity, or heavy metals). Control over water and nutrient uptake has improved. The topic has received positive reviews (Raviv et al. 2019; Jones Jr 2014). Crops grown utilizing soilless culture are frequently cultivated in troughs, bags, or containers to facilitate effective management of the root zone (Fig. 6). In their list of typical nutrient sources for soilless cultivation, Raviv et al. (2019) mention raw irrigation water, fertilizers that are frequently incorporated into a substrate, substrate components, and a provision to modify the pH of the substrate. A thorough description of management techniques for soilless culture systems is given by Nelson (2012).

Nutrient-film technique (NFT), deep-water culture (DWC; also known as deep-flow method, raft, raceway, or floating hydroponics, among other names), and aggregate culture are hydroponic systems that are frequently employed in UA (Gómez et al. 2019). Crops grown in slanted troughs with a thin film of nutrient solution flowing over the roots (either constantly or sporadically) constitute NFT. Roots are continuously submerged in a nutritional solution in DWC systems. In aggregate culture, crops are grown in containers or on substrates that have been bagged, with drip systems used to apply nutritional solutions. For leafy greens and herbs, NFT and DWC systems are frequently employed. Aggregate culture is recommended for long-term fruiting crops including strawberry (*Fragaria x ananassa*), cucumber (*Cucumis sativus*), sweet pepper (*Capsicum annuum*), and tomato (*Lycopersicon esculentum*) (Gómez et al. 2019).

For soilless culture, substrate selection is a critical. Primary substrate components consists of >40% of the substrate volume. They are organic materials with low bulk density and high water-holding capacity like peatmoss and coconut coir fiber (Argo and Fisher 2002; Gómez et al. 2019). On the other hand, secondary components that consist of <40% substrate volume include expanded minerals like perlite, vermiculite, clays, sand, and composts that increase drainage and cation exchange capacity to increase aeration and nutrient retention (see Raviv et al. 2019 for a review).

1.2.2 Optimal Growth Conditions

The CEA sector struggles to attain economic viability due to ineffective microclimate and rootzone-environment management and excessive prices. Microclimate control, comprising light, temperature, ventilation, CO_2, and humidity, is crucial for producing uniform, high-quantity, and high-quality crops (Ojo and Zahid 2022). The focus of the most recent 10 years' research has been on the establishment of intelligent systems in CEA facilities, such as nutrient solution management for hydroponic farms and cloud-based microenvironment monitoring and control systems (Michael et al. 2021). According to Monteiro et al. (2018), artificial intelligence (AI) algorithms have also opened up new possibilities for intelligent predictions and self-learning. A subset of machine learning called deep learning (DL), which has a large presence in many contemporary technologies, has attracted a lot of interest in recent years.

In order to automate watering in vertical stack farms and microclimate control, computer vision and deep learning algorithms have been used (Ruscio et al. 2019). This has made it easier for growers to carry out quantitative assessments for high-level decision-making. A tiny indoor farm of less than 1500 ft^2 requires three personnel to complete manual CEA, which is labor-intensive. However, intelligent automation may be able to overcome these issues employing optical sensors coupled with DL-based prediction models (Namuduri et al. 2020). Several sensors, including cameras and LiDAR, are used to detect targets (Mendez et al. 2021).

1.2.3 Optimal Growth Environment and Automation

The crop quality and yield can be impacted by a number of indoor circumstances (Gibson 2018; Engler and Krarti 2021). The reported literature indicates four primary elements as being essential to creating ideal indoor growing settings:

- Temperature
- Humidity and transpiration
- Chemical balances
- Photosynthetic photon flux (PPF)

Temperature Temperature influences the timing of plant growth events such as maturation, flowering, and fruiting, and seeding is temperature-influenced in most plants (Kozai et al. 2019). For example, warmer temperatures speed up the process until flowering occurs at ideal levels. Below this threshold temperature, flowering progresses slowly and eventually stops completely at the ceiling temperature (Engler and Krarti 2021). Stressing plants at the end of their life is standard procedure for all flowering and fruiting plants. Stresses are modulated to mimic the challenges that plants face in the wild before they die, including imposing drought conditions, reducing temperature and nitrogen levels. Graamans et al. (2018) estimated the growth rate of lettuce and found that the optimum temperature for photosynthesis is between 20 and 25 °C, the optimum for respiration is between 30 and 35 °C, and the optimum dry matter production is between 16 and 17 °C (Graamans et al. 2018). LEDs are commonly used in CEA applications. It emits far less far-infrared radiation and is more energy efficient than traditional high-pressure sodium lamps used in greenhouses. Therefore, LEDs can help keep plants at the right temperature (Kozai et al. 2019).

Humidity and Transpiration Plant transpiration is hampered by the high relative humidity at the CEA facility. Vapor pressure deficit (VPD) is used to determine how much water can be contained in the air around a leaf, depending on its surface characteristics and a given temperature. The ideal VPD range for many plants is between 0.8 kPa and 0.95 kPa (Kozai et al. 2019). Reduced VPD prevents transpiration, which leads to water storage by the plant, promotes the growth of fungus, and finally reduces output (Linker et al. 2011). Yet, higher VPD needs higher water consumption, potential loads, and heating, ventilation, and air conditioning (HVAC) requirements. Plant stomata have the capacity to completely close, stopping transpiration (Engler and Krarti 2021). Stomata, which are openings in the plant wall, are used for respiration. Stomata can detect changes in light, temperature, humidity, and CO_2 concentration. The amount of water transpired is influenced by the root-shoot ratio, VPD, leaf area, and surface characteristics (Kozai et al. 2019; Bramley et al. 2022). Deep roots enable a plant to store more water for transpiration by the shoot. In plants, larger leaves often absorb more water than smaller leaves. Plants with thick cuticles, thick cell walls, sunken stomata, or hairs can reduce the rate of transpiration in order to raise the boundary layer between the stomata and the sensible heat of the flowing air (Passioura and Angus 2010). It is assured that CO_2 and water vapor will diffuse into the plant's leaves by maintaining a horizontal airflow rate of 0.3–0.5 m/s. The ideal airflow for some plants, such as tomato seedlings, is 0.7 m/s, but generally speaking, airflows up to 1.0 m/s can unduly stress the plant. Additionally, natural convection caused by ventilation can stop overheating in the top rows of a CEA plant (Kozai et al. 2019).

Chemical Balances Improved rates of nutrient intake, photosynthetic assimilation, and product nutritional value are all strongly associated with CO_2 enrichment (Vanhove et al. 2011). Due to cost constraints, persistently gloomy weather, or high ventilation rates in hot regions, CO_2 enrichment might only be practical for a small number of CEA sites (Li et al. 2018a, b). The production of biomass and amino acids in lettuce is said to be enhanced by a CO_2 concentration of 1000 ppm, monochromatic LED, and appropriate nutrient distribution (Miyagi et al. 2017). Increasing nitrogen concentrations in recirculating hydroponic systems from 100 mg/L to 400 mg/L increases lettuce yields by 0.8 kg m^{-2} in the fall, 2.6 kg m^{-2} in the winter, and 2.3 kg m^{-2} in the spring (Djidonou and Leskovar 2019). For CEA facilities, tracer gases are utilized to measure air exchange rates. N_2O or SF_6 are frequently utilized tracer gases in construction sites and CEA facilities. Moreover, CO_2 cannot be utilized in CEA facilities, since it can be absorbed by plants, despite the fact that it is employed as a tracer gas in other sectors. These gases' resulting energy balance can be used to forecast the right ventilation rates, which would save operational expenses. The use of H_2O as a tracer gas is now the subject of research (Engler and Krarti 2021).

Photosynthetic Photon Flux (PPF) The photoperiod, or duration of the night, which characterizes the growth season for a specific latitude, determines flowering. While exposed to light, plants absorb CO_2; when it is dark, they retain it. The level of CO_2 within a CEA facility is impacted by this pattern naturally (Li et al. 2018a, b). For a number of reasons, LEDs are preferable to incandescent, fluorescent, and HID bulbs. According to Graamans et al. (2018), LEDs installed in plant factories are often set at 52%, with the remaining 48% of power being distributed as sensible heat to aid in plants' evapotranspiration. The suggested growing parameters for CEA facilities are available Table 2.

Automation Automation in CEA or protected agriculture can be achieved through the implementation of the Internet of Things (IoT) (Shi et al. 2019a, b). A network of physical items that are equipped with sensors, software, and other technologies is known as the Internet of Things (IoT). These "things" are able to share real-time data with other linked devices and systems through networks because they are connected to the Internet.

Table 2 Ideal indoor conditions for different crops

Crop	Temperature (°C)	RH	VPD	Lighting density	photoperiod	CO_2 concentration	Cultivation period (days)
Lettuce	25 °C (day); 22 °C (night)	60–70%	0.85–0.95	200 PPF	16 h	800–1200 ppm	30
Tomato	25–31 °C	75–85% vegetative; 65–75% for flowering	0.65–0.8 vegetative; 0.85–1.0 flowering	900 PPF	12 h	1000 ppm	90
Herbs	20 °C (day); 15 °C (night)	70–80%	0.6 (day); 0.45 (night)	67% re at 620 nm; 33% blue light at 450 nm	12 h	800–1200 ppm	50–90 depending on variety
Micro-greens	21 °C (day); 17 °C (night)	80%	0.5 (day); 0.39 (night0	85% red, 15% blue; 300–600 PPF depending on variety	16 h	500–800 ppm	7–21 depending on variety
Shiitake mushrooms	12.8–24 °C	85% early on. Can drop to 60% after 3 days of pinning, 80% is ideal	0.6–0.8	55–100 PPF	12 h	800 ppm (6–10 air changes per hours to keep CO_2 levels low)	50–60
Cannabis	22–30 °C	65–75% vegetative; 60–70% flowering; 55–65% stressing	0.80–0.95 vegetative; 1.0–1.15 flowering; 1.16–1.35 stressing; 0.65 (day); 0.45 (night0	400–600 PPF	18 h veg; 12 h flowering	800–1100 ppm	70–80
Cucumbers	21–25 °C (day); 20 °C (night)	75–80%	0.65 (day); 0.45 (night)	100 PPF early stages; 250 PPF mature; 85% red, 15% blue	16 h	450–600 ppm	55–65
Peppers	21 °C (day); 17 °C (night);16 °C (flowering)	55–65%	3–5 seeding; 1.0 (day); 0.75 (night 0 vegetative; 1.0 (day); 0.0 (night)	430–500 PPF	18 h seed; 14 h veg	450–500 ppm	50–65

Crop	Temperature (°C)	RH	VPD	Lighting density	photoperiod	CO_2 concentration	Cultivation period (days)
Strawberries	30 °C (day); 15 °C night)	65% (day); 100% (night)	1.0 (day); 0.0 (night)	300 PPF	16 h	1000 ppm	90
Blueberry transplants	23 °C average	90%	0.28	100–200 PPF	16 h	1500 ppm	30–40
Rice	23–27 °C	70%	1.05 (day); 0.85 (night)	800 PPF	12 h	400 ppm	100

Adopted from: Engler and Krarti (2021). doi: https://doi.org/10.1016/j.rser.2021.110786

With the development of agricultural sensor, wireless communication, cloud computing, machine learning, and big data technologies, IoT technology has grown and is progressively being promoted and used in the field of protected agriculture (Kamilaris and Prenafeta-Boldú 2018). It is playing an important role in many areas of protected agriculture due to its capacity to help farmers check soil quality, climatic change, and the health of animals and plants (Shi et al. 2019a, b). In the event that environmental variables alter above the predetermined threshold, IoT will automatically send an alert message to the administrator demanding that the hidden threat be eliminated. Additionally, according to Liu et al. (2018), it has the capacity to alter environmental factors like temperature, humidity, carbon dioxide concentration, and illumination in real time.

Additionally, the IoT system's cameras can capture crop diseases and insect pests in the greenhouse in real time, helping farmers to spot problems and put preventative measures into place (Ma et al. 2015). GPS, radio frequency identification (RFID), and other location-based sensors enable tracking and visual monitoring of produce during storage and transportation. Supermarket managers use their computer or smartphone to monitor and forecast product status and demand in order to get things on the shelves. Users and customers can obtain details on the variety, origin, processing, and other features of agricultural products by utilizing a QR code, barcode, etc. With the use of IoT for protected agriculture, a rural community may be constructed that is knowledgeable, connected, advanced, and adaptable. Cheap embedded devices can improve how people engage with the physical world. For further information on IoT, read the section on technology in DA. Big data, cloud computing, and edge computing can all provide insightful analysis and information that can be used to make decisions (Shi et al. 2019a, b; Quy et al. 2022).

1.3 Challenges Facing Food Production and Food Supply Chain

The food sector is crucial in providing the fundamentals and needs to support a range of human behaviors and activities (Cooper and Ellram 1993). In order for the food to reach the ultimate consumers by the due date, it must be stored, delivered, and retailed after it has been produced or harvested. According to reports, around 1.3 billion tons (or about one-third) of the food produced each year is abandoned or wasted (Manning et al. 2006). Around 1 billion tons of food are wasted each year, with two-thirds of that occurring in the supply chain during harvest, shipping, and storage (Fritz and Schiefer 2008). Consider fruit and vegetables as an example. Due to inefficient and ineffective food supply chain management (FSCM), 492 million tons of such perishable food were wasted globally in 2011 (Gustavsson et al. 2011). FSCM is important to save our food as a result (see Zhong et al. 2017 for a review).

The food supply chain has quickly evolved in recent decades, spreading internationally and engaging many more partners, making the supply chain longer and more sophisticated than before. Today's consumers expect exotic delicacies, fresh on their plates, year-round. As if things weren't already challenging enough, the multiyear COVID pandemic shutdown in 2020 put even more strain on supply chains by closing down numerous restaurant and food service supply chains and raising the stakes for retail chains and direct-to-consumer food delivery (Huang et al. 2021).

1.3.1 Blockchain Technology

Blockchain technology, a sophisticated database system, permits open information exchange inside a business network. In a blockchain database, data is held in blocks that are linked together in a chain. The data is still constant in time, since the chain cannot be deleted or changed without network agreement. You can set up an unchangeable or immutable ledger using blockchain technology to manage orders, payments, accounts, and other transactions. The system's built-in capabilities, which also prevent unauthorized transaction submissions, make it possible to see these transactions as a whole.

1.3.2 e-Commerce Software

A stand-alone program or software suite called e-commerce software gives the ability to sell your goods and services online. The front end, which is your website, makes it simple for customers to make purchases, while the back end allows you to streamline all of your procedures from inventory to sales.

Each style of e-commerce software is available and can be customized to meet your objectives and financial constraints. Although it's not a rule, the sort of e-commerce website software you use usually depends on the size of your company. Software-as-a-Service, Platform-as-a-Service, or an on-premise platform that gives you control over the server and software used to offer your e-commerce website are all options for your e-commerce needs.

1.4 Climate Smart Agriculture

Climate change has imposed several adversaries to the planet ecosystem through erratic environmental fluctuations in temperature, rain pattern, and drought occurrence (IPCC 2018). The continuous changing scenario not only disturbs the crop growth and production but also affects the food security and the incidence of diseases (Chakraborty and Newton 2011). It has been unequivocally demonstrated that the climate change has set an impact on all the pathogen, host, and plant environment (Singh et al. 2023). Since agricultural productivity is crucially affected by plant diseases, the fluctuating climatic environment has led to different disease related modalities, such as distribution pattern, resurgence, widespread infestation, and new pathotypes (Velásquez et al. 2018). Cases like intense Ascochyta blight in chickpea occurred due to infrequent late rainfall resulting in yield and quality losses (Addisu et al. 2023), and the shift in rainfall pattern due to an El Nino event has damaged lentil crop due to rust infestation in Ethiopia (Pathak et al. 2018).

Since the dawn of agriculture, there have been technological developments, which have paved the way for improvement of crop plants and refining the crop cultivation and management. Plant breeding has witnessed genetic and agronomic interventions to enhance the pace and accuracy of plant selection (Wijerathna-Yapa and Pathirana 2022). Keeping in view of the population progression, declining land resources, and climate vagaries, there is a need to develop selection methods with more accuracy and precision. The advancement of artificial intelligence (AI) in the past decade has offered great potential to augment the climate smart agriculture. AI technology through the use of high-throughput genomics and phenomics methods can quicken the course of breeding new plant varieties (Khan et al. 2022; Harfouche et al. 2019). The machine learning tools have found their application in marker-assisted selection, genomic prediction, and genomic selection (Esposito et al. 2020; Reinoso-Peláez et al. 2022). The tools including ML, deep learning, and predictive analysis can help in the analysis of complex, huge agricultural datasets to extract useful information about traits, and their associations of plant responses to stress conditions (Tong and Nikoloski 2021; Crane-Droesch 2018). Genomic technologies together with high-throughput phenotyping provide the trait related information to researchers to guide and notify the breeding methods to adopt for climate-smart breeding (Marsh et al. 2021). AI plays a vital role in integrating and handling the huge data by conducting association studies to identify genomic targets associated with disease response traits (Khan et al. 2022). Breeders can use the data for management of crop plants for their adaption to stresses and introgression through the use of genomic selection or genome editing tools (Harfouche et al. 2019).

Plant diseases inflict severe losses on plant productivity and affect global food security. It has been demonstrated that the changing climatic factors worsen the conditions for resurgence of plant and crop diseases. This warrants the need of a greater understanding of the changing climate effects on crop plants in a spatial and temporal manner under realistic field scenario. The intervention of information technologies such as the Internet of Things (IoT), remote sensing, unmanned aerial vehicles, and artificial intelligence has revolutionized the agriculture (Gao et al. 2020). These digital technologies have been pivotal in generating huge amount of data to aid the understanding of crop breeding for several applications, such as prediction of yield, weed and pest/disease detection and forecast, risk management, food safety, and spoilage inhibition. Kreuze et al. (2022) suggested the use of image detection from smartphones or unmanned aerial vehicles for monitoring of pest and disease and data handling for modeling, predictions, and forecasting regarding climate change in root, tuber crops, and banana.

The deep learning tools have also found their place in agriculture, for weather forecast (Schultz et al. 2021). Neural networks are regularly used in the context of plant diseases, such as epidemiology or remote sensing (Zhang et al. 2005;

Selvaraj et al. 2019). In case of powdery mildew disease, UV-B light has shown good application for disease management in grapes and strawberry (Onofre et al. 2021; Meyer et al. 2021). Application of pesticides on crops like grapes can be very well done using robotics systems (Oberti et al. 2016). Disease phenotyping often plays a crucial role in field grown plants, for example, in potato in the context of potato blight, efficient phenomics-assisted screening has been used for disease resistance (Gold et al. 2020). The deep learning and machine learning are also used to precisely categorize breeding germplasm for resistance to potato late blight (Gold et al. 2020), *Rice hoja blanca virus* (Delgado et al. 2019), and banana Xanthomonas wilt (Selvaraj et al. 2020). There have been several studies indicating that it is possible to go for early, nondestructive prediction of the onset of disease based on primary symptoms such as mild and small lesions by using imaging spectroscopy (Gold 2021).

1.5 Technologies in DA

The phrase "Internet of Things" was first coined in 1999 by computer scientist Kevin Ashton. While working at Procter & Gamble, Ashton promoted the use of radio frequency identification (RFID) chips to track products as they move through a supply chain. A five-layer IoT architecture was created by Shi et al. (2019a, b) based on the realities of protected agriculture and the expertise of other academics. In Fig. 7, these levels are succinctly proposed as a five-layer system.

- Perception layer: This layer is made up of various sensors, terminal devices, farm machinery, wireless sensor networks (WSN), RFID tags and readers, etc. Common sensors include machines, wireless sensor networks (WSN), RFID tags and readers, and other objects. Common sensors include those that collect data on the environment, plants and animals and other agriculturally related sensors. These sensors can offer temperature, humidity, and wind speed data to agriculture. Data on variables, including temperature, humidity, wind speed, plant diseases, insect infestations, and animal vital signs, can all be collected with these sensors. Information has been acquired about plant diseases, insect pests, and animal vital signs. The gathered data is simply analyzed by the embedded device and uploaded to a higher layer through the network for additional processing and analysis.
- Network layer: The infrastructure of the Internet of Things is made up of a converged network that consists of the Internet and various other communication networks. The transmission medium network is made up of the Internet and other communication networks. For the transmission, the medium can be either wired technology, such as CAN bus and RS485 bus, or wireless technology, such as Bluetooth, LoRa, and NB-IoT, as well as wireless technology, such as Zigbee. Agricultural data is also transmitted across the network layer using Bluetooth, LoRa, Zigbee, and NB-IoT. The network layer not only transmits

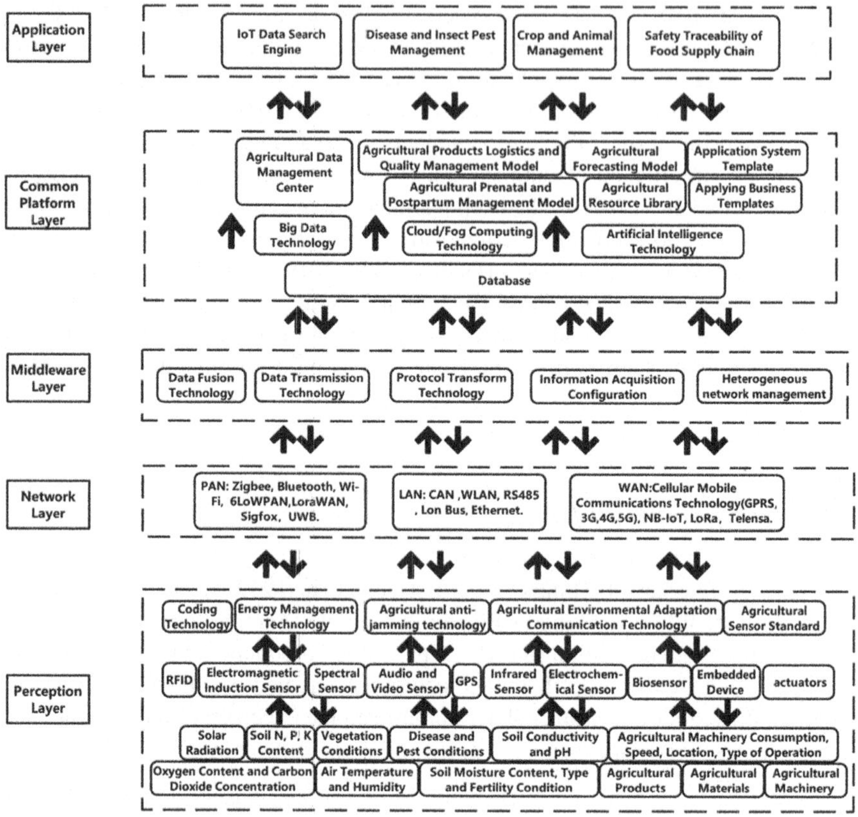

Fig. 7 Structure of IoT in protected agriculture. (After Shi et al. 2019a, b; courtesy: Sensors; doi: https://doi.org/10.3390/s19081833)

different kinds of related information gathered by the perception layer to the higher layer, but it also sends control agricultural related information gathered by the perception layer to the higher layer and commands from the application layer to the perception layer, causing the related network layer devices to act appropriately.

- Middleware layer: IoT may provide a range of services to fit a range of devices. Because each device's technical requirements (CPU, power source, communication module, and system) are unique from the others, heterogeneity issues can occur. Different devices are unable to connect to and communicate with one another as a result. The middleware layer's aggregation, filtering, and processing cause heterogeneity issues. The middleware layer collects, filters, and processes data from IoT devices, greatly lowering processing time and cost while providing developers with a more flexible tool to build their applications. It also simplifies the processes for introducing new hardware and software, facilitating its faster integration with existing systems and boosting IoT compatibility.

- Common platform layer: The organization, decision-making, summary, and statistics of agricultural data, as well as the creation of diagnostic analysis, forecasting, and early warning systems, are all responsibilities of the common platform layer. Machine learning, big data, edge computing, cloud computing, fog computing, diagnostic reasoning, and early warning and prediction are all part of this layer. An algorithm, extra commonly used core processing technologies, and its business model are all included in this layer.
- Application layer: The value and utility of the Internet of Things are most clearly seen at here, the highest level of the architecture. This layer includes a number of intelligent platforms or systems for environmental monitoring and control of plants and animals, early warning and management of diseases and insect pests, and traceability of the safety of agricultural products. These systems can all improve production efficiency and save money and time.

1.5.1 Crucial Technologies of IoT

Sensor Technology In order to collect data about the environment, plants, and animals, sensors are crucial and one of the technological barriers in the development of the Internet of Things (Shi et al. 2019a, b). Around 6000 research and production groups, including well-known companies like Honeywell, Foxboro, ENDEVCO, Bell & Howell, and Solartron, are now working on sensor research, representing more than 40 different countries. The three most often used types of agricultural sensors are physical property type sensors, biosensors, and micro-electromechanical system (MEMS) sensors. The majority of temperature, humidity, and gas sensors fall under the category of physical property sensors, which convert signals by physically altering the material's sensitivity. The biosensor (Li et al. 2018a, b) is primarily used to detect pesticide residue, heavy metal ions, antibiotic residue, and toxic gas and includes enzyme sensors (Zheng et al. 2015), microbial sensors, adaptive sensors (Jiao et al. 2018), etc. It transmits information based on the organism's reaction to the outside environment. The MEMS sensor is a standout among the most recent research and development efforts in the area of dependable, affordable, and compact sensors (Negara et al. 2014). There could be hundreds, thousands, or even millions of nodes in a sensor network. The cost of each node needs to be kept to around $1 in order for the sensor network to be practicable; however, it is now as high as $80 (Shi et al. 2019a, b).

Data Transmission Technology When compared to conventional transmission technologies like fieldbus, wireless communication technology offers advantages, including inexpensive construction and maintenance costs, low-power consumption, and great extensibility. In order to develop their WSN for environmental monitoring (Kumar and Hancke 2014), autonomous irrigation (Rajalakshmi and Mahalakshmi 2016), and remote control (Revathi and Sivakumaran 2016), the majority of scientists, enterprises, and producers currently employ it. The heterogeneity of the IoT has been slightly increased as a result of businesses and research groups developing their own wireless devices. Additionally, interference between wireless signals from several protocols that use the same band, such as Bluetooth, Wi-Fi, and ZigBee, is possible (Čolaković and Hadžialić 2018). Given its high power consumption and quick connection, Wi-Fi is a viable option for the deployment of sensor networks at fixed locations. Since Bluetooth has a small communication range, exceptional security, and high power consumption, it is perfect for short-term, close-range networking. ZigBee offers the advantages of low consumption, low cost, and self-organization, because each node can serve as a relay station for data transmission between close-by nodes. As a result, it makes for the ideal long-distance, large-range sensor networking and enables simple coverage expansion.

WSN The WSN is a multi-hop self-organizing network system created via wireless communication in order to cooperatively sense, gather, and process various data about the observed item in the network coverage area (Srbinovska et al. 2015; Ferentinos et al. 2017). It is made up of a number of sensor nodes, the majority of which are battery-operated. It can be divided into terrestrial WSN and wireless subterranean sensor networks (WUSN). Lower frequency wireless solutions are preferred for agricultural sensors, which are often buried in the ground, because of WUSN's low attenuation. In comparison to terrestrial WSN, WUSN also consumes more energy and has larger antennas (Ojha et al. 2015). IoT may no longer require a mesh-style WSN with power-based routing, where one node forwards packets of other nodes, as low-power wide-area network (LPWAN) technology develops.

Cloud Computing Cloud computing is the on-demand provision of computer system resources, particularly data storage (in the form of cloud storage) and processing power, without the user's active involvement. Cloud computing is a result of distributed computing, parallel computing, and network computing. A variety of hardware, infrastructure, platform, software, and storage services are offered for IoT applications via this Internet-based computing system. A system for dynamically assigning, deploying, monitoring, and reallocating pools of virtualized computing and storage resources is at the heart of it (Hashem et al. 2015). This system enables users to access compute, data storage, and platform services that adhere to quality-of-service criteria. This will have a significant impact on the expansion of IoT in agriculture. First, cloud computing has made it possible for farmers to store text, pictures, videos, and other types of agricultural data using inexpensive data storage services, which has considerably reduced the cost of storage for agricultural businesses (Nativi et al. 2015). Second, relying on farmers' technical expertise to make decisions using this raw data is challenging. Cloud computing is the only technology that can support intelligent large-scale data processing systems (Ferrández-Pastor et al. 2016). Third, using cloud computing can create a safe environment for developing different IoT applications, such as monitoring agricultural activities (Botta et al. 2016).

Edge Computing Edge computing, as defined by Satyanarayanan (2017), is a new computing model that makes advantage of calculations at the network's edge. Any computer and network resources between the data source and the cloud computing center path are referred to as the edge of edge computing. Cloud services are represented by the edge's downlink data, IoT services are represented by the edge's uplink data, and both are represented by the edge's uplink data. Edge hormone, which shifts some of the computing activities to the network edge device, can improve data transmission performance, guarantee real-time processing, and lower the computational load on the cloud computing center. Because processing occurs close to the source rather than in the cloud, edge computing also provides greater data security (Shi et al. 2019a, b).

Machine Learning A sophisticated method known as machine learning (ML) allows computers to learn new knowledge, continuously improve their performance, and reach perfection. Theoretical, algorithmic, and practical advances in machine learning have been made recently (Biamonte et al. 2017), and it has been combined with other agricultural technologies to optimize crop output while reducing input costs (Shi et al. 2019a, b). The main machine learning methods include naive Bayes, discriminant analysis, K-nearest neighbor, support vector machines (SVM), K-means clustering, fuzzy clustering, gaussian mixture models, artificial neural networks (ANN), deep learning (Ojo and Zahid 2022), decision tree algorithm, and others (Edwards-Murphy et al. 2016). A theoretical framework for agricultural decision-making is provided by ML, which can make accurate predictions, reveal the internal linkages between jumbled, modelless, and complex agricultural data and discover these relationships. Machine learning technologies are useful for intelligent irrigation planning, crop breeding, disease detection, pest and disease prediction, and agricultural expert systems (Russell and Norvig 2018). For instance, historical farming data may be examined using machine learning technology, along with crop productivity under varied climatic conditions and the inheritance of a particular phenotype. Furthermore, by utilizing ML technology, it is feasible to look at association rules and then develop a probability model to identify the genes that are most likely to be involved in the expression of a particular desired trait in the plant (Montesinos-López et al. 2019). This can help the breeding specialist create a breeding experiment that will be effective. The method used three processes to identify the maturity of a single intact tomato using machine learning: pixel-based segmentation, blob-based segmentation, and individual fruit detection. Using criteria including color, shape, texture, and size, decision trees were built in the first two steps and then utilized to segment photos. The different fruit of each tomato was finally automatically identified using the X-means clustering technique. Their method has a precision of 0.88 and a recall of 0.80, per the results of the tomato detection picture test (Kyosuke et al. 2014).

Big Data Protected agriculture generates millions of dynamic, intricate, and geographical data points, including soil databases, greenhouse environment data, animal vaccination records, and government investment data. Contrary to relational data structures, which logically express themselves using two-dimensional tables, agricultural data is more unstructured and contains many hypermedia elements, including expert experience, knowledge, and agricultural models in the form of text, charts, pictures, animations, and voice/video. The four characteristics that best sum

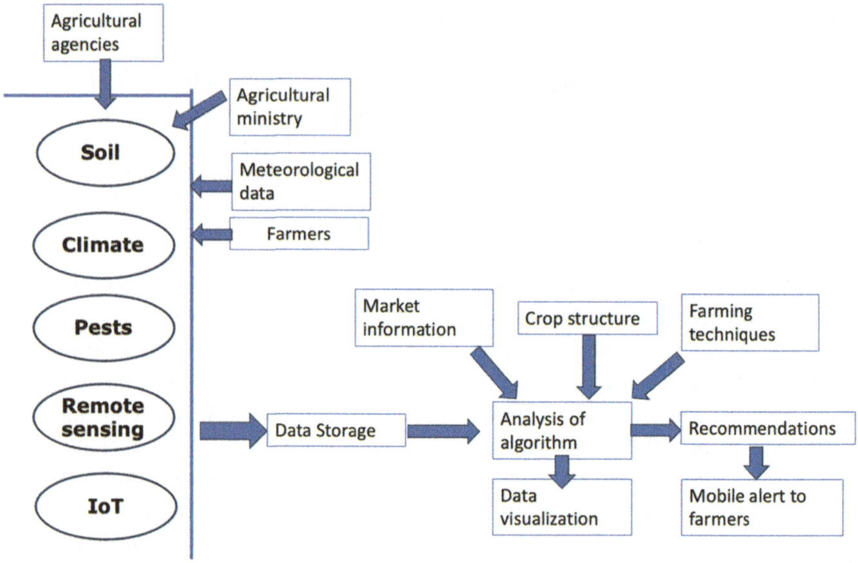

Fig. 8 Big data technology

up how "big" these data are volume, velocity, diversity, and honesty (Zhou et al. 2016). Big data technology can find new knowledge, discover hidden connections within a data collection, and provide data support for subsequent processes. This is done by employing information mining and other techniques. The methods that are most frequently used to deal with big data technology are image processing, modeling and simulation, machine learning, statistical analysis, and geographic information systems (GIS) (Kamilaris et al. 2017) (Fig. 8).

1.5.2 IoT and Plant Management

By using greenhouse technology, which is partially uncontrolled by the natural environment and promotes the intense and efficient use of agricultural resources, protected agriculture, as opposed to open-field farming, offers a more favorable and manageable environment for crop growth. Numerous studies have shown that building and testing various monitoring and control systems to alter greenhouse environmental parameters, like air temperature and humidity, light intensity, and CO_2 concentration, are both technically feasible and economically viable (Sreekantha and Kavya 2017). At the early phases of IoT development, the environmental data are simply processed and frequently provided in sheet and plot form (Mat et al. 2016).

Fig. 9 Conceptual image of IoT-based agricultural solutions

With the development of cloud computing, ML, etc., IoT solutions may easily achieve smart data processing and analysis at low cost and in a straightforward manner (Elijah et al. 2018). Deng et al. (2018) built a closed-loop control system in a factory that makes salad-growing plants based on the kinetic model. Zamora-Izquierdo et al. (2019) developed a low-cost smart agricultural Internet of Things infrastructure based on edge and cloud computing for soilless culture greenhouses. There were three parts to the platform: local, edge, and cloud. While the edge component handled primary management responsibilities and might improve the stability of these systems, the local component dealt with data collecting and automatic control via cyber-physical systems. Data analyses were performed by the cloud component. When compared to a standard open control, the platform conserved more than 30% more water (Liao et al. 2017). According to Zamora-Izquierdo et al. (2019), an online watering system for hydroponic greenhouse crops increased water and fertilizer use efficiency by 100%. Liao et al. installed an IoT-based system in an orchid greenhouse to monitor environmental factors and the growth status of Phalaenopsis. The suggested method might provide high spatiotemporal resolution quantitative data to flower growers and aid in the future improvement of phalaenopsis farming practices (Katsoulas et al. 2017). For a conceptual representation of IoT-based agricultural solutions, see Fig. 9.

Crop growth is greatly threatened by diseases and insect pests, and conventional technology and chemical prevention have several drawbacks and harmful effects (Larsen et al. 2019). Because of the development of IoT, crop disease and pest control now have more intelligent and effective solutions. Numerous IoT sensor types may collect information about location, greenhouse environment state, crop development, and pest situation anywhere in real time, helping farmers to keep an eye on agricultural pests and diseases. Following transmission to cloud data centers, the raw data and photos are processed and evaluated using a range of models and algorithms based on different diseases and pests (Pixia and Xiangdong 2013). The following services are often provided to farms by these cloud computing facilities: disease or pest detection, disaster warning and warning of approaching calamities, and expert system-recommended governance activities. The diagnosis and early warning of agricultural illnesses, as well as online monitoring, should therefore be the main areas of future research.

The source of all IoT data is sensing. The agri-food industry produces a significant amount of heterogeneous datasets with the help of many IoT devices, both in terms of content, structure, and storage type (Lokers et al. 2016). According to Ahmed et al. (2019), big data frequently demonstrates heterogeneity, variety, unstructuredness, noise, and excessive redundancy. Such enormous datasets require sophisticated methods for data curation and storage, as well as time-consuming statistical methods and programing models to extract relevant data. The knowledge required to understand the state of the (agri-food) system is produced through the preprocessing and conditioning of raw data. By employing sophisticated algorithms, observing the system's performance in respect to the desired outcomes, and allowing the system to make independent localized judgments and take the necessary actions, a system can be created capable of doing so. An IoT system is deemed "intelligent" when it reaches this level of independence, which permits autonomy in sensing, decision-making, and actuation (Misra et al. 2022).

1.5.3 AI in Digital Agriculture

The imitation of human intelligence functions by machines, especially computer systems, is artificial intelligence. Vendors have been rushing to highlight how AI is used in their goods and services as AI buzz has grown. Frequently, what they classify as AI is just a part of the technology, like machine learning. For the creation and training of machine learning algorithms, AI requires a foundation of specialized hardware and software. Python, R, Java, C^{++}, and Julia all offer characteristics that are well-liked by AI engineers, yet no one programing language is exclusively associated with AI.

In commercial IT, the phrases artificial intelligence (AI), machine learning (ML), and deep learning (DL) are frequently used interchangeably (van Dijk et al. 2021) (Fig. 10a). However, there are differences. The 1950s saw the invention of the term "AI," which describes devices that mimic human intelligence. As new technologies are created, it encompasses a set of skills that is constantly changing. Machine learning and deep learning are examples of technologies that fall under the category of AI (Madakam et al. 2022). With the aid of machine learning, software programs may predict outcomes more accurately without having to be expressly programed to do so. In order to forecast new output values, machine learning algorithms use historical data as input. The availability of big datasets for training increased the effectiveness of this strategy significantly. Deep learning, a branch of machine

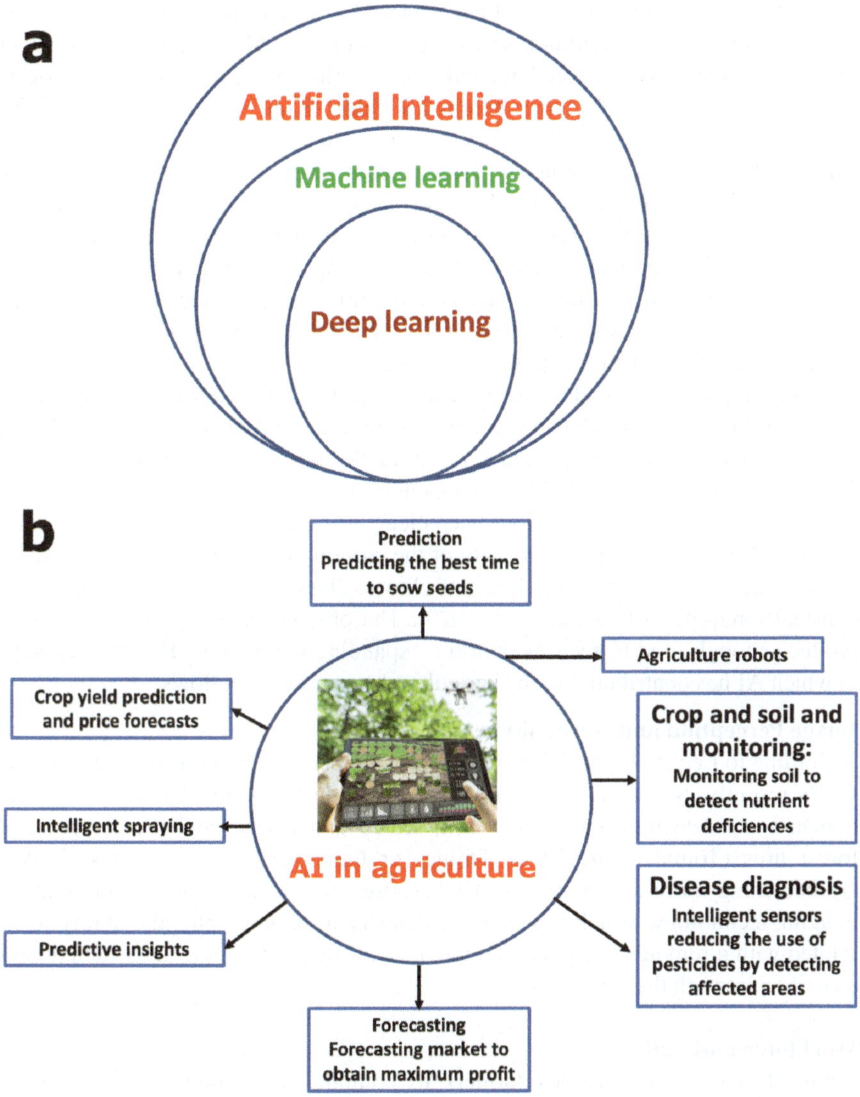

Fig. 10 (**a**) The phrases artificial intelligence (AI), machine learning (ML), and deep learning (DL) are frequently used interchangeably. (**b**) AI-based technologies assisting to increase efficiency across all fields

learning, is based on our knowledge of the anatomy of the human brain. Recent developments in AI, such as self-driving cars and ChatGPT, are underpinned by deep learning's usage of artificial neural networks' structure.

In addition to managing the challenges faced by various industries, including the various fields in the agricultural sector, such as crop yield, irrigation, soil content sensing, crop monitoring, weeding, and crop establishment, AI-based technologies also help to increase efficiency across all fields (Kim et al. 2008) (Fig. 10b). In order to supply high-value AI applications in the aforementioned industry, agricultural robots are constructed (Talaviya et al. 2020). The agricultural industry is experiencing a problem as a result of the rising worldwide population. AI has the ability to provide a crucial remedy. AI-based technical advancements have allowed farmers to increase output while using less input, improve output quality, and ensure a quicker go-to-market for the produced crops. Farmers were using 75 million linked devices in 2020 (Talaviya et al. 2020). The typical farm is anticipated to produce an average of 4.1 million data points per day by 2050.

Over the past few decades, the agriculture production systems have had a great deal of difficulty due to changes in the climate, rising production costs, declining water supplies for irrigation, and an overall decline in farm labor (Jung et al. 2021). In addition, the COVID-19 pandemic poses a threat to the disruption of supply chains and food production. Such elements pose a risk to the environment's sustainability as well as the continuity of the current and future food supply chain. To keep ahead of the ongoing effects of climate change, significant inventions are constantly required (Talaviya et al. 2020). The obvious challenge here is how to produce enough food to feed the world's expanding population. The various ways in which AI has contributed in the agricultural sector are as follows:

Image Perception and Recognition

According to Lee et al. (2017), there has been an increase in interest in autonomous UAVs recently. Some of these applications include recognition and surveillance, human body detection and geolocation, search and rescue, and the detection of forest fires (Tomic et al. 2012). Drones or unmanned aerial vehicles (UAVs) are becoming more and more popular because of their adaptability and amazing imaging technology, which ranges from delivery to photography, the ability to be piloted with a remote controller, and the devices' dexterity in the air, which allows us to do a lot with these devices.

Workforce and Skills

Artificial intelligence enables farmers to compile vast amounts of data from public and government websites, analyze it all, and give farmers answers to many ambiguous problems (Panpatte 2018). It also gives us a smarter way of irrigation, which increases the farmers' yield. A combination of technology and biological talents will be used in farming in the near future as a result of artificial intelligence, which will not only improve quality for all farmers but also reduce their losses and workloads. According to the UN, by 2050, two-thirds of the world's population would be living in cities, necessitating a reduction in the load on farmers (Talaviya et al. 2020). AI in agriculture can be used to automate many operations, reduce risks, and give farmers with relatively simple and effective farming.

Increase the Output

Variety selection and seed quality determine the maximum performance level for all plants (Ferguson et al. 1991). Emerging technologies have aided in crop selection and even improved the selection of hybrid seed options that are most suited to farmer demands. It has been implemented by studying how the seeds react to varied weather conditions and soil kinds. Plant diseases can be reduced by gathering this information. We can now meet market trends, yearly outcomes, and customer needs, allowing farmers to maximize agricultural returns more efficiently.

Farmers' Chatbots

The conversational virtual assistants that automate conversations with users are known as chatbots. With the use of machine learning and artificial intelligence-powered chatbots, we can now understand natural language and communicate with users more personally. Agriculture has made use of this facility by supporting the farmers in receiving answers to their unanswered queries, for offering them counsel, and for providing other recommendations as well. They are mostly equipped for retail, travel, and media.

Machines that are used on farms to hoe and harvest crops, perform weeding, use drones to spray weeds and pesticides, and gadgets used in automatic milking are a few examples of AI-based agricultural technologies (Ryan et al. 2021). Robotics have assisted in an 80% reduction in the amount of herbicides sprayed on crops (Revanth 2019). According to studies, this optimization can reduce pesticide and herbicide costs by 90% while also protecting the environment from the negative consequences of chemical use (Revanth 2019). Drone-captured images of crops can be utilized for a variety of purposes, including nutrient deficiency monitoring, farm animal health monitoring, and agricultural cultivation optimization (Marvin et al. 2021).

On the basis of a given dataset, machine learning (ML) creates algorithms that learn to carry out particular tasks. It is a branch of artificial intelligence that is extensively employed in both academia and business. Between supervised and uncontrolled learning, there are significant differences. A predictive model is improved through supervised learning by setting its parameters to perform well on labeled training data, which consists of inputs and known outcomes. The generated models can then forecast new test data that hasn't yet been seen. On the other hand, unsupervised learning looks for patterns in unlabeled data. It is more difficult to quantify the performance of an unsupervised model compared to supervised methods (van Dijk et al. 2021).

1.5.4 DL, Genomics and Breeding

As was previously stated, there are two basic categories of ML problems: supervised and unsupervised. The goal of supervised learning is to create a model that associates predictors with target variables, such as histone marks, such as DNA sequences. Target variables might be either continuous (regression) or categorical (classification). The prediction of regulatory and nonregulatory regions in the maize genome (Mejia-Guerra and Buckler 2019), the prediction of mRNA expression levels (Washburn et al. 2019), sequence tagging in rice (Do et al. 2018), plant stress phenotyping (Ghosal et al. 2018), and the prediction of macronutrient deficiencies in tomatoes (Tran et al. 2019) are a few examples of supervised learning applications. The issue becomes unsupervised if there is no information about the outcome in the data collection (Wang et al. 2020).

In order to solve complicated biological challenges, deep learning has been utilized in the fields of genomics, transcriptomics, proteomics, metabolomics, and systems biology (Xu and Jackson 2019). Numerous studies demonstrated that DNA shape significantly influences the specificity of transcription factor (TF) DNA-binding (Lai et al. 2019). Chromatin accessibility assays (like MNase-seq, DNase-seq, and FAIRE) and other genomic assays (such microarray and RNA-seq expression) are only a few of the many data types that are available. The same is true for transcription factor (TF) binding, which can be studied using ChIP-seq data, gene expression profiles, DNA affinity purification sequencing (DAP-seq), and ampDAP-seq, which uses amplified and consequently demethylated DNA as substrates and histone modifications (Zampieri et al. 2019).

Several deep learning techniques were created to model TF DNA-binding specificity and analyze these enormous datasets (Wang et al. 2020). Several deep learning-based techniques have been developed to predict in vivo TF binding. For instance, DeepBind can learn several motifs to forecast the binding sites of proteins that bind DNA and RNA (Alipanahi et al. 2015). Cell-specific TF binding is predicted by TFImpute (Qin and Feng 2017). In DeepSEA (Zhou and Troyanskaya 2015), DeFine (Wang et al. 2018), and DFIM (Greenside et al. 2018), the impacts of functional noncoding variations were assessed. DRNApred was created (Yan and Kurgan 2017) to distinguish between residues that bind to DNA and those that bind to RNA.

It is difficult to pinpoint the important genomic regulatory regions in species like maize, which have a large number of repeated elements and broad intergenic areas. In order to overcome these difficulties, techniques like k-mer grammars, which are based on natural language processing, have been employed to precisely and cheaply annotate regulatory areas in maize lines. Modeling transcription factor binding locations has benefited significantly from machine learning techniques. Several facets of plant biology have shown the effectiveness of machine learning models. For better in vivo transcription binding sites (TFBSs) prediction, they can be trained using several types of sequencing data, either separately or in combination, and they can also further integrate additional data, such as DNase I hypersensitivity data.

1.6 Remote Sensing Technologies

The agricultural community now has a diagnostic tool thanks to remote sensing (RS) technology that may serve as an early warning system. This enables quick action to stop any problems before they spread widely and negatively impact crop productivity. The agricultural community now has access to a variety of RS options as a result of recent advancements in sensor technologies, data management, and data analytics. However, the agriculture business has not yet fully utilized RS technologies due to knowledge gaps about their sufficiency, suitability, and techno-economic viability. The use of RS technologies in agricultural production has increased significantly over the past 20 years, while use of unmanned aerial systems (UASs) has increased significantly since 2015. The region that produced the most research articles concerning UASs was Europe (34% of the total), followed by the USA (20%) and China (11%) (Khanal et al. 2020). Prior RS research tended to concentrate more on soil moisture and crop health monitoring during the growing season and less on issues like soil compaction, subsurface drainage, and crop grain quality monitoring.

Modern technology have always been used by agricultural research experts as they look for new methods to incorporate them into agricultural systems. Dynamic crop simulation models have proven helpful tools for integrating various agriculture system components and enabling us to investigate how those components operate within the system. Because of its ability to utilize huge data, which is now more readily available through the use of unmanned aircraft systems (UASs), it is currently attracting a lot of attention within the agriculture disciplines (Jung et al. 2021). By enabling advanced analytics for managing agricultural systems, UAS offers a previously unheard-of-chance to increase production systems' resilience and efficiency (Lezoche et al. 2020).

1.7 Precision Agriculture Technologies for Crop Production

Precision agriculture (PA) enables the agro-management by using advanced technology sensor and analysis tools. PA employs a huge volume of data and information to progress the use of agricultural resources, yields, and the quality of crops (Singh et al. 2020) and drought-related decisions in agriculture (Rhee and Im 2017). The changing weather and its effect on ecosystem threaten crop production and food security for the present and future generations. Machine learning approaches have been applied for the management of agri-related factors such as water availability, soil fertility, environment and diseases/pests (Priya and Ramesh 2019). Smart, digital agriculture can also benefit from the integration of the IoT devices, smart systems, and sensors to enable farmer's agri-practices (Chehri et al. 2020). Among the PA applications, remote sensors, GPS, GIS, and yield maps are among the most in use (Cisternas et al. 2020). Other tools that have shown great interest for PA include UAVs and WSNs for diverse functions including aerial crop monitoring and smart spraying tasks (Radoglou-Grammatikis et al. 2020).

PA ensembles a huge amount of information about the crop status or crop health in the growing season at high spatial resolution. Independently of the data source, the most crucial objective of PA is to provide support to farmers in managing their farming practices. Several agro-related variables, such as soil condition, plant health, fertilizer and pesticide effect, irrigation, and crop yield, have to be efficiently managed to realize higher yield and better crop growth under natural and environmentally challenging conditions (Abdullahi and Sheriff 2017). Monitoring all the above with precision is important for rational use of farming resources and their management (Wu et al. 2022). Remote sensing methods like satellite- and UAV-based hyperspectral imaging offer solutions as biophysical indicator maps during the various stages of crop growth cycle and seasons (Bégué et al. 2018; Wu et al. 2022) besides soil and plant health. Other tools like AI and ML have also been useful in precision agriculture for prediction and appraisal of crop yield, detection of diseases, and weeds (Liakos et al. 2018) (Figs. 11 and 12).

1.8 Conclusion and Recommendations

The use of big data in food production, along with the implementation of the Internet of Things (IoT), blockchain technology, artificial intelligence (AI), machine learning, cloud computing, as well as unmanned aerial vehicles (UAVs), and

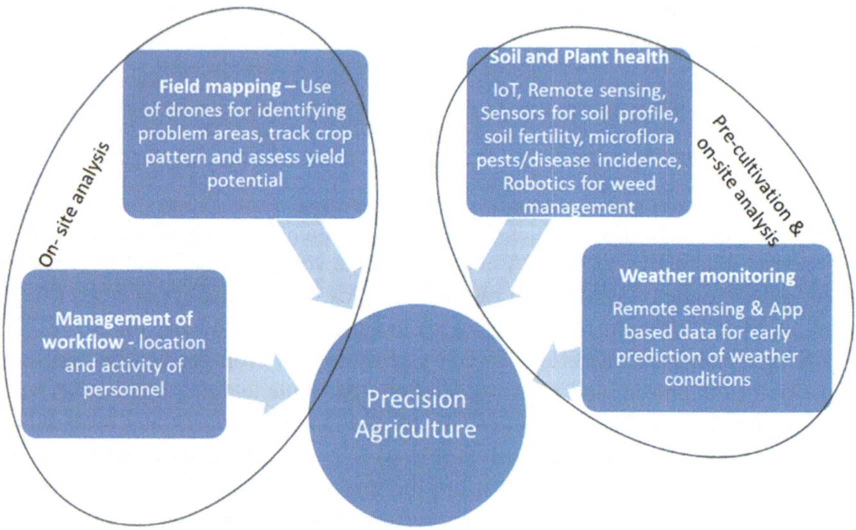

Fig. 11 Diverse applications of precision agriculture

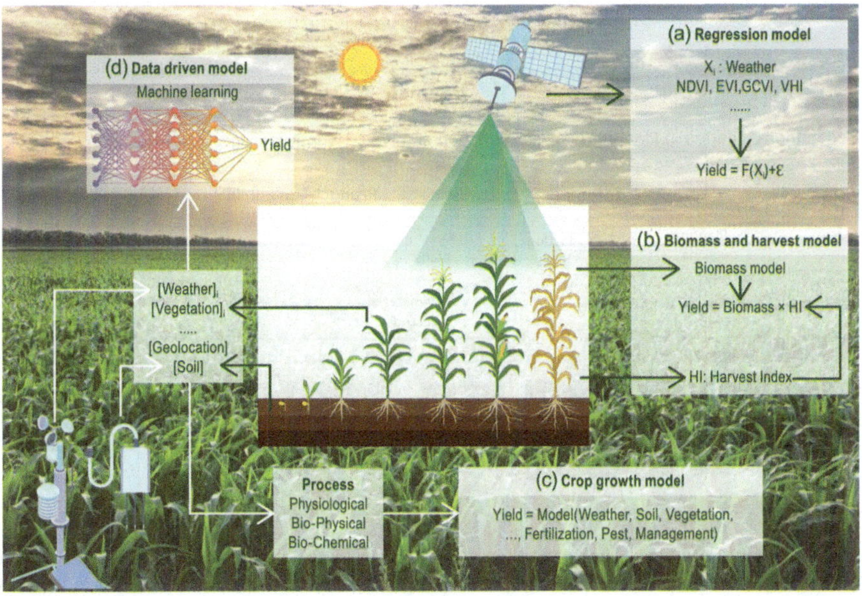

Fig. 12 Crop yield prediction methods. (**a**) regression method; (**b**) biomass and harvest index; (**c**) crop growth model; (**d**) data-driven models. (Courtesy: Wu et al. 2022; doi: https://doi.org/10.1093/nsr/nwac290)

robotics, is referred to as framework of digital agriculture. The components of the digital agriculture framework are as follows:

- Basic information databases pertaining to agriculture: These databases include essential information about farmland, genetic resources, weather patterns, social and economic contexts, etc. that is pertinent to agricultural activities.
- A method for acquiring data that can be used to update databases and keep track of agricultural activities in real time (or almost real time). This system is made up of digital data collectors that are tasked with collecting information from aerial or satellite-based sensors, above- and below-ground sensors, and data on the weather, plants, and soil.
- Digital network transmission system: This system is a sort of media that enables the distribution of commands and the gathering of data.
- System for central processing in order to control the functioning of digital agricultural machinery, cyber physical system (CPS) assesses all the information amassed and develops feasible judgments using GIS, agricultural models, and expert systems.
- Digitized agricultural machinery (DAM): This category comprises tools for harvesting, seeding, and managing fertilizer and water. As digital agricultural machinery performs CPS commands and returns processing results either directly

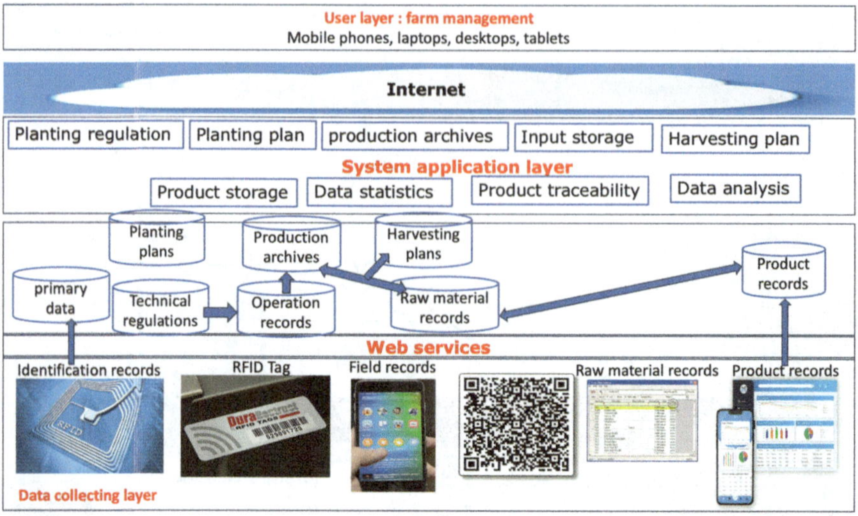

Fig. 13 Framework of digital agriculture (Radio frequency identification (RFID) refers to a wireless system comprised of two components: tags and readers. The reader is a device that has one or more antennas that emit radio waves and receive signals back from the RFID tag)

or through a real-time (quasi real-time) information collecting system, it uses digital networks, GPS, and GIS to assist it (see Rijswijk et al. 2021 for details).

The framework for digital agriculture is shown in Fig. 13. Each component is connected by a common data interface. A computerized agricultural system that uses core information databases to set the planting schedule for a year also monitors crop growth vigor and provides data on soil structure, water content, disease, weather, and other important elements. Digital agriculture technology is used to carry out a series of operations, such as planting, controlling water or fertilizer, harvesting, and sending the data back to CPS. CPS does thorough information analysis before making decisions. The whole analysis' report is then produced by CPS. The interconnected development of each component is underlined in digital agriculture. The foundation for digital agriculture can only be laid when all the parts are perfectly connected and advance at the same time. The phrase "digital agriculture" cannot be used to describe a single element or a group of elements that are developing separately.

References

Abdullahi H, Sheriff R (2017) Case study to investigate the adoption of precision agriculture in Nigeria using simple analysis to determine variability on a maize plantation. J Agric Econ Rural Dev 3(3):279–292

Addisu S, Fininsa C, Bekeko Z (2023) Distribution of chickpea (Cicer arietinum L.) Ascochyta blight (Didymella rabiei) and analyses of factors affecting disease epidemics in Central Ethiopia. Eur J Plant Pathol 2023. https://doi.org/10.1007/s10658-023-02672-5

Ahmed M, Choudhury S, Al-Turjman F (2019) Big data analytics for intelligent internet of things. In: Artificial intelligence in IoT. Springer, pp 107–127

Alipanahi B, Delong A, Weirauch MT, Frey BJ (2015) Predicting the sequence specificities of DNA- and RNA-binding proteins by deep learning. Nat Biotechnol 33:831–838

Alm E et al (2016) Digitizing The Netherlands: how The Netherlands can drive and benefit from an accelerated digitized economy in Europe. Boston Consulting Group

Argo WR, Fisher PR (2002) Understanding pH management for container-grown crops. Meister Publishing, Willoughby

Avgoustaki DD, Xydis G (2020) Indoor vertical farming in the urban nexus context: business growth and resource savings. Sustainability 12:1965. https://doi.org/10.3390/su12051965

Bacco M et al (2019) The digitisation of agriculture: a survey of research activities on smart farming. Array 3–4. https://doi.org/10.1016/j.array.2019.100009

Basso B, Antle J (2020) Digital agriculture to design sustainable agricultural systems. Nat Sustain 3:254–256. https://doi.org/10.1038/s41893-020-0510-0

Beacham AM (2019) Vertical farming: a summary of approaches to growing skywards. J Hortic Sci Biotechnol 94:277–283. https://doi.org/10.1080/14620316.2019.1574214

Bégué A, Arvor D, Bellon B, Betbeder J, De Abelleyra DPD, Ferraz R, Lebourgeois V, Lelong C, Simões MR, Verón S (2018) Remote sensing and cropping practices: a review. Remote Sens 10(1):99. https://doi.org/10.3390/rs10010099

Benis K, Reinhart C, Ferrão P (2017) Development of a simulation-based decision support workflow for the implementation of Building-Integrated Agriculture (BIA) in urban contexts. J Clean Prod 147:589–602. https://doi.org/10.1016/j.jclepro.2017.01.130

Bertoglio R et al (2021) The digital agricultural revolution: a bibliometric analysis literature review digital object identifier. https://doi.org/10.1109/ACCESS.2021.3115258

Biamonte J, Wittek P, Pancotti N, Rebentrost P, Wiebe N, Lloyd S (2017) Quantum machine learning. Nature 549:195–202

Botta A, de Donato W, Persico V, Pescapé A (2016) Integration of Cloud computing and internet of things: a survey. Futur Gener Comput Syst 56:684–700

Bramley H, Ranawana SRWMCJK, Palta JA, Stefanova K, Siddique KHM (2022) Transpirational leaf cooling effect did not contribute equally to biomass retention in wheat genotypes under high temperature. Plan Theory 11:2174. https://doi.org/10.3390/plants11162174

Carmassi G, Incrocci L, Maggini R, Malorgio F, Tognoni F, Pardossi A (2007) An aggregated model for water requirements of greenhouse tomato grown in closed rockwool culture with saline water. Agric Water Manag 88:73–82

Chakraborty S, Newton AC (2011) Climate change, plant diseases and food security: an overview. Plant Pathol 60:2–14. https://doi.org/10.4060/cb4769en

Chehri A, Chaibi H, Saadane R, Hakem N, Wahbi M (2020) A framework of optimizing the deployment of IoT for precision agriculture industry. Proc Comput Sci 176:2414–2422

Chugh G, Siddique KHM, Solaiman ZM (2021) Nanobiotechnology for agriculture: smart technology for combating nutrient deficiencies with nanotoxicity challenges. Sustainability 13:1781. https://doi.org/10.3390/su13041781

Cisternas I, Velásquez I, Caro A, Rodríguez A (2020) Systematic literature review of implementations of precision agriculture. Comput Electron Agric 176(2020):105626. https://doi.org/10.1016/j.compag.2020.105626

Cohen AR, Chen G, Berger EM, Warrier S, Lan G, Grubert E, Dellaert F, Chen Y (2022) Dynamically controlled environment agriculture: integrating machine learning and mechanistic and physiological models for sustainable food cultivation. ACS ES &T Eng 3:3–19. https://doi.org/10.1021/acsestengg.1c00269

Čolaković A, Hadžialić M (2018) Internet of Things (IoT): a review of enabling technologies, challenges, and open research issues. Comput Netw 144:17–39

Cooper MC, Ellram LM (1993) Characteristics of supply chain management and the implications for purchasing and logistics strategy. Int J Logist Manag 4(2):13–24
Cordell D, Neset TSS, Prior T (2012) The phosphorus mass balance: identifying "hotspots" in the food system as a roadmap to phosphorus security. Curr Opin Biotechnol 23:839–845
Crane-Droesch A (2018) Machine learning methods for crop yield prediction and climate change impact assessment in agriculture. Environ Res Lett 13:114003
da Silveira F, Amaral FG (2022) Agriculture 4.0. In: Zhang Q (ed) Encyclopedia of smart agriculture technologies. Springer, Cham. https://doi.org/10.1007/978-3-030-89123-7_207-1
Dawson CJ, Hilton J (2011) Fertiliser availability in a resource-limited world: production and recycling of nitrogen and phosphorus. Food Policy 36:S14–S22
Delgado C, Benitez H, Cruz M, Selvaraj M (2019) Digital disease phenotyping. In: Proceedings of the IGARSS 2019 – 2019 IEEE international geoscience and remote sensing symposium, New York, pp 5702–5705
Deng X, Dou Y, Hu D (2018) Robust closed-loop control of vegetable production in plant factory. Comput Electron Agric 155:244–250
Djidonou D, Leskovar DI (2019) Seasonal changes in growth, nitrogen nutrition, and yield of hydroponic lettuce. Hort Sci 54:76–85. https://doi.org/10.21273/hortsci13567-18
Do H, Than K, Larmande P (2018) Evaluating named-entity recognition approaches in plant molecular biology. https://doi.org/10.1101/360966
Eaves J, Eaves S (2018) Comparing the profitability of a greenhouse to a vertical farm in Quebec. Can J Agric Econ 66:43–54. https://doi.org/10.1111/cjag.12161
Ebi KL, Vanos J, Baldwin JW, Bell JE, Hondula DM, Errett NA, Hayes K, Reid CE, Saha S, Spector J, Berry P (2021) Extreme weather and climate change: population health and health system implications. Annu Rev Public Health 42:293–315. https://doi.org/10.1146/annurev-publhealth-012420-105026
Edwards-Murphy F, Magno M, Whelan PM, O'Halloran J, Popovici EM (2016) b+ WSN: Smart beehive with preliminary decision tree analysis for agriculture and honey bee health monitoring. Comput Electron Agric 124:211–219
Elijah O, Rahman TA, Orikumhi I, Leow CY, Hindia MN (2018) An overview of internet of things (IoT) and data analytics in agriculture: benefits and challenges. IEEE Internet Things J 5:3758–3773
Engler N, Krarti M (2021) Review of energy efficiency in controlled environment agriculture. Renew Sust Energ Rev 141(2021):110786. https://doi.org/10.1016/j.rser.2021.110786
Esposito S, Carputo D, Cardi T, Tripodi P (2020) Applications and trends of machine learning in genomics and phenomics for next-generation breeding. Plan Theory 9:34
FAO (2011) Looking Ahead in World Food and Agriculture: Perspectives to:2050
FAO (2017) Information and communication technology (ICT) in agriculture: a report to the G20 agricultural deputies. FAO, Rome
FAO (2018) The state of food security and nutrition in the world: building resilience for peace and food security. FAO, Rome. http://www.fao.org/news/story/en/item/1152031/icode/
FAO (2019a) The state of the world's biodiversity for food and agriculture. In: Bélanger J, Pilling D (eds) FAO commission on genetic resources for food and agriculture assessments. FAO, Rome, p 572
FAO (2019b) FAO's work on climate change. The Food and Agriculture Organization of the United Nations (FAO), Rome
FAO (2021) The impact of disasters and crises on agriculture and food security. FAO, Rome. https://doi.org/10.4060/cb3673en
Ferguson JM, Keys RD, McLaughlin FW, Warren JM (1991) Seed and seed quality. AG-North Carolina Agricultural Extension Service, North Carolina State University (USA)
Ferentinos KP, Katsoulas N, Tzounis A, Bartzanas T, Kittas C (2017) Wireless sensor networks for greenhouse climate and plant condition assessment. Biosyst Eng 153:70–81
Ferrández-Pastor FJ, García-Chamizo JM, Nieto-Hidalgo M, Mora-Pascual J, Mora-Martínez J (2016) Developing ubiquitous sensor network platform using internet of things: application in precision agriculture. Sensors 16:1141

Fritz M, Schiefer G (2008) Food chain management for sustainable food system development: a European research agenda. Agribusiness 24(4):440–452

Gao D, Sun Q, Hu B, Zhang S (2020) A framework for agricultural pest and disease monitoring based on internet-of-things and unmanned aerial vehicles. Sensors 20(5):1487. https://doi.org/10.3390/s20051487

Ghosal S, Blystone D, Singh AK, Ganapathysubramanian B, Singh A, Sarkar S (2018) An explainable deep machine vision framework for plant stress phenotyping. Proc Natl Acad Sci U S A 115:4613–4618

Gibson T (2018) Room to grow. ASEE Prism 27:26–31. https://doi.org/10.2307/26820026

Goedde L, Katz J, Menard A, Revellat J (2020) Agriculture's connected future: how technology can yield new growth. McKinsey Global Publishing, p 10

Gold KM (2021) Plant disease sensing: studying plant-pathogen interactions at scale. mSystems 6:e01228–e01221

Gold KM, Townsend PA, Herrmann I, Gevens AJ (2020) Investigating potato late blight physiological differences across potato cultivars with spectroscopy and machine learning. Plant Sci 295:110316

Gómez C et al (2019) Controlled environment food production for urban agriculture. Hort Sci 54:1448–1458. https://doi.org/10.21273/HORTSCI14073-19

Graamans L, Baeza E, van den Dobbelsteen A, Tsafaras I, Stanghellini C (2018) Plant factories versus greenhouses: comparison of resource use efficiency. Agric Syst 2018(160):31–43. https://doi.org/10.1016/j.agsy.2017.11.003

Greenside P, Shimko T, Fordyce P, Kundaje A (2018) Discovering epistatic feature interactions from neural network models of regulatory DNA sequences. Bioinformatics 34:i629–i637

Gustavsson J, Cederberg C, Sonesson U, Van Otterdijk R, Meybeck A (2011) Global food losses and food waste. Food and Agriculture Organization of the United Nations, Rome

Harfouche AL, Jacobson DA, Kainer D, Romero JC, Harfouche AH, Mugnozza GS, Moshelion M, Tuskan GA, Keurentjes JJB, Altman A (2019) Accelerating climate resilient plant breeding by applying next-generation artificial intelligence. Trends Biotechnol 37:1217–1235

Hashem IAT, Yaqoob I, Anuar NB, Mokhtar S, Gani A, Ullah Khan S (2015) The rise of "big data" on cloud computing: review and open research issues. Inf Syst 47:98–115

Hatfield JL, Prueger JH (2015) Temperature extremes: effect on plant growth and development. Weather Clim Extremes 10:4–10

Hatfield JL, Boote KJ, Kimball BA, Ziska LH, Izaurralde RC, Ort D, Thomson AM, Wolfe DW (2011) Climate impacts on agriculture: implications for crop production. Agron J 103:351–370. https://doi.org/10.2134/agronj2010.0303

Himesh S, Rao EVSP, Gouda KC et al (2018) Digital revolution and big data: a new revolution in agriculture. CABI Rev. https://doi.org/10.1079/PAVSNNR201813021

Hodges CN, Groh JE, Johnson AW (1968) Controlled-environment agriculture for coastal desert areas. Proc Natl Agric Plast Conf 8:58–68

Huang, Y., Li, J., Qi, Y., and Shi, V. 2021. Predicting the impacts of the COVID-19 pandemic on food supply chains and their sustainability: a simulation study. Discret Dyn Nat Soc 2021, 7109432, 9 doi: https://doi.org/10.1155/2021/7109432

Iddio E, Wang L, Thomas Y, McMorrow G, Denzer A (2019) Energy efficient operation and modeling for greenhouses: a literature review. Renew Sust Energ Rev 117. https://doi.org/10.1016/j.rser.2019.109480

ILOSTAT (2019) Employment database. International Labour Organization, Geneva. Data retrieved May 2019

Ingram J, Maye D (2020) What are the implications of digitalisation for agricultural knowledge? Front Sustain Food Syst 4:66. https://doi.org/10.3389/fsufs.2020.00066

IPCC (2018) Annex IV: Expert Reviewers of the IPCC Special Report on Global Warming of 1.5 °C. In V. Masson-Delmotte, P. Zhai, H.-O. Pörtner, D. Roberts, J. Skea, P. R. Shukla, A. Pirani, W. Moufouma-Okia, C. Péan, R. Pidcock, S. Connors, J. B. R. Matthews, Y. Chen, X. Zhou, M. I. Gomis, E. Lonnoy, T. Maycock, M. Tignor, & T. Waterfield (Eds.), Global Warming of 1.5 °C. An IPCC Special Report on the Impacts of Global Warming of

1.5 °C above Pre-Industrial Levels and Related Global Greenhouse Gas Emission Pathways, in the Context of Strengthening the Global Response to the Threat of Climate Change, Sustainable Development, and Efforts to Eradicate Poverty (51 p.). Geneva, Switzerland: World Meteorological Organization.

Javaid M, Haleem A, Singh RP, Suman R (2022) Enhancing smart farming through the applications of agriculture 4.0 technologies. Int J Intell Netw 3:150–164

Jiao Y, Fu J, Hou W, Shi Z, Guo Y, Xia S, Yang Q, Li F (2018) Homogeneous electrochemical aptasensor based on a dual amplification strategy for sensitive detection of profenofos residues. New J Chem 42:14642–14647

Jones JB Jr (2014) Complete guide for growing plants hydroponically. CRC Press, Boca Raton

Jung J, Maeda M, Chang A, Bhandari M, Ashapure A, Landivar-Bowles J (2021) The potential of remote sensing and artificial intelligence as tools to improve the resilience of agriculture production systems. Curr Opin Biotechnol 70:15–22. https://doi.org/10.1016/j.copbio.2020.09.003

Kamilaris A, Prenafeta-Boldú FX (2018) Deep learning in agriculture: a survey. Comput Electron Agric 147:70–90

Kamilaris A, Kartakoullis A, Prenafeta-Boldú FX (2017) A review on the practice of big data analysis in agriculture. Comput Electron Agric 143:23–37

Katsoulas N, Bartzanas T, Kittas C (2017) Online professional irrigation scheduling system for greenhouse crops. Acta Hortic 1154:221–228

Khan MHU, Wang S, Wang J, Ahmar S, Saeed S, Khan SU, Xu X, Chen H, Bhat JA, Feng X (2022) Applications of artificial intelligence in climate-resilient smart-crop breeding. Int J Mol Sci 23:11156. https://doi.org/10.3390/ijms231911156

Khanal S, Fulton JP, Shearer S, Ozkan E (2020) Remote sensing in agriculture – accomplishments, limitations, and opportunities. Remote Sens 12:3783. https://doi.org/10.3390/rs12223783

Kim DJ, Ferrin DL, Rao HR (2008) A trust-based consumer decision-making model in electronic commerce: the role of trust, perceived risk, and their antecedents. Decis support syst 44:544–564

Klerkx L, Jakku E, Labarthe P (2019) A review of social science on digital agriculture, smart farming and agriculture 4.0: new contributions and a future research agenda. NJAS Wageningen J Life Sci 90–91:100315. https://doi.org/10.1016/j.njas.2019.100315

Kozai T (2019) Towards sustainable plant factories with artificial lighting (PFALs) for achieving SDGs. Int J Agric Biol Eng 12:28–37

Kozai T, Niu G, Takagaki M (2019) Plant factory: an indoor vertical farming system for efficient quality food production, 2nd edn. Amsterdam University Press, Amsterdam. https://doi.org/10.1016/C2018-0-00969-X

Kreuze et al. (2022) Innovative Digital Technologies to Monitor and Control Pest and Disease Threats in Root, Tuber, and Banana (RT&B) Cropping Systems: Progress and Prospects. In: G. Thiele et al. (eds) Root, Tuber and Banana Food System Innovations. https://doi.org/10.1007/978-3-030-92022-7_9 Springer

Kumar A, Hancke GP (2014) Energy efficient environment monitoring system based on the IEEE 802.15.4 standard for low cost requirements. IEEE Sensors J 14:2557–2566

Kyosuke Y, Wei G, Yosuke Y, Seishi N (2014) On plant detection of intact tomato fruits using image analysis and machine learning methods. Sensors 14:12191–12206

Lai X, Stigliani A, Vachon G, Carles C, Smaczniak C, Zubieta C, Kaufmann K, Parcy F (2019) Building transcription factor binding site models to understand gene regulation in plants. Mol Plant 12:743–763

Lajoie-O'Malley A, Bronson K, van der Burg S, Klerkx L (2020) The future(s) of digital agriculture and sustainable food systems: an analysis of high-level policy documents. Ecosyst Serv 45:101183. https://doi.org/10.1016/j.ecoser.2020.101183

Larsen AE, Patton M, Martin EA (2019) High highs and low lows: elucidating striking seasonal variability in pesticide use and its environmental implications. Sci Total Environ 651:828–837. https://doi.org/10.1016/j.scitotenv.2018.09.206

Lee J, Wang J, Crandall D, Sabanovic S, Fox G (2017) Real-time, cloud-based object detection for unmanned aerial vehicles. In: 2017 first IEEE international conference on robotic computing (IRC). https://doi.org/10.1109/irc.2017.77

Lezoche M, Hernandez JE, Diaz MDEA, Panetto H, Kacprzyk J (2020) Agri-food 4.0: a survey of the supply chains and technologies for the future agriculture. Comput Ind 117:103187

Li Y, Ding Y, Li D, Miao Z (2018a) Automatic carbon di oxide enrichment strategies in the greenhouse: a review. Biosyst Eng 171:101–119. https://doi.org/10.1016/j.biosystemseng.2018.04.018

Li F, Wang X, Sun X, Guo Y, Zhao W (2018b) A dual-signal amplification strategy for kanamycin based on ordered mesoporous carbon-chitosan/gold nanoparticles-streptavidin and ferrocene labelled DNA. Anal Chim Acta 1033:185–192

Li HM, Wang XC, Zhao XF, Qi Y (2021) Understanding systemic risk induced by climate change. Adv Clim Chang Res 12:384–394. https://doi.org/10.1016/j.accre.2021.05.006

Liakos KG, Busato P, Moshou D, Pearson S, Bochtis D (2018) Machine learning in agriculture: a review. Sensors 18(8):2674. https://doi.org/10.3390/s18082674

Liao MS, Chen SF, Chou CY, Chen HY, Yeh SH, Chang YC, Jiang JA (2017) On precisely relating the growth of Phalaenopsis leaves to greenhouse environmental factors by using an IoT-based monitoring system. Comput Electron Agric 136:125–139

Linker R, Kacira M, Arbel A (2011) Robust climate control of a greenhouse equipped with variable-speed fans and a variable-pressure fogging system. Biosyst Eng 110:153–167. https://doi.org/10.1016/j.biosystemseng.2011.07.010

Lioutas ED, Charatsari C, De Rosa M (2021) Digitalization of agriculture: a way to solve the food problem or a trolley dilemma?

Liu P, Li B, Yu X (2018) Intelligent control system of cucumber production in the greenhouse based on internet of things. In: Proceedings of the international conference on cloud computing and security, Haikou, China, 8–10 June 2018, pp 395–406

Lokers R, Knapen R, Janssen S, van Randen Y, Jansen J (2016) Analysis of Big Data technologies for use in agro-environmental science. Environ Model Softw 84:494–504

Lowder SK, Skoet J, Raney T (2016) The number, size and distribution of farms, smallholder farms, and family farms worldwide. World Dev 86:16–29

Ma J, Li X, Wen H, Fu Z, Zhang L (2015) A key frame extraction method for processing greenhouse vegetables production monitoring video. Comput Electron Agric 111:92–102

MacPherson J, Voglhuber-Slavinsky A, Olbrisch M, Schöbel P, Dönitz E, Mouratiadou I, Helming K (2022) Future agricultural systems and the role of digitalization for achieving sustainability goals. A review. Agron Sustain Dev 42:70. https://doi.org/10.1007/s13593-022-00792-6

Madakam S, Uchiya T, Mark S, Lurie Y (2022) Artificial intelligence, machine learning and deep learning (literature: review and metrics). Asia-Pac J Manag Res Innov 18:7–23. https://doi.org/10.1177/2319510X221136682

Manning L, Baines R, Chadd S (2006) Quality assurance models in the food supply chain. Br Food J 108(2):91–104

Marsh JI, Hu H, Gill M, Batley J, Edwards D (2021) Crop breeding for a changing climate: integrating phenomics and genomics with bioinformatics. Theor Appl Genet 134:1677–1690

Marvin HJP, Bouzembrak Y, Fels-Klerx HJ, Kempenaar C, Veerkamp R, Chauhan A et al (2021) Digitalisation and artificial intelligence for sustainable food systems. Trends Food Sci Technol 120:344–348. https://doi.org/10.1016/j.tifs.2022.01.020

Mat I, Kassim MRM, Harun AN, Yusoff IM (2016) IoT in precision agriculture applications using wireless moisture sensor network. In: Proceedings of the 2016 IEEE conference on open systems (ICOS), Langkawi, Malaysia, 10–12 October 2016, pp 24–29

Maurel VB, Brossard L, Garcia F, Mitton N, Termier A (2022) Agriculture and digital technology: getting the most out of digital technology to contribute to the transition to sustainable agriculture and food systems, pp 1–185

Mehrabi Z, McDowell MJ, Ricciardi V et al (2021) The global divide in data-driven farming. Nat Sustain 4:154–160. https://doi.org/10.1038/s41893-020-00631-0

Mejia-Guerra MK, Buckler ES (2019) A k-mer grammar analysis to uncover maize regulatory architecture. BMC Plant Biol 19:103

Mendez J, Molina M, Rodriguez N, Cuellar MP, Morales DP (2021) Camera-LiDAR multi-level sensor fusion for target detection at the network edge. Sensors 21:3992. https://doi.org/10.3390/s21123992

Meyer P, Van de Poel B, De Coninck B (2021) UV-B light and its application potential to reduce disease and pest incidence in crops. Hort Res 8(1):194. https://doi.org/10.1038/s41438-021-00629-5

Michael G, Tay F, Then Y (2021) Development of automated monitoring system for hydroponics vertical farming. J Phys Conf 1844:012024

Misra NN, Dixit Y, Al-Mallahi A, Bhullar MS, Upadhyay R, Martynenko A (2022) IoT, big data, and artificial intelligence in agriculture and food industry. IEEE Internet Things J 9:6305–6324. https://doi.org/10.1109/JIOT.2020.2998584

Miyagi A, Uchimiya H, Kawai-Yamada M (2017) Synergistic effects of light quality, carbon dioxide and nutrients on metabolite compositions of head lettuce under artificial growth conditions mimicking a plant factory. Food Chem 218:561–568. https://doi.org/10.1016/j.foodchem.2016.09.102

Monteiro J, Barata J, Veloso M, Veloso L, Nunes J (2018) Towards sustainable digital twins for vertical farming. In: Proceedings of the 2018 thirteenth international conference on digital information management (ICDIM), Berlin, Germany, 24–26 September 2018, pp 234–239

Montesinos-López OA, Martín-Vallejo J, Crossa J, Gianola D, Hernández-Suárez CM, Montesinos-López A, Juliana P, Singh R (2019) A benchmarking between deep learning, support vector machine and Bayesian threshold best linear unbiased prediction for predicting ordinal traits in plant breeding. G3 Genes Genomes Genet 9:601–618

Namuduri S, Narayanan BN, Davuluru VSP, Burton L, Bhansali S (2020) Review – deep learning methods for sensor based predictive maintenance and future perspectives for electrochemical sensors. J Electrochem Soc 167. https://doi.org/10.1149/1945-7111/ab67a8

Nativi S, Mazzetti P, Santoro M, Papeschi F, Craglia M, Ochiai O (2015) Big data challenges in building the global earth observation system of systems. Environ Model Softw 68:1–26

Negara C, Vieth K-U, Lafontaine M, Freund M (2014) Automatic fruit sorting by non-destructive determination of quality parameters using visible-near infrared to improve wine quality: II. Regression analysis. NIR News 25:4–6

Nelson PV (2012) Greenhouse operation and management, 7th edn. Prentice Hall, Upper Saddle River

Oberti R, Marchi M, Tirelli P, Calcante A, Iriti M et al (2016) Selective spraying of grapevines for disease control using a modular agricultural robot. Biosyst Eng 146:203–215

Ojha T, Misra S, Raghuwanshi NS (2015) Wireless sensor networks for agriculture: the state-of-the-art in practice and future challenges. Comput Electron Agric 118:66–84

Ojo MO, Zahid A (2022) Deep learning in controlled environment agriculture: a review of recent advancements, challenges and prospects. Sensors 22:7965. https://doi.org/10.3390/s22207965

Onofre RB, Gadoury DM, Stensvand A, Bierma A, Rea MS, Peres NA (2021) Use of ultraviolet light to suppress powdery mildew in strawberry fruit production fields. Plant Dis 105:2402–2409

Panpatte DG (2018) Artificial intelligence in agriculture: An emerging era of research. Intutional Science, Canada, pp 1–8

Park YS, Egilmez G, Kucukvar M (2016) Energy and end-point impact assessment of agricultural and food production in the United States: a supply chain-linked ecologically-based life cycle assessment. Ecol Indic 62:117–137

Passioura JB, Angus JF (2010) Improving productivity of crops in water-limited environments. Adv Agron 2010(162):37–75. https://doi.org/10.1016/S0065-2113(10)06002-5

Pathak R, Singh SK, Tak A, Gehlot P (2018) Impact of climate change on host, pathogen and plant disease adaptation regime: a review. Biosci Biotech Res Asia 15(3). https://doi.org/10.13005/bbra/2658

Pauschinger D, Klauser FR (2022) The introduction of digital technologies into agriculture: space, materiality and the public–private interacting forms of authority and expertise. J Rural Stud 91:217–227
Pixia D, Xiangdong W (2013) Recognition of greenhouse cucumber disease based on image processing technology. Open J Appl Sci 3:27
Priya R, Ramesh D (2019) ML based sustainable precision agriculture: a future generation perspective. Sustain Comput Inf Syst 28:100439. https://doi.org/10.1016/j.suscom.2020.100439
Protopop I, Shanoyan A (2016) Big data and smallholder farmers: big data applications in the agri-food supply chain in developing countries. Int Food Agribus Manag Rev 19. https://doi.org/10.22004/ag.econ.240705
Qin Q, Feng J (2017) Imputation for transcription factor binding predictions based on deep learning. PLoS Comput Biol 13:e1005403
Quy VK, Hau NV, Anh DV, Quy NM, Ban NT, Lanza S, Randazzo G, Muzirafuti A (2022) IoT-enabled smart agriculture: architecture, applications, and challenges. Appl Sci 12:3396. https:/ /doi.org/10.3390/app12073396
Radoglou-Grammatikis P, Sarigiannidis P, Lagkas T, Moscholios I (2020) A compilation of UAV applications for precision agriculture. Comput Netw 172:107148. https://doi.org/10.1016/j.comnet.2020.107148
Rajalakshmi P, Mahalakshmi SD (2016) IOT based crop-field monitoring and irrigation automation. In: Proceedings of the 2016 10th international conference on intelligent systems and control (ISCO), Coimbatore, India, 7–8 January 2016, pp 1–6
Raviv M, Lieth JH, Bar-Tal A (2019) Soilless culture: theory and practice. Elsevier, San Diego
Reinoso-Peláez EL, Gianola D, González-Recio O (2022) Genome-enabled prediction methods based on machine learning. In: Ahmadi N, Bartholomé J (eds) Genomic prediction of complex traits; methods in molecular biology. Humana, New York
Revanth (2019) Toward future farming: how artificial intelligence is transforming the agriculture industry. WIPRO. https://www.wipro.com/holmes/towards-future-farming-how-artificial-intelligence-is-transforming-the-agriculture-industry/
Revathi S, Sivakumaran N (2016) Fuzzy based temperature control of greenhouse. IFAC-PapersOnLine 49:549–554
Rhee J, Im J (2017) Meteorological drought forecasting for ungauged areas based on machine learning: using long-range climate forecast and remote sensing data. Agric For Meteorol 237–238:105–122
Rijswijk K, Klerkx L, Bacco M, Bartolini F, Bulten E, Debruyne L, Dessein J, Scotti I, Brunori G (2021) Digital transformation of agriculture and rural areas: a socio-cyber-physical system framework to support responsibilisation. J Rural Stud 85:79–90. https://doi.org/10.1016/j.jrurstud.2021.05.003
Rose DC, Chilvers J (2018) Agriculture 4.0: broadening responsible innovation in an era of smart farming. Front Sustain Food Syst 2:87. https://doi.org/10.3389/fsufs.2018.00087
Rotz S et al (2019) The politics of digital agricultural technologies: a preliminary review. Sociol Ruralis 59:203–229
Ruscio F, Paoletti P, Thomas J, Myers P, Fichera S (2019) Low-cost monitoring system for hydroponic urban vertical farms. Int J Agric Biosyst Eng 13:267–271
Russell SJ, Norvig P (2018) Artificial intelligence: a modern approach. Pearson Education Limited, New York
Rutten CJ et al (2013) Invited review: sensors to support health management on dairy farms. J Dairy Sci 96:1928–1952
Ryan M, van der Burg S, Bogaardt M-J (2021) Identifying key ethical debates for autonomous robots in agri-food: a research agenda. AI and Ethics 1−15. https://doi.org/10.1007/s43681-021-00104-w
Satyanarayanan M (2017) The emergence of edge computing. Computer 50:30–39
Schultz M, Betancourt C, Gong B, Kleinert F, Langguth M (2021) Can deep learning beat numerical weather prediction? Philos Trans R Soc A 379:20200097

Selvaraj MG, Vergara A, Ruiz H, Safari N, Elayabalan S (2019) AI-powered banana diseases and pest detection. Plant Methods 15:92
Selvaraj MG et al (2020) Detection of banana plants and their major diseases through aerial images and machine learning methods: a case study in DR Congo and Republic of Benin. ISPRS J Photogramm Remote Sens 169:110–124
Shepherd M, Turner JA, Small B, Wheeler D (2018) Priorities for science to overcome hurdles thwarting the full promise of the 'digital agriculture' revolution. J Sci Food Agric 100(14):5083–5092. https://doi.org/10.1002/jsfa.9346
Shi X, An X, Zhao Q, Liu H, Xia L, Sun X, Guo Y (2019a) State-of-the-art internet of things in protected agriculture. Sensors 19:1833. https://doi.org/10.3390/s19081833
Shi W, Cao J, Zhang Q, Li Y, Xu L (2019b) Edge computing: vision and challenges. IEEE Internet Things 3:637–646
Silberbush M, Ben-Asher J, Ephrath JEE (2005) Model for nutrient and water flow and their uptake by plants grown in a soilless culture. Plant Soil 271:309–319
Singh P, Pandey PC, Petropoulos GP, Pavlides A, Srivastava PK, Koutsias N, Deng KAK, Bao Y (2020) Hyperspectral remote sensing in precision agriculture: present status, challenges, and future trends. In: Pandey PC, Srivastava PK, Balzter H, Bhattacharya B, Petropoulos GP (eds) Earth observation-hyperspectral remote sensing. Elsevier, pp 121–146
Singh BK, Delgado-Baquerizo M, Egidi E (2023) Climate change impacts on plant pathogens, food security and paths forward. Nat Rev Microbiol. https://doi.org/10.1038/s41579-023-00900-7
Specht K et al (2013) Urban agriculture of the future: an overview of sustainability aspects of food production in and on buildings. Agric Hum Value 31:33–51. https://doi.org/10.1007/s10460-013-9448-4
Springmann M et al (2018) Options for keeping the food system within environmental limits. Nature 562:519–525. https://doi.org/10.1038/s41586-018-0594-0
Srbinovska M, Gavrovski C, Dimcev V, Krkoleva A, Borozan V (2015) Environmental parameters monitoring in precision agriculture using wireless sensor networks. J Clean Prod 88:297–307
Sreekantha DK, Kavya AM (2017) Agricultural crop monitoring using IOT – a study. In: Proceedings of the 2017 11th international conference on intelligent systems and control (ISCO), Coimbatore, India, 5–6 January 2017, pp 134–139
Talaviya TT, Shah D, Patel N, Yagnik H, Shah M (2020) Implementation of artificial intelligence in agriculture for optimisation of irrigation and application of pesticides and herbicides. Artif Intell Agric 4:58–73. https://doi.org/10.1016/j.aiia.2020.04.002
Tomic T, Schmid K, Lutz P, Domel A, Kassecker M, Mair E, Grixa IL, Ruess F, Suppa M, Burschka D (2012) Toward a fully autonomous UAV: research platform for indoor and outdoor urban search and rescue. IEEE Robot Automat Mag 19:46–56. https://doi.org/10.1109/mra.2012.2206473
Tong H, Nikoloski Z (2021) Machine learning approaches for crop improvement: leveraging phenotypic and genotypic big data. J Plant Physiol 257:153354
Tong Z et al (2016) A case study of air quality above an urban roof top vegetable farm. Environ Pollut 208:256–260. https://doi.org/10.1016/j.envpol.2015.07.006
Tran T-T, Choi J-W, Le T-T, Kim J-W (2019) A comparative study of deep CNN in forecasting and classifying the macronutrient deficiencies on development of tomato plant. Appl Sci 9:1601
Trendov NM, Varas S, Zeng M (2019) Digital technologies in agriculture and rural areas. FAO Publication
UNESCO (2019) Rapport mondial des Nations Unies sur la mise en valeur des ressources en eau 2019: ne laisser personne pour compte. Disponible le 20-04-2021 sur https://unesdoc.unesco.org/ark:/48223/pf0000367303_fre.locale=fr
van Delden SH et al (2021) Current status and future challenges in implementing and upscaling vertical farming systems. Nat Food 2:944–956. https://doi.org/10.1038/s43016-021-00402-w
van Dijk ADJ, Kootstra G, Kruijer W, de Ridder D (2021) Machine learning in plant science and plant breeding. iScience 24:2021. https://doi.org/10.1016/j.isci.2020.101890

Van Ginkel SW, Igou T, Chen Y (2017) Energy, water and nutrient impacts of California-grown vegetables compared to controlled environmental agriculture Systems in Atlanta, GA. Resour Conserv Recycl 122:319–325

Vanhove W, Van Damme P, Meert N (2011) Factors determining yield and quality of illicit indoor cannabis (Cannabis spp.) production. Forensic Sci Int 212:158–163. https://doi.org/10.1016/j.forsciint.2011.06.006

Velásquez AC, Castroverde CDM, He SY (2018) Plant-pathogen warfare under changing climate conditions. Curr Biol 28(10):R619–R634. https://doi.org/10.1016/j.cub.2018.03.054

Wang M, Tai C, Weinan E, Wei L (2018) DeFine: deep convolutional neural networks accurately quantify intensities of transcription factor-DNA binding and facilitate evaluation of functional non-coding variants. Nucleic Acids Res 2018(46):e69

Wang H, Cimen E, Singh N, Buckler E (2020) Deep learning for plant genomics and crop improvement. Curr Opin Plant Biol 54:34–41

Washburn JD, Mejia-Guerra MK, Ramstein G, Kremling KA, Valluru R, Buckler ES, Wang H (2019) Evolutionarily informed deep learning methods for predicting relative transcript abundance from DNA sequence. Proc Natl Acad Sci U S A 116:5542–5549

Wijerathna-Yapa A, Pathirana R (2022) Sustainable agro-food systems for addressing climate change and food security. Agriculture 12:1554. https://doi.org/10.3390/agriculture12101554

Wolfert S et al (2017) Big data in smart farming – a review. Agric Syst 153:69–80

Wu B, Zhang M, Zeng H, Tian F, Potgieter AB, Qin X, Yan N, Chang S, Zhao Y, Dong Q, Boken V, Plotnikov D, Guo H, Wu F, Zhao H, Deronde B, Tits L, Loupian E (2022) Challenges and opportunities in remote sensing-based crop monitoring: a review. Natl Sci Rev 10(4):nwac290. https://doi.org/10.1093/nsr/nwac290

Xu C, Jackson SA (2019) Machine learning and complex biological data. Genome Biol 20:76

Yan J, Kurgan L (2017) DRNApred, fast sequence-based method that accurately predicts and discriminates DNA- and RNA-binding residues. Nucleic Acids Res 45:e84

Zamora-Izquierdo MA, Santa J, Martínez JA, Martínez V, Skarmeta AF (2019) Smart farming IoT platform based on edge and cloud computing. Biosyst Eng 177:4–17

Zampieri G, Vijayakumar S, Yaneske E, Angione C (2019) Machine and deep learning meet genome-scale metabolic modeling. PLoS Comput Biol 15:e1007084

Zhang Z et al (2005) Remote sensing and spatial statistical analysis to predict the distribution of Oncomelania hupensis in the marshlands of China. Acta Tropica 96:205–212. https://doi.org/10.1016/j.actatropica.2005.07.027

Zheng Y, Liu Z, Jing Y, Jie L, Zhan H (2015) An acetylcholinesterase biosensor based on ionic liquid functionalized graphene–gelatin-modified electrode for sensitive detection of pesticides. Sens Actuators B Chem 210:389–397

Zhong R, Xu X, Wang L (2017) Food supply chain management: systems, implementations, and future research. Ind Manag Data Syst 117:2085–2114. https://doi.org/10.1108/IMDS-09-2016-0391

Zhou J, Troyanskaya OG (2015) Predicting effects of noncoding variants with deep learning-based sequence model. Nat Methods 12:931–934

Zhou K, Fu C, Yang S (2016) Big data driven smart energy management: from big data to big insights. Renew Sust Energ Rev 56:215–225

Zhuang Y et al (2022) Economies of scale in constructing plant factories with artificial lighting and the economic viability of crop production. Front Plant Sci 13:992194. https://doi.org/10.3389/fpls.2022.992194

Agriculture and Food Security in the Era of Climate Change

Karla Gasparini, Diego Dias Rafael, Lázaro Eustáquio Pereira Peres, Dimas Mendes Ribeiro, and Agustin Zsögön

Abstract Climate change is an ongoing threat worldwide, concerning food security in developing countries but also affecting crop productivity even in well-developed regions. These continuous changes in the climate have a multidimensional and complex impact on food availability and population health, leading to an urge for a science-based approach that can simultaneously take advantage of the new imposed environmental conditions for food productivity and security.

In this context, elevated atmospheric CO_2 (eCO_2) arises as a flagship in climate change conditions, and despite showing a positive influence on the photosynthesis rate of many C_3 species, the C_4 species response is relatively small, also occasioning a decrease in proteins, vitamins, and micronutrients content in both metabolisms under certain conditions, reducing nutritional quality. Temperature oscillation also influences crop productivity with complex interactions through ambient CO_2 concentration, water availability, and nutrient availability. In the concern of temperature, high day temperature (HDT) and high night temperature (HNT) affect productivity in different ways, making it detrimental to understand how and which crops are affected by each or both temperature variations and in which developmental stage crops are most affected. Furthermore, crop improvement and smart land management are crucial to alleviate the ubiquitous climate change events.

Keywords Carbon dioxide · Food security · High temperature · Uncertain climate

---

K. Gasparini · L. E. P. Peres
Departamento de Ciências Biológicas, Escola Superior de Agricultura "Luiz de Queiroz", Universidade de São Paulo, Piracicaba, SP, Brazil
e-mail: lazaro.peres@usp.br

D. D. Rafael · D. M. Ribeiro · A. Zsögön (✉)
Departamento de Biologia Vegetal, Universidade Federal de Viçosa, Vicosa, MG, Brazil
e-mail: dimas.ribeiro@ufv.br; agustin.zsogon@ufv.br

P. M. Priyadarshan et al. (eds.), *Digital Agriculture*,
https://doi.org/10.1007/978-3-031-43548-5_2

1 Introduction

The transition to agriculture and sedentary food production is closely linked to climate events. The switch from gathering food in the wild to farming was probably triggered by climate constraints. This long period during the Pleistocene was characterized by progressively colder and dryer weather, marked by extreme climate events (Alley 2000). The concomitant origin of sedentary, farming societies in as many as ten geographically independent areas of the world coincides with the start of the currently ongoing interglacial cycle (the Holocene) around 12,000 years ago. The onset of relatively warmer temperatures and the increase in local rainfall likely played a role in the appearance of the first agricultural societies (Ferrio et al. 2011). Since then, stable climate has been the norm, and a new glacial cycle is not expected for the next 50,000 years (Ganopolski et al. 2016). Notably, climate anomalies like the Iron Age Cold Epoch (900–300 BCE), the Roman Warm Period (250 BCE to 400 CE), and the Little Ice Age (1550–1700 CE) led to disruptions in food supply and alterations in demographic trends (Bevan et al. 2017). However, these events were localized phenomena, in contrast to the unprecedented global increase in temperature starting in the early twentieth century (Neukom et al. 2019). This novel climatic pattern threatens the sustainable intensification of agriculture required to support the growing population in the coming decades.

The impact of climate change on extant crops could be compared to the novel conditions experienced by early crops when radiating from their respective centers of origin (Fig. 1). There are many examples, of which soybean (*Glycine max*) is probably one of the most representative, where the latitudinal range of a crop

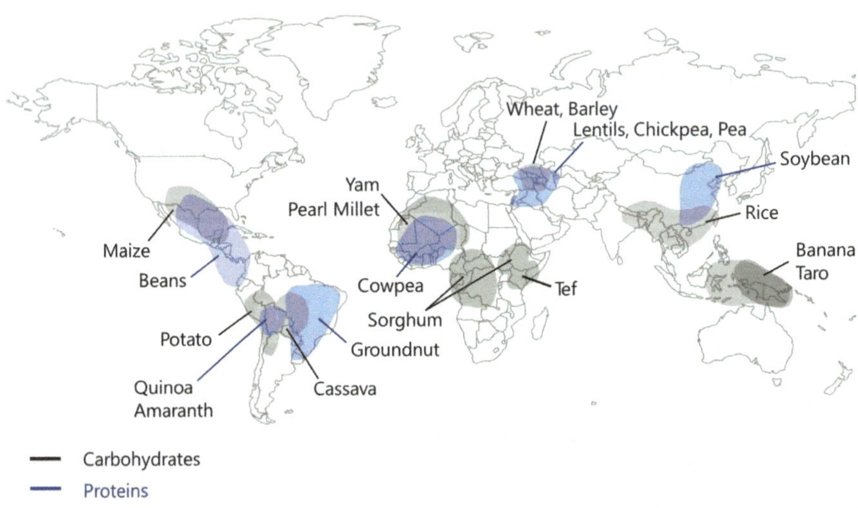

Fig. 1 Centers of origin of crops responsible for the main sources of protein and carbohydrate worldwide

has been expanded via genetic changes in photoperiodic response, duration of growth cycle, and time to maturity. Crops of the Compositae family like lettuce (*Lactuca sativa*) were also selected to avoid precocious flowering in tropical environments. However, the classic example in this regard is the transfer of potato (*Solanum tuberosum*) to Europe from South America, which was selected to initiate tuberization in long-day conditions, as opposed to its natural short-day tuberization response. Similarly, another native South American species, the tomato (*Solanum lycopersicum*), suffered a profound alteration in its circadian clock machinery and a reduction of heterostyly to adapt to the more extreme oscillations in photoperiod and the lack of natural pollinators in its new environment in Europe, respectively.

2 Elevated CO_2 and Its Impacts on Food Security

As the world population continues to increase, crop production must intensify proportionally to ensure food security in the coming decades, while remaining sustainable by reducing its environmental impact. However, climate change poses a serious challenge to achieve these goals (Giller et al. 2021). Crop growth and yield depends on a combination of factors such as plant genotype, temperature, precipitation, sunlight, nutrient availability, and atmospheric CO_2 concentration (Sharon and Siobhan 2016). In this context, elevated atmospheric CO_2 (eCO_2) has the potential to positively alter the rate of photosynthesis for many C3 species, which may lead to increased growth and crop yield (Dong et al. 2019; Poorter et al. 2022) (Fig. 2). On the other hand, the response of C4 crops such as maize (*Zea mays*) and sorghum (*Sorghum bicolor*) to eCO_2 exposure is expected to be relatively small compared to C3 crops like rice (*Oryza* spp.) and wheat (*Triticum aestivum*)

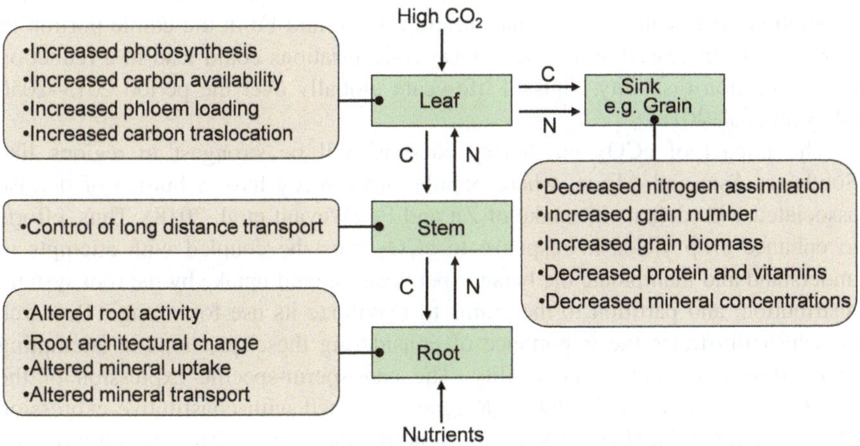

Fig. 2 Schematic model summarizing the effects of high CO_2 on crop physiology and yield

(Leakey et al. 2006). Despite improving C3 crop yields, eCO_2 in isolation (i.e., without concomitant alterations in air relative humidity or temperature) decreases the concentrations of protein, vitamins, and micronutrients essential for humans in edible parts of crops, negatively impacting food security (Myers et al. 2014; Zhu et al. 2018) (Fig. 2). Such losses in nutritional quality represent an extra challenge for agriculture to provide enough nutrition for a population that is rapidly expanding (Nelson et al. 2018). Moreover, crop responses to increases in atmospheric CO_2 are related to nutrient availability (Jin et al. 2019). In other words, the additional carbon acquired through photosynthesis in response to eCO_2 can only result in increased crop yield if plant nutrition is adequate. These deficiencies of plants grown at eCO_2 in using photosynthetic carbon gain under nutrient-limited conditions are of serious concern. It is apparent that the eCO_2 effect on crop productivity may be reduced in low-income countries, where the availability of fertilizers is a limiting factor in agricultural production. Hence, countries that depend on agriculture for a large share of their income are at risk of more food insecurity. In this context, a deeper understanding of how eCO_2 regulates crop yield and nutritional quality is required to ensure food security over the coming decades.

In response to eCO_2, cereal crops increase grain number and grain biomass, which is often associated with reduction in nutritional quality of crops (Dong et al. 2019). There are sufficient data to indicate that eCO_2 leads to a decrease on concentrations of Zn and Fe in staple crops like wheat, rice, potato, and legumes (Loladze 2014; Myers et al. 2014; Zhu et al. 2018). The fertilization effects of eCO_2 associated with incidence of climate impacts on grain mineral concentrations are projected to decrease the global availability of Zn by 14.6% and Fe by 13.6% in 2050 (Beach et al. 2019). The losses in grain mineral concentrations in response to eCO_2 may be attributed to the lower absorption and/or translocation to grains as well as yield dilution and concentration effect (Ujiie et al. 2019; Jin et al. 2019) (Fig. 2). In this context, the effect of eCO_2 on concentrations of Zn and Fe may therefore cause a nutritional deficit in these key nutrients for a large segment of the world's population. The reduction on concentration of Zn and Fe in the edible portion of crops due to increased atmospheric CO_2 concentrations could lead to a reduction of 125 million disability-adjusted life-years globally over the period 2015–2050 (Weyant et al. 2018).

The impact of eCO_2 on dietary patterns will be strongest in regions like Southeast Asia and Africa, where populations already have a burden of disease associated with deficits in intake of Zn and Fe (Weyant et al. 2018). Thus, efforts to enhance crop yields in response to eCO_2 must be coupled with attempts to understand and manipulate the balance between mineral uptake by the root system, distribution, and partition to the grains to maximize its use for storage. Research on wheat illustrates the importance of considering these questions in attempting to increase grain nutritional quality. The endosperm-specific expression of the *VACUOLAR IRON TRANSPORTER* gene combined with constitutive expression of the *NICOTIANAMINE SYNTHASE* gene increased grain Zn concentration and altered the redistribution of Fe within the grain, which led to an increase in Fe in wheat flour (Harrington et al. 2022). However, for the full potential for wheat improvement to be realized, this focus on grain nutritional quality must be accompanied by increased understanding of how climate change could affect grain mineral concentrations. This consideration raises a multitude of new and complex questions about the integration of carbon assimilation, quantity, and nutritional quality of crops.

Protein concentration of cereals (the ratio of grain protein amount to grain yield) is an important trait affecting the market value and nutritional value of grain (Geyer et al. 2022). The predicted changes in atmospheric CO_2 concentration alone can increase the total amount of protein in grain of C3 crops such as rice and wheat, but also decrease its concentration (Myers et al. 2014). Although the precise mechanism behind this remarkable effect remains uncertain, decreased protein concentration under eCO_2 conditions can be attributed to higher starch accumulation and lower assimilation of nitrate into organic nitrogen compounds (Bloom et al. 2014) (Fig. 2). It is also likely that eCO_2 decreases the concentration of available N in the soil, contributing to the lower N concentration in vegetative tissues and probably reduction grain protein concentration (Jin et al. 2019). The effects of increases in atmospheric CO_2 are thus predicted to decrease global availability of dietary protein by 4.1% (Loladze 2014), which can disproportionately affect countries that already have high levels of nutritional deficiency. Additionally, the negative effect on protein concentration under eCO_2 could lead to a decrease on S availability for human because plant proteins are the important source of the S-containing amino acid methionine (Tcherkez et al. 2020). Thus, breeders and biotechnologists need to identify plant traits that can be targeted to improve nutritional quality of crops in relation to increasing atmospheric CO_2 concentration.

It should also be contemplated that atmospheric CO_2 concentration is not just the source of carbon for photosynthetic organisms, but a long-wave-radiation trapping gas, with consequences for global temperature and precipitation patterns, climatic variables that affect yields and nutritional quality of crops. The interactions between eCO_2 and other variables (e.g., temperature and precipitation patterns) lead to effects on agricultural productivity and global nutrient availability that are not easily predictable from the studies of the individual components. This has major consequences for discussion of how, and to what extent, yields and nutritional quality of crops can be optimized in a changing environment.

Taken together, research into growth and yield regulatory processes under influence of eCO_2 conditions indicates that the increase in CO_2 alone may improve the energy efficiency of plant metabolism of C3 crops and thus more fixed carbon could be allocated to grain, increasing yield. These responses, however, have the unintended effect of reducing grain nutritional quality. The challenge is to understand how the grain nutritional quality is coordinated with the availability of photosynthate under eCO_2 environments at the levels of single cells and whole plants. Moreover, the regulation of gene expression and signaling cascades that regulate many mineral transporters in response to eCO_2 conditions remain to be elicited. These are steps toward learning how an increase in the levels of photosynthetic carbon modifies plant carbon-to-nutrient ratios, which in turn may lead to the development of a sustainable production under eCO_2 conditions. However, one needs to be cautious that there are other biophysical conditions, especially temperature and precipitation, interacting with eCO_2 and the nutritional value of the crop. This reinforces the complexity of developing models to predict the impact of CO_2 conditions on global food security in the context of climate change.

3 Temperature Changes

The global average surface temperature increased 0.6 °C during the twentieth century and, according to the most recent forecasts, is expected to increase 2.6 °C by the end of the century, compared to the preindustrial era. Increased temperature and more frequent heatwaves will have a strong impact on agriculture in tropical regions but also in some temperate countries (Fig. 2). Globally, 31% of agricultural areas are considered as "high risk" of heat stress in the twenty-first century. Climate risks could thus lead to food shortages, massive migrations, and other societal disruptions. The full impact of temperature increase on crops is an area of intense ongoing research, as the final effects depend on complex interactions with ambient CO_2 concentration and water and nutrient availability (Moore et al. 2021). Particularly worrisome is the increasing occurrence of simultaneous stresses, for instance, high temperature and drought. Water scarcity will impinge strongly on agricultural output. Climate models project that rising temperatures will lead to changes in rainfall patterns that exacerbate existing trends, that is, dry regions will get drier and wet areas will become wetter (Bathiany et al. 2018). The Mediterranean basin, for instance, is particularly susceptible to drought, so a large share of the agricultural output in countries of Southern Europe and North Africa is expected to be affected. Entire agriculture-based industries, like wine production in Spain, could be disrupted (Droulia and Charalampopoulos 2021).

Many open questions are still the subject of intense research to provide new knowledge that can help mitigate the effects of temperature extremes. First, the physiological impact of a steady increase in temperature will differ from that of discrete, extreme temperature events (e.g., heat waves, unseasonable frosts). What are the genetic networks controlling the responses to each one and how much overlap (if any) is there between them?

As mentioned above, increasing temperatures resulting from climate change drastically impact crop production around the globe. High temperatures affect crop yields by direct and indirect effects, causing water stress through reduction of soil water and increased atmospheric water demand (Lobell et al. 2013), leading to stomatal closure to avoid dehydration thereby impairing CO_2 uptake, and enhanced root growth, both causing reduction of shoot biomass. Considering the same crop, optimum temperatures differ at different growth stages, and changes in temperature conditions can happen at any developmental stage at field conditions.

High day temperature (HDT) refers to higher than optimum temperatures during daytime for crop development. In rice (*Oryza sativa* L.), HDT during vegetative stage affect tiller formation and continuous stress exposure during boot stage impacts directly on spikelet meristem differentiation, and during reproductive stage increases spikelet fertility, and continuous stress exposure during seed development impacts grain weight (Xu et al. 2020). Photosynthesis is the rate-limiting factor preferentially inhibited by HDT, with great decrease in the photoassimilate production due to great reduction in leaf carbohydrate content due to photorespiration (Dusenge et al. 2019). HDT also affects molecular pathways with the purpose to avoid, scape, and tolerate stressful conditions. For example, *EXTRA GLUME* 1 (EG1) encodes a predominantly mitochondria-localized lipase that functions upstream of floral identity genes in rice (*OsMADS1*, *OsMADS6*, and *OsG1*) to promote floral development sturdiness under HDT (Zhang et al. 2016). Tomato (*Solanum lycopersicum* L.) is one of the main crops in which yield losses have been massively reported when heat stress takes place during the reproductive phase. Tomato fruit number per truss and fruit weight is directly affected by HDT, ranging from a few days (when pollen development or fruit set is disturbed) to a whole developmental period (Sato et al. 2006). In potato (*Solanum tuberosum*), HDT decreased tuber yield (~18.1%) by reducing photoassimilates, which was probably attributed to decreased photosynthetic efficiency through a feedback inhibition (Kim and Lee 2019). Moreover, night temperature appears to be increasing at a faster pace than day's causing harmful effect on crop growth, development, and yields due to a reduced diurnal temperature range (Bahuguna and Jagadish 2015).

High night temperature (HNT) occurs when there is an uneven temperature increase, with larger increase of night's compared to day temperatures (Schaarschmidt et al. 2021). Reduction of grain yield was reported after HNT exposure, and it seems that disturbed translocation of photoassimilates was the main cause (Wu et al. 2017), also affecting pollen viability in rice (Yang et al. 2017) and decreased spikelet fertility, grains per spike, grain size, and quicker grain filler period in wheat (Narayanan et al. 2016). Quality parameters were also altered after HNT, as grain length, grain width, and grain area. All together shows that HNT has more deleterious effect on grain quality compared to HDT (Fahad et al. 2016; Schaarschmidt et al. 2021), although the impact on yield decrease and quality is directly related with the HNT tolerance of the species. This can be assumed since HNT affects gene regulation, metabolic pathways, and hormone metabolism. Glaubitz et al. (2017) performed transcriptomic and metabolomic analysis on leaves from six rice cultivars under HNT. An overlap of six significantly differentially expressed genes was pinpointed in five cultivars and were all upregulated, encoding proteins involved in transcription regulation (helix-loop-helix proteins), signal transduction (protein kinase), protein-protein interactions (TIFY domain containing protein), and biosynthesis of polyphenols (phenylalanine ammonia-lyase). Metabolites profile revealed involvement of 4-amino-butanoic acid (GABA) signaling, providing a link to the TCA cycle in sensitive cultivars and of myo-inositol as precursor for inositol phosphates also linking jasmonates signaling to the HNT response mainly in tolerant cultivars. In potato, during tuber initiation, HNT delayed tuber development, thus altering tuber mass distribution by reducing the yield proportion (~53.7%) and lowering early harvest index (16.1%), causing yield loss (~17.2%) without photosynthesis damage (Kim and Lee 2019).

Considering the aforementioned detrimental effects of heat stress on crop production, some points need to be elucidated: (1) how and which crops are differentially affected by HDT and HNT stresses, (2) the anatomical-molecular-physiological mechanisms related with yield consistency and impairment (tolerance and susceptibility), (3) which developmental stages are most affected by heat stress for each crop, and (4) benefits and challenges in the development of new heat tolerant varieties throughout molecular pathways (Xu et al. 2020).

4 Adapting Agriculture to Uncertain Climate

Even though climate is hard to predict, the current consensus from many independent studies (Lehmann and Rillig 2014; Mazdiyasni and AghaKouchak 2015; Bigot et al. 2018; Anderson and Song 2020; Grossiord et al. 2020; Zandalinas et al. 2021) indicates that the mean surface temperature will increase steadily over the current century. The latest IPCC report states that at least half of the increase in global mean temperature between 1951 and 2010 has been likely caused by anthropogenic greenhouse gases: CO_2 levels have risen from 250 ppm to over 400 ppm over the period. Agriculture itself has led to considerable detrimental effects: the destruction of tropical forests releases a trillion tons of carbon per year, an eighth of all anthropogenic CO_2 emissions (Friedlingstein et al. 2010). It is anticipated that, if unchecked, global warming will lead to altered distribution of rainfalls, exacerbating flooding in some areas and drought in others. Expected adverse effects on crop growth include decreased seed germination, increased incidence of plant disease, and herbivory (Lobell and Gourdji 2012; Taiz 2013; Wheeler and Von Braun 2013). Climate change models have furthermore suggested the increased incidence of extreme climatic events (Otto 2015), which are likely to have devastating impact on crop yields.

Climate extremes, such as drought or heat stress, can lead to harvest failures and threaten the livelihood of agricultural producers and the food security of communities. Improving the understanding of their impacts on maize production is crucial to enhance the resilience of the global food system. Climate factors, including mean climate and climate extremes, explain 16–39% of the variance of yield anomalies (YA), with 10–31% of the explained variance attributable to climate conditions. YA related more closely with temperature extremes than with precipitation-related factors (Vogel et al. 2019). The forecast for future scenarios is a loss of climatic suitability for maize in sub-Saharan Africa and Latin America regions but accompanied by an expansion in the northern hemisphere, particularly in Europe. The relative change in climatically suitable areas for future maize production was estimated for the top five producers. Production in 2050 is expected to increase 8% for the USA and 4% for China and to decrease 5% for Brazil, 2% for Argentina, and 11% for Mexico. The incidence of low temperature and waterlogging, presently common in Europe and Asia, is projected to diminish, whereas heat stress in Africa and drought stress in South America are projected to increase (Ramirez-Cabral et al. 2017).

In 2010, FAO introduced the concept of "climate-smart agriculture" to cope with future threats to food security and climate change. One of the key drivers of "climate-responsible" intensification of agriculture is diversification. However, conservation of agro-biodiversity is not an end in itself. Conservation must be strongly linked to utilization, either actual or potential.

Changing highly engrained dietary habits is probably more challenging than breeding new crops and creating resilient agricultural systems (Fanzo et al. 2013). However, past experiences show that it is possible through a combination of policy and individual endeavor. As recently as 300 years ago, European peasants were reluctant to grow potatoes for a variety of reasons including superstition, resemblance with poisonous nightshade, or simply taste preferences (McNeill 1999). Today, Europe is responsible for 30% of the total production of potato worldwide, and Germany, France, the Netherlands, and Poland are among the top 10 world producers. The first commercial orchard of kiwifruit (*Actinidia deliciosa*) was established in New Zealand in the 1930s. Today, the total world production is well over four million tonnes per year and could expand and diversify through the exploitation of closely related species: *A. arguta* (already grown in low scale in Europe and in the USA), *A. kolomikta* (high in vitamin C and adapted to colder areas), or *A. eriantha* (high in vitamin C) (Ferguson 2013).

Genomic analyses are widening to capture the large-scale range of ecological variation of crops. They now include wild species, landraces, and cultivars, and they aim at identifying relevant genetic signatures for valuable agronomic traits. This is a fundamental first step, which in an ideal pipeline should be followed by physiological characterization and agronomic field assays.

References

Alley RB (2000) The Younger Dryas cold interval as viewed from central Greenland. Quat Sci Rev 19:213–226

Anderson JT, Song BH (2020) Plant adaptation to climate change—where are we? J Syst Evol 58:533. https://doi.org/10.1111/jse.12649

Bahuguna RN, Jagadish KSV (2015) Temperature regulation of plant phenological development. Environ Exp Bot 111:83–90. https://doi.org/10.1016/j.envexpbot.2014.10.007

Bathiany S, Dakos V, Scheffer M, Lenton TM (2018) Climate models predict increasing temperature variability in poor countries. Sci Adv 4:eaar580. https://www.science.org/doi/10.1126/sciadv.aar5809

Beach RH, Sulser TB, Crimmins A, Cenacchi N, Cole J, Fukagawa NK, Mason-D'Croz D, Myers S, Sarofim MC, Smith M, Ziska LH (2019) Combining the effects of increased atmospheric carbon dioxide on protein, iron, and zinc availability and projected climate change on global diets: a modelling study. Lancet Planet Health 3:e307–e317. https://doi.org/10.1016/S2542-5196(19)30094-4

Bevan A, Colledge S, Fuller D, Fyfe R, Shennan S, Stevens C (2017) Holocene fluctuations in human population demonstrate repeated links to food production and climate. Proc Natl Acad Sci U S A 114:E10524. https://doi.org/10.1073/pnas.1709190114

Bigot S, Buges J, Gilly L, Jacques C, Le Boulch P, Berger M, Delcros P, Domergue JB, Koehl A, Ley-Ngardigal B, Tran Van Canh L, Couée I (2018) Pivotal roles of environmental sensing and signaling mechanisms in plant responses to climate change. Glob Chang Biol 24:5573. https://doi.org/10.1111/gcb.14433

Bloom AJ, Burger M, Kimball BA, Pinter PJ (2014) Nitrate assimilation is inhibited by elevated CO2 in field-grown wheat. Nat Clim Chang 4:477–480. https://doi.org/10.1038/nclimate2183

Dong J, Grylls S, Hunt J, Armstrong R, Delhaize E, Tang C (2019) Elevated CO2 (free-air CO2 enrichment) increases grain yield of aluminium-resistant but not aluminium-sensitive wheat (Triticum aestivum) grown in an acid soil. Ann Bot 123:461. https://doi.org/10.1093/aob/mcy171

Droulia F, Charalampopoulos I (2021) Future climate change impacts on European viticulture: a review on recent scientific advances. Atmosphere 12(4):495. https://doi.org/10.3390/atmos12040495

Dusenge ME, Duarte AG, Way DA (2019) Plant carbon metabolism and climate change: elevated CO2 and temperature impacts on photosynthesis, photorespiration and respiration. New Phytol 221:32–49. https://doi.org/10.1111/nph.15283

Fahad S, Hussain S, Saud S, Hassan S, Chauhan BS, Khan F, Ihsan MZ, Ullah A, Wu C, Bajwa AA, Alharby H, Amanullah, Nasim W, Shahzad B, Tanveer M, Huang J (2016) Responses of rapid viscoanalyzer profile and other rice grain qualities to exogenously applied plant growth regulators under high day and high night temperatures. PLoS One 11:e0159590. https://doi.org/10.1371/journal.pone.0159590

Fanzo J, Hunter D, Borelli T, Mattei F (2013) Diversifying food and diets: using agricultural biodiversity to improve nutrition and health. Routledge, London

Ferguson AR (2013) Kiwifruit. The wild and the cultivated plants. Adv Food Nutr Res 68:15–32

Ferrio JP, Voltas J, Araus JL (2011) Global change and the origins of agriculture. In: Crop stress management and global climate change. CAB International, Wallingford, pp 1–14

Friedlingstein P, Houghton RA, Marland G, Hackler J, Boden TA, Conway TJ, Canadell JG, Raupach MR, Ciais P, Le Quéré C (2010) Update on CO2 emissions. Nat Geosci 3:811. https://doi.org/10.1038/ngeo1022

Ganopolski A, Winkelmann R, Schellnhuber HJ (2016) Critical insolation-CO2 relation for diagnosing past and future glacial inception. Nature 529:200. https://doi.org/10.1038/nature16494

Geyer M, Mohler V, Hartl L (2022) Genetics of the inverse relationship between grain yield and grain protein content in common wheat. Plan Theory 11:2146. https://doi.org/10.3390/plants11162146

Giller KE, Delaune T, Silva JV et al (2021) The future of farming: who will produce our food? Food Secur 13:1073–1099. https://doi.org/10.1007/s12571-021-01184-6

Glaubitz U, Li X, Schaedel S, Erban A, Sulpice R, Kopka J, Zuther E (2017) Integrated analysis of rice transcriptomic and metabolomic responses to elevated night temperatures identifies sensitivity- and tolerance-related profiles. Plant Cell Environ 40:121–137. https://doi.org/10.1111/pce.12850

Grossiord C, Ulrich DEM, Vilagrosa A (2020) Controls of the hydraulic safety-efficiency trade-off. Tree Physiol 40:573. https://doi.org/10.1093/TREEPHYS/TPAA013

Harrington SA, Connorton JM, Nyangoma NIM, McNelly R, Morgan YML, Aslam MF, Sharp PA, Johnson AAT, Uauy C, Balk J (2022) A two-gene strategy increases iron and zinc concentrations in wheat flour, improving mineral bioaccessibility. Plant Physiol 29:kiac499. https://doi.org/10.1093/plphys/kiac499

Jin J, Armstrong R, Tang C (2019) Impact of elevated CO_2 on grain nutrient concentration varies with crops and soils – a long-term FACE study. Sci Total Environ 651:2641–2647. https://doi.org/10.1016/j.scitotenv.2018.10.170

Kim Y-U, Lee B-W (2019) Differential mechanisms of potato yield loss induced by high day and night temperatures during tuber initiation and bulking: photosynthesis and tuber growth. Front Plant Sci 10:300. https://doi.org/10.3389/fpls.2019.00300

Leakey ADB, Uribelarrea M, Ainsworth EA et al (2006) Photosynthesis, productivity, and yield of maize are not affected by open-air elevation of CO_2 concentration in the absence of drought. Plant Physiol 140:779–790

Lehmann J, Rillig M (2014) Distinguishing variability from uncertainty. Nat Clim Chang 4:153. https://doi.org/10.1038/nclimate2133

Lobell DB, Gourdji SM (2012) The influence of climate change on global crop productivity. Plant Physiol 160:1686. https://doi.org/10.1104/pp.112.208298

Lobell DB et al (2013) The critical role of extreme heat for maize production in the United States. Nat Clim Chang 3:497–501. https://doi.org/10.1038/nclimate1832

Loladze I (2014) Hidden shift of the ionome of plants exposed to elevated CO_2 depletes minerals at the base of human nutrition. eLife 3:e02245. https://doi.org/10.7554/eLife.02245

Mazdiyasni O, AghaKouchak A (2015) Substantial increase in concurrent droughts and heatwaves in the United States. Proc Natl Acad Sci U S A 112:11484. https://doi.org/10.1073/pnas.1422945112

McNeill W (1999) The glance of the eye: Heidegger, Aristotle, and the ends of theory. State University of New York Press, Albany

Moore CE, Meacham-Hensold K, Lemonnier P, Slattery RA, Benjamin C, Bernacchi CJ, Lawson T, Cavanagh AP (2021) The effect of increasing temperature on crop photosynthesis: from enzymes to ecosystems. J Exp Bot 72:2822. https://doi.org/10.1093/jxb/erab090

Myers SS, Zanobetti A, Kloog I, Huybers P, Leakey AD, Bloom AJ, Carlisle E, Dietterich LH, Fitzgerald G, Hasegawa T, Holbrook NM, Nelson RL, Ottman MJ, Raboy V, Sakai H, Sartor KA, Schwartz J, Seneweera S, Tausz M, Usui Y (2014) Increasing CO_2 threatens human nutrition. Nature 510:139–142. https://doi.org/10.1038/nature13179

Narayanan S, Tamura PJ, Roth MR, Prasad PVV, Welti R (2016) Wheat leaf lipids during heat stress: I. High day and night temperatures result in major lipid alterations. Plant Cell Environ 39:787. https://doi.org/10.1111/pce.12649

Nelson G, Bogard J, Lividini K et al (2018) Income growth and climate change effects on global nutrition security to mid-century. Nat Sustain 1:773–781. https://doi.org/10.1038/s41893-018-0192-z

Neukom R, Steiger N, Gómez-Navarro JJ, Wang J, Werner JP (2019) No evidence for globally coherent warm and cold periods over the preindustrial Common Era. Nature 571:550. https://doi.org/10.1038/s41586-019-1401-2

Otto FEL (2015) Climate change: attribution of extreme weather. Nat Geosci 8:581. https://doi.org/10.1038/ngeo2484

Poorter H, Knopf O, Wright IJ, Temme AA, Hogewoning SW, Graf A, Cernusak LA, Pons TL (2022) A meta-analysis of responses of C3 plants to atmospheric CO2: dose–response curves for 85 traits ranging from the molecular to the whole-plant level. New Phytol 233:1560–1596

Ramirez-Cabral NY, Kumar L, Shabani F (2017) Global alterations in areas of suitability for maize production from climate change and using a mechanistic species distribution model (CLIMEX). Sci Rep 7:5910. https://doi.org/10.1038/s41598-017-05804-0

Sato S, Kamiyama M, Iwata T, Makita N, Furukawa H, Ikeda H (2006) Moderate increase of mean daily temperature adversely affects fruit set of *Lycopersicon esculentum* by disrupting specific physiological processes in male reproductive development. Ann Bot 97(5):731–738. https://doi.org/10.1093/aob/mcl037

Schaarschmidt S, Lawas LMF, Kopka J, Jagadish SVK, Zuther E (2021) Physiological and molecular attributes contribute to high night temperature tolerance in cereals. Plant Cell Environ 44:2034. https://doi.org/10.1111/pce.14055

Sharon BG, Siobhan MB (2016) Plant developmental responses to climate change. Dev Biol 419:64–77. https://doi.org/10.1016/j.ydbio.2016.07.023

Taiz L (2013) Agriculture, plant physiology, and human population growth: past, present, and future. Theor Exp Plant Physiol 25(3):167–181. https://doi.org/10.1590/s2197-00252013000300001

Tcherkez G, Ben Mariem S, Larraya L, García-Mina JM, Zamarreño AM, Paradela A, Cui J, Badeck FW, Meza D, Rizza F, Bunce J, Han X, Tausz-Posch S, Cattivelli L, Fangmeier A, Aranjuelo I (2020) Elevated CO2 has concurrent effects on leaf and grain metabolism but minimal effects on yield in wheat. J Exp Bot 71:5990. https://doi.org/10.1093/jxb/eraa330

Ujiie K, Ishimaru K, Hirotsu N, Nagasaka S, Miyakoshi Y, Ota M, Tokida T, Sakai H, Usui Y, Ono K, Kobayashi K, Nakano H, Yoshinaga S, Kashiwagi T, Magoshi J (2019) How elevated CO2 affects our nutrition in rice, and how we can deal with it. PLoS One 14:e0212840. https://doi.org/10.1371/journal.pone.0212840

Vogel E, Donat MG, Alexander LV, Meinshausen M, Ray DK, Karoly D, Meinshausen N, Frieler K (2019) The effects of climate extremes on global agricultural yields. Environ Res Lett 14:054010. https://doi.org/10.1088/1748-9326/ab154b

Weyant C, Brandeau ML, Burke M, Lobell DB, Bendavid E, Basu S (2018) Anticipated burden and mitigation of carbon dioxide-induced nutritional deficiencies and related diseases: a simulation modeling study. PLoS Med 15:e1002586. https://doi.org/10.1371/journal.pmed.1002586

Wheeler T, Von Braun J (2013) Climate change impacts on global food security. Science (1979) 341:508–513. https://doi.org/10.1126/science.1239402

Wu C, Cui K, Wang W, Li Q, Fahad S, Hu Q, Huang J, Nie L, Mohapatra PK, Peng S (2017) Heat-induced cytokinin transportation and degradation are associated with reduced panicle cytokinin expression and fewer spikelets per panicle in rice. Front Plant Sci 8:371. https://doi.org/10.3389/fpls.2017.00371

Xu J, Henry A, Sreenivasulu N (2020) Rice yield formation under high day and night temperatures – a prerequisite to ensure future food security. Plant Cell Environ 43:1595–1608. https://doi.org/10.1111/pce.13748

Yang Z, Zhang Z, Zhang T, Fahad S, Cui K, Nie L, Peng S, Huang J (2017) The effect of season-long temperature increases on rice cultivars grown in the central and southern regions of China. Front Plant Sci 8:1908. https://doi.org/10.3389/fpls.2017.01908

Zandalinas SI, Fritschi FB, Mittler R (2021) Global warming, climate change, and environmental pollution: recipe for a multifactorial stress combination disaster. Trends Plant Sci 26:588. https://doi.org/10.1016/j.tplants.2021.02.011

Zhang B, Wu S, Zhang Y, Xu T, Guo F, Tang H, Li X, Wang P, Qian W, Xue Y (2016) A high temperature-dependent mitochondrial lipase EXTRA GLUME1 promotes floral phenotypic robustness against temperature fluctuation in rice (*Oryza sativa* L.). PLoS Genet 12:e1006152. https://doi.org/10.1371/journal.pgen.1006152

Zhu C, Kobayashi K, Loladze I, Zhu J, Jiang Q, Xu X, Liu G, Seneweera S, Ebi KL, Drewnowski A, Fukagawa NK, Ziska LH (2018) Carbon dioxide (CO_2) levels this century will alter the protein, micronutrients, and vitamin content of rice grains with potential health consequences for the poorest rice-dependent countries. Sci Adv 4:eaaq1012. https://doi.org/10.1126/sciadv.aaq1012

Part I
Vertical Farming and Nurseries (Both Controlled and Uncontrolled Environments)

Soilless Smart Agriculture Systems for Future Climate

Rajiv Ranjan Singh and Anirban Jyoti Hati

Abstract Global warming will have a negative impact on agricultural land in underdeveloped nations, as the land warms up more rapidly and easily than water. By the 2080s, the demand for food is predicted to triple due to an increase in global population and affluence. Climate smart agriculture (CSA) is an integrated method of managing landscapes that address the interrelated problems of food security and climate change. Smart agriculture systems use sensors and monitoring tools to gather information on variables such as temperature, humidity, water levels, and fertilizer levels. Robotic systems for planting, harvesting, and weeding can also be part of smart agriculture systems, such as automated watering and fertilizing systems. Soilless smart agriculture technologies can be carried out in a controlled setting and are more resistant to adverse weather, reducing the carbon footprint of food production. A controlled environment, like a greenhouse, can be used for year-round production and shelter from harsh weather. Soilless agriculture is an adaptation to climate change, as it is more resistant to adverse weather and uses less water than conventional agriculture. Urban agriculture is becoming increasingly important, as people are relocating to cities and demand for food production is rising. It is important to consider the environmental and social impact of these methods, such as energy consumption for a controlled environment, and ensure they are sustainable in the long run. In this chapter, we have summarized the methodologies and enabling technologies for indoor soilless smart agriculture systems (ISSAS) considering both global and Indian scenarios.

Keywords Climate change · Soilless agriculture · Enabling technologies

R. R. Singh (✉)
Department of Electronics and Communication Engineering, Presidency University, Bengaluru, Karnataka, India
e-mail: rajivranjansingh@presidencyuniversity.in

A. J. Hati
Department of Education and Training, National Institute of Electronics and Information Technology, Kolkata, India
e-mail: anirban.jyotihati@presidencyuniversity.in

P. M. Priyadarshan et al. (eds.), *Digital Agriculture*,
https://doi.org/10.1007/978-3-031-43548-5_3

1 Introduction

Global warming will have a negative impact on agricultural land in underdeveloped nations compared to industrialized nations, since the land warms up more rapidly and easily than water. Crops can be impacted by climate change in two ways: (1) increased soil evaporation and (2) higher temperatures that alter plants' ability to absorb and utilize moisture. By the 2080s, the demand for food is predicted to triple due to an increase in global population and affluence (Cline 2008). As a result, the supply and demand equation is unstable, and climate change would make it significantly worse.

The need for a larger emphasis on climate change adaptation in agriculture is becoming more urgent. Few studies evaluate the adoption rates and propensity for the effectiveness of potential response tactics. A broader risk-management framework that takes climate unpredictability and market dynamics into account will be needed to support further adaptation activities. Science must change as well by continuously evaluating the need for new research and improving managerial techniques (Howden et al. 2007).

Additionally, climate change can lead to increase in migration, poverty, etc. Therefore, the concept of climate smart agriculture (CSA) has been conceptualized, which is an integrated method of managing landscapes that addresses the interrelated problems of food security and climate change. It includes farming, raising cattle, managing forests, and managing fisheries. The World Bank (Cline 2008) has specified three outcomes of CSA: (1) increased productivity by producing more and better food to enhance earnings and improve nutrition security, especially for the 75% of the world's poor who reside in rural regions and mostly depend on agriculture; (2) improved resilience in terms of lower susceptibility to pests, diseases, drought, and other climate-related hazards and shocks and increase ability to adapt and develop in the face of longer-term pressures, such shortened seasons and unpredictable weather patterns; (3) lessened emissions by striving for lower emissions per calorie or kilogram of food produced, limiting agricultural deforestation, and finding techniques to remove carbon dioxide from the environment.

Despite the fact that CSA is marketed as a multidisciplinary idea, persistent biases toward scientific and technical challenges still influence how researchers view CSA on a worldwide scale. To find CSA solutions, there is enough technical advice and scientific support, but the literature on social, management, and economic issues is underdeveloped. In particular, there is a shortage of research to support better coherence, coordination, and integration of the CSA pillars in the areas of gender, markets, broader landscape features, and decision-making. The many CSA pillars have significant overlaps and divergences. Trade-offs would be necessary if multiple orientations were pursued, and these trade-offs might favor one CSA pillar over another (Chandra et al. 2018).

Smallholder farmers can benefit most from optimal combinations of adaptation and mitigation initiatives by contributing to their socioeconomic development.

Development has had a significant influence on CSA discussions (Chandra et al. 2018).

The food and agriculture industry both contributes significantly to climate change and is particularly susceptible to its worst effects. Complex and contentious political processes are at play as new governance agendas are implemented, and much is at stake. This unique forum brings together a collection of presentations that highlight three overlapping themes that are at the heart of these disputes in order to unravel these concerns (Clapp et al. 2018).

The confluence of food and agriculture with climate change has offered a forum for discussing new solutions and rehashing old debates. We may observe how debates in agrarian studies over land rights, control over agricultural technologies, access to them, governance of fisheries and marine resources, trade liberalization, and food sovereignty are once more at the forefront (Clapp et al. 2018).

Eleven case studies are used to examine scaling-up strategies based on value chains and private sector involvement, agro-advisory services, and policy engagement (Westermann et al. 2018). The case studies highlighted several challenges: estimating the costs and benefits of different scaling activities, integrating knowledge across multiple levels, and addressing equity issues. Results showed that these different strategies exhibit different characteristics. One is the issue of estimating the costs and benefits of different scaling activities. While it may be envisaged that strategies for scaling up based on value chains, ICT/agro-advisory services and policy engagement could be highly cost-effective, more rigorous information is needed, and this warrants further work. A second challenge is that of integrating knowledge across multiple levels. This is not only just the challenge of moving from successful small-scale projects to informing and implementing policy with broad reach; it also requires devolving action from national levels to local levels (or scaling down) to ensure that interventions are appropriately contextualized and locally viable. The third challenge is that of addressing equity considerations in scaling up CSA interventions (Westermann et al. 2018).

The use of technology to increase the productivity and efficiency of agricultural activities is referred to as “smart agriculture systems.” Sensors and monitoring tools are frequently used in these systems to gather information on variables, including temperature, humidity, water levels, and fertilizer levels. The growing conditions are then adjusted in real-time to maximize plant development and production using the data collected. Robotic systems for planting, harvesting, and weeding can also be a part of smart agriculture systems, as can automated watering and fertilizing systems (Oliveira et al. 2021).

Future food production could be efficient and sustainable thanks to soilless smart agriculture technologies, especially in light of the world’s changing climate (Banerjee et al. 2022). Soilless smart agricultural systems have the potential to be a crucial instrument for coping with and lessening the effects of climate change in terms of the future climate. The demand for food will rise as the world’s population expands. The key benefits from soilless agriculture could be envisaged as below:

Climate Adaptation The demand for food will rise as the world's population continues to rise. Additionally, it is anticipated that climate change would bring about more extreme weather events like droughts and floods, which could be detrimental to traditional agriculture (Altieri et al. 2015). However, soilless agriculture methods can be carried out in a controlled setting and are often more resistant to adverse weather. For instance, indoor hydroponics and aeroponics operations can shield plants from harsh weather conditions, including heatwaves, cold snaps, and heavy rain (Rayhana et al. 2020).

Efficiency in Water Use Conventional agriculture can use a lot of water. This can be a serious issue in locations where water is already in short supply. However, soilless agriculture uses a lot less water than conventional agriculture (Eigenbrod and Gruda 2015). For instance, only the water that is absorbed by the plants is wasted in hydroponic systems where the water is recycled. In arid areas, where traditional agriculture would be impossible, soilless agriculture technologies can be used (Schröder and Lieth 2002).

Reduced Carbon Footprint Food production's carbon impact is a significant environmental concern. A large amount of the world's greenhouse gas emissions is caused by traditional agriculture (Bozchalui et al. 2015). However, soilless agriculture methods utilize substantially fewer chemical inputs than conventional agriculture, such as fertilizers and pesticides, which can assist in lowering the carbon footprint of food production (Eigenbrod and Gruda 2015). The ability to precisely manage the nutrient levels, pH, and water supply for the plants, resulting in higher development and yields, is one of the key benefits of soilless agriculture (Lakhiar et al. 2018).

Urban Agriculture People are relocating to cities in greater numbers as urbanization continues to rise. The demand for food production in urban areas is rising along with the population. In urban locations, where traditional agriculture is not feasible, it is possible to grow food using methods of soilless agriculture, such as vertical farming (Goldstein 2018). This can give urban people with fresh, locally grown vegetables while lowering the carbon footprint of food transportation (Goodman and Minner 2019). A controlled environment, like a greenhouse, can also be used for soilless agriculture, enabling year-round production and shelter from harsh weather (Goodman and Minner 2019; Rayhana et al. 2020).

Additionally, it is also important to consider the environmental and social impact of these methods, for example, energy consumption for controlled environment, and to ensure that they are sustainable in the long run.

According to the United Nations, by 2030, India is expected to have 1.5 billion people, making it the most populated nation on earth (UN DESA 2022). India's population is expanding quickly, and it is anticipated that demand for food will rise sharply over the next few years. But there are also serious obstacles to India's food security. Around 20% of children under the age of five are underweight, and 14.5% of the population is undernourished. India's agriculture is anticipated to be significantly impacted by climate change. According to predictions made by the

Intergovernmental Panel on Climate Change (IPCC), India would likely experience an increase in the frequency and severity of extreme weather events like droughts and floods as a result of climate change (Anderson et al. 2020).

In order to understand the impacts of climate change and the requirements for food production in developing countries in the context of soilless agriculture techniques, we have taken India into consideration. India is experiencing a severe water problem in terms of its water supplies. India's water storage has fallen to a disconcertingly low 28% of its maximum capacity, according to the Central Water Commission (Sikka et al. 2022). The ancient agricultural practices in India are up against several difficulties due to the growing population and water shortage. In this situation, soilless agriculture techniques, like hydroponics, aeroponics, aquaponics, and vertical farming, have a critical role to play in supplying food and coping with climate change. But the use of soilless farming techniques in India is still in its infancy, and there are still a number of obstacles to overcome, including a lack of knowledge, a lack of technical know-how, and high start-up expenses.

This chapter reviews the research of researchers from various parts of the world related to vertical farming, greenhouse farming, precision farming, climate control, fertilizer optimization, crop planning, soilless agriculture methods, agriculture in LED light, etc., taking into account both global and Indian scenarios. It then discusses the components and enabling technologies of indoor soilless smart agriculture systems (ISSAS) and their related challenges.

2 Soilless Smart Agriculture Systems (SSAS)

Soilless smart agriculture systems (SSAS) can be both indoors and outdoors, depending on the type of system and the crops being grown. However, indoor systems are more commonly used for SSAS due to their many advantages, such as climate control, better disease and pest management, and year-round production. On the other hand, outdoor SSAS are typically used for larger-scale agricultural operations such as field crops or orchards. Outdoor SSAS are often referred to as precision agriculture, as they use data and technology to optimize crop yields and minimize environmental impact. Hydroponics, aeroponics, aquaponics, and vertical farming are some of the popularly known methods of soilless agriculture techniques that have the potential to significantly influence future climatic conditions. Hydroponics, aeroponics, and aquaponics are the three soilless agriculture methods that could be used in vertical farming, but they are not exclusive to it. The primary advantage of vertical farming is that it allows for efficient use of space, making it ideal for urban areas where land is limited. Figure 1 illustrates three popular soilless agriculture methods like hydroponics, aquaponics, and aeroponics, which are recognized as viable alternatives to traditional farming worldwide.

Broadly, in some literature "smart indoor factories" and "smart indoor farms" have been used to refer to any indoor growing system that incorporates some level of technology or automation, while "soilless smart agriculture systems" specifically

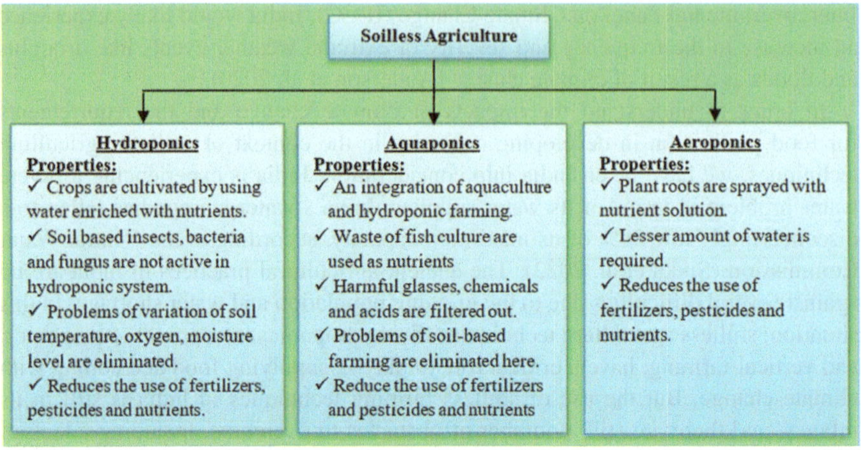

Fig. 1 Features of three popular soilless farming methods: hydroponics, aquaponics, and aeroponics

refers to growing systems that do not use soil and rely on advanced technology and data analytics to optimize plant growth. In order to make it more precise to the readers and to have a clear understanding of the growing systems being discussed, we propose to use the term "indoor soilless smart agriculture systems (ISSAS)" henceforth in this chapter, which highlights the environmental benefits of soilless growing techniques and precise control over growing conditions, as well as the potential for promoting sustainable agriculture. Finally, it emphasizes the innovative nature of technology and its potential for transforming the way we grow food. We will now concentrate on various ISSAS approaches and technology in the discussions that follow.

2.1 Hydroponics, Aquaponics, and Aeroponics

Hydroponics uses mineral nutrient solutions in water to grow plants without using soil. As a means of effective and sustainable food production, it is gaining popularity. Plant growing beds and a reservoir of plant nutrient solution make up the majority of hydroponics systems. Compared to conventional gardening techniques, this soilless farming method produces more while using about 20 times less water (AlShrouf 2017). Romeo et al. (2018) justified that in comparison to traditional open field farms and greenhouse cultivations, the hydroponic farm performs better. Vertical hydroponic farming can outperform the two conventional kinds of agriculture if the source of the electrical input is carbon neutral, such as wind energy.

Sharma et al. (2018) discussed various hydroponic structures, such as wick, ebb and flow, drip, deep water culture and nutrient film technique (NFT) systems, their operations, benefits and limitations, performance of different crops, and water conservation and found that NFT technique has been used commercially for successful production of leafy as well as other vegetables with 70–90% savings of water. In another study, Sambo et al. (2019) state that soilless cultivation requires specific knowledge and skills to manage aspects such as NO3− management and crop quality increase. New technologies such as nanoparticles and PGPRs are being studied, but better knowledge of the processes underpinning the acquisition of nutrients and their allocation in the different tissues is essential. A decoupling of hardware component management from software components will require a service center specialized in smart agriculture.

Hydroponic systems have advantages over field culture systems, such as reuse of water, ease in controlling external factors, and a reduction in traditional farming practices, but have limitations such as high setup cost, rapid pathogen spread, and a need for specialized management knowledge (Lee and Lee 2015). Low-cost techniques are essential for successful implementation of commercial hydroponic technology, which should also try optimization techniques to reduce plant diseases and enhance food quality and quantity.

Aquaponics combines hydroponics and aquaculture (fish farming), producing fish and vegetables in a closed-loop water system. By utilizing the nutrients from the fish waste as a source for nutrients, the plants help to purify the water for the fish. It creates a closed-loop ecosystem by utilizing the waste from one component as a resource for another. A mix of aquaculture and hydroponics, aquaponics collects nutrients from an aquaculture tank rather than from an outside source. Due to their ability to reuse water resources, aquaponics systems use 90% less water than conventional techniques. In order to successfully adopt aquaponics, producers need to start with catfish and then shift to a high-value fish species for niche markets (Bosma et al. 2017). It was concluded that by producing 1250 kg fish, 6000 kg lettuce, and 300 kg tomato per year would have a net-benefit-cost Ratio of 1.3 after 20 years.

Aeroponics involves misting plant roots with a nutrient solution. As the plants grow in an atmosphere of air or mist, their roots are suspended in the air. Aeroponics, which is regarded as a more advanced form of hydroponics, is known for its high yields and effective use of water and fertilizer. Plants in aeroponics systems are suspended in the air, and nutrients are delivered to the roots of the plants using a spray system. When compared to conventional systems, the systems use 95% less water and take up less space. In an aeroponics system, plants exhibit a rapid growth. If the supply of water and nutrients is managed while taking the plants' needs into account, aeroponics systems are inexpensive and offer higher growth rates.

Aeroponics is an innovative and appropriate technology that has the potential to cultivate plants in large quantities, tree saplings associated with soil microorganisms, and reforestation of degraded land in humid regions. It is an indoor horticulture practice that reduces labor cost, consumes less water, fertilizer usage, pesticide and herbicides usage, and maximize plant yield by 45% to 75% (Lakhiar et al. 2018).

The system is an environmentally friendly and economically efficient plant growing system that requires a high level of proficiency and advanced equipment to operate and control.

The evaluation, assessment, and utilization of aeroponics system for commercial plant developing purpose should focus on root research, nutrient concentration, plant spacing, and pest/disease control. Artificial lighting should be used to grow the plant. Aeroponics is a highly specialized cultivation system that can be used in developing countries of the Third World to accommodate intensive food production in areas without fresh water and fertile soils. Future research will focus on understanding why aeroponic cultivation is more productive than hydroponic or soil cultivation, understanding root developmental architecture, understanding the relationship between aeroponic fertilization and daily cycles, identifying aerosol generation technology, and establishing experimental and analytical frameworks for comparison of vertical farming technologies (Eldridge et al. 2020).

The aforementioned three soilless agricultural methods, when combined with smart sensing and control, will increase output while using fewer resources in indoor conditions, proving the necessity of CSAs or indoor soilless smart agriculture systems (ISSAS). Table 1 summarizes some notable research works on three popular ISSAS: hydroponics, aquaponics, and aeroponics.

2.2 *Vertical Farming*

Vertical farming and nurseries (both controlled and uncontrolled environments) for agro-climate regulation through minimal dependence on external input and reduced land footprint. Vertical farming involves growing crops in vertically stacked layers, usually in a controlled environment such as a greenhouse or a warehouse. This type of ISSAS is becoming increasingly popular, because it allows for high-density crop production and efficient use of space, making it ideal for urban areas where land is limited. Some examples of indoor ISSAS are hydroponics, aeroponics, and aquaponics.

Vertical farming is a concept that encompasses a range of technologies and methods used to grow crops in a vertical arrangement. This can be done using hydroponic, aeroponic, or other soilless techniques, but it can also be done using traditional soil-based methods. The primary advantage of vertical farming is that it allows for efficient use of space, making it ideal for urban areas where land is limited.

With this technique, plants are grown in a controlled environment in layers. This makes it possible for plants to grow at a considerably higher density than in conventional horizontal farming. The exact control of the growing conditions made possible by the controlled environment also results in increased yields and less consumption of water and other resources.

Indoor vertical farming is a growing field, with several types of vertical construction, big rooms, little containers, and huge greenhouse farms. Kalantari et al.

Table 1 Summary of notable research works on three popular ISSAS: hydroponics, aquaponics, and aeroponics

Article	Broad area	Methodology adopted	SSAS type/ adoptability	Technological bene-fits/drawbacks
Nalwade (2017)	Hydroponics farming techniques	Automated water delivery and required pH and electrical conductivity (EC) maintenance. When using the root-dipping technique, plants are immersed in the manure mixture. It is used once and then replaced, as opposed to circulating manure blend	Yes/majorly indoors	Automatic maintenance of pH and EC
Nishimura et al. (2017)	Sensor design for hydroponics farming	A new hardware module senses and measures water level and nutrient concentration	Yes/majorly indoors	Measurement accuracy is impacted by the sensor cable's instability in water
Kaewwiset and Yooyativong (2017) and Fuangthong and Pramokchon (2018)	Maintaining EC and pH of hydroponics solution	Fuzzy logic and linear regression algorithms are utilized to calculate the amount of nitric acid needed to fill the hydroponics reservoir and maintain the desired EC and pH levels	Yes/majorly indoors	The accuracy for regulating pH and EC using linear regression is 95% and 80.8%, respectively
Eridani et al. (2018)	Automatic nutrition level controlling of hydroponics solution	Proximity sensor for detecting water level, total dissolved solids (TDS) sensor for measurement of EC of nutrient solution. Automatic nutrient controlling using nutrient film technique	Yes/majorly indoors	TDS sensor gives 97.8% accuracy
Kyaw and Ng (2017)	Smart aquaponics system	Through a cloud server, the processing unit is connected to mobile and Web applications for the control of water quality, light intensity, and fish feeding	Yes/majorly indoors	The user can remotely control the parameters
Lopes et al. (2017)	Fish farming: automatic biomass estimation	The structured light vision system, which utilizes a camera and laser, is used to create 3D models of fish	Yes/majorly indoors	Estimating fish growth is possible
Idris and Sani (2012) and Sani et al. (2017)	Monitoring and control of aeroponics farming system	This system regulates the delivery of nutrients, the caliber of the growing medium, pH, temperature, and humidity. In order to improve the efficiency of resource utilization, an ultrasonic mist producer and fan are employed for spraying after a set period of time	Yes/majorly indoors	Efficient utilization of the available water and nutrients

(2017) found that an automated plant factory equipped with optimized light emitting diode (LED) lighting, renewable energy sources, smart water management systems, crop planning and management systems, artificial climate control systems, soil and fertilizer management systems, and smart data collection and management systems can significantly impact the agriculture sector. Vertical greenery systems are indoor agricultural systems integrated with vertical buildings that reduce average energy consumption in buildings and contribute to the sustainable growth of populous cities by generating fresh air and reducing the temperature of the environment (Singh et al. 2017). Suparwoko and Taufani's (2017) performed analysis of the green building concept for Sleman, Indonesia, found that this innovative approach to urban farming not only boosts agricultural productivity but also lessens the shortage of arable land.

2.3 *Other Soilless Methods*

Some of the other soilless techniques are listed below, albeit they are outside the purview of this chapter, in addition to the characteristics of common indoor soilless smart agricultural systems shown in Fig. 2.

Substrate-based Systems Supporting the roots of the plants using a solid medium, such as peat moss, perlite, or rockwool.

Drip Irrigation Using a network of tubes and emitters to provide water and nutrients to plant roots is known as drip irrigation.

Ebb and Flow Ebb and flow is the process of flooding and draining a growth tray with nutritional solution utilizing a series of pumps and timers.

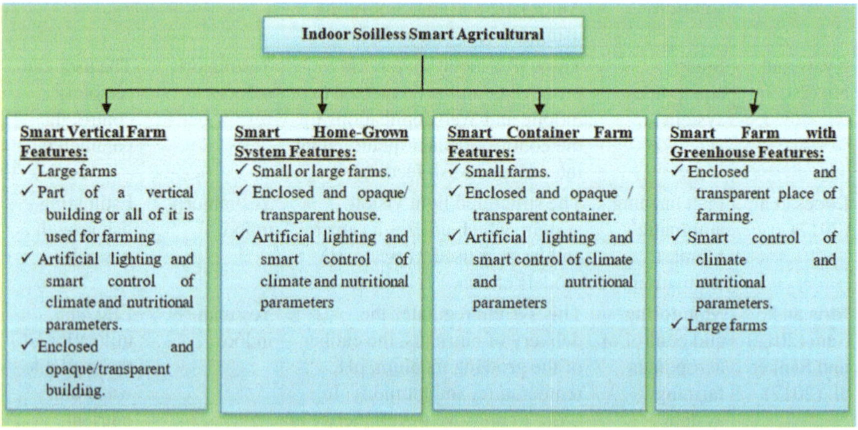

Fig. 2 Features of Prevalent Indoor Soilless Smart Agricultural Systems

NFT (Nutrient Film Technique) Utilizing a stream of water that is shallow and contains all the dissolved nutrients necessary for plant growth is known as NFT (nutrient film technique).

3 Indoor Soilless Smart Agriculture Systems (ISSAS): Methodologies

Some of the key methodologies used in ISSAS include artificial climate control, crop planning, plant disease detection, artificial lighting, and smart nutrition management. These techniques are designed to optimize plant growth and maximize crop yields while minimizing environmental impact. Artificial climate control involves creating and maintaining the ideal temperature, humidity, and other environmental conditions for plant growth. Crop planning involves using data and analytics to plan the timing and location of crop planting and harvesting, based on factors, such as weather patterns and market demand. Plant disease detection involves using sensors and other technology to detect and diagnose plant diseases early, allowing for prompt treatment and prevention of crop losses. Artificial lighting can be used to supplement natural light in indoor ISSAS, providing the optimal light spectrum and intensity for plant growth. Finally, smart nutrition management involves carefully monitoring and adjusting the nutrient levels in the growing medium to ensure that plants receive the right balance of essential nutrients for optimal growth.

There are a few other methodologies that are commonly used in soilless smart agriculture systems, including:

Automated Irrigation Systems These systems use sensors and software to monitor soil moisture levels and automatically adjust watering schedules to ensure that plants receive the right amount of water.

Remote Monitoring and Control This involves using sensors and cameras to monitor plant growth and environmental conditions and remotely controlling various aspects of the growing environment, such as temperature and lighting.

Data Analytics and Machine Learning By collecting and analyzing large amounts of data on plant growth, environmental conditions, and other factors, farmers can use machine learning algorithms to optimize crop production and minimize resource inputs.

Integrated Pest Management This involves using a combination of biological, chemical, and cultural methods to manage pests and diseases in an environmentally sustainable way.

Overall, these methodologies are essential for achieving the full potential of soilless smart agriculture systems. By integrating these technologies and techniques, farmers can produce high-quality crops with higher yields, lower resource inputs, and reduced environmental impact but with maximum efficiency, making them a promising technology for the future of agriculture.

3.1 Artificial Climate Control

Climate smart agriculture (CSA) technologies are being used to cope with harsh biophysical conditions, such as flood, drought, soil erosion, heavy precipitation, etc. ISSAS do not experience adverse weather conditions, so CSA technologies can be used to regulate the indoor climate (Morton et al. 2017; Khatri-Chhetri et al. 2017; Mwongera et al. 2017). Popa and Ciocarlie (2011) created a distributed smart indoor climate control system that connects data-gathering nodes, servers, clients, and actuators over the Internet. Microclimatic factors are controlled using a variety of techniques, such as adaptive control of outdoor climate, proportional integral derivative (PID)-based control, fuzzy logic-based control, artificial neural network (ANN)-based management system, and neuro-fuzzy-based control (Ardabili et al. 2016; Afram et al. 2017).

The ideal indoor environmental state is achieved by monitoring and maintaining key factors in a predetermined range (Wicaksono et al. 2018). A self-tuning PID controller has been used to keep temperature and humidity within a preset range (Heidari and Khodadadi 2017; Janprom et al. 2017). A fuzzy immune PID controller provides greater dynamic performance (Revathi et al. 2017). To simulate the heating requirements of greenhouses, Ahamed et al. (2018) suggested a quasi-steady state thermal model.

Indoor farming is characterized by heat transmission by conduction and convection, air exchange, heat exchange through the floor and perimeter, and evapotranspiration. The nature of greenhouse ventilation rate and other microclimatic factors is nonlinear and non-affine, so fuzzy logic systems are used to simulate the system's unknowable dynamics and monitor the system's output parameters. Indoor environments can also use a dynamic climate model of greenhouses to calculate the climatic state (Su and Xu 2015; Taki et al. 2016).

The Kalman filter eliminates sensor noise and processes noise to reduce inaccuracy and smooths a climate control system's control signals (Shi et al. 2012). Particle swarm optimization (PSO)-based nonlinear model predictive control (MPC) algorithms can maximize the objective function while using the least amount of energy (Zou et al. 2010). Multi-objective evolutionary algorithms (MOEAs) seek for control signals in the solution space (Member 2010).

Researchers have used thermal modelling to better understand how different internal designs and building materials affect a building's microclimate (Kisilewicz 2015). Phase change materials (PCM) are often used in light-weight buildings to prevent an abrupt change in the outer environment from having an impact on the indoor climate (Li et al. 2015a, b).

Green wall planting is an economical technique that serves as both an air filter and a cooling insulator. Buildings use green wall vegetation, micro pot plants, or pocket plants as three types of green wall planting to protect the inside environment from the effects of external heat (Lee and Chuang 2017). Plant walls improve indoor microclimate by lowering particulates and stabilizing carbon dioxide levels (Liu et al. 2018). Natural nighttime ventilation in desert regions lowers the energy needed for cooling, and researchers have created a thermo-aerodynamic numerical model of natural night ventilation that may also be used for indoor farming (Hamdani et al. 2017). The design and installation of sensing and actuating components, air conditioning, ventilation, thermal insulation, and the best automation algorithms are difficult, but ISSAS can be monitored and controlled effectively by gathering pertinent farm data from distant areas. Table 2 summarizes some of the notable research works on enabling technologies for artificial climate control.

3.2 Crop Planning

Crops grown by farmers include food crops (rice, wheat, maize, pulses, vegetables, fruits, etc.), plantation crops (cotton, coffee, tea, cocoa, oil seeds), horticulture crops (fruits, vegetables, spices, beverages, nuts, etc.), forage crops (barley, grass, alfalfa, etc.), and manure crops (beans, red clover, lupin, winter tare, etc.). These crops are grown in various seasons based on the availability of irrigation water, land, weather, and fertilizer use. An optimized cropping pattern that takes into account all the relevant elements would help to increase productivity (Saranya and Amudha 2017).

Indoor farming is only possible indoors due to soil type, nutrients, water resources, fertilizers, pesticides, harvesting techniques, and economic profitability. Machine learning approaches make it easier to find a cropping pattern while taking into account all the relevant constraints (Kumar et al. 2015).

Machine learning techniques include artificial neural networks (ANN), information fuzzy networks (IFN), decision trees, regression analysis, clustering techniques, principal component analysis (PCA), Bayesian belief networks, time series analysis, Markov chain models, etc. ANN is a type of supervised learning technology that makes predictions about the future based on training models created from training data. IFN, a supervised learning algorithm, builds a fuzzy network. Regression analysis uses statistical techniques to determine the relationship between various variables. Clustering is an unsupervised machine learning process that divides the dataset into groups. PCA identifies uncorrelated variables, and Bayesian networks express conditional dependencies of variables using a graphical and probabilistic model. A probabilistic mathematical model called the Markov chain model determines output based on prior knowledge (Mishra et al. 2016).

Cropping pattern prediction is an optimization issue that is subject to several restrictions. Piecewise genetic algorithm (PWGA) is used to identify the best solution for the crop pattern and water allocation problem. To prevent a water disaster, crop patterns must be optimized based on water resource availability. Particle swarm optimization (PSO), simulated annealing (SA), and other meta-

Table 2 Summary of notable research works on enabling technologies for artificial climate control

Article	Broad area	Adopted methodology	SSAS type/ adoptability	Technological bene-fits/drawbacks
Popa and Ciocarlie (2011)	Distributed smart system for monitoring and control of indoor temperature	LPC 2148 microcontroller based on AR7TDMI-S-based core, ENC28J60 Ethernet controller communicating with microcontroller using IEEE 802.3 compliant SPI (serial peripheral interface) interface	Indoor. Adoptable in all forms of indoor farming	Fast data transfers and affordable infrastructure
Wicaksono et al. (2018)	Smart temperature control of poultry farm	Temperature and humidity sensor, WSN and IEEE 802.15.4 protocol for communication	Broiler poultry farms. Adoptable in all forms of indoor farming	Low error percentage in sensing. 1.51% is the highest error value
Heidari and Khodadadi (2017) and Revathi et al. (2017)	Climate control of green house	Fuzzy logic-based proportional integral derivative controller used for the purpose of actuation	Greenhouse. Adoptable in all forms of indoor farming	The fuzzy logic-based controller has self-tuning capabilities
Ahamed et al. (2018) and Taki et al. (2016)	A study of heating requirements and energy consumption of greenhouses	Heat transfer model considering heat loss due to plant evapotranspiration and environmental heat gain etc.	Greenhouse. Adoptable in all forms of indoor farming	Applicable for commercial greenhouses
Su et al. (2016)	Climate control and dealing with actuator saturation problem	Fuzzy logic system (FLS) for estimation of unknown nonlinear parameters of the control system	Greenhouse. Adoptable in all forms of indoor farming	Successfully tested to estimate the ventilation rate
Su and Xu (2015)	Simulation of greenhouse climate model	Modeling of convection, condensation, ventilation, transpiration, photosynthesis, respiration, etc., using algebraic fitting technique	Greenhouse. Adoptable in all forms of indoor farming	Temperature, humidity and carbon dioxide content can be predicted

(continued)

Table 2 (continued)

Article	Broad area	Adopted methodology	SSAS type/ adoptability	Technological bene-fits/drawbacks
Shi et al. (2012)	Climate control of greenhouses	Extended Kalman filter algorithm is used to estimate the control states and filter out the noises	Greenhouse. Adoptable in all forms of indoor farming	Useful for control systems with nonlinear system dynamics
Zou et al. (2010)	Green house climate control	Internal temperature is controlled using particle swam optimization (PSO)with the help of solar radiation, wind speed, outside temperature, ventilation, etc. parameters	Greenhouse. Adoptable in all forms of indoor farming	Energy consumption is reduced
Kisilewicz (2015), Li et al. (2015a, b), Lee and Chuang (2017) and Liu et al. (2018)	Controlling of indoor climate for buildings	EnergyPlus software for building energy simulation, lightweight buildings constructed with phase change material (PCM) to reduce the room temperature and heat flux inside the room, smart plant wall, etc.	Vertical buildings. Adoptable mainly in the vertical farm buildings	Contribute to reduce urban heat effect

heuristic algorithms can be employed to tackle the multi crop planning (MCP) problem (Bou-Fakhreddine et al. 2016). The best cropping strategy for managing the water resource can be determined using a multi-objective fuzzy stochastic model based on GA (Dutta et al. 2016). When four popular evolutionary algorithms (EA) are compared to find the best crop pattern when there is a limited amount of normal and sufficient water resources, PSO, DE, and EP perform better than GA (Pant et al. 2010).

The Lingo software tool was used to develop an optimum crop pattern for various seasons in a case study of the Rajolibanda Diversion Scheme area in Mahabubnagar, Andhra Pradesh, India (Rani 2012). Trials were conducted on a variety of crops to determine the pattern that yielded the greatest profit. Linear programming was used to optimize the cropping pattern in three locations of Egypt to maximize the annual profit while controlling the limitations of the available water and land resources (Osama et al. 2017). In India's Karnataka state's Markandeya command region, the cropping strategy was adjusted using linear programming to achieve the greatest profit to make the best use of irrigation water (Shreedhar et al. 2015; Chowdhury and Chakrabarty 2015).

The most popular methods for predicting agricultural yield are fuzzy logic, multiple linear regression, artificial neural networks, and adaptive neuro-fuzzy inference system (ANFIS). The accuracy of the projected outcome is examined using RMSE, mean square error (MSE), and correlation coefficient approaches. ANFIS is more accurate than other approaches due to its ability to take into account all relevant internal and external elements (Yuvaraj and Dolui 2021). Machine learning and optimization methods have been used to optimize the cropping pattern of ISSAS. Designing an effective decision-making module can help produce an optimal cropping pattern, boosting productivity while guaranteeing the viability of the farm. Some of the notable research works on enabling technologies for crop planning have been summarized in Table 3.

3.3 Detection of Plant Diseases

Plant diseases are caused by the host plant being vulnerable to a specific disease or illness, the presence of plant diseases in the host plant, and the environment, which helps plant pathogens thrive and produce spores. To increase output, ISSAS produce artificial climates to promote the development and spread of plant disease spores.

Intelligent computer vision-based periodic monitoring of plants can lead to early detection of plant illnesses, allowing for the earliest implementation of curative therapies. Plants can be monitored using cameras installed on robotic platforms or incorporated into other systems, and sick plants can be identified by looking at the photos. This can be accomplished using both RGB and NIR (near-infrared spectroscopy)-based cameras. In comparison to NIR-based detection, the performance of the RGB camera is better (Schor et al. 2016). An integrated system that combines machine vision and the Internet of Things can be used to detect crop infections early and apply prompt cures (Tanmayee 2017). Drones, also known as unmanned aerial vehicles (UAVs), have been adopted for smart agricultural tasks. Using UAV for plant disease diagnosis and other tasks in an indoor agricultural environment would be an intriguing idea (Castelao Tetila et al. 2017).

The research on plant diseases and the algorithms used to detect them is covered in this part. One example is the intelligent classification of damage in sugarcane billets and correlation of it with sugarcane germination using computer vision technology (Alencastre-Miranda et al. 2018). Infections, like powdery mildew (PM) and tomato spotted wilt virus (TSWV), have been determined using algorithms like principal component analysis (PCA), neural network, support vector machines (SVM), etc. Plants with PM and TSWV infections can be found with a high degree of accuracy using PCA. Spots on the leaves of field crops, forages, and vegetables are caused by the fungus *Septoria*, while wheat is impacted by yellow rust. SVM-based classification algorithms are more effective than artificial neural network (ANN) at detecting *Septoria* and yellow rust (Han et al. 2015).

Baquero et al. (2015) used the nearest neighbor algorithm to identify the six prevalent diseases of tomato plants, including early blight, chlorosis, sooty molds, powdery mildew, necrosis, and white fly. Early blight damages stems, fruits, and leaves and causes defoliation and sunscald. Color descriptors such as CSD, CLD, and SCD are used to identify regions of interest (ROIs). A 1-NN classifier is used to distinguish between healthy and diseased plants, and image segmentation is used to extract pertinent characteristics (Molina et al. 2015). Careful picture segmentation

Table 3 Summary of notable research works on enabling technologies for crop planning

Article	Broad area	Adopted methodology	SSAS type/ adoptability	Technological bene-fits/drawbacks
Saranya and Amudha (2017)	Crop planning optimization research and review	Crop planning based on various factors, i.e., irrigation, land, labor, soil, climate, transportation, fertilizers and pesticides, weed, etc.	Outdoor. Crop planning based on labor, soil, transportation, fertilizers, pesticides, and weed are relevant for indoor farming	Role of bioinspired optimization algorithm discussed
Kumar et al. (2015)	Crop selection to maximize crop yield	Proposed crop selection method (CSM) algorithm based on predicted yield, sowing time and days of plantation	Outdoor. Crop planning based on soil type, water density, and crop type is relevant for indoor farming.	Overall crop yield rate is increased
Ghasemi et al. (2016)	Crop pattern optimization	Piecewise genetic algorithm (PWGA) is used to find the optimal crop pattern. A ground water model is used to solve the water allocation problem	Outdoor. Water allocation solutions can the adopted in the scope of indoor farming	Piecewise genetic algorithm (PWGA) deals with nonlinearity and handles large number of variables involved
Bou-Fakhreddine et al. (2016), Pant et al. (2010), Rani (2012), Osama et al. (2017), Shreedhar et al. (2015) and Chowdhury and Chakrabarty (2015)	Crop planning under deficit irrigation situation	Simulated annealing (SA), particle swam optimization (PSO), and linear programming (LP) are used to maximize profit when water supply is not enough. Evolutionary algorithm, such as genetic algorithms (GA), particle swarm optimization (PSO), differential evolution (DE), and evolutionary programming (EP), is used for optimization	Outdoor. Adoptable in all forms of indoor farming	Agriculture with low water availability
Dutta et al. (2016)	Optimization of crop pattern subjected to total supplied water in an agricultural farm	Genetic algorithm and fuzzy stochastic programming are used in this process	Outdoor. Adoptable in all forms of indoor farming	Increase of irrigated area with fixed water supply

is required to identify the pertinent image segment required to classify the plant diseases (Singh and Misra 2017). An optimization approach like the genetic algorithm (GA) is used to eliminate duplicate features and complexity (Ghyar and Birajdar 2018). Researchers have proposed a convolutional neural network-based method to classify an image dataset of 3750 images into six classes, i.e., healthy plant, early blight, late blight, yellow leaf curl virus, spider mite damage, and bacterial spot (Golhani et al. 2018; Bhatt et al. 2017). This method uses the learning capabilities of neural networks (NN) to make it one of the successful classifiers of hyper spectral images. A fuzzy logic-based classification algorithm has been effectively tested to distinguish between iron-deficient or infected strawberry leaves and healthy strawberry leaves, which mimics the abilities of experienced farmers to categorize sick crops (Ghyar and Birajdar 2018).

A Web-based tool was used to identify diseased pomegranates (Bhange and Hingoliwala 2015). The collected photos were used to extract features based on color, morphology, and color coherence vector. The training dataset was first clustered using the K-means approach before being fed into the support vector machine (SVM). Many bacterial and fungal diseases, including bacterial blight, fruit spot, fruit rot, leaf spot, etc., can infect pomegranate plants. Noise filtering strategy would improve classification accuracy. The rotating kernel transform (RKT) features, its modified versions, or other directional features correctly reflect the picture information, since the input images of leaves, fruits, and stems include edge information and directional statistics (Ullagaddi and Raju 2017). Mobile image capture is a low-cost and low-energy method of taking pictures with mobile phones, but its inability to capture fine details present in an image (Prasad et al. 2014).

Plant diseases are caused by plant pathogens' ability to survive in favorable environmental circumstances, so it is important to understand how environmental factors and plant diseases are related. Beta regression models can be used to determine the relationship between environmental factors, such as temperature, humidity, leaf wetness, etc., and plant infection (Shivling et al. 2016). A successful model will aid in the prediction of various plant diseases and notify farmers to consider essential treatments. Computer vision approaches are used to enhance the utility of decision-making modules and IoT infrastructure. A number of research works have been summarized in Table 4 on enabling technologies for plant disease detection.

3.4 Artificial Lighting

Researchers have conducted tests to determine a different energy source for indoor farming environments, particularly for vertical farming infrastructure. The primary artificial lighting sources are fluorescent lamps, high-intensity discharge (HID) lamps, and light emitting diodes (LED). LEDs are less expensive, produce less heat, and offer the highest levels of photosynthetically active radiation (PAR) efficiency (Darko et al. 2014). Olle and Virile (2013) investigated how green vegetables and a

Table 4 Summary of notable research works on enabling technologies for plant disease detection

Article	Broad area	Name of disease	Adopted methodology	SSAS type/adoptability	Technological benefits/drawbacks
Schor et al. (2016)	Smart disease detection of plants	Powdery mildew (PM) and tomato spotted wilt virus (TSWV)	For the detection of PM, principal component analysis (PCA) and leaf condition classification (LCC) are used. For TSWV, PCA-based analysis and coefficient of variation have been used	Greenhouse. Adoptable in all forms of indoor farming	For the detection of PM, PCA gives an accuracy of 95.2%, and LCC gives an accuracy of 64.3%. For the detection of TSWV, PCA gives an accuracy of 90%, and coefficient of variation gives an accuracy of around 85%
Castelao Tetila et al. (2017)	Soybean disease detection	Soybean foliar diseases	Images are captures using unmanned aerial vehicle (UAV). Simple linear iterative clustering (SLIC) methods are used for segmentation of super pixel	Outdoor. Adoptable in all forms of indoor farming	Detection accuracy of 98.34% (at height between 1 and 2 meter)
Alencastre-Miranda et al. (2018)	Robotics in sugarcane cultivation	Billet damage detection of sugarcane	Computer vision and robotics	Outdoor. Sugarcane is not generally cultivated in indoor conditions. Computer vision and robotics technique can be adopted	Around 90% accuracy
Han et al. (2015)	Automatic crop disease detection	Wheat disease: *Septoria* and yellow rust	Support vector machine (SVM) and artificial neural network (ANN)	Outdoor. Wheat is not preferred at indoor condition, but SVM and ANN can be used for detection of other diseases	For SVM, the accuracy to detect *Septoria* is 70% and yellow rust is 95%. By using ANN, the accuracy for detecting *Septoria* is 72% and for yellow rust is 53%
Molina et al. (2015)	Automatic disease detection of tomato crops	Early blight infection	Color descriptors such as color structure descriptor (CSD), color layout descriptor (CLD), and scalable color descriptor (SCD) are used in the process	Outdoor. Adoptable in all forms of indoor farming	The maximum accuracy is around 99% using CSD descriptor

(continued)

Table 4 (continued)

Article	Broad area	Name of disease	Adopted methodology	SSAS type/adoptability	Technological benefits/drawbacks
Ghyar and Birajdar (2018)	Detection of rice leaf diseases	Leaf blast and brown spot detection	Genetic algorithm for feature selection and ANN and SVM is used for classification purpose	Outdoor. Rice is not preferred at indoor condition, but SVM and ANN can be used for detection of other diseases	For ANN, accuracy is 87.5%, and for SVM, accuracy is 92.5%
Kiani and Mamedov (2017)	Soft computing technique detection of plant infection	Iron deficiency and fungal disease of strawberry leaves	Fuzzy logic-based classifier is used to avoid computing complexity	Outdoor. Adoptable in all forms of indoor farming	Accuracy is above 92%
Bhatt et al. (2017)	Tomato crop health analysis	Early blight, late blight, yellow leaf curl virus, spider mite, damage, bacterial spot	Convolutional neural network (CNN) is used for classification	Offline image analysis. Adoptable in all forms of indoor farming	Accuracy is 99.7%
Bhange and Hingoliwala (2015)	Pomegranate disease detection	Bacterial blight disease	K-means for clustering and SVM for classification purpose	Offline image analysis. Adoptable in all forms of indoor farming	Accuracy is approximately 82%
Ullagaddi and Raju (2017)	Disease recognition of mango crop	Anthracnose disease	Modified rotation kernel transformation based directional feature and artificial neural network based disease identification	Offline image analysis. Adoptable in all forms of indoor farming	Maximum accuracy 98%

few other plants responded to artificial LED light in terms of metabolism, growth, and photosynthesis. The results showed that plants need red and blue light for photosynthesis, with far red light having greater effects on photomorphogenetic processes and plant growth. Plants typically respond physiologically to the colors green and yellow, but the red and blue portions of efficient light spectrum for artificial farming are larger (Urrestarazu 2018).

The experiment was carried out in four different LED lighting environments to observe the effects of LED irradiance on tomato plantlets. Results showed that the maximum photosynthetic rate was recorded under lighting conditions with a red to blue ratio of 10:1, but the highest growth in plant height was observed under conditions with 100% red LED (Naznin and Lefsrud 2014). Another experiment with *Brassica chinensis* showed that continuous light therapy outperformed pulsed light treatment (Harun et al. 2016). The basic metrics that are recorded and compared to gauge the overall growth of the plants are leaf count, plant height, fresh weight, dry weight, moisture content, and chlorophyll content.

Pepper plants exhibit improved morphology when environmental elements, such as temperature, carbon dioxide level, humidity, water cycle, and photosynthetic photon flux density value (PPFD), are regulated (Liang et al. 2018). The Osaka Prefecture University in Japan used an artificial hybrid LED light source (i.e., mixtures of red, blue, white, and far infrared rays) and report their optimal pulse width modulation duty cycle, light intensity, and frequency of luminance (Sugano 2015). Hop crops are typically radiation-sensitive, so a specially created LED lighting system with two channels (red and blue) may manage the radiant flux of the channels to create a supportive atmosphere (Tavares et al. 2018). Commercial LED lighting modules implement a control mechanism to optimize electrical energy consumption and boost photosynthesis rate (Almeida et al. 2014). Exposure to red and blue LEDs causes the enrichment of carotenoids and chlorophyll a and b, which increases photosynthesis (Wojciechowska et al. 2013).

Artificial illumination combined with continuous lighting can improve photosynthesis rates and overall plant growth. Only the most useful portion of the light spectrum should be used, and pulsed red and blue lighting produces the best photosynthetic rate. More research is needed to develop novel systems that maximize the use of light energy while saving electrical energy. In Table 5, we summarize some notable research works on enabling technologies for artificial lighting.

3.5 Smart Nutrition Management

In the context of soilless smart agriculture systems in an indoor setup, smart nutrition management involves using technology and data to optimize plant growth and nutrient uptake in the absence of traditional soil-based growing methods. This can include using hydroponic or aeroponic systems to grow plants in nutrient-rich

Table 5 Summary of notable research works on enabling technologies for artificial lighting

Article	Broad area	Adopted methodology	SSAS type/adoptability	Technological benefits/drawbacks
Darko et al. (2014)	Photosynthesis and plant metabolism in artificial light	Reviews all possible artificial light sources, effect on plant growth, and metabolism with changing light intensity, quality etc.	Indoor. Adoptable in all forms of indoor farms	Improved plant growth and better production
Olle and Viršile (2013)	LED lighting in greenhouse	Explores how LED lighting effects on photosynthesis, nutrition values, production, and growth of different green vegetables	Greenhouse. Adoptable in all forms of indoor farms	Improved plant growth and better production
Urrestarazu (2018)	Artificial lighting in agriculture	Explores measurement of spectrum, quality, colors, photoperiod, photo chrome of artificial light, and their effects to plants	Indoor. Adoptable in all forms of indoor farms	Only effective part of light spectrum can be used for better growth and photosynthesis rate
Naznin and Lefsrud (2014)	Effect of LED irradiance on photosynthesis	Conducted experiment to observe photosynthesis rate under four light treatments with different ratios of red to blue light	Indoor. Adoptable in all forms of indoor farms	Only effective part of light spectrum can be used for better growth and photosynthesis rate
Harun et al. (2016)	Remote monitoring of plant growth in LED plant factory	Observed plant growth parameters in continuous and pulse lighting condition. WSN used for monitoring purpose	Indoor. Adoptable in all forms of indoor farms	Automated LED control to get better growth rate
Kárász and Kopják (2017)	Study of different LED driving methods	Reviewed LED source with different ratio of red, blue, and white light at different wavelength, intensity, duty cycle. Corresponding sensing elements are also discussed	Indoor. Adoptable in all forms of indoor farms	Comparative study helps to understand the difference among different LED driving methods
Sugano (2015)	Key technologies required for plant factory with artificial lighting	Robotic technology for screening of seedlings, transport system with smart control and wireless communication system, artificial lighting and ventilation system, data processing facility, building construction with insulation from external heat	Large-scale indoor farms. Adoptable in all forms of indoor farms	The plant factory can produce around 5000 heads of lettuce per day

Tavares et al. (2018)	Hop (*Humulus lupulus*) cultivation in two-channel LED lighting system	Two-channel LED lighting array of red and blue lights are used. Photo-electro thermal (PET) is used to estimate radiant flux of lighting system and calculate the operating points	Indoor. Adoptable in all forms of indoor farms	Hops are generally grown at low latitude outdoor location. Lighting needs of hops are artificially created in this experiment.
Almeida et al. (2014)	LED lighting system to support plant physiology	Design of an electronic system so that controlled current can be sent to LED strings. A spectroradiometer Labsphere CDS610 (350–1000 nm) with SpectraSuite utility equipped with photosynthetically active radiation (PAR) plug-in is used for PAR radiation measurement	Indoor. Adoptable in all forms of indoor farms	Efficient and low-cost design
Harun et al. (2013)	Artificial lighting for *Brassica chinensis* farming	12 h of continuous lighting system and pulse lighting are compared. Red and blue LED channels (ratio 16:4) are used as source of light	Indoor. Adoptable in all forms of indoor farms	Pulse lighting treatment gives 291% higher photosynthesis rate
Wojciechowska et al. (2013)	Impact of LED lighting on *Valerianella locusta*	Effect of white and red plus blue LED lamp on photosynthesis rate, transpiration, photosynthesis pigment content, etc. is measured	Indoor. Adoptable in all forms of indoor farms	Red plus blue lighting condition gives the best performance

water or mist, as well as using sensors and automation to monitor and adjust nutrient levels in real time.

One advantage of soilless smart agriculture systems is that they allow for greater control over plant nutrition, since growers can precisely monitor and adjust nutrient levels to meet the specific needs of each crop. This can help to minimize nutrient waste and reduce the environmental impact of agriculture while also producing higher yields and healthier plants. In addition, smart agriculture systems can help to reduce labor costs and improve efficiency, since growers can use data and automation to optimize growing conditions and minimize the risk of crop failure.

Overall, smart nutrition management is a critical component of soilless smart agriculture systems in an indoor setup, since it allows growers to optimize plant growth and nutrient uptake in the absence of traditional soil-based growing methods. By using technology and data to monitor and adjust nutrient levels in real time, growers can produce healthier plants, higher yields, and more sustainable agriculture practices.

Nitrogen, phosphorus, and potassium are essential for plant growth, while secondary nutrients such as sulfur, calcium, and magnesium are needed (Gruhn et al. 2000). To determine soil fertility, pH, electrical conductivity, organic carbon, primary and secondary nutrients, soil texture, density, water-retention capacity, etc. can be measured (Kumar et al. 2017).

The management of plant nutrition includes the balanced and ideal application of fertilizer. Imam et al. have used an integrated artificial neural network (ANN) and bidirectional improved particle swarm optimization to optimize the fertilizer dose (Cholissodin et al. 2017). To teach farmers how to utilize fertilizers most effectively, OFRA created the fertilizer optimization tool (FOT) for 65 agroecological zones (AEZs) and 14 crops (Macharia et al. 2016). The farm output would increase if intelligent indoor farms included smart nutrition management.

4 Enabling Technologies for Indoor Soilless Smart Agriculture Systems

Indoor soilless smart agriculture systems are a promising solution for sustainable and efficient food production, using sensors, automation, and artificial intelligence to monitor and optimize plant growth in a controlled environment. This chapter will explore the latest developments in enabling technologies for these systems, including their benefits, challenges, and potential applications.

4.1 ISSAS Generic Architecture

A typical indoor soilless smart agriculture system (ISSAS), as shown in Fig. 3a, might include a farming bed connected to a number of sensing components to

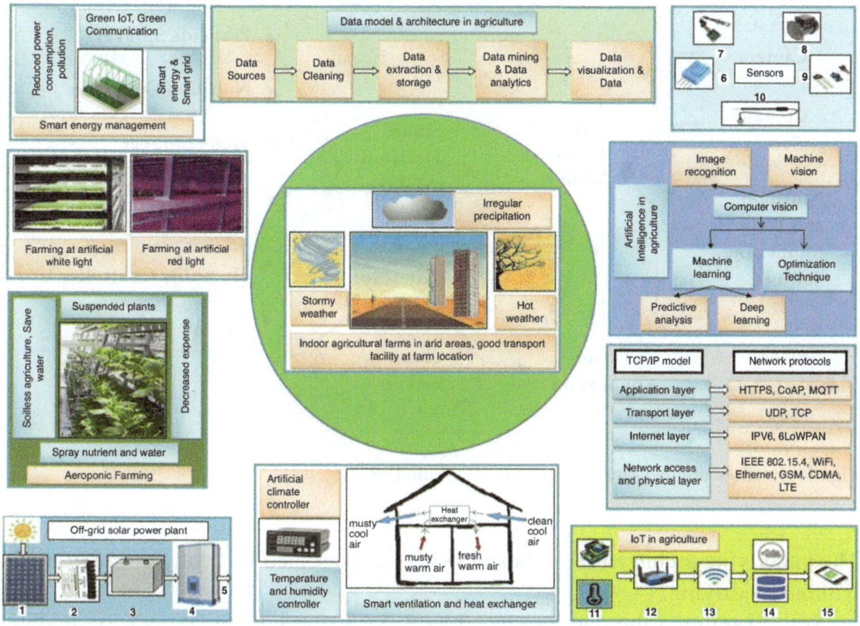

Fig. 3 Typical architecture of a standalone indoor soilless smart agriculture systems. (1: solar panel, 2: charge controller, 3: battery, 4: inverter, 5: external load, 6: temperature and humidity sensor, 7: pH sensor, 8: airflow sensor, 9: soil moisture sensor, 10: electrical conductivity sensor, 11: sensors and devices, 12: gateway, 13: wide-area network, 14: data storage, processing, and cloud server, 15: user application)

gather pertinent data (such as information about the moisture condition of the soil, lighting conditions, water level, temperature, and humidity), as well as short-range and long-range communication channels for relaying the data to processing components for local processing and to cloud computing platforms for long-term analysis, respectively. The lighting, temperature, humidity, pH, water levels, etc. are all controlled by an advanced decision-making module. For quick visualization and actuation, a mobile device with a mobile application is also employed.

The core components of ISSAS are thought to be data management and analytics. As shown in Fig. 4 for a typical aeroponic SSAS, data is gathered, prepared, analyzed, and then provided to the predictive model to determine the next course of action. In this method, the necessary nutrients, water, etc. are misted onto the plant roots. A crucial component of the ISSAS is artificial smart lighting. In order to boost the production of plants, a formula of precise spectrum and intensity of lighting is created using a combination of LED lights. The plants flourish more than they would in a traditional agricultural farm thanks to careful micro- and macronutrient feeding. The typical life cycle of many dangerous pests is disrupted by the smart regulation of microclimatic features and controlled smart growth methods, resulting in a higher yield. In terms of energy requirements, it is a standalone solar off-grid system.

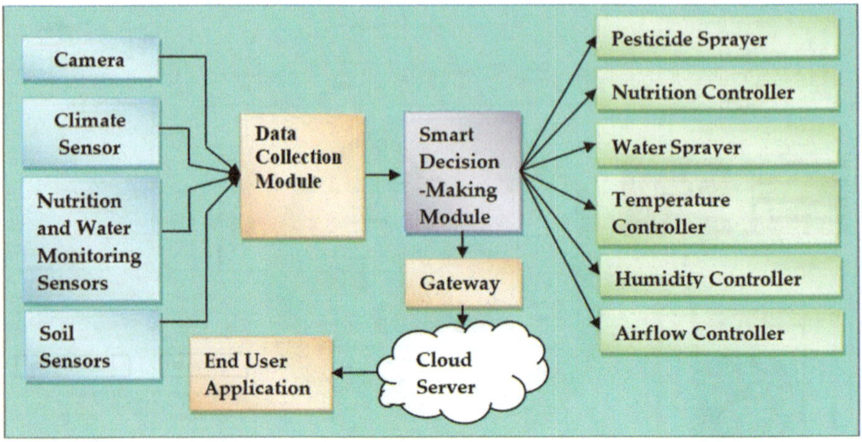

Fig. 4 Dataflow architecture of a typical aeroponic ISSAS

The underlying architecture of these ISSAS can be scaled vertically or horizontally without modifying.

After examining the data produced by ISSAS, the intelligent decision-making module makes intelligent decisions and starts the actuation process. Its architecture, which is capable of learning from prior data, is shown in Fig. 5. It has access to both original sensor-generated data and supplementary databases for the soil, climate, and other pertinent domains. In most cases, obtained data are not ordered, thus preprocessing is done on them before characteristics are extracted. A forecasting model is constructed and optimized using extracted information. By evaluating real-time data gathered from monitoring climate conditions, soil, plant nutrition, plant development, and plant health-related aspects, the validated forecasting model aids in making dynamic and intelligent judgments.

Wireless sensor networks installed in the farm assist in the monitoring process. Primary data, or sensor data, are gathered periodically. Plant growth and related factors are recorded using camera sensors. The gathered data provides an understanding of the current stage of the plant life cycle, which is then fed into the forecasting module to produce optimum actuations. The data is afterwards saved in local and cloud servers. Algorithms based on artificial intelligence are crucial for data analysis and forecasting model construction. The architecture also gains a particular capability for remotely controlling sensing and actuating devices, thanks to cloud-based control. The end-user application provides users with only the most pertinent and important indicators. The graphical user interface offered by the Web server enables the user to visualize the state of the plant or factory and also provides alerts in the event of a technical issue. As part of the analysis of the data, the user's contribution is also a crucial factor.

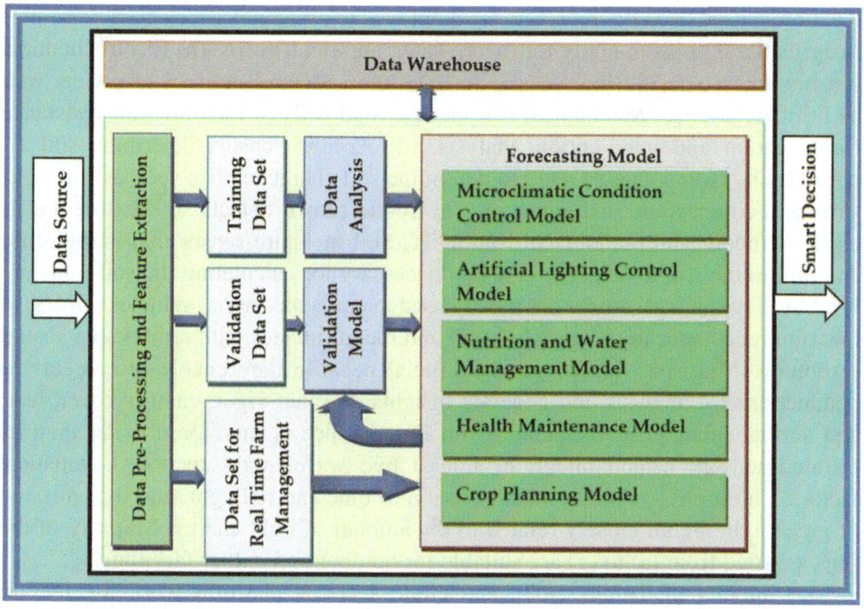

Fig. 5 Architecture of smart decision-making module

4.2 *Internet of Things*

One of the most relevant definitions of Internet of Things (IoT) have been given by Gubbi et al. (2013) as "Interconnection of sensing and actuating devices providing the ability to share information across platforms through a unified framework, developing a common operating picture for enabling innovative applications. This is achieved by seamless ubiquitous sensing, data analytics and information representation with Cloud computing as the unifying framework." The Internet of Things (IoT) offers a wide range of uses, including smart agriculture, smart environments, personal and home monitoring, and enterprise. Four primary elements make up the IoT ecosystem in smart agriculture: IoT devices, communication technologies, Internet, data storage, and processing (Elijah et al. 2018).

ISSAS use a data importation frontend, software module for administration and decision-making, and a cloud-based actuation module (Tan 2016; O'Grady and O'Hare 2017). Using any one of the following communication protocol standards: IEEE 802.15.4 (low-rate Wireless personal area network), IEEE 802.11 (wireless local area network standard for Wi-Fi communication), IEEE 802.15.1 (wireless personal A), a centralized wireless sensor network-based monitoring system collects temperature, humidity, light, pressure, leaf area index, and other necessary data for data collection and importation at predetermined intervals (Buratti et al. 2009).

In order to support farmers during the life cycle of crops, as discussed by Maddikunta et al. (2021) among the sensing elements are the following: (a) Smart location sensors and GPS receivers are employed to pinpoint various places and sites in agricultural fields to apply fertilizer, water, and treat weeds. (b) Electrochemical sensors detect specific ions in soil to determine pH and nutrient levels, as well as fertilizer use. (c) Mechanical sensors use load cells to measure soil resistance for irrigation and intervention analysis. (d) Airflow sensors determine soil air permeability, which can be used to determine soil characteristics such as soil type, structure, compaction, and signature. (e) Sound sensor detects soil texture and is used for indoor and outdoor cultivation. (f) Soil moisture sensor that is dielectric determines the soil's dielectric constant necessary for calculating the soil moisture level. And (g) optical sensors are mostly used in unmanned aerial vehicles (UAVs) to measure reflectance in the near-infrared and record images with remarkable spatial resolution. Multispectral sensors are crucial because they enable researchers to conduct precise analysis and generate insights on plant vigor, canopy cover, leaf, and several other plant elements. Crop fluorescence is monitored using thermal infrared sensors, which integrate at least two wavelengths to assess statistical factors. Chlorophyll content, the absorption of blue and red light, and the emission of green light are all closely related to the amount of light energy. Majority of the sensors listed from (a) to (g) are suitable for outdoor and indoor farming.

The Internet of Things (IoT) and cloud services platform work together to provide Web services for connected ICT components (Karim et al. 2017), such as a Google Web Toolkit-based greenhouse monitoring and management system (Wang et al. 2018). Researchers have applied IoT to indoor farming, such as creating a remote-controlled water delivery system based on the state of the plants' soil moisture (Bin Ismail and Thamrin 2018).

IoT has the potential to revolutionize conventional farming, but its cost, adoption of long-range communication protocols, cost, and other issues will prevent it from being economically viable. Cost is the biggest challenge for ISSAS due to their constrained space. Summary of research and enabling technologies related to IoT in indoor farming is listed Table 6.

4.3 Big Data and Data Modeling

Big Data is data with high volume, velocity, and variety, and one of its primary sources is wireless sensor networks. It is used in farm management through connected processes such as data collection, storage, transport, transformation, analytics, and marketing. Cloud-based data warehouses are popular, because they are quick to access, are inexpensive, and don't require farms to buy any gear.

The Hadoop Distributed File System (HDFS) is a distributed file system is known for its excellent fault tolerance performance and ease of installation on inexpensive hardware (Wolfert et al. 2017). Precision agricultural systems by taking intelligent decisions by using Global Positioning Systems (GPS), data-gathering sensors, contemporary communication technologies, variable rate technology, geo-mapping, and automated machineries have revolutionized agriculture. Big data is

Table 6 Summary of notable research works on enabling technologies for IoT

Article	Broad area	Adopted methodology	SSAS type/ adoptability	Technological Benefits/ Drawbacks
Tan (2016)	Smart decision support system for precision agriculture	A cloud-based software architecture	Precision agriculture/yes, indoor farms	Application specific smart decision
Pahuja et al. (2013)	Monitoring and control of climate using wireless sensor network	Fuzzy logic-based controller, IEEE 802.15.4- and XMesh-based networking, RS-485-based actuator, and customized application software	Green house/ yes, indoor farms	Online monitoring and control facility
Akkaş and Sokullu (2017)	Agricultural monitoring system	IEEE 802.15.4 compliant 2.4 GHz MicaZ mote modules for low-power WSN, MIB 250 service support platform data analysis and management	Green house/yes, indoor farms	Monitoring from remote location
Li et al. (2015)	Design of leaf area sensor	Leaf area index estimation using WSN and computer vision	Outdoor condition/yes, indoor farms	Estimation of growth of the plants
Wang et al. (2018)	Monitoring and control of environment	Software architecture based on Google Web toolkit	Green house/yes, indoor farms	User can access using Android app

essential for ensuring knowledge and information continue to move through the agricultural value chain (AVC) (Pham and Stack 2018). The agricultural decision-making process is defined by a hierarchy of facts, information, knowledge, and wisdom, with wisdom at the top of the list (Lokers et al. 2016).

Wireless sensor networks (WSNs), remote sensing (RS) technology, and unmanned aerial vehicles (UAVs) are used to gather information about various spatial and temporal variables of agricultural fields (Zhang et al. 2018a, b). WSNs-based data collection can be implemented in ISSAS, and Web services architectures, such as SOAP and REST, are used to communicate between multiple applications across the World Wide Web (Vitolo et al. 2015).

The agriculture sector generates a vast volume of data from numerous sources, making it important to model it to gain a better understanding (Rodriguez et al. 2017). The higher-order singular value decomposition (HOSVD) is a method suggested by researchers from around the world to extract the core value by removing undesirable data dimensions (Sabarina and Priya 2015).

Researchers use data analysis and modelling tools like AgBiz Logic and TOA-MD to quantify the economic, social, environmental, and other effects associated with the agricultural farm. To project the effects of climate change on the agricultural sector, the Agricultural Model Intercomparison and Improvement Project (AgMIP) models crop, economic, and climatic data (Antle et al. 2017). Wolfert et al. (2017) explored a variety of issues related to managing Big Data in smart

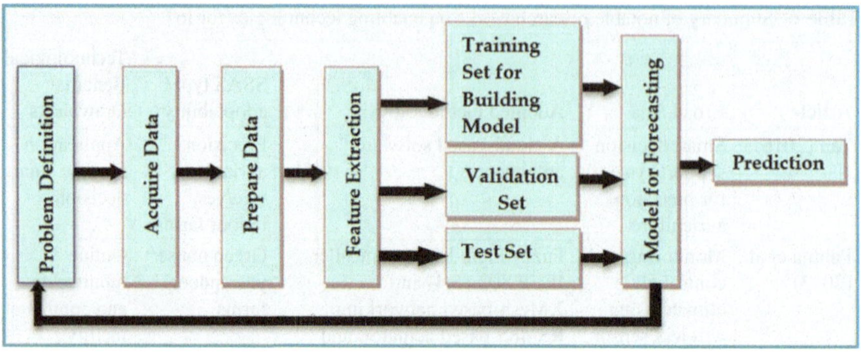

Fig. 6 Process flow of data analysis using machine learning

farming, including data ownership, data quality, intelligent processing and analytics, sustainable integration of Big Data sources, alluring business models, and platform openness.

ISSAS are designed to maximize the benefits for small and medium-scale farmers by providing intelligent processing of high-quality data, identification of data sources, platform independence, business models, and security. IoT and Big Data can work together to provide a smart model that makes use of all the information collected due to their shared characteristics (Capalbo et al. 2017). Summary of research and enabling technologies related to big data and data modeling in ISSAS is listed Table 7.

4.4 *Machine Learning in Smart Agriculture*

Machine learning is a subset of artificial intelligence (AI). It constructs a forecasting model to anticipate future results by learning from the patterns in the data already available. Raw data is gathered from a variety of sources and preprocessed before being split into training, testing, and validation sets. A forecasting model is created using the attributes taken from historical data. Machine learning is a useful technique in today's data-centric smart agriculture for analysis of yield prediction, crop health condition monitoring, water control management, soil management, etc. (Liakos et al. 2018). In machine learning, the flow of data analysis and various processes have been presented in Fig. 6.

Data on climate characteristics (Veenadhari et al. 2014), soil quality (Cunha et al. 2018), and production from prior years (Shakoor et al. 2017; Rahman et al. 2014) must be gathered to forecast agricultural productivity in intelligent indoor farms. This can be done using soil data, interior microclimate data, and photographs of plants at various phases of their life cycles.

Table 7 Summary of notable research works on enabling technologies for big data

Article	Broad area	Adopted methodology	SSAS type/ adoptability	Technological benefits/drawbacks
Rani et al. (2017) and Zhang et al. (2018a, b)	Big data collection through WSNs	Big data efficient gathering (BDEG) algorithm for energy efficient data collection. WSNs and unmanned aerial vehicles (UAVs) are used for data collection	Any WSN with high density of nodes Adoptable for WSN used in indoor farms	Saving time and energy when collecting big amounts of data
Wolfert et al. (2017), Pham and Stack (2018), Lokers et al. (2016), Vitolo et al. (2015), Rodriguez et al. (2017), Sabarina and Priya (2015), Antle et al. (2017) and Capalbo et al. (2017)	Data in agricultural science	Data chain: refers to sequence of activities (i.e., data capture, storage, transfer, transformation, analytics, and marketing) to manage data for farm management Agriculture value chain (AVC): used to explain the connection among different stakeholders LIAISE (linking impact assessment instruments to sustainability expertise) toolkit: impact assessment tool for the new policy RPy2: interface to R programming language from python programming language PostgreSQL: free and open-source relational database management system PyWPS: server implementation of Web processing service using python programming language IHOSVD: data dimensionality reduction algorithm AgBiz: analytical tool for agricultural business	Conventional agriculture Adoptable for large-scale indoor agricultural farms	Enhanced, well-organized, and effective data modelling and knowledge inference from agricultural farms

Machine learning techniques are used in crop health monitoring, plant phenotyping, and soil fertility grading. SVM, Bayesian network, neural network, regression, and other techniques are used to anticipate crop pests (Kim et al. 2014). Images from a farm field are taken using cameras and evaluated using a machine learning algorithm to determine the nutritional shortage in plants (Merchant et al. 2018; Shah et al. 2018). DeepPheno is a concept that uses deep learning to examine the phenology of plants, and images and sensor data are analyzed in stress phenotyping (Yalcin 2018). The quality of agricultural output is largely determined by our understanding of the soil, so machine learning approaches are applied in many applications. In some of the interrelated domains associated to ISSAS, such as weed identification (Zhang et al. 2018a, b), soil sensor design (Luciani et al. 2019), production quality assessment (Chokey and Jain 2019), etc., machine learning may also be used.

Machine learning algorithms support decision-making by analyzing data from sensors for crop suggestion, yield prediction, disease detection, and control mechanisms for artificial lighting, nutrition management, climate control, and optimal water use. ISSAS aid in the monitoring and gathering of data related to the plant life cycle and activate the required control mechanisms. Research in these areas has to be expanded to increase crop output. Summary of research and enabling technologies related to machine learning in ISSAS is listed Table 8.

4.5 Plant Phenotyping

Plant phenotyping is the process of measuring and analyzing plant traits, or phenotypes, in order to better understand plant growth, development, and response to environmental factors. There are several categories of plant phenotyping, including morphological, physiological, and molecular phenotyping.

Morphological phenotyping involves measuring physical characteristics of plants, such as leaf size, stem diameter, and root length. Physiological phenotyping involves measuring the function and activity of different plant organs and systems, such as photosynthesis, water use efficiency, and nutrient uptake. Molecular phenotyping involves analyzing the expression and activity of specific genes and proteins within plant cells.

In order to effectively measure and analyze plant phenotypes, researchers use a variety of parameters and modeling approaches. Some common parameters used in plant phenotyping include growth rate, biomass accumulation, and nutrient content. Researchers may also use imaging techniques, such as fluorescence microscopy or hyperspectral imaging, to visualize and quantify plant traits at a high resolution.

In addition, researchers may use modeling approaches, such as mathematical models or machine learning algorithms, to better understand and predict plant growth and development. These models can help to identify key factors that influence plant phenotypes, as well as predict how plants will respond to different environmental conditions and stressors. A correlation of various steps consisting of

Table 8 Summary of notable research works on enabling technologies for machine learning

Article	Broad area	Adopted methodology	SSAS type/adoptability	Technological benefits/drawbacks
Liakos et al. (2018)	A review of machine learning in agriculture	This study discusses the use of machine learning algorithms for agricultural, water, soil, weed, and plant health management	Adoptable in all forms of indoor farms	Various algorithms and their efficiencies are discussed
Veenadhari et al. (2014)	Crop yield forecasting from climatic condition	Decision tree C4.5 algorithm	Outdoor farming. This research cannot be utilized directly in indoor farms as they have controlled microclimates	Overall accuracy 82%. It is applicable for different crops
Cunha et al. (2018)	Crop yield forecasting by using precipitation and soil quality data	Deep neural network based model	Outdoor farming. Only prediction based on soil quality data is relevant in indoor environment	Good prediction even with fewer data
Rahman et al. (2014)	Crop yield forecasting in Bangladesh from climatic condition	K-means clustering, neural network, regression tree, ensemble learning, linear regression, generalized linear regression (GLM)	Outdoor farming. This research cannot be utilized directly in indoor farms as they have controlled microclimates	Applicable for a number of crops and even various models give the same result
Merchant et al. (2018) and Shah et al. (2018)	Identification of nutrient deficiency of plants	Image processing and machine learning techniques	Adoptable in all forms of indoor farms	Advance detection of plant diseases
Yahata et al. (2017) and Yalcin (2018)	Recognition of plant phenology	Simple linear iterative clustering (SLIC) for image segmentation, convolutional neural network for classification of different phonological stages of plant	Adoptable in all forms of indoor farms	Gives a better insight of relation between productivity, plant health, and environmental condition
Rahman et al. (2019)	Classification of soil and suggestion of crop	K-nearest neighbor, Gaussian support vector machine, bagged trees	Adoptable in all forms of indoor farms	Average accuracy 92.93%
Jia et al. (2010)	Grading of soil fertility	Transfer learning algorithm based on Bayesian network	Adoptable in all forms of indoor farms	Due to transfer learning approach similar problems can be solved
Luciani et al. (2019)	Designing of a contactless soil moisture sensor	Sensing based on microwave spectroscopy technique, partial least square regression for analysis of sensing data	Adoptable in all forms of indoor farms	Machine learning algorithm improves sensing performance

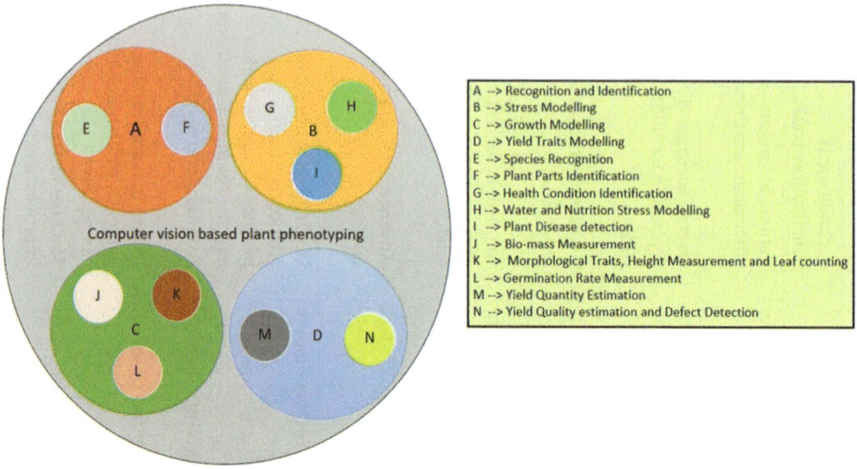

Fig. 7 Computer vision-based plant phenotyping

recognition and identification, modelling of stress, growth, and yield traits needed in computer vision-based plant phenotyping has been presented in Fig. 7.

Overall, plant phenotyping is a critical tool for understanding plant biology and developing more efficient and sustainable agricultural practices. By measuring and analyzing plant phenotypes, researchers can better understand how plants respond to different environmental factors, as well as identify traits that are important for crop yield, quality, and resistance to pests and diseases. Table 9 elaborates the plant phenotyping categories, parameters, and modelling approaches employed in computer vision-based plant phenotyping.

5 Challenges for Indian Farmers

India remains an agriculture-based country. About 53% of its total workforce, which is approximately 243 million citizens, is employed in agricultural sector. As shown in Table 10, India is one of the primary contributors in world's agricultural production.

According to recent statistics though almost half of the workforce works in farming sector, contribution of Indian agriculture to its gross domestic product (GDP) has reduced from 54% in 1950–1951 to 15.4% in 2015–2016 (Fig. 8). The agricultural yield, which is defined as the production per unit land area, is lower in India compared to other primary agricultural producer such as the USA, China, and Brazil (Deshpande 2017). However, several factors are responsible, which are adversely affecting the yield and productivity of Indian agriculture. Some of the primary reasons are as follows.

Table 9 Plant phenotyping: categories, parameters, and modelling approaches

Sl No	Category	Task details: exploring phenotyping parameters	Computer vision-based approach
1	Pre-phenotyping task	Plant species recognition	Image classification
2	Plant growth and development	Plant organ counting	Semantic segmentation and object detection
		Modelling of plant morphological changes	Object detection, semantic segmentation, and regression
		Dynamic modelling of plant height	Object detection and regression method
		Understandings of root architectural traits	Semantic segmentation method
		Determining plant imbibition and germination rates	Object detection, semantic segmentation, and regression
		Plant biomass identification	Semantic segmentation method and regression method
3	Plant stress phenotyping	Health condition identification	Image classification and semantic segmentation
		Plant disease detection	Semantic segmentation
		Water stress identification	Object detection and semantic segmentation method
		Nitrogen stress identification	Object detection and semantic segmentation method
		Root health condition identification	Image classification and semantic segmentation
4	Plant yield rate and post harvesting	Flowering rate and time	Object detection and regression
		Yield traits identification	Object detection, semantic segmentation, and regression
		Determination of chemical composition of fruit and vegetable	Regression method
		Detection of defect in fruits and vegetables	Object detection and semantic segmentation method

Table 10 India's contribution to world's agricultural production

Sl. No	Crops	Percentage contribution of total production
1	Pulse	25
2	Rice	22
3	Wheat	13
4	Cotton	25

5.1 Small Land Holding

In the past few decade, marginal (<1 hectare) and small land (in between 1 and 2 hectare) holding have significantly increased in India. From 1971 to 2011, marginal

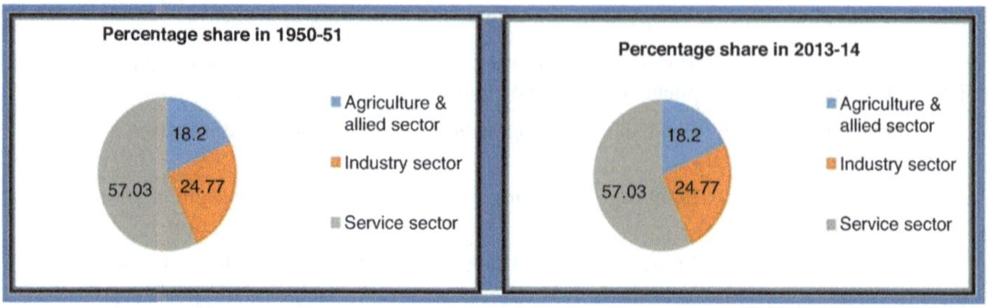

Fig. 8 Sector-wise percentage share of India's GDP in 1950–1951 and in 2013–2014

land holdings have increased 2.58 times (Deshpande 2017). Recently, it has been reported that more than 80% of Indian farmers are marginal or small land holders (Gopalakrishnan and Thorat 2015). As most of the farmers with small and marginal land holding do not have any legal lease agreement, they are not eligible for insurance, subsidies, and beneficial govt. schemes.

5.2 *Scarcity of Water*

In India, only 40.6% of the food grains are cultivated with the help of irrigation water (ADB Report 2016). Irrigation water uses almost 83% of the total available water of India. As per the prediction of researchers agricultural sector used 688 Billion Cubic Meter (BCM) of water in 2010, which will increase to 1072 BCM by 2050 (Sonekar 2017). Due to inefficient use of irrigation water, low availability of per capita water resources and too much dependency on rainwater Indian agriculture faces a scarcity of water throughout the year.

5.3 *Natural Disasters*

Every year natural calamities like floods, draught, landslides, storms, and hails cause heavy losses to crops (Gupta et al. 2020). The poor farmers with no access to banking insurance system face irreparable damages.

5.4 *Quality of Soil*

Due to increase of food production and repeated use of agricultural lands, nutrient level and water level of soil have decreased, affecting the growth and production of the plants. One of the effects of soil degradation is soil erosion, which directly affects the agricultural production (Gupta 2013).

5.5 Improper Use of Fertilizers

Due to lack of knowledge, farmers use fertilizers in improper ratio, which leads to declined soil fertility and loss in crop production. Nitrogen (N), Phosphorus (P) and Potassium (K) are the major nutrient elements that are required for the crops. Indian farmers fail to maintain the recommended ratio of NPK while using fertilizers. This leads to loss of soil fertility.

5.6 Imbalanced Use of Pesticides

Imbalanced and unregulated use of pesticide is harmful for agricultural products. Due to absence of any effective regulatory authority to control the manufacture, purchase, and sell of pesticides in India, low-quality pesticides are present in the market. Due to lack of knowledge, farmers use them without following any pest-management system causing damage to the production of the crops.

5.7 Lack of Good Quality Seeds

Small- and marginal-scale farmers cannot afford high-quality seed. There is also limited access to good-quality seed and necessary research innovations for betterment of seed qualities.

5.8 Lack of Smart Machineries

Agricultural machineries are used to reduce human labor in agriculture. Machines are mainly used in threshing, harvesting, and irrigation activities. Most of the small and marginal farmers have not yet adopted automated techniques of farming due to economic reason, which reduces the agricultural production in India.

5.9 Poor Postharvesting Activities

Due to poor transportation, packaging, and storage facilities, food are wasted at different stages of post harvesting activities.

5.10 Absence of Minimum Support Price and Price Deficiency System

Minimum support price (MSP) is the price at which govt. buys crops produced from the farmers. In price deficiency system, govt. compensates the farmers in case market price of crops falls below the MSP. Though NITI Aayog has recommended price deficiency system for Indian farmers, presently no such existing system is there in place (Deshpande 2017).

6 ISSAS: The Game Changer in Global and Indian Perspective

The USA, Japan, and some of the countries of Middle East and Europe have embraced indoor farming as a consistent and sustainable source of food supplier, making it one of the fastest growing industries in urban areas. Smart indoor farming techniques as of today have not been adopted in India on a larger scale. If simultaneously implemented in India with its conventional outdoor farming, the overall agricultural production will increase, providing the required food security to its citizen in the upcoming years.

The reasons for which indoor farming has the potential to become a new dimension of Indian agriculture are as follows:

6.1 Efficient Supplier of Food

ISSAS are efficient producer of food and uses lower amount of land compared to outdoor farming. Farming in smart indoor environment is sustainable as it uses water, nutrients, and human labor in an optimized way (Al-Kodmany 2018).

6.2 To Deal with Climate Change in India

In past few decades, the world has witnessed the effect of climate change in agricultural production. Many of the Indian states have witnessed climate change of varied nature. The last century's summer monsoon rainfall in India showed no significant trend, with three subdivisions showing a decreasing trend and eight subdivisions showing significant increasing trends (Venkateswarlu and Rao 2013). Projected change in temperature in India is shown in Fig. 9a. The conventional farming lands are one of the main sources of greenhouse gasses, which cause global warming and climate change. The primary greenhouse gasses, which are generated during farming are methane (CH_4), nitrous oxide (N_2O), and carbon-dioxide (CO_2). Table 11 enlists the main sources of these greenhouse gasses (Pathak et al. 2014).

Climate change directly affects the yield and agricultural production (Shah and Srivastava 2017), photosynthesis rate, and fertility of the farming land (Karmakar et al. 2016). Sometimes natural disasters, like floods, drought, etc., are caused by climate changes (Mall et al. 2007). As indoor farms grow crops in closed environment, climate change and natural disasters do not affect directly to their

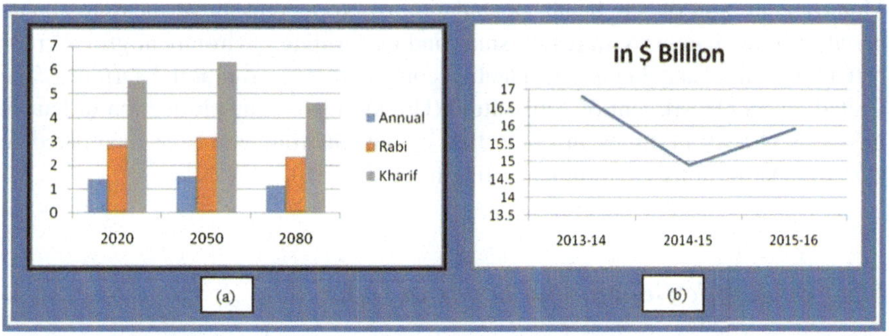

Fig. 9 (**a**) Projected change in mean temperature (in °C) in India in the upcoming years during annual, Rabi, and Kharif seasons; (**b**) increase in cost involvement for importing food grain in India. (Source: Venkateswarlu and Rao 2013)

Table 11 Primary greenhouse gasses and their sources

Sl. No.	Greenhouse gasses	Sources in the farming field
1	CH_4	Microbial decomposition of organic matter, agricultural field submerged in water, organic manure, and crop residues while getting burnt
2	CO_2	Biological decomposition of organic matters present in soil, burning of agricultural residues, agricultural operations which uses fuel
3	NO_2	Aerobic microbial oxidation of ammonium nitrate, fertilizers, manure, sewage sludge, minerals containing nitrogen.

production. If plants are grown in smart and climate-controlled environment, it also avoids generation of greenhouse gasses.

6.3 An Answer to Changing Demographic Pattern of India

According to the estimation of United Nations, by 2050 about 80% of the world's population would be living in cities and urban areas. India is an exception to this, as in India the figure is 55% (Agarwal and Sinha 2017). To meet the increasing demand of food, Indian cities need to grow food in indoor environment. This will help to deal with the threat on Indian food security concern in the upcoming years.

6.4 Quality and Quantity Come Together

In ISSAS the crops are grown in controlled environment, where pesticides, nutrients, fertilizers, water, and other resources are used efficiently to get the optimal output. Moreover, in case of any health degradation of crops, preventive measures could be taken immediately. As a result, smart indoor farming environment grows crops better in quality and higher in quantity compared to conventional farming. The United States Department of Agriculture (USDA) reports the production of lettuce increases almost 11 times in controlled indoor environment when compared to conventional farming environment (Higgins et al. 2016).

6.5 Economic Benefit and Scope in India

The govt. of India provides food grains, such as wheat, rice and coarse cereals, etc., at subsidized price to almost 68% of its population, so that all citizens get enough access to food. Moreover, to meet the demand of the food of its people, the agricultural imports have increased in India over past few decades (Deshpande 2017). The agricultural import statistics of India shows rice, wheat, pulses, and other cereals are among the major food grains, which are imported regularly from other countries. Figure 9b represents the cost involvement for importing this food grains.

Indoor farming has the potential to reduce Indian agricultural import significantly. Though all the crops cannot be grown at indoor environment, according to researchers, a wide varieties of greens, hops, strawberries, vine crops, flowers, herbs, micro greens, vegetables, fruits, cannabis, commodities, forestry seedlings, etc. can be grown indoor (Higgins et al. 2016).

As most of the ISSAS are situated in urban and city areas, the crops thus produced can directly be sold to the local market, reducing the transport cost. This may save the overall costs up to 60% (Al-Kodmany 2018).

6.6 *To Maintain Better Balance in Ecosystem*

In Brazil, 1,812,992 sq. km of farmland has been converted to farmland in the past 50 years (Al-Kodmany 2018). To avoid deforestation in India and maintain the ecological balance, a new source of food for the people of urban and city area is required. Controlled indoor farming can restore the biodiversity in urban and city areas.

7 Discussion and Concluding Remarks

About 1.26 billion of Indian population is suffering from nutritional and health challenges. Approximately 38.7% of the children and 15% of the total population are reported to be malnourished. The International Food Policy Research Institute (IFPRI) in its report of Global Hunger Index, 2018, has ranked India 103 out of 119 countries (Grebmer et al. 2018). These indicate strong presence of hunger and undernourishment in India. Moreover, with the current growth rate, India's population will reach 1.6 billion by 2050, generating more requirement of food (Ritchie et al. 2018). Even in global scenario due to increase of human population, the consumptions of food and biofuel have also increased, resulting in the increase in the demand for agricultural production by 60% to 110% (Ray et al. 2013).

As reflected in the Fig. 10, net area sown has remained almost same in India over a period of approximately 50 years. Indoor farming can help to grow Indian agricul-

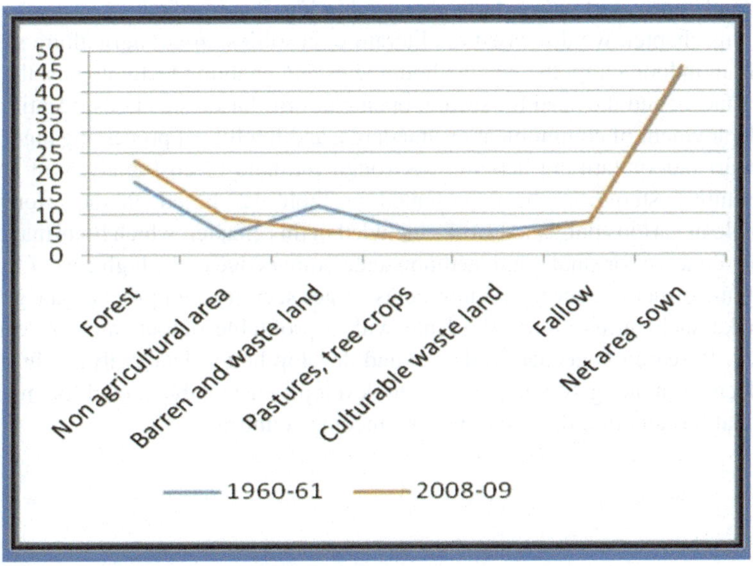

Fig. 10 Land use categories of India and changes in its percentage share over the years

ture further. The nonarable lands, the lands which have been declared as wastelands due to their climatic constraints or anthropogenic limitations, abandoned buildings, closed rooms, garden areas, and rooftops can be used for indoor framing. In India, 114.01 million hectare lands are degraded and wastelands (Balasubramanian 2016). A part of these wastelands can be used to build up cost-efficient infrastructure for indoor farming. Hence, India has a huge potential to take ISSAS forward if its research scopes are explored by the researchers.

Indoor soilless smart agriculture system development and deployment is a new area of research in India. There exist several research opportunities in this field that will add new dimensions to it to make it more efficient and sustainable. ISSAS require much attention on how it can be made more cost-effective so that poor farmers can afford this technique with minimum investment. An improved and secure architecture with a better hardware software ecosystem and better interoperability technique would increase the efficiency of the smart management system. Researchers also need to explore how self-learning capabilities can be incorporated in the architecture with the help of machine learning and other data analysis algorithms. With the self-learning capabilities, it can learn from plant life cycle data and initiate smart actuation accordingly. Creation of artificial climate in the indoor environment is another challenge in this field. Optimized use of water, nutrition, energy, and other resources to avoid resource crisis is an important dimension of research exploration. The challenges of energy crisis need to be addressed using energy harvesting techniques. Finally, how a standalone ISSAS can provide a sustainable solution and how it can be implemented simultaneously with conventional farming are the research questions that need to be answered. Considering the above requirements and the feasibility analysis, the architecture of a standalone smart indoor farm presented in Fig. 3 is justified.

In this chapter, we discussed the literature on soilless smart agriculture systems mostly in indoor setup, the key methodologies and enabling technologies, the challenges faced, and the need for secure, open platform-based standards, identification and deployment of communication standards, and intelligent processing algorithms for smart indoor farming activities. A rough estimate of cost analysis of soilless agriculture systems has been presented in Table 12, based on the information available in various published works included in this chapter, which hints that cost is a major concern for small- and medium-scale farmers due to the high cost of IoT and cloud infrastructure, energy requirements, water scarcity, and proper management. To make such farms more affordable with a profitable output, more research is needed. Research is needed to design and develop better data analysis algorithms and decision-making systems in order to develop a sustainable model for small and marginal farmers in India and other low-income countries.

Table 12 Cost analysis of soilless agriculture systems

Heading level	Globally	Indian
Hydroponics	A small-scale home hydroponics system can cost around \$500–\$1000, while a commercial-scale hydroponics system can cost around \$50,000–\$5,00,000 or more	A small-scale home hydroponics system can cost around INR 30,000–INR 50,000, while a commercial-scale hydroponics system can cost around INR 5,00,000–INR 50,00,000 or more
Aeroponics	A small-scale home aeroponics system can cost around \$1000–\$2000, while a commercial-scale aeroponics system can cost around \$100,000–\$1,000,000 or more	A small-scale home aeroponics system can cost around INR 50,000–INR 1,00,000, while a commercial-scale aeroponics system can cost around INR 10,00,000–INR 1,00,00,000 or more
Aquaponics	A small-scale home aquaponics system can cost around \$2000–\$5000, while a commercial-scale aquaponics system can cost around \$50,000–\$5,00,000 or more	A small-scale home aquaponics system can cost around INR 1,00,000–INR 2,50,000, while a commercial-scale aquaponics system can cost around INR 25,00,000–INR 2,50,00,000 or more
Vertical farming	A small-scale vertical farm can cost around \$500,000–\$1,000,000, while a commercial-scale vertical farm can cost around \$10,000,000–\$50,000,000 or more	A small-scale vertical farm can cost around INR 3,50,00,000–INR 7,00,00,000, while a commercial-scale vertical farm can cost around INR 70,00,00,000–INR 3,50,00,00,000 or more

References

ADB Report (2016) Agriculture and natural resources sector program in India: a background paper and desk review. https://www.adb.org/sites/default/files/linked-documents/12-India-Agriculture-and-Natural-Resources.pdf. Online accessed 30 Aug 2018

Afram A, Janabi-Sharifi F, Fung AS, Raahemifar K (2017) Artificial neural network (ANN) based model predictive control (MPC) and optimization of HVAC systems: a state of the art review and case study of a residential HVAC system. Energ Buildings 141:96–113

Agarwal HP, Sinha R (2017) Urban farming-A sustainable model for Indian cities. Int J Emerg Technol 8(1):236–242

Ahamed MS, Guo H, Tanino K (2018) A quasi-steady state model for predicting the heating requirements of conventional greenhouses in cold regions. Inf Process Agric 5:33–46. https://doi.org/10.1016/j.inpa.2017.12.003

Akkaş MA, Sokullu R (2017) An IoT-based greenhouse monitoring system with Micaz motes. Procedia Comput Sci 113:603–608. https://doi.org/10.1016/j.procs.2017.08.300

Al-Kodmany K (2018) The vertical farm: a review of developments and implications for the vertical city. Buildings 8. https://doi.org/10.3390/buildings8020024

Alencastre-Miranda M, Davidson JR, Johnson RM, Waguespack H, Krebs HI (2018) Robotics for sugarcane cultivation: analysis of billet quality using computer vision. IEEE Robot Autom Lett 3:3828–3835. https://doi.org/10.1109/LRA.2018.2856999

Almeida CC, Almeida PS, Monteiro NRC, Pinto MF, Braga HAC (2014) LED-based electronic system to support plant physiology experiments. IEEE Int Symp Ind Electron 531–536. https://doi.org/10.1109/ISIE.2014.6864669

AlShrouf A (2017) Hydroponics, aeroponic and aquaponic as compared with conventional farming. Am Sci Res J Eng Technol Sci 27(1):247–255

Altieri MA, Nicholls CI, Henao A, Lana MA (2015) Agroecology and the design of climate change-resilient farming systems. Agron Sustain Dev 35(3):869–890

Anderson R, Bayer PE, Edwards D (2020) Climate change and the need for agricultural adaptation. Curr Opin Plant Biol 56:197–202

Antle JM, Jones JW, Rosenzweig CE (2017) Next generation agricultural system data, models and knowledge products: introduction. Agric Syst 155:186–190. https://doi.org/10.1016/j.agsy.2016.09.003

Ardabili SF, Mahmoudi A, Gundoshmian TM (2016) Modeling and simulation controlling system of HVAC using fuzzy and predictive (radial basis function, RBF) controllers. J Build Eng 6:301–308

Balasubramanian A (2016) The wastelands in India. https://doi.org/10.13140/RG.2.2.35824.87045

Banerjee A, Paul K, Varshney A, Nandru R, Badhwar R, Sapre A, Dasgupta S (2022) Soilless indoor smart agriculture as an emerging enabler technology for food and nutrition security amidst climate change. In: Plant nutrition and food security in the era of climate change. Academic Press, pp 179–225

Baquero D, Molina J, Gil R, Bojacá C, Franco H, Gómez F (2015) An image retrieval system for tomato disease assessment. 2014 19th Symp Image Signal Process Artif Vision STSIVA 2014:1–5. https://doi.org/10.1109/STSIVA.2014.7010156

Bhange M, Hingoliwala HA (2015) Smart farming: pomegranate disease detection using image processing. Procedia Comput Sci 58:280–288. https://doi.org/10.1016/j.procs.2015.08.022

Bhatt P, Sarangi S, Pappula S (2017) Comparison of CNN models for application in crop health assessment with participatory sensing. GHTC 2017 – IEEE Glob Humanit Technol Conf Proc 2017-Janua:1–7. https://doi.org/10.1109/GHTC.2017.8239295

Bin Ismail MIH, Thamrin NM (2018) IoT implementation for indoor vertical farming watering system. 2017 Int Conf Electr Electron Syst Eng ICEESE 2017 2018-Janua:89–94. https://doi.org/10.1109/ICEESE.2017.8298388

Bosma RH, Lacambra L, Landstra Y, Perini C, Poulie J, Schwaner MJ, Yin Y (2017) The financial feasibility of producing fish and vegetables through aquaponics. Aquac Eng 78:146–154

Bou-Fakhreddine B, Abou-Chakra S, Mougharbel I, Faye A, Pollet Y (2016) Optimal multi-crop planning implemented under deficit irrigation. Proc 18th Mediterr Electrotech Conf Intell Effic TechnolServ Citizen, MELECON 2016:18–20. https://doi.org/10.1109/MELCON.2016.7495480

Bozchalui MC, Cañizares CA, Bhattacharya K (2015) Optimal energy management of greenhouses in smart grids. IEEE Trans Smart Grid 6(2):827–835. https://doi.org/10.1109/TSG.2014.2372812

Buratti C, Conti A, Dardari D, Verdone R (2009) An overview on wireless sensor networks technology and evolution. Sensors 9:6869–6896. https://doi.org/10.3390/s90906869

Capalbo SM, Antle JM, Seavert C (2017) Next generation data systems and knowledge products to support agricultural producers and science-based policy decision making. Agric Syst 155:191–199. https://doi.org/10.1016/j.agsy.2016.10.009

Castelao Tetila E, Brandoli Machado B, Belete NADS, Guimaraes DA, Pistori H (2017) Identification of soybean foliar diseases using unmanned aerial vehicle images. IEEE Geosci Remote Sens Lett 14:2190–2194. https://doi.org/10.1109/LGRS.2017.2743715

Chandra A, McNamara KE, Dargusch P (2018) Climate-smart agriculture: perspectives and framings. Clim Pol 18(4):526–541

Chokey T, Jain S (2019) Quality assessment of crops using machine learning techniques. Proc – 2019 Amity Int Conf Artif Intell AICAI 2019:259–263. https://doi.org/10.1109/AICAI.2019.8701294

Cholissodin I, Dewi C, Surbakti EE (2017) Integrated ANN and Bidirectional improved PSO for optimization of fertilizer dose on Palawija plants. In: Proceeding – 2016 2nd Int Conf Sci Inf Technol ICSITech 2016 Inf Sci Green Soc Environ, pp 193–197. https://doi.org/10.1109/ICSITech.2016.7852632

Chowdhury MA, Chakrabarty D (2015) Optimal Crop yield under limited water availability – a linear programming approach. J Basic Appl Eng Res 2:892–895

Clapp J, Newell P, Brent ZW (2018) The global political economy of climate change, agriculture and food systems. J Peasant Stud 45(1):80–88

Cline WR (2008) Global warming and agriculture. Finance Dev 45(1):23–27

Cunha RLF, Silva B, Netto MAS (2018) A scalable machine learning system for pre-season agriculture yield forecast. Proc IEEE 14th Int Conf eScience e-Science 2018:423–430. https://doi.org/10.1109/eScience.2018.00131

Darko E, Heydarizadeh P, Schoefs B, Sabzalian MR (2014) Photosynthesis under artificial light: the shift in primary and secondary metabolism. Philos Trans R Soc B Biol Sci 369. https://doi.org/10.1098/rstb.2013.0243

Deshpande T (2017) State of agriculture in India. PRS Legislative Research 53(8):6–7

Dutta S, Sahoo BC, Mishra R, Acharya S (2016) Fuzzy stochastic genetic algorithm for obtaining optimum crops pattern and water balance in a farm. Water Resour Manag 30:4097–4123. https://doi.org/10.1007/s11269-016-1406-7

Eigenbrod C, Gruda N (2015) Urban vegetable for food security in cities. A review. Agron Sustain Dev 35:483–498

Elijah O, Rahman TA, Orikumhi I, Leow CY, Hindia MN (2018) An overview of internet of things (IoT) and data analytics in agriculture: benefits and challenges. IEEE Internet Things J 5:3758–3773. https://doi.org/10.1109/JIOT.2018.2844296

Eldridge BM, Manzoni LR, Graham CA, Rodgers B, Farmer JR, Dodd AN (2020) Getting to the roots of aeroponic indoor farming. New Phytol 228(4):1183–1192

Eridani D, Wardhani O, Widianto ED (2018) Designing and implementing the arduino-based nutrition feeding automation system of a prototype scaled nutrient film technique (NFT) hydroponics using total dissolved solids (TDS) sensor. In: Proc 2017 4th Int Conf Inf Technol Comput Electr Eng ICITACEE 2017 2018-Janua, pp 170–175. https://doi.org/10.1109/ICITACEE.2017.8257697

Fuangthong M, Pramokchon P (2018) Automatic control of electrical conductivity and PH using fuzzy logic for hydroponics system. 3rd Int Conf Digit Arts Media Technol ICDAMT 2018:65–70. https://doi.org/10.1109/ICDAMT.2018.8376497

Ghasemi MM, Karamouz M, Shui LT (2016) Farm-based cropping pattern optimization and conjunctive use planning using piece-wise genetic algorithm (PWGA): a case study. Model Earth Syst Environ 2:1–12. https://doi.org/10.1007/s40808-016-0076-z

Ghyar BS, Birajdar GK (2018) Computer vision based approach to detect rice leaf diseases using texture and color descriptors. Proc Int Conf Inven Comput Informatics, ICICI 2017:1074–1078. https://doi.org/10.1109/ICICI.2017.8365305

Goldstein H (2018) The green promise of vertical farms. IEEE Spectr 55:50–55

Golhani K, Balasundram SK, Vadamalai G, Pradhan B (2018) A review of neural networks in plant disease detection using hyperspectral data. Inf Process Agric 5:354–371. https://doi.org/10.1016/j.inpa.2018.05.002

Goodman W, Minner J (2019) Will the urban agricultural revolution be vertical and soilless? A case study of controlled environment agriculture in New York City. Land Use Policy 83:160–173

Gopalakrishnan R, Thorat Y (2015) What India can do differently in agriculture. Sarthak Krishi Yojana. http://www.tata.com

von Grebmer K, Bernstein J, Hammond L, Patterson F, Sonntag A, Klaus L, Fahlbusch J, Towey O, Foley C, Gitter S, Ekstrom K, Fritschel H (2018) 2018 global hunger index: forced migration and hunger. Welthungerhilfe and Concern Worldwide, Bonn and Dublin. Available at https://www.globalhungerindex.org/pdf/en/2018.pdf

Gruhn P, Goletti F, Yudelman M (2000) Integrated nutrient management, soil fertility, and sustainable agriculture: current issues and future challenges. Food, Agriculture, and the Environment Discussion Paper

Gubbi J, Buyya R, Marusic S, Palaniswami M (2013) Internet of Things (IoT): a vision, architectural elements, and future directions. Futur Gener Comput Syst 29:1645–1660. https://doi.org/10.1016/j.future.2013.01.010

Gupta P (2013) Soil and its economic implications in India. Int J Res Dev Pharm Life Sci 2(6):650–666

Gupta AK, Hodom S, Singh S, Srivastava R, Chary GR, Sehgal VK, Ray S, Singh KK, Pathak H, Krishnan P, Attri SD (2020) National Agriculture Disaster Management Plan. Dept. of Agriculture, Cooperation and Farmers Welfare, Ministry of Agriculture & Farmers Welfare, Govt. of India, New Delhi, p 348

Hamdani M, Bekkouche SMA, Cherier MK, Benouaz T, Belarbi R (2017) Study on effects of window with an external shutters for natural ventilation for buildings in hot climates. Proc 2016 Int Renew Sustain Energy Conf IRSEC 2016:780–784. https://doi.org/10.1109/IRSEC.2016.7983905

Han L, Haleem MS, Taylor M (2015) A novel computer vision-based approach to automatic detection and severity assessment of crop diseases. Proc 2015 Sci Inf Conf SAI 2015:638–644. https://doi.org/10.1109/SAI.2015.7237209

Harun AN, Ahmad R, Mohamed N (2016) WSN application in LED plant factory using continuous lighting (CL) method. ICOS 2015 – 2015 IEEE Conf Open Syst:56–61. https://doi.org/10.1109/ICOS.2015.7377278

Harun AN, Ani NN, Ahmad R, Azmi NS (2013) Red and blue LED with pulse lighting control treatment for brassica chinensis in indoor farming. 2013 IEEE Conf Open Syst ICOS 2013:231–236. https://doi.org/10.1109/ICOS.2013.6735080

Heidari M, Khodadadi H (2017) Climate control of an agricultural greenhouse by using fuzzy logic self-tuning PID approach. ICAC 2017 – 2017 23rd IEEE Int Conf Autom Comput Addressing Glob Challenges through Autom Comput:7–8. https://doi.org/10.23919/IConAC.2017.8082074

Higgins C, Gordon-Smith H, Schreiber K, Blume A, Grosbard A, Ross D, Mattson N, Goodman W, Cabo D, Williams C, Keane Graham M, Wall Media B, Wakefield S, Koschei J, Krieb K (2016) Agrylist: State of Indoor Farming. Available online: https://www.bayer.com/sites/default/files/stateofindoorfarming-report-2017.pdf. Accessed on 5 Mar 2023

Howden SM, Soussana JF, Tubiello FN, Chhetri N, Dunlop M, Meinke H (2007) Adapting agriculture to climate change. Proc Natl Acad Sci 104(50):19691–19696

Idris I, Sani MI (2012) Monitoring and control of aeroponic growing system for potato production. Proc 2012 IEEE Conf Control Syst Ind Informatics, ICCSII 2012:120–125. https://doi.org/10.1109/CCSII.2012.6470485

Janprom K, Wangnippanto S, Permpoonsinsup W (2017) Embedded control system with PID controller for comfortable room. 2017 Int Electr Eng Congr iEECON 2017:8–10. https://doi.org/10.1109/IEECON.2017.8075728

Jia HY, Chen J, Yu HL, Liu DY (2010) Soil fertility grading with Bayesian Network transfer learning. 2010 Int Conf Mach Learn Cybern ICMLC 2010 3:1159–1163. https://doi.org/10.1109/ICMLC.2010.5580915

Kaewwiset T, Yooyativong T (2017) Electrical conductivity and pH adjusting system for hydroponics by using linear regression. ECTI-CON 2017 2017 14th Int Conf Electr Eng Comput Telecommun Inf Technol:761–764. https://doi.org/10.1109/ECTICon.2017.8096350

Kalantari F, Mohd Tahir O, Mahmoudi Lahijani A, Kalantari S (2017) A review of vertical farming technology: a guide for implementation of building integrated agriculture in cities. Adv Eng Forum 24:76–91. https://doi.org/10.4028/www.scientific.net/aef.24.76

Kárász C, Kopják J (2017) Comparative study about LED driving methods and feedback system for indoor plant cultivation. CINTI 2016 17th IEEE Int Symp Comput Intell Informatics Proc:219–224. https://doi.org/10.1109/CINTI.2016.7846407

Karim F, Karim F, Frihida A (2017) Monitoring system using web of things in precision agriculture. Procedia Comput Sci 110:402–409. https://doi.org/10.1016/j.procs.2017.06.083

Karmakar R, Das I, Dutta D, Rakshit A (2016) Potential effects of climate change on soil properties: a review. Sci Int 4(2):51–73

Khatri-Chhetri A, Aggarwal PK, Joshi PK, Vyas S (2017) Farmers' prioritization of climate-smart agriculture (CSA) technologies. Agric Syst 151:184–191. https://doi.org/10.1016/j.agsy.2016.10.005

Kiani E, Mamedov T (2017) Identification of plant disease infection using soft-computing: application to modern botany. Procedia Comput Sci 120:893–900. https://doi.org/10.1016/j.procs.2017.11.323

Kim YH, Yoo SJ, Gu YH, Lim JH, Han D, Baik SW (2014) Crop pests prediction method using regression and machine learning technology: survey. IERI Procedia 6:52–56. https://doi.org/10.1016/j.ieri.2014.03.009

Kisilewicz T (2015) Passive control of indoor climate conditions in low energy buildings. Energy Procedia 78:49–54. https://doi.org/10.1016/j.egypro.2015.11.113

Kumar R, Singh MP, Kumar P, Singh JP (2015) Crop Selection Method to maximize crop yield rate using machine learning technique. 2015 Int Conf Smart Technol Manag Comput Commun Control Energy Mater ICSTM 2015 Proc:138–145. https://doi.org/10.1109/ICSTM.2015.7225403

Kumar RA, Aslam MKM, Raj VPJ, Radhakrishnan T, Kumar KS, Manojkumar TK (2017) A statistical analysis of soil fertility of Thrissur district, Kerala. Proc 2016 Int Conf Data Sci Eng ICDSE 2016:7–11. https://doi.org/10.1109/ICDSE.2016.7823953

Kyaw TY, Ng AK (2017) Smart aquaponics system for urban farming. Energy Procedia 143:342–347. https://doi.org/10.1016/j.egypro.2017.12.694

Lakhiar IA, Jianmin G, Syed TN, Chandio FA, Buttar NA, Qureshi WA (2018) Monitoring and control systems in agriculture using intelligent sensor techniques: a review of the aeroponic system. J Sens 2018:1–18

Lee J, Chuang IT (2017) Living green Shell: urban micro-vertical farm. Proc. 2017 IEEE Int. Conf. Appl. Syst. Innov Appl Syst Innov Mod Technol ICASI 2017:1087–1090. https://doi.org/10.1109/ICASI.2017.7988181

Lee S, Lee J (2015) Beneficial bacteria and fungi in hydroponic systems: types and characteristics of hydroponic food production methods. Sci Hortic 195:206–215

Li X, Liu Q, Yang R, Zhang H, Zhang J, Cai E (2015a) The design and implementation of the leaf area index sensor. Sensors 15(3):6250–6269

Li Y, Wang Y, Meng X, Wang M, Long E (2015b) Research on indoor thermal environment improvement of lightweight building integrated with phase change material under different climate conditions. Procedia Eng 121:1628–1634. https://doi.org/10.1016/j.proeng.2015.09.193

Liakos KG, Busato P, Moshou D, Pearson S, Bochtis D (2018) Machine learning in agriculture: a review. Sensors (Switzerland) 18:1–29. https://doi.org/10.3390/s18082674

Liang L, Tian H, Ning P (2018) Artificial light LED planting system design. 2017 14th China Int Forum Solid State Light. Int Forum Wide Bandgap Semicond. China, SSLChina IFWS 2017 2018-Janua:88–90. https://doi.org/10.1109/IFWS.2017.8245981

Liu Y, Akram Hassan K, Karlsson M, Weister O, Gong S (2018) Active Plant Wall for green indoor climate based on cloud and internet of things. IEEE Access 6:33631–33644. https://doi.org/10.1109/ACCESS.2018.2847440

Lokers R, Knapen R, Janssen S, van Randen Y, Jansen J (2016) Analysis of big data technologies for use in agro-environmental science. Environ Model Softw 84:494–504. https://doi.org/10.1016/j.envsoft.2016.07.017

Lopes F, Silva H, Almeida JM, Pinho C, Silva E (2017) Fish farming autonomous calibration system. Ocean. 2017 – Aberdeen 2017-Octob:1–6. https://doi.org/10.1109/OCEANSE.2017.8084565

Luciani G, Siboni M, Crescentini M, Romani A, Tartagni M, Berardinelli A, Ragni L (2019) When machine learning boosts sensing performance: a compact and contactless soil moisture sensor example. Proc 2018 Int Conf Sens Networks Signal Process SNSP 2018:88–91. https://doi.org/10.1109/SNSP.2018.00026

Macharia M, Oduor G, Kayuki C, Rware H (2016) Fertilizer use optimization approach: an innovation to increase agricultural profitability for African farmers. Afr J Agric Res 11(38):3587–3597

Maddikunta PKR, Hakak S, Alazab M, Bhattacharya S, Gadekallu TR, Khan WZ, Pham QV (2021) Unmanned aerial vehicles in smart agriculture: applications, requirements, and challenges. IEEE Sensors J 21(16):17608–17619

Member LX (2010) 1. Department of Control Science and Engineering, Tongji University, Shanghai, 200092, China 2. School of Information Engineering, Zhejiang Forestry University, Lin'an City, Zhejiang Province, 311300, China. Control 63–67.

Merchant M, Paradkar V, Khanna M, Gokhale S (2018) Mango leaf deficiency detection using digital image processing and machine learning. 2018 3rd Int Conf Converg Technol I2CT 2018:6–8. https://doi.org/10.1109/I2CT.2018.8529755

Mishra S, Mishra D, Santra GH (2016) Applications of machine learning techniques in agricultural crop production: a review paper. Indian J Sci Technol 9. https://doi.org/10.17485/ijst/2016/v9i38/95032

Molina JF, Gil R, Bojacá C, Gómez F, Franco H (2015) Automatic detection of early blight infection on tomato crops using a color based classification strategy. 2014 19th Symp Image Signal Process Artif Vision STSIVA 2014 1:1–5. https://doi.org/10.1109/STSIVA.2014.7010166

Morton LW, McGuire JM, Cast AD (2017) A good farmer pays attention to the weather. Clim Risk Manag 15:18–31. https://doi.org/10.1016/j.crm.2016.09.002

Mwongera C, Shikuku KM, Twyman J, Läderach P, Ampaire E, Van Asten P, Twomlow S, Winowiecki LA (2017) Climate smart agriculture rapid appraisal (CSA-RA): a tool for prioritizing context-specific climate smart agriculture technologies. Agric Syst 151:192–203. https://doi.org/10.1016/j.agsy.2016.05.009

Nalwade R (2017) Hydroponics farming. pp 645–650

Naznin MT, Lefsrud MG (2014) Impact of LED irradiance on plant photosynthesis and action spectrum of plantlet. Opt Photonics Inf Process VIII 9216:921602. https://doi.org/10.1117/12.2061236

Nishimura T, Okuyama Y, Matsushita A, Ikeda H, Satoh A (2017) A compact hardware design of a sensor module for hydroponics. 2017 IEEE 6th Glob Conf Consum Electron GCCE 2017 2017-Janua:1–4. https://doi.org/10.1109/GCCE.2017.8229255

O'Grady MJ, O'Hare GMP (2017) Modelling the smart farm. Inf Process Agric 4:179–187. https://doi.org/10.1016/j.inpa.2017.05.001

Oliveira LF, Moreira AP, Silva MF (2021) Advances in agriculture robotics: a state-of-the-art review and challenges ahead. Robotics 10(2):52

Olle M, Viršile A (2013) The effects of light-emitting diode lighting on greenhouse plant growth and quality. Agric Food Sci 22:223–234

Osama S, Elkholy M, Kansoh RM (2017) Optimization of the cropping pattern in Egypt. Alex Eng J 56:557–566. https://doi.org/10.1016/j.aej.2017.04.015

Pahuja R, Verma HK, Uddin M (2013) A wireless sensor network for greenhouse climate control. IEEE Pervasive Comput 12:49–58. https://doi.org/10.1109/MPRV.2013.26

Pant M, Thangaraj R, Rani D, Abraham A, Srivastava DK (2010) Estimation of optimal crop plan using nature inspired metaheuristics. World J Model Simul 6:97–109

Pathak H, Bhatia A, Jain N (2014) Greenhouse gas emission from Indian agriculture: trends, mitigation and policy needs. Indian Agricultural Research Institute, New Delhi, p 39

Pham X, Stack M (2018) How data analytics is transforming agriculture. Bus Horiz 61:125–133. https://doi.org/10.1016/j.bushor.2017.09.011

Popa M, Ciocarlie H (2011) Distributed intelligent climate system for indoor locations. 9th IEEE Int Symp Appl Mach Intell Informatics SAMI 2011 Proc:291–296. https://doi.org/10.1109/SAMI.2011.5738892

Prasad S, Peddoju SK, Ghosh D (2014) Energy efficient mobile vision system for plant leaf disease identification. IEEE Wirel Commun Netw Conf WCNC 4:3314–3319. https://doi.org/10.1109/WCNC.2014.6953083

Rahman MM, Haq N, Rahman RM (2014) Machine learning facilitated Rice prediction in Bangladesh. Proc 2014 Annu Glob Online Conf Inf Comput Technol GOCICT 2014:1–4. https://doi.org/10.1109/GOCICT.2014.9

Rahman SAZ, Mitra KC, Islam SMM (2019) Soil classification using machine learning methods and crop suggestion based on soil series. 2018 21st Int Conf Comput Inf Technol ICCIT 2018:1–4. https://doi.org/10.1109/ICCITECHN.2018.8631943

Rani S, Ahmed SH, Talwar R, Malhotra J (2017) Can sensors collect big data? An energy-efficient big data gathering algorithm for a WSN. IEEE Trans Ind Informatics 13:1961–1968. https://doi.org/10.1109/TII.2017.2656899

Rani YR (2012) Multi Objective Crop Planning For Optimal Benefits 2:279–287

Ray DK, Mueller ND, West PC, Foley JA (2013) Yield trends are insufficient to double global crop production by 2050. PLoS One 8. https://doi.org/10.1371/journal.pone.0066428

Rayhana R, Xiao G, Liu Z (2020) Internet of things empowered smart greenhouse farming. IEEE J Radio Freq Identif 4(3):195–211

Revathi S, Radhakrishnan TK, Sivakumaran N (2017) Climate control in greenhouse using intelligent control algorithms. Proc Am Control Conf:887–892. https://doi.org/10.23919/ACC.2017.7963065

Ritchie H, Reay D, Higgins P (2018) Sustainable food security in India—domestic production and macronutrient availability. PLoS One 13:1–17. https://doi.org/10.1371/journal.pone.0193766

Rodriguez D, de Voil P, Rufino MC, Odendo M, van Wijk MT (2017) To mulch or to munch? Big modelling of big data. Agric Syst 153:32–42. https://doi.org/10.1016/j.agsy.2017.01.010

Romeo D, Vea EB, Thomsen M (2018) Environmental impacts of urban hydroponics in Europe: a case study in Lyon. Procedia Cirp 69:540–545

Sabarina K, Priya N (2015) Lowering data dimensionality in big data for the benefit of precision agriculture. Procedia Comput Sci 48:548–554. https://doi.org/10.1016/j.procs.2015.04.134

Sambo P, Nicoletto C, Giro A, Pii Y, Valentinuzzi F, Mimmo T et al (2019) Hydroponic solutions for soilless production systems: issues and opportunities in a smart agriculture perspective. Front Plant Sci 10:923

Sani MI, Siregar S, Kumiawan AP, Jauhari R, Mandalahi CN (2017) Web-based monitoring and control system for aeroponics growing chamber. ICCEREC 2016 – Int Conf Control Electron Renew Energy Commun 2016 Conf Proc:162–168. https://doi.org/10.1109/ICCEREC.2016.7814977

Saranya S, Amudha T (2017) Crop planning optimization research – a detailed investigation. 2016 IEEE Int Conf Adv Comput Appl ICACA 2016:202–208. https://doi.org/10.1109/ICACA.2016.7887951

Schor N, Bechar A, Ignat T, Dombrovsky A, Elad Y, Berman S (2016) Robotic disease detection in greenhouses: combined detection of powdery mildew and tomato spotted wilt virus. IEEE Robot Autom Lett 1:354–360. https://doi.org/10.1109/LRA.2016.2518214

Schröder FG, Lieth JH (2002) Irrigation control in hydroponics. In: Hydroponic production of vegetables and ornamentals, pp 263–298

Shah R, Srivastava R (2017) Effect of global warming on Indian agriculture. Sustain Environ 2(4):366

Shah A, Gupta P, Ajgar YM (2018) Macro-nutrient deficiency identification in plants using image processing and machine learning. 2018 3rd Int Conf Converg Technol I2CT 2018:1–4. https://doi.org/10.1109/I2CT.2018.8529789

Shakoor MT, Rahman K, Rayta SN, Chakrabarty A (2017) Agricultural production output prediction using Supervised Machine Learning techniques. 2017 1st Int Conf Next Gener Comput Appl NextComp 2017:182–187. https://doi.org/10.1109/NEXTCOMP.2017.8016196

Sharma N, Acharya S, Kumar K, Singh N, Chaurasia OP (2018) Hydroponics as an advanced technique for vegetable production: an overview. J Soil Water Conserv 17(4):364–371

Shi P, Luan X, Liu F, Karimi HR (2012) Kalman filtering on greenhouse climate control. Chinese Control Conf CCC:779–784

Shivling VD, Goap A, Ghanshyam C, Sharma SK, Arora D, Kumar S, Kumar R (2016) A real time computational and statistical model (with high availability) of early warning for plant protection and pest control for crops (exp. Kutki). 2015 IEEE Int Conf Comput Graph Vis Inf Secur CGVIS 2015:22–26. https://doi.org/10.1109/CGVIS.2015.7449886

Shreedhar R, Hiremath CG, Shetty GG (2015) Optimization of cropping pattern using linear programming model for Markandeya command. Area 6:1311–1326

Sikka AK, Alam MF, Mandave V (2022) Agricultural water management practices to improve the climate resilience of irrigated agriculture in India. Irrig Drain 71:7–26

Singh PK, Jain S, Mathur A, Kumar Y (2017) An analysis on the potentials of Vertical Greenery System (VGS) in context to the application viewpoint. 4th IEEE Int Conf Signal Process Comput Control ISPCC 2017 2017-Janua:632–636. https://doi.org/10.1109/ISPCC.2017.8269754
Singh V, Misra AK (2017) Detection of plant leaf diseases using image segmentation and soft computing techniques. Inf Process Agric 4:41–49. https://doi.org/10.1016/j.inpa.2016.10.005
Sonekar BL (2017) Problem of water stress in Indian agriculture. J Rural Develop Rev 1(1)
Su Y, Xu L (2015) A greenhouse climate model for control design. 2015 IEEE 15th Int Conf Environ Electr Eng EEEIC 2015 – Conf Proc:47–53. https://doi.org/10.1109/EEEIC.2015.7165318
Su Y, Xu L, Li D (2016) Adaptive fuzzy control of a class of MIMO nonlinear system with actuator saturation for greenhouse climate control problem. IEEE Trans Autom Sci Eng 13:772–788. https://doi.org/10.1109/TASE.2015.2392161
Sugano M (2015) Elemental technologies for realizing a fully-controlled artificial light-type plant factory. 2015 12th Int Conf Expo Emerg Technol a Smarter World, CEWIT 2015:1–5. https://doi.org/10.1109/CEWIT.2015.7338169
Suparwoko, Taufani B (2017) Urban farming construction model on the vertical building envelope to support the green buildings development in Sleman. Indonesia Procedia Eng 171:258–264. https://doi.org/10.1016/j.proeng.2017.01.333
Taki M, Ajabshirchi Y, Ranjbar SF, Rohani A, Matloobi M (2016) Modeling and experimental validation of heat transfer and energy consumption in an innovative greenhouse structure. Inf Process Agric 3:157–174. https://doi.org/10.1016/j.inpa.2016.06.002
Tan L (2016) Cloud-based decision support and automation for precision agriculture in orchards. IFAC-PapersOnLine 49:330–335. https://doi.org/10.1016/j.ifacol.2016.10.061
Tanmayee P (2017) Rice crop monitoring system-A lot based machine vision approach. 2017 Int Conf Nextgen Electron Technol Silicon to Software, ICNETS2 2017:26–29. https://doi.org/10.1109/ICNETS2.2017.8067890
Tavares PL, Guimaraes IAB, Braga HAC, Bender VC, Almeida PS (2018) LED system with independent red and blue channels employing radiant flux estimation and indirect flux control for greenhouse hop cultivation. 14th Brazilian Power Electron Conf COBEP 2017 2018-Janua:1–9. https://doi.org/10.1109/COBEP.2017.8257403
The World Bank: Climate Smart Agriculture. https://www.worldbank.org/en/topic/climate-smart-agriculture
Ullagaddi SB, Raju SV (2017) Disease recognition in Mango crop using modified rotational kernel transform features. 2017 4th Int Conf Adv Comput Commun Syst ICACCS 2017. https://doi.org/10.1109/ICACCS.2017.8014610
United Nations Department of Economic and Social Affairs, Population Division (2022). World Population Prospects 2022: Summary of Results. UN DESA/POP/2022/TR/NO. 3 https://www.un.org/development/desa/pd/sites/www.un.org.development.desa.pd/files/wpp2022_summary_of_results.pdf
Urrestarazu M (2018) Technical article artificial lighting in agriculture January 2018 Silvia Bures, Miguel Urrestarazu Gavilán and Stiina Kotiranta
Veenadhari S, Misra B, Singh CD (2014) Machine learning approach for forecasting crop yield based on climatic parameters. 2014 Int Conf Comput Commun Informatics Ushering Technol Tomorrow, Today, ICCCI 2014:1–5. https://doi.org/10.1109/ICCCI.2014.6921718
Venkateswarlu B, Rao VUM (2013) Climate change and its impact on Indian agriculture. Climate Change Modeling, Mitigation, and Adaptation:419–453
Vitolo C, Elkhatib Y, Reusser D, Macleod CJA, Buytaert W (2015) Web technologies for environmental big data. Environ Model Softw 63:185–198. https://doi.org/10.1016/j.envsoft.2014.10.007
Wang J, Zhou J, Gu R, Chen M, Li P (2018) Manage system for internet of things of greenhouse based on GWT. Inf Process Agric 5:269–278. https://doi.org/10.1016/j.inpa.2018.01.002
Westermann O, Förch W, Thornton P, Körner J, Cramer L, Campbell B (2018) Scaling up agricultural interventions: case studies of climate-smart agriculture. Agric Syst 165:283–293
Wicaksono D, Perdana D, Mayasari R (2018) Design and analysis automatic temperature control in the broiler poultry farm based on wireless sensor network. In: Proc 2017 2nd Int Conf Inf

Technol Inf Syst Electr Eng ICITISEE 2017 2018-Janua, pp 450–455. https://doi.org/10.1109/ICITISEE.2017.8285549

Wojciechowska R, Kołton A, Długosz-Grochowska O, Żupnik M, Grzesiak W (2013) The effect of LED lighting on photosynthetic parameters and weight of lamb's lettuce (Valerianella locusta). Folia Hortic 25:41–47. https://doi.org/10.2478/fhort-2013-0005

Wolfert S, Ge L, Verdouw C, Bogaardt MJ (2017) Big data in smart farming – a review. Agric Syst 153:69–80. https://doi.org/10.1016/j.agsy.2017.01.023

Yahata S, Onishi T, Yamaguchi K, Ozawa S, Kitazono J, Ohkawa T, Yoshida T, Murakami N, Tsuji H (2017) A hybrid machine learning approach to automatic plant phenotyping for smart agriculture. Proc Int Jt Conf Neural Networks 2017-May:1787–1793. https://doi.org/10.1109/IJCNN.2017.7966067

Yalcin H (2018) Phenology recognition using deep learning: DeepPheno. 26th IEEE Signal Process Commun Appl Conf SIU 2018:1–4. https://doi.org/10.1109/SIU.2018.8404165

Yuvaraj RM, Dolui B (2021) Statistical and machine intelligence based model for landslide susceptibility mapping of Nilgiri district in India. Environ Chall 5:100211

Zhang W, Hansen MF, Volonakis TN, Smith M, Smith L, Wilson J, Ralston G, Broadbent L, Wright G (2018a) Broad-leaf weed detection in pasture. 2018 3rd IEEE Int Conf Image Vis Comput ICIVC 2018:101–105. https://doi.org/10.1109/ICIVC.2018.8492831

Zhang Y, Chen D, Wang S, Tian L (2018b) A promising trend for field information collection: an air-ground multi-sensor monitoring system. Inf Process Agric 5:224–233. https://doi.org/10.1016/j.inpa.2018.02.002

Zou Q, Ji A, Zhang S, Shi M, Luo Y (2010) Model predictive control based on particle swarm optimization of greenhouse climate for saving energy consumption. 2010 World Autom Congr WAC 2010:123–128

Intelligent Nutrient Controlling System for Precision Urban Agriculture

Nico Surantha and Vito Vincentdo

Abstract Urban agriculture has gained significant attention in recent years due to its potential to address various challenges, such as food security, urbanization, and climate change. However, urban farming method requires special treatment for controlling the water temperature, water level, and acidity (pH) of nutrient solutions. The emergence of the Internet of Things (IoT) has enabled the integration of sensors and devices with the physical world, leading to the emergence of intelligent systems that can be applied in various domains, including urban agriculture. Integrating IoT technologies with urban agriculture makes it possible to create intelligent systems that can monitor and control different aspects of the production process in real-time. In this chapter, we conduct a review about intelligent nutrient-controlling system for precision urban agriculture. Specifically, this chapter discusses about the latest development of intelligent system, the IoT architecture, and the future challenge of intelligent nutrient-controlling system.

Keywords Urban agriculture · Internet of Things · Nutrient control systems · Hydroponics · Artificial intelligence

1 Introduction

Urban agriculture, defined as the practice of growing food in urban areas, has gained significant attention in recent years due to its potential to address various challenges,

N. Surantha (✉)
Department of Electrical, Electronic and Communication Engineering, Faculty of Engineering, Tokyo City University, Tokyo, Japan

Computer Science Department, BINUS Graduate Program - Master of Computer Science Bina Nusantara University, Jakarta, Indonesia
e-mail: nico@tcu.ac.jp

V. Vincentdo
Computer Science Department, BINUS Graduate Program – Master of Computer Science Bina Nusantara University, Jakarta, Indonesia

P. M. Priyadarshan et al. (eds.), *Digital Agriculture*,
https://doi.org/10.1007/978-3-031-43548-5_4

such as food security, urbanization, and climate change. However, the efficiency of urban agriculture depends on several factors, including the use of advanced technologies that can optimize production processes, reduce resource consumption, and improve crop yields (D'Ostuni et al. 2022). The proliferation of the Internet of Things (IoT) has enabled the integration of sensors and devices with the physical world, leading to the emergence of intelligent systems that can be applied in various domains, including urban agriculture. Integrating IoT technologies with urban agriculture makes it possible to create intelligent systems that can monitor and control different aspects of the production process in real time (Herman 2020). In this context, intelligent IoT systems can enhance the efficiency and sustainability of urban agriculture by providing real-time data and insights for decision-making, automating tasks, and optimizing resource usage (Xu et al. 2022).

In recent years, there has been an increasing interest in the application of IoT in urban agriculture, and numerous studies have been conducted to explore the potential benefits of this technology. Researchers have investigated various aspects of IoT technology in urban agriculture, including sensors, data analytics, and automation. One study was conducted by Herman et al. (2019) explored the use of IoT technology in hydroponic systems to monitor and control nutrient and water levels. Ouafiq et al. (2021) developed an intelligent system for urban agriculture that combines IoT, big data, and artificial intelligence to optimize resource use and improve crop yield. Stevens et al. (2018) proposed a smart agricultural tool called as MicroCEA that can be controlled via a mobile application. Parameters monitored in the hydroponic system created are LED lights, air humidity, CO_2 level in the air, air temperature, pH level in water, and EC level in water.

Another study by Vianny et al. (2022) investigated the use of IoT in precision irrigation systems to optimize water usage and reduce waste. In addition to improving resource management, IoT can also help farmers detect and prevent diseases in their crops. A study by Cruz et al. (2022) used IoT technology to monitor the growth of strawberries and detect early signs of disease. Similarly, Puengsungwan et al. (2020) used IoT sensors to detect plant stress caused by environmental factors such as temperature and humidity.

IoT technology can also facilitate the integration of urban agriculture into the food supply chain. A study by Onwude et al. (2020) used IoT sensors to monitor the freshness of produce during transportation, while another study by Kamble et al. (2020) used IoT technology to track the origin and quality of vegetables in urban farms. However, the implementation of IoT in urban agriculture also presents challenges such as the cost of sensors and data management. A study by Podder et al. (2021) explored the use of edge computing to reduce the cost and improve the efficiency of IoT in urban agriculture. Another study by Chaganti et al. (2022) investigated the use of blockchain technology to improve the security and transparency of data in IoT systems. These studies demonstrate that integrating IoT technology in urban agriculture can revolutionize the field and promote sustainable and efficient agriculture practices. They demonstrated IoT technology's various benefits and challenges in urban agriculture, and further research is needed to realize its potential fully.

This paper presents an overview of the potential benefits of integrating IoT technologies with urban agriculture for nutrient controlling system. Urban farming method requires special treatment for controlling the water temperature, water level, and acidity (pH) of nutrient solutions. Nutritional solutions for hydroponic systems are aqueous solutions containing inorganic ions, especially from salts which are important elements for plants which are tall (Trejo-Téllez and Gómez-Merino 2012). Plants need frequent watering and fertilization (Charumathi et al. 2017). To be able to produce plants that are good in the harvest period, these treatments and regular must be done every day. The checks carried out include checking the water content in the installation, the nutrients contained, the dose of the pH, the temperature and humidity of the air, etc., which must meet the specified standards. If one of these elements does not meet the right dose, the plant will not grow as expected. Therefore, regular checks must be done every day. Due to the need for regular checks, the hydroponic method becomes inefficient because it requires a long time and high costs for maintenance (Lochan Mishra and Jain 2015). This also impacts on the selling price of hydroponic plants; the plants become more expensive. While hydroponic method is a solution to the problem of limited land, it also requires complicated care, making it not efficient for agriculture.

This paper is organized as follow. The literature review of latest development in an intelligent system for nutrient-controlling systems is discussed in Sect. 2. The general IoT system architecture is discussed in Sect. 3. The future challenge of research and implementation of intelligent nutrient-controlling system is discussed in Sect. 4. Finally, the conclusion is presented in Sect. 5.

2 Intelligent Nutrient Control Systems

Intelligent nutrient control systems have been increasingly applied in urban agriculture to improve plant growth and yield while reducing waste and environmental impact. These systems utilize sensors and automation to monitor and regulate nutrient levels in the soil or hydroponic solutions, while data analytics and machine learning algorithms are used to optimize nutrient delivery and minimize resource usage. In this section, some research on intelligent nutrient control system in urban farming is discussed.

Herman et al. (2020) proposed a hydroculture system that is monitored using sensors and controlled by a microcontroller especially 8266 and actuators. The sensors used include pH, electrical conductivity, humidity, and temperature levels to see the current conditions in the hydroculture. The data from sensor then analyzed with Sugeno fuzzy logic method to automatically regulate the water and nutrient pump. The results of the study had significant differences in leaf width and plant height in lettuce and bok choy plants.

Mehra et al. (2018) proposed the implementation of deep neural networks (DNN) in deep water culture (DWC) hydroponic. The DWC hydroponic technique is the most straightforward hydroponic technique. It only uses a water reservoir, and the plants are directly on top of the water reservoir. The input parameters for the DNN are PPM, water level, temperature, light intensity, and humidity. The input parameters are fitted with models that have been trained in the cloud and will provide a classification of actions to regulate the hydroponic system environment. The system's output can only classify which actuator needs to be turned on or off.

Alipio et al. (2019) used the nutrient film technique (NFT) hydroponic systems. NFT is a hydroponic technique that continuously circulates dissolved nutrients from the water reservoir to the growing media using a pump. The dissolved nutrients are flowed through the gutter and pass through all the roots of the plants. The study uses a Bayesian network that acts as the system's brain to automatically regulate the water reservoir's pH and EC. pH, EC, humidity, light intensity, and water temperature are the Bayesian network input parameters. The Bayesian network processes the sensors' data to give the proper action needed to regulate the hydroponic system environment. The detected data and respective output are sent to the cloud so the user can monitor it.

Adidrana et al. (2019) proposed an NFT (nutrient film technique) hydroponic nutrition control system using the KNN method and IoT. This control system expected to provide accurate calculation results to command the microcontroller to turn on or off the nutrition controllers more than one at a time, such as pH down, pH up, AB nutrition, and filter pump. KNN (k-nearest neighbor) algorithm uses for predicting the classification of nutrient conditions, so the system can provide information on nutrition conditions to the user. pH and TDS values controlled using pH (up and down) solution, nutrients (A and B) to increase the TDS value, and nutrient filter to reduce the TDS value obtained from the pH sensor and TDS sensor.

Atmaja et al. proposed a multistep fuzzy logic method for NFT hydroponics system in making decisions for parameter adjustments in the hydroponic system. The multistep fuzzy logic is proposed to be able activating relay within the same time. After the relay was activated at the same time, it is possible there are some calibrations needed to tune the mixed solution to add a difference into the hydroponics main system. The calculation result data is sent via the ESP8266 and NRF24L01 modules. With the results of the evaluation of the multistep fuzzy logic method, it is in accordance with the expectations of the created fuzzy rule. From the real-time data transmission method, the success of sending data is 30% from the ESP82166 and 75% of the NRF24L01 with a shortage of the NRF24L01 data loss. For the relay, activation can be accommodated with dynamic programming. As for multistep fuzzy logic, hydroponics was tested to reach optimal water condition for kale crops, resulting in an average 12.8 iterations of calibration from condition where researches add water only from the start.

From the all the research that has been explained in this section, the researchers tried to address the difficulty in fostering urban farming that requires a precision water and nutrition intake. From their research results, it is evident that machine learning or deep learning method can be used to analyze the data from sensor and automatically regulate the nutrient pump. Therefore, the plants can receive nutritional intake according to their needs and will grow more optimally (Table 1).

Table 1 Intelligent nutrient control systems

No	Publication	Proposed technique	Results
1	Herman et al. (2020)	Sugeno fuzzy logic to control pH, nutrient, and temperature	Proposed system shows better plant growth in terms of length and width of the leaves and plant's height
2	Mehra et al. (2018)	Deep neural networks (DNN) in deep water culture (DWC) hydroponic	Plant growth in hydroponics is far better in terms of height compared to the traditional soil growth
3	Alipio et al. (2019)	Bayesian network to detect and regulate humidity, sunlight, water temperature, pH level, and electrical conductivity	The prediction model obtained 84.53% accuracy after model validation, and the yielded crops on the automatic control was 66.67% higher than the manual control
4	Adidrana et al. (2019)	KNN (k-nearest neighbor) algorithm to predict the classification of nutrient conditions	Achieves 93.3% accuracy
5	Atmaja et al. (2022)	Multistep fuzzy logic method for NFT hydroponics system	To reach optimal water condition for kale crops resulting in average 12.8 iterations calibration from condition where researches add water only from the start

3 IoT System Architecture

This section discusses the general architecture of an intelligent nutrient control system. Generally, as shown in Fig. 1, the system consists of three sections: the sensor layer, the actuator layer, and the data processing layer. A detail explanation of each layer is presented in Sects. 3.1, 3.2, and 3.3.

3.1 Sensor Layer

Intelligent nutrient control systems use sensors to monitor and control nutrient levels in plants or aquaculture systems. The sensors commonly used include pH sensors, electrical conductivity (EC) sensors, dissolved oxygen (DO) sensors, temperature sensors, soil moisture sensors, and nutrient sensors, such as ammonium, nitrate, and potassium sensors.

pH sensors are commonly used in intelligent nutrient control systems for urban farming to monitor and maintain the acidity or alkalinity of the nutrient solution (Herman 2019; Adidrana and Surantha 2019). In hydroponic or aeroponic systems,

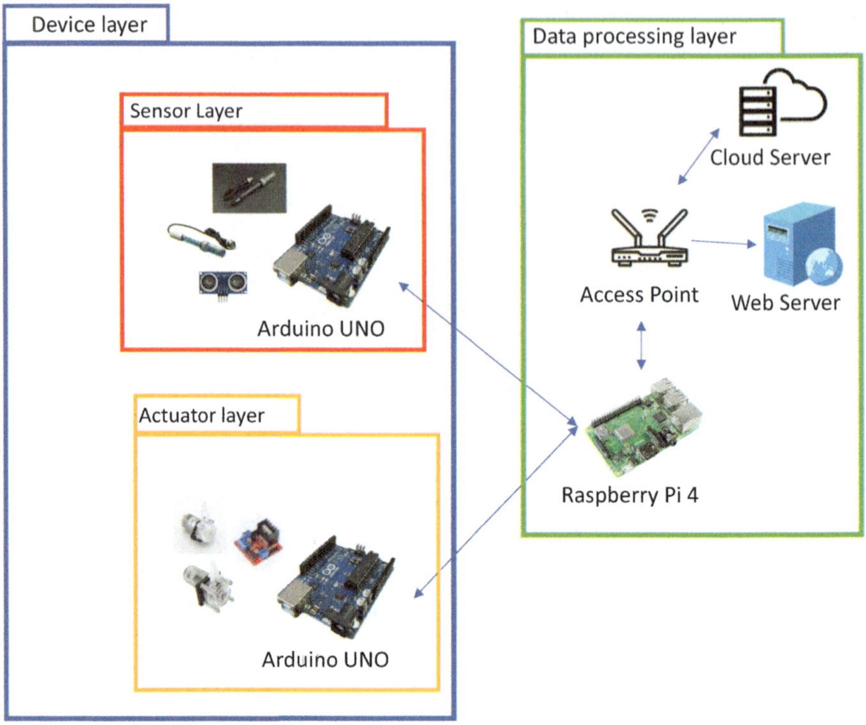

Fig. 1 General System Architecture

plants receive nutrients directly from a nutrient solution rather than from soil, so it is important to ensure that the pH of the solution is within the optimal range for plant growth. The optimal pH range depends on the plant species and can range from around 5.5 to 6.5 for most leafy greens and to around 6.5 to 7.5 for tomatoes and cucumbers (Goddek et al. 2020). pH sensors can be used to continuously monitor the pH of the nutrient solution and provide real-time feedback to an automated nutrient control system. The system can adjust the pH by adding acid or base solutions to maintain the desired pH range. This helps to ensure that the plants have access to the nutrients they need and can grow optimally. pH sensors can also be used to diagnose problems such as nutrient imbalances, which can cause the pH to drift outside of the optimal range. Overall, pH sensors are an essential component of an intelligent nutrient control system for urban farming to ensure optimal plant growth and health.

EC sensors are commonly used in intelligent nutrient control systems for urban farming to measure the concentration of nutrients in the hydroponic solution (Yolanda et al. 2016). In urban farming, where space is often limited, hydroponic systems are used to grow plants without soil in a nutrient-rich water solution. The EC sensor measures the electrical conductivity of the solution, which is directly related to the concentration of dissolved salts in the solution. By measuring the EC, the system can determine the nutrient concentration of the solution and make adjustments to ensure that the plants are receiving the proper nutrients for optimal growth. In addition to nutrient monitoring, EC sensors can also be used to monitor the overall health of the hydroponic system (Lochan Mishra et al. 2007). For example, if the EC is too high, it may indicate that there is a buildup of salts in the solution, which can be harmful to the plants. Similarly, if the EC is too low, it may indicate that the plants are not receiving enough nutrients.

DO sensors measure the amount of oxygen dissolved in the nutrient solution, which is critical to the health of the plants (Deepthi et al. 2021). If the oxygen level is too low, it can lead to root rot, which can kill the plants. If the oxygen level is too high, it can create an environment that promotes the growth of harmful bacteria. DO sensors can detect changes in oxygen levels and alert the system to adjust the oxygen supply to maintain optimal levels (Kyaw and Ng 2017). This ensures that the plants receive the right amount of oxygen to grow and remain healthy. Overall, DO sensors play an important role in maintaining the health and productivity of plants in urban farming systems.

Temperature sensors are an essential component of intelligent nutrient control systems in urban farming (Joseph Balinado 2016). These sensors measure the temperature of the nutrient solution, which is critical for plant growth and health. Temperature affects plant metabolism, nutrient uptake, and the growth rate of plants. For example, if the temperature is too high, it can lead to lower oxygen levels in the nutrient solution, which can harm plant roots. On the other hand, if the temperature is too low, it can slow down plant growth and reduce nutrient uptake. By monitoring the temperature of the nutrient solution, the intelligent nutrient control system can adjust other parameters such as pH and nutrient levels to optimize plant growth and health. Additionally, temperature sensors can also be used to monitor the temperature in the growing environment (Alipio et al. 2017), which is important for controlling the microclimate and preventing heat stress or cold damage to plants. Overall, temperature sensors play a critical role in maintaining optimal growing conditions and maximizing yield in urban farming systems.

Nutrient sensors such as ammonium, nitrate, and potassium sensors can be used in intelligent nutrient control systems for urban farming to monitor the nutrient levels in hydroponic or aeroponic systems (John and Mahalingam 2021). These sensors can detect the concentration of specific nutrients in the solution and provide real-time data that can be used to adjust the nutrient levels. For example, ammonium sensors can detect the concentration of ammonium ions in the solution, which is important for plants as a source of nitrogen. Nitrate sensors can detect the concentration of nitrate ions in the solution, which is also important for plants as a source of nitrogen. Potassium sensors can detect the concentration of potassium ions in the solution, which is essential for plant growth and development (Silva et al. 2022). By using these nutrient sensors in combination with other sensors such as pH sensors and EC sensors, an intelligent nutrient control system can adjust the nutrient levels in real-time to ensure that the plants are getting the optimal amount of nutrients for growth and development. This can lead to increased yield and improved quality of produce in urban farming systems.

Finally, soil moisture sensors are commonly used in intelligent nutrient control systems for urban farming to help optimize plant growth and nutrient uptake (Hostalrich et al. 2022). These sensors provide real-time data on the moisture levels in the soil, allowing farmers to adjust their watering schedule and fertilizer application to meet the needs of the plants. Integrating soil moisture sensors with other sensors, such as pH, EC, DO, temperature, and nutrient sensors, can create a more comprehensive system for intelligent nutrient control. By combining data from multiple sensors, farmers can better understand the overall health of their plants and adjust their nutrient levels accordingly (Janani et al. 2022). The summary of sensor used in intelligent nutrient control system is presented in Table 2.

3.2 Actuator Layer

Actuators are an important component of an intelligent nutrient control system for urban farming. They are used to control the delivery of nutrients, water, and other inputs to plants in hydroponic or aeroponic systems. The three main types of actuators used in these systems are pumps, solenoid valves, and dosing systems.

Pumps are commonly used in nutrient control systems to deliver nutrient solutions to plants (Safira et al. 2022). Peristaltic pumps are often used because they are precise and have a low risk of contamination. The system can control them

Table 2 Sensor used in intelligent nutrient control system

No.	Sensor type	Function	References
1	pH sensors	Measure acidity or alkalinity of nutrient solution	Herman (2020), Adidrana and Surantha (2019) and Yolanda et al. (2016)
2	EC sensors	Measure concentration of nutrients in the solution	Adidrana and Surantha (2019), Yolanda et al. (2016) and Lochan Mishra et al. (2007)
3	DO sensors	Measure oxygen concentration in the water	Deepthi et al. (2021) and Kyaw and Ng (2017)
4	Temperature sensors	Measure the temperature of nutrient solution and temperature of plant environment	Adidrana and Surantha (2019), Joseph Balinado (2016) and Alipio et al. (2017)
5	Specific nutrient sensors (ammonium, nitrate, and potassium)	Measure levels of specific nutrients in the solution	John and Mahalingam (2021) and Silva et al. (2022)
6	Soil moisture sensors	Measure moisture levels in the soil	Hostalrich et al. (2022) and Janani et al. (2022)

to deliver precise amounts of nutrient solution to the plants based on the real-time data collected from the sensors (Rico 2020).

Solenoid valves are used to control the flow of water and nutrient solutions in hydroponic and aeroponic systems (Xu et al. 2020). These valves can be controlled electronically, allowing for precise control of the amount of solution delivered to each plant. They are often used in combination with pumps to deliver nutrient solutions to plants (Iswanto and Ma'Arif 2020).

Dosing systems are used to deliver precise amounts of nutrients to the plants. They can be used to mix and deliver nutrient solutions based on the real-time data collected from the sensors (Lennard and Ward 2019). Some dosing systems are automated, allowing for precise control of the nutrient delivery to each plant (Hosseini et al. 2021).

In summary, pumps, solenoid valves, and dosing systems are the primary types of actuators used in intelligent nutrient control systems for urban farming. These actuators allow for precise control of the delivery of nutrients, water, and other inputs to plants based on real-time data collected from sensors. By using these actuators, urban farmers can optimize plant growth and development, leading to increased yields and improved quality of produce. The summary of actuators used in intelligent nutrient control system is presented in Table 3.

3.3 Data Processing Layer

Intelligent nutrient control systems for urban farming often incorporate microcontrollers and cloud computing technology to automate and remotely monitor the growing environment. Microcontrollers and cloud computing are used for data analytics process and data storage for the plant monitoring systems.

Microcontrollers such as Arduino (Ibrahim et al. 2015) and Raspberry Pi (Crisnapati et al. 2017) can be used in intelligent nutrient control systems for urban farming. These microcontrollers can be programmed to read data from various sensors, such as pH, EC, dissolved oxygen, and nutrient sensors, and adjust the nutrient levels in real-time based on the data. The microcontrollers can also be used to control other components in the system such as pumps, valves, and lights. This allows

Table 3 The summary of actuator used in intelligent nutrient control system

No.	Sensor type	Function	References
1	Pumps	Deliver nutrient solutions to plants	Safira et al. (2022) and Rico (2020)
2	Solenoid valves	Control the flow of water and nutrient solutions	Xu et al. (2020) and Iswanto and Ma'Arif (2020)
3	Dosing systems	To mix and deliver nutrient solutions based on the real-time data collected from the sensors	Lennard and Ward (2019) and Hosseini et al. (2021)

for precise and automated control of the nutrient solution, leading to increased yield and improved produce quality in urban farming systems. Additionally, these microcontrollers are cost-effective and easily accessible, making them a popular choice for small-scale urban farming operations.

FPGAs (Field Programmable Gate Arrays) can also be used in intelligent nutrient control systems for urban farming to process the data from various sensors and control the nutrient delivery system (Oukaira et al. 2021). FPGAs are also highly customizable and can be reprogrammed to accommodate system changes or add new sensors or control functions. Additionally, FPGAs can operate at high speeds with low latency, making them ideal for real-time control in urban farming systems where quick response times are essential (Kumar et al. 2020).

Cloud computing technology enables the remote monitoring and control of the growing environment from a smartphone, tablet, or computer. This allows farmers to monitor and adjust the growing environment from anywhere, at any time, which is especially important for urban farming where space and time are often limited. Cloud computing can also be used to store and analyze data from sensors, providing insights into the performance of the growing system and allowing for continuous optimization.

Microsoft Azure is an example of a cloud computing platform used for intelligent nutrient control systems. Microsoft Azure offers IoT Hub and Time Series Insights, which allow farmers to connect and monitor sensors in real time and analyze historical data to make informed decisions about nutrient control (Rahul et al. 2022). The platform also offers machine learning tools, which can be used to optimize nutrient control and predict plant growth based on historical data. Another example of a cloud computing platform used for intelligent nutrient control systems is AWS IoT. AWS IoT offers a suite of services, including IoT Core, which enables farmers to connect sensors and devices, and IoT Analytics, which provides real-time analysis of sensor data (Ponnusamy et al. 2021; Philimon et al. 2022). The platform also offers machine learning tools such as SageMaker, which can be used to predict plant growth and optimize nutrient levels based on historical data (Shaif 2021). Additionally, AWS Greengrass allows for local compute and analytics at the edge, enabling farmers to quickly respond to changes in nutrient levels in real time (Tawalbeh et al. 2020).

Blynk and Growlink are two examples of software platforms that can be used to create intelligent nutrient control systems for urban farming. Blynk is an IoT (Internet of Things) platform that allows users to build custom apps to control and monitor various devices, including sensors and actuators (Herman 2020). Growlink, on the other hand, is a software platform specifically designed for agriculture and hydroponics systems (Srivastava and Das 2022). By integrating Blynk or Growlink with various sensors such as pH, EC, temperature, and nutrient sensors, an intelligent nutrient control system can be created for urban farming. These systems can monitor and adjust nutrient levels in real time, based on the data collected from the sensors. The platforms also allow for remote monitoring and control of the system, which can save time and increase efficiency for urban farmers. Overall, Blynk and Growlink offer user-friendly and customizable solutions for creating

Table 4 Data processing layer

No	Technology	Example of technology	References
1	Microcontroller/ edge device	Arduino	Joseph Balinado (2016) and Ibrahim et al. (2015)
		Raspberry-Pi	Atmaja and Surantha (2022) and Crisnapati et al. (2017)
		FPGA	Oukaira et al. (2021) and Kumar et al. (2020)
2	Cloud computing platform	Microsoft Azure: IoT hubs and time-series insight	Rahul et al. (2022)
		AWS IoT: IoT Core, SageMaker, AWS Greengrass	Ponnusamy et al. (2021), Philimon et al. (2022), Shaif (2021), and Tawalbeh et al. (2020)
		Blynk	Herman (2020)
		Growlink	Srivastava and Das (2022)

intelligent nutrient control systems for urban farming. By using these platforms in combination with various sensors, urban farmers can optimize the growth and quality of their crops while also saving time and resources.

In combination, microcontrollers and cloud computing technology provide a powerful tool for intelligent nutrient control in urban farming. By automating nutrient delivery and environmental control and remotely monitoring and adjusting the growing environment, farmers can optimize plant growth and productivity while reducing labor costs and environmental impact. Additionally, the use of cloud computing allows for real-time analysis and optimization of the growing system, leading to increased efficiency and improved crop yield (Table 4).

4 Future Challenges

The future of intelligent nutrient control systems for urban farming faces several challenges. In this section, we identify several potential challenges for research in this field.

4.1 Integration of Sensors and Controls System

One of the key challenges is integrating multiple sensors and control systems to optimize nutrient delivery to plants. With increasing technological advancements, more sensors and control systems are becoming available, making it difficult to select the most effective and efficient systems for a particular urban farming environment (Hostalrich et al. 2022). Therefore, there is a need to develop integrated systems that can work together seamlessly to optimize plant growth.

On the other hand, as more sensors and actuators are added to the system, the interactions between them become more complicated, making it difficult to optimize nutrient delivery to the plants (Herman 2020). For example, a system that includes sensors for measuring pH, EC, ammonium, nitrate, and potassium levels and actuators for adjusting the nutrient delivery system can generate a vast amount of data that needs to be processed in real time. The system needs to be able to analyze this data and make decisions about when and how to adjust the nutrient delivery system to maintain optimal nutrient levels. In addition, there may be interactions between the sensors and actuators that need to be considered. For example, changes in pH levels can affect the availability of certain nutrients to the plants, which may require adjustments to the nutrient delivery system (Herman 2019). This requires a system that can integrate the data from multiple sensors and actuators and make decisions based on the overall state of the system.

4.2 Data Analytics and Machine Learning for Predicting Nutrient Control

There is a need for more advanced data analytics and machine learning algorithms to analyze the vast amounts of data generated by the sensors in real time. This requires developing advanced algorithms that can identify trends and patterns in the data to make more accurate predictions about plant growth and nutrient requirements (Mehra et al. 2018). Another challenge is the need to develop algorithms that can accurately predict plant growth and nutrient requirements. This requires training the algorithms with large amounts of data from a variety of environmental conditions, plant species, and nutrient solutions (Deren et al. 2021). The algorithms need to be able to identify patterns and trends in the data to make accurate predictions. Additionally, machine learning algorithms need to be able to adapt to changes in the system, such as the addition of new sensors or changes in the nutrient solution. Developing algorithms that can learn and adapt to changes in the system is essential for maintaining the optimal nutrient delivery to the plants (Atmaja and Surantha 2022).

4.3 Sustainable and Eco-Friendly Nutrient Delivery Systems

Another challenge is the development of more sustainable and eco-friendly nutrient delivery systems. One challenge is reducing water consumption, as many nutrient delivery systems require large amounts of water. This can be achieved by using recirculating systems that reuse water or by using systems that capture and reuse rainwater or other alternative water sources (Bisaga et al. 2019). Implementing these systems can significantly reduce water consumption and improve the sustainability of the nutrient delivery system.

Another challenge is reducing energy consumption, as many nutrient delivery systems require energy to pump and circulate the nutrient solution. Using solar or wind power to generate energy for the system can reduce the environmental impact of the system and lower energy costs. Additionally, optimizing the timing and frequency of nutrient delivery can reduce the amount of energy required to operate the system.

Another challenge is minimizing waste, as many nutrient delivery systems generate nutrient-rich runoff that can be harmful to the environment if not properly managed. Implementing systems that capture and reuse this runoff can significantly reduce waste and improve the sustainability of the nutrient delivery system.

Finally, there is a need to develop nutrient solutions that are more sustainable and environmentally friendly. Currently, many nutrient solutions use synthetic fertilizers that can be harmful to the environment (Havlin 2020). Developing more sustainable and organic nutrient solutions can improve the nutrient delivery system's sustainability and the quality of the produce grown in the system.

4.4 More Affordable and Accessible Intelligent Nutrient Control Systems

Finally, there is a need to develop intelligent nutrient control systems that are more affordable and accessible to urban farmers. One challenge is the cost of the sensors and control systems used in these systems. Many of the sensors and control systems can be expensive, making them inaccessible to small-scale urban farmers with limited resources. Therefore, there is a need to develop more affordable sensors and control systems that can be easily integrated into urban farming environments.

Another challenge is the need for specialized knowledge to operate these systems. Many of the current systems require advanced technical knowledge to operate, which can be a barrier for small-scale urban farmers who may not have the necessary skills or training (Atmaja and Surantha 2022). Therefore, there is a need to develop more user-friendly systems that are easy to operate and require minimal technical knowledge.

Additionally, there is a need to develop open-source systems that can be customized and modified by urban farmers to meet their specific needs. Open-source systems can be more affordable and accessible because they allow users to modify and customize the system using readily available and low-cost components.

Finally, there is a need to develop training and support programs for urban farmers to help them implement and operate these systems. Many urban farmers may not have the necessary technical knowledge or experience to operate these systems, so providing training and support can help ensure the success of the system and improve the accessibility of intelligent nutrient control systems for urban farming.

5 Conclusion

The proliferation of IoT and AI leads to the emergence of intelligent system that can be used to improve various aspect in urban farming. One of the main application of IoT in urban farming is for the automatic nutrient controlling system. Urban farming method requires special treatment for controlling the water temperature, water level, and acidity (pH) of nutrient solutions. The intelligent system will help the beginner urban farmer grow the plants optimally. In this chapter, we have discussed some of the algorithms developed for intelligent nutrient controlling. We have also discussed the general architecture of IoT system and its detail component. Finally, we discuss the potential challenge of research and implementation of intelligent nutrient systems in society. With this study, hopefully there are more research to be done to improve the feasibility of intelligent nutrient system in urban farming.

References

Adidrana D, Surantha N (2019) Hydroponic nutrient control system based on internet of things and K-nearest neighbors. In: 2019 international conference on computer, control, informatics and its applications: emerging trends in big data and artificial intelligence, IC3INA 2019

Alipio MI, Dela Cruz AEM, Doria JDA, Fruto RMS (2017) A smart hydroponics farming system using exact inference in Bayesian network. In: 2017 IEEE 6th global conference on consumer electronics GCCE 2017. Janua(Gcce), pp 1–5

Alipio MI, Dela Cruz AEM, Doria JDA, Fruto RMS (2019) On the design of nutrient film technique hydroponics farm for smart agriculture. Eng Agric Environ Food (December 2018):0–1

Atmaja P, Surantha N (2022) Smart hydroponic based on nutrient film technique and multistep fuzzy logic. Int J Electr Comput Eng 12(3):3146–3157

Bisaga I, Parikh P, Loggia C (2019) Challenges and opportunities for sustainable urban farming in South African low-income settlements: a case study in Durban. Sustainability 11(20)

Chaganti R, Varadarajan V, Gorantla VS, Gadekallu TR, Ravi V (2022) Blockchain-based cloud-enabled security monitoring using internet of things in smart agriculture. Fut Internet 14(9):1–20

Charumathi S, Kaviya RM, Manisha R, Dhivya P (2017) Optimization and control of hydroponics agriculture using IOT. Asian J Appl Sci Technol 1(2):96–98

Crisnapati PN, Wardana INK, Aryanto IKAA, Hermawan A (2017) Hommons: hydroponic management and monitoring system for an IOT based NFT farm using web technology. In: 2017 5th international conference on cyber and IT service management CITSM 2017

Cruz M, Mafra S, Teixeira E, Figueiredo F (2022) Smart strawberry farming using edge computing and IoT. Sensors 22(15)

D'Ostuni M, Trombadore A, Orsini F. Food in the city: enabling integrated solutions for urban agriculture and food production in buildings. 2022 231–254.. https://link.springer.com/chapter/10.1007/978-3-030-68556-0_8

Deepthi AS, Niranjanaa A, Hari Haran A, Austin Joel A, Dhinakaran SB, Monisha Thangam K et al (2021) Integrated smart system for urban farming. In: 2021 international conference on advancements in electrical, electronics, communication, computing and automation ICAECA 2021

Deren L, Wenbo Y, Zhenfeng S (2021) Smart city based on digital twins. Comput Urban Sci 1(1): 1–11

Goddek S, Joyce A, Kotzen B, Burnell GM (2020) Correction to: aquaponics food production systems. Aquapon Food Prod Syst:1

Havlin JL (2020) Soil: fertility and nutrient management. Landsc L Capacit [Internet]:251–265. Available from: https://www.taylorfrancis.com/chapters/edit/10.1201/9780429445552-34/soil-fertility-nutrient-management-john-havlin

Herman SN (2019) Intelligent monitoring and controlling system for hydroponics precision agriculture. In: 2019 7th international conference on information and communication technology, ICoICT 2019

Herman SN (2020) Smart hydroculture control system based on IoT and fuzzy logic. Int J Innov Comput Inf Control 16(1):207–221

Hosseini H, Mozafari V, Roosta HR, Shirani H, van de Vlasakker PCH, Farhangi M (2021) Nutrient use in vertical farming: optimal electrical conductivity of nutrient solution for growth of lettuce and basil in hydroponic cultivation. Horticulturae 7(9)

Hostalrich D, Pelegri-Sebastia J, Sogorb T, Pellicer V (2022) Intelligent management of hydroponic systems based on IoT for agrifood processes. J Sensors 2022

Ibrahim MNR, Solahudin M, Widodo S (2015) Control system for nutrient solution of nutrient film technique using fuzzy logic. TELKOMNIKA (Telecommunication Comput Electron Control [Internet]) 13(4):1281. Available from: http://journal.uad.ac.id/index.php/TELKOMNIKA/article/view/2113

Iswanto MP, Ma'Arif A (2020) Nutrient film technique for automatic hydroponic system based on arduino. In: Proceeding – 2020 2nd international conference on industrial electrical and electronics, ICIEE 2020, pp 84–86

Janani GM, Santhiya N, Vigneshwari G, Manuka A, Kumar K (2022) Automatic indoor hydroponic plant grow pot using Arduino. In: Proceedings of 2nd international conference on artificial intelligence and smart energy, ICAIS, pp 1614–1618

John J, Mahalingam PMPR (2021) Automated fish feed detection in IoT based aquaponics system. In: 8th international conference on smart computing and communications. Artificial Intelligence(AI) driven applications for a smart world, ICSCC 2021, pp 286–290

Joseph Balinado FC (2016) Fuzzy logic controller implementation to an arduino-based solar-powered aquaponics system prototype. J Eng Comput Stud [Internet] 3(2) Available from: http://lpulaguna.edu.ph/wp-content/uploads/2016/10/Fuzzy-Logic-Controller-Implementation-to-an-Arduino-Based-Solar-Powered-Aquaponics-System-Prototype.pdf

Kamble SS, Gunasekaran A, Sharma R (2020) Modeling the blockchain enabled traceability in agriculture supply chain. Int J Inf Manag 52:101967

Kumar PM, Kumar TGS, Aravindhan R, Sairam D, Moorthi S (2020) Design and development of an embedded framework to test soil for urban terrace farming. In: Proceedings of 3rd international conference on emerging technologies in computer engineering: machine learning and Internet OF Things, ICETCE 2020, pp 41–45

Kyaw TY, Ng AK (2017) Smart aquaponics system for urban farming. Energy Procedia [Internet] 143:342–347. https://doi.org/10.1016/j.egypro.2017.12.694

Lennard W, Ward J (2019) A comparison of plant growth rates between an NFT hydroponic system and an NFT aquaponic system. Horticulturae 5(2)

Lochan Mishra R, Jain P (2015) Design and implementation of automatic hydroponics system using ARM processor. Int J Adv Res Electr Electron Instrum Eng 4(8):6935–6940

Lochan Mishra R, Jain P, Student ME (2007) Design and implementation of automatic hydroponics system using ARM processor. Int J Adv Res Electr Electron Instrum Eng (An ISO [Internet]) 3297. Available from: http://ijareeie.com/upload/2015/august/16_Design.pdf

Maria Manuel Vianny D, John A, Kumar Mohan S, Sarlan A, Adimoolam AA (2022) Water optimization technique for precision irrigation system using IoT and machine learning. Sustain Energy Technol Assess 52:102307

Mehra M, Saxena S, Sankaranarayanan S, Tom RJ, Veeramanikandan M (2018) IoT based hydroponics system using deep neural networks. Comput Electron Agric 155(October):473–486

Onwude DI, Chen G, Eke-Emezie N, Kabutey A, Khaled AY, Sturm B (2020) Recent advances in reducing food losses in the supply chain of fresh agricultural produce. PRO 8(11):1–31

Ouafiq EM, Elrharras A, Mehdary A, Chehri A, Saadane R, Wahbi M (2021) IoT in smart farming analytics, big data based architecture [Internet]. Vol. 189, Smart innovation, systems and technologies. Springer, Singapore, pp 269–279. https://doi.org/10.1007/978-981-15-5784-2_22

Oukaira A, Benelhaouare AZ, Kengne E, Lakhssassi A (2021) Fpga-embedded smart monitoring system for irrigation decisions based on soil moisture and temperature sensors. Agronomy 11(9)

Philimon SP, Tsun Tee MK, Yan Lo CC, Hung Ho DC, Yan Chieng JZ, Bee Lian Soon RM (2022) Proposing a foundational IoT smart planter concept towards reducing carbon footprint in Malaysia via gamification. In: Proceedings – 2022 international conference on computer and drone applications, IConDA 2022, pp 34–39

Podder AK, Al Bukhari A, Islam S, Mia S, Mohammed MA, Kumar NM et al (2021) IoT based smart agrotech system for verification of urban farming parameters. Microprocess Microsyst [Internet] 82(December 2020):104025. https://doi.org/10.1016/j.micpro.2021.104025

Ponnusamy V, Natarajan S, Ramasamy N, Clement C, Rajalingam P, Mitsunori M (2021) An IoT-enabled augmented reality framework for plant disease detection. Rev d'Intelligence Artif 35(3):185–192

Puengsungwan S, Jirasereeamornkul K (2020) IoT Based root stress Detection for lettuce culture using infrared leaf temperature sensor and light intensity sensor. Wirel Pers Commun [Internet] 115(4):3215–3233. https://doi.org/10.1007/s11277-020-07219-z

Rahul, Rauniyar K, Monika A, Khan JA (2022) Application of IoT and AI in the development of smart cities. In: Smart Cities [Internet], pp 181–196. Available from: https://www.taylorfrancis.com/chapters/edit/10.1201/9781003287186-7/application-iot-ai-development-smart-cities-rahul-kritesh-rauniyar-monika-javed-ahmad-khan

Rico ALJ (2020) Automated pH monitoring and controlling System for hydroponics under greenhouse condition. ARPN J Eng Appl Sci 15(2):523–528

Safira MR, Lim MW, Chua WS (2022) Design of control system for water quality monitoring system for hydroponics application. IOP Conf Ser Mater Sci Eng 1257(1):012027

Shaif A. Predictive maintenance in smart agriculture using machine learning: a novel algorithm for drift fault detection in hydroponic sensors. 2021

Silva AF, Löfkvist K, Gilbertsson M, Van Os E, Franken G, Balendonck J et al (2022) Hydroponics monitoring through UV-vis spectroscopy and artificial intelligence: quantification of nitrogen, phosphorous and potassium 88

Srivastava A, Das DK (2022) A comprehensive review on the application of Internet of Thing (IoT) in smart agriculture [Internet]. Vol. 122, wireless personal communications. Springer, pp 1807–1837. https://doi.org/10.1007/s11277-021-08970-7

Stevens JD, Shaikh T (2018) MicroCEA: developing a personal urban smart farming device. In: 2nd international conference on smart grid and smart cities, ICSGSC 2018, pp 49–56

Tawalbeh L, Muheidat F, Tawalbeh M, Quwaider M (2020) Applied sciences IoT privacy and security: challenges and solutions. Mdpi:1–17

Trejo-Téllez LI, Gómez-Merino FC (2012) Nutrient solutions for hydroponic systems. Hydroponics – a stand methodology for plant. Biol Res:1–22

Xu K, Kitazumi Y, Kano K, Shirai O (2020) Automatic management of nutrient solution for hydroponics-construction of multi-ion stat. Anal Sci [Internet] 36(9):1141–1144. Available from: https://link.springer.com/article/10.2116/analsci.20A002

Xu J, Gu B, Tian G (2022) Review of agricultural IoT technology. Artif Intell Agric 1(6):10–22

Yolanda D, Hindersah H, Hadiatna F, Triawan MA (2016) Implementation of real-time fuzzy logic control for NFT-based hydroponic system on Internet of Things environment. In: Proceedings of the 6th international conference on system engineering and technology (ICSET) [Internet]. IEEE, pp 153–159. Available from: http://ieeexplore.ieee.org/document/7849641/

Vertical Farming of Medicinal Plants

Françoise Bafort and M. Haïssam Jijakli

Abstract Medicinal plants have been used in traditional medicine, health food supplements, rituals, and for health care purposes for thousands of years. According to the Food and Agriculture Organization of the United Nations (FAO), the worldwide production of medicinal and aromatic plants is estimated to be 330 million tons for a total area of 77 million ha. Nowadays, the sector of medicinal plants is subject to inconstancy, and issues about the yield, quality, and efficacy of plant extracts have been reported. The present review describes the current status of medicinal plants worldwide, including a detailed description of the sector in France. The suitability of vertical farming for the production of medicinal plants is discussed, and its advantages and drawbacks are presented. Indoor cultivation in a controlled environment requires appropriate adjustment of abiotic factors to optimize biomass and secondary metabolite contents. Light quantity and quality, nutrient solution, temperature, and CO_2 concentration are presented in relation with their impact on plants and on the production of the targeted phytocompound. A case-study on the technic feasibility and economic viability of producing a plant-based drug in a vertical container is presented, including plant cultivation and drug extraction steps. Based on the costs related directly to the production activity, it provides a rapid estimate of the direct production cost of each step. The largest contributor to cultivation costs is labor, averaging 48%, followed by energy (20%) and investment cost (20%). The largest contributor to extraction and purification costs is the operating and maintenance cost of equipment (47%), followed by energy cost (31%) and labor cost (16%). The largest contributor to the whole plant-based drug production process, from plant cultivation to drug production, is the research and development cost (98–67%), followed by cultivation and extraction costs (1–24%) and drug manufacturing costs (1–8%), depending on the number of containers, i.e., on the productivity of the cultivation and extraction steps.

F. Bafort (✉) · M. H. Jijakli
Integrated and Urban Plant Pathology Laboratory, Gembloux Agro-Bio Tech, Liège University, Gembloux, Belgium
e-mail: francoise.bafort@uliege.be

P. M. Priyadarshan et al. (eds.), *Digital Agriculture*,
https://doi.org/10.1007/978-3-031-43548-5_5

Keywords Vertical farming · Economic viability · Medicinal plant · Indoor cultivation · Controlled-environment cultivation · Quality of medicinal plants

1 Introduction: Vertical Farming and Medicinal Plants

1.1 Current Challenges of Agriculture

Agriculture currently faces many challenges and difficulties in terms of environmental performances. Although the Organization for Economic Cooperation and Development (OECD) points out that agricultural nitrogen and phosphorous nutrient surpluses in the OECD countries steadily declined between 1990 and 2009, farmed soils still contain an average surplus of 63 kg/ha of nitrogen and 6 kg/ha of phosphorous. These levels remain very high as to their potential to cause surface, groundwater and coastal water pollutions (OECD 2013). In most European member states, agriculture is responsible for over a third of the total nutrient discharge into surface and coastal waters (OECD 2013). Agriculture is also the major user of pesticides, with 70% of the mean pesticide sales in OECD countries related to agriculture. As a result, agricultural soils are major reservoirs of pesticides that affect soil microbial communities and represent sources of water and air pollutants (Tao et al. 2008; OECD 2013; Hvězdová et al. 2018; Dou et al. 2020a, b). Almost a third of OECD member countries is affected by moderate to severe water-related soil erosion, while far fewer countries are suffering from wind-related erosion (OECD 2013). Erosion due to agricultural practices can be mainly attributed to continued cultivation on fragile and marginal soils, overgrazing of pasture, or unsuited farming and tillage practices (Bullock 2005; OECD 2013; Gebrehiwot 2022; Hassan et al. 2022). The mean energy consumption related to agriculture between 2008 and 2010 was low – 1.6% – but the sector is vulnerable to changes in crude oil prices, and sensitive to dramatic changes (OECD 2013). Although the OECD indicates that the agriculture sector reduced its water withdrawals over the past decade, agriculture remains a major user of water accounting for an average 44% of total water withdrawals (OECD 2013). Biodiversity as measured from farmland bird populations has been declining continuously in almost all countries over the 1990–2010 period (OECD 2013). The main reason is the considerable use of land and water resources on which wild species are highly dependent (OECD 2013). Agricultural intensification in recent decades has resulted in reduced crop diversity and losses of plant species (Storkey et al. 2012; Meyer et al. 2013; Abeli et al. 2022). Figure 1 shows the pressure exerted by agriculture in several sectors.

The value of primary agriculture round the world can be partly understood by looking at trade statistics from worldwide databases. Although trade data are never complete and products are categorized differently, they give a global picture of the importance of primary agriculture and the share of medicinal plants within primary agriculture. Customs nomenclature referring to primary agriculture includes several codes: 07 “Edible vegetables and certain roots and tubers,” 08 “Edible fruit and nuts;

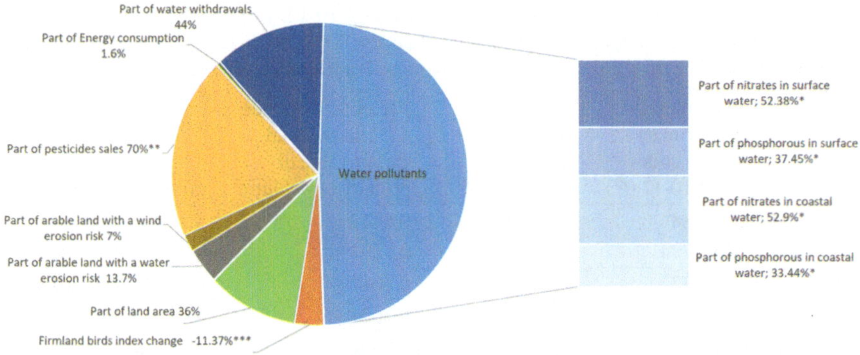

Fig. 1 Overview of the pressure exerted by agriculture in several sectors: water pollutants, energy consumption, water withdrawal, pesticide sales, land area, biodiversity, and water and wind erosion risk (OECD 2013). Each sector is represented on a 100% basis, in which the share of primary agriculture is indicated *Mean based on a limited number of OECD countries **The number of OECD countries monitoring pesticides in water systems is limited. However, data for Austria reveals that the development of pesticide sales is closely related to level of pesticides detected in surface waters ***Mean average annual percentage change

peel of citrus fruit or melons," 09 "Coffee, tea, maté and spices," 10 "Cereals," and 12 "Oil seeds and oleaginous fruits; miscellaneous grains, seeds and fruit; industrial or medicinal plants; straw and fodder," Table 1 shows the exported value of primary agriculture and the relative significance of nomenclature 1211 corresponding to "Plants and parts of plants, incl. seeds and fruits, of a kind used primarily in perfumery, medicaments or for insecticidal, fungicidal or similar purposes, fresh or dried, whether or not cut, crushed or powdered," It shows that the export value of medicinal plants represents a small percentage of primary agriculture (about 0.7%) but is constantly growing.

In 2019, the total exported quantities of code 1211 represented 731,606 tons and 2,892,682 euros. The exported value of code 1211 *per* ton was about ten times higher than those of rice (code 1006), cereals (code 1001), and potato (code 0701) and about three times higher than that of tomatoes (code 0702), showing that this category has a high added value.

1.2 The Current Status of Medicinal Plants

The Current Status of Medicinal Plants Round the World

Medicinal plants, including medicinal herbs, have long been used round the world. The use of medicinal plants is one of the oldest forms of treatment, coming from ancestral and empirical uses, that still plays a significant role in Africa and Asia (World Health Assembly 2003). The World Health Organization (WHO) reports that at least half of the world population do not receive the healthcare services they

Table 1 World total export value (€) of codes 07-08-09-10-12, world total export value (€) of code 1211, percentage of code 1211 in the total of codes 07-08-09-10-12, and annual export value growth of code 1211 from 2018 to 2021 (International Trade Center, no date)

Year	World total export value of primary agriculture codes 07-08-09-10-12 (€)	World total export value of code 1211 (€)	Percentage of code 1211 in the total export value of primary agriculture	Annual export growth of code 1211
2021	483,564,488	3,310,966	0.6847%	5.26% (2020–2021)
2020	431,281,779	3,145,490	0.7293%	8.74% (2019–2020)
2019	410,466,532	2,892,682	0.7047%	6.15% (2018–2019)
2018	387,958,090	2,725,045	0.7024%	–

Codes: 07, "Edible vegetables and certain roots and tubers"; 08, "Edible fruit and nuts; peel of citrus fruit or melons"; 09, "Coffee, tea, maté and spices"; 10, "Cereals"; 12, "Oil seeds and oleaginous fruits; miscellaneous grains, seeds and fruit; industrial or medicinal plants; straw and fodder"; 1211, "Plants and parts of plants, incl. seeds and fruits, of a kind used primarily in perfumery, medicaments or for insecticidal, fungicidal or similar purposes, fresh or dried, whether or not cut, crushed or powdered"

need and that about 80% are using traditional medicines to meet their healthcare needs (World Health Assembly 2003; World Health Organization 2022). One way of understanding the importance of the medicinal plant market at the level of a country is to look at the number of national research institutes dedicated to traditional and complementary medicines, which are fully or partially funded by the governments and indicate strong national policy support. The WHO report on traditional and complementary medicine shows that the highest number of countries reporting a national research institute were in the South-East Asia Region (64%), followed by the African Region (62%), the Eastern Mediterranean Region (48%), the Western Pacific Region (33%), the Region of the Americas (26%), and the European Region (21%) (Fig. 2) (WHO 2019). The regions with the highest percentage correspond to countries, where medicinal plants strongly belong to the traditional healthcare system (Pan et al. 2014; Howes et al. 2020).

The trade database shows that India and China are the major providers of medicinal plants round the world with 24.1 and 10.6% of the total export value in 2021 (Fig. 3), followed by Germany (6%), the USA (4.4%), Egypt (4.3%), Canada (4.1%), Spain (2.8%), Poland (2.6%), Korea (2.3%), and Mexico, Vietnam, France, all three at 1.8%. All other countries are below 1.8% and represent 33% of the total world exports.

Current Status of the Medicinal Plant Sector in Europe

In Europe, medicinal and aromatic plants are cultivated on more than 200,000 ha, most of which are located in France (52,000 ha), Poland (30,000 ha), Spain (27,800 ha), Bulgaria (16,800 ha), Germany (13,000 ha), Croatia (8500 ha), the Czech Republic (7225 ha), Italy (7191 ha), Greece (6800 ha), and Austria (4136 ha) (EIP-AGRI 2020). The export market of customs code 1211 in Europe in 2021 was dominated by Germany (26%), Spain (12%), Poland (12%), and France (8%),

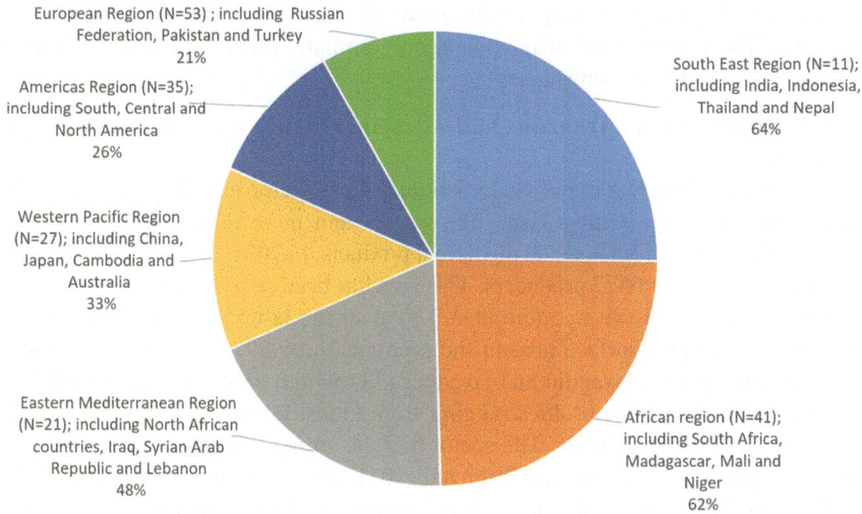

Fig. 2 Percentages of national research institutes for traditional and complementary medicines or herbal medicines in six regions of the planet. Each region is represented by N countries. The percentages represent the numbers of countries having a national research institute in a specific region. (WHO, 2019)

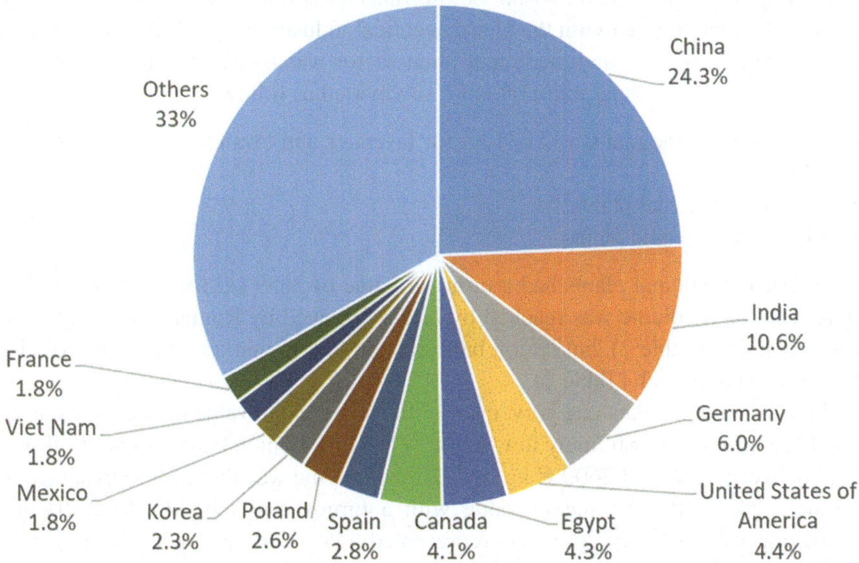

Fig. 3 Main exporter countries of medicinal plants in 2021. Export values in 2021 expressed as percentages, according to the trade database from the International Trade Center (no date). Export values were calculated from custom nomenclature 1211 "Plants and parts of plants, incl. seeds and fruits, of a kind used primarily in perfumery, medicaments or for insecticidal, fungicidal or similar purposes, fresh or dried, whether or not cut, crushed or powdered"

followed by Austria (5%), Italy and Bulgaria (4% each) (International Trade Center, no date). The current state of medicinal and aromatic plants in France is further studied in the following subchapter.

The French Perfume, Aromatic, and Medicinal Plant Sector (FranceAgriMer 2020, 2021)

The French perfume, aromatic, and medicinal plant sector includes the cultivation and regular picking of more than 300 species and more than 1000 products of marketed perfume, aromatic, and medicinal products. In 2021, this sector covered an area of 67,513 ha for 6527 producers. This area has been constantly increasing since the 2000s and has grown by more than 32% over the last 5 years. Perfume plants represent the largest surface area in the sector with 37,897 ha in 2021 and three predominant species: lavandin and lavender (33,094 ha) and clary sage (3400 ha). The farms have multiple profiles ranging from industrial cultivation to very small farms in disadvantaged areas. The sector had the strongest surface area growth in absolute value (>33%) between 2017 and 2021. Aromatic plants were grown on 9644 ha in 2021. The main species were coriander, parsley, thyme, fennel, mint, dill, tarragon, marjoram, oregano, basil, rosemary, and chives. This sector strongly grew (>66%) between 2017 and 2021. The medicinal plant sector includes the largest number of species (more than 150 species including poppy, chamomile, milk thistle, lemon balm, etc.). Its surfaces decreased by 4% to 19,972 ha in 2021 compared to 2020 (20,712 ha), but altogether increased by 19% between 2017 and 2021. Poppy (*Papaver somniferum* var. *nigrum*) and ginkgo biloba are exclusively produced under contract with the pharmaceutical industry. In 2021, the total area of the perfume, aromatic, and medicinal plant sector was 67,513 ha and represented less than 1% of French agricultural land, subdivided as follows:

- 56% for perfume plants (37,234 ha for lavender and lavandin areas, i.e., nearly 49%)
- 30% for medicinal plants
- 14.3% for aromatic plants

In 2020, medicinal plants had a turnover value of 3659 k€ for a volume of 385 tons. The main volume was reached by birch, followed by Roman chamomile and rose geranium (Table 2), but lemon balm ranked first in market value, followed by beech wood and birch (Table 2).

In the medicinal plant category, most of the commercial value relates to essential oils (355 k€) for an extremely low volume (154 kg). Lemon balm essential oil was sold between 2000 and 2600 €/kg in 2020, and thyme was the most representative aromatic plant (PA) in market value with a turnover of 630 k€. In 2020, the marketing value of "dry" products represented 44% of all aromatic plants. It was 29% for fresh products (including frozen ones) and 25% for essential oils (Fig. 4). The market shares of dried plants predominated over all medicinal plants with 36%, followed by "fresh/frozen" processed plants (31%). The market share of essential oils was 27%, and the remaining 6% included hydrolates, oily macerates, and stabilized extracts (Fig. 4). This shows that dried plants had the main market

Table 2 Major medicinal plants produced in France. Surfaces (ha), market volumes (t), market values (k€), and main uses

	2021 Surfaces (ha)	2020 Market volume (tons)	2020 Market value (k€)	Main form
Betula s.l.	/	21	202	Concentrated bud macerates, sap, traditional health syrup
Chamaemelum nobile	362 (2017)	14	60	Essential oil, tea, floral water, health food supplement, extract
Pelargonium 'rosat'	/	14	/	Essential oil, mother tincture, tea, concentrated bud macerates
Filipendula ulmaria	/	10	88	Tea, health food supplement, hydroalcoholic extract
Aloysia citrodora	/	9	100	Tea, essential oil, extract, mother tincture
Centaurea cyanus	22 (2017)	9	55	Floral water, tea, extract, hydrolat
Calendula officinalis	/	8	/	Extract, tea, hydroalcoholic extract, mother tincture, oily macerate
Melissa officinalis	260	8	404	Essential oil, health food supplement, extract, tea, hydrolat
Arnica montana	/	3–10 (harvest)	66	Mother tincture, oily macerate, extract, vegetable oil
Leontopodium alpinum	/	/	78	Flower extract, flower essence
Vitis vinifera	/	/	87	Health food supplement, water extract
Gentiana lutea	/	1600 (harvest-2017)	91	Extract, tea, mother tincture, health food supplement
Fagus s.l.	/	/	300	Concentrated bud macerates
Ribes nigrum	603 (perfume and essential oil)	30 (bud-harvest-perfume and essential plant) 60 (leaf-harvest-medicinal plant)	1443 (perfume and essential plant)	Concentrated bud macerates, health food supplement, extract, mother-tincture, macerate
Thymus	955 (aromatic plant)	328 (aromatic plant)	630 (aromatic plant)	Essential oil, healthy traditional syrup, concentrated bud macerates, health food supplement

(continued)

Table 2 (continued)

	2021 Surfaces (ha)	2020 Market volume (tons)	2020 Market value (k€)	Main form
Salvia rosmarinus	204 (aromatic plant)	60 (aromatic plant)	153 (aromatic plant)	Essential oil, concentrated bud macerates, health food supplement, extract, tea
Silybum marianum	300	/	/	Health food supplement, extract
Cynara cardunculus	250	/	/	Health food supplement, tea, extract
Angelica archangelica	179	/	/	Mother-tincture, extract, health food supplement, essential oil
Plantago afra L.	74	/	/	Seed, health food supplement
Lavandula L.	33,094	140 (Lavander – essential oil) 2000 (Lavandin – essential oil)	/	Essential oil, health food supplement
Papaver	10,000 (estimation)	/	/	License with a pharmaceutical company – derivatives for the production of alkaloids

Source: FranceAgriMer (2020, 2021)

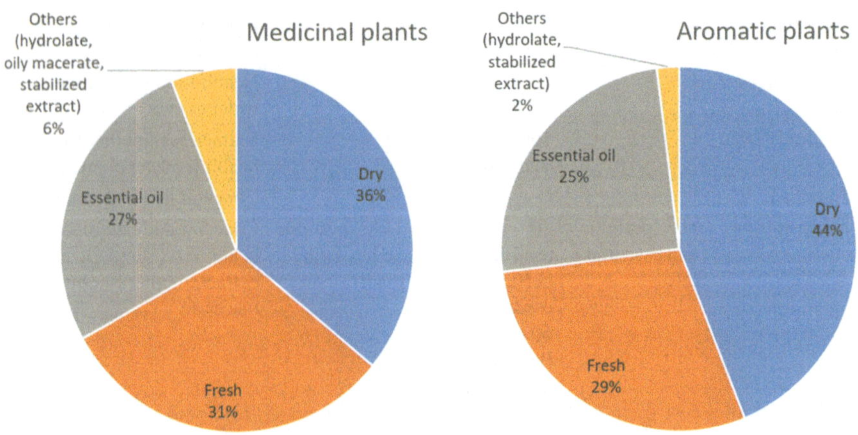

Fig. 4 Share of the marketing value of medicinal plants and aromatic plants according to the type of processing in France in 2020. (FranceAgriMer 2021)

value in both the medicinal and aromatic plant markets, followed by fresh plants and essential oils.

The use of medicinal plants is increasing in industrialized countries; the percentages of the population that had used a plant-based medicine at least once were 70% in Canada, 49% in France, 48% in Australia, 42% in the United States of America, and 31% in Belgium (World Health Assembly 2003). The global market value of herbal products is predicted to grow to US$ 5 trillion by 2050 (Pan et al. 2014). The increasing demand for medicinal plants has serious consequences such as overharvesting, quality inconsistencies, and uncertain efficacy (World Health Assembly, 2003; Howes et al. 2020; Zobayed 2020; Singh et al. 2022).

Overharvesting of medicinal plants has a major impact on biodiversity; for example, (1) Asian *Taxus brevifolia* Nutt., *T. chinensis*, *T. mairei*, and *T. contorta* Giff. populations harvested for paclitaxel extraction have undergone significant population reductions, (2) *Encephalartos woodii Sander* is extinct in the wild, (3) about 80% of Ethiopian medicinal plants are harvested from the wild with serious threats on their preservation, or (4) *Arnica montana L.* has been overexploited in Europe for its anti-inflammatory properties and shows decreasing populations; it is now included in the red list of several European countries (Balabanova and Vitkova 2010; Howes et al. 2020; Vera et al. 2020).

Medicinal plants are mainly harvested from wild plants (Zobayed 2020). Under field cultivation, some methods have had a negative impact on the environment. For example, field cultivation of *Panax ginseng* Meyer in Asia led to deforestation and soil microbial diversity losses in farmlands, which in turn brought about serious soil-borne diseases affecting the quality and yield of *P. ginseng* (Tong et al. 2021). The quality of medicinal plants is subject to inconstancies, and issues about the quality and efficacy of plant extracts have been reported (World Health Assembly 2003). Outdoor plants are exposed to variations of their growing conditions in water content, temperature, light characteristics (photoperiod, intensity, ozone and UV radiation), and soil characteristics. All these parameters vary according to the season, annual climate changes, and location and impact the plant contents in specific metabolites. Moreover, open-field harvesting is often seasonal and conditions the annual yield. Issues related to quality, efficacy, microbial and pollutant contamination, and contamination with misidentified plant species are often reported (World Health Assembly 2003; Zobayed 2020).

1.3 General Interest of Vertical Farming

Vertical farming consists in growing vegetables in vertically/horizontally stacked layers made of hydroponic or aeroponic soilless crop units mounted in (1) an indoor closed production system with artificial light, where environmental factors (airflow, temperature, CO_2, humidity and nutrients) are completely controlled, or (2) a greenhouse with vertically stacked layers, in semi-closed production systems, possibly adding artificial light to natural sunlight.

Vertical farming could contribute to answer some challenges of outdoor agriculture:

- First of all, *yields* in vertical farming are widely described as being significantly higher than in conventional agriculture, because they combine three factors: (1) the yield per square meter is increased thanks to a reduced land footprint resulting from the vertical succession of crop production units, (2) the photosynthetic rate is better as a result of a constant and ideal combination of environmental factors, and (3) production is possible all year round (Banerjee and Adenaeuer 2014; Barbosa et al. 2015; Avgoustaki and Xydis 2020a, b). The yield depends on the number of plants *per* square meter and on the maximizing of the vertical indoor space, which implies plants no taller than 30 cm, such as leafy greens, herbs, transplants, and medicinal plants (Kozai and Niu 2020).
- Secondly, *water use* is significantly lowered, because plants are grown hydroponically, irrigation water is supplied in a closed loop, and drought events are absent – climate is stable (Barbosa et al. 2015; Benke and Tomkins 2017; Graamans et al. 2017, 2018; Kalantari et al. 2018; Avgoustaki and Xydis 2020a).
- Thirdly, *pesticide use* is dramatically lowered, because exposure to the outdoor environment is reduced, although the risk of pest contamination cannot be completely excluded (Cowan et al. 2022). Moreover, if a pest appears, it is likely to spread exponentially because of the interconnected irrigation system and the high plant density.
- Fourthly, nitrogen and phosphorous *nutrient losses* in soil and aquatic sources are reduced, because the nutrient solutions are recirculated in a closed-loop system (Cowan et al. 2022). However, the recycling of the nutrient solution is not complete: nutrient imbalance gradually appears, and the nutrient solution has to be replaced unless a dynamically managed system is used (Silberbush and Ben-Asher 2001; Zeidler et al. 2017; Michael et al. 2021; Cowan et al. 2022).
- Finally, *farmland use* is reduced because crop production is soilless, the crop system is multilayered and can be implemented in urban areas and hostile places, such as desert, tundra, polluted and cold regions (Cowan et al. 2022).

However, several challenges are reported for vertical farming:

- Vertical farming *requires energy*, hence a carbon footprint. More electricity is required than in open-field and greenhouse farming; these high energy expenses are mainly linked to lighting and air and hydric management (Zeidler et al. 2017; Graamans et al. 2018; Sparks and Stwalley III 2018; Avgoustaki and Xydis 2020a; Bafort et al. 2022; Cowan et al. 2022).
- Other difficulties are the *global cost* to start vertical farming, linked to high start-up costs, high property costs in urban areas, high labor requirements, and the low market price of leafy-green crops challenging its viability (Zeidler et al. 2017; Bafort et al. 2022).
- The use of *mineral nutrients* has a big impact on soil resources and ecology. Other nutrient sources should be considered. Organic nutrient sources are often described, e.g., manures, bulky organic manures, or organic fertilizers. Most organic nutrient sources, including waste materials, have widely varying compositions and often only a low concentration of variably available nutrients and need to be processed before use (Szekely and Jijakli 2022).

- *Plastic* is largely used in hydroponics: the materials used for hydroponic culture (nutrient film technique, ebb-and-flow systems, deep water systems, aeroponics systems, and drip irrigation systems) are mainly plasticware. Efforts to decrease the use of plastic in hydroponics materials should be done.
- Rockwool is mainly used as a *substrate* in hydroponics. However, it has low durability as it has to be discarded after one or two cultivation cycles and requires high energy during its manufacturing process (Bar-Tal et al. 2019). To increase the durability of rockwool, its reuse has been developed as raw material for horticultural and insulation applications and in brick production in European countries, but this reuse network is not well developed yet (Bar-Tal et al. 2019). Clay beads are characterized by a very good long-term stability that allows for their reuse. Reuse induces increased costs because workforce and water are needed to rinse and clean the clay beads. Coco fiber is natural and recyclable, but its use in deep-water systems causes filtering problems, because coco fibers are degraded rapidly, so that more labor work needed to clean the filtering system very regularly (Bafort et al. 2022). As a consequence, the use of ecological hydroponic media should be emphasized.

Figure 5 summarizes the main challenges of outdoor farming and vertical farming.

1.4 Interest of Vertical Farming for Growing Medicinal Plants

The economic viability of leafy vegetable cultivation in indoor vertical farms with artificial lighting is complex, in particular on the European market because of their low market price, and high start-up, energy, and labor costs. In the United States, only 50% of container farms and 27% of indoor vertical farms reported operating profitability after 7 years of existence (Agrilyst 2017). Several studies on leafy greens in container farms reported that production costs were too high for them to be viable (Sparks and Stwalley III 2018; Debusschere and Boekhout 2021; Bafort et al. 2022). The selling prices in a simulated multilayer vertical farm – two layers containing four levels of lettuce each and two layers containing 18 rows of tomato each – were calculated to be 5.81 €/kg for an annual yield of 810 tons for lettuce and 9.94 €/kg for an annual yield of 215 tons for tomato, making profitability impossible (Zeidler et al. 2017). In Europe, several cases of bankruptcy of vertical farms have been reported, confirming the difficulty for vertical farming to be economically feasible (Sijmonsma 2019; VerticalFarmDaily.com 2021; Perreau 2022). Diversification by cultivating high-added-value plants, such as medicinal plants, could be less challenging economically. The economic approach of vertical farming of medicinal plants is discussed in Sect. 3.

Vertical farming is particularly suitable for producing medicinal plants. The stability of the environment makes it possible to increase stable and predictable yields and provide a stable quality with regular and high concentrations in phy-

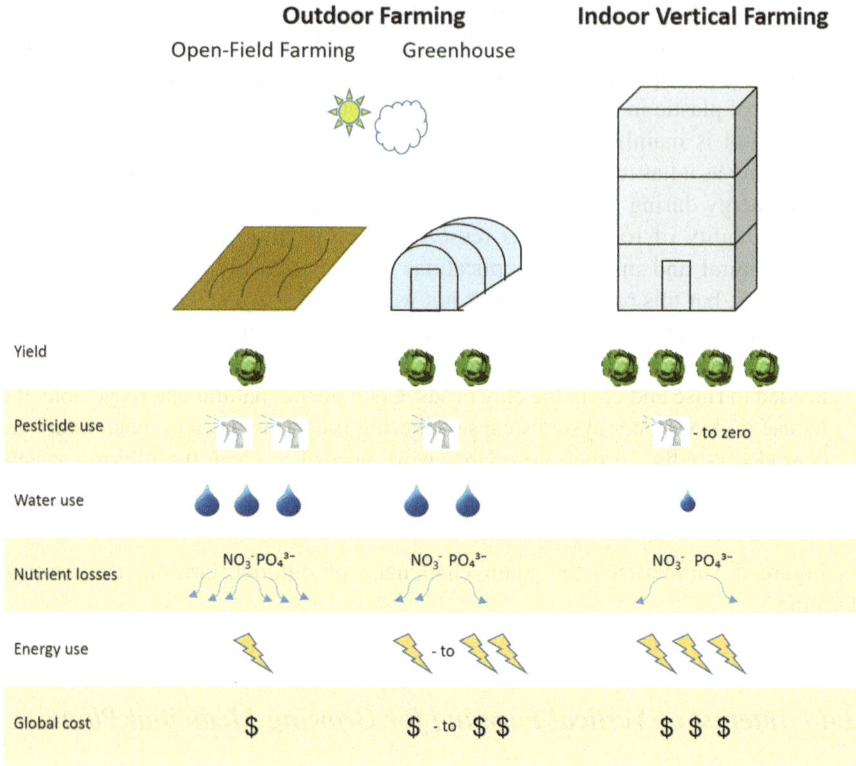

Fig. 5 Schematic overview of the current challenges of traditional farming and vertical farming. Please note that many forms of greenhouses exist, from plastic tunnels to fully automated greenhouses with complementary lighting devices

tochemicals, without soil contamination by microbes or pollutants (Goto 2012; Zobayed 2020). However, high biomass is contradictory with high concentrations in secondary metabolites, and a combination of these two criteria both important to reach economic viability is difficult to reach. Biomass increases are obtained by an ideal combination of abiotic factors – the most important variables are light, the water status, and the CO_2 concentration – so that photosynthesis is promoted and the production of primary metabolites such as starch and sucrose is promoted. Primary metabolites (lipids, proteins, and carbohydrates) are critical for plant growth and development. Plant growth is closely related to photosynthesis and respiration, and more than 90% of the crop biomass is derived from photosynthesis (Yamori 2020). Based on primary metabolites, plants metabolize various molecules with complex structural compositions called secondary metabolites (Naik and Al-Khayri 2016; Twaij and Hasan 2022). When plants encounter abiotic or biotic stresses, secondary metabolites are synthesized to communicate and act as a defense mechanism (Naik and Al-Khayri 2016; Dadhich et al. 2022). Plant secondary metabolites are usually

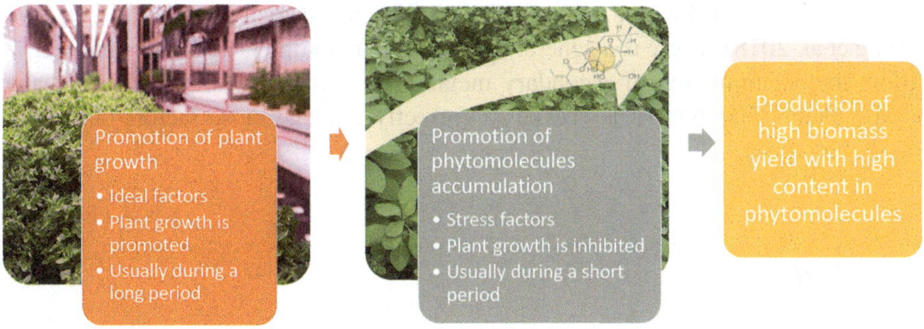

Fig. 6 Schematic overview of a two-step production principle allowing medicinal crop production with a significant biomass yield and an increased content in phytomolecules. The plant illustrated in the figure is *Euphorbia peplus* for its ingenol-mebutate content

classified in four major groups: (1) phenolics; (2) terpenes, saponins, and steroids; (3) nitrogen-containing compounds (such as alkaloids); and (4) glycosides (Hussein and El-Anssary 2018; Twaij and Hasan 2022). Following their specific presence and concentration, they characterize the medicinal property of the plant and its interest for the healthcare and pharmaceutic sector. However, the stress-induced enhancement of secondary metabolites alters plant development and growth (Itoh 2018; Dadhich et al. 2022). The enhancement of biomass is antagonistic with the enhancement of secondary metabolite production. Therefore, a dynamic two-step production of medicinal compounds has been proposed (Itoh 2018; Zobayed 2020) (Fig. 6).

2 Abiotic Factors Affecting the Quality of a Medicinal Crop

The environmental factors that play a role on plant photosynthesis and respiration also have an impact on plant growth and the accumulation of crop biomass. Ensuring the best environmental factors in a closed and controlled environment allows for a stable, maximized yield of high-quality plants, while stressing them may reallocate carbon to secondary metabolite production. Secondary metabolites are described as nonessential molecules for plant growth and biomass accumulation but are crucial for their interaction and adaptation to environmental fluctuations. Producing secondary metabolites is costly for plants because it requires primary metabolites, enzymes, cofactors, and energy. Secondary metabolites do not all have the same cost: terpenoids require less photosynthetically-produced carbon than alkaloids do (Gulmon and Mooney 1986; Cipollini et al. 2017). Plants' environments are usually classified in three main categories: (1) adverse biotic factors, such as fungi, bacteria, viruses, herbivores, and competing plants; (2) favorable biotic factors, such as symbiotic microorganisms, pollinators, seed dispersers, and plant-to-plant communication; and (3) abiotic factors, such as light, water availability, minerals availability, soil fertilization, temperature, and, in closed environment, the CO_2 level (Yang et al. 2018). There is a general consensus that abiotic factors can significantly affect the accumulation of secondary metabolites and in turn the medicinal value of the plant. Therefore, it is crucial to correctly manage those factors during the production process.

2.1 Light

Light affects plants in two ways – as an energy source and as an information medium (Dou and Niu 2020). The energy of light is transmitted by the photons, and about 10% of sunlight are converted into chemical energy – carbohydrates – through photosynthesis, while the remaining 90% are converted into heat energy (Dou and Niu 2020). The absorption of light for photosynthesis is initiated by photosynthetic pigments – chlorophylls and carotenoids; chlorophylls strongly absorb red and blue light, and carotenoids strongly absorb blue light (Yamori 2020). Chloroplasts and whole leaves absorb most of the light, including green light (Yamori 2020). Plant photoreceptors measure the light composition variations and trigger plant responses independently from photosynthesis, as in photoperiodism and photomorphogenesis, and regulate the expression of genes associated with cell division and enlargement (Dou and Niu 2020). Five classes of photoreceptors have been described. They allow plants to perceive a broad spectrum of light from ultraviolet to far-red wavelengths and to regulate multiple physiological and metabolomic responses (Fig. 8).

2.1.1 Effect of the Quantity of Light

In controlled environments, artificial light is usually constant without the seasonal variation in intensity, duration, and spectrum of natural sunlight to which plant growth is subjected under natural conditions. The daily light integral (DLI) describes the total amount of photosynthetically active photons that are delivered to a specific area over a 24-hour period; it usually has a linear relationship with crop yield in controlled environments (Dou et al. 2018). The effects of three DLI levels of 8.64, 14.4, and 28.8 mol m^{-2} d^{-1} under a 16-h photoperiod were tested on the shoot biomass and the accumulation of a diterpene – ingenol-mebutate – by the medicinal plant *Euphorbia peplus* (Bafort et al. 2022). Increasing DLIs had a positive effect on yield, with shoot fresh biomass rises of 111% and 212% compared to the values obtained with a DLI of 8.64 mol m^{-2} d^{-1} (Fig. 7). The same trend was observed for shoot dry biomass. The calculated positive correlation was relatively low. It was attributed to the low homogeneity of the yield, which varied dramatically with the position of the plant in the vertical container, especially under the lowest DLI. In the same study, the content in ingenol-mebutate of *E. peplus* was not modified with the DLI level.

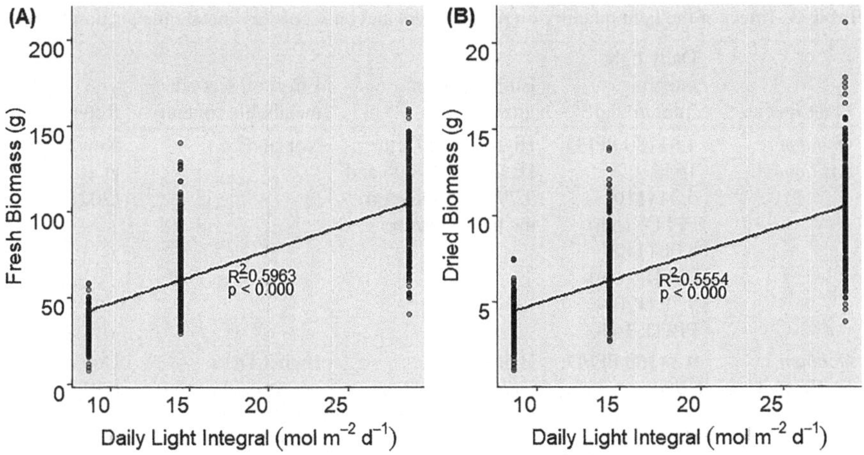

Fig. 7 Correlations and p-values between the shoot fresh biomass (**a**), the shoot dried biomass (**b**), and the daily light integral (DLI) of *Euphorbia peplus* grown at different DLI levels (8.64, 14.4, and 28.8 mol m^{-2} d^{-1}) in a vertical container farm for 47 days. (Bafort et al. 2022)

In a completely closed and controlled environment, the DLI is modulated in two ways: (1) by adjusting the light intensity and (2) by adjusting the photoperiod.

Several studies have addressed the *role of the DLI* on yields and secondary metabolite contents by acting on the light intensity (photosynthetic photon flux density; PPFD) or on the photoperiod, or on both (Table 3). Basil (*Ocimum basilicum* L.) and lettuce (*Lactuca sativa*) are ideal crops for vertical farming, because they are well adapted to closed controlled and soilless environments, have short cultivation cycles and a limited height. Therefore, they have been extensively studied to determine the effect of environmental factors. The plant biomass is generally proportional to the DLI within a certain range (Dou and Niu 2020; Xu et al. 2021). Increasing the DLI increased the shoot fresh biomass yield of *O. basilicum* L., because of a higher photosynthetic rate and a linear accumulation of anthocyanins, phenols, and flavonoids *per* plant (Dou et al. 2018). However, the positive biomass correlation of basil with the DLI was also found cultivar dependent. For example, the Ararat variety had the largest weight at a DLI of 6.34, while the Yerevan sapphire variety reached its best yield at a DLI of 9.79 (Kondrat'Ev et al. 2021). A fixed DLI of 12.9 did not cause the yield of basil to vary, whatever the photoperiod-PPFD combination (Dou and Niu 2020). Red perilla shoot dry weight increased with the DLI but not in a linear manner, because light utilization efficiency decreased with increased PPFD (Yoshida et al. 2022). Anthocyanins accumulated *per* dry weight unit with higher DLI, but the essential oil perillaldehyde did not (Yoshida et al. 2022).

The effect of an *extended photoperiod* has been studied. Compared to DLIs of 5.8, 8.6, and 11.6, basil and lettuce growth were improved under a DLI of 14.4 corresponding to a PPFD of 250 μmol m^{-2} s^{-1} under a 16 h photoperiod, and so were the

Table 3 Effect of the light quantity on plant growth and on secondary metabolite accumulation

Plant species	Daily light integral (mol m^{-2} d^{-1})	Effect on plant growth	Effect on secondary metabolite content	Reference
Ocimum basilicum L.	4.61 (80 PPFD; 16 h) 6.34 (110 PPFD; 16 h) 8.06 (140 PPFD; 16 h) 9.79 (170 PPFD; 16 h)	Highest yield with DLI = 6.34, 8.06 and 9.79, depending on the basil cultivar	Not studied	Kondrat'Ev et al. (2021)
Ocimum basilicum L.	9.3 (160 PPFD; 16 h) 11.5 (200 PPFD; 16 h) 12.9 (224 PPFD; 16 h) 16.5 (290 PPFD; 16 h) 17.8 (310 PPFD; 16 h)	Higher DLIs increased yield, but no significant differences in yield between DLIs of 12.9, 16.5 and 17.8	Higher DLIs increased the total anthocyanin, phenolic and flavonoid contents *per* plant	Dou et al. (2018)
Ocimum basilicum L.	12.9 ((298 PPFD; 12 h) 12.9 (256 PPFD; 14 h) 12.9 (224 PPFD; 16 h) 12.9 (199 PPFD; 14 h) 12.9 (179 PPFD; 20 h)	No yield differences between photoperiod and PPFD variation with a fixed DLI of 12.9	Not studied	Dou and Niu (2020)
Ocimum basilicum L. *Lactuca sativa* L	5.8 (100 PPFD; 16 h) 8.6 (150 PPFD; 16 h) 11.5 (200 PPFD; 16 h) 14.4 (250 PPFD; 16 h) 17.3 (300 PPFD; 16 h)	Highest yield with DLI = 14.4	Higher antioxidant capacity, phenolics and flavonoids in *L. sativa* at DLI = 14.4	Pennisi et al. (2020)
Lactuca sativa L.	8.64 (150 PPFD;16 h) 8.64 (200 PPFD; 12 h) 9.04 (2x3h at 100 PPFD and 6 h at 300 PPFD)	Better yield obtained with longer photoperiod and multi-segment light intensity	Not studied	Mao et al. (2019)

(continued)

Table 3 (continued)

Plant species	Daily light integral (mol m^{-2} d^{-1})	Effect on plant growth	Effect on secondary metabolite content	Reference
Perilla frutescens L.	2.88 (50 PPFD; 16 h) 5.76 (100 PPFD; 16 h) 11.52 (200 PPFD; 16 h) 23.04 (400 PPFD; 16 h)	Increased yield with increased DLI	Perillaldehyde content *per* unit of dry weight similar whatever the DLI. Anthocyanin content *per* unit of dry weight increased with DLI	Yoshida et al. (2022)
Catharanthus roseus (L.)	4.32 (75 PPFD; 16 h) 8.64 (150 PPFD; 16 h) 17.28 (300 PPFD; 16 h) 34.56 (600 PPFD; 16 h)	Best fresh total leaf weight obtained with DLI = 17.28	Highest vindoline and catharanthine contents with DLI = 8.64	Fukuyama et al. (2015)
Ophiorrhiza pumila	2.16 (50 PPFD; 12 h) 2.9 (100 PPFD; 8 h) 4.32 (100 PPFD; 12 h) 5.8 (100 PPFD; 16 h) 6.48 (150 PPFD; 12 h)	Best yield with DLI = 5.8	Highest camptothecin content with DLI = 5.8	Lee et al. (2020)
Tropaeolum majus L.	17.3 (300 PPFD; 16 h) 17.3 (200 PPFD; 24 h) 25.9 (300 PPFD; 24 h) 34.6 (400 PPFD; 24 h)	Linear increase in total biomass with DLI. At same DLI (17.3), better shoot yield with increased photoperiod.	Antioxidant capacity and total phenolic content increased with increased DLI	Xu et al. (2021)
Stevia rebaudiana	7.2 (249 PPFD; 8 h) 7.2 (165 PPFD; 12 h) 7.2 (125 PPFD; 16 h) 7.2 (125 PPFD; 16 h intermittent)	Highest yield with constant longer photoperiod	Highest yield of stevioside and rebaudioside A *per* plant under 16 h photoperiod but higher rebaudioside A concentration under 8 h photoperiod	Rengasamy et al. (2022)

(continued)

Table 3 (continued)

Plant species	Daily light integral (mol m^{-2} d^{-1})	Effect on plant growth	Effect on secondary metabolite content	Reference
Nasturtium officinale L	11.52 (266 PPFD; 12 h) 11.52 (200 PPFD; 16 h) 11.52 (160 PPFD; 20 h) 11.52 (133 PPFD; 24 h)	Highest yield under 20 h photoperiod	Highest total glucosinolate content *per* plant shoot under 20 h photoperiod	Lam et al. (2021)
Amaranthus tricolor, Brassica oleracea var. viridis, Ocimum basilicum	14 (250.8 PPFD; 16 h) 14 (166.6 PPFD; 24 h) 21 (376.9 PPFD; 16 h) 21 (247.6 PPFD; 24 h)	Highest yield under DLI = 21 with constant lighting	High DLI with constant lighting and high DLI Increased *A. tricolor* and *B. oleracea var. viridis* phenolic, anthocyanin and antioxidant contents. Unaffected secondary metabolite concentrations in basil	Lanoue et al. (2022)

water, energy, and light use efficiencies (Pennisi et al. 2020). Secondary metabolites also accumulated in lettuce at a DLI of 14.4 (Pennisi et al. 2020). An extended photoperiod (16 h) under a low light intensity (PPFD = 100 μmol m^{-2} s^{-1}) promoted chlorophyll accumulation and improved the root/shoot ratio, helping lettuce to absorb enough light energy and improve its growth under low light conditions (Mao et al. 2019). Lettuce increased its photosynthetic capacity significantly under multi-segment lighting, which simulated circadian rhythms and resulted in an increased yield (Mao et al. 2019).

Shade plants such as *Ophiorrhiza pumila* have a low saturation point and showed better biomass yield and camptothecin accumulation under a low PPFD (100 μmol m^{-2} s^{-1}) and a long photoperiod (16 h) (Lee et al. 2020a, b). Mid-shade plants such as *Catharanthus roseus* showed an increased yield up to a certain level of DLI (17.28), but a higher DLI led to the inhibition of growth (Fukuyama et al. 2015). In the same plant, vindoline and catharanthine accumulation were greatest under a lower DLI (8.64) (Fukuyama et al. 2015). An extended photoperiod strategy can also be well adapted to tropical countries, where natural weather conditions and the day-neutral photoperiod restrict field growth of some plants. For example, stevia plant productivity and quality were enhanced under a long and constant photoperiod (16 h) at a low light intensity (PPFD = 125 μmol m^{-2} s^{-1}) (Rengasamy et al. 2022).

The effect of *continuous lighting* (24 h) has also been studied. *Tropaeolum majus* L. showed a linear increase in dry weight with the DLI under continuous lighting with DLIs ranging from 17.3 (PPFD = 200 μmol m^{-2} s^{-1}) to 34.6 mol m^{-2} d^{-1} (PPFD = 400 μmol m^{-2} s^{-1}) (Xu et al. 2021). The increased yield resulted in reversible photoinhibition during plant growth and in an adaptive process to protect the photosynthetic apparatus from light stress (Xu et al. 2021). With a fixed DLI of 17.3, secondary metabolite production was increased under continuous lighting compared to a higher light intensity and a shorter photoperiod (Xu et al. 2021). Continuous lighting and a higher DLI – hence higher light intensities – maintained the secondary metabolite content (Xu et al. 2021). The productivity and quality of four microgreens were tested under two DLIs and constant lighting or a long photoperiod (16 h) (Lanoue et al. 2022). For each fixed DLI, the yield was better under constant lighting and maximized at the highest DLI (Lanoue et al. 2022). Interestingly, higher energy-use-efficiencies of lighting were observed under constant light, and a reduced electricity cost *per* unit of fresh biomass was measured (Lanoue et al. 2022). The nutritional quality of amaranth and collard greens was also improved at high DLIs, without or with constant lighting, and unchanged in basil (Lanoue et al. 2022). However, constant lighting and a low DLI – i.e., a low light intensity – can impact plant growth negatively. *Nasturtium officinale* L. growth was decreased under constant lighting and a low light intensity (133 μmol m^{-2} s^{-1}) because of reduced net photosynthesis and stomatal conductance (Lam et al. 2021). On the contrary, the total glucosinolate concentrations were highest in those conditions, but the total glucosinolate content *per* shoot dry weight was reduced, because of the markedly reduced biomass (Lam et al. 2021). Continuous lighting can also induce negative effects on sensitive plants, e.g., leaf chlorosis, growth inhibition, and leaf necrosis that may result from photo-oxidative damage (Xu et al. 2021). The hypothesis is that continuous-lighting-tolerant plants have high antioxidant contents that protect them (Xu et al. 2021). For example, continuous lighting induced higher chlorogenic acid content in lettuce plants that could protect them against high levels of reactive oxidative species generated by physiological stresses (Shimomura et al. 2020). On the contrary, basil growth under continuous lighting induced physiological stress, such as chlorosis, stunting, and leaf necrosis (Sipos et al. 2021).

2.1.2 Effect of the Quality of Light: Spectral Quality and UV Radiation

Spectral Quality The quality of light is perceived by photoreceptors, whose reaction to light quality is species-specific. Therefore, the effect of light quality should be considered separately for each plant species (Dou and Niu 2020; Karimi et al. 2022). Light quality influences plant growth and the synthesis of bioactive compounds (Yang et al. 2018; Dou and Niu 2020).

In general, *red (R)* and *blue (B) lights* are the most commonly used spectra in indoor cultivation, because they correspond to the absorption peaks of chlorophylls

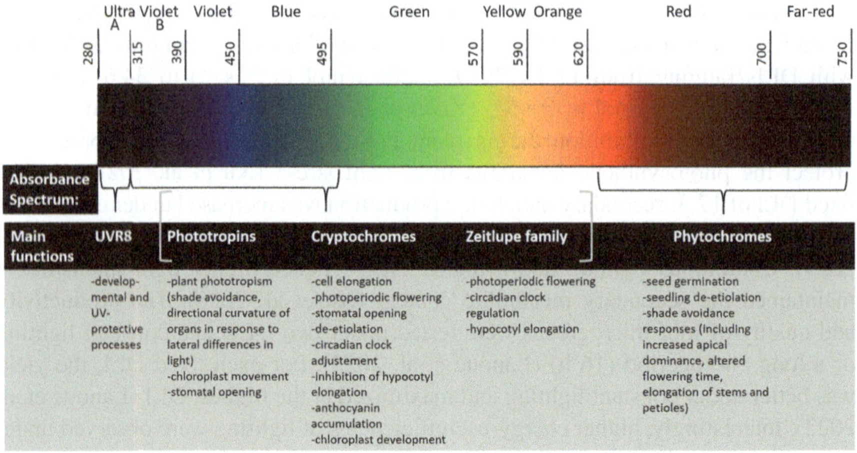

Fig. 8 Representation of the main absorbance spectra and main functions of the five photoreceptor classes. (From Christie et al., 2015; Dou, Niu and Gu, 2019; Appolloni et al. 2022; Paradiso and Proietti 2022). *UVR8* UV resistance locus 8

and to the main plant functions, as showed in Fig. 8 (Dou and Niu 2020; Appolloni et al. 2022). Combined R&B lights are more efficient than monochromatic blue or red lights for plant growth, which can induce physiological disorders in some plant species (Dou and Niu 2020). Full-spectrum *white light-emitting diodes (LEDs)* have recently been found efficient in indoor culture; they supply a full spectrum that optimizes plant growth (Dou and Niu 2020). *Green light* is not fully absorbed by chlorophyll and has long been considered less effective than red and blue lights in promoting plant growth (Paradiso and Proietti 2022). Nowadays, it is admitted that green light penetrates deeper into the plant canopy and may promote better photosynthesis in the whole canopy (Paradiso and Proietti 2022). Inclusion of green light in dichromatic red and blue LEDs impacted plant growth differently depending on its proportion (Orlando et al., 2022a, b). A high proportion of green light (25–44%) generated opposite responses to blue- or red-light-induced effects and negatively affected sweet basil and microgreen quality (Kim et al. 2005; Zhang and Folta 2012; Dou et al. 2019, 2020a, b). A low proportion of supplemental green light – under 10% – to red and blue spectra did not affect the fresh or dried biomass of several microgreens (Ying et al. 2020; Orlando et al. 2022a, b). However, 12–24% green light addition to red and blue lights positively affected the biomass of lettuce and kale and induced secondary metabolite accumulation in *Crocus sativus* and lettuce plants (Kim et al. 2005; Bian et al. 2016; Meng et al. 2019; Orlando et al. 2022a, b).

The effect of light spectra on the accumulation of *phenol* metabolites in medicinal plants has been investigated. Combined blue (38%) + red (62%) lights and combined blue (38%) + green (12%) + red (50%) lights have been tested on *C. sativus* tepal biomass and bioactive metabolite accumulation and compared with those obtained under greenhouse cultivation (Orlando et al. 2022a, b). The inclusion of green light increased the total flavonoid content and the biomass remained unaffected as compared to the greenhouse production. Therefore, cultivation under LEDs may positively valorize *C. sativus* by-products. Blue LEDs, red LEDs, combined blue (70%) + red (30%) LEDs, and white LEDs have been tested on the growth and the phenolic compound production of *Dracocephalum forrestii* shoots (Weremczuk-Jeżyna et al. 2021). The best biomass values, shoot propagation, and secondary metabolite production were obtained under blue LEDs. The enhancement of the antioxidant capacity was positively correlated with the maximum total polyphenolic acid content. Blue, red, and white LEDs were tested on the roots, stems, and leaves of *Scutellaria baicalensis* seedlings for 2 weeks (Yeo et al. 2021). The roots treated with white LEDs showed increased concentrations of the flavonoids baicalin, baicalein, and wogonin and reduced concentrations of carbohydrates, suggesting the need for energy to enhance the biosynthesis of phenolic compounds. The effects of monochromatic red, blue, and green LEDs, several dichromatic red (60–90%) and blue (40–10%) LEDs, and several trichromatic red (50–90%), green (10%), and blue (40–0%) LEDs were tested on the growth and bioactive compound biosynthesis of *Crepidiastrum denticulatum* (Park et al. 2020a, b). The total phenolic content was similar among all treatments, but the antioxidant capacity and dry weight *per* shoot were increased under the trichromatic red (80%) + green (10%) + blue (10%) LEDs. The addition of far-red light to dichromatic blue (20%) and red (80%) LEDs was tested on the growth and phenolic content of *C. denticulatum* (Bae et al. 2017). Growth was increased under supplemental far-red irradiation, while the phenolic content *per* unit dry weight remained unaffected by the different light treatments.

Several light combinations have been tested on the accumulation of bioactive *terpene/terpenoid* compounds produced by medicinal plants. Six light treatments – monochromatic red and blue LEDs and dichromatic red (80–20%) and blue (20–80%) LEDs – were tested on *Hypericum perforatum* (Karimi et al. 2022). The plants under the monochromatic red light showed an increased accumulation of foliage, higher flower and root fresh and dry weights, and an increased percentage of hypericin, pseudohypericin, and hyperforin in their flowers *per* square meter. Red light stimulated the expression of genes related to *H. perforatum* flowering. Enhanced accumulation of artemisinin and artemisinic acid and other terpenoids in *Artemisia annua* and increased fresh leaf weight were measured under white and blue spectra (Sankhuan et al. 2022). Moreover, crude extracts under the same light treatment showed improved antimalarial anti-*Plasmodium falciparum* activity compared to crude extract under monochromatic red light treatment and greenhouse cultivation. Red light treatment decreased the level of terpenoid production and induced distinct phytochemical profiles.

The effect of the light spectrum on *alkaloid* accumulation has been studied in medicinal plants. Several light spectra were applied on embryogenic *Fritillaria cirrhosa* D. Don calluses for 3 months to measure their effect on growth and alkaloid production (Chen et al. 2020). Monochromatic red, blue, and far-red, warm, and cold white lights and various combinations of red, blue, green, and far-red treatments induced differential development and growth of *F. cirrhosa*. The maximum fresh weight was obtained under monochromatic red light, and the highest contents in peimisine, peiminine, and peimine were recorded under the monochromatic red light and infrared light. *Picea abies* seedlings were exposed to white light with 12% or 45% added blue light (Kivimäenpää et al. 2021). The spectra with the highest blue light content decreased the alkaloid, terpene, and terpenoid concentrations in needles, although the contents in total flavonoids and acetophenones were increased. Growth and the carbohydrate and pigment contents were unaffected, suggesting carbon reallocation from alkaloid and terpenoid synthesis to flavonoid synthesis as a response to increased blue light.

UV Radiation UV radiation induced multiple responses ranging from slowed down photosynthesis to increased DNA repair, defense mechanisms, and specialized metabolite production (Vanhaelewyn et al. 2020). Reactive species (ROS) in response to UV-B radiation cause DNA damage, affect the plant metabolism, and generate defense mechanisms such as the production of ROS-scavenging enzymes and antioxidant compounds (Park et al., 2020a, b). Specialized metabolites are synthesized, thanks to the reallocation of carbon toward the production of phenolics (e.g., flavones, flavonols, anthocyanins), alkaloids, carotenoids, and glucosinolates (Vanhaelewyn et al. 2020). Supplemental UV-B radiation typically decreases biomass; therefore, using this light stress needs fine-tuning to achieve both good yield and enhanced bioactive metabolites (Dou and Niu 2020).

UV-B radiation has been tested on *C. denticulatum* growth and its biosynthesis of total carotenoids, phenolics, and terpenes (Park et al. 2020a, b). High-energy UV-B light reduced the chlorophyll content and several sesquiterpene contents and increased the total carotenoid, phenolic, and hydroxycinnamic acid contents, while it decreased *C. denticulatum* growth. Moderate energy levels of UV-B radiation (0.1 and 0.25 W m^{-2}) increased the antioxidant capacity, the total hydroxycinnamic acid content, and several sesquiterpenes without inhibiting growth and were considered as a eustress (Park et al. 2020a, b). The effect UV-B light on the terpene content of *Panax ginseng* C.A. Meyer has been tested (Choi et al. 2022). A low-energy dose of 0.1 W m^{-2} for 1, 2, or 3 hours during the preharvest days did not modify the total ginsenoside content. Several spectra – monochromatic blue and red and red with high energy (5 W m^{-2}) UV-A – were tested on the growth and the alkaloid vinblastine content of *C. roseus* for 7 days (Fukuyama et al. 2017). The total leaf dry weight was unaffected whatever the spectrum, while the vinblastine content *per* dry weight unit was significantly increased after 3 days of UV-A treatment and highest after 7 days of UV-A treatment. The effect of several UV-A energy levels combined with red light revealed a positive correlation with the UV-A energy levels on the leaf vinblastine content and a negative correlation on the leaf vindoline and catharanthine contents (Fukuyama et al. 2017).

2.1.3 Light Combined with Others Factors

The growth of a plant depends on many abiotic factors. The plant's response may differ when a single factor or several additional environmental factors are studied. Therefore, checking the effects of multiple factors is an interesting approach. A classic approach is the one-factor-at-a-time (OFAT) design, which makes only one factor vary while the other variables are kept constant. Some limitations are that the interactions between factors cannot be estimated, and the risk of obtaining a false optimum is high when more than two factors are considered (Czitrom 1999). Another method is the design of experiments (DOE), for example, the response surface methodology or the Box-Behnken experimental design, which search for the factor level combination that gives the best answer (i.e., yield, content in phytomolecules). In this case, multiple factors can be modified together, the interactions among factors are estimated, and the response is optimized (Czitrom 1999).

Several studies have addressed the effect of multiple factors on plant growth and phytomolecule production. Growth and bioactive metabolite production by red and green *Perilla* were tested by making three levels of electrical conductivity (EC) and three levels of PPFD vary (Lu et al. 2017). The concentration of perillaldehyde – a terpene – was not affected by EC or light intensity in red perilla, but the content in rosmarinic acid – a phenol – was highest under the highest light intensity and the lowest EC and decreased significantly when EC was increased. The shoot dry weight was promoted by higher light intensities under mid and high EC. In green perilla, the shoot dry weight increased with PPFD and EC, the perillaldehyde and rosmarinic acid concentrations decreased with increased EC, and rosmarinic acid was promoted by higher PPFD. Yield, anthocyanins, and soluble sugars were measured in *Brassica rapa* var. Chinensis under several light intensities and nitrogen concentrations (Hao et al. 2020). The yield was enhanced by the combination of a moderate PPFD and a moderate nitrogen concentration, but anthocyanins were optimized under high PPFD and nitrogen, and soluble sugars were promoted by the lowest nitrogen concentration. This shows how difficult it is to obtain a unique optimum for all parameters taken together. The optimal light intensity, temperature, and nutrients for *H. perforatum* L. accumulation of bioactive compounds were investigated (Kuo et al. 2020). Hyperforin and rutin were significantly affected by the light intensity and temperature, but the nutrient concentration had little effect. Melatonin seemed to be unaffected by the environmental factors considered in the study. The leaf biomass was enhanced with light intensity, temperature, and nutrients. Based on the response surface methodology, the optimal conditions for the yield of each specific metabolite were calculated.

2.1.4 Toward Sustainability of the Use of the Light Resource

The need for more sustainable agriculture is important in indoor cultivation systems where energy consumption is one of the major drawbacks. Artificial lighting represents a major share of the energy requirements. The energy and light use efficiencies are two ways of measuring the energy costs of crop production in indoor systems. Energy use efficiency (EUE) is expressed in grams of biomass produced *per* kWh, and light use efficiency (LUE) is expressed in grams of biomass produced by light integral. Both units are useful to find the optimal response of plant growth to light intensity, and using them can show if higher light intensity – and higher energy requirements – can bring enough yield gain to be expressed as increased light and energy use efficiencies. The technological evolution of artificial lights has already improved the EUE of lettuce cultivated under LED light (EUE = 40.6 g kWh^{-1}) compared with lettuce cultivated under fluorescent lamps (EUE = 15.9 g kWh^{-1}) (Zhang et al. 2018). With further technological developments and societal demand, next-generation LEDs will improve energy supply and will allow for improved sustainability. Moreover, the use of the right spectral composition can improve EUE, as showed for indoor lettuce and basil cultivation (Pennisi et al. 2019a, b).

2.2 *Nutrient Solutions*

Nutrient solutions in soilless crop cultivation have to bring all the nutrients necessary for plant growth. Nutrients are described as essential macroelements and microelements, i.e., nutrients that cannot be replaced by another element, whose absence induces deficiency symptoms. They are directly involved in the plant metabolism (Tsukagoshi and Shinohara 2020). The nine macroelements are used in relatively large amounts, and the eight microelements are required in small amounts. Three macro-nutrients – carbon, oxygen, and hydrogen – are supplied from atmospheric carbon dioxide and water and are not included in fertilizers. However, enough dissolved oxygen has to be present in water for root respiration, generally brought by air pumps or agitation of the nutrient solution. The remaining macro-nutrients are nitrogen, phosphorous, potassium, calcium, magnesium, and sulfur. Micronutrients are iron, boron, manganese, copper, zinc, molybdenum, chlorine, and nickel. The main functions of each element are well-known and summarized in Fig. 9 (Tsukagoshi and Shinohara 2020).

Typical formulas have been developed and commercialized for soilless application and exist in a ready-to-use form. However, nutrient compositions should be ideally tested according to the plant type, its growth stage, the substrate type, and the targeted quality (Tsukagoshi and Shinohara 2020). Several ways of studying the effect of nutrition on plant growth and secondary metabolite accumulation are available. We selected four methodologies among them.

1. *Tailor-made nutrient recipes* have been developed and tested. Nitrogen, potassium, and phosphorous supplies were modulated on two medicinal plants – *Lavandula angustifolia* and *Mentha spicata* – to assess the yield and quality of essential oils (Chrysargyris and Tzortzakis 2021). Lower camphor and higher carvone contents were measured in *L. angustifolia* under nitrogen levels above

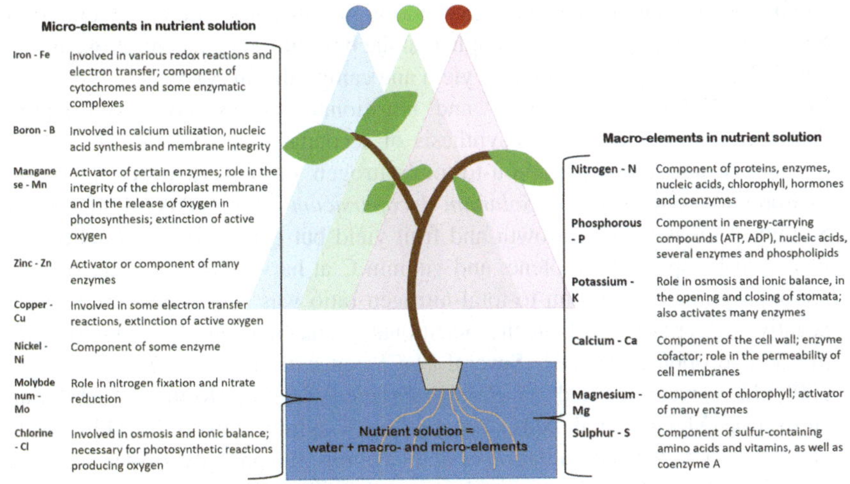

Fig. 9 Schematic representation of an indoor soilless plant and the macro- and microelements to be added to water to form a nutrient solution. The main functions of the nutrients are indicated (Tsukagoshi and Shinohara 2020). The macronutrients carbon, hydrogen, and oxygen are supplied by atmospheric carbon dioxide and water

200 mg L^{-1}, both indicating increased oil quality. The carvone and limonene contents of *M. spicata* were more sensitive to the nitrogen and potassium contents than to the phosphorous content. A home-made nutrient solution was tested on the growth, antioxidant level, and chicoric acid contents of *C. denticulatum* (Park et al. 2016). Increased EC increased *C. denticulatum* biomass, total phenolic content, chicoric acid content, and antioxidant capacity.

2. *Testing several concentrations of typical formulas* (e.g., Hoagland, Otsuka composition, commercial fertilizers) by making EC or application rates vary. Several concentrations of a ready-to-use fertilizer solution were tested on the growth and alkaloid content of *Mitragyna speciosa* (Zhang et al. 2020). Growth was promoted by increasing amounts of fertilizer, while the alkaloid concentrations were highly variable. Lower and medium fertilizer rates promoted the accumulation of several alkaloids, suggesting that nitrogen was reallocated to secondary metabolite synthesis. The yield, total phenolic content, and antioxidant capacity of *O. basilicum* L. were measured under several EC levels (Ren et al. 2022). Biomass was increased by medium to high EC, while the total phenolic content and antioxidant capacity were increased at low EC. Two-step cultivation was successfully applied, consisting in a first, long step under medium-high EC that promoted a good yield of sweet basil, followed by a second, short step just before harvest, when water (no fertilizer) or a low EC promoted total phenolic accumulation and the antioxidant capacity. The effects of the nutrient formula concentration and root temperature were tested on *Ophiorrhiza pumila* growth and camptothecin accumulation (Lee et al. 2020a, b). Growth and the

camptothecin content were best at a mid-high nutrient solution concentration. Several root temperatures were applied at the best nutrient concentration, among which 20 °C gave the optimum in yield and camptothecin content.

3. *Applying NaCl stress*. Salinity and nutritional stresses have been largely described to modulate the biosynthesis of secondary metabolites. The impact of salinity and the ammonium-to-total-nitrogen ratio were tested in closed hydroponic cultivation of *Solanum lycopersicum* (Tzortzakis et al. 2022). Salinity decreased plant growth and fruit yield but enhanced fruit quality, and increased lycopene, β-carotene, and vitamin C at harvesting or during storage. An appropriate ammonium-to-total-nitrogen ratio was suggested to reduce the negative effects of NaCl on the nutritional status of plants by regulating the pH in hydroponic systems. Several NaCl concentrations – 1.7, 25, 50, and 100 mM – were applied on *Reichardia picroides* (L.) Roth in hydroponic cultivation (Maggini et al. 2021). After 6 weeks, salinity above 1.7 mM induced a decreased yield but accumulation of anthocyanins, flavonol glycosides, and total phenols and improved the antioxidant capacity. The effects of increasing NaCl concentrations (1–40 mM) were tested in hydroponic and aquaponic cultivation systems of the drug-type *Cannabis sativa* L. during the flowering period (Yep et al. 2020). The cannabinoid contents decreased linearly with increasing NaCl concentrations in both systems. Decreased yields in hydroponic systems have been observed from NaCl concentrations above 5 mM. Forty mM was phytotoxic in hydroponics, but not in aquaponics, suggesting a potential NaCl tolerance induced by aquaponics. The impacts of salinity, calcium chloride – that may alleviate salt stress – and successive harvests were tested on two *O. basilicum* L. cultivars (Ciriello et al. 2022). Moderate salinity in the presence or absence of calcium chloride and high salinity in the presence of calcium chloride showed improved nutritional quality with improved phenol concentrations and reduced nitrate levels without affecting the eucalyptol content. In the green cultivar, the yield decreased with increased salinity. Successive harvests increased the phenol and vitamin C concentrations but reduced the eucalyptol content. The impact of nutrient deficiency and salinity was tested on the soilless greenhouse cultivation of the halophyte *Crithmum maritimum* (Castillo et al. 2022). Increasing salinity induced reduced foliar accumulation of several terpenes and total lipids, while nutrient deficiency increased the concentrations of some polyphenols. Salt stresses were applied in soilless greenhouse cultivation of *Schizonepeta tenuifolia* Briq. (Zhou et al. 2018). Salt treatments positively modulated the density of total glandular trichomes on both leaf sides, while their relative contents in pulegone, other monoterpenes, and sesquiterpene decreased significantly. On the other hand, ketones, alkanes, and esters increased significantly in glandular trichomes with increasing salt stress.
4. *Adding a plant-growth-promoting rhizobacterium or a natural bioactive compound*. The impact of mineral nutrient supply (S or N) and rhizobacterium inoculation on two *O. basilicum* L. cultivars was investigated (Kolega et al. 2020). Fortified nutrient solutions positively impacted the fresh biomass of both cultivars, while inoculation with *Azospirillum brasilense* did not promote growth.

Metabolomics analyses revealed that rhizobacterium inoculation modulated the accumulation of more than 400 secondary metabolites, e.g., terpenoids, phenols, alkaloids, and phenylpropanoids. The primary metabolism was also influenced, with changes in the metabolism of fatty acids, carbohydrates, and amino acids. However, the observed responses were rather cultivar-dependent than following a generalized modification of the phytochemical profile. The effects of natural bioactive products (NBP) – two from fermented plant extracts and microorganisms; microorganisms; bioactive substances extracted from *Ecklonia maxima* – on the growth and ginsenoside content of vertically and aeroponically cultivated *P. ginseng* were tested (Kim et al. 2012). The effects on the root and leaf ginsenoside content were treatment- and location- (upper or lower layer) dependent. A biostimulant made of a plant-derived protein hydrolysate and saline conditions were tested on soilless greenhouse production of *L. sativa* L. (Lucini et al. 2015). Salt stress decreased the shoot and root dry biomass of lettuce, but application of a biostimulant under salt stress increased fresh yield, dry biomass, improved the plant nitrogen metabolism, and delayed photoinhibition as compared to plants under salinity stress. Root and leaf application of the biostimulant under salt stress induced changes in sterol and terpene composition.

Sustainable Nutrients in Vertical Farming
Mineral fertilizers are mainly used in hydroponics nutrient solutions. However, exploiting these resources contributes to land degradation, water contamination, excessive energy consumption, and air pollution (Szekely and Jijakli 2022). In a perspective of sustainability and to meet the challenges of agriculture and climate change worldwide, alternatives should be developed. The organic form of hydroponics (called bioponics) recycles organic waste into a nutrient solution. Several studies have showed positive effects on plant disease mitigation and crop quality, notably with higher health-promoting compounds and/or lower nitrate levels in leafy vegetables (Szekely and Jijakli 2022).

2.3 Temperature

Temperature stress induces many changes in the physiological, biochemical, and metabolic processes and alters the production of bioactive compounds (Fig. 10). Crops with cold or heat tolerance mechanisms better cope with temperature stress (Hasanuzzaman et al. 2013). At low chilling temperature, enzymatic activities are slowed down. In leaves, the balance between light harvesting by photosystem II (PSII) and light utilization through metabolic enzymatic activity is disrupted, leading to photoinhibition and decreased photosynthetic activity (Miura and Furumoto 2013). The reduced activities of antioxidant enzymes result in the accumulation of reactive oxygen species (ROS) (Hasanuzzaman et al. 2013). Adaptive mechanisms have been described, such as promotion of the cyclic electron flow, regulation of energy distribution, antioxidant activity inititation, and accumulation of osmotic

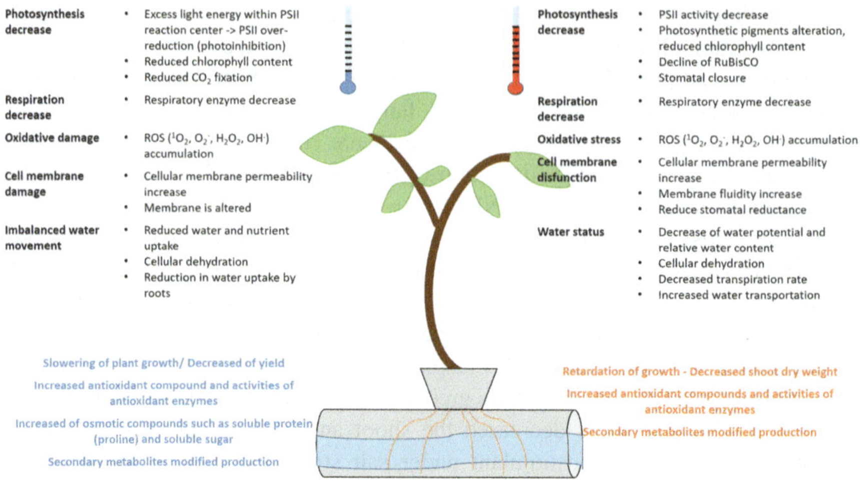

Fig. 10 Representation of the main physiological modifications induced by cold or heat stress in a hydroponically cultivated plant

regulators (soluble sugars and soluble proteins such like proline and betaine) (Li et al. 2022). Under heat stress, the efficiency of photosynthesis declines, because PSII activity, Rubisco activity, the photosynthetic pigment content, and the carbon fixation capacity are reduced (Zhao et al. 2020). Other physiological changes occur like altered cell membrane thermostability or oxidative damage (Zhao et al. 2020). Plants accumulate antioxidants (proline, glutathione, ascorbate, carotenoids), and the activity of antioxidant enzymes is increased (Hasanuzzaman et al. 2013). Another adaptive response may be a reduced chlorophyll content, as this decreases the energy absorption linked to chlorophyll energy absorption and lowers leaf heating (Mesa et al. 2022).

The response of *Paspalum wettsteinii* under heat stress treatments has been investigated (Zhao et al. 2022). A metabolic analysis revealed that biosynthesis of flavonoids and anthocyanins was both up- and downregulated under heat stress. Heat and cold stresses were applied on *S. lycopersicum* L. (Mesa et al. 2022). Heat stress decreased plant productivity and increased tocochromanols in the leaves and ascorbic acid in the fruit. The effect of short low/high temperature treatments on the root zone of *Coriandrum sativum* L. have been studied (Nguyen et al. 2020). Short temperature treatments reduced fresh biomass, while carotenoids, phenolics, chlorogenic acid, ascorbic acid, and the antioxidant capacity of the plants were enhanced under the extreme temperature treatments (15 °C or 35 °C) for 6 days.

Terpene emission is generally controlled by temperature (Staudt and Bertin 1998; Tarvainen et al. 2005; Ibrahim et al. 2010; Yang et al. 2018). Augmenting night temperature increased the terpene content of *Betula pendula* and *Populus tremula* (Ibrahim et al. 2010). The influence of antagonistic successive stresses – cold (4 °C, 30 min)/heat (from 25 °C to 60 °C within 5 min) and heat (from

25 °C to 60 °C within 5 min)/cold (4 °C, 30 min) – on *O. basilicum* L. and *Salvia officinalis* L. was tested (Copolovici et al. 2022). Terpene emissions were enhanced in plants under successive stresses as compared to the control plants, while the phenolic and flavonoid contents remained unaffected. The impact of temperature stress on *H. perforatum* was investigated (Zobayed et al. 2005). The shoot contents in hypericin, pseudohypericin, and hyperforin increased with high temperature (35 °C). The effects of temperature on the growth and terpene production of *Platycodon grandiflorum* A. DC in soil and soilless culture systems were measured (Nguyen et al. 2022). Fresh weight was highest under soilless cultivation conditions at 20 °C, and the shoot contents in platycodin D3, polygalcin D, and total saponin were optimized at 20 °C and 25 °C.

The increase in metabolite production should be calculated along with CO_2 emissions if temperature is increased or decreased in order to improve sustainability and lower environmental costs.

2.4 CO_2 Level

Increased levels of CO_2 induced increased photosynthesis, mainly due to increased Rubisco activity, which is not saturated at current atmospheric CO_2 concentrations. Increased photosynthesis results in better growth and yield. The photosynthetic rate, the transpiration rate, stomatal conductance, and the leaf, stem, and root carbon contents of *Withania somnifera* (a medicinal plant native to India) increased significantly in elevated CO_2 conditions, and dry weight increased too (Sharma et al. 2018). Elevated CO_2 levels improve water use efficiency and mitigate the negative effects of drought stress (Li et al. 2018). Cucumber seedlings under drought stress conditions and increased CO_2 levels had a higher leaf water content, regulated the cell osmotic pressure by accumulating carbohydrates, and accumulated secondary metabolites (Li et al. 2018). Several studies were conducted under CO_2 enrichment to increase the medicinal properties of *Labisia pumila*, a medicinal plant found in the Indochinese Peninsula. The total phenolic and flavonoid contents increased under high CO_2, together with a reduced chlorophyll content (Ibrahim and Jaafar 2011a). The enhanced secondary metabolite content could be due to reallocation of phenylalanine from protein synthesis to secondary metabolite production (Ibrahim and Jaafar 2011b). Under 1200 μmol mol^{-1} of CO_2 enrichment, increased nitrogen fertilization reduced the total phenolic and flavonoid contents (Ibrahim and Jaafar 2011b, 2017).

Combined light intensities and CO_2 levels have been investigated. The cumulated values of secondary metabolites and antioxidant activity were observed at the lowest light intensity (PPFD = 225 μmol m^{-2} s^{-1}) and the highest CO_2 level (1200 μmol mol^{-1}) (Ibrahim et al. 2014). The cytotoxicity of *L. pumila* variety *alata* leaf extract toward cancer cells was strongest under elevated CO_2 (1200 μmol mol^{-1}) and low light intensity (PPFD = 300 μmol m^{-2} s^{-1}), and the concentrations of different phenolics and flavonoids, the total phenolic, flavonoid, and saponin contents were highest (Karimi et al. 2016). Some medicinal plants showed a positive correlation of their secondary metabolite content with light intensity and the CO_2 concentration. For example, *H. perforatum* L. (a herb native to Europe and West Asia) showed increased hypericin and pseudohypericin contents under a high CO_2 level and increased light intensity (Mosaleeyanon et al. 2005).

The combination of temperature and CO_2 concentrations has been investigated. *Gynostemma pentaphyllum* (a herbal drug that grows in Asian countries) showed increased biomass but a reduced total antioxidant capacity and reduced levels of antioxidant compounds when cultivated under elevated CO_2 and increased temperature (Chang et al. 2016).

The impact of CO_2 enrichment seems to be species- as well as growth-stage-specific. If CO_2 is increased, one should check that it is well absorbed by the plants and that all the other conditions are optimal for the growth of the plant.

3 Economic Approach of Vertical Farming of Medicinal Plants

Vertical farms are shortly defined as multilayer soilless crop production systems including various ways of producing vegetables. "Vertical" refers to layers that can be vertically or horizontally mounted and to crop production systems that can be installed in closed or semi-closed structures. Semi-closed systems are typically greenhouses with sunlight that can be supplemented with artificial light. Indoor vertical farms are closed systems defined as plant factories using artificial lighting (PFALs), e.g., a container or a closed building. PFALs are controlled systems and are ideal for producing medicinal plants because the system ensures stable high standards, constant quality, and constant quantity. However, PFALs use intensive technology and are expensive because of expensive facilities and high energy and labor costs. As discussed previously, vertical farming of leafy greens is economically tricky in Europe, mainly due to high investment, energy, and labor costs combined with low market prices for such commodities. Economic studies on vertical farm construction, operation, and viability are lacking (Baumont de Oliveira et al. 2022). Most economic feasibility studies are based on hypothetical case studies and horticultural crop predictions, and none of them deals with medicinal plants. Cultivating high-added-value plants is assumed to be less economically challenging. However, no studies have been carried out on the whole process of making medicinal plant, from the indoor growing to the final product. The complete production scheme of medicinal plants depends on the form and application of the final product (Figs. 11 and 12). A medicinal plant product can be under various forms depending on its use, e.g., infusion, decoction, paste, poultice, multi-metabolites extract, or powder. The end-user's choice involves a more or less complex production pattern. The cultivation process has to target the yield and the metabolite content through a fine-tuning of abiotic factors. If the final product is used fresh, the postharvest

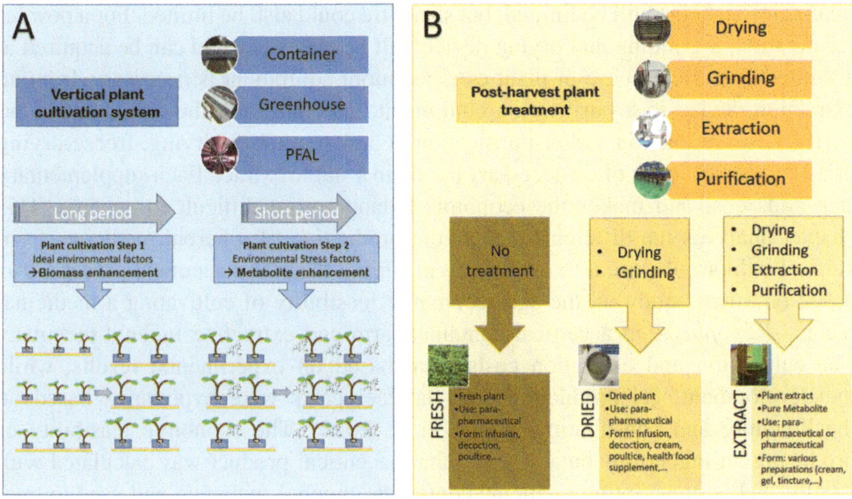

Fig. 11 Schematic representation of the two-step cultivation process of vertical medicinal plant farming (**a**) and the possible postharvest plant treatments (**b**)

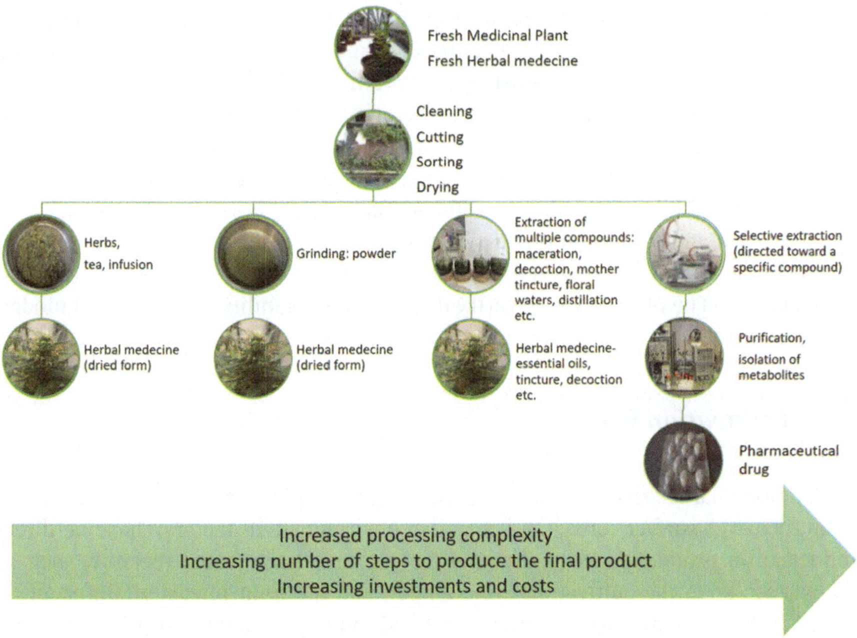

Fig. 12 Schematic representation of the various forms of herbal medicine

treatments and cost will be limited, but shelf-life could also be limited. For a powder formulation, a grinding and drying device will be necessary and can be acquired at a limited cost. In the case of plant extracts, more equipment is necessary, from an extraction device to a purification equipment, depending on the extract type and purity. After extraction and/or purification, a new treatment (drying, freeze-drying or dielectric drying) is often necessary to obtain a stable extract. Each supplementary step adds cost and makes the economic balance more difficult to achieve. This chapter analyzes the different steps of the production of a herbal medicine, from vertical indoor cultivation to extraction, including the pharmaceutical process. It is based on a case study on the agro-economic feasibility of cultivating a medicinal plant – *E. peplus* – in a vertical container farm and extracting ingenol-mebutate. The cultivation and extraction costs were based on experimental results, while the development, gel production, and flat fees costs were hypotheses based on the literature and consultation (Bafort et al. 2022). The economic feasibility of producing an ingenol-mebutate-based pharmaceutical product was calculated with Picato® gel, a prescription medicine containing ingenol-mebutate and used to treat skin actinic keratosis. Data on another medicinal plant – *Artemisia annua* – is also discussed (Bafort et al. 2023). The cost price is an economic term that refers to all the costs supported by a company to produce goods or a service. The sum has to include direct costs and indirect costs. Indirect costs are expenses that are not directly linked to the production of the product or service (advertising, rental of premises, salaries, etc.). Different calculation approaches exist, based on variable cost prices, direct cost prices, coefficient methods, and activity-based costing (Niessen and Chanteux 2005). Therefore, a company that offers different products and services has to choose the right analysis in order to understand how much a service or a product costs. In the paper, all the costs are related directly to production. The case-study is useful to forecast an economical evaluation of (i) cultivation and extraction process and (ii) pharmaceutical drug production. The forecast calculation for the pharmaceutical market is based on assumptions and general costs. The objective is to verify the economic viability of this type of model.

3.1 Cultivation Cost

Cultivation characteristics, such as the plant species, plant biomass, culture length, plant density, surface area, and specific environmental factors, have a direct influence on production costs. Several factors have been tested recently, such as the surface area, the cultivation cycle length, and the light intensity (Bafort et al. 2022). The production cost is strongly related to the productivity of the cultivation system, which can be described in different ways. Annual biomass – fresh or dried – is one of them and can also be described partly by the mean fresh or dried biomass *per* plant. The productivity of a cultivation system varies according to the following factors:

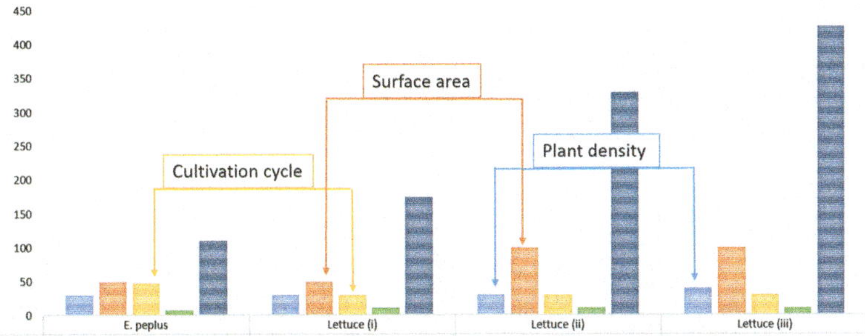

	E. peplus	Lettuce (i)	Lettuce (ii)	Lettuce (iii)
Density (plant/m2)	30	30	30	40
Surface area (m2)	50	50	100	100
Cultivation cycle (days)	49	30	30	30
Number of cycle/year	7	11	11	10
Annual production (0.1 kg.yr)	111	175	329	425

Fig. 13 Influence of plant density, surface area, and the length of the cultivation cycle on the annual production yield. First column and bars, typical cultivation conditions of *E. peplus* in a vertical container. Second to fourth columns and bars, simulation of Romaine lettuce cultivation in a vertical container with a shorter cultivation cycle (i), an increased surface area (ii), and an increased density of plants (iii)

(i) The *cultivation cycle*. For a fixed biomass *per* plant, Fig. 13 shows how the annual biomass output depends on the cultivation cycle. If the cycle is short, more cycles can be achieved *per* year and productivity is increased. The cycle can be shortened by modifying abiotic factors. For example, reducing vegetative growth of hemp (*C. sativa*) by modifying the photoperiod shortens the cultivation cycle. Working with cuttings or in-vitro propagated plants instead of seeds can also make the cycle shorter. Container farming of the medicinal plant *E. peplus*, which has a cycle of 48.5 days, allows 7.2 cycles *per* year, taking the time needed for harvesting and cleaning into account, and gives an output of 1106 kg *per* year. Cultivating Romaine lettuce – a crop with a shorter culture cycle (30 days) – increased the number of cycles *per* year and increased annual biomass to 1745 kg.

(ii) The *surface area*. Small cultivation surface units decrease productivity. Figure 1 shows that for a same crop and under identical environmental conditions, doubling the cultivating area enhanced annual productivity by 88.5%, from 1745 kg to 3285 kg.

(iii) The *plant density*. The plant density is a way of increasing the productivity of a crop system. It can be improved by a specific design/improvement of the production area. For example, cultivation on vertically stacked layers can improve the plant density for some species, especially small plant. It will also need light to be placed not only above the cultivation tray but also surrounding the crop production layer. A greater plant density of lettuce from 30 to 40 plants *per* m^2 resulted in a 29.5% increase to reach 4255 kg of lettuce *per* year (Fig. 13).

(iv) The *biomass per plant*. Higher biomass results in higher annual productivity and higher output of the medicinal product (dried leaves, infusion bags,

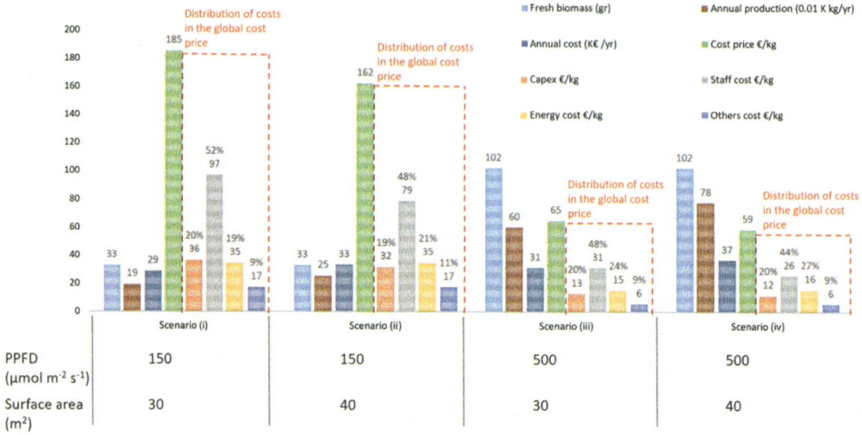

Fig. 14 Fresh shoot biomass *per* plant, annual output, CapEx, OpEx (subdivided in staff cost, energy cost, and other costs), total cost (CapEx + OpEx), and cost price of *Euphorbia peplus* cultivation in a vertical container farm following several scenarios. The dotted box includes CapEx and OpEx, the sum of which corresponds to the production cost price; the relative percentage of each cost in the cost price is indicated above each bar. Scenario (i), cultivation under a PPFD of 150 $\mu mol^{-2}\ s^{-1}$ with a surface area of 30 m^2; scenario (ii), cultivation under a PPFD of 150 $\mu mol^{-2}\ s^{-1}$ with a surface area of 40 m^2; scenario (iii), cultivation under a PPFD of 500 $\mu mol^{-2}\ s^{-1}$ with a surface area of 30 m^2; and scenario (iv), cultivation under a PPFD of 500 $\mu mol^{-2}\ s^{-1}$ with a surface area of 40 m^2. (Bafort et al. 2022)

poultice, etc.). The biomass *per* plant can be optimized by cultivation under optimized environmental factors (Fig. 14) or by breeding or selecting high-biomass varieties.

If all three factors – cultivation cycle length, surface area, and plant density – are upgraded from the initial crop system, the annual production could sharply rise by 284% as the yield could increase from 1106 to 4255 kg (Fig. 13).

Production costs are influenced by several parameters, among others the productivity of the cultivation system. Costs are divided in two items: (1) capital expenditures (CapEx), long-term investment (e.g., equipment, property, buildings), and (2) operating expenditures (OpEx), daily expenses necessary to keep the business operational (e.g., labor, energy, water consumption, nutrients, seeds). Figure 14 represents the production cost of the medicinal weed *E. peplus* in a vertical indoor hydroponic container and shows the relationship between productivity and the production cost (Bafort et al. 2022). At a low light intensity, *E. peplus* growth was not optimized, and the mean biomass *per* plant reached 33 g. This resulted in a low annual productivity, and costs were distributed across a small volume of production. When the mean biomass *per* plant and the surface area increased, through modification of the cultivation process, productivity increased too. If the induced costs (e.g., for structural modifications, more powerful LEDs) increased moderately, the production cost *per* kg of plant decreased. Figure 14 shows that if the surface area is increased by 33% by placing an additional layer in the container under the same light intensity (scenarios (i) and (ii)), CapEx and OpEx increase by 14%. As productivity is increased by 31.2%, the production cost is cut by 23€ *per* kg. In scenario (iii), the environmental factors have been modified: an increased light intensity results in a significant 209% rise of the mean biomass *per* plant as compared with scenario (i). Although the total cost is higher due to increased energy consumption and investment in upgraded LEDs, the total costs increase by only 9%, whereas annual production is increased by 212%, hence a 65% decrease of the production cost. In this optimized plant environment, if the surface area dedicated to production is increased by 10 m^2 (scenario iv), the production cost is reduced by 68% as compared to scenario (i). This shows the importance of optimizing the technical cultural itinerary to maximize productivity, as investment, labor, and energy costs are important.

The production cost of the vertical farming of another medicinal plant – *Artemisia annua* L. – in a modified shipping container has been calculated (Fig. 15) (Bafort et al. 2023). *A. annua* is an annual herb native to Asia. It has been used in traditional Asian medicine for treating and preventing fever and chills for many centuries and has been widely used for treating malaria (Kim et al. 2015). Again, the production cost of 1 kg of *A. annua* is closely related to the productivity of the horticultural process, and the production cost can be significantly reduced if the optimization of productivity does not increase costs too much. Higher-intensity LED

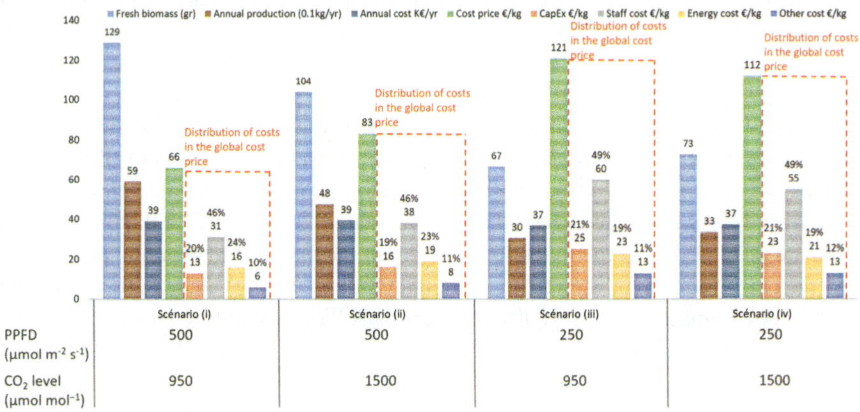

Fig. 15 Fresh shoot biomass *per* plant, CapEx, OpEx (subdivided in staff cost, energy cost, and other costs), annual fresh shoot biomass output, total cost (CapEx + OpEx), and cost price of *Artemisia annua* cultivation in a vertical *container f*arm following several scenarios. The dotted box includes CapEx and OpEx, the sum of which corresponds to the production cost price; the relative percentage of each cost in the cost price is indicated above each bar. Scenario (i), cultivation under a PPFD of 500 $\mu mol^{-2}\ s^{-1}$ with a CO_2 concentration of 950 $\mu mol^{-2}\ s^{-1}$; scenario (ii), cultivation under a PPFD of 500 $\mu mol^{-2}\ s^{-1}$ with a CO_2 concentration of 1500 $\mu mol^{-2}\ s^{-1}$; scenario (iii), cultivation under a PPFD of 250 $\mu mol^{-2}\ s^{-1}$ with a CO_2 concentration of 950 $\mu mol^{-2}\ s^{-1}$; and scenario (iv), cultivation under a PPFD of 500 $\mu mol^{-2}\ s^{-1}$ with a CO_2 concentration of 1500 $\mu mol^{-2}\ s^{-1}$

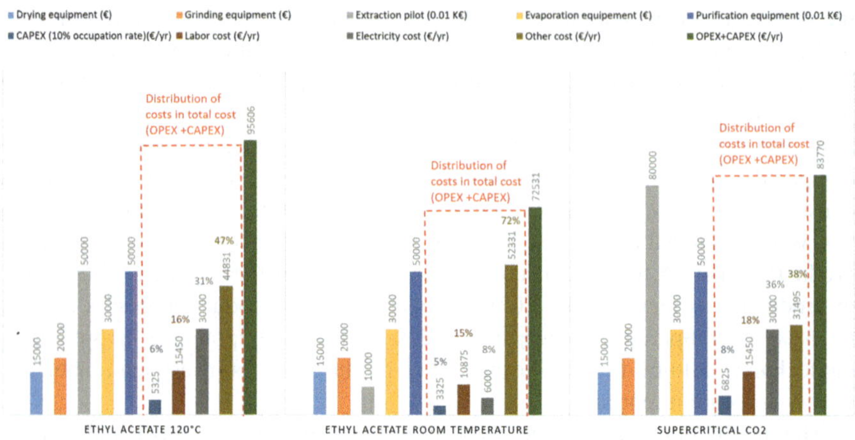

Fig. 16 Evaluation of the costs of extracting a diterpene from *Euphorbia peplus*. Estimated costs of various facilities (drying, grinding, extraction pilot, evaporation, and purification), CapEx *per* year at a 10% occupation rate and on a 20-year depreciation basis, OpEx *per* year (labor cost, electricity cost, and other costs), and total expenses *per* year (CapEx + OpEx). The dotted box includes CapEx and OpEx, the sum of which corresponds to the total cost (CapEx + OpEx); the relative percentage of each cost in the total cost is indicated above each bar. (Bafort et al. 2022)

lamps increased the total cost by 7%, but their use increased productivity by 56.4% so that the production cost *per* kg decreased significantly. Under the same light intensity (PPFD = 500 μmol^{-2} s^{-1}), increasing the CO_2 concentration induced a negative stress on *A. annua* and the mean fresh shoot biomass decreased, hence a higher production cost *per* kg (Fig. 16).

The cultivation cost is a vital economic piece of data, and its calculation is a key step for establishing the breakeven point of a product and a coherent selling price. It does not include distribution, marketing, or storage costs. If the medicinal plant is used fresh, such costs have to be added to calculate the cost price of the product and evaluate profitability. If the plant has to be dried, the drying and grinding process, or any necessary additional step (e.g., cleaning, cutting, sorting), have to be calculated in the same manner as for the cultivation cost (OpEx and CapEx).

3.2 *Extraction Cost*

The cultivation of a medicinal plant is the first step of its production. Depending on the final use of the plant (Fig. 11), the next steps after harvest can range from drying to the manufacturing of a pharmaceutical specialty. The cost of terpene extraction from the shoot biomass of a medicinal plant after vertical container cultivation has been studied recently (Bafort et al. 2022). Cultivation of *E. peplus* in a container farm resulted in a yearly output of 776 kg of fresh shoot biomass (Fig. 14). This output was divided into several batches of 103 kg each, representing a very low load for industrial drying, grinding, extraction, evaporation, and purification devices, which can handle much more biomass. To take the low level of occupation of the devices into account, the occupation rate of the drying, extraction, and purification devices was set to 10% with a depreciation rate of 20 years. Three extraction methods were evaluated (ethyl acetate at 120 °C, ethyl acetate at room temperature, and supercritical CO_2), and their respective costs were calculated (Fig. 14). The investment costs were similar for the drying, grinding, evaporation, and purification facilities. However, the extraction method represented different investment costs depending on the extraction technique. The cost was higher for supercritical CO_2 extraction, and lower for ethyl acetate extraction at room temperature, which induced the highest and lowest CapEx, respectively. The OpEx differed depending on the extraction method. The method generating the highest operational cost was the "ethyl acetate at 120 °C" method, followed by the "supercritical CO_2" method, and finally the "ethyl acetate at room temperature" method. The distribution of costs differed between the extraction methods. With ethyl acetate at 120 °C, the largest contributor to annual cost was the operating and maintenance cost of equipment ("other costs" – 47%), followed by energy cost (31%) and labor cost (16%), while CapEx only represented 6%. With supercritical CO_2, the main costs were the "other costs" (38%) together with electricity cost (36%), and with ethyl acetate at room temperature, the major cost was "other costs" (72%), while electricity and labor costs and CapEx were much lower (15–5%).

The production cost of a metabolite depends on the CapEx and OpEx of the cultivation and extraction processes, but it is also strongly dependent on the extraction yield (Fig. 17). Although extraction by ethyl acetate at 120 °C generated the highest OpEx and CapEx, it also gave a significantly higher yield; as a result, this diterpene extraction method was the cheapest, with a cost of 37.8 € *per* mg. The increased yield allowed reducing production costs by 34% and 19.5% as compared to the "ethyl acetate at room temperature" and "supercritical CO_2" methods, respectively. This shows that production costs are related to multiple factors and how important it is to evaluate the expenses and yield of each step when producing medicinal plants and extracts. In the case study of ingenol-mebutate production from *E. peplus*, the concentration of this metabolite in the plant was low (about 60–70 mg *per* kg of plant shoot) (Bafort et al. 2022). The selection of the appropriate cultivation method (i.e., a high light intensity and an increased surface area) increased the extraction yield as compared to other studies (Hohmann et al. 2000). However, the plant content in ingenol-mebutate is constitutively low, so that the extraction yield remained low too. Increasing the content in a specific metabolite by appropriate abiotic factors such as high-temperature stress during the cultivation process could increase the content in terpene and ultimately the extraction yield (see also Chap. 2). Another possible way of reducing the cost of the production process of a metabolite is to increase the surface area to augment the cultivation yield or productivity.

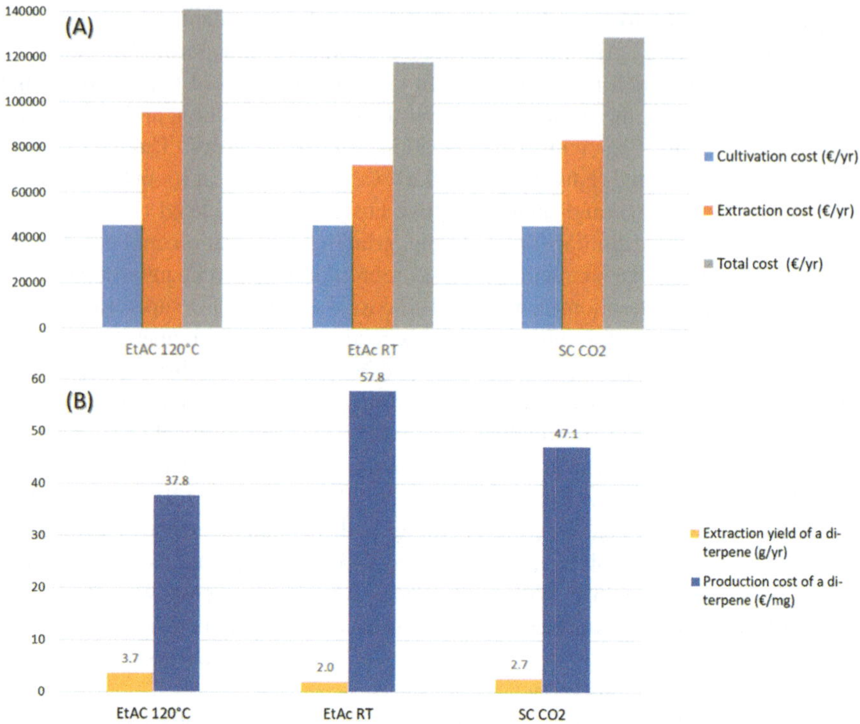

Fig. 17 Evaluation of the production cost of a diterpene extracted from a medicinal plant cultivated in a vertical container farm following three extraction methods. (A) Cultivation, extraction, and total cost following three extraction methods. EtAC 120 °C, ethyl acetate at 120 °C; EtAc RT, ethyl acetate at room temperature; SC CO_2, supercritical CO_2. (B) Extraction yield and production cost *per* mg of a diterpene following three extraction methods: EtAC 120 °C, ethyl acetate at 120 °C; EtAc RT, ethyl acetate at room temperature; SC CO_2, supercritical CO_2. (Bafort et al. 2022)

3.3 *Pharmaceutical Drug Production Cost*

Medicinal plants have secondary metabolites that can be of interest for pharmaceutical applications as purified molecules. In this case, the use of the metabolite in a pharmaceutical drug has to be approved by an official agency like the European Medicines Agency. Approval requires significant development costs showing the safety and efficacy of the drug and includes preclinical and clinical studies. Estimating the average cost of developing a drug is difficult. It largely varies according to studies, from US$ 92.0 million to US$ 884 million and even US$ 1395 million (Morgan et al. 2011; DiMasi et al. 2016). Moreover, the clinical costs of drug development vary, depending on the treatment category. They range from US$ 312 million for analgesics/anesthetics to US$ 448 million for anti-infective drugs (Morgan et al. 2011). Therefore, pharmaceutical use requires far more investment than traditional para-pharmaceutical use (e.g., extracts (decoction, infusion, poultice, etc.)). The extract will also need a pharmaceutical-grade certification and will have to be manufactured in a "Good Manufactory Practices"-certified factory. Cultivating a medicinal plant in a vertical indoor farm is particularly suited for pharmaceutical or high-grade standard quality, because the process is completely controlled and ensures large, regular, and predictable quantities and constant high-quality metabolites. Moreover, the pharmaceutical use of the crop will give a higher added value to the metabolite. Few studies have investigated the entire cost of processing a medicinal plant from cultivation to the final pharmaceutical drug. The economic feasibility of producing a medicinal molecule was calculated from *E. peplus* annual biomass yield and ingenol-mebutate extraction yield (Bafort et al. 2022), based on a prescription medicine containing ingenol-mebutate and used to treat precancerous skin lesions. Figure 18 shows the output, CapEx and OpEx of *E. peplus* cultivation in a 40-m^2 vertical container farm under high light intensity producing 776 kg of fresh shoot crop *per* year, from which 3.73 gr of ingenol-mebutate *per* year are extracted with ethyl acetate at 120 °C. This process gave an output of 0.56 M€ with the selling of pharmaceutical gels containing 0.015% and 0.05% of the metabolite. The development costs were estimated to be 300 M€ (15 M€ *per* year) allocated over the term of a 20-year patent. Compared with this very high investment cost, other OpEx appeared as a very low load: 0.14 M€ for the cultivation and extraction costs and 0.12 M€ for gel manufacturing and flat

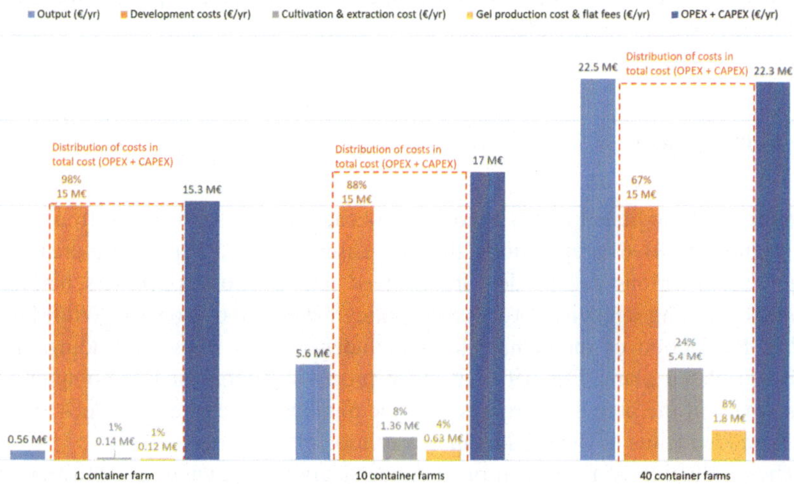

Fig. 18 Simulation of the production costs of a pharmaceutical gel based on ingenol-mebutate extracted from vertical container farming of *Euphorbia peplus*. Vertical cultivation in one container, 10 containers, or 40 containers and generated outputs, OpEx and CapEx. OpEx are subdivided in development costs, cultivation and extraction costs, gel production costs, and flat fees. The dotted box includes CapEx and OpEx, the sum of which corresponds to the total production cost; the relative percentage of each cost in the total cost is indicated above each bar. (Bafort et al. 2022)

fees. The total cost – the sum of CapEx (pharmaceutical manufacturing building) and OpEx – reached 15.3 M€ for 0.56 M€ of output. Therefore, the return time on investment for a total annual cost was 27 years. By multiplying the cultivation yield by 10 (by acquiring 10 vertical container farms), the extraction yield would be multiplied by 10, with a total of 37.3 g of ingenol-mebutate manufactured *per* year. This would raise the output to 5.6 M€ *per* year. The CapEx and OpEx costs, except the development costs, would also be increased and would reach an annual total of 17 M€. Therefore, the return time on investment for a total annual cost would be 3 years. Forty vertical container farms would be needed to reach a return time on investment of 1 year, with an output of 22.5 M€ and a total annual cost of about 22.3 M€, without being sure that the demand would absorb such a production. When looking at the distribution of costs of the whole plant-based drug production process from plant cultivation to drug production, the largest contributor is the R&D cost (98–67%), followed by cultivation and extraction costs (1–24%) and drug manufacturing costs (1–8%), depending on the number of containers, i.e., on the productivity of the cultivation and extraction steps.

Although the simulation of the profitability of the pharmaceutical gel showed that economic feasibility was difficult to reach, some factors could rapidly increase the profitability of ingenol-mebutate production. The improvement of the ingenol-mebutate content in the plant by a more specific and adapted cultivation process would increase the extraction yield rapidly. Furthermore, upcoming new plant factory designs with increased growing surfaces and planting densities together with digital agriculture will reduce the CapEx and OpEx and the cost *per* kg of crop, and profitability will be less challenging.

4 Conclusion

The sector of medicinal plants is complex because many forms exist, from freshly cut plants to dried preparations through essential oils, macerates, creams, or poultice, and various stakeholders are involved among whom consumers, herbalists, retailers, funding agencies, processors, policymakers, and growers (WildMapsFit 2020). Production of medicinal plants includes various steps, depending on the final use of the plant (Fig. 19) that make processing more or less complex. The complexity of the process increases with the number of steps, and so does the cost, but the added value of the product increases too.

The production of medicinal plants under a controlled environment offers new opportunities (WildMapsFit 2020; Zobayed 2020):

- A greater number of botanically reliable products free of misidentified plants.
- A product uncontaminated by pollutants, pesticides, and microbes.
- A stable source of guaranteed raw material.
- Uniform and optimized biochemical profiles.
- Quantity and quality are predictable and guaranteed.

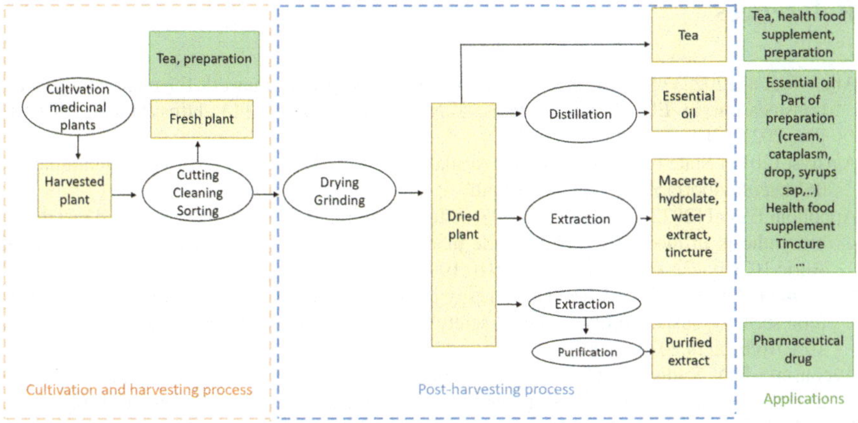

Fig. 19 Manufacturing processes of medicinal plants. (Adapted from EIP-AGRI 2020)

- Relationships between producers and purchasers are enhanced, based on stable and predictable production.
- Controlled postharvest handling.
- Quality control can be more easily implemented in such structures.
- Product certification or labeling.

Growers may in turn consider medicinal plant cultivation safer and more profitable than traditional crops. However, growers need to carefully calculate the economic viability of these production systems.

The demand for sustainability from consumers and regulators is increasing. Actors in the value chain must respond to consumer expectations, e.g., raw material sourcing, traceability, quality regulation, efficiency, and safety, while considering sustainability in the cultivation process (WildMapsFit 2020).

Five factors of the crop cultivation process under a controlled environment need to be optimized: (1) productivity (fresh or dried biomass *per* year) has to be maximized, (2) the plant content in metabolites of interest has to be maximized, (3) yields of postharvest processes (drying, extraction, purification, etc.) have to be maximized, (4) the sustainability of the process (life cycle assessment, energy use efficiency, light use efficiency, water use efficiency) has to be maximized, and (5) costs have to be minimized. The cultivation practices need a fine-tuning of environmental factors that should be specific to the crop and the metabolite of interest as plant responses to abiotic factors are mainly species-specific. The use of controlled environment cultivation systems can facilitate the development of safe, steady cultures in quality and quantity and high-quality (para)pharmaceutical plant products extracted from medicinal plants.

Funding This research work was funded by the European Union and the Walloon Region with the European Funds for Regional Development 2014–2020 within the framework of the VERDIR Tropical Plant Factory program (OptiBiomasse project) (Form number: 814687–481490).

References

Abeli T et al (2022) On farm plant reintroduction: a decision framework for plant conservation translocation in EU agro-ecosystems. J Nat Conserv 65:126113. https://doi.org/10.1016/j.jnc.2021.126113

Agrilyst (2017) State of indoor farming. Available at: https://artemisag.com/wp-content/uploads/06/stateofindoorfarming-report-2017.pdf

Appolloni E et al (2022) Beyond vegetables: effects of indoor LED light on specialized metabolite biosynthesis in medicinal and aromatic plants, edible flowers, and microgreens. J Sci Food Agric 102(2):472–487. https://doi.org/10.1002/jsfa.11513

Avgoustaki DD, Xydis G (2020a) How energy innovation in indoor vertical farming can improve food security, sustainability, and food safety? Adv Food Secur Sustain 5:1–51. https://doi.org/10.1016/bs.af2s.2020.08.002

Avgoustaki DD, Xydis G (2020b) Indoor vertical farming in the urban nexus context: business growth and resource savings. Sustainability 12(5). https://doi.org/10.3390/su12051965

Bae J-H, Park S-Y, Oh M-M (2017) Supplemental irradiation with far-red light-emitting diodes improves growth and phenolic contents in Crepidiastrum denticulatum in a plant factory with artificial lighting. Hortic Environ Biotechnol 58(4):357–366. https://doi.org/10.1007/s13580-017-0331-x

Bafort F et al (2022) The agro-economic feasibility of growing the medicinal plant Euphorbia peplus in a modified vertical hydroponic shipping container. Horticulturae 8. https://doi.org/10.3390/horticulturae8030256

Bafort F, Bouhadada A, Ancion N, Crutzen N, Haïssam JM (2023) Optimization of hydroponic production of the medicinal plants Euphorbia peplus and Artemisia annua in container. Acta Hortic 1369:171–178. https://doi.org/10.17660/ActaHortic.2023.1369.21

Balabanova V, Vitkova A (2010) Peculiarities in ontogenesis of arnica Montana L. in Bulgaria. Comptes Rendus de L'Academie Bulgare des Sciences 63(9):1301–1306. https://www.scopus.com/inward/record.uri?eid=2-s2.0-77958000334&partnerID=40&md5=2aae6bafab20e676a2e523f3c3643fee

Banerjee C, Adenaeuer L (2014) Up, up and away! The economics of vertical farming. J Agric Stud 2(1):40–60

Bar-Tal A et al (2019) Chapter 7 – inorganic and synthetic organic components of soilless culture and potting mixtures. In: Raviv M, Lieth JH, Bar-Tal A (eds) Soilless culture, 2nd edn. Elsevier, Boston, pp 259–301. https://doi.org/10.1016/B978-0-444-63696-6.00007-4

Barbosa GL et al (2015) Comparison of land, water, and energy requirements of lettuce grown using hydroponic vs. conventional agricultural methods. Int J Environ Res Public Health 12(6):6879–6891. https://doi.org/10.3390/ijerph120606879

Baumont de Oliveira FJ, Ferson S, Dyer RAD (2022) How high is high enough? Assessing financial risk for vertical farms using imprecise probability. Sustainability 14. https://doi.org/10.3390/su14095676

Benke K, Tomkins B (2017) Future food-production systems: vertical farming and controlled-environment agriculture. Sustain: Sci Pract Policy 13(1):13–26. https://doi.org/10.1080/15487733.2017.1394054

Bian Z-H et al (2016) Continuous light from red, blue, and green light-emitting diodes reduces nitrate content and enhances phytochemical concentrations and antioxidant capacity in lettuce. J Amer Soc Hort Sci 141(2):186–195. https://doi.org/10.21273/JASHS.141.2.186

Bullock P (2005) Climate change impacts. In: Hillel D (ed) Encyclopedia of soils in the environment. Elsevier, Oxford, pp 254–262. https://doi.org/10.1016/B0-12-348530-4/00089-8

Castillo JM et al (2022) Interactive effects between salinity and nutrient deficiency on biomass production and bio-active compounds accumulation in the halophyte Crithmum maritimum. Sci Hortic 301:111136. https://doi.org/10.1016/j.scienta.2022.111136

Chang J-D et al (2016) Effects of elevated CO_2 and temperature on Gynostemma pentaphyllum physiology and bioactive compounds. J Plant Physiol 196–197:41–52. https://doi.org/10.1016/j.jplph.2016.02.020
Chen C-C et al (2020) Led lights affecting morphogenesis and isosteroidal alkaloid contents in fritillaria cirrhosa d. Don – an important chinese medicinal herb. Plan Theory 9(10):1–15. https://doi.org/10.3390/plants9101351
Choi J et al (2022) Effect of far-red and UV-B light on the growth and ginsenoside content of ginseng (Panax ginseng C. A. Meyer) sprouts aeroponically grown in plant factories. Hortic Environ Biotechnol 63(1):77–87. https://doi.org/10.1007/s13580-021-00380-9
Chrysargyris A, Tzortzakis N (2021) Tailor-made hydroponic recipes for the production of high quality medicinal and aromatic plants. Acta Hortic 1321:87–94. https://doi.org/10.17660/ActaHortic.2021.1321.11
Cipollini D, Walters D, Voelckel C (2017) Costs of resistance in plants: from theory to evidence. Annual Plant Rev Online:263–307. https://doi.org/10.1002/9781119312994.apr0512
Ciriello M et al (2022) Differential response to NaCl osmotic stress in sequentially harvested hydroponic red and green basil and the role of calcium. Front Plant Sci 13. https://doi.org/10.3389/fpls.2022.799213
Copolovici L et al (2022) Antagonist temperature variation affects the photosynthetic parameters and secondary metabolites of Ocimum basilicum L. and Salvia officinalis L. Plan Theory 11(14). https://doi.org/10.3390/plants11141806
Cowan N et al (2022) CEA systems: the means to achieve future food security and environmental sustainability? Front Sustain Food Syst 6. https://doi.org/10.3389/fsufs.2022.891256
Czitrom V (1999) One-factor-at-a-time versus designed experiments. The Am Statist. [American Statistical Association, Taylor & Francis, Ltd.] 53(2):126–131. https://doi.org/10.2307/2685731
Dadhich A et al (2022) Unveiling the interaction of divergent abiotic stresses and their consequences in terms of bacosides in Bacopa monnieri (L.) Wettst. J Appl Res Med Aromat Plants 31. https://doi.org/10.1016/j.jarmap.2022.100423
Debusschere T, Boekhout R (2021) When will vertical farming become profitable? CEA 40 Conference, Tech 4.0, Vertical Farming Daily. Available at: https://www.verticalfarmdailycom/article/9321424/when-will-vertical-farming-become-profitable/. Accessed 6 Aug 2021
DiMasi JA, Grabowski HG, Hansen RW (2016) Innovation in the pharmaceutical industry: new estimates of R&D costs. J Health Econ 47:20–33. https://doi.org/10.1016/j.jhealeco.2016.01.012
Dou H et al (2018) Responses of sweet basil to different daily light integrals in photosynthesis, morphology, yield, and nutritional quality. Hort Science Horts 53(4):496–503. https://doi.org/10.21273/HORTSCI12785-17. Washington, DC: American Society for Horticultural Science
Dou H et al (2020a) Morphological and physiological responses in basil and brassica species to different proportions of red, blue, and green wavelengths in indoor vertical farming. J Amer Soc Hort Sci. Washington, DC: American Society for Horticultural Science 145(4):267–278. https://doi.org/10.21273/JASHS04927-20
Dou H, Niu G (2020) Chapter 9 – plant responses to light. In: Kozai T, Niu G, Takagaki M (eds) Plant factory, 2nd edn. Academic Press, pp 153–166. https://doi.org/10.1016/B978-0-12-816691-8.00009-1
Dou H, Niu G, Gu M (2019) Photosynthesis, morphology, yield, and phytochemical accumulation in basil plants influenced by substituting green light for partial red and/or blue light. Hort Science 54(10):1769–1776. https://doi.org/10.21273/HORTSCI14282-19
Dou R et al (2020b) Contamination of pyrethroids and atrazine in greenhouse and open-field agricultural soils in China. Sci Total Environ 701:134916. https://doi.org/10.1016/j.scitotenv.2019.134916
EIP-AGRI (2020) Plant-based medicinal and cosmetic group. EIP-AGRI, Agriculture and Innovation. Available at: https://ec.europa.eu/eip/agriculture/en/publications/eip-agri-focus-group-plant-based-medicinal-and

FranceAgriMer (2020) L'estimation des externalités produites par la filière des PPAM sur des territoires donnés- juin 2020, November 2020. Available at: https://www.franceagrimer.fr/fam/content/download/65461/document/EtudeExternalitesPPAM_juin 2020_Com.pdf?version=2

FranceAgriMer (2021) Marchés des plantes à parfum, aromatiques et médicinales – Panorama 2020, November 2021. Available at: https://www.franceagrimer.fr/content/download/67749/document/20211212_MARCHE_PPAM_2020.pdf. Accessed 2 January 2023

Fukuyama T et al (2017) Effects of ultraviolet a supplemented with red light irradiation on vinblastine production in catharanthus roseus. Environ Control Biol 55(2):65–69. https://doi.org/10.2525/ecb.55.65

Fukuyama T, Ohashi-Kaneko K, Watanabe H (2015) Estimation of optimal red light intensity for production of the pharmaceutical drug components, vindoline and catharanthine, contained in Catharanthus roseus (L.) G. Don. Environ Control Biol 53(4):217–220. https://doi.org/10.2525/ecb.53.217

Gebrehiwot K (2022) In: Jhariya MK et al (eds) Chapter 3 – soil management for food security. Elsevier, pp 61–71. https://doi.org/10.1016/B978-0-12-822976-7.00029-6

Goto E (2012) Plant production in a closed plant factory with artificial lighting. Acta Hortic:37–49. https://doi.org/10.17660/ActaHortic.2012.956.2

Graamans L et al (2017) Plant factories; crop transpiration and energy balance. Agr Syst 153:138–147. https://doi.org/10.1016/j.agsy.2017.01.003

Graamans L et al (2018) Plant factories versus greenhouses: comparison of resource use efficiency. Agr Syst 160:31–43. https://doi.org/10.1016/j.agsy.2017.11.003

Gulmon SL, Mooney HA (1986) Costs of defense and their effects on plant productivity. On the economy of plant form and function: proceedings of the Sixth Maria Moors Cabot Symposium, Evolutionary Constraints on Primary Productivity, Adaptive Patterns of Energy Capture in Plants, Harvard Forest, August 1983

Hao X et al (2020) An optimization model of light intensity and nitrogen concentration coupled with yield and quality. Plant Growth Regul 92(2):319–331. https://doi.org/10.1007/s10725-020-00641-0

Hasanuzzaman M, Nahar K, Fujita M (2013) Extreme temperature responses, oxidative stress and antioxidant defense in plants. In: Vahdati K, Leslie C (eds) Abiotic stress. IntechOpen, Rijeka. https://doi.org/10.5772/54833

Hassan W et al (2022) Improved and sustainable agroecosystem, food security and environmental resilience through zero tillage with emphasis on soils of temperate and subtropical climate regions: a review. Int Soil Water Conserv Res 10(3):530–545. https://doi.org/10.1016/j.iswcr.2022.01.005

Hohmann J et al (2000) Diterpenoids from Euphorbia peplus. Planta Med 66(3):291–294. https://doi.org/10.1055/s-2000-8568

Howes M-JR et al (2020) Molecules from nature: reconciling biodiversity conservation and global healthcare imperatives for sustainable use of medicinal plants and fungi. Plants, People, Planet. John Wiley & Sons, Ltd 2(5):463–481. https://doi.org/10.1002/ppp3.10138

Hussein RA, El-Anssary AA (2018) In: Builders PF (ed) Plants secondary metabolites: the key drivers of the pharmacological actions of medicinal plants. IntechOpen, Rijeka, p Ch. 2. https://doi.org/10.5772/intechopen.76139

Hvězdová M et al (2018) Currently and recently used pesticides in central European arable soils. Sci Total Environ 613–614:361–370. https://doi.org/10.1016/j.scitotenv.2017.09.049

Ibrahim MA et al (2010) Elevation of night-time temperature increases terpenoid emissions from Betula pendula and Populus tremula. J Exp Bot 61(6):1583–1595. https://doi.org/10.1093/jxb/erq034

Ibrahim MH et al (2014) Allocation of secondary metabolites, photosynthetic capacity, and antioxidant activity of Kacip Fatimah (Labisia pumila benth) in response to CO_2 and light intensity. Sci World J 2014:1. https://doi.org/10.1155/2014/360290

Ibrahim MH, Jaafar HZE (2011a) Enhancement of leaf gas exchange and primary metabolites under carbon dioxide enrichment up-regulates the production of secondary metabo-

lites in labisia pumila seedlings. Molecules 16(5):3761–3777. https://doi.org/10.3390/molecules16053761
Ibrahim MH, Jaafar HZE (2011b) Involvement of carbohydrate, protein and phenylanine ammonia lyase in up-regulation of secondary metabolites in labisia pumila under various CO_2 and N_2 levels. Molecules 16(5):4172–4190. https://doi.org/10.3390/molecules16054172
Ibrahim MH, Jaafar HZE (2017) Impact of carbon dioxide and increasing nitrogen levels on growth, photosynthetic capacity, carbohydrates and secondary metabolites in kacip fatimah (labisia pumila). Annual Res Rev Biol 18(5):1. https://doi.org/10.9734/ARRB/2017/36540
International Trade Center (n.d.) ITC – trade map. Available at: https://www.trademap.org/. Accessed 3 Jan 2023
Itoh H (2018) Functional plants. Plant Factory Using Artificial Light: Adapting to Environmental Disruption and Clues to Agricultural Innovations, pp 143–154. https://doi.org/10.1016/B978-0-12-813973-8.00013-0
Kalantari F et al (2018) Opportunities and challenges in sustainability of vertical farming: a review. J Landsc Ecol, Berlin: Sciendo 11(1):35–60. https://doi.org/10.1515/jlecol-2017-0016
Karimi E, Jaafar HZE, Ghasemzadeh A (2016) Chemical composition, antioxidant and anticancer potential of Labisia pumila variety alata under CO_2 enrichment. NJAS – Wagening J Life Sci 78:85–91. https://doi.org/10.1016/j.njas.2016.05.002
Karimi M, Ahmadi N, Ebrahimi M (2022) Photoreceptor regulation of Hypericum perforatum L. (cv. Topas) flowering under different light spectrums in the controlled environment system. Environ Exp Bot 196:104797. https://doi.org/10.1016/j.envexpbot.2022.104797
Kim G-S et al (2012) Effects of natural bioactive products on the growth and Ginsenoside contents of Panax ginseng cultured in an aeroponic system. J Ginseng Res 36(4):430–441. https://doi.org/10.5142/jgr.2012.36.4.430
Kim H-H et al (2005) Light-emitting diodes as an illumination source for plants: a review of research at Kennedy Space Center. Habitation 10(2):71–78. https://doi.org/10.3727/154296605774791232
Kim W-S et al (2015) Anti-inflammatory, antioxidant and antimicrobial effects of artemisinin extracts from Artemisia annua L. Korean J Physiol Pharmacol 19(1):21–27. https://doi.org/10.4196/kjpp.2015.19.1.21
Kivimäenpää M et al (2021) Changes in light spectra modify secondary compound concentrations and BVOC emissions of Norway spruce seedlings. Can J For Res 51(9):1218–1229. https://doi.org/10.1139/cjfr-2020-0120
Kolega S et al (2020) Nutraceutical profiles of two hydroponically grown sweet basil cultivars as affected by the composition of the nutrient solution and the inoculation with Azospirillum brasilense. Front Plant Sci 11. https://doi.org/10.3389/fpls.2020.596000
Kondrat'Ev VM et al (2021) Influence of LED lighting power on basil (Ocimum basilicum L.). IOP Conference Series: Earth and Environmental Science. https://doi.org/10.1088/1755-1315/723/3/032079
Kozai T, Niu G (2020) Chapter 1 - introduction. In: Kozai T, Niu G, Takagaki M (eds) Plant factory, 2nd edn. Academic Press, pp 3–6. https://doi.org/10.1016/B978-0-12-816691-8.00001-7
Kuo C-H et al (2020) Optimization of light intensity, temperature, and nutrients to enhance the bioactive content of Hyperforin and Rutin in St. John's Wort. Molecules 25(18). https://doi.org/10.3390/molecules25184256
Lam VP, Choi J, Park J (2021) Enhancing growth and glucosinolate accumulation in watercress (Nasturtium officinale l.) by regulating light intensity and photoperiod in plant factories. Agri 11(8). https://doi.org/10.3390/agriculture11080723
Lanoue J et al (2022) Continuous lighting can improve yield and reduce energy costs while increasing or maintaining nutritional contents of microgreens. Front Plant Sci 13. https://doi.org/10.3389/fpls.2022.983222
Lee J-Y, Hiyama M et al (2020a) Effects of concentration and temperature of nutrient solution on growth and camptothecin accumulation of ophiorrhiza pumila. Plan Theory 9(6):1–14. https://doi.org/10.3390/plants9060793

Lee J-Y, Shimano A et al (2020b) Effects of photosynthetic photon flux density and light period on growth and camptothecin accumulation of ophiorrhiza pumila under controlled environments. J Agric Meteorol 76(4):180–187. https://doi.org/10.2480/agrmet.D-20-00026

Li M et al (2018) Metabolomics analysis reveals that elevated atmospheric CO_2 alleviates drought stress in cucumber seedling leaves. Anal Biochem 559:71–85. https://doi.org/10.1016/j.ab.2018.08.020

Li X et al (2022) The protective effect of cold acclimation on the low temperature stress of the lotus (Nelumbo nucifera). Hortic Sci 49:29. https://doi.org/10.17221/62/2020-HORTSCI

Lu N et al (2017) Growth and accumulation of secondary metabolites in perilla as affected by photosynthetic photon flux density and electrical conductivity of the nutrient solution. Front Plant Sci 8. https://doi.org/10.3389/fpls.2017.00708

Lucini L et al (2015) The effect of a plant-derived biostimulant on metabolic profiling and crop performance of lettuce grown under saline conditions. Sci Hortic 182:124–133. https://doi.org/10.1016/j.scienta.2014.11.022

Maggini R et al (2021) Effects of NaCl on hydroponic cultivation of Reichardia picroides (L.) Roth. Agronomy 11(11). https://doi.org/10.3390/agronomy11112352

Mao H et al (2019) Both multi-segment light intensity and extended photoperiod lighting strategies, with the same daily light integral, promoted Lactuca sativa L. Growth and photosynthesis. Agronomy 9(12). https://doi.org/10.3390/agronomy9120857

Meng Q, Kelly N, Runkle ES (2019) Substituting green or far-red radiation for blue radiation induces shade avoidance and promotes growth in lettuce and kale. Environ Exp Bot 162:383–391. https://doi.org/10.1016/j.envexpbot.2019.03.016

Mesa T et al (2022) Differential physiological response to heat and cold stress of tomato plants and its implication on fruit quality. J Plant Physiol 268:153581. https://doi.org/10.1016/j.jplph.2021.153581

Meyer S et al (2013) Dramatic losses of specialist arable plants in Central Germany since the 1950s/60s – a cross-regional analysis. Divers Distrib. John Wiley & Sons, Ltd 19(9):1175–1187. https://doi.org/10.1111/ddi.12102

Michael GW, Tay FS, Then YL (2021) Development of automated monitoring system for hydroponics vertical farming. J Phys Conf Series 1844(1):12024. https://doi.org/10.1088/1742-6596/1844/1/012024. IOP Publishing

Miura K, Furumoto T (2013) Cold signaling and cold response in plants. Int J Mol Sci 14(3):5312–5337. https://doi.org/10.3390/ijms14035312

Morgan S et al (2011) The cost of drug development: a systematic review. Health Policy 100(1):4–17. https://doi.org/10.1016/j.healthpol.2010.12.002

Mosaleeyanon K et al (2005) Relationships between net photosynthetic rate and secondary metabolite contents in St. John's Wort. Plant Sci 169(3):523–531. https://doi.org/10.1016/j.plantsci.2005.05.002

Naik PM, Al-Khayri JM (2016) Impact of abiotic elicitors on in vitro production of plant secondary metabolites: a review. J Adv Res Biotechnol 1(2):1–7

Nguyen DTP et al (2020) Short-term root-zone temperature treatment enhanced the accumulation of secondary metabolites of hydroponic coriander (Coriandrum sativum L.) grown in a plant factory. Agronomy 10(3). https://doi.org/10.3390/agronomy10030413

Nguyen TKL et al (2022) Optimization of cultivation type and temperature for the production of balloon flower (Platycodon grandiflorum A. DC) sprouts in a plant factory with artificial lighting. Horticulturae 8(4). https://doi.org/10.3390/horticulturae8040315

Niessen W, Chanteux A (2005) Les tableaux de bord et business plan. Edited by Editions des Chambres de Commerce et d'Industrie de Wallonie. Edipro

OECD (2013) OECD compendium of agri-environmental indicators. https://doi.org/10.1787/9789264186217-en

Orlando M, Trivellini A, Puccinelli M et al (2022a) Increasing the functional quality of Crocus sativus L. by-product (tepals) by controlling spectral composition. Hortic Environ Biotechnol 63(3):363–373. https://doi.org/10.1007/s13580-021-00407-1

Orlando M, Trivellini A, Incrocci L et al (2022b) The inclusion of green light in a red and blue light background impact the growth and functional quality of vegetable and flower microgreen species. Horticulturae 8. https://doi.org/10.3390/horticulturae8030217

Pan S-Y et al (2014) Historical perspective of traditional indigenous medical practices: the current renaissance and conservation of herbal resources. Evid Based Complement Alternat Med 2014:525340. https://doi.org/10.1155/2014/525340

Paradiso R, Proietti S (2022) Light-quality manipulation to control plant growth and photomorphogenesis in greenhouse horticulture: the state of the art and the opportunities of modern LED systems. J Plant Growth Regul 41(2):742–780. https://doi.org/10.1007/s00344-021-10337-y

Park S-Y et al (2016) Evaluating the effects of a newly developed nutrient solution on growth, antioxidants, and chicoric acid contents in Crepidiastrum denticulatum. Hortic Environ Biotechnol 57(5):478–486. https://doi.org/10.1007/s13580-016-1060-2

Park S-Y et al (2020a) Physiologic and metabolic changes in crepidiastrum denticulatum according to different energy levels of UV-B radiation. Int J Mol Sci 21(19):1–16. https://doi.org/10.3390/ijms21197134

Park S-Y, Bae J-H, Oh M-M (2020b) Manipulating light quality to promote shoot growth and bioactive compound biosynthesis of Crepidiastrum denticulatum (Houtt.) Pak & Kawano cultivated in plant factories. J Appl Res Med Aromat Plants 16:100237. https://doi.org/10.1016/j.jarmap.2019.100237

Pennisi G, Orsini F et al (2019a) Resource use efficiency of indoor lettuce (Lactuca sativa L.) cultivation as affected by red: blue ratio provided by LED lighting. Sci Rep 9(1):1–11. https://doi.org/10.1038/s41598-019-50783-z

Pennisi G, Blasioli S et al (2019b) Unraveling the Role of Red:Blue LED Lights on Resource Use Efficiency and Nutritional Properties of Indoor Grown Sweet Basil. Front Plant Sci 10. https://doi.org/10.3389/fpls.2019.00305

Pennisi G et al (2020) Optimal light intensity for sustainable water and energy use in indoor cultivation of lettuce and basil under red and blue LEDs. Sci Hortic 272:109508. https://doi.org/10.1016/j.scienta.2020.109508

Perreau C (2022) La ferme urbaine Agricool placée en redressement judiciaire. LesEchos. Available at: https://www.lesechos.fr/start-up/impact/la-ferme-urbaine-agricool-placee-en-redressement-judiciaire-1395891

Ren X et al (2022) Optimization of the yield, total phenolic content, and antioxidant capacity of basil by controlling the electrical conductivity of the nutrient solution. Horticulturae 8(3). https://doi.org/10.3390/horticulturae8030216

Rengasamy N et al (2022) Artificial lighting photoperiod manipulation approach to improve productivity and energy use efficacies of plant factory cultivated Stevia rebaudiana. Agronomy 12(8). https://doi.org/10.3390/agronomy12081787

Sankhuan D et al (2022) Variation in terpenoids in leaves of Artemisia annua grown under different LED spectra resulting in diverse antimalarial activities against plasmodium falciparum. BMC Plant Biol 22(1):128. https://doi.org/10.1186/s12870-022-03528-6

Sharma R et al (2018) 'Adaptive physiological response, carbon partitioning, and biomass production of Withania somnifera (L.) Dunal grown under elevated CO_2 regimes. Biotech 8:267. https://doi.org/10.1007/s13205-018-1292-1

Shimomura M et al (2020) Continuous blue lighting and elevated carbon dioxide concentration rapidly increase chlorogenic acid content in young lettuce plants. Sci Hortic 272:109550. https://doi.org/10.1016/j.scienta.2020.109550

Sijmonsma A (2019) Swedish vertical farming company 'Plantagon international' declares bankruptcy, Hortidaily.com. Available at: https://www.hortidaily.com/article/9075157/swedish-vertical-farming-company-plantagon-international-bankrupt/

Silberbush M, Ben-Asher J (2001) Simulation study of nutrient uptake by plants from soilless cultures as affected by salinity buildup and transpiration. Plant and Soil 233(1):59–69. https://doi.org/10.1023/A:1010382321883

Singh PA et al (2022) An overview of some important deliberations to promote medicinal plants cultivation. J Appl Res Med Aromat Plants 31:100400. https://doi.org/10.1016/j.jarmap.2022.100400

Sipos L et al (2021) Optimization of basil (Ocimum basilicum L.) production in LED light environments – a review. Sci Hortic 289:110486. https://doi.org/10.1016/j.scienta.2021.110486

Sparks RE, Stwalley RM III (2018) Design and testing of a modified hydroponic shipping container system for urban food production. Int J Appl Agric Sci 4(4):93–102. Science Publishing Group

Staudt M, Bertin N (1998) Light and temperature dependence of the emission of cyclic and acyclic monoterpenes from holm oak (Quercus ilex L.) leaves. Plant Cell Environ 21(4):385–395. https://doi.org/10.1046/j.1365-3040.1998.00288.x

Storkey J et al (2012) The impact of agricultural intensification and land-use change on the European arable flora. Proc R Soc B Biol Sci 279(1732):1421–1429. https://doi.org/10.1098/rspb.2011.1686

Szekely I, Jijakli MH (2022) Bioponics as a promising approach to sustainable agriculture: a review of the main methods for producing organic nutrient solution for hydroponics. Watermark 14. https://doi.org/10.3390/w14233975

Tao S et al (2008) Organochlorine pesticides contaminated surface soil as reemission source in the Haihe plain, China. Environ Sci Technol 42(22):8395–8400. https://doi.org/10.1021/es8019676

Tarvainen V et al (2005) Temperature and light dependence of the VOC emissions of Scots pine. Atmos Chem Phys 5(4):989–998. https://doi.org/10.5194/acp-5-989-2005

Tong A-Z et al (2021) Diversity and composition of the Panax ginseng rhizosphere microbiome in various cultivation modesand ages. BMC Microbiol 21(1):18. https://doi.org/10.1186/s12866-020-02081-2

Tsukagoshi S, Shinohara Y (2020) Chapter 14 – nutrition and nutrient uptake in soilless culture systems. In: Kozai T, Niu G, Takagaki M (eds) Plant factory, 2nd edn. Academic Press, pp 221–229. https://doi.org/10.1016/B978-0-12-816691-8.00014-5

Twaij BM, Hasan MN (2022) 'Bioactive secondary metabolites from plant sources: types, synthesis, and their therapeutic uses. Int J Plant Biol 13:4–14. https://doi.org/10.3390/ijpb13010003

Tzortzakis N et al (2022) Ammonium to total nitrogen ratio interactive effects with salinity application on Solanum lycopersicum growth, physiology, and fruit storage in a closed hydroponic system. Agronomy 12(2). https://doi.org/10.3390/agronomy12020386

Vanhaelewyn L et al (2020) Ultraviolet radiation from a plant perspective: the plant-microorganism context. Front Plant Sci 11. https://doi.org/10.3389/fpls.2020.597642

Vera M et al (2020) Living at the edge: population differentiation in endangered Arnica Montana from NW Iberian Peninsula. Plant Syst Evol 306(2). https://doi.org/10.1007/s00606-020-01673-9

VerticalFarmDaily.com (2021) Not possible to market our vertically grown vegetables in a financially attractive way, VerticalFarmDaily. Available at: https://www.verticalfarmdaily.com/article/9349889/not-possible-to-market-our-vertically-grown-vegetables-in-a-financially-attractive-way/. Accessed 31 Aug 2021

Weremczuk-Jeżyna I et al (2021) The protective function and modification of secondary metabolite accumulation in response to light stress in dracocephalum forrestii shoots. Int J Mol Sci 22(15). https://doi.org/10.3390/ijms22157965

WildMapsFit (2020) European medicinal and aromatic plant (MAP) farming. Processing and Training Alliance. Available at: https://wildmapsfit.eu/wp-content/uploads/2022/04/D2-Market-Analysis-of-Wild-MAP.pdf

World Health Assembly, 56 (2003) Traditional medicine: report by the secretariat. Geneva PP - Geneva: World Health Organization. Available at: https://apps.who.int/iris/handle/10665/78244

World Health Organization (2019) WHO global report on traditional and complementary medicine 2019. World Health Organization, Geneva PP – Geneva. Available at: https://apps.who.int/iris/handle/10665/312342

World Health Organization (2022) Universal health coverage. Available at: https://www.who.int/health-topics/universal-health-coverage#tab=tab_1. Accessed 21 Oct 2022

Xu W et al (2021) Continuous lighting and high daily light integral enhance yield and quality of mass-produced nasturtium (Tropaeolum majus l.) in plant factories. Plan Theory 10(6). https://doi.org/10.3390/plants10061203

Yamori W (2020) Chapter 12 – photosynthesis and respiration. In: Kozai T, Niu G, Takagaki M (eds) Plant factory, 2nd edn. Academic Press, pp 197–206. https://doi.org/10.1016/B978-0-12-816691-8.00012-1

Yang L et al (2018) Response of plant secondary metabolites to environmental factors. Molecules 23(4). https://doi.org/10.3390/molecules23040762

Yeo H-J et al (2021) Metabolic analysis of root, stem, and leaf of scutellaria baicalensis plantlets treated with different led lights. Plan Theory 10(5). https://doi.org/10.3390/plants10050940

Yep B, Gale NV, Zheng Y (2020) Aquaponic and hydroponic solutions modulate NaCl-induced stress in drug-type Cannabis sativa L. Front Plant Sci 11. https://doi.org/10.3389/fpls.2020.01169

Ying Q, Kong Y, Zheng Y (2020) Growth and appearance quality of four microgreen species under light-emitting diode lights with different spectral combinations. HortScience. Washington, DC: American Society for Horticultural Science 55(9):1399–1405. https://doi.org/10.21273/HORTSCI14925-20

Yoshida H et al (2022) Changes in the levels of bioactive compounds in red perilla at various leaf positions and ages under different photosynthetic photon flux densities. Acta Hortic:179–186. https://doi.org/10.17660/ActaHortic.2022.1337.24

Zeidler C, Schubert D, Vrakking V (2017) Vertical farm 2.0: designing an economically feasible vertical farm – a combined European endeavor for sustainable urban agriculture

Zhang M et al (2020) Effects of nutrient fertility on growth and alkaloidal content in Mitragyna speciosa (Kratom). Front Plant Sci 11. https://doi.org/10.3389/fpls.2020.597696

Zhang T, Folta KM (2012) Green light signaling and adaptive response. Plant Signal Behav. Taylor & Francis 7(1):75–78. https://doi.org/10.4161/psb.7.1.18635

Zhang X et al (2018) Effects of environment lighting on the growth, photosynthesis, and quality of hydroponic lettuce in a plant factory. Int J Agric Biol Eng 11(2):33–40

Zhao J et al (2020) Plant responses to heat stress: physiology, transcription, noncoding RNAs, and epigenetics. Int J Mol Sci 22(1). https://doi.org/10.3390/ijms22010117

Zhao X et al (2022) Differential physiological, transcriptomic, and metabolomic responses of Paspalum wettsteinii under high-temperature stress. Front Plant Sci 13. https://doi.org/10.3389/fpls.2022.865608

Zhou Y et al (2018) Effects of salt stress on plant growth, antioxidant capacity, glandular trichome density, and volatile exudates of schizonepeta tenuifolia briq. Int J Mol Sci 19(1). https://doi.org/10.3390/ijms19010252

Zobayed S (2020) Chapter 17 – medicinal components. In: Kozai T, Niu G, Takagaki M (eds) Plant factory, 2nd edn. Academic Press, pp 245–250. https://doi.org/10.1016/B978-0-12-816691-8.00017-0

Zobayed SMA, Afreen F, Kozai T (2005) Temperature stress can alter the photosynthetic efficiency and secondary metabolite concentrations in St. John's wort. Plant Physiol Biochem 43(10–11):977–984. https://doi.org/10.1016/j.plaphy.2005.07.013

Vertical Farms for Future Cities

Kheir Al-Kodmany

Abstract The Food and Agriculture Organization of the United Nations forecasts that by 2050 the global population will grow by nearly 2 billion persons. Consequently, we must sustainably produce 70% more food (United Nations, Department of Economic and Social Affairs. https://www.un.org/development/desa/en/news/population/2018-revision-of-world-urbanization-prospects.html). However, water supply and arable lands are shrinking. In recent years, the impacts of the vicious pandemic and climate change manifested by weather extremes have hurt agriculture and the entire food production systems. Further, "food miles," referring to the distance that food travels from the place of production to the plate, is becoming an alarming problem. This chapter examines the potential of the vertical farm (VF) to support food security. It also discusses the challenges it faces.

Keywords Food production · Carbon emissions · Climate change · Water resources · Food quality · Crop yields · Space efficiency

1 Introduction

1.1 Goals and Scope of the Study

The goal of this chapter is to enlighten about recent developments in VF. It attempts to answer basic questions, including:

- What is a VF?
- Why should we integrate VF into our cities?
- What are the VF methods and technologies?

K. Al-Kodmany (✉)
Department of Urban Planning and Policy (UPP), College of Urban Planning and Public Affairs (CUPPA), the University of Illinois at Chicago, Chicago, IL, USA
e-mail: kheir@uic.edu

P. M. Priyadarshan et al. (eds.), *Digital Agriculture*,
https://doi.org/10.1007/978-3-031-43548-5_6

- What are the salient VF projects?
- What are the VF implications for future cities?

1.2 What Is a VF?

The VF engages the vertical plane in growing plants and vegetation to optimize food production in a limited indoor space. Like libraries that stack books on shelves instead of spreading them on floors to save space, the VF does the same for agriculture. It stacks growing beds along tall technology-supported structures instead of spreading them over the ground, reaching maximum compactness and reducing footprint (Kah et al. 2019; Armanda et al. 2019). The VF utilizes specialized cultivation methods (hydroponics, aeroponics, and aquaponics) and advanced technologies (artificial intelligence, LEDs, and robots) to enhance the cultivation environment, improve food quality, and increase yields. It is suited to producing leafy and microgreens because they feature a high harvest index, fast growth rate, low photosynthetic energy demand, and compact shape. New VFs have demonstrated staggering capacities of growing thousands of crops in just a few hundred square feet (Al-Kodmany 2018). They occur in new or retrofitted buildings of various sizes and heights. Therefore, vertical farming is an environmentally friendly method to produce quality food with less space by engaging technology and the vertical dimension.

1.3 Why VF?

1.3.1 Food Security

Food insecurity is becoming an acute problem. Over the coming decades, an expanding global population, a changing climate, environmental stressors, and rising food costs will substantially impact food security. While the increased urban population is placing a great demand on food, agronomists, ecologists, and geologists warn of soaring shortages of cropland. Indeed, the sprawling fringes of suburban developments continue taking over more farmland. Creative solutions and urban policies are urgently needed, including options for water conservation, land use efficiencies, and food production. Simply, as the food demand will be greater than the supply, our planet is growing hungrier for solutions. The VF offers a creative solution that merges food production and consumption in the same place to produce fresh food locally while reducing transportation and saving the environment (Armanda et al. 2019; Al-Kodmany 2018; Edmondson et al. 2020).

1.3.2 Climate Change

Climate change is a severe threat to food security. It has already decreased arable land. Manifesting in horrific events, such as storms, flooding, hurricane, and drought, it has damaged valuable agricultural production (Okeke et al. 2022). For instance, the 2011 drought in the USA damaged grain crops with a value estimated at $110 billion (Al-Kodmany 2018; Edmondson et al. 2020). Similarly, heat waves in California have resulted in significant loss of crops. Further, traditional farming demands enormous fossil fuels to conduct agricultural activities. The travel distances of food from production (farms) to consumption (cities) or the “food miles” have increased significantly. On average, food travels 1500 miles from the farm field to the consumer’s plate (Okeke et al. 2022; Llorach-Massana et al. 2016). Transporting food counts for 0.4 tons of carbon dioxide emissions per household yearly (Edmondson et al. 2020). Regrettably, the increased greenhouse gas emissions from food transport and fossil fuels-based agricultural activities have exacerbated climate change (Fig. 1).

1.3.3 Urban Space and Density

Urban agriculture suffers from finding space for farming. As the urban population grows, demand for urban increases, and it becomes difficult to find land in urban areas for urban agricultural activities. Further, land prices have been increasing,

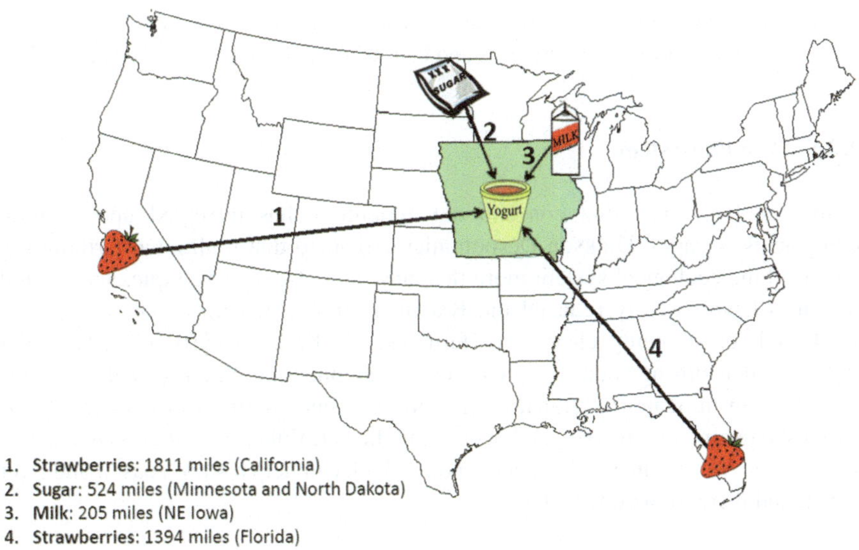

Fig. 1 Map showing the distances that the essential ingredients of a small strawberry yogurt can travel. (Adapted from Edmondson et al. 2020)

making it unfordable for farming. VF may offer a solution by maximizing agricultural work on a little lot. Harnessing the vertical dimension increases production many folds.

VF facilitates compact urban agriculture, which supports compact urban living, a core element of sustainability. VF frees land to house more urban population, services, and infrastructure. Researchers have critiqued urban agriculture for decreasing density, entailing longer commutes and travel time, and more significant fuel costs and carbon emissions. They explained that the increased gas utilization rising from moving a small percentage of farmland into urban areas would create an extra 1.77 tons of CO_2 per household yearly (Engler and Krarti 2021).

1.3.4 Human and Environmental Health

Traditional farming inflicts harm on human health and the natural environment. The World Health Organization explains that 50% of the world's farms use raw animal waste as fertilizer, which may contain diseases transmitted to crops. Traditional farms use pesticides and herbicides, which create polluting agricultural runoff. They cause erosion, contaminate soil, and generate excessive wastewater. When leftover fertilizer washes into water bodies (e.g., oceans, rivers, streams), a high concentration of nutrients is developed (called eutrophication), which could disturb the ecological equilibrium. Further, traditional farming uses far more water than high-tech VF, about one-tenth of that used in conventional agriculture, by offering precision irrigation and efficient scheduling. Agricultural activities use excessive freshwater – in most regions of the world, over 70% of freshwater is used for agriculture – competing with urban areas. The water crisis may worsen as climate change triggers warmer temperatures and causes more droughts (Okeke et al. 2022).

1.3.5 The Ecosystem

Some scholars argue that conventional agriculture has infringed upon natural ecosystems for ages. Dickson Despommier explained that traditional farming has damaged the ecological system more than anything else. For example, agricultural activities have severely reduced the Brazilian rainforest, with about two million hardwood forests being cleared for farmland (Al-Kodmany 2018). Despommier indicated that infringement on ecosystems is augmenting climate change. In this way, VF can mitigate traditional agricultural influence on the world's ecosystems by reestablishing biodiversity and decreasing the harmful effects of climate change. Further, the VF eliminates fertilizer runoff, which can help restore coastal and river water, and increase wild fish stock.

1.3.6 Economics

Proponents of the VF argue that as technologies improve, its food prices will drop. Indeed, new VFs are embracing sophisticated technologies, automated systems, robots, artificial intelligence (AI), and advanced data models to offer competitive prices. Advanced VF will generate greater yields many folds, making it affordable to larger populations. Simultaneously, the soaring expenses of conventional farming rapidly reduce the cost gap. For example, when VFs are placed strategically in urban areas, they will sell products directly to the consumer, decreasing transportation costs and eliminating the middleman. In addition, VF can generate local employment and support the local economy. Abandoned urban buildings and disused warehouses can be converted into VFs to supply healthy food in neighborhoods where fresh produce is scarce.

2 VF Methods

Researchers have been advancing environmentally friendly methods of food production. The following section highlights three main VF methods: hydroponics, aeroponics, and aquaponics.

2.1 Hydroponics

Hydroponics is a method of growing plants in water containing nutrients without soil. The term stems from the Greek words *hydro* and *ponos*, meaning "water doing labor" or "water works." The hydroponics technique involves planting a seed in a tiny cub of sponge, and when the delicate roots poke after a week, it is transplanted into water-filled tanks containing a nutritious liquid with chemical fertilizers. Besides, oxygen and sunshine (or artificial light) are the only ingredients needed. The soilless hydroponics method can eliminate soil-related cultivation problems, such as bacteria that grow in soil, fungus, and insects. It is also low maintenance since it disengages weeding, tilling, kneeling, and dirt removal. The hydroponic method is less labor-intensive because it involves less space (Engler and Krarti 2021). It could also be cleaner than traditional methods, for it does not contain animal excreta. Furthermore, it offers an easier way to control nutrient levels, pH balance, oxygen level, moisture, and microorganisms. Therefore, the hydroponic method may result in higher-quality crops (Engler and Krarti 2021).

2.2 Aeroponics

Aeroponics is a soilless method that relies on air to deliver a high-pressured, nutrient-rich mist to the plant's roots, which are suspended in the air. Aeroponics means "working air" and stems from the Greek words for air, "aer," and labor, "ponos." Therefore, aeroponics builds off that of hydroponic systems, in which exposed roots are held in a soilless growing medium. However, aeroponics does not require containers or grow trays to hold water because it uses nutritious mist instead of water. Further, the "misted" system delivers extra oxygen to roots, resulting in faster growth. Like hydroponics, the aeroponics method eliminates soil-related cultivation problems and is free of fertilizers or pesticides (Engler and Krarti 2021; Khan et al. 2020). Also, aeroponics does not need hydroponic tanks and uses much less water than hydroponics. Since it uses a minimal amount of water (95% less water than conventional farming), it is an efficient way of growing plants. Overall, the aeroponic method substantially saves water and space, making it superior to traditional farming practices.

2.3 Aquaponics

Aquaponics is a farming method that integrates an aquatic environment (where aquatic animals like snails and fishes live) into a hydroponics environment where plants grow. The combined system achieves symbiosis by using the nutrient-rich waste from fish tanks as a fertilizer for the hydroponic production beds. Interestingly, while the plant roots filter the water for the fish, the fish provides fertilizer for the plants. As such, the hydroponic beds act as biofilters that remove acids, gases, and chemicals, such as phosphates, nitrates, and ammonia, from the water. Concurrently, the gravel beds provide habitats for nitrifying bacteria, augment nutrient cycling, and filter water. Consequently, the freshly cleansed water is recirculated into the fish tanks (Fig. 2). As such, aquaponics reduces or eliminates the need for chemicals and artificial fertilizers. It also offers two unique products: fresh vegetables and fish simultaneously (Benis and Ferrão 2018; Khot and Mueller 2019; Sipos et al. 2020).

Table 1 summarizes the aforementioned three methods.

3 Vertical Farm Projects

Vertical farming is sprouting rapidly. The following section highlights a dozen projects (Armanda et al. 2019; Sipos et al. 2020; Angotti 2015; Abbasi et al. 2022) in different parts of the world.

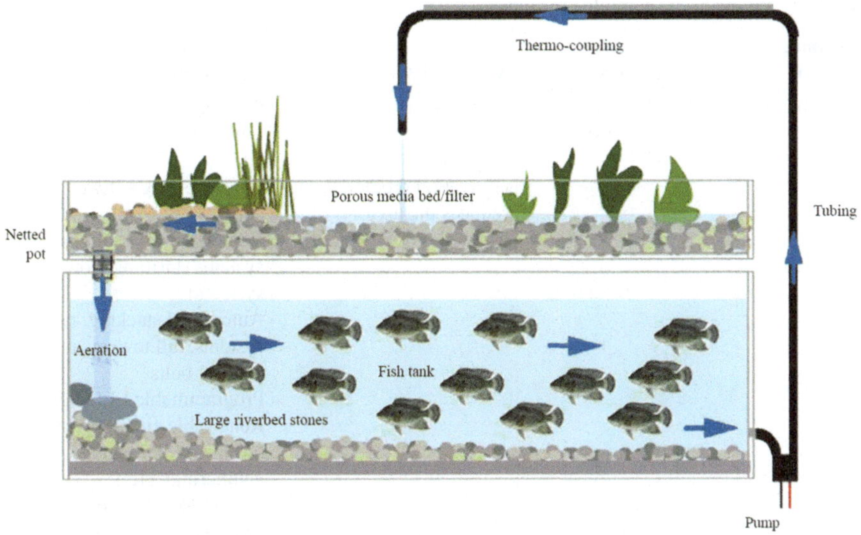

Fig. 2 Aquaponic method. (Adapted from Engler and Krarti 2021)

3.1 Sky Greens, Singapore

Singapore is a crowded small island with over five million inhabitants. With only 250 acres of farmland, it generates only 7% of its food need. The remaining need is provided by importing food, ensuring high transportation costs. Consequently, Singapore has pioneered VF. One of its first commercial VF projects is Sky Greens. The ten-year-old project is a three-story building that contains translucent greenhouses to grow tropical leafy vegetables (e.g., Chinese cabbage, lettuce, xiao bai cai, spinach cai, bayam, cai xin, gai lan, kangkong, and nai bai) with a rate of ½ ton of fresh veggies daily. It uses various growing media, including soil-based and soilless hydroponics. Sky Gardens produce high-quality fresh food at competitive prices. In addition, it offers educational programs to expose students and residents to VF (Al-Kodmany 2018).

3.2 Green Spirit Farms, New Buffalo, Michigan, USA

Green Spirit Farms (GSF) company started with a modest building of about 3716 m^2 (40,000 ft^2). It aimed at providing nongenetically modified organism (GMO) foods of greater demand (e.g., Brussel sprouts, lettuce, kale, arugula, peppers, basil, spinach, tomatoes, stevia, strawberries) at reasonable prices. The company has grown and opened VFs in Philadelphia, East Benton (Pennsylvania), Atlanta, the

Table 1 VF methods (compiled by author)

Farming method	Key characteristics	Major benefits	Common/applicable technologies
Hydroponics	Soilless-based, uses water as the growing medium	Fosters quick plant growth; decreases even eliminates soil-related cultivation problems; reduces the use of pesticides or fertilizers.	Computerized systems Laptops, cell phones, and tablets Food growing software and apps Remote control software and systems (farming-from-afar systems) Automated stacking, racking systems, tall towers, and moving belts Programmable LED lighting Renewable energy (wind turbines, solar panels, geothermal, etc.) Closed-loop systems, anaerobic digesters Programmable nutrient systems Water recirculating and recycling systems Climate control, HVAC systems Insect-killing systems Robots Rainwater collectors
Aeroponics	A variant of hydroponics involves spraying plant's roots with mist or nutrient solutions.	In addition to the benefits mentioned above, aeroponics requires less water.	
Aquaponics	It integrates aquaculture (fish farming) with hydroponics.	It creates symbiotic relationships between the plants and the fish by using the nutrient-rich waste from fish tanks to "fertigate" hydroponics production beds. The hydroponic bed cleans water for fish habitat.	

UK, and Canada. The East Benton is an extensive VF that contains 1715 vertical growing stations. It produces leafy vegetables, herbs, tomatoes, and peppers, the equivalent of 200 acres of farmland yearly (Al-Kodmany 2018).

3.3 *FarmedHere, Illinois, USA*

Founded in 2011, FarmedHere is a company that has three locations in Illinois: Englewood, Flanagan, and Bedford Park. Given the generational demands for healthy and organic foods, the company has flourished, supplying 6% or more of the Chicagoland's demand for premium green and culinary herbs. The company's product is spreading in several grocery stores, including The Green Grocer, Whole Foods Market, Mariano's Fresh Market, Trader Joe's, and Meijer. Hyped as one of the largest VF in America, Bedford Park's VF is about 8361 m2 (90,000 ft^2), followed by Flanagan (929 m^2 (10,000 ft^2)) and Englewood (371 m^2 (4000 ft^2)). Bedford Park VF uses aquaponics and aeroponics systems and produces about 136,078 kg (300,000 lb) of $453{,}59^2$ kg of chemical, herbicide, and pesticide-free leafy greens yearly (Khot and Mueller 2019; Sipos et al. 2020).

3.4 *The Plant, Chicago, Illinois, USA*

The four-story, 8686 m^2 (93,500 ft^2) VF is a retrofitted warehouse. Aiming for zero energy, it uses an anaerobic digester that converts food waste into biogas that powers, heats, and cools the facility. Daily, the anaerobic digester catches the methane from tons of food waste and burns it to produce electricity and heat (Orsini et al. 2020). Completed in 2016, The Plant uses the facility as a food business incubator, research lab, and educational facility. It produces greens, mushrooms, and kombucha tea. The Plant VF closed-loop system works as follows. The anaerobic digester turns organic materials into biogas, which is channeled into a turbine generator that generates power. Kombucha tea brewery makes CO_2 to the plants, while plants make oxygen to the kombucha tea brewery. Plants clean the water for the fish, while fish waste functions as fertilizer for plants. Sludge generated by the digester becomes algae duckweed that feeds the fish. Further, the turbine makes steam piped to the commercial kitchen, brewery, and entire building for heating and cooling (Al-Kodmany 2018). Notably, the kitchen generates kombucha tea, fish, fresh vegetables, food, and beer with no waste (Fig. 3).

3.5 *Green Girls, Memphis, Tennessee, USA*

Green Girls VF supplies local restaurants with year-round fresh, healthy food. The 60,000-ft^2 facility responds to restaurants' desire for microgreens that give meals intense flavor, texture, vivid color, and pizzazz. The goal of Green Girls is to make microgreens affordable, given their high market prices. The facility uses an automated hydroponics system, reducing laborers to only two. The system is efficient in using water; it uses one-tenth of what conventional farming uses. The

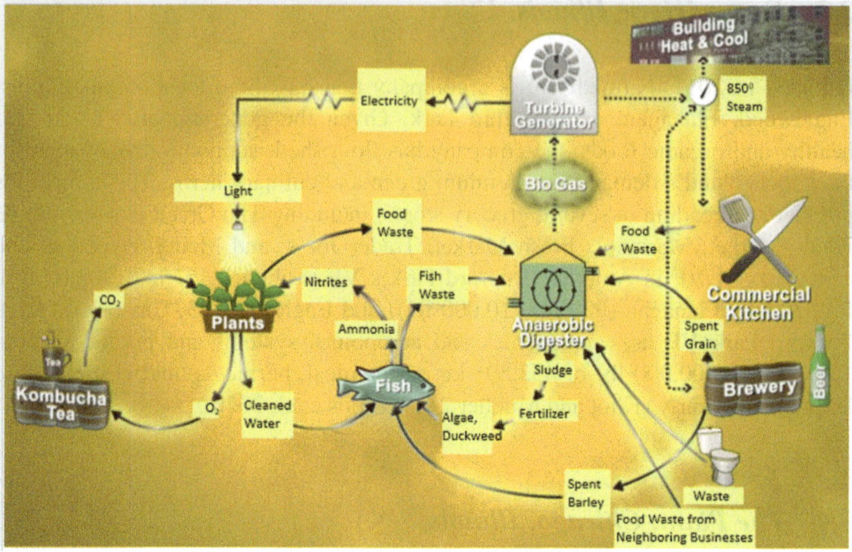

Fig. 3 The plant's anaerobic digester system. (Al-Kodmany 2018)

facility also uses LED lighting (Armanda et al. 2019; Engler and Krarti 2021). LED reduces the light's seven waves to the essential two lights (red and blue) for photosynthesis, which entail energy saving (Al-Kodmany 2018).

3.6 Gotham Greens, Brooklyn, New York, USA

Gotham Greens is a 1394 m^2 facility that sits atop a two-story building. Constructed in 2011, it uses controlled-environment agriculture (CEA) that enables high efficiency, with a rate of eight times of a traditional farm of the same size. Gotham Greens grows 80–100 tons of lettuce, salad greens, and herbs. It uses thermal insulation, double-glazing, natural ventilation, high-efficiency pumps and fans, and on-site solar photovoltaics to reduce energy consumption. Its hydroponic system also uses water efficiently (Al-Kodmany 2018; Llorach-Massana et al. 2016; Abbasi et al. 2022).

3.7 China National Cereals, Oils and Foodstuffs Corporation, Beijing, China

Completed in 2015 and with an area of 80,000 m^2, it is one of the largest VF in China. It features advanced hydroponic systems, temperature control, and artificial lighting, and produces fresh, pesticide-free food at affordable prices. The VF idea is suitable for China, which faces rapid urbanization. It is expected that by 2035 more than one billion people will be living in urbanized areas (Abbasi et al. 2022).

3.8 Vertical Urban Farm, Romainville, France

It is a seven-story VF building made mainly from sustainable materials such as wood. It uses natural light solely, saving energy. In addition to commercial purposes, it is an educational facility that educates residents and students about vertical farming. The building's ground floor contains a restaurant, shops, and a community garden (Abbasi et al. 2022).

3.9 Pasona Headquarters, Tokyo, Japan

Located in Tokyo, Japan, and designed by Kono Designs, Pasona Headquarters is a nine-story building that refurbished a 50-year-old building. The project was completed in 2010. The building integrates a rooftop garden and urban farming facilities that allow employees to grow and harvest their food at work. Interior spaces contain plants, fruits, vegetables, and rice. Interior partitions integrate lemon and passion fruit trees, tomato vines dangle from the ceiling; and beans sprout under benches. The building has a double-skin green facade with flowers and orange trees planted on small balconies. Outside, the office block is draped in green foliage. Ducts, pipes, and vertical shafts were relocated to the building's perimeter to increase the ceiling's height and to accommodate a climate control system that monitors humidity, temperature, and airflow to ensure the comfort and health of employees and greeneries (Armanda et al. 2019; Al-Kodmany 2018; Engler and Krarti 2021).

3.10 Kameoka Plant, Kameoka, Kyoto, Japan

Spread Company (one of Japan's largest vertical farming companies) established the Kameoka Plant in 2007. It is a 2787 m^2 (30,000 ft^2) hydroponic indoor environment with 5295 m^2 (57,000 ft^2) of vertical grow space that produces a variety of lettuces safe from the nearby Fukushima nuclear plant. It is a nonautomated vertical farm that can deliver 21,000 heads of lettuce daily. This large-scale operation brought the yield rate to 97%, and the facility became profitable in 2013. Lately, the company has upgraded the facility by adding a highlight-efficient water filtering system and an environmental control system that monitors the temperature, humidity, CO_2 levels, and light sources. Spread also plans to make tasks like raising seedlings, replanting, and harvesting accomplished by machines and artificial intelligence (Monteagudo et al. 2020; Orsini et al. 2020).

3.11 Techno Farm Keihanna, Keihanna, Kyoto, Japan

Spread Company completed this project in 2018 and is considered one of its most advanced facilities. Located in Keihanna Technopolis, it is one of the world's most automated vertical farms and utilizes the next-generation food production system Techno Farm™. Its automated cultivation system can produce 30,000 heads of lettuce daily and makes four kinds of leaf lettuce without pesticides. Inside the building, vegetation trays are stacked one after another, and a robotic arm performs planting. The irrigation and harvesting of this "AI farm" are almost entirely handled by robotic arms. White and purple specialized LED lights alternate to assist batches of crops in completing photosynthesis without interruption for 24 h. The facility recycles 90% of its water. With increased automation, it cuts down 50% of labor costs. Spread has incorporated more rigid standards for hygiene control of the cultivation environment and aims to gain the international certification of food safety standard "FSSC22000" (Al-Kodmany 2018).

3.12 PlantLab, Den Bosch, Holland, The Netherlands

PlantLab is a Dutch indoor farming pioneer. In 2010, it completed its earliest facilities in Holland. It is a three-story underground vertical farm. It uses advanced LED technology that calibrates light composition and intensity to precise needs, while eliminating the sunlight wavelengths that prevent plant growth. The farm features an automated system that monitors and controls several variables, including light intensity, light color, irrigation, nutritional value, humidity, CO_2, air velocity, and air temperature. The high-tech farm reduces water use by 90% and produces a yield three times the amount of the average greenhouse. In 2020, PlantLab opened a new VF in Indianapolis, Indiana (Al-Kodmany 2018; Benis and Ferrão 2018; Abbasi et al. 2022).

4 Discussion

4.1 VF Benefits

The VF has the potential to support food security in our cities. It offers a sustainable, safe food source. The VF is needed as the urban population increases, and we continue to face food shortages, increases in transportation costs, and climate change. The increasing fuel costs, water shortages, and shrinking arable land make a case for the VF. The hydroponic and aeroponic methods are very efficient in using water as their irrigation systems target the plant roots, and the controlled environment reduces evaporation. Some VFs even collect and recycle the water condensed within the controlled environment. The VF may also use recycling wastewater systems (grey or black) and harness rainwater. This closed-cycle approach decreases water consumption by 90–98%. Further, it has the added advantage of preventing nutrients and fertilizers from harming the land or being washed in rivers and streams (Pasha and Akash 2020).

Overall, the VF can offer a sustained food production paradigm that supplies crops year-round without interruption caused by climate change, seasons, or adverse natural events (e.g., floods, drought, hurricanes). Crop production is protected from seasonal weather patterns that are highly vulnerable to disruption due to our challenging climate. Countries facing extreme climatic and agricultural conditions may find the VF a helpful solution. For example, some Middle Eastern countries (e.g., United Arab Emirates, Saudi Arabia, Kuwait, Oman, Qatar, and Bahrain) face three significant challenges to traditional agriculture, including hot climate, water scarcity, and infertile soil. Similarly, North European countries (e.g., Denmark, Finland, Ireland, Norway, Sweden, Iceland, and the United Kingdom) face challenges of little sunlight and freezing temperatures that damage crops. In a VF, temperature, water, and lighting can be enhanced to eliminate climatic risks and improve production rates. Also, the soil is not an issue because it is not the prime cultivation medium (Benis and Ferrão 2018; Walker and Buhler 2020).

The VF could be useful in countries that import a significant portion of their foods (some of the Middle Eastern and North European countries mentioned earlier). For example, recently, Dubai opened Emirates Crop One. With over 330,000 ft^2 and the capacity to produce two million pounds of leafy greens annually, it is one of the world's largest VFs. The facility is located near Al Maktoum International Airport at Dubai World Central and its major clients are airlines (Hall 2020). On their flights, passengers will eat leafy greens, including arugula, lettuce, spinach, and mixed salad greens. This facility is Crop One's second VF after the one in Millis, Massachusetts.

Some literature suggests that the VF can consolidate some 700 acres of farmland into a big-box retail store. We can harvest 365 days a year and shorten the growth cycle to about ten days for many of the products, which is nearly a 700 increase in yield while saving a million gallons of water weekly and using 1% of land compared to traditional farming. MIRA's facility near Tokyo can generate yields 50–100 times greater than conventional crop farms. It uses AI and an extensive vertical and automated racking system to optimize space utilization. The VF space efficiency explains why it is spreading rapidly in countries like Japan and Singapore, where land is scarce (Abeliotis et al. 2016; Duncan et al. 2016).

In addition to giving greater yields per space unit, VF features a faster production cycle. For example, the time needed to grow lettuce in a VF is about one-half of that in a traditional farm (Pasha and Akash 2020). Additionally, with an automated system, the products are cleaner than that produced by conventional farms as they are not touched by a human hand. The product is clean enough that it does not need to be washed. There are no bugs, no pesticides, and no bird waste on it.

Further, the prices of VF produce are not affected by weather conditions as in conventional farming. Grand schemes, like the one proposed by Studio NAB, could even see the vertical farming concept broadened to include fish and honey production while reconnecting consumers with the food production process and establishing sustainable jobs for the surrounding community. Today, unhealthy food dominates people's diets. On average, people consume one-third of what they need of healthy food. The VF product offers high-nutrition food (Guineé et al. 2017).

Additionally, VF's high-tech, computer-based environment can make farming fun. Hence, the practice has enticed a technology-savvy younger generation, grooming a new breed of farmers. Further, VF offers the impetus for developing innovative agricultural technologies. Finally, the VF could reconnect city dwellers with nature by engaging in farming activity. In summary, the VF supports sustainability's three pillars, social, economic, and environmental, as illustrated in Table 2.

4.2 Challenges

With benefits come some challenges of the VF. Constructing VFs continues to be more expensive than building outdoor farms. The production costs have been rendered to be high due to high power consumption, expensive technology, and unaffordable startup costs. Replacing sunlight with artificial ones continues to require substantial power. Energy prices have been increasing. For example, recently, energy prices in the EU increased by nearly 58%. Two years ago, European VF spent around 25% of their operational costs on power, but that has increased to 40% (Trouwborst et al. 2016).

Also, with high power consumption, the VF may entail high carbon emissions, increasing its footprint. As such, the claim of reducing carbon emission via reducing food miles is offset by the high carbon emission resulting from the utility of lots of power. However, some VFs have been attempting to use renewable energy, such as solar power, to reduce reliance on fossil fuel-generated power. Future LED lighting will further decrease power use (Hall 2020).

Another challenge concerns finding employees with proper education, skills, and expertise. This problem may ease as educational systems adapt to new needs and demands. Finally, the cost and availability of land for vertical farming in cities can prove challenging. In response, many VFs find their homes in repurposed shipping containers, former factories, and disused warehouses (Orsini et al. 2020).

Further, most VFs grow leafy salad vegetables, e.g., shoots, herbs, and microgreens, because they produce fast under LEDs and have a brief shelf life and extra price point. However, with recent inflation, consumers might skip pricey VF herbs for cheaper choices. The inflation issue has been manifesting in the European food

Table 2 VF supports sustainability's three pillars, including social, economic, and environmental (compiled by the author)

#	Benefit	Environmental	Social	Economic
1	Decreasing food miles (travel distances)	Decreasing air pollution	Enhancing air quality, which improves the environment and people's health People receive "fresher" local food	Decrease energy consumption, packaging, and fuel to transport food
2	By using high-tech irrigation methods and recycling systems, VF reduces water consumption for food production	Reducing surface water runoff of traditional farms	Making potable water available to more people	Reduce costs
3	Recycling organic waste	Save the environment by reducing needed landfills	Improve food quality and, subsequently, consumers' health	Turn waste into an asset
4	Generating local jobs	Employees will work nearby, decreasing their travel and ecological footprint	Create a local community of workers and connections with farmers	Support the domestic economy and local employment
5	Reducing the use of fertilizers, herbicides, and pesticides	Improve the environmental well-being	Improve food quality and, subsequently, consumers' health	Minimize costs
6	Improve productivity	Needs less space	Reduce laborious work, and save time to do productive and socially rewarding activities	Offer greater yields
7	Avoid crop losses due to floods, droughts, hurricanes, overexposure to the sun, and inclement weather	Reduce environmental damage and required cleanups of farms after damage	Improve food security	Avoiding economic loss
8	Control product/produce regardless of seasons	Produce food regardless to season	Increase accessibility year-round and improve response to population demand	Fuel economic activities year-round
9	Using renewable energy	Reducing fossil fuel	Improve air quality	Reduce costs
10	Bringing nature closer to the city	Increase biodiversity	Enhance the health and psychological well-being	Generate local jobs

(continued)

Table 2 (continued)

#	Benefit	Environmental	Social	Economic
11	Promoting science and green technology	Green technology reduces harm to the urban and natural environments	Encourage seeking higher education and modern skills	Offers new jobs in bioengineering, biochemistry, biotechnology, construction, and research and development
12	Decreasing traditional farming activities and practices	Preserving the natural ecological system	Improve the health of citizens	Saving money required to correct environmental damage
13	Repurposing dilapidated buildings	Enhance the environment Remove eye sores and stigma from neighborhoods	Create opportunities for social interaction	Revive economy

market and the VF product may face competition from harvests that are grown in traditional farms or greenhouses (Abbasi et al. 2022; Carvalho and Folta 2014).

4.3 *Future Technologies and Data Models*

Increasingly advanced technologies are likely to make VF a more efficient method of food production. For example, LED technology has been improving while prices are dropping. Further, automation and use of robots and artificial intelligence (AI) will likely increase efficiency. Likewise, data modeling will better connect VF with the marketplace. For example, with data modeling, VF owners can accurately predict the output of each crop every day, year-round. The controlled microclimate environment and automation process help to do so. The amount of production can be scaled based on the market demand for each crop. The VF will increase production if the need increases for a particular crop. Conventionally, this has been a severe problem. Market demand may not match supply and using high-tech and data models may help solve the problem (Benis and Ferrão 2018; Forchino et al. 2017).

4.4 *Education and Consumer Behavior*

People should be educated about food systems. They should be aware of "food miles" and learn about the carbon footprint of the different foods we consume. Overall, food that travels by airplane has a much greater footprint than food that is shipped because an airplane produces more carbon emissions per pound. Research explains, "Food that flies can generate more than one hundred times the carbon emissions per kilometer of food that travels by ship." For example, if I eat an avocado flown to the UK from Mexico, its transport emissions are much higher than if I eat a banana shipped from Colombia (Forchino et al. 2017).

Furthermore, people should abandon bad habits of wasting food and overconsumption to decrease demand for food production and travel. Avoiding food from going to waste is one of the most straightforward and decisive actions to save money and lower climate change footprint by decreasing greenhouse gas (GHG) emissions and conserving natural resources. Most humans do not realize how much food they throw away daily — from uneaten remnants to ruined produce to portions of fruits and vegetables that could be consumed or repurposed. A third of all food in the United States is wasted. In 2019, the EPA estimated that 96% of households' wasted food went to landfills, combustion plants, or the sewer system. To reduce wasting food, people should be educated about the benefits of preventing wasted foods and the ways to do it by learning about shopping tips, storage ideas, cooking and preparation instructions, etc. (Kobayashi et al. 2014).

Likewise, people may develop the habits of eating food that is in season and local and reduce consumption of refrigerated food as they demand cooling, reducing a significant source of carbon emissions. Also, seasonal and local foods often taste better than imported ones.

4.5 Will the VF Help the Poor Population?

Most, if not all, VF projects are happening in well-to-do countries, while developing countries continue to suffer from maximum food insecurity. Unfortunately, developing countries lack the financial resources, technologies, and expertise to build VFs. As such, VF applications may support food security in places where they are already better off than other countries. The VF model may empower the already powerful nations and leave the poor behind. In other words, it is likely to enlarge the gap in quality of life between the poor and rich countries (Armanda et al. 2019; Al-Kodmany 2018).

5 Conclusion

Food insecurity is a rising global problem. VF bears the substantial potential to supply quality food grown in a controlled and clean environment without pesticides and with minimal water. It can provide food year-round, closer to cities' inhabitants (reducing food miles), and with marginal waste. It will become more needed as climate change predominates and available farmland per capita declines. However, VF faces challenges, mainly economics. The required construction and operational costs, primarily energy costs, are significant. Also, it continues to offer limited choices of crops and does not serve the poor population. Nevertheless, as technologies improve production and increase yields, it is hoped that the costs will drop and its products will reach a wider segment of population. While VF still represents a tiny portion of the global food production industry, its benefits to our ever-expanding population could tilt the farming landscape by 90 degrees in the future. Innovation and automation will drive down costs, and future VF advances will likely make it mainstream to feed the increasing urban population.

References

Abbasi R, Martinez P, Ahmad R (2022) An ontology model to represent aquaponics 4.0 system's knowledge. Inf Process Agric 9(4):514–532, ISSN 2214-3173. https://doi.org/10.1016/j.inpa.2021.12.001

Abeliotis K, Barla SA, Detsis V, Malindretos G (2016) Life cycle assessment of carnation production in Greece. J Clean Prod 112(2016):32–38. https://doi.org/10.1016/j.jclepro.2015.06.018

Al-Kodmany K (2018) The vertical farm: a review of developments and implications for the vertical city. Buildings 8:24. https://doi.org/10.3390/buildings8020024

Angotti T (2015) Urban agriculture: long-term strategy or impossible dream?. Lessons from Prospect farm in Brooklyn. Public Health 129:336. https://doi.org/10.1016/j.puhe.2014.12.008

Armanda DT, Guinée JB, Tukker A (2019, ISSN 2211-9124) The second green revolution: Innovative urban agriculture's contribution to food security and sustainability – a review. Glob Food Sec 22:13–24. https://doi.org/10.1016/j.gfs.2019.08.002

Benis K, Ferrão P (2018) Commercial farming within the urban built environment – taking stock of an evolving field in northern countries. Glob Food Sec 17(2018):30–37. https://doi.org/10.1016/j.gfs.2018.03.005

Carvalho SD, Folta KM (2014) Environmentally modified organisms – expanding genetic potential with light. Crit Rev Plant Sci 33:486–508. https://doi.org/10.1080/07352689.2014.929929

Duncan DW, Collins A, Fuhrman NE, Knauft DA, Berle DC (2016) The impacts of a school garden program on urban middle school youth. J Agric Educ 57:174–185. https://doi.org/10.5032/jae.2016.04174

Edmondson JL, Cunningham H, Densley Tingley DO et al (2020) The hidden potential of urban horticulture. Nat Food 1:155–159. https://doi.org/10.1038/s43016-020-0045-6

Engler N, Krarti M (2021) Review of energy efficiency in controlled environment agriculture. Renew Sustain Energy Rev 141:110786, ISSN 1364-0321. https://doi.org/10.1016/j.rser.2021.110786

Forchino AA, Lourguioui H, Brigolin D, Pastres R (2017) Aquaponics and sustainability: the comparison of two different aquaponic 2 techniques using the life cycle assessment (LCA). Aquacult Eng 77:80–88. https://doi.org/10.1016/j.aquaeng.2017.03.002

Guineé JB, Heijungs R, Vijver MG, Peijnenburg WJGM (2017) Setting the stage for debating the roles of risk assessment and life-cycle assessment of engineered nanomaterials. Nat Nanotechnol 12:727. https://doi.org/10.1038/NNANO.2017.135

Hall C (2020) Crop One, Emirates open 'world's largest vertical farm' in Dubai. TechCrunch, July 19. https://techcrunch.com/2022/07/19/crop-one-emirate-worlds-largest-vertical-farm-in-dubai/%23: :text=Crop%20One%20Holdings%20and%20Emirates%20Flight%20Catering%20announced%20this%20week,Airport%20at%20Dubai%20World%20Central

Kah M, Tufenkji N, White JC (2019) Nano-enabled strategies to enhance crop nutrition and protection. Nat Nanotechnol 14:532–540. https://doi.org/10.1038/s41565-019-0439-5

Khan MM, Akram MT, Janke R, Qadri RWK, Al-Sadi AM, Farooque AA (2020) Urban horticulture for food secure cities through and beyond COVID-19. Sustainability 12:9592. https://doi.org/10.3390/su12229592

Khot RA, Mueller F (2019) "Human-food interaction", foundations and trends®. Hum–Comput Interact 12(4):238–415. https://doi.org/10.1561/1100000074

Kobayashi M, Kanto T, Fujikawa T et al (2014) Supplemental UV radiation controls rose powdery mildew disease under the greenhouse conditions. Environ Control Biol 51:157–163. https://doi.org/10.2525/ecb.51.157

Llorach-Massana P, Peña J, Rieradevall J, Montero JI (2016) LCA & LCCA of a PCM application to control root zone temperatures of hydroponic crops in comparison with conventional root zone heating systems. Renew Energy 85:1079–1089. https://doi.org/10.1016/j.renene.2015.07.064

Monteagudo A, Kiss T, Mayer M, Casas AM, Igartua E, Karsai I (2020) Genetic diversity in developmental responses to light spectral quality in barley (Hordeum vulgare L). 20:1471–2229. https://doi.org/10.1186/s12870-020-02416-1

Okeke ES, Okoye CO, Atakpa EO, Ita RE, Nyaruaba R, Mgbechidinma CL, Akan OD (2022, ISSN 0921-3449) Microplastics in agroecosystems-impacts on ecosystem functions and food chain. Resour Conserv Recycl 177:105961. https://doi.org/10.1016/j.resconrec.2021.105961

Orsini F, Pennisi G, Zulfiqar F, Gianquinto G (2020) Sustainable use of resources in plant factories with artificial lighting (PFALs). Eur J Hortic Sci 85(16114426):297. https://doi.org/10.17660/eJHS.2020/85.5.1

Pasha N, Akash MH (2020) Impact of artificial lights in cultivation & generate electricity besides agriculture: feasibility & constraints analysis in Bangladesh. CONECCT. 978-1-7281-6828-9, 1. https://doi.org/10.1109/CONECCT50063.2020.9198365

Sipos L, Boros IF, Csambalik L, Székely G, Jung A, Balázs L (2020, ISSN 0304-4238) Horticultural lighting system optimization: a review. Sci Hortic 273:109631. https://doi.org/10.1016/j.scienta.2020.109631

Suthaparan A, Torre S, Stensvand A et al (2010) Specific light-emitting diodes can suppress sporulation of Podosphaera pannosa on greenhouse roses. Plant Discov 94:1105–1110. https://doi.org/10.1094/PDIS-94-9-1105

Trouwborst G, Hogewoning SW, van Kooten O et al (2016) Plasticity of photosynthesis after the 'red light syndrome' in cucumber. Environ Exp Bot 121:75–82. https://doi.org/10.1016/j.envexpbot.2015.05.002

United Nations, Department of Economic and Social Affairs (n.d.) https://www.un.org/development/desa/en/news/population/2018-revision-of-world-urbanization-prospects.html

Walker LP, Buhler D (2020) Catalyzing holistic agriculture innovation through industrial biotechnology. Indus Biotechnol 16(1550–9087):189. https://doi.org/10.1089/ind.2020.29222.lpw

Part II
IoT (Internet of Things) in Agriculture for Improved Farm Use Efficiency, Plant and Soil Management

Remote Sensing in Precision Agriculture

U. Surendran, K. Ch. V. Nagakumar, and Manoj P. Samuel

Abstract Agriculture plays a vital role in feeding the world's growing population despite facing challenges such as dwindling arable land, water scarcity, changing climatic conditions, and the need for sustainable resource management. To address these challenges and to optimize agricultural productivity, the integration of remote sensing technologies has emerged as a transformative approach within the realm of precision agriculture. Remote sensing, encompassing satellite imagery, drones, and ground-based sensors, provides invaluable data and insights for informed decision-making, resource allocation, and yield optimization. This chapter explores the significance of remote sensing applications in modern agriculture. Satellite imagery, acquired at various spatial and temporal scales, allows farmers, agronomists, and researchers to monitor crop health, identify areas of stress, and assess the impact of environmental factors. Drones equipped with high-resolution cameras and multispectral sensors enable localized data collection, facilitating detailed field-level analysis. Ground-based sensors complement these technologies by providing real-time data on soil moisture, nutrient levels, and weather conditions. The integration of remote sensing data with geographic information systems (GIS) and data analytics tools empowers stakeholders to make precise interventions, leading to reduced resource wastage and increased efficiency. Through the identification of variability within fields, growers can implement site-specific management strategies, tailoring irrigation, fertilization, and pest control practices to the unique needs of each area. This targeted approach not only maximizes crop yield but also minimizes the environmental impact of agricultural operations. Furthermore, remote sensing fosters early detection of disease outbreaks, pest infestations, and nutrient deficiencies. Timely interventions based on accurate and up-to-date information result in improved crop health and reduced reliance on chemical inputs. Additionally, remote sensing assists in monitoring land-use changes, assessing soil erosion, and promoting sustainable land management practices. To conclude, remote sensing applications are revolutionizing agriculture by enabling precise and data-driven

U. Surendran (✉) · K. Ch. V. Nagakumar · M. P. Samuel
CWRDM, Kozhikode, Kerala, India
e-mail: suren@cwrdm.org

P. M. Priyadarshan et al. (eds.), *Digital Agriculture*,
https://doi.org/10.1007/978-3-031-43548-5_7

decision-making. By harnessing the power of satellite imagery, drones, and ground-based sensors, the agricultural sector can achieve enhanced productivity, resource efficiency, and environmental sustainability.

Keywords Remote sensing · Precision agriculture · Satellite imagery · Drones · Sensors · GIS · Agricultural productivity · Resource efficiency · Sustainability

1 Introduction

Remote sensing, which is both an art and a science, is the process of learning about the characteristics of the Earth's surface without physically touching it (visible, infrared, and microwaves). Remote sensing systems are capable of providing consistent, synoptic, multitemporal, and multispectral coverage (spatial and temporal resolution) and play an important role in providing variety of information. Satellites, planes, drones, and other advanced aerial technologies along with use of various sensors are able to perceive the reflected energy from the surface of the Earth. The advantage of this is that the data may be collected from areas which are not accessible. The principal component of this technology is the source of energy, which helps to illuminate the target. In general, this is based on the energy emitted or reflected from the target that will be captured, processed, analyzed and then use that information for the required applications. Each target responds differently to these wavelength regions thus helping in distinguishing different features like vegetation, soil, water, and other similar features. Electromagnetic radiation (EMR) is how the energy is present.

The energy which moves in a harmonic wave pattern with the light velocity is known as electromagnetic energy. This EMR (energy) contains both magnetic and electrical field. In remote sensing, the two important characteristics used are wavelength and frequency of EMR. The length of a wave cycle, or the distance within a cycle between any two points, is referred to as the wavelength. Greek letter lambda (λ) is used to denote it. Typically, it is expressed in micrometer (mm, 10^{-6} m) or as nanometer (nm, 10^{-9} m). Frequency denotes the number of wave crests that passes at a specific point in the specific time frame. It is expressed in hertz (Hz). The EMR spectrum ranges from nanometers to kilometers. Further, these units are divided as spectral bands. A typical EMR with different regions are depicted in Fig. 1.

For more precise and accurate resource monitoring, combine remote sensing with on-the-ground observations. In a nutshell, the method of remote sensing can help in monitoring the Earth's surface features by offering cost-effective geographical, temporal, synoptic, and repetitive data on the Earth's surface (Justice et al. 2002). Based on the signal source, remote sensing is classified into two types, that is, (a) active and (b) passive. In the case of active remote sensing, the sensors will make use of their own energy for collecting the data (RADAR, LiDAR, and SONAR technologies); with respect to passive remote sensing, the sensors will use the

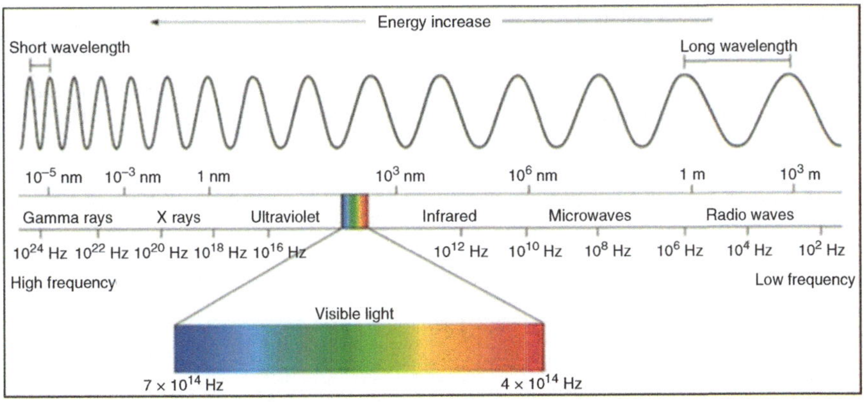

Fig. 1 Electromagnetic spectrum

energy from the sun, which is reflected. In recent decades, remote sensing is being used in a large number of field-level applications and helps to address the issues related to climate change, food security, land sustainability and other environmental issues, and many more. The extensive use and adoption of remotely sensed data and geospatial datasets is a result of GPS technologies, smartphones, and the mapping services provided by numerous mapping portals.

Three forms of resolution are utilized in remote sensing: spectral, temporal, and spatial. The term "spatial resolution" describes the region that is the measure of the smallest object or feature that the sensor can detect, or the area imaged for the instantaneous field of view (IFOV) in the area of interest (ground) by the sensor, or the area denoted by each pixel. The length of time required to collect the datasets and return back to the same area (location) is known as the temporal resolution. This is dependent in significant part on the sensor platform's orbital properties. For example, the Landsat satellite revolves the globe in every 16 days, whereas SPOT in every three days; hence, the same area can be revisited in the gap of 16 days in the case of Landsat and three days in the case of SPOT. The capacity of the sensor to specify narrower wavelength ranges is known as spectral resolution. The wavelength range for a given band is narrower as the spectral resolution is finer. For instance, the visible portion of the energy spectrum is recorded by band 1 of the Landsat TM sensor between 0.45 and 0.52 m. The terms "coarse spectral resolution" and "fine spectral resolution" describe the width of the EMF spectrum's intervals, respectively. For example, the panchromatic SPOT sensor is having a coarser spectral resolution, since it records the wavelength interval between 0.51 and 0.73 μm in EMR. The capacity of an imaging system to distinguish very minute variations in energy is known as its radiometric resolution. A sensor's sensitivity to detect minute variations in energy emitted or reflected increases with the radiometric resolution of the sensor. Table 1 shows few satellites and their resolution as examples.

Table 1 Few satellites and their resolution data

Satellite	Sensor	Spectral bands		Resolution			Launch Date
				Spatial m	Radiometric Bit	Temporal day	
Landsat 4	MSS	1	0.5–0.6 G	30	6	18	1982
		2	0.6–0.7 R				
		3	0.7–0.8 NIR				
		4	0.8–1.1 NIR				
Landsat 5	TM	1	0.45–0.52 B	30 120-TIR	8	16	1984
		2	0.52–0.6 G				
		3	0.6–0.7 R				
		4	0.76–0.90 NIR				
		5	1.55–1.75 SWIR1				
		6	10.4–12.5 TIR				
		7	2.08–2.35 SWIR2				
Landsat 7	ETM+	TIR 61 and 62		60	8	816	1999
IRS 1D	LISS III	B2	0.52–0.59 G	23.5	7	24	1997
		B3	0.62–0.68 R				
		B4	0.77–0.86 NIR				
		B5	1.55–1.70 SWIR				
	PAN	0.5–0.75		5.8	6	5	
	WIFS	B3	0.62–0.68 R	188	7	5	
		B4	0.77–0.86 NIR				
IRS P6	LISS III			23.5	7	24	2003
	AWIFS	B2	0.52–0.59 G	56	10	5	
		B3	0.62–0.68 R				
		B4	0.77–0.86 NIR				
		B5	1.55–1.70 SWIR				
	LISS IV	B2	0.52–0.59 G	5.8	7		
		B3	0.62–0.68 R				
		B4	0.77–0.86 NIR				

2 A Global Perspective on Remote Sensing (RS) Technologies in Agriculture

In current scenario, remote sensing (RS) plays avital role in variety of applications in agriculture across the globe. Remote sensing research on crop canopies has revealed important information about the crop's various agronomic characteristics. On July 23, 1972, NASA launched the Earth Resources Technology Satellite (ERTS), subsequently known as Landsat 1, which marked the beginning of the use of remote sensing in agriculture. At the early stages of remote sensing in agriculture, researchers were mainly focusing on the use of data for differentiation of land-use/land cover types to identify different crops. However, in recent years, the use has been diversified for understanding different kinds of stress and also assess the plant biophysical properties including its health and crop productivity. After Landsat 1, a series of Landsat satellites (Landsat 2–9) were launched in a continuous stream to deliver high-quality photos, which are crucial to agriculture. France and India both deployed the SPOT 1 and IRS-1A satellites in 1986 and 1988 for use in agriculture and natural resource management, respectively. Remote sensing is the preferred technology for monitoring the crop over conventional method, since it has the ability to provide recurrent information in a nondestructive way and will provide valuable information for precise agricultural applications (Adhikary et al. 2022). Satellite-based remote sensing in agriculture is used for the estimation of crop area, forecasting, and production assessment of agricultural crops by different researchers. These RS satellites are helpful in characterizing the crop yield based on biophysical characters of crops, yield forecasting/estimation, crop phenological information, detection of stress situations (abiotic and biotic), characterization of soils for their properties, mapping of problem soils, crop and soil suitability, etc. Details of few Indian satellites used in agriculture applications are listed in Table 2.

From a historical point of view, the spatial and temporal resolution of remote sensing data used in agriculture has improved throughout the years. The optical sensors' resolution saw a significant upgrade, allowing for the measurement of more bands, including narrow bands. Numerous vegetation indices have been developed as a result, and automatic detection of invalid pixels such as clouds and shadows has been improved. The use of active sensors, such as SAR, which provides measurements independent of clouds, improving the regularity of image availability, or satellite sun-induced fluorescence, which could provide insightful data on the effectiveness of the photosynthetic process, was considered a significant shift in agriculture applications.

Despite the rapid advancement of RS technologies, the majority of their use in the agricultural sector has been limited to specialized research, and very few applications have been created for farmers' activities or for the end users. This is partially explained by the fact that the majority of RS applications in agriculture have only been carried out by RS professionals who lack field agronomy expertise. Basic understanding of RS usage should be widely disseminated among the many

Table 2 Remote sensing satellites of Indian Space Research Organization involved in agriculture applications

Satellite type	Satellite	Objectives
Multispectral imaging satellite	Resourcesat-2 and Resourcesat-2A	Multispectral satellite images for forecast of crop area and productivity, land, water, and natural resource inventory and management, and support on disaster management-related activities
Cartography satellite	Cartosat-1	High-resolution images for DEM (digital elevation model mapping),generation of cartographic maps, mapping of drainage and irrigation networks, contour and topographic maps
Radar imaging	RISAT-1	This will be helpful for imaging during monsoon seasons. It is for *Kharif crop* (June to November) during southwest and *Rabi crop* northeast monsoon seasons. Besides, this is being used for flood and natural disaster management
Meteorological forecasting	Kalpana-1	Comprehensive weather status reporting and forecasting
Meteorological observation	INSAT-3D and INSAT-3DR	Meteorological observations including vertical (temperature and humidity) atmosphere weather forecasting and disaster warning

agricultural disciplines in order to make the implementation of satellite RS more practicable on agricultural fields.

In order to successfully apply spatial and temporal basic informative layers to a variety of fields, including flood plain mapping, hydrological modeling, surface energy flux, urban development, land-use changes, crop growth monitoring, and stress detection, remote sensing in conjunction with GIS is highly advantageous (Kingra et al. 2016). Nanosatellites and other spacecraft launched after 2000, including SuperView-1 (2018), GeoEye-1 (2008), Pleiades-1A (2011), WorldView-3 (2014), and SkySat-2 (2014), capture multispectral images with a daily or sub-daily revisit period at a high spatial resolution of about 2 m. The development of narrow-band or hyperspectral sensors and an improvement in the spatial resolution of sensors installed on aircraft or satellites have advanced the use of remote sensing techniques and facilitated a more thorough investigation of crop classification. Agriculture monitoring has become less expensive and more effective because of the use of various types of sensors that deliver trustworthy data in a timely manner for a small fraction of the cost of conventional data collection methods. The future of agriculture is likely to be altered by the unprecedented affordability of UAVs.

3 Remote Sensing and Precision Agriculture

Temperature and moisture sensors, robots, GPS technology, and aerial images and geospatial data, hyperspectral sensing, and many more cutting-edge technologies must all be integrated into the current agricultural environment. It should be coupled with advanced farming practices like precision agriculture. Such novel approaches help the farmers toward accessing effective production strategies, banking and financial services, and real-time market information.

Precision agriculture (PA) is an agricultural management technique that makes use of technology to guarantee that the land and crops receive the precise care needed to maximize their yield. This agricultural management method is also known as site-specific crop management (SSCM) because it is based on the precise specifications and location. Precision agriculture is centered on information technology, with the framework being supported by (SSCM), geospatial techniques (RS and GIS) like GPS-based soil sampling, drones, robotics, sensors, and telematics. The development of wireless sensor network-based applications for precision agriculture enables growth in efficiency of water use and fertilizer use efficiencies, maximizing the yield and profitability while reducing the negative impacts on the environment, in agricultural systems. The technological advancement is providing real-time information from the fields to farmers and will provide a solid platform for them to adjust strategies at any time, and even their scientific advancement is being linked to the user departments/universities/experts to provide real-time solutions to the problems. Precision agriculture is a cutting-edge approach that gathers data, processes, and analyses spatially and temporally before combining it with other data to make management decisions about the calculated variability of crops, soils, and climate, among other things, for better resource input use efficiency, crop productivity, quality, and profitability, as well as sustainability of agricultural production. Precision agriculture combines cutting-edge data and methods in the decision-making process to improve crop output while minimizing water and fertilizer losses and maximizing resource use. PA is utilized in all aspects of agriculture, including horticulture, fishery, pasture, and dairy–livestock management.

Since the 1970s, satellite devices have been widely used for PA. There has been a sharp rise in the use of remote sensing technology for PA throughout the last decades due to the rapid development of this technology. Different sensing components used in the PA has been listed in Fig. 2. The use of remote sensing has been encouraged in numerous PA applications, including crop monitoring, irrigation management, nutrient application, disease and pest control, and yield prediction, as a result of the accessibility of high-resolution (spatial, spectral, and temporal) satellite pictures. A number of remote sensing-based PA technologies have already been adopted by commercial farmers such as the variable fertilizer rate application system by using handheld sensors such as the GreenSeeker and the Crop Circle, which are based on remote sensing. Since they are affordable and flexible, unmanned aerial vehicles (UAVs) have significantly increased their use over the past 10 years due to

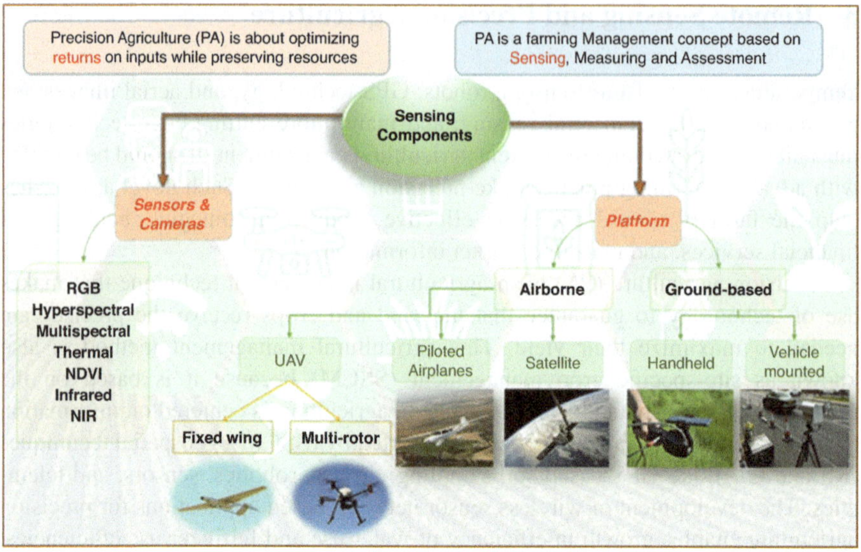

Fig. 2 Image showing the components of PA sensing components

their ability to acquire the high-resolution (cm-scale) images for PA purposes that are required. Several researchers and academicians are investigating cutting-edge methods to store, process, and analyze satellite data due to the availability of a large volume of satellite data, and they are experimenting with cutting-edge concepts like cloud computing and machine learning. It is essential to look into and build a user-friendly yet dependable workflow for the real-time application of remote sensing given the complexity of the image processing process and the amount of technical knowledge and ability required to be able to apply remote sensing in PA. The creation of precise yet user-friendly systems will probably lead to a wider adoption of remote sensing technologies in both commercial and noncommercial PA applications.

4 Linking Remote Sensing Observations to Interest Variables

Field data and remote sensing data can be integrated to estimate agriculture variables over vast areas. These estimations' accuracy, for instance, depends on how effectively the field data can be correlated with satellite images and how well agricultural areas can be distinguished. Since it can be challenging to differentiate agricultural regions from other land cover classes in practice, estimates may be skewed as a result. For quite a while, it has been suggested to use remote sensing as a technique for calculating agricultural yield on a locality basis.

4.1 Remote Sensing and Production Agriculture

Since there is uncertainty in using remotely sensed data to calculate crop yield (mostly due to an indirect relationship between remotely sensed data and crop state variables), crop models may be employed in conjunction with remote sensing. There are several techniques to incorporate remotely sensed data into models and various types of crop models (statistical, deterministic, and semiempirical) (forcing, recalibration, statistical correction). Understanding crop response to management strategies and environmental stresses for PA requires understanding of spatial variability in agricultural production. Crop biophysical indicators acquired from remote sensing, also known as vegetation indices, have a positive association with crop production and biomass measurements, suggesting a possibility for use in yield estimation.

4.2 Pre Season Planning

The cost of production, suitability of the land for the crop, marketability, management, accessibility of the market, and availability of inputs all affect crop selection. It is possible to locate water sources and construct land suitability using satellite imagery analysis. This satellite and drone data can be used to determine general trends of land use, estimate crop growth, and more. By highlighting regions that need attention for planting development and probable disease outbreaks, remote sensing technology enhances crop production operations. Crop yield predictions, global monitoring of other agricultural processes, and tracking of significant changes in land usage can all be done via remote sensing. Agriculture can employ remote sensing to map the characteristics of soil to find areas with high productivity and hospitable soil. With the use of this data, farmers can determine which soil is ideal for which kind of crop. Remote sensing also aids in the monitoring of droughts. In order to mitigate drought, monitoring is crucial. Observing trends in weather patterns is necessary to determine how long droughts will persist and which places are at risk. The availability of sensors that offer real-time information on rainfall, soil moisture, streamflow, and groundwater levels has made drought monitoring easier. Flood mapping as well as its regular checking is accomplished by field surveys, remote sensing, or a combination of both. Additionally, remote sensing can be utilized to locate any potential development-blocking features like streams, marshes, and stones. When done properly, it gives a true picture of the land use and usable space.

4.3 Topography Mapping

Topography is the primary component impacting productivity and soil nutrient concentration. As a result, slope can significantly limit production, especially in undrained or steep areas that have been eroded. Depending on whether the year is rainy or dry, flow buildup can also have a significant impact on yield. Topographical land features in some areas can account for as much as a mid-double-digit percentage of yield variability. Because topography impacts how water moves across a place, agriculture is impacted by it. For instance, swiftly flowing downhill water may remove soil nutrients or leave crops too dry for optimum growth (depending on what type of crops are grown). Slow-moving water can provide moisture to plants for a longer period of time, but because it drags soil particles away, it also hastens erosion. Additionally, topography has an impact on agricultural output by regulating the amount of sunlight that reaches plants at various locations throughout an area. Flat areas frequently receive more sunlight than mountainous ones since there are fewer obstacles in the way of the sun's rays that could prevent them from reaching leaves. A region's topography significantly affects the types of agricultural activities that can be done there. It directly influences the amount of rainfall that different parts of a country will receive as well as the amount of water that is available to water crops. The ideal machinery and equipment for a farm can depend on the topography. For instance, farmers must use their own body power to complete tasks because hillsides are sometimes too steep for tractors. The farm's labor requirements are influenced by topography. During the planting or harvesting season, moving about will not require much effort if the land is level and smooth, but if it is steep or uneven, movement will require more effort. The impact of topography on the amount of rain any region receives annually is another factor in the importance of topography in agriculture. More clouds are forming above adjacent mountains, dumping their water into those locations, making them wetter than other nearby areas; therefore, there will be more rain there than in a place with flat land. A digital elevation model (DEM) is a depiction of the topographic surface of the Earth devoid of any surface features like trees. Several sources, including topographic maps and high-resolution LiDAR data, are used to construct DEMs. Moreover, detailed analysis of DEM maps can reveal the slope and elevation of that area.

4.4 Subsurface Tile Drain Mapping

Buried perforated pipes used for subsurface drainage are designed to catch water below the surface of the Earth and send it to an outlet. Since the subsurface pipeline was built using clay or concrete slabs up to the 1970s, subsurface drainage is frequently referred to as "tile" drainage. Subsurface tile drainage has been frequently employed in agricultural fields in the Midwest to remove surplus water from the soil using perforated tubes that are buried beneath the surface. Although it is essential for enabling agricultural activities in moist yet productive areas, this system also has a significant impact on the dynamics of water and nutrients, as well as the water quality in this area. Topographic depressions and tillage on the soil's surface prevent standard optical image processing from reliably capturing the precise positions of subsurface tile drainage structures. These difficulties are overcome by using thermal imaging to locate a subsurface drainage pipe. The hypothesis states that the unique spatial distribution of soil moisture produced by tile drains may be the cause of the change in surface soil temperature between places close to and far from drainage pipes.

4.5 Artificial Intelligence and RS

The application of artificial intelligence (AI), machine learning (ML), and computer vision (CV) has been the most recent breakthrough for extracting precise and accurate data over wide areas from satellite imagery (CV). In several domains where gathering vast amounts of image data is necessary for pattern recognition and the development of computer-based algorithms, AI and ML models have achieved remarkable success. AI and ML can assist the user in comprehending the data gathered in order to quickly find answers to the specific issue at hand. The study of broad areas of interest, item classification, detecting and monitoring land use, data fusion, cloud removal, and spectral analysis of ecological changes from satellite photography can all be improved by AI. Deep learning and neural networks using computer vision models can help AI with data collecting, processing, and interpretation so that data users can manage data more quickly and effectively. There is pressure for increased agricultural production and accurate crop condition information globally due to increasing population. The world's agricultural resources must be managed more effectively to accomplish these goals, especially in developing nations. AI, ML, and CV algorithms are used to extract spectral evaluated data from high-resolution, multispectral satellite photos. This data is then converted into management solutions for improved crop health and production targets. By leveraging image data gathered by satellites, fixed-wing aircraft, or unmanned aerial vehicles, artificial intelligence (AI) and geographic information systems (GIS) applications can assist farmers in doing crop forecasts and managing their agricultural production (UAV). To diagnose crop stress, waterlogging, control production yields, and grade trees, this data is gathered and processed to produce the NDVI and many other vegetation indicators. For people farmers and the agricultural industries, the capacity of AI and GIS to evaluate and visualize agricultural landscapes and workflows has proven to be quite useful.

ICT in agriculture helps farmers in many nations by giving them crucial information on planting, crop protection, and enhancing soil fertility, allowing them to increase agricultural productivity. With the use of weather-related advisories and notifications, farmers may better plan for unpredictable occurrences like floods, droughts, and even pest and disease outbreaks, minimizing crop loss. The Water Resources Information System developed by the states like Kerala and Karnataka help to get all details of weather, surface water, groundwater, reservoir operations, canal flow, soil moisture, water quality, and all other relevant and required information on water in the state on a single platform.

At a different scale, the application of IT in modern agriculture has also profoundly changed farming and agriculture in industrialized nations. Big Data, Cloud Computing, ML, Deep Learning, Hierarchical Systems, and Internet of Things (IoT) have all had a significant impact on how effective current procedures are. Many farmers in the USA and Europe operate their farms remotely by employing drones, sensing technology, and other tools that collect essential information on the soil, air, crop health, and weather.

One of the collaborative research projects called FATIMA (FArming Tools for external nutrient Inputs and water MAnagement) uses Earth observation data to monitor and manage agricultural resources more effectively and efficiently. The European Commission provided funding for the project through its Horizon 2020 program for research and innovation. There are nine active participating nations in this multinational effort. In the experiment, pilot plots of several crops that have been traditionally grown were observed using satellite data from Landsat 8 and Sentinel-2 sources. Each harvest during the course of the project's 3 years produced fresh information regarding anticipating crop water requirements (CWR) and crop yield unpredictability.

Open ET is another example of using modern spatial tools for direct use of farmers. It offers conveniently available satellite-based evapotranspiration (ET) estimations for better water management in the Western United States. On the website openetdata.org, users can browse ET data in two different ways: at the field scale for millions of distinct fields or at the original quarter-acre resolution of the satellite data.

4.6 Remote Sensing and Weather Forecasting

A detector is placed far from a target while using remote sensing. The sensor might be a part of a radar or satellite system that keeps an eye on the weather and the ocean, or it might be attached to unmanned aerial vehicles. Images and data from distant sensors are used for weather monitoring and forecasting at all scales. For monitoring on coming fronts and storms (such as hurricanes and blizzards), imaging water (such as seas, lakes, rivers, soil moisture, vapor in the air, clouds, and snow cover), and calculating runoff and flood potential from thawing, remote sensing is used.

Sensors collect a wide range of data from clouds because snow and dense clouds appear in photos as a bright white color. Without clouds in the sky, the land and sea appear dark gray in the photographs. The sun is the only source of visible light; hence, throughout the night when there is no sunlight, there are no visible satellite photographs available. The temperature reduction with altitude is thus measured by the recorded radiation images, and it may be deduced that high clouds are cooler because of less radiation. The radiations emitted by the atmospheric water vapor are captured by sensors. This is because satellites do not receive radiation from low clouds. The amount of water vapor in the atmosphere typically determines how much radiation is detected at the remote sensor. High humidity makes the sensor have bright shades. Typically, weather radars measure Doppler winds and rain reflectivity. Weather radars can show variations in rainfall intensity because they capture photographs more frequently than meteorological satellites. In places that are prone to rain, they efficiently monitor any change in rainfall intensity. Quick scat satellites are specialized satellites that use remote sensors to measure scattered microwave signals from ocean waves and reduce near-surface wind speed and direction. This can be used to find oceanic cyclones. Weather radars have remote sensors with a unique, higher resolution than other satellites. As a result, they reveal very precise information about weather variations, particularly when it comes to rainfall. It can also calculate the direction of the wind and the temperature of the atmosphere as measured from space, enhancing weather predictions.

4.7 Soil Moisture and Temperature Mapping

The data made with microwave, SAR, optical, or thermal infrared (TIR) sensors are commonly used to determine soil moisture and temperature (Barrett and Petropoulos 2012; Kerr et al. 2010). Moran et al. (2004) compared optical, and microwave attempts for measuring surface soil moisture.

4.8 Soil Compaction Assessment

Soil compaction has a detrimental influence on soil health and agricultural productivity by reducing soil porosity, hydraulic conductivity, and nutrient availability. Farmers can reduce in-field compaction and related agriculture losses by improving their understanding of the temporal and spatial extents of soil compaction in a field. Traditionally, soil compaction has been measured using cone penetrometers, a laborious, time-consuming, and incomplete method due to the discrete nature of the data collected. Traditionally, soil compaction has been measured using cone penetrometers, a laborious, time-consuming, and incomplete method due to the discrete nature of the data collected. A study conducted by Kulkarni et al. (2010) examined the effect of soil compaction on canopy spectral reflectance and cotton yields in Arkansas using hyperspectral data. Using green normalized difference vegetation index (GNDVI) imagery, soil compaction can be assessed by comparing the green and near-infrared spectrum bands. Despite this, cone penetrometer data and NIR data failed to show a strong correlation, making compacted soil difficult to identify. There has been little research into the spatial aspect of soil compaction throughout the year and its impact on soil hydraulic properties. There is currently no widely used method for measuring mechanical qualities in a field in order to estimate soil compaction. To assess soil compaction, more research into RS-based assessments of soil properties and their integration with cutting-edge machine learning algorithms is needed.

4.9 Crop Emergence and Density

Crop emergence is the initial sign of crop success. Crop planting date, soil moisture, soil temperature, seed variety, and other elements all have an impact on crop emergence. Remote sensing data has been used to map the land surface phenology (LSP) at different spatial resolutions ranging from a few hundred meters to a few meters. The product green-up dates (or the start of the season) can be linked to crop emergence dates. According to earlier studies, crop emergence dates are often used to determine when remote sensing green-up dates are identified. Depending on the algorithm, the detection lag can range from a few days to a few weeks (Gao et al. 2017). Since 2001, LSP measures with a 500 m spatial resolution, including green-up dates, have been made available through the Moderate Resolution Imaging Spectroradiometer (MODIS) land cover dynamics data package (MCD12Q2). The within-season emergence (WISE) technique was developed and validated across the Beltsville Agricultural Research Center (BARC) experimental fields in Beltsville, MD, during the 2019 growing season, utilizing the Vegetation and Environment monitoring on a New Micro-Satellite (VENμS) time series (5-m, 2-day revisit). Findings demonstrate that, 2 weeks following crop emergence, WISE is capable of accurately detecting early crop growth stages at the subfield scale. The dates of remote sensing green-up were typically 4–5 days after crop emergence (Gao et al. 2020).

4.10 Monitoring Crops for Yield Optimizations and Crop Yield Forecasting

To develop effective management plans for fieldwork and/or remedies, crop growth and yield must be monitored in order to understand how the crop responds to the environment and agronomic practices. Leaf area index (LAI) and biomass are both significant measures of the growth and health of a crop. LAI and biomass have been calculated using RS data for a number of crops, including row crops, orchards, and vine crops. Such research often uses a set of reference data to construct a regression or machine-learning-based approach to estimate LAI and/or biomass for a target field. These reference data can include measured LAI and accompanying vegetation indices. Yue et al. (2017) used a variety of spectral indicators along with observed plant height to estimate biomass ($R^2 = 0.74$) in several irrigation and nutrient treatment plots for winter wheat produced in China. There are two methods that have historically been used to estimate crop yields using RS data. Initially, crop models are used to estimate crop yield and biomass using biophysical factors acquired from remotely sensed data, such as the leaf area index. Second, connections between crop parameters/indices derived from remote sensing (e.g., NDVI, LAI) and observed crop yield and biomass are created in a typical agricultural field using statistical methods (e.g., regression). To map crop yield at a target crop field, one could utilize the developed regression model or empirical connection. Crop modeling is a data-intensive technique that needs a lot of data for the model's input parameters, weather data, and yield and biomass data. To investigate the connection between maize yield and biomass and spectral indicators obtained at the V12 stage, Maresma et al. (2016) used a regression-based method. Wide dynamic range vegetation index (WDRVI) and the red-based NDVI were found to have the strongest correlations with grain yields across a variety of fertilizer application rates. An improved estimate of crop biomass and yield is anticipated to result from spatial mapping of crop biophysical characteristics or indices over the course of a growing season as opposed to a single season-long snapshot.

4.11 Agriculture for Ecosystem Services

Agriculture ecosystems are vital to human health because they give us food, fodder, bioenergy, and medicines. These systems rely on the ecosystem services provided by natural ecosystems, including pollination, biological pest control, preservation of soil fertility and structure, nutrient cycling, and hydrological processes. Assessments indicate that the importance of these ecological services to agriculture is significant and frequently under appreciated. On-farm management techniques can considerably enhance the ecosystem services provided by agriculture. In order to manage for increased provisioning services, farmers typically utilize inputs and management techniques to raise yields. Yet, same strategies can also enhance pollination, biological pest control, soil fertility and structure, water regulation, and biodiversity support. Pollinators or natural enemies may be able to obtain the resources they need from the agroecosystem's habitat management (Tscharntke et al. 2005). Several studies have shown that perennial vegetation is essential for maintaining biodiversity in general and beneficial creatures in particular (e.g., Perfecto and Vandermeer 2008). There is proof that management techniques that emphasize crop diversification through the use of polycultures, cover crops, crop rotations, and agroforestry can decrease the amount of insect pests that specialize on a certain crop while simultaneously giving natural enemies a haven and an alternative prey source (Andow 1991). Similar strategies, such as minimal pesticide usage, no-till systems, and crop rotations with mass-flowering crops, may be advantageous for natural pollinators.

A range of farming techniques can successfully cut or balance out agricultural greenhouse gas (GHG) emissions. Animal waste emissions can be considerably reduced with good manure management. Agricultural output can reduce its CO_2 emissions by half by adopting biological nitrogen fixing by legumes rather than synthetic nitrogen fertilizers (Drinkwater and Snapp 2007). Perennialization and legume intensification in agroecosystems change internal cycling processes and increase N usage efficiency through the recoupling mechanisms discussed above. These circumstances allow for the reduction of chronic excess inorganic N additions, which are currently common, which lowers NOx and N_2O emissions. Agriculture must boost soil's potential to absorb and store carbon, known as carbon sequestration, in order to reduce greenhouse gas emissions (Lal 2008a, b). The net flux of CO_2 between the land and the atmosphere is determined by the balance between carbon gains from plant growth and soil carbon sequestration and carbon losses from changing land use and land management practices. Crop rotations and cover crops, as well as soil conservation methods like conservation tillage and no-till agriculture, can particularly help reduce the decomposition of subsurface carbon. In general, water management and erosion control can sustain soil organic carbon (Lal 2008a). Soil carbon sequestration provides extra ecosystem advantages to agriculture by maintaining soil structure and fertility, strengthening soil quality, increasing the use efficiency of agronomic inputs, and improving water quality by filtering and denaturing contaminants (Lal 2008b; Smith et al. 2008).Crops can be grown on agricultural land in order to produce biofuel. Particularly, cellulosic biofuels have the potential to substitute for some fossil fuels and cut greenhouse gas emissions (Smith et al. 2008). While using fossil fuels raises the atmospheric concentration of carbon, correctly managed bioenergy crops reduce this by recycling carbon. Carbon is released into the atmosphere when bioenergy feedstocks are burned, but plants also absorb carbon as they develop. The use of solar energy instead of fossil fuels to generate electricity has the potential to reduce CO_2, N_2O, and NOx emission.

Ecosystem services provided by agricultural systems are essential for human well-being. Other ecosystem services, such as those that facilitate provisioning and regulation, are also produced and used by them. Agroecosystems can maximize their provisioning services at the price of other ecosystem services, but with good management, these trade-offs can be considerably reduced or even eliminated. Techniques for managing agriculture are crucial for maximizing environmental benefits and minimizing adverse consequences. Climate change will make these problems worse despite recent advances in our ability to assess the worth of diverse ecosystem services related to agriculture and to consider ways to reduce trade-offs and increase synergies. Future research must address these difficulties within geographically and temporally defined frameworks.

4.12 Nitrogen Stress Monitoring

Application of fertilizer must be timely and appropriate to enhance crop growth and yields while minimizing environmental harm from nutrient losses to groundwater and surface water. In tractor-mounted systems, remote sensors are frequently fitted prior to the spray boom. In these systems, nitrogen (N) application rates are established based on estimated vegetation indices (such as NDVI). The nutrient applicator/spreader receives these application rates and uses them to apply fertilizer in real time. Using a variety of techniques, the measured vegetation indexes are converted into the proper N application rates. By comparing the measured vegetation indices in the target field with the reference vegetation index, which is often measured in a fertilized (N-rich) plot or strip that is representative of the target field, the nitrogen (N) application rates are frequently determined. In order to determine the in-season N requirements for many crops based on vegetation indices, numerous fertilizer rate calculation algorithms, including the nitrogen fertilizer optimization algorithm (Raun et al. 2005), have been developed and successfully integrated in these commercially available sensors (Scharf et al. 2011). Notwithstanding its commercialization, farmers continue to use variable rate N-management technology based on proximal remote sensing at a low rate in many agricultural firms. Lack of strong data indicating the substantial economic benefits (crop yield and/or profitability), particularly in commercial farm settings, is preventing the broad use of these technologies (i.e., vast fields). To advance these remote sensing-based technologies and maximize their advantages, research is being conducted with UAVs and other remote sensors for a variety of crops in various climatic regions. Maresma et al. (2016) used images from a UAV to examine the suitability of several vegetation indicators and crop height in determining in-season fertilizer application rates for maize grown in Spain.

4.13 Crop Disease Monitoring

Diseases have the ability to drastically lower farm income and agricultural productivity. A plant disease's geographic distribution and early detection can be used to stop the disease's spread and reduce output losses. The method of disease identification known as field scouting is labor-intensive, time-consuming, and prone to human error. Remote sensing could be used to efficiently monitor the condition, particularly in the early stages of infection development when it may be difficult to recognize the symptoms of sickness with field scouting. Many techniques, including RGB, multispectral, hyperspectral, thermal, and fluorescence imaging, have been used to detect illnesses in a variety of crops. In Italy, Di Gennaro et al. (2016) discovered a strong link between NDVI produced from UAV imageries and grapevine leaf stripe disease. Citrus canker may be identified with 96% accuracy even in its early stages of growth by Abdulridha et al. (2019) using a machine learning method and vegetation indices obtained from hyperspectral UAV photos. When compared to commonly used vegetative indices (e.g., NDVI), the development of disease-specific spectral disease indicators (SDI) might increase the accuracy of disease diagnosis and distinction in actual field settings (Mahlein et al. 2013). Despite the fact that classification study into plant diseases has been conducted, additional effort is needed to develop disease identification techniques that are more accurate, automated, and reproducible in a range of environmental and field circumstances.

4.14 Weed Identification and Classification

Traditional weed control methods involve consistent herbicide spraying, which is inefficient and increases the risk of off-site chemical losses. Herbicide can be applied at a variable rate as needed to improve treatment effectiveness, lower input costs, and reduce environmental contamination. For site-specific weed treatment, remote sensing has frequently been employed to map weed patches in agricultural areas. Specific spectral signature and other phenological or morphological characteristics of weeds distinguish them from crop plants. Over the past few years, categorizing images for weed mapping has shown to be a very accurate and effective process using machine learning approaches. For weed mapping, supervised and unsupervised classification are the two main types of image classification algorithms that are typically utilized. Despite the fact that each method has advantages and disadvantages of its own, supervised categorization takes more time and physical labor.

UAVs have emerged as the most popular remote sensing platform for weed mapping and management because of their capacity to produce the 5 cm-scale resolution images necessary for weed detection and mapping. Huang et al. (2018) achieved up to 90% accuracy when mapping weeds in a Chinese rice field using the fully convolutional network method. Partel et al. (2019) developed a target weed sprayer for ground-sensor-based weed detection using deep learning neural network approach, which delivered 71% application accuracy in trial fields in Florida, USA. Given the technical knowledge needed to use sophisticated software and the intricate application processes, commercial adoption of these technologies is still difficult.

4.15 Grain Quality Assessment

To boost the sustainability of grain production at the regional level, there is a need for operational, dependable systems for crop monitoring during the whole growing season as well as for techniques for early yield and quality prediction of winter wheat grain. Using satellite data of the seasonal dynamics of the vegetation index NDVI, crop physiological states and crop size are evaluated. In a study, Eroshenko et al. (2020) investigated the correlation between remote sensing data and indices of winter wheat quality for the Stavropol region. According to the analysis of the data, there is a maximum correlation coefficient of 0.83 between NDVI and the amount of grains in the second and third classes, with a minus sign denoting the grain creation phase. The winter wheat harvests' vegetative index NDVI and quality indicators for the Stavropol region are most strongly correlated throughout the period between 10 and 22 calendar weeks.

4.16 Crop Residue Assessment

By preserving a protective mulch on the soil's surface and assisting in the reduction of soil temperature, erosion, nutrient loss, and evaporation, crop residue cover (CRC) maintenance can greatly enhance the environmental performance of cropping systems. Remote sensing methods can be used to locate crop residue, which can be used to monitor the application of conservation tillage strategies. Broad spectrum contrasts between shortwave infrared (SWIR) and near-infrared (NIR) reflectance as well as narrow contrasts looking at cellulose absorption in the SWIR have been utilized to quantify CRC using multispectral and hyperspectral data. The development of a viable operational use of remote sensing to map CRC and tillage intensity, however, still confronts hurdles. The need for scene-specific calibration, the impact of residue and soil moisture content on spectral features, the variety of residue and soil characteristics, and interference from vegetation are a few of these difficulties. Furthermore, there is a wide range of capabilities for nearby, airborne, and spaceborne sensors. Remote sensing of non-photosynthetic vegetation has the potential to improve rangeland management, our knowledge of vegetation dynamics, and the monitoring of carbon fluxes in the larger agricultural landscape. However, there are still scientific challenges that must be overcome before it can be used effectively in practice.

4.17 Agriculture with Remote Sensing: Possibilities, Limits, and Problems

Nearly every element of PA, from soil preparation to harvesting, has potential implications for remote sensing. PA has undergone a paradigm shift as a result of the widespread use of high-spatial-resolution, multi temporal satellite data, inexpensive UAVs, and widely accessible ground-based proximity sensors. Many cutting-edge methodologies have been utilized to examine the potential uses of remote sensing in PA, including empirical, regression, and different machine learning techniques. Similar to this, a number of vegetation indices, including disease control, weed mapping, variable fertilizer management, irrigation scheduling, and yield forecasting, have been developed and assessed for their potential to help PA operations. There are numerous challenges that must be overcome before remote sensing technology may be widely used in both commercial and non-commercial agriculture.

Even though the majority of satellite data are freely accessible, processing them for practical purposes may necessitate substantial technological know-how and skill. For instance, software experts and specialized skills are needed for image pre- and post-processing. Furthermore, many PA procedures, like the management of weeds and diseases, call for data with fine spatial resolution (cm-scale) and high spectral and temporal (e.g., daily) precision. Most satellite data that is made available to the public does not adhere to these standards. Additionally, many satellite photos might not be acceptable for usage on cloudy days or when there is irregular or changeable irradiance from the sun.

Users and farmers may be required to pay for high-resolution satellite data, which might be prohibitively expensive, particularly for small farms. For small farm operations, a low-cost alternative may be provided by photos captured by UAVs (Candiago et al. 2015). The usage of UAVs and tractor-mounted sensors necessitates expert operators (such as drone licensing) and requires the use of specialized software for data analysis (Ali et al. 2017). The cutting-edge sensors deployed on some of the most recent satellite launches and unmanned aerial vehicles (UAVs) produce hyperspectral images that contain a wealth of data on crop biophysical parameters. However, these sensors (UAVs) are pricey, and picture processing is complex (Khanal et al. 2018). It is necessary to research and develop cutting-edge information and communication technologies, as well as chemometric and spectrum decomposition approaches, in order to produce and supply the useful information needed for PA applications. At the scale necessary for many PA applications, machine learning and other artificial intelligence techniques can provide spatially and temporally continuous information from real-time satellite data. (Reichstein et al. 2019). Such AI techniques can be complemented by hybrid methods, which use the information from physically based models to build methodologies helpful in PA decision-making (Weiss et al. 2020).

Technology is now a crucial part of any commercial farm as agriculture evolution. New precision agriculture businesses are enabling farmers to maximize harvests by automating the control of every crop farming variable, including moisture levels, pest stress, soil conditions, and microclimates. By providing more exact techniques for planting and producing crops, precision agriculture aids farmers in increasing production and reducing costs. Precision agriculture-focused businesses have a fantastic chance to grow. The younger generation of farmers is lured to businesses that are faster and more flexible and that meticulously maximize agricultural productivity.

India being a country leading in space technology has access to large quantity of such accurate geospatial data. Apart from this, we should also have field-level primary data. These data should be scouted, processed, and analyzed in such a way that some user-friendly products and deliverables are developed and made available to the farmers. These products can be in the form of a DSS, expert system, mobile app, Web-based tool, url, GUI, or anything. In the end, it should make it possible for farmers and agribusinesses to carefully monitor crop cultivation inputs and practices, utilize natural resources and agrochemicals more effectively, and promptly adjust to changing environmental conditions. The Internet of Things (IoT), for example, has several uses in agriculture, ranging from tracing a product's origin, its environmental impact, and its storage settings along the supply chain to real-time monitoring of soil, plant, and animal health using in situ sensors. According to predictions, the Internet of Things (IoT) could transform into the "Internet of Action" by 2030. In the near future, traditional labor-intensive farming will be replaced by completely automated farming, where sensors and equipment based on built-in AI and data analytics capabilities are capable of self-optimizing and beginning activities on their own, without much human participation. The self-managed precision farming systems with agrobots will also come to a reality soon.

Although much research has been done on the use of remote sensing in PA, there are not many approaches or frameworks that have been shown to be trustworthy, reproducible, and effective over a wide range of meteorological, soil, crop, and management scenarios. The geographical, spectral, temporal picture resolution, atmospheric, climatic, and weather conditions, crop and field traits (such growth stage, land cover), and the analytic approach all have an impact on how accurate remote sensing systems are perfoming at the field level. For instance, there is a lot of uncertainty in PA decision-making since the precision of surface energy balancing methodologies for ET estimate varies greatly over time and space. To fully grasp the spatio temporal nature of uncertainty in calculating ET, soil moisture, disease stress, and other crop factors, more research is required. Crop condition and responses to site factors (such as soil and topography), management, and stresses (biotic and abiotic) are reflected in a crop's spectral signature (e.g., diseases, weeds, nutrient, and water stress). In the real world, where numerous biotic and abiotic factors affect crop response or circumstances, a disease identification technique that performs well in a lab context may not perform as well. It is essential to investigate and develop a straightforward and reliable workflow for image preprocessing, analysis, and application in real time given the complexity of image processing methods and the amount of technical skill and ability required for application. There are still many challenges and constraints in the development of tools and frameworks that can facilitate end users' access to satellite data for real-time applications. Remote sensing data will probably be used more frequently in commercial and non commercial PA activities as a result of the development of precise, user-friendly technologies. However, it should be coupled with financial inclusion and risk management for farmers supported by better capacity building, empowerment, extension, and advisory services.

References

Abdulridha J, Ampatzidis Y, Kakarla SC, Roberts P (2019) Detection of target spot and bacterial spot diseases in tomato using UAV-based and benchtop-based hyperspectral imaging techniques. Precis Agric 21:955–978

Adhikary S, Biswas B, Naskar MK, Mukherjee B, Singh AP, Atta K (2022. [cited 2022 Dec 25]. Available from: https://www.intechopen.com/online-first/83417) Remote sensing for agricultural applications. In: Eyvaz APM, Albahnasawi A, Gürbulak E, Tekbaş M (eds) Arid environment – perspectives, challenges and management [working title] [internet]. IntechOpen, London. https://doi.org/10.5772/intechopen.106876

Ali MM, Al-Ani A, Eamus D, Tan DKY (2017) Leaf nitrogen determination using non-destructive techniques—a review. J Plant Nut 40:928–953

Andow DA (1991) Vegetational diversity and arthropod population response. Annu Rev Entomol 36:561–586

Barrett BW, Petropoulos GP (2012) Satellite remote sensing of surface soil moisture. In: Petropoulos GP (ed) Remote sensing of energy fluxes and soil moisture content. CRC Press, Boca Raton, pp 85–120

Candiago S, Remondino F, De Giglio M, Dubbini M, Gattelli M (2015) Evaluating multispectral images and vegetation indices for precision farming applications from UAV images. Remote Sens 7:4026–4047

Di Gennaro SF, Battiston E, Di Marco S, Facini O, Matese A, Nocentini M, Palliotti A, Mugnai L (2016) Unmanned Aerial Vehicle (UAV)-based remote sensing to monitor grapevine leaf stripe disease within a vineyard affected by esca complex. Phytopathol Mediterr 55:262–275

Drinkwater L, E.&Snapp S. S. (2007) Nutrients in agroecosystems: re-thinking the management paradigm. Adv Agron 92:163–186

Eroshenko F, Storchak I, Engovatova I, Likhovid N (2020) Use of remote sensing earth data for regional assessment of winter wheat grain quality. InterCarto InterGIS 26:240–251. https://doi.org/10.35595/2414-9179-2020-3-26-240-251

Gao F, Anderson MC, Zhang X, Yang Z, Alfieri JG, Kustas WP, Mueller R, Johnson DM, Prueger JH (2017) Toward mapping crop progress at field scales through fusion of Landsat and MODIS imagery. Remote Sens Environ 188:9–25

Gao F, Anderson M, Daughtry C, Karnieli A, Hively D, Kustas W (2020) A within-season approach for detecting early growth stages in corn and soybean using high temporal and spatial resolution imagery. Remote Sens Environ 242:111752

Huang Y, Chen Z, Yu T, Huang X, Gu X (2018) Agricultural remote sensing big data: management and applications. J Integr Agric 7:1915–1931

Justice CO, Townshend JRG, Vermata EF, Masuoka E, Wolfe RE, Saleons N, Ray DP, Morisette JT (2002) An overview of MODIS land data processing and product status. Remote Sens Environ 83:3–15

Kerr YH, Waldteufel P, Wigneron JP, Delwart S, Cabot F, Boutin J, Mecklenburg S (2010) The SMOS Mission: new tool for monitoring key elements of the global water cycle. Proc IEEE 98(5):666–687

Khanal S, Fulton J, Klopfenstein A, Douridas N, Shearer S (2018) Integration of high resolution remotely sensed data and machine learning techniques for spatial prediction of soil properties and corn yield. Comput Electron Agric 153:213–225

Kingra PK, Majumder D, Singh SP (2016) Application of remote sensing and Gis in agriculture and natural resource management under changing climatic conditions. Agric Res J 53:295–302. https://doi.org/10.5958/2395-146X.2016.00058.2

Kulkarni SS, Bajwa SG, Huitink G (2010) Investigation of the effects of soil compaction in cotton. Am Soc Agric Biol Eng 53:667–674

Lal R (2008a) Soil carbon stocks under present and future climate with specific reference to European ecoregions. Nutr Cycl Agroecosyst 81:113–127

Lal R (2008b) Sequestration of atmospheric CO_2 in global carbon pools. Energy Environ Sci 1:86–100

Mahlein A-K, Rumpf T, Welke P, Dehne H-W, Plümer L, Steiner U, Oerke E-C (2013) Development of spectral indices for detecting and identifying plant diseases. Remote Sens Environ 128:21–30

Maresma Á, Ariza M, Martínez E, Lloveras J, Martínez-Casasnovas JA (2016) Analysis of vegetation indices to determine nitrogen application and yield prediction in maize (*Zea mays* L.) from a standard UAV service. Remote Sens *8*:973

Moran MS, Peters CD, Watts JM, McElroy S (2004) Estimating soil moisture at the watershed scale with satellite-based radar and land surface models. Can J Remote Sens 30(5):805–826

Partel V, Kakarla SC, Ampatzidis Y (2019) Development and evaluation of a low-cost and smart technology for precision weed management utilizing artificial intelligence. Comput Electron Agric 157:339–350

Perfecto I, Vandermeer J (2008) Biodiversity conservation in tropical agroecosystems: a new conservation paradigm. Year Ecol Conserv Biol 1134:173–200

Raun WR, Solie JB, Stone ML, Martin KL, Freeman KW, Mullen RW, Zhang H, Schepers JS, Johnson GV (2005) Optical sensor-based algorithm for crop nitrogen fertilization. Commun Soil Sci Plant Anal 36:2759–2781

Reichstein M, Camps-Valls G, Stevens B, Jung M, Denzler J, Carvalhais N, Prabhat. (2019) Deep learning and process understanding for data-driven earth system science. Nature 566:195–204

Scharf PC, Shannon DK, Palm HL, Sudduth KA, Drummond ST, Kitchen NR, Mueller LJ, Hubbard VC, Oliveira LF (2011) Sensor-based nitrogen applications out-performed producer-chosen rates for corn in on-farm demonstrations. Agron J 103:1684–1691

Smith P et al (2008) Greenhouse gas mitigation in agriculture. Philos Trans R Soc B 363:789–813

Tscharntke T, Klein AM, Kruess A, Steffan-Dewenter I, Thies C (2005) Landscape perspectives on agricultural intensification and biodiversity: ecosystem service management. Ecol Lett 8:857–874

Weiss M, Jacob F, Duveillerc G (2020) Remote sensing for agricultural applications: a meta-review. Remote Sens Environ 236:111402

Yue J, Yang G, Li C, Li Z, Wang Y, Feng H, Xu B (2017) Estimation of winter wheat above-ground biomass using unmanned aerial vehicle-based snapshot hyperspectral sensor and crop height improved models. Remote Sens 9:708

Sensing Climate Change Through Earth Observations: Perspectives at Global and National Level

Girish S. Pujar, Alok Taori, Abhishek Chakraborty, and Tarik Mitran

Abstract The window to mitigate and adapt to climate change is closing very fast for humanity. Unless drastic measures are introduced and implemented with unambiguous policy oversight, the Earth is going to witness irreversible damage to biophysical systems to the peril of life on this planet. Emphatic arguments on access of climate alerts to everyone, by the United Nations, highlight the need to build a comprehensive system observing the trinity of the ocean, land and atmosphere. Earth observation systems comprising remote sensing in increasingly innovative interactions of energy and matter offer unprecedented scope of watching phenomenon across spatial scales. Coupling these observations with substantiated process models gives insights into future scenarios at reasonable levels of confidence. Current review attempts to comprehend remote sensing systems in place, for observing atmosphere followed by ocean and land phenomenon as well as the information systems, in Indian context, designed for application by varied users. Data and information derived by range of sensors on board Indian satellites are discussed, and latest understanding of vulnerability patterns in varied land cover contexts such as snow, water, crop, and forest is summarized for benefit of decisions across hierarchy of managing natural resources. However, a wide variety of phenomenon linked to climate change impact which could not be covered does

G. S. Pujar (✉)
Rural Development and Watershed Monitoring Division, Remote Sensing Application Area
National Remote Sensing Centre (ISRO), Hyderabad, India
e-mail: pujar@nrsc.gov.in

A. Taori
Atmospheric Science Division, Earth and Climate Sciences Area, NRSC (ISRO), Hyderabad, India

A. Chakraborty
Agroecosystem and Modeling Division, Agricultural Sciences & Applications Group, Remote Sensing Applications Area, NRSC(ISRO), Hyderabad, India

T. Mitran
Soils & Land Resources Assessment Division, Soil Resources & Land Use Mapping Group, Remote Sensing Application Area, NRSC (ISRO), Hyderabad, India

P. M. Priyadarshan et al. (eds.), *Digital Agriculture*,
https://doi.org/10.1007/978-3-031-43548-5_8

not exclude the scope of applying Earth observation technology far and wide. Global and national perspectives, hence, deliberated herewith correspond to the generic know-how of the current technological scope and application to mostly operational needs.

1 Introduction

Landmark report released by Intergovernmental Panel on Climate Change (IPCC) during February 2022 has clearly declared that the window to mitigate climate change is fast closing and stated "any further delay in concerted anticipatory global action on adaptation and mitigation will miss a brief and rapidly closing window of opportunity to secure a livable and sustainable future for all". Further to it, report on mitigation (IPCC 2022) says that by 2030 it is possible to halve the emissions. Changing climate is marked by wide variety of global manifestations that can threaten life and diversity on Earth by inducing extreme droughts, rising sea levels, disappearing snow cover and ice caps, increased wildfires, heatwaves, storms, insect outbreaks, reducing farm yields, decreased accessibility to water and increased conflicts over natural resources. It is also coupled with indirect effects such as physical and mental health impacts, destroyed infrastructure, mass population displacement and possible widespread hunger. As per United Nations' news release, greenhouse gas emissions generated by human activity have increased since 2010 'across all major sectors globally' and global community is on pathway to warming of more than double the 1.5 °C limit that was agreed in Paris in 2015. Such a status warrants application of state-of-the-art sensors to detect, monitor, model and virtualize the geo-intelligence in space and time to develop alerts for every citizen in a manner that is easily connectable and extant. Conventional observations from ground network provide information up to a certain degree of detail but fail to provide spatial scale, which logically leads to application of space-based sensors to collect critical information at desired frequency (Cracknell and Cracknell 2001).

Climate change has been defined by the UN Framework Convention on Climate Change (UNFCCC) as 'a change of climate which is attributed directly or indirectly to human activity that alters the composition of the global atmosphere and which is in addition to natural climate variability observed over comparable time periods'. The twin problems of climate change and associated extreme weather events are of grave concern under the present scenario of global warming. It is, however, not a new phenomenon as the presence of ice age was conclusively proven in the 1840s. Natural processes like changes in the solar output, Earth's orbit and volcanic eruptions may cause climate change. But over the past century, there has been a noticeable acceleration in climate change through global warming caused by the accumulation of greenhouse gases in the atmosphere due to anthropogenic activities. Failure to keep climate change in check can have a grim fallout, as witnessed ~250 million years ago through the mass extinction of nearly 96% of marine and

70% of terrestrial species due to runaway climate change triggered possibly by volcanic gas emissions.

The fourth assessment report of the IPCC estimated an increase of 0.74 °C in global mean temperatures from 1906 to 2005. Global mean surface temperatures in 2018 are higher than pre-industrial (1850) levels by nearly 1 °C. The year 2019 is likely to be recorded as the hottest year globally with record-high maximum temperatures recorded over India and Europe. Compared against pre-industrial levels (circa 1850–1880), there has been an increase in global surface temperature by 0.85 °C, increase in ocean acidity by 26% and rise in global sea levels by 3.3 mm $year^{-1}$. The Antarctic ice sheet has decreased by 152 km^3 between 2002 and 2006, and the Arctic sea ice area is decreasing by 2.7% per decade. The top 700 m ocean temperature has risen by 0.302 °F since 1969 resulting in thermal expansion, sea level rise and inundation of low-lying coastal areas. Climate change has overall had a negative effect on crop productivity and increasing frequency and intensity of extreme weather events such as heatwaves, heavy precipitation, cyclones and lightning. The elevation in ocean temperature and acidity are imposing adverse effects on marine ecosystem, driving the extinction of marine species and triggering frequent occurrence of storms.

Most of the climate-related information has been conventionally derived from point-based measurements and they often sense spatial spread across entire Earth only as samples at best. Remote sensing offers scope to build continuous surfaces related to climate processes and enable continuous observations. Processes especially related to carbo dioxide, clouds, oceans, surface and air temperature, vegetation phenology and water surfaces determine the way climate change will be managed in future. Satellites serve an extremely pivotal role in this (Canadian Space Agency 2022).

Diversity of information regarding ability to observe climate processes and concomitant changes using Earth observation approaches is humongous. Conveying the order, essence, utility and stakes inherent to this body of knowledge (NASA n.d.) is of paramount importance as of now, since mitigation and adaptation measures require content that is clear and decipherable. Across 1,20,000 articles published on climate change since 1960s, work on analysis of satellite-derived sea surface temperature in tandem with in situ observations (Reynolds 2002) stands out as eighth among the top ten most cited climate change articles (McSweeney 2015) Here, an attempt is made to comprehend the latest innovations in operation and proposed in near term horizon across the globe to observe climate and its change. Special emphasis will be laid on to the Indian scenario of observation and inputs to adaptation and mitigation.

1.1 Scale of Vulnerability

The US National Oceanic and Atmospheric Administration's Geophysical Fluid Dynamics Laboratory applied several ocean–atmosphere coupled models to predict the cumulative effect of GHG emissions from different population, economic and energy use projections that may affect Earth and created future scenarios called 'Representative Concentration Pathways (RCPs)' similar to earlier scenarios known as SRES level (Special Report on Emission Scenarios). RCPs are known by the radiative forcing characteristic of the global concentration of GHGs prevalent, such as for instance.

RCP 4.5 indicates retention of 4.5 watts of energy by Earth's atmosphere per sq.m. In this context, the Coupled Model Intercomparison Project (CMIP) is a collaborative framework designed to improve knowledge of climate change and facilitates concentration pathway modelling. Overall, the vulnerability of a development pathway is clearly brought out by these indicators, and pathways include scope to bring policy modulation to reduce the emissions (Fig. 1).

Different RCPs are characterized by features and assumptions related to emissions due to policy-driven actions. RCP 2.6 being a very low future emissions pathway has carbon dioxide remaining constant until early of this century, followed by a decline to turn negative by 2100. Assumption for the pathway is that fossil fuel witnesses sharp decline, more biofuel-derived from crop land and reduced methane emission by 40%. Low to moderate pathway of RCP 4.5 assumes slight increase in carbon emissions until mid-century, then declines, with stabilized methane emissions. Here, large-scale reforestation coupled with sharp decline in energy use is assumed, coupled with reduction in size of agricultural land to increase yield and lowered meat consumption with stricter climate policies. Very high future emission pathway of RCP 8.5 points to three times higher emission of carbon dioxide than

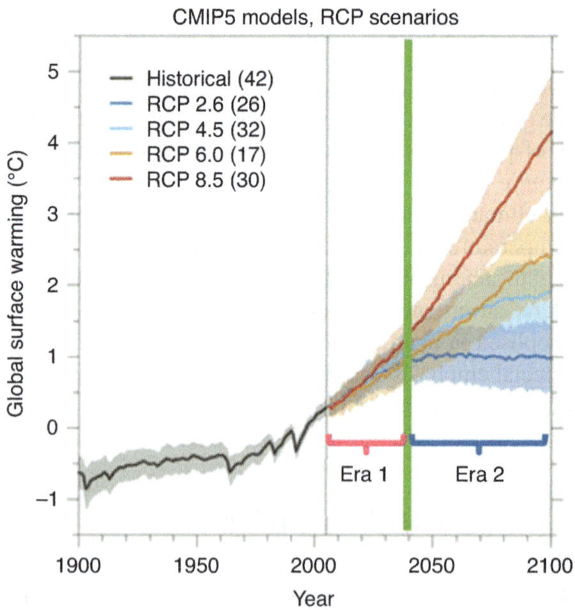

Fig. 1 Intergovernmental Panel on Climate Change (IPCC) graph of future temperature change under alternative greenhouse gas emission scenarios. (McCarl et al. 2016)

present with large increase in methane output. Fossil fuels dominate over uptake of renewables with least implementation of strict climate policy (Climate information n.d.).

Extremities projected include four different scenarios of warming at 1.5, 2, 3 and 4 °C at 2100 with its impacts on nature (Fig. 2) and major Earth processes in relation with RCPs. Mildest of all with only 1.5 degree (as proposed in the Paris Agreement) will in fact has potential to induce minimum two months of drought and wildfires, suffered crop yields such as rice, maize, wheat and soybean with rising sea level

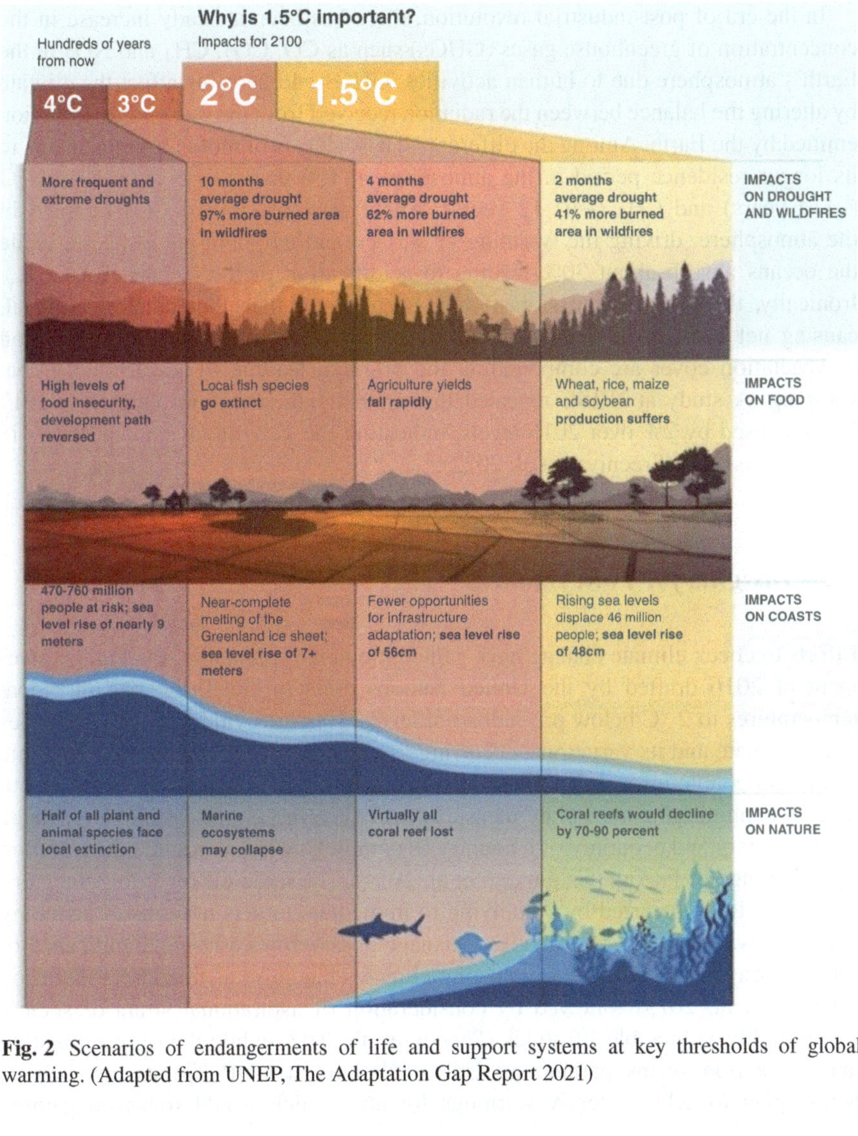

Fig. 2 Scenarios of endangerments of life and support systems at key thresholds of global warming. (Adapted from UNEP, The Adaptation Gap Report 2021)

. Sea level is likely to rise by 48 cm and coral reefs would decline by about 70–90%. As the warming goes higher to 2 degrees, wildfires may increase by 62%, with average droughts occurring 4 months a year and most of coral reefs being lost, with a 56 cm rise in the sea level. Other severe scenarios of global warming of 3 and 4 degrees at much distant future at century scale would be witnessing 10 and more months of drought, with local fish going extinct and sea would be rising at 7–9 m. Marine ecosystems would be in total collapse and half of all plant and animal species face local extinction. Hence, controlling warming around 1.5 degree is critical to the survival of most of marine and terrestrial biodiversity, food systems as well as coastal habitation and economy.

In the era of post-industrial revolution, there has been a steady increase in the concentration of greenhouse gases (GHGs) such as CO, CO_2, CH_4 and NO_2 in the Earth's atmosphere due to human activities. GHGs and aerosols affect the climate by altering the balance between the radiation received from the Sun and the radiation emitted by the Earth. Among the different GHGs, CO_2 is of more importance due to its longer residence period in the atmosphere (~150 years) as compared to N_2O (~114 years) and CH_4 (~10–12 years). About 40% of emitted CO_2 retained in the atmosphere, driving the warming of surface and adjoining atmosphere, while the oceans absorb about 30% leading to acidification of the oceans (ESA n.d.). Ironically, the rates of emission of GHGs are faster than their natural removal, causing net accumulation in the atmosphere. Change in land use and reduction of vegetation cover are compounding the problem leading to accelerated global warming. A study at NRSC revealed that annual Indian carbon budget for 2017 has increased by 2% over 2016 levels, indicating the key role of human beings in carbon emissions (Sreenivas et al. 2022).

1.2 *Insights for Policymakers*

Efforts to check climate change have gained momentum globally; the Paris Agreement of 2016 drafted by the United Nations aims to check the rise in global temperatures to 2 °C below pre-industrial levels. It is thus vital to continue studies of the climate and its variations, occurring as a result of natural and anthropogenic factors, to enable informed decision-making on mitigation/adaptation measures to face a changing climate. Policy measures to reduce the impact of climate change on the society and economy of a country as complex as India require consideration of wide range of factors (Chaturvedi et al. 2014). The trade-off regarding the gains and losses to be incurred by complying to immediate targets in terms of economy and poverty alleviation is difficult to manage at governance level. Significance of focusing on the co-benefits of managing climate change so as to address inclusion (Dubash et al. 2013) followed by consideration of aspirational strata of society living highly vulnerable lifestyles (Pandey et al. 2018) while devising mitigation and adaptation seems paramount. United Nations at COP 27 has unveiled an action plan to achieve 'early warnings for all', which would focus on gaining disaster risk knowledge, observations and forecasting, preparedness and response and communication of early warnings to all. In view of such ambitious global initiative, it is certainly essential that expertise involved in policymaking and its implementation needs to have well-informed choices for assessing vulnerability and adapt to it. India's instances of disaster preparedness and handling several hydro-climatological extremes is of high quality and needs to be upscaled further using state-of-the-art technology such as space-based approach. The simplicity and alacrity of information availability as early warning and impact event should determine how fast the policy managers handle the climate change-related aspects deftly. Mere labyrinth of terminologies and hidden inferences should not defer the usage of information, since scientific penetration in India is yet to reach satisfactory levels. However, the hopeful part is that advent of information technology can deliver cascade of content in understandable manner through open-source tools such as ISRO Bhuvan.

2 Earth Observation Scenario in India

2.1 National-Level Datasets

2.1.1 Climate Science and Information System

Towards describing global warming and associated climate change, the Global Climate Observing System (GCOS) of the WMO (World Meteorological Organization) has identified 54 key parameters or essential climate variables (ECVs), necessary for characterizing the Earth's climate, encompassing the domains of land, atmosphere and ocean. These parameters can objectively quantify the changes of the Earth's ecosystems with space and time. Realizing the gravity of the situation and to address the lack of an accurate climate quality database from the Indian perspective, the Indian Space Research Organization (ISRO) has established 'National Information System for Climate and Environment Studies (NICES)' at the National Remote Sensing Centre (NRSC), Hyderabad, in 2012 to generate long-term, consistent and accurate database using satellite data (NRSC 2022). NICES (Fig. 3) is a multi-institutional endeavour from which currently 64 bio-/geophysical parameters are being generated and freely disseminated to stakeholders through web-enabled services. The objectives of NICES are the establishment of appropriate observational network, acquisition and processing of international and national missions' data, generation of spatially and temporally blended climate products, establishment of supporting infrastructure and services and effective dissemination of data for scientific utilization of data towards impact assessment, adaptation, vulnerability and mitigation strategy. The historic database will help understand the impacts of climate variability on ecology and to quantify how different ecosystems have historically responded to climate change as well as the uncertainty in projected biophysical impact on biosphere and humanity.

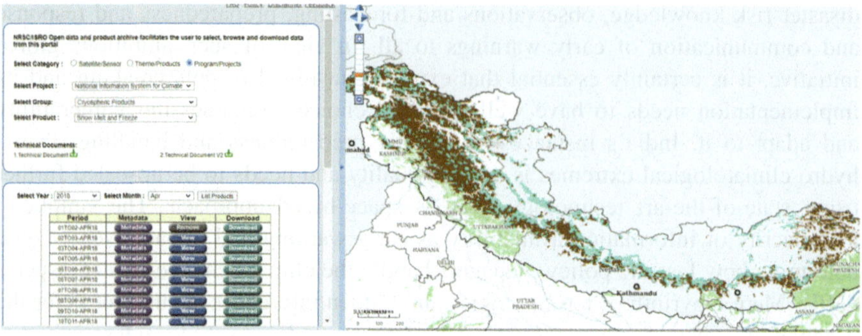

Fig. 3 Dissemination link interface for snow and glacial melt data download datewise, from Bhuvan NICES portal

Endeavour also includes regional workshops to create awareness about the climate change and the innovative use of the climate database being generated at NRSC. Interest has been created among academia and different stakeholders to contribute in the building up of Indian national database on climate change studies. The collaboration in the form of projects will enable academic interface to share the domain expertise towards developing better understanding about Earth system sciences, new algorithms, calibration, validation and establishment of new sensor network across the nation. The dedicated team of the Earth and Climate Science Area (ECSA) of NRSC is engaged in studying the climate and its variability, empirical/process-based methods for estimation of climate variability impacts across multiple spatial–temporal scales, including emerging Earth observation technologies and distributed sensor networks, methods for forecasting impacts of climate variability on biosphere response and translating forecasts into useful decision support for the farmers and policymakers.

2.1.2 Climate Science Activities

As part of climate science-related experiments and campaigns, regular participation of the centre in major national and international cruises in deep oceans is assured, to collect the ocean biological, physical and chemical parameters which are otherwise absent or availability is restricted. This helps to understand anthropogenic impacts better. Climate science activities contribute for improved and value-enhanced datasets under NICES in the long run.

Key areas of research findings cover areas related to measurement of carbon dioxide, methane, clouds, aerosols, cloud top temperature, land and ocean productivity and sea level. The 35th Indian Scientific Expedition to Antarctica in 2016 witnessed CO_2 levels exceeding 400 ppm for the first time since international efforts to monitor CO_2 levels gained momentum. Airborne measurements over different parts of the country help towards understanding the variation in greenhouse gases and aerosol parameters with height. State-of-the-art observatory for continuous measurements of greenhouse gases, aerosols, radiation and other meteorological since 2014 have been yielding reliable records to support climate change analysis and help validate satellite-based observations.

Reduction in cloudiness during the past decade, over a belt extending from the Arabian Sea, southern and eastern parts of India to the Bay of Bengal and Northeast India, compared to increases up to 10% over the north-western parts has been observed in a preliminary study using satellite data. Declining cloud cover over the east coast is of concern in view of weakening of the summer monsoon and lower frequency of monsoonal depression over the Bay of Bengal, which may lead to lower precipitation. Results also indicated a decrease in planetary albedo following an increase in soil moisture and precipitable water vapour. The planetary albedo increases with increase in atmospheric aerosol concentrations. Remote sensing data over the past decade reveals that atmospheric aerosols are dominated by scattering type of aerosols which reflect solar radiation and thus cool the surface. An increasing trend in the amount of scattering aerosols over India was noticed during post-monsoon and winter, while negative trends are seen in pre-monsoon and summer monsoon.

Apart from aerosols, cloud top temperature assumes key role improving the regional climate change modelling, since it helps to characterize the cloud physical process. A mixture of light-absorbing and light-scattering aerosols contribute to atmospheric solar heating and surface cooling. The sum of the two climate forcing terms – the net aerosol forcing effect – is thought to be negative and may have masked as much as half of the global warming attributed to the recent rapid rise in greenhouse gases. However, the aerosol forcing effect remains largely underestimated (Ramanathan et al. 2007). On the other hand, it is also gradually being understood that transport determines trends in the aerosols (Prijith et al. 2018).

Under the National Carbon Project (NCP) within the ambit of the ISRO Geosphere Biosphere Programme (IGBP), the GHGs dynamics have been studied and budget was estimated by means of a robust observational network across the country for continuous measurement of CO_2. The findings reveal that during biomass burning, CO_2 and CH_4 have increased by ~2 and ~0.06%, respectively. Climate variability has been driving an increasing trend in vegetation cover and net primary productivity over crop/grass land-dominated northwestern and south-central Peninsular India and a decreasing trend over forested regions of the North-East and Western Ghats. The Indian terrestrial ecosystem is currently acting as a net sink for CO_2 at an average rate of 20 $TgCyr^{-1}$. Carbon dioxide levels in the surface layer of the atmosphere are increasing at one-third the rate of increase in surface fluxes, implying most part of surface emissions is transported out from the surface layer of the atmosphere over India.

Remote sensing of the ocean has observed distinct phytoplankton bloom and its seasonality in the northern Indian Ocean. The oceans cover more than 70% of the Earth's surface and act as a major sink for atmospheric CO_2 through the process of carbon sequestration by diverse microscopic organisms, called phytoplankton, which form the base of the food chain and play a crucial role in climate regulation and influence food security and ocean productivity. Initiatives were taken up to identify the carbon sinks and sources along the Indian coast which include small and important ecosystems such as Chilika Lake. Further, an annual increase in regional sea levels between 2 mm and 8 mm has been detected in the southern and northern tropical Indian Ocean using satellite gravity data. Changes in sea levels are influencing changes in shoreline along the Indian coast over the decades.

2.1.3 Datasets in NICES

This system focuses on serving range of products as essential climate variable (ECV) as well as geophysical product for the purpose of climate modelling and monitoring. An essential climate variable (ECV) is a physical, chemical or biological variable or a group of linked variables that critically contributes to the characterization of Earth's climate (GCOS 2021). As per latest GCOS (Global Climate Observing System) norms, ECVs for three domains related to climate change are listed as being realized under NICES (Table 1). ECV may be sourced from the sensor or as a modelled product using assimilation of information from non-satellite sources as well as a determinant parameter from satellite data. The products are from domains of terrestrial, oceanographic and atmospheric sensors. Terrestrial products are derived in majority from IRS datasets from Resourcesat LISS III or AWiFS sensors and gridded for appropriate resolutions of 5 km and higher. Atmospheric products are derived from sensors on board Indian satellites such as INSAT 3D, 3DR, Kalpana as well as Suomi NPP, AVHRR and AURA. Oceanography-related products are either sourced or modelled. Data from both indigenous sensors such as Oceansat, SARAL and international sensors such as TMI, MODIS etc. Totally there are seven terrestrial ECVs and 29 geophysical products available, while ocean products consist 12 ECVs and 22 geophysical products in NICES. Atmospheric products consist of 3 ECVs and 2 geophysical products, respectively (Table 2).

2.1.4 Features of Products

Surface soil moisture plays key role in water and energy cycles and determines land atmosphere feedback. It a key component in the carbon dioxide exchange. This ECV is retrieved using brightness temperature data from Advanced Microwave Scanning Radiometer 2 (AMSR 2, on board GCOM-W-1) collected at 13.30 local time using 10.65 and 36.5 G Hz channels. Using Land Parameter Retrieval Model surface soil moisture at 25 km resolution is prepared and disseminated once in 2 days for entire country. Soil carbon is generated at 5×5 km grid using spatial modelling approach involving data on soil types, land use and agroclimatic subregion. Net sown area as fractional product over same grid size is derived using multitemporal data for three seasons mapped using monthly data composites and has utility in mesoscale models.

Table 1 List of global-level essential climate variables as per global climate observation system standards

Global ECVs as per GCOS-240 (2021 report)					
Atmosphere	14	Ocean	16	Terrestrial	19
Upper atmosphere		Surface ocean physics		Cryopshere	
	Lightening		Sea surface temperature		Ice sheets and ice shelves
	Wind speed and direction		Sea ice		Glaciers
	Upper air water vapour		Salinity		Snow
Atmospheric composition			Sea level		Permafrost
	Aerosols		Stress	Hydrosphere	
	Precursors of aerosols/ozone		Heat flux		Groundwater
	Ozone		Currents		River discharge
	Clouds		Sea state		Lakes
	CO2/CH4/GHGs	Ocean biochemistry		Biosphere	
Surface atmosphere			Inorganic carbon		Aboveground biomass
	Precipitation		Nutrients		Fire
	Surface radiation budget		Nitrous oxide		Land cover
	Surface temperature		Oxygen		Soil carbon
	Surface pressure		Transient tracers		Leaf area index
	Wind speed and direction		Ocean colour		Albedo
	Surface water vapour	Biological/ecosystems			Fraction of APAR
			Marine habitats		Land surface temperature
			Plankton		Evaporation from land
					Soil moisture
				Anthroposphere	Anthropogenic GHG fluxes
					Anthropogenic water use

Table 2 ECVs available with NICES (NRSC-ISRO) for user community

Terrestrial products			
Sl no.	**Geophysical dataset**	**Satellite/sensor**	**Resolution**
ECV			
1.1	Land-use land cover (MM5 compatible)	Resourcesat-2/AWiFS	30″/2′/5′
1.2	Land-use land cover (WRF compatible)	Resourcesat-2/AWiFS	30″/2′/5′
2	Mean organic soil carbon density	Resourcesat-2/AWiFS	5 km
3	Surface soil moisture	Aqua AMSR-E and GCOM-W1/AMSR2	0.25°
4	Snow cover fraction	Resourcesat-2/AWiFS	$3' \times 3'$
5	Average annual Forest fire density	Aqua and Terra/MODIS	5 km
6	Surface water bodies fraction	Resourcesat-2,2A/AWiFS	$3' \times 3'$
Geophysical products			
1	Albedo	Oceansat-2/OCM-II	1 km
3	NDVI		
2.1	NDVI	Oceansat-2/OCM-II	8 km
2.2	NDVI	Oceansat-2/OCM-II	1 km
2.3	Filtered NDVI	Oceansat-2/OCM-II	1 km
3	Vegetation fraction	Oceansat-2/OCM-II	1 km
4	Soil		
4.1	Mean inorganic soil carbon density	Resourcesat-2/AWiFS	5 km
4.2	Fraction soil depth	Resourcesat-2/AWiFS	5 km
4.3	Fraction soil textural class	Resourcesat-2/AWiFS	5 km
5	Land degradation (3 layers)		
5.1	Fraction water erosion	Resourcesat-2/LISS-III	5 km
5.2	Fraction wind erosion	Resourcesat-2/LISS-III	5 km
5.3	Fraction salt-affected	Resourcesat-2/LISS-III	5 km
6	Forest fire		
6.1	St. dev. of average annual forest fire density	Aqua and Terra/MODIS	5 km
6.2	Length of fire period	Aqua and Terra/MODIS	5 km
7	Forest cover fraction	SOI/Landsat MMS and TM/Resourcesat-2/AWiFS	5 km
8	Forest types	Resourcesat-2/AWiFS	5 km
9	Net sown area		
9.1	Fractional net sown area	Resourcesat-2/AWiFS	5 km
9.2	Fractional kharif sown area	Resourcesat-2/AWiFS	5 km
9.3	Fractional rabi sown area	Resourcesat-2/AWiFS	5 km
9.4	Fractional fallow area	Resourcesa-2/AWiFS	5 km
15	Snow melt and freeze	Oceansat-2/OSCAT	2.225 km
16	Snow cover fraction	Resourcesat-2/AWiFS	$3' \times 3'$
17	Himalayan glacial lakes and water bodies	Resourcesat-2/AWiFS	1:250 k
18	Snow melt and freeze	Oceansat-2/OSCAT	2.225 km
19	Snow albedo	Resourcesat-2/AWiFS	250 m

(continued)

Table 2 (continued)

Atmospheric products			
	ECV		
1	Cloud fraction	Kalpana/VHRR and INSAT-3D imager	0.25° × 0.2°
2	Cloud top temperature	INSAT-3D	0.5° × 0.5°
3	Lightning	Ground network	0.1° × 0.1°
	Geophysical products		
1	Derived tropospheric ozone	OMI and MLS/Aura	1° × 1°
2	Planetary boundary layer height	Suomi NPP/CrIS	0.25° × .25°
Ocean products			
Sl no.	**Geophysical products**	**Satellite/sensor**	**Resolution**
	ECV		
1	Ocean surface winds		
1.1	Ocean surface winds	OSCAT/ScatSat-1	0.5°
1.2	Ocean surface winds	OSCAT/ScatSat-1	0.25°
2	Wind stress		
2.1	Wind stress	OSCAT/ScatSat-1	0.5°
2.2	Wind stress	OSCAT/ScatSat-1	0.25°
3	Ocean surface currents	Altika and ScatSat-1	0.25°
4	Ocean chlorophyll		
4.1	Chlorophyll concentration (OC2 algorithm) (N.Indian Ocean, NInOc)	Oceansat-2/OCM II	1 km
4.2	Chlorophyll concentration (OC4 algorithm) (NInOc)	Oceansat-2/OCM II	1 km
4.3	Chlorophyll concentration (OC2 algorithm)	Oceansat-2/OCM II	4 km
4.4	Chlorophyll concentration (OC4 algorithm)	Oceansat-2/OCM II	4 km
	Geophysical products		
1	Wind curl		
1.1	Wind curl	OSCAT/ScatSat-1	0.5°
1.2	Wind curl	OSCAT/ScatSat-1	0.25°
2	Sea level pressure	Oceansat-2/OSCAT and ScatSat-1	0.5°
3	Ekman currents	OSCAT/ScatSat-1	0.25°
4	Sea surface height anomaly	SARAL/Altika	0.25°
5	Geostrophic currents	SARAL/Altika	0.25°
6	Eddy kinetic energy (EKE)	Altimeter SSHA (AVISO)	0.25°
7	Monthly mean sea level anomaly (MMSLA)	Altimeter SSHA (AVISO)	1°
8	Diffuse attenuation coefficient		
8.1	Diffuse attenuation coefficient at 490 nm (KD490)	Oceansat-2/OCM II	1 km
8.2	Diffuse attenuation coefficient at 490 nm (KD490)	Oceansat-2/OCM II	4 km

(continued)

Table 2 (continued)

9	Total alkalinity (TA)	Aquarius and MODIS	0.25°
10	Dissolved inorganic carbon (DIC)	Aquarius and MODIS	0.25°
11	Co-tidal map		
11.1	K1O1 co-tidal map	Model derived	2′
11.2	M2S2 co-tidal map	Model derived	2′
12	Ocn heat content (OHC) and ocn. Mean temp (OMT) at different depths	TMI/AMSR-2 SST and Altimeter SSHA	0.25°
13	Tropical cyclone heat potential	TMI/AMSR-2 SST and Altimeter SSHA	0.25°
14	Ocean heat content of 700 m layer	TMI/AMSR-2 SST and Altimeter SSHA	0.25°
15	Tropical cyclone heat potential forecast	Model derived	0.5°
16	Depth of 26 degree isotherm	Model derived	0.5°

Snow melt and freeze is a key parameter indicating several energy and water fluxes in the system in the form of delivering and removing heat as well as in run-off estimations (Fig. 3). Scatterometer data for different satellites is used to derive this parameter. Scatterometers operate on the principle of active microwave scanning using rotating sensors.

Among ocean-related parameters, ocean surface winds drive heat exchange and momentum at ocean–atmosphere interface as well as provide key forcing of ocean circulation responsible for global carbon transport. Daily global gridded wind fields from two-day composite are generated using ascending and descending pass data from Oceansat scatterometer (OSCAT and SCATSAT). Ocean colour is key product indicating the photosynthetic potential of ocean and hence carbon sequestration efficiency. Chlorophyll-a is deduced from reflectance from OCM sensor on board Indian satellite Oceansat. It also has great application in potential fishing zone, blue economy. Daily ocean mean temperature and heat content is another key set of ECV, measured as kilojoules per sq. cm, which has significant bearing on climate change understanding since climate dynamics and interior thermodynamics are linked to it. In NRSC, this parameter is being modelled since 1998 till date using neural network techniques involving sea surface height anomaly and sea surface temperature from Tropical Rainfall Measuring Mission's microwave imager. Monthly sea level anomaly is a key product from sensors on board TOPEX/Poseidon, ERS-1 or 2, Envisat, Jason-1 and 2, HY-2 and SARAL/AltiKa and conveys the area of ocean water sinking and upwelling, connected to influences of climate change. Total alkalinity is buffering capacity of oceans, which acts as natural feedback of changing ocean pH, in turn connected to carbon dioxide enrichment of oceans. Dissolved inorganic carbon (DIC), an important sink of atmospheric CO2, in the form of carbonates and bicarbonates, is strong parameter and linked to gross primary production. Sea surface salinity from Aquarius and drifting buoys and sea surface temperature and chlorophyll from MODIS-A are used to generate these two products.

Cloud top temperature (CTT) and lightening are two critical variables derived using satellite and ground sensor network, respectively. Small change in abundance and distribution of clouds can change climate more than global factors of climate change. Continuous monitoring of clouds alone ensures their better representation in models. Data from INSAT 3D using thermal IR emission during day and night is used to prepare CTT. It has both research and operational utility. Lightening seems to indicate extreme events increasingly witnessed due to climate change and also due to production of nitrogen oxide that controls ozone formation strongly. This is detected using network of sensors having 50% overlap of event detection using algorithmic mapping (Taori et al. 2022, 2023). INSAT 3D imaging has been employed to characterize the diurnal variation of cloud top temperature (CTT) and delineation of cloud mask involving validation against radiosonde observations as well as inter-comparison against MODIS and CALIOP (Cloud-Aerosol Lidar with Orthogonal Polarization) derived products. Algorithmic retrieval provided about 85% accuracy of estimation of cloud presence, wherein cloud detection algorithm employed nine different tests, in accordance with solar illumination, satellite angle and surface type conditions to generate pixel-resolution cloud mask (Lima et al. 2019).

2.2 *MOSDAC (Meteorological and Oceanographic Satellite Data Archival Centre)*

Information system built as web enabled data dissemination system from the suite of Indian remote sensing satellites observing atmosphere and ocean processes through algorithmic retrieval. The Meteorological and Oceanographic Satellite Data Archival Centre (MOSDAC) is a data centre of Space Applications Centre (SAC) and has facility for satellite data reception, processing, analysis and dissemination. MOSDAC is operationally supplying Earth observation data from Indian meteorology and oceanography satellites, to cater to national and international research requirements. Degree of extant information served has high utilitarian value especially for extreme weather events as well as regular Earth system dynamics at global scale. It deals with information on cold waves, heatwaves, heavy rain, lightning and state of seas along with solar and wind energy related content. MOSDAC-LIVE is a web-enabled data and information visualization and analysis system of MOSDAC, SAC/ISRO. MOSDAC LIVE provides access to satellite data products and information products derived from satellite and model forecast in near real-time basis. Details of the missions in orbit brief the major satellite systems in place.

(i) INSTA 3D: Payload consists of an imager, sounder and communication sensors. Imager provides imaging of the Earth disc from geostationary altitude. This uses six spectral ranges of one visible (0.52–0.72 micrometres) and five infrared; 1.55–1.70 micrometer (SWIR), 3.80–4.00 micrometer

(MIR), 6.50–7.00 micrometer (water vapour), 10.2–11.2 micrometer (TIR-1) and 11.5–12.5 micrometer (TIR-2) bands. The ground resolution at the sub-satellite point is nominally 1 km × 1 km for visible and SWIR bands, 4 km × 4 km for one MIR and both TIR bands and 8 km × 8 km for WV band. It is an improved design of VHRR/2 (very-high-resolution radiometer) heritage instrument flown on the Kalpana-1 and INSAT-3A missions.

(ii) INSAT 3DR: Continuation mission of INSAT 3D.

(iii) Kalpana-1: Launched in 2002, this satellite has a very-high-resolution radiometer (VHRR)/2 which is a modified version of the VHRR heritage imagers flown on INSAT-2A, 2B and 2E. Sensors measures in in VIS, water vapour and TIR bands providing a spatial resolution of 2 km in VIS band and 8 km for the rest. It is an indigenously developed sensor.

(iv) INSAT-3A: Towards meteorological observation, it has a three-channel very-high-resolution radiometer (VHRR) with 2 km resolution (VIS) and 8 km in thermal infrared and water vapour bands. In addition, a charge-coupled device camera operates in the visible and shortwave infrared bands providing a spatial resolution of 1 km as well. Apart from this, it has many communication channels meant for search and rescue operations, linking to stand-alone beacons and others.

(v) Megha-Tropiques: This is an Indo–French joint mission for studying the water cycle and energy exchanges in the tropics, to understand the life cycle of convective systems. This provides information on condensed water in clouds, water vapour in the atmosphere, precipitation and evaporation using sensors Microwave Analysis and Detection of Rain and Atmospheric Structures (MADRAS), an imaging radiometer, Sounder for Probing Vertical Profiles of Humidity (SAPHIR), Scanner for Radiation Budget (ScaRaB) and Radio Occultation Sounder of Atmosphere (ROSA).

(vi) SARAL/AltiKa: This is a joint CNES/ISRO system, which is part of global altimetry system and participates to the precise and accurate observations of ocean circulation and sea surface elevation. It is a Ka-band altimeter with enhanced bandwidth with improved resolution than Envisat satellite at vertical resolution of 0.3 m.

(vii) Oceansat-2 and -3: Oceansat-2 and -3 are launched in 2009 and 2022, respectively. They measure ocean colour through an ocean colour monitor sensor in eight spectral channels at 360 m spatial resolution and help in understanding ocean productivity. Other sensors are Ku-band pencil beam scatterometer (SCAT) developed by ISRO and ROSA developed by the Italian Space Agency. Scatterometer provides a global ocean coverage and wind vector retrieval with a revisit time of 2 days. ROSA is a scientific payload for understanding ionosphere.

(viii) ScatSat-1: This is a scatterometer system providing wind vectors using Ku-band pencil beam scatterometer. It is an active microwave radar operating at 13.515 GHz providing a ground resolution cell of size 25 × 25 km. In vertical–vertical polarization (VV), it covers 920 km circle for scanning yielding 1840 km swath, while in horizontal– horizontal (HH) polarization, it covers 700 km circle with 1400 km swath.

Wide-ranging satellite imaging/senor-based information has been disseminated as tabulated in the following, keeping in view of the dynamic requirements of national and regional community based on satellite imaging and modelled products. Several of the products (Table 3) represented are from experimental level modelling processes and hence carry a disclaimer regarding their value in claims related to loss of life and property during the extreme event. However, the comprehensive record available on the portal makes it relevant for the alerts as societal requirements with regard to daily, seasonal, annual and episodic extremes and trends as well. The range addressed forecasts, nowcasts, alerts as well as ocean and meteorological applications.

3 Global Framework of Earth Observation in Measuring the Climate Processes

3.1 Remote Sensing of Atmospheric Components

Atmospheric components that are in focus to understand as the system of elements connected with climate change comprise of measurement of temperature, rainfall, wind, composition of gases, clouds and aerosols. Since alteration of gases in the composition causes the entire essence of warming potential, the theme assumes primacy in causation and hence demands higher order of innovation in observing the subtlety at higher precision. Modelling efforts to represent the long-term interaction and relations to land surface processes require precision measurements of atmospheric processes. Earliest sensors observing atmosphere involved observing the phenomena that were physically discernible using earliest sensors capable of basis interaction of matter and light, especially from geostationary orbits, since it offered continuous watch capacity. Advances in sensor technology and complex interactions of atmospheric composition with narrower bandwidths of electromagnetic spectrum made intricate imaging and sounding possible to bring about unprecedented patterns. Sensitive active and passive microwave observations have made measuring winds possible, while innovations in translating interactions of various atmospheric chemistry elements with smallest windows of light at dimensions of Angstrom level could image gases such as carbon dioxide, nitrogen oxides and many other gases precisely.

3.1.1 Precipitation

Towards rainfall measurement, the Tropical Rainfall Measurement Mission has played a pivotal role in developing rainfall patterns using combination of active and passive microwave imaging deployed in non-sun-synchronous orbit at different parts of the day, by NASA and JAXA since 1997. This mission also involved

Table 3 Details of climate and weather services provided through MOSDAC

Forecast	City weather	Cold waves	Heatwaves	Heavy rain	Lighting
	Cloudiness	Every 48 hrs monitoring	Every 48 hrs	Past 48 h	Every 4 days forecast
	Wind speed and directions	Minimum surface temp*	Maximum surface temp		
	Rainfall	Change in extreme cold	Change in extreme heat		
	Temperature				
	Relative humidity				
	Surface pressure deviation from 1000 hPa				
Forecast	**Monsoon**	**Sea state**	**Solar wind**		
	Trend in rainfall from 2017–2022 using NCAR NCEP community atmosphere model (CAM) and global initial condition data	Sea state forecast (120 hrs) in terms of several parameters ****	Forecast of Windspeed and solar insolation for 3 days Wind as speed and degree of direction		
Nowcast	**Cloudburst**	**Heavy rain**	**Alerts**	**India**	**SE Asian countries**
	Cloudburst potential alerts over Western Himalayan region for 6 hrs, using the in-house model Nowcasting of ExTreme orographic RAinfall events (NETRA) of SAC, ISRO	Satellite-based heavy rainfall alerts over India and surrounding regions, using the INSAT series of satellite		Cloudburst/cyclone/heavy rain, animations of visible and thermal infrared imagery of three days cloud pattern movement over each state and region	

Current events	**AWS time series**	**Heavy rain**	**MOSDAC live**		
	Automatic weather stations data				
Met applications	**Urja**	**Varsha**	**Vayu**	**Soil moisture**	**Scorpio**
	Solar energy	Rainfall-related parameters	INSAT-3D fog	Soil moisture (CuM/CuM)	Satellite-based cyclone observation and prediction over Indian Ocean**
			INSAT-3D aerosol optical depth	Soil wetness index	Historical records of all cyclones till 2013
			Parameters shown below #	Trends for about one month	
Ocean applications	**Oil spill**	**Ocean subsurface fields**	**Safe search**	**Ocean eye**	**Eddy currents**
	Lagrangian coherent structure cores (LCS -cores) based oil spill dynamics	Density anomalies of subsurface	Risk prediction at selected beach for three days	Ocean surface current	Oceanic eddy parameter information prepared for 24 years (1993–2016) over bay of Bengal using AVISO***
	LCS		Rip current risk	Surface winds	Eddy track
	Stretching direction		Wave period	Wave height	Eddy point
	LCS core		Wave height	Sea level pressure	
	FTLE (finite-time Lyapunov exponent)		Wave direction	Heat content anomaly	

(continued)

Table 3 (continued)

Forecast	City weather	Cold waves	Heatwaves	Heavy rain	Lighting
	****Sea state**		**#Vayu**		
	Mixed layer depth		INSAT-3R thermal infrared 1 count providing		
	Swell height		Air quality index		
	Sea level anomaly		PM25		
	Sea surface salinity		PM10		
	Sea surface temperature		NO_2		
	TCHP anomaly		SO_2		
	Mean wave period		Ozone		
	Significant wave height		Carbon monoxide		
			Animation wind over the region		

*Based on WRF model
**Cyclogenesis, track, intensity and landfall of cyclones, ship avoidance region, storm surge, coastal inundation, etc., and satellite-based cyclone information are disseminated
**AVISO (archiving, validation and interpretation of satellite oceanographic data) is a service set up by CNES to process, archive and distribute data and products from altimetry satellite missions. Merged and gridded satellite altimeter product of sea surface height (SSH) anomaly at 7-day interval having special resolution of 0.25 deg used in the current database

integration of information from other rainfall measuring satellites as well. Sensors on board included precipitation radar, microwave imager, VNIR scanner along with lightening imaging sensor and system to measure cloud and earth's radiant energy. Precipitation radar measuring at 2 mm wavelength provided three-dimensional images of the clouds. Satellite provided near real-time monitoring of hurricanes and accurate estimates of rainfall accumulation over time. Harnessing the legacy of rainfall measurement mission, Global Precipitation Measurement Mission was launched in 2004 following the decommissioning of TRMM. GPM has only two sensors DPR (www.earthdata.nasa.gov/learn/articles/trmm-to-gpm) and GMI (GPM Microwave Imager). Dual-frequency precipitation radar (DPR) provides 3D profiles and intensity estimates of precipitation ranging from rain to snow employing dual frequency radar. Microwave Image (GMI) has additional frequency range than TRMM (4 more channels than 9 of TRMM) that allows measurement of precipitation intensity and type through all cloud layers using wider swath. GPM covers data approximately between 65 degree north and south latitudes, while TRMM collected between 35 degree limits and allows tracking of storms as they form in tropics and move to middle and high latitudes. GPM has a system called core observatory that calibrates data from constellation of other climate observing satellites by setting up a reference, from among 13 channels to each of these. Apart from the instrumentation, the initiative of IMERG (Integrated Multi-Satellite Retrievals for GPM) is focusing on harmonizing data retrievals from different satellites in to one dataset, especially matching TRMM data with GPM data, to build time series till 1998, through a consistent algorithm.

Extreme weather event, as per IPCC guidelines, would normally be as rare as or rarer than the tenth or 90th percentile of a probability density function estimated from observations and may vary from place to place depending upon the resolution of observation. A study emphasized the potential of the ERDS IMERG half-hourly early run data, working at the global scale with a spatial resolution of $0.1° \times 0.1°$ (a satellite precipitation measurement) as the input for a near real-time extreme rainfall detection system. There has been an attempt to improve the extreme rainfall detection system using GPM IMERG data by employing varied aggregation intervals of rainfall events from 12 to 96 h aggregation. An aggregation at 24 h interval ensures a probability of detection (defined as the number of hits divided by the total number of observed events) greater than 80% (Mazzoglio et al. 2019).

Tropical Rainfall Measurement Mission (TRMM) and Global Precipitation Mission (GPM) microwave sensors stand out as the advanced sensors providing three-dimensional estimate of rainfall systems across tropical belt using microwave imaging in various windows and look angles. A consistent long-term data records from both the sensors with a 13-month common operational period using WindSat, a polarimetric microwave radiometer. Such a record helps to ensure a consistent long-term precipitation record (Chen and Jones 2018). Indian summer monsoon is a unique phenomenon of Earth's climate system and critical for the availability of freshwater for drinking and irrigation, agricultural production, power generation, water resources management and the overall economy of the country. Warming of the climate as evident from observations on increasing global average air temperature (Jones and Moberg 2003) has considerable bearing on the Indian summer monsoon in terms of spatio-temporal distribution of precipitation (New et al. 2001).

Changes in spatio-temporal pattern of rainfall and rainy days over the monsoon month/season were estimated using daily gridded rainfall data for the last 35 years (1971–2005) by Das et al. (2014). Positive trends of both rainfall and rainy days were found over the southern region of the Indian peninsula, covering coastal Andhra Pradesh and Rayalaseema. Marathwada, south interior Karnataka, Telangana, Madhya Maharashtra. Significant negative trends in case of rainfall as well as rainy days during the monsoon season were found in the west coast (Kerala, coastal Karnataka), the eastern region (Jharkhand, Arunachal Pradesh) and western desert region (east and west Rajasthan). A significant decrease in rainfall either in monsoon months or season without any significant changes in the rainy days was reported in the northeastern region of India covering the sub-Himalayan West Bengal, Assam and Nagaland–Manipur–Mizoram–Tripura. On the other hand, significant negative trend of rainy days either over the monsoon months or season (negative changes in the rainfall distribution) was observed in the north and central regions of India covering Punjab, Haryana, west and east Uttar Pradesh, west and east Madhya Pradesh, Gujarat and Orissa. Statistically significant increasing trend of rainfall with decreasing trend of rainy days, indicating higher probability of high-intensity rainfall and flash floods, was reported in Uttarakhand and Himachal Pradesh and Jammu and Kashmir region.

3.1.2 Greenhouse Gases

India's mean surface air temperature has increased significantly by about 0.4 °C over the past century. Carbon emissions from the energy sector amount to 71 MT a year, equivalent to all other sectors combined. From land-use data, a marginal net sequestration of 5.25 million tonnes of carbon occurred during 1986. Following the IPCC guidelines, methane emissions from rice and livestock are estimated at 17.4 and 12.8 Tg yr^{-1}, respectively. According to recent climate model projections, India may experience a further rise in temperature of 1 °C by the year 2050, about four times the rate of warming experienced over the past 100 years. About 70% of the electricity generation in India is from coal-based power stations. Altering this dependence significantly to reduce emissions would imply a substantial change in the present energy policy of India. There is great potential for improving energy efficiency and conservation. The adoption of cleaner coal technologies should be considered, as must the development of renewable, non-conventional energy sources. In all cases, serious institutional barriers and resource limitations need to be addressed. The scope for carbon sequestration is limited by land availability and other factors. It is argued that any response to global warming must be located firmly in the framework of sustainable development. India's population growth, urbanization trends, patterns of income distribution and increasing industrial production lead to increasing waste generation. Inappropriate waste management results in emission

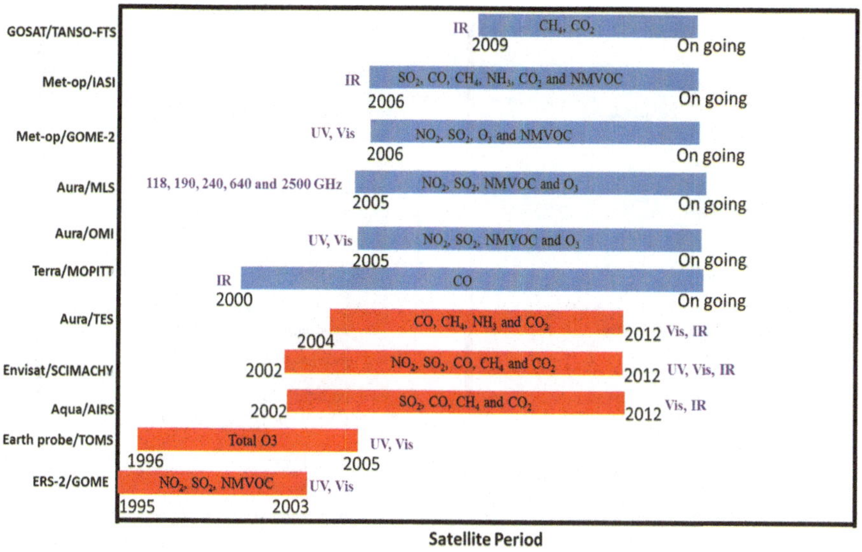

Fig. 4 Availability of space-based information for GHG monitoring

of greenhouse gases (GHG) constituting methane and nitrous oxide, contributing to global warming (Fig. 4).

Globalization and liberalization policies of the government in the 90s have increased the number of road vehicles nearly 92.6% from 1980–81 to 2003–04. These vehicles mainly consume nonrenewable fossil fuels and are a major contributor of greenhouse gases, particularly CO_2 emission. This paper focuses on the statewide road transport emissions (CO_2, CH_4, CO, NOx, N_2O, SO_2, PM and HC), using region specific mass emission factors for each type of vehicles. The country-level emissions (CO_2, CH_4, CO, NOx, N_2O, SO_2 and NMVOC) are calculated for railways, shipping and airway, based on fuel types. In India, transport sector emits an estimated 258.10 Tg of CO_2, of which 94.5% was contributed by road transport (2003–04). Among all the states and union territories, Maharashtra's contribution is the largest, 28.85 Tg (11.8%) of CO_2, followed by Tamil Nadu 26.41 Tg (10.8%), Gujarat 23.31 Tg (9.6%), Uttar Pradesh 17.42 Tg (7.1%), Rajasthan 15.17 Tg (6.22%) and Karnataka 15.09 Tg (6.19%). These six states account for 51.8% of the CO_2 emissions from road transport (Fig. 5).

IPCC 2006 model estimated GHG emissions from waste sector across India considering a gross domestic product growth rate of 6.5% as 70.13 million tones CO_2 eq in the year 2011, which is expected to rise 1.60 times by the year 2031. Emission mitigation options for waste sectors including diversion of organic waste from landfills towards treatment options, diversion of wastewater from domestic and commercial sectors towards sewer and further capturing and utilizing methane from landfills and effluent treatment units indicate a potential to lower the emissions to around 78.75 million tonnes CO_2 eq in year 2031. There is an urgent need to

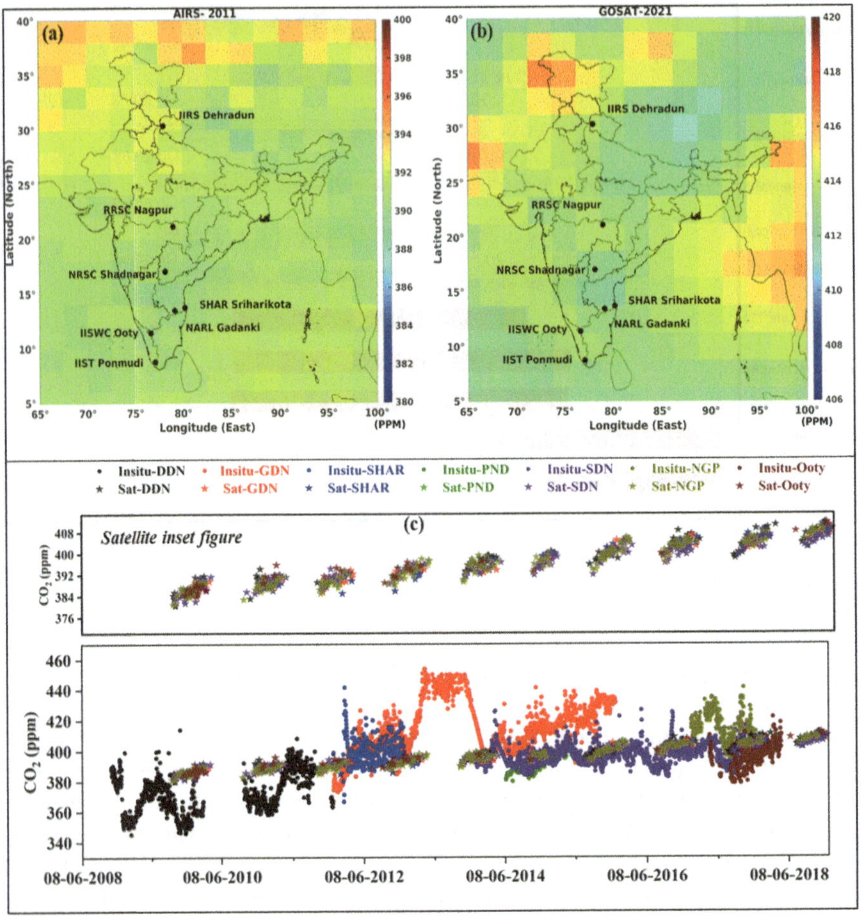

Fig. 5 Spatio-temporal variability of CO_2 indicating the near continuous increase. Locations of ground-based monitoring of CO_2 by NRSC, ISRO, are filled circles in map

apply appropriate policy, political will, financial resources, capacity building and indigenous technology to reduce impact of our activities on global warming.

The 'Emissions Gap Report 2022: The Closing Window' released ahead of the UN Climate Change Conference (COP 27) in Egypt said the international community is still falling far short of the Paris goals, with no credible pathway to limiting global temperature rise to 1.5 °C in place. To address climate change, countries adopted the Paris Agreement in 2015 to limit global temperature rise in this century to well below 2 °C, preferably to 1.5 °C, compared to pre-industrial levels. The report states that world average per capita GHG emissions including land use, land-use change and forestry were 6.3 tCO_2e in 2020. The USA remains far above this level at 14 tCO_2e, followed by 13 tCO_2e in the Russian Federation, 9.7 tCO_2e in China, about 7.5 tCO_2e in Brazil and Indonesia and 7.2 tCO_2e in the European Union. India remains far below the world average at 2.4 tCO_2e. On average, least developed countries emit 2.3 tCO_2e per capita annually.

3.1.3 Surface Temperature

A global dataset of air temperature derived from integration of remote sensing data from MODIS and global station data illustrates the strength of remote sensing to overcome the limitation of the data surfaces prepared using only station data, which at times might represent several discrepancies (Hooker et al. 2018). Study used geographically weighted and climate space weighted regression approaches, where LST from MODIS is weighted using more than 3253 records between 2003 and 2012 and Worldclim surfaces equivalent to MODIS data, respectively. Regressions are estimated using open-source approach and error part is provided along with the coefficients to provide continuous surfaces across land continuum.

Land surface temperature and emissivity determine total long-wave radiation quantity from Earth surface, indicating climate variability, land cover change and energy balance between land and atmosphere. A long-term and consistent Earth system data record is essential for such parameter. Products based on MODIS (MOD21) and VIIRS (VNP21) sensors have overcome the issues in accuracy and consistency using temperature emissivity separation, as continuity with respect to earlier existing MYD21 and VNP21 products at 0.5 K temperature level and only 1–2% difference of magnitude with respect to land validation sites using quartz sand and grasslands (Hulley et al. 2017).

MODIS-based land surface temperature product was used to detect the hottest place on Earth at Lut Desert in Iran with a recorded temperature of 70.7 °C in 2018 and diurnal variability has been studied. Improvements in estimation methods and high spatial resolution brought in clarity of estimations (Azarderakhsh et al. 2020).

The spatial patterns of temporal trends in temperature and its extremes have been analysed over the homogeneous temperature regions of India using daily gridded temperature data for the period of 1969–2005 (Chakraborty et al. 2017). The study reported a general warming trend over India with notable spatio-temporal variations in terms of magnitude and direction. The magnitude and spatial extent of increasing trend (0.02–0.04 °C $year^{-1}$) of minimum temperature was found to be higher than that of maximum temperature (0.01–0.02 °C $year^{-1}$), and it is more pronounced during winter and pre-monsoon season. Dry and arid northwest region of India showed consistent positive trends of minimum temperature. The southern peninsula region of India was found to have significant positive trend of maximum temperature during the cooler months (November, December and January). Significant negative trend of minimum temperature over the eastern part of India was found during monsoon months, whereas same observations were made for maximum temperature over the northwest, north central and northeast regions.

3.1.4 Heatwaves

Heatwave is a condition of weather when local temperatures cross 40 °C in plains, 37 °C in coastal areas and 30 °C in hills according to the Indian Meteorological Division. Heatwave is declared on the day temperature crosses 4.5–6.4 degrees above normal. Severe heatwave sets in when rise is more than 6.4 degrees. The year 2022 recorded 280 heatwave days from March 11 to March 18 across various states of India. Exceptionally unusual early heatwaves that swept India and Pakistan in 2022 (Fig. 6) were made 30 times more likely due to direct impact of climate change (World Weather Attribution Network) through an analysis based on observations

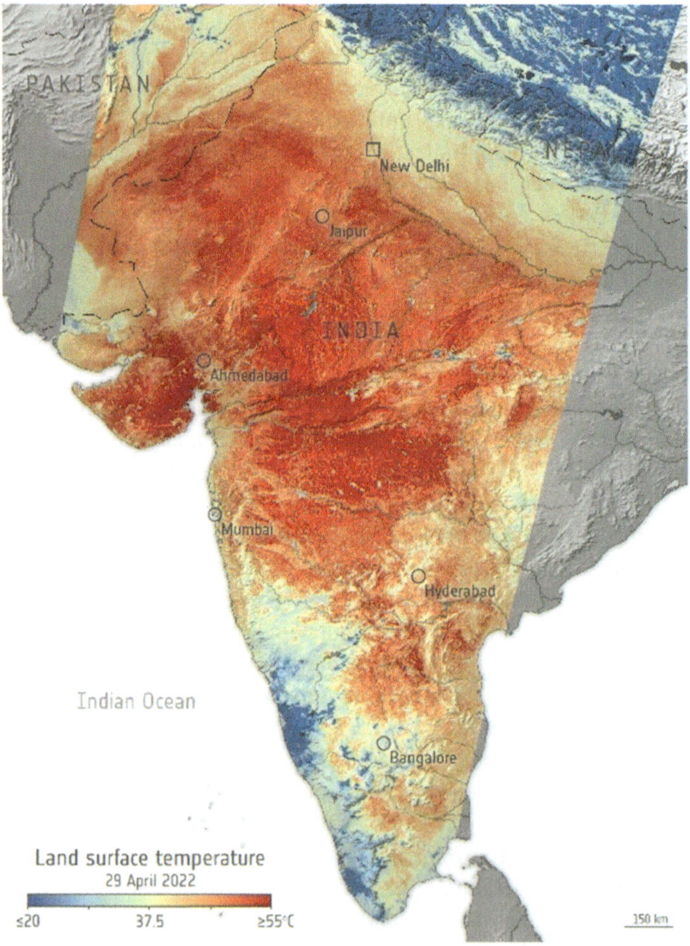

Fig. 6 Land surface temperature-based heatwave condition map of the Indian region on April 29, 2022, based on Sentinel-3 SLSTR (www.esa.int/ESA_Multimedia/Missions/Sentinel-3/(result_type)/images)

from 20 models. Review of more than 400 peer-reviewed studies analysing weather extremes across the globe have asserted that of the 152 extreme heat events assessed by scientists, 93% found that climate change made the event or trend more likely or more severe (www.carbonbrief.org/mapped-how-climate-change-affects-extreme-weather-around-the-world).

Remotely sensed land surface temperature estimates are highly sensitive to characteristics of the measuring instrument. Spatial sampling as resolution and viewing geometries involving oblique or nadir view as well as to the algorithms and auxiliary data used in the retrievals affect the observations. Notably, assumptions about surface emissivity affect the estimates. Ability of land surface temperature (LST) retrieved from the Spinning Enhanced Visible and InfraRed Imager (SEVIRI) on board Meteosat Second Generation (MSG) to monitor heat extremes is harnessed for monitoring approach relying on monthly anomalies performed as departures from the median and the monthly number of hot days (NHD). Both of these factors were computed for satellite LST derived from MSG and MODIS, and for 2 m air temperature (T2m) from ERA5 reanalysis, using as threshold the 90th percentiles (Gouveia et al. 2022). Results of the study highlighted the suitability of MSG LST to study heat extremes alone or combined with dry and bright conditions. It prompts the potential of other climate data records from geostationary satellites to characterize these climate extremes, which may become norm in future events.

Sentinel-3 with the sensor SLSTR (Sea and Land Surface Temperature Radiometer) is designed to retrieve global coverage sea and land surface skin temperatures (with zero bias and an uncertainty of ± 0.3 K) for a $5^\circ \times 5^\circ$ latitude–longitude area. The temporal stability of measurements is 0.1 K/decade and has dual-view capability of both oblique and nadir views, with respective swaths of 1400 km and 740 kms. Sensor observes Earth in visible, SWIR, MWIR and thermal IR regions. Imaging of Indian region heatwave condition, as illustrated on ESA portal, provides direct surface temperature as contrasted to the conventional air temperature and is an unprecedented depiction pointing to the enormity of the event. The uniform high range temperatures beyond 45° points to the severity of the event and occurred close to highest temperature record during March, in the past 120 years of air temperature measurement. Such an image sums up every aspect of alarm needed for all stakeholders alike.

3.1.5 Snow, Ice and Glaciers

Snow cover is one among the most sensitive parameters to changing climate since warm summers in upper latitudes and higher altitudes in India are inducing snowmelt in an increasing manner (Singh et al. 2022). Globally, spectacular retreat of glaciers is witnessed across all major physiographic regions characterized by terrains of frozen snow. Nothing illustrates the intensity of climate change than the increasing reduction in sea ice in June in the Arctic region. It has dropped by almost 18% per decade over the last 30 years. Analysis revealed that Arctic is warming about three times faster than other regions. Arctic sea ice apparently has hit its tenth lowest wintertime extent (Candanosa 2022). In February 2022, the Antarctic sea ice also has dropped to record low minimum extent. MODIS remote sensing data of snow cover from 2009 to 2014 were used to estimate snow cover percent in study conducted for Lidder watershed in upper Jhelum catchment. Supporting digital elevator model from ASTER sensor for the intended catchment was also used in the study. Snow melt run-off model was used to model and estimate streamflow in the snow regions daily. Measured degree day on a parameter day affects the discharge on the next day, while a critical temperature can trigger run-off from instant rain or snowfall. With the help of modelling, it was shown that a 2 °C temperature increase caused a 53% rise in catchment discharge (Kumar et al. 2022).

Satellite remote sensing from optical and microwave sensors (MODIS and EnviSat ASAR) has been assimilated with meteorological point measurement from stations, numerical weather predictions in spatial and temporal scales. Snow maps from optical and microwave imaging reveal systematic differences which need to be compensated appropriately for use in snowmelt models (SRM). Intermittent availability of satellite images required that prognostic equations were applied to predict the daily snow cover extent to update the model. Snow characterization is done using normalized difference snow index that harnesses strong decline of reflectance of snow (1628–1652 nm) in middle-infrared spectral region to discriminate snow from most other natural surfaces and dense water clouds. MODIS band 3 (459–479 nm) performs better to differentiate shadow zones than band 4 (545–565 nm) while calculating NDSI at 500 m spatial resolution. Clouds which are to be segregated precisely need spectral bands such as thermal emission in band 31 (10.78–11.28 mm) and 35(13.78–14.08 mm) for high level clouds formations. Emission in band 31 discounted by emission/reflectance in band 21 (329–3.99 mm) for detecting low and medium level water clouds. Small convective clouds at low elevation missed by other classifiers can be detected by ratio of band 1 (620–670 mm) and band 6 (1628–1652 nm) (Nagler et al. 2008).

During extended cloudy periods, optical data needs to be augmented with synthetic aperture radar (SAR) data, A SAR is a multimode C-band SAR system operating on systems such as European Space Agency's Envisat that can provide spatial resolution of 30 m to 150 m depending on look angle configuration. SAR data can be used to monitor snow thawing since totally dry snow does not return any energy to sensor unless it has begun to melt. Apart from this, steep mountain slope suffer from layover effect and hence need to be compensated using reliable extrapolations from cover from similar altitudes, derived from DEMs. On the other hand, all slopes illuminated at local incidence angle of <17 were excluded as they fail to form clear signals of cover. This comprehensive study managed successfully to predict short-term snow hydrology based on the assembling of data from various sources, which enables better understanding of impacts of climate change.

Most of Himalayan glaciers have been reported to have rates similar to glaciers in other parts of the globe, with an exception of stability of mass gain in Karakoram range (Bolch et al. 2012). Diversity of climate conditions and terrain extremities within region may make projections of glacier dynamics speculative. Gains and losses of mass of glaciers as response function to climate change as manifestation in length and area are not easy to interpret as climate-related phenomenon. Satellite images with extensive seasonal snow or maps from them can be a serious source of uncertainty in glacial study. But due to several factors such as inaccessible terrain, political situations, in situ measurements of glacier being not fully possible, remote sensing provides a meaningful substitution albeit partially (Gaddam et al. 2022).

Detailed analysis of 12 glaciers in Alakananda basin in the Himalayas using satellite-based inventory from 1968 to 2020 coupled with limited sampling of selected glacier has revealed critical facts about deglaciation and loss of spread. Images from 1968 to 2020 using Corona and Sentinel-2A satellite coupled with Landsat 7 and 8 data for intervening period were employed. Corona satellite employed film return technique wherein the physical film thrown from satellite in orbit was captured by aircraft during parachuted decent in atmosphere during the 1960s. This high-resolution (about 5 m) panchromatic data was geometrically rectified using Cartosat-1 ortho corrected data as well as digital elevation model. Snout and boundary position of glaciers weas validated using real-time kinematic GPS having an accuracy of ~1 cm (Remya et al. 2022). From 1968, the number of glaciers increased from 98 to 116 over 52-year period, while glacier area reduced from 742 ($\pm$ 44.4) sq km to 683 ($\pm$ 47.8) sq. km with annual average recession of 11.75 ($\pm$ 1.6) m over the entire basin. Significant deglaciation and fragmentation observed are augmented by increase in winter time temperature of 0.03 °C.

Glacier thickness plays a major role in understanding future sea level rise by virtue of mass of water they hold and possible ablation. Flow models are used to estimate the thickness involving inputs from microwave remote sensing. Ice thickness of HMA (High-Mountain Asia) glaciers covering states of Himachal, Uttarakhand, Bhutan, Sikkim and Arunachal was estimated using DInSAR (Differential Interferometry Synthetic Aperture Radar) approach (Nela et al. 2023). Two satellite passes separated by 14 days for ALOS-2/PALSAR sensor L-band backscatter data were processed to calculate the phase difference. Phase differences provide the velocity of the surface ice elements later computed into thickness by integrating into laminar flow law, which revealed about 100 m as mean thickness of ice over the entire study region. Retreat of Himalayan glaciers in (Satopanth and Bhagirath) India has been studied using IRS data and CORONA images of 1968 in association with MOD11A2-derived land surface temperature pattern for the period of 2000–2020 indicating significant warming trend. Significant negative trend in snow cover was witnessed. Image analysis clearly indicated retreat of 23.5 m and 18.2 m from 1968 to 2017 in Satopanth and Bhagirath glaciers of Mana Basin in India. These two glaciers are converging to a single point. Studies show that in the altitude range of 3200–5600 m amsl signals of warming has been distinct in this part of Himalayas and there's a possible impact on glacial retreat (Thapliyal et al. 2023) (Fig. 7).

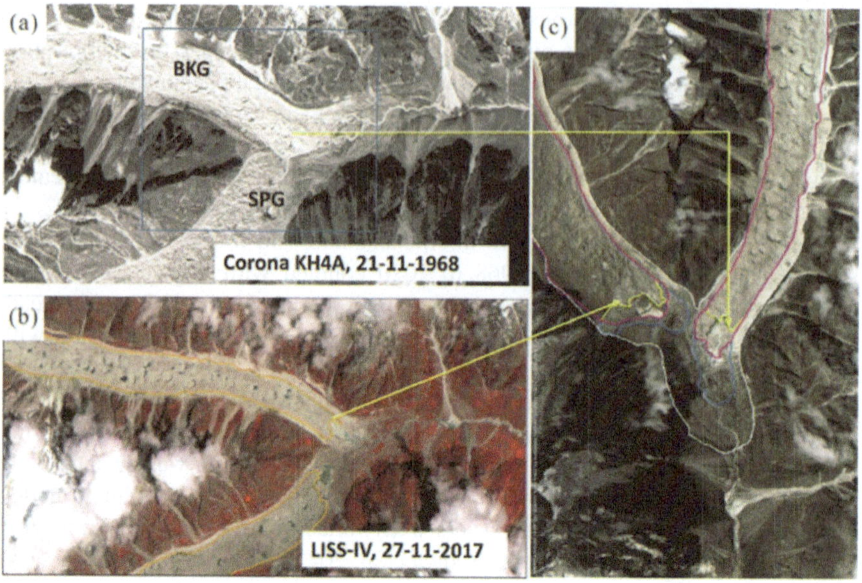

Fig. 7 . Temporal trends in Satopanth glacier (SPG) and Bhagirathi Kharak glacier (BKG) juncture as observed in 1968 (Corona KH4 of 21 Nov '68) (**a**) followed by observation in 2017 (Resourcesat-2, LISS IV of Nov. 21, 2017) (**b**). High-resolution Google Earth image (**c**) depicts clear recession of snouts at confluence

3.2 Climate Change and Oceans

Top five metres of ocean stores as much as energy of entire atmosphere and this has led to ocean warming which in turn has threatened habitat of coral mortally, melted sea and nearby land ice as well as led to increase in sea level. Oceans have absorbed 90% of the heat generated in recent decades by anthropogenic causes of global warming. Very critical impacts of warming will be linked to ocean processes, since they contribute large surface of Earth and influence all the geophysical exchanges with atmosphere and land. Global trend on ocean heat illustrated here summarizes the severity of the situation in terms of rise in warmth of ocean (Fig. 8).

3.2.1 Sea Surface Temperature

Sea surface temperature is an important factor or physical variable that helps to understand global warming as a pivotal parameter. SST enables understanding, quantification and prediction of complex interactions between ocean and atmosphere, since the energy movement patterns once the heat is absorbed by ocean trigger several climate-related events (Li et al. 2001). Daily SST maps for operational systems and climate modelling are normative now as a matured and

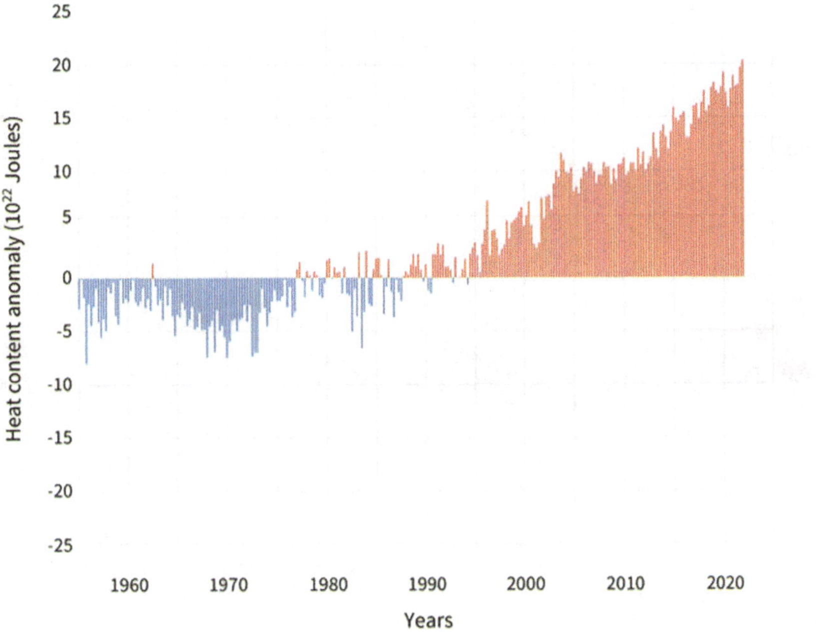

Fig. 8 Ocean heat above the average: seasonal (three-month) heat energy in the top half-mile of the ocean compared to the 1955–2006 average. Heat content in the global ocean has been consistently above-average (red bars) since the mid-1990s. This graph is based on data (0–700 m) from the NCEI Ocean Heat Content product collection (Lindsey and Dahlman 2020)

sustained information service for wide ranging stakeholders (Fig. 9). Data streams of global networks such as Group for High Resolution SST and CEOS SST Virtual Constellation are harmonized through steps of sharing, indexation, processing, quality control, analysis and documentation to provide the products. A combination of low Earth orbit or geostationary thermal or near-infrared sensors or microwave imaging sensors along with in situ data from moored or drifting buoys, Argo floats are combined for comprehensive information as spatial datasets (O'Carrol et al. 2019) An exhaustive review of five decades (Minnet et al. 2019) of remote sensing sea surface temperature provides exhaustive insights into instruments, orbital platforms, data analysis approaches and way forward about this key area of climate change research.

Climate data records which are essential for scientific communities towards many applications have been produced using integration of satellite data, model outputs and ground measurements. Satellite-derived radiances are evaluated for clear sky and sea water pixel for initial exclusion of false signals, by involving both observation and modelled rulesets about clear sky radiance and sea skin temperature effects. Skin SST and uncertainty estimated to provide quality flag was followed by estimation of daily depth SST. This estimation was converted into single sensor gridding, followed by multisensory analysis and then creation of gap-filled SST for

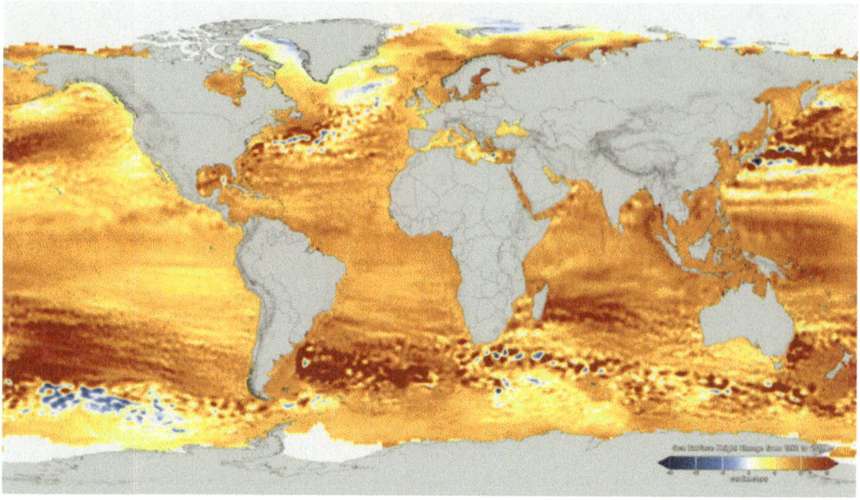

Fig. 9 Total sea level change between 1992 and 2019, based on data collected from the TOPEX/Poseidon, Jason-1, Jason-2 and Jason-3 satellites. Orange to red indicates 5 to 15 cm level increase and blues loss of height up to 15 cm (sealevel.jpl.nasa.gov/resources/1258/27-year-sea-level-rise-topexjason)

global level. In this approach, results from development of methods as well as testing were applied to get parameters for processing the reflectance data (Merchant et al. 2019). Further, sea surface temperature (SST) and upper ocean heat content (OHC, upper 700 m) in the tropical Indian Ocean underwent rapid warming during 1950–2015, with the SSTs showing an average warming of about 1 °C. The SST and OHC trends are very likely to continue in the future, under different emission scenarios. Climate models project a rise in tropical Indian Ocean SST by 1.2–1.6 °C and 1.6–2.7 °C in the near (2040–2069) and far (2070–2099) futures. Indian Ocean warming seems to have resulted in decreasing trend in oxygen (O_2) concentrations in the tropical Indian Ocean and declining trends in pH and marine phytoplankton over the western Indian Ocean. The observed trends in O_2, pH and marine phytoplankton are projected to increase in the future with continued GHG emissions (Roxy 2020).

Climate data record of SST over 35 years from 1981 to 2016 was developed from 4×10^{12} satellite measurements of thermal infrared radiance (TIR). Each pixel that represented SST estimates ranged between 1 and 45 sq. km depending upon the source being local pixel or GAC (global area coverage) pixel. And the good quality observations were 13 in number per each sq. km in this long-term analysis. TIR measurements were collected by two series of sensors, viz. 11 AVHRRs and 3 ATSR (along-track scanning radiometer). SST as skin temperature determines the air–sea fluxes and controls the radiative cooling of the ocean as well as humidity of the air in contact with air–sea interface. Depending on the satellite passes at local time, data quality can vary which needs adjustment that influences the skin-to-depth temperatures (Merchant et al. 2019).

Tropical Indian Ocean is warming quite rapidly in comparison with the rest of tropical oceans (Roxy et al. 2015) and Arabian sea is warming since 1990s (D'Mello and Kumar 2018). Decline of oil sardine fishery in Southwest India (Shetye et al. 2019) and coral bleaching and mortality in Lakshadweep Archipelago (Vineetha et al. 2018) point towards the negative ecological impact prevalent. Impact of rise in SST is evident as threat to thermosensitive reef building corals. Marine heatwaves caused by El Niño–Southern Oscillation has caused bleaching and mortality in tropical Indo-Pacific regions. Study using NOAA's Coral Reef Watch satellite-based alert data on SST was tested for its efficacy as proxy for coral deaths. Parameters such as bleaching threshold (BT), positive SST anomaly (PA) and degree heating weeks (DHW) were calculated to assess thermal stress (Arora et al. 2022) for the period of 2010–2019 which clearly brought out the massive mortality in the Gulf of Kachchh and Malvan in Gujarat and Maharashtra in India, respectively. Kachchh region experienced alert level 2 status (DHW > 8 °C, very warm sea) in 2020, while Malvan in 2010.

3.2.2 Sea Surface Height

Global mean sea level is a critical indicator of global warming and sea level rise. Satellite altimetry is the method of deriving the sea level and anomalies of sea surface heights. Sea surface height (SSH) and sea surface temperature (SST) are two most key indicators related to warming studies. Sea surface height (SSH) is the height of the sea surface above a reference ellipsoid. This is the direct product of satellite altimetry (Subrahmanyam and Robinson 2000; Vignudelli et al. 2019). Sea surface height values are provided along the satellites' ground tracks or at regular grids interpolated from the values determined along the satellite tracks. An important usage of SSH is to derive the SSH anomaly which is the difference between the long-term average for different regions of the ocean and what is actually observed by satellites. Anomalies (SSHA and SSTA) in these patterns over decade or similar scale clearly define the vulnerability of global geophysical systems to warming. Between 1900 and 1990, it was analysed that sea level rose between 1.2 mm and 1.7 mm per year on average. Alarmingly, it rose to about 3.2 mm/year by 2000, and the rate in 2016 is estimated at 3.4 mm per year (Kopp et al. 2016) revealing that seas rose about 14 cm, from 1900 to 2000. In the absence of human-induced warming, sea levels would have remained at somewhere between a 3 cm fall and a 7 cm rise as per modelled estimates. Over 470–760 million residents in coastal cities will be inundated if a warming of 4 degree rise happens.

Instruments to measure sea level rise exploit (Fig. 10) the principle of precise radars to measure backscattered signals from the ocean's surface to determine the height of the ocean. Sensitivity of instruments is so high that even a level difference of 5 mm can be measured from a height of 10 km. The system is supported by laser station to provide precise position of the satellite using differential measurements as well as radiometer to measure water vapour level which influences the accuracy. Analysis involving geoid height at local point with satellite altitude and ocean

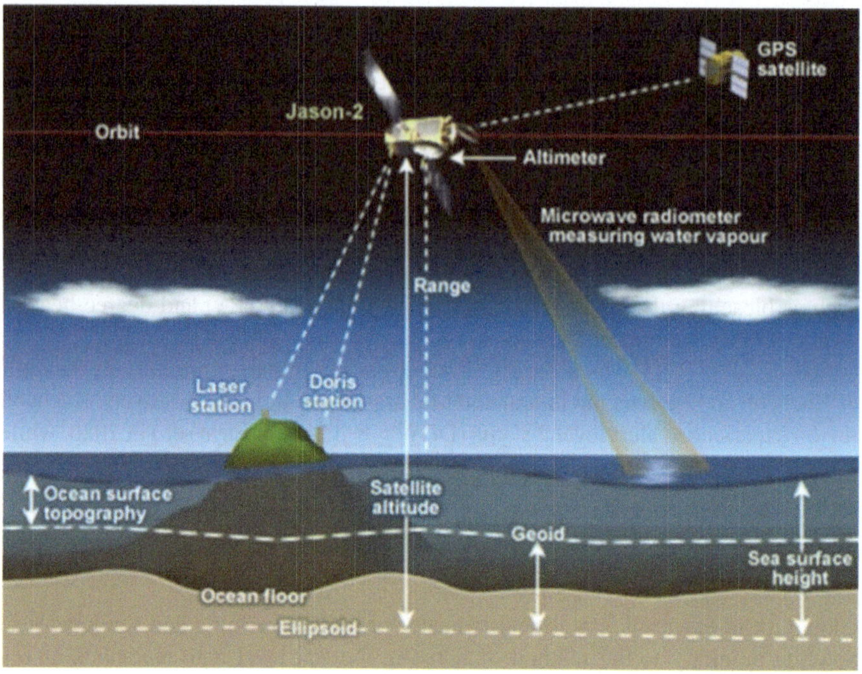

Fig. 10 Schematic showing the principle of satellite altimetry

surface topography provides sea surface height. (ocean.si.edu/through-time/ancient-seas/sea-level-rise). It involves study of radar return waveform analysis which can be influenced by the range of dynamic conditions prevalent during data acquisition. It involves complex method to correct humidity of troposphere, geometry of wave as well the geostrophic current-induced instability of the surface. Geostrophic currents are concurrently measured and included as part of the data model resulting from the measurements based on observations from moorings. Geostrophic currents are oceanic currents in which the pressure gradient force is balanced by the Coriolis effect. Direction of geostrophic flow is parallel to the isobars, with the high pressure to the right of the flow in the Northern Hemisphere and the high pressure to the left in the Southern Hemisphere (en.wikipedia.org/wiki/Geostrophic_current).

SARAL/AltiKa is an innovative joint Indo–French backscatter sensor operating a first time Ka-band radar altimetry mission took the advantage of smaller footprint of about 100 sq. km compared with earlier footprints of 300 sq. km (Jason Series). Such a sensor has potential to provide much better assessment of sea surface height near coasts along with better spatial resolution due to 40 Hz frequency than earlier 20 Hz (Verron et al. 2018). Significance of near coastal altimetry is critical due to the fact that physics and chemistry of these areas determine the impact on the coastal ecosystems as well as economy. Another advanced sensor called SWOT (Surface Water and Ocean Topography) is also in pipeline to measure the topography using the principle of interferometry in Ka-band using along-track and cross-track scans so as to provide enhanced spatial resolution of the observations of SSH. This will be a technological leap in the sense of providing high-resolution topography far better than any existing sensor.

Analysis of satellite-derived SSHA, sea surface temperature (SST) and ocean reanalysis data in the tropical Indian Ocean reveals that patterns of SSHA, SST, ocean temperature, upper ocean heat content (UOHC) and propagations of Kelvin and Rossby waves differ during strong and weak monsoon years thereby modulates the regional climatic processes (Bulusu and Robinson 2000; Rao et al. 2010, and references therein). It has been noticed that during strong monsoons positive SSH, SST and UOHC anomalies develop over large parts of northern Indian Ocean, whereas during weak monsoons much of the northern Indian Ocean is covered with negative anomalies. These patterns can be used as a standard tool for evaluating the performance of coupled and ocean models in simulating and forecasting strong and weak monsoons by climate modellers. Moreover, the rainfall pattern of central India is found to be significantly correlated with SSHA over the regions Arabian Sea and West central Indian Ocean and Bay of Bengal where SSHA is positively large during strong monsoons and in contrast weak monsoon with negative SSHA.

On the other hand, the relationships between Indian Ocean SST and Asian monsoon rainfall have been subject of many studies (Rao et al. 1988; Li et al. 2001; Vibhute et al. 2020). Studies suggest Indian monsoon rainfall has significant positive correlations with the Indian Ocean SST and moisture flux transport in the preceding winter and spring. The effect of this SST influence is quite different from the remote forcing of the Indian monsoon rainfall by the Eastern Pacific SST, which is more dominant on the El Niño–Southern Oscillation (ENSO, 3–7 year) timescale. More specifically, the SST anomalies are found to significantly correlate with the seasonal June to September over the Indian region (Rao et al. 1988). For the first time, it has been observed that heavy or deficient rainfall years over the Indian subcontinent are associated with large-scale coherent changes in SST over the northern Indian Ocean. Further, it has been observed that the correlation between SST and seasonal monsoonal rainfall undergoes changes in sign from significantly positive with pre-monsoon SST to negative over the post-monsoon months. Therefore, all these suggest a strong connection between the SSH, SST and Indian monsoon with the deployment of numerous Argos in recent times, and it is expected that our understanding of these complex air–sea interactions will become much better with strong implications in Earth system models.

4 Observing Climate Change Impacts on Vegetation and Soil

4.1 Agriculture

Global agricultural productivity has dropped due to climate change by 21% since 1961; for some regions like Africa, Latin America and Caribbean region, it is much higher, while at the same time agriculture has both feedback and impact relations with climate change and hence can cause warming too. Greenhouse gas emissions from global agriculture was estimated to be 700 million metric tons in 2018 (eos.com/blog/remote-sensing-to-face-ag-risks-due-to-climate-change). Estimation of changes of agriculturally relevant growing season parameters across the globe comprising of start of season and length of growing season in the primary regions of rainfed agriculture for 26 years showed district patterns. Study used AVHRR NDVI dataset containing 15-day maximum value composites at 8 km resolution for July 1981 to December 2006. Weather data also has been employed to understand their anomalies due to climate factors. Weather parameters were gridded to provide the accumulated growing degree days (AGDD) and humidity data derived from global land data assimilation system (GLADS) in the growing season analysis. This GLADS was 3-hourly gridded meteorological data synthesized by assimilating ground-based, remote sensing and surface climate reanalysis data (Brown et al. 2012). Growing degree days computed by subtracting base temperature (5 c) from average daily temperature were summed up over the 18-month period to create AGDD and accumulated relative humidity (ARHUM). Validation of the patterns with field data was accomplished using crop statistics from the USA and Europe for major crops. Analysis of impact of phenological variation on production at the country level was enabled through this step. Analysis across global contexts demonstrated increasing importance of moisture conditions necessary for crops and other vegetation to harness the desirable higher temperature and growing seasons. Significant correlations were recorded in the study between the peak position measured in growing degree days and relative humidity with rainfed cereal production. This in turn indicated continued vulnerability of the agricultural system to local climate (Brown et al. 2012).

Crops require certain amount of heat units in terms of growing degree days (GDDs), to reach different stages of growth. Warming climate has significantly changed the seasonal GDD patterns across India. Significantly high positive trend (2.3° days year^{-1}) of *kharif* degree days was found over the northwestern region. Significant positive trend of kharif degree days (1.2–1.8° days year^{-1}) with moderate magnitude was observed over north central, northeast Indian Peninsula, East Coast and West Coast. On the other hand, significantly high positive trend of *rabi* season degree days (2–2.8° days year^{-1}) was observed over northwest, East Coast and West Coast, whereas moderate positive trends (1.5–1.7° days year^{-1}) were observed over northeast and Indian Peninsular region. Such large increase in the GDDs during the two major crop-growing seasons, i.e. *kharif* and *rabi*, has significant ramifications on the crop phenology, crop duration, crop water use efficiency, dynamics of pest and diseases and crop yield.

Substantial changes are observed in the annual frequency of occurrence of temperature extremes due to the warming trend over India. Cold extremes have been found to be decreasing significantly, whereas occurrences of hot extremes have increased significantly across India. Significant negative trend in occurrences of cold nights (−0.27 to −0.51 days year^{-1}) was reported over large contiguous area of north and northeastern part covering Western Himalayas, northwest, north central, northeast and West Coast. On the other hand, hot nights showed an increasing trend (0.3–0.9 days year^{-1}) over Western Himalayas, northwest and northeast. The annual frequencies of cold days showed a decreasing trend (−0.4 to −0.6 days year^{-1}) over southern part of India (IP, EC, WC), while hot days did not show any significant trend.

A comprehensive review of drought index used in monitoring meteorological, agricultural, hydrological and socioeconomic drought using database from Google Scholar, Scopus and ScienceDirect revealed presence of 111 drought indices of which 67 were devised using remote sensing data. Considering remote sensing-based drought indices, 90% are employed for agricultural drought monitoring and 10% for hydrological and meteorological drought monitoring. Advances in satellite technologies have been responsible for accelerated design of new drought indices and satellite observations have replaced traditional location specific data with acknowledged success. It was found that PDSI, SPI and NDVI indices were most popular in drought monitoring and had global representations. Integrated indices using both remote sensing and ancillary data have proved to be nonreliable. Scaled Drought Severity Index (SDSI) uses derivatives of vegetation, precipitation and temperature such as VCI, PCI and TCL combined on a scale of 0 to 1 vegetation–soil water deficit has distinction of combining parameters of precipitation soil moisture and potential evapotranspiration which can identify drought with grater accessory (Alahacoon and Edirisinghe 2022).

Both the frequency and intensity of extreme weather and climate events in last decades have increased worldwide, causing unprecedented losses (Halsnæs et al. 2018). Damage to agricultural crop due to increased extreme weather events is an important aspect of applying remote sensing for post event damage detection (Sosa et al. 2021) and possible recoveries in early crop stages. Hail, squalls and flood affect the crop intensely and damage substantial fractions due to reasons ranging from foliage destruction, rupture of major plant parts, damaged reproduction stages and total lodging of crops. Such events need to be assessed over large areas since insuring the crops against damage is an important economic activity involved in crop management. Both optical and microwave energies have been exploited at various resolutions to assess the damage. C-band microwave images at appropriate resolutions on board satellites such as Sentinel-1, RISAT and RADARSAT have ability to image during cloud-masked conditions and provide distinction of damaged parts clearly. The SAR sensors then measure amplitude and phase of wavelength coming back from surface. In a hailstorm impact study in Iowa, in 2016 and 2017 were measured using VH (vertical–horizontal combination of polarization

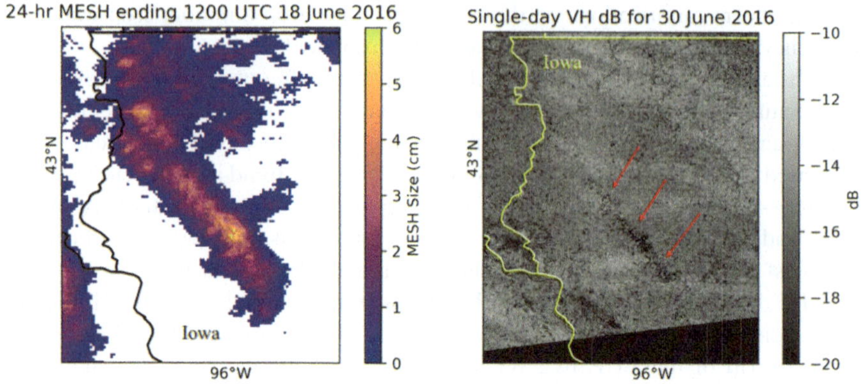

Fig. 11 24 hr maximum estimated size of hail ending at 1200 UTC on June 18, 2016 (left), and hail damage swath in Northwest Iowa on June 30, 2016, Sentinel-1 VH dB image

for forward and backscatter energy) combination of backscatter, which provided clear indication (Fig. 11) of the damaged crop area (Bell et al. 2018). The backscatter signals observed as time series data over cropping seasons indicated slight difference in dB values for damaged and intact crops, meaning instance based comparison is better.

In a study exploring capability of Sentinel-2 optical imagery within a cloud computing platform, eight indices referring to both plant health and water content using various band combinations of near-infrared, shortwave infrared, green and red spectral regions were applied to assess the crop damage (Ha et al. 2022). Temporal profiles that helped derive area under curve for the performances of spectral indices over eight dates of imaging helped to estimate the damage, especially overcoming cloud-related issues. Apart from area under curve, simple differences in vegetation indices (NDVI and NDWI) between pre- and post event have also been quite effective.

Machine learning-based method was standardized for applying microwave and spectral indices to evaluate the hailstorm damage so as to adopt an unsupervised approach. Dual Polarization SAR Vegetation Index and Normalized Pigment Chlorophyll Ratio Index were found to be most sensitive to changes by damage. Time series and rates of change of these indices were used to derive variables in k-means method to develop homogenous damage zones, with a resultant accuracy of 87% (Sosa et al. 2021).

Recent climate change impacts may aggravate crop production loss from frequent flooding. Damage of crops due to flood hazard has caused 57% of all the damages, between 2003 and 2013, induced by extreme events. A comprehensive review of flood-induced damage (Rahman and Di 2020) indicates three major categories of flood-based damage such as flood intensity-based approach, crop condition-based approach and model-based loss assessment. Remote sensing data from optical and microwave sensor play a vital role in each phase of flood crop damage assessment, which depends on information such as extent, depth and duration of flood followed by condition of the crop affected using vegetation indices. In applying models of flooding, availability of good-quality digital elevation model, derived from laser, optical wavelengths or microwave backscatter as direct or interferometric, that has total hydrological compliance, is critical to assess the vulnerable pockets. Out of 60 studies assessed, 43 studies were conducted post 2011, which may indicate both event intensification and improved technology access.

One of the most important early indicators of the impact of climate change on ecosystem is the changes in crop phenology, i.e. recurring pattern of crop growth and development (White et al. 1997). Changes in crop phenology alters the global carbon water and nitrogen cycles, crop production, pollination window and diseases/pest distribution, leading to broad impacts on terrestrial ecosystems and human societies (Penuelas and Fiella 2001). Therefore, phenological study has recently become an important focus for ecological and climatic research (Menzel et al. 2001; Cleland et al. 2007; Chakraborty et al. 2014). Satellite-based study can reveal broad-scale phenological trends that would be difficult, if not impossible, to detect from the ground. Long-term satellite data, with proper standardization and calibration across the sensors, could provide continuous phenological information over large temporal range, with low cost even over inaccessible regions.

Chakraborty et al. (2014) used satellite-based NDVI of 25 years (1982–2006) and detected significant shift in different crop phenology metrics, i.e. start of the growing season (SGS), seasonal NDVI amplitude (AMP) and seasonally integrated NDVI (SiNDVI), during kharif season (June to October) over Indian subcontinent. Pre-occurrence of the SGS (0.1–0.7 days/year) was reported over large contiguous areas of Punjab, Haryana, West Uttar Pradesh, Marathwada, Vidarbha and Madhya Maharashtra, whereas delay in the SGS (0.9–1.6 days/year) was found in Rayalaseema, Coastal Andhra Pradesh, Bihar, Gangetic West Bengal and sub-Himalayan West Bengal. Significant greening trend (increased SiNDVI) along with increase in the seasonal amplitude (AMP) was reported over Punjab, Haryana, West and East Uttar Pradesh, West and East Rajasthan, West and East Madhya Pradesh, Bihar, sub-Himalayan West Bengal, Sourashtra and Kutch and Rayalaseema. On the other hand, Marathwada and Vidarbha showed increase in SiNDVI along with decrease in the AMP implying increase of the length of the growing period. Significant browning trends were reported in most of the south and eastern part of India covering Tamil Nadu, South Interior Karnataka, Coastal Andhra Pradesh, Madhya Maharashtra, Gujarat, Chhattisgarh, Jharkhand and Gangetic West Bengal. Such changes in the crop phenology may be driven by climatic or anthropogenic factors and can lead to the changes in crop calendar, cropping pattern, crop type, net sown area, etc.

4.2 Climate Change Impact on Soils

Soils provide ecosystem services and are essential to plant life as well as for sustainable agriculture. The relations between the atmosphere and soils in a climate change scenario are essential to comprehend altered climate and its possible influence on soils. The changing climate may influence soil through alteration in soil moisture conditions, soil properties, specifically soil organic matter dynamics, soil temperature, CO2 levels, soil erosion, nutrients and alkalinity, soil organisms, etc. Hence, the study on the impact of climate changes on soil properties and the process needs to be more detailed. Many researchers have assessed impact of climate change on soil erosion in India. Gupta (2015) reported that due to climate changes soil annual loss may increase by 25.64 and 20.33% (in the A2 climate change scenario) and 25.3 and 23.38% (under the B2 climate change scenario), respectively, in 2050 and 2080 in the Indian Himalayan regions. Mondal et al. (2015) simulated the impact of climate change on future soil erosion over Narmada River basin of India and reported that average soil erosion would be 15.5 and 105.8% in the year 2050 and 2080, respectively. Besides, numerous studies have assessed the effects of climate change on the stocks, dynamics and distribution of soil organic carbon and have forecasted trends under various climate change scenarios, ranging from regional to global scale (Banger et al. 2015). The knowledge of how climate change affects soil carbon in India, however, is based on various process-oriented models (CENTURY, RothC, etc.) and digital mapping techniques (Falloon et al. 2007; Bhattacharyya et al. 2007; Banger et al. 2015; Mitran et al. 2018). These models work well when combined with GCMs to calculate the effects of climate change. By combining the RothC model with the HadCM3LC climate change forecast, Falloon et al. (2007) evaluated the effects of climate change on carbon storage in India and found that soil carbon stocks would decline by 0.11 Pg from the baseline value of 8.62 Pg in 1860 by the end of the twenty-first century. In a study over the Indo–Gangetic plains of India, Bhattacharyya et al. (2007) similarly noted a declining trend in SOC stocks using Global Environment Facility Soil Organic Carbon (GEFSOC) modelling in conjunction with the empirical Intergovernmental Panel on Climate Change (IPCC) technique. According to Gupta (2015), using the CENTURY model and baseline data from 2010, the Indian Himalayan region's soil C content will decrease by 11.6–19.2% (in the A2 climate change scenario) and 9.62–16.9% (under the B2 climate change scenario) by 2099 as a result of climate change. In order to forecast SOC changes in the semi-arid region of India, Mitran et al. (2018) employed satellite-derived indices and a geostatistical technique. They anticipated a decrease in soil carbon stock. Jain and Mitran (2020) anticipated a decline in total SOC stocks, ranging between 1.12–4.93 and 0.45–4.49 Tg, respectively, by 2050 and 2070 utilizing remote sensing-based indices and geostatistical modelling methods over a semi-arid region of India.

4.3 Natural Vegetation

Scale at which the forests are turning vulnerable, due to desiccating forces of drying climate during summers, across the globe is alarming and unprecedented. Recent spate of forest fire especially in temperate forests of the Northern Hemisphere is threatening all downstream and downwind ecosystems including hitherto valuable human habitation. Vast extents of forest landscapes dried up and experienced infernos that threatened biodiversity as well as contributed massive amount of greenhouse gases to already precariously tipped global concentrations. Apart from this, extreme droughts in otherwise moist and wet habitats such as Amazon forests have induced forest mortality of a magnitude not even simulated by best models available. Resilience of global forests, which cover about 30% of Earth's surface, is under severe threat especially in tropical, arid and temperate forests, with boreal forests being exception (Forzieri et al. 2022).

Long-term vegetation responses using remote sensing reflectance on board global-scale sensors provide clear indication of the ecosystem behaviour such as regime shifts after perturbations in terms of leaf area index and species composition of net primary productivity (Schefferet et al. 2001; Nes et al. 2016). Global gross primary productivity (GPP) and net primary productivity (NPP) products from 1981 to 2018 were estimated using multisource data, viz. fraction of absorbed photosynthetically active radiation (FPAR) and leaf area index (LAI) data from the global land surface satellite (GLASS) dataset and the light use efficiency (LUE) providing 0.05 degree GPP product. Average NPP declined in Asia and Amazon tropical forests and increased in African tropical rainforest due to mainly the climate change. Multisource analysis compared better than MOD17 and showed improved letter component.

Clear understanding of climate change and carbon cycle long-term understanding of the vegetation productivity involving time series datasets are crucial. Gross primary productivity (GPP) and net primary productivity (NPP) products at 0.05° resolution (approximately 5 km grid) starting from 1981 till 2018 (Fig. 12) estimated using a special NPP algorithm named MuSyQ (multisource data synergized

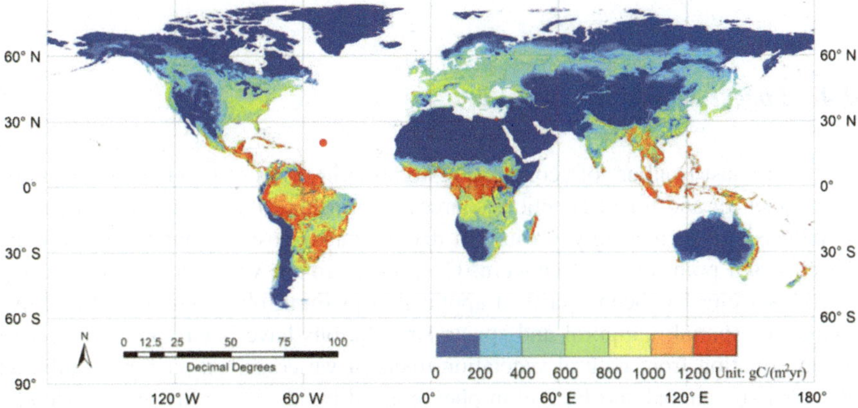

Fig. 12 Global spatial pattern of the mean annual NPP from 1981 to 2018 derived from MuSyQ model that incorporates satellite-based inputs and field observations. (Wang et al. 2021)

quantitative). This model was based on the fraction of absorbed photosynthetically active radiation (FPAR) and leaf area index (LAI) data from the global land surface satellite (GLASS) dataset, the light use efficiency (LUE) from the parameterization approach with the clearness index (CI), the ERA-Interim meteorological data and other environmental factors. As a method integrated modelling performed better than standalone satellite-based product of MOD17 GPP product (MODIS based) and FLUXNET GPP (computed from flux tower network-based observation data) for major forest types such as evergreen broadleaf, deciduous broadleaf, wetlands, woody savannah, dense shrub land and crops. While RMSE of both modelled and MOD17 products remained similar at 214.6 gC/Sqm year, the relationship (R-square) was far higher in modelling (0.81 > 0.55), indicating the strength of integrated approach in assessment of GPP which is central to climate change assessment. Study brought out that NPP declined significantly in Asia and Amazon tropical rainforests and increased significantly in African tropical rainforest. Global NPP has shown a significant increasing trend, with an annual growth rate of 0.10 PgC/year over the past 38 years. However, contribution of tropical rainforest NPP of Amazon, Africa and Asia to the global NPP dropped significantly, wherein except African forests other two forest regions witnessed total decline of NPP (Wang et al. 2021). Greening and browning trends of vegetation for entire Indian region and their responses to climatic (rainfall, temperature and others) and non-climatic (cropping area, irrigated area, fertilizer use) drivers have been studied for a span of 35 years (1981–2015) for India (Parida et al. 2020) using 8 km bimonthly GIMMS-based NDVI3g data along with precipitation (0.25°), temperature (1°), monthly solar radiation (0.5°), soil moisture (0.5°) and seasonal crop statistics. Analysis based on Theil–Sen trends showed that 47% of the nation showed prominent large-scale greening, while in south peninsula warming trends have caused reduction in greening trends (both in kharif and rabi seasons) beyond year 2000. Vegetation over the Himalayas and Northeast India revealed a browning trend that seems to be related to temperature-induced moisture stress (Fig. 13).

4.4 *Forest Mortality*

Forests are last frontiers of ecological restoration in terms of habitats, biodiversity, water security and forest genetic resources for Earth's future. Mortality of forests as induced by increasingly warm and dry climate across temperate and tropical systems is a point of great concern. Observing remote vegetation patterns using remote sensing has been a critical application in Earth observation, and multiple sources of spatial, temporal and spectral resolutions have been applied to derive the information from natural vegetation tracts of varied crown packing exhibiting characteristic annual and long-term phenological trends. Mortality experienced by forests at an unprecedented scale as well as stunting of trees being experienced across Europe, for instance (Newburger 2020), highlights the scale of vulnerability and hence needs state-of-the-art remote sensing observations for land surfaces.

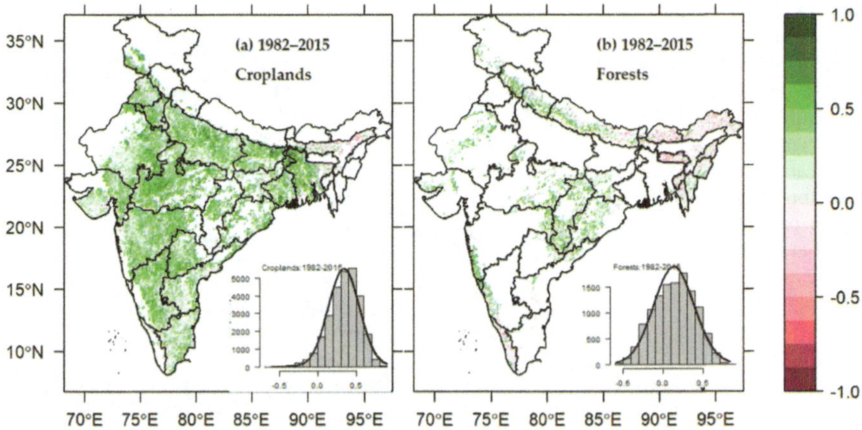

Fig. 13 Grid point Mann–Kendall test for NDVI trends (Tau) for the whole period of 1982–2015 based on annual means of NDVI3g data. Trends are statistically significant at $p < 0.1$ when Tau is above $\pm$ 0.10. Croplands and forests are presented in (**a**, **b**) along with corresponding histograms

Dried and dead forests are turning vulnerable to forest fire and often are bordered by populations already under attack by pests and diseases. It was reported that carbon dioxide storage that is lost because of invasive insects killing the trees every year in forests is equivalent to the emissions from five million vehicles (Fei et al. 2019); this forest age and height are getting reduced gradually and will continue to happen in future.

Excessive forest mortality in Europe has been linked to drought using long-term canopy mortality maps from 1987 to 2016 by using relationships of mortality to water availability. Integrated climatic water balance from March to July fell below -1.6 standard deviation of its long-term average (Fig. 14). At continental level, about 500,000 ha of excess mortality was reported, which seems to be the precursor of further drought-related mortality (Senf et al. 2020).

Remote sensing-based vulnerability to drought in recent past was correlated with long-term record from dendrochronological analysis, using tree ring growth (ring width index, RWI) in southern Sierra Nevada Mountains, near the epicentre of drought severity and mortality associated with the 2012–2015 California drought and concurrent outbreak of western pine beetle (Keen et al. 2022). It was analysed that widespread mortality was presaged by five decades (from time span ranging between 1900 and 2016) of increasing sensitivity of both tree growth and $\Delta^{13}C$ to Palmer Drought Severity Index (PDSI). The sensitivity in fact constitutes early warning signal for mortality caused by direct and indirect effects of drought. Normalized Difference Moisture Index (NDMI) from Landsat between 2012 and 2016 dry seasons was employed for sampling location of trees for stable isotope measurements. Drier sites showed increasing sensitivity of RWI to PDSI over the last century as well as higher mortality rates associated with drought level event

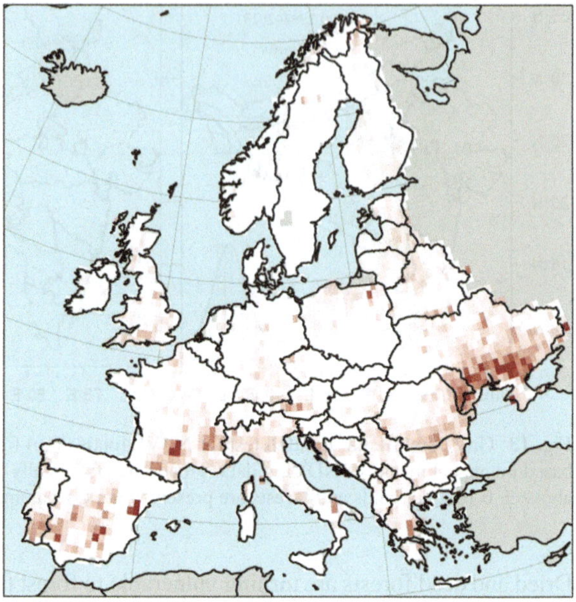

Fig. 14 Percent of the total forest canopy mortality attributable to drought-related excess forest canopy mortality between 1987 and 2016 across Europe. Value of 30 means that 30% of the total forest canopy mortality in this particular grid cell. (Senf et al. 2020)

compared to wetter sites. Forest responses to continued climate warming can be forecasted employing forest modelling using remote sensing and dendrochronology.

However, as far as climate change-driven forest loss is concerned, remote sensing tools still have limitations to detect the diffuse and gradual tree mortality (Hartmann et al. 2018) in contrast to the gregarious loss that can be confidently detected in case of land-use changes in forests and wildfire events. Though ground measurements sufficiently evidence the mortality due to climate, reflectance-based spatial models are difficult to be derived with higher confidence. Large-scale forest mortalities recorded in Thuringia, Germany, in 2018–19 were characterized using soil moisture index using satellite-based measurements, which revealed worst drought documented in the last 70 years. Mortality rates of Scots pine increased tenfold from <0.1% in 2018 to almost 1% to 2019. Exceptional drought could be assessed using images of September 2019 over more than 50% of the region (Hartmann et al. 2022).

Natural vegetation systems are exposed to high stresses and degradation due to anthropogenic activities and climatic changes. It is translated into reduction in forest vigour, degradation and deforestation leading to loss of carbon stocks, biodiversity, ecosystem services and livelihoods of dependent people (Midha and Mathur 2010). Multi-temporal long-term satellite data can be used to retrieve vegetation phenology, and the subtle changes in forest vigour can be assessed by analysing changes in its seasonal greenness. Recently, long-term MODIS NDVI data has been used to assess the spatial patterns of significant negative trend of seasonal greenness over the different forest types of India particularly its hotspots in the core forest areas (Chakraborty et al. 2018). Lion's share of these negative changes in greenness were found to be in tropical moist deciduous (2.06 m ha) followed by tropical dry deciduous (1.4 m ha) forest. Interestingly, nearly 80% of these changes took place in the core forest areas, which seems to be alarming. The states of Odisha, Chhattisgarh, Madhya Pradesh, Telangana and Uttarakhand were found to be hotspots of these negative changes in deciduous forest. The study has also identified significant negative changes of seasonal greenness over the large protected areas such as national park, wildlife sanctuary an conservation reserve for prioritization of biodiversity conservation and climate change mitigation programmes.

4.5 *Forest Fire*

Though wildfires were normal processes earlier, current spate of blazes raging across Europe, North America and other continents are certainly abnormal and seem to have origins due to excessive heat and dry conditions prevalent. Biodiversity loss, air pollution and habitat damage are clear fallouts of such extreme events. As per United Nation Environment Programme, global warming and land-use change are projected to increase extreme fires by up to 14 to 50% by the end of century, with increases of up to 14% by 2030, 30% by the end of 2050 and 50% by the end of the century. Remote sensing of forest fire is an exhaustively researched and understood subject and is made possible from earliest days of visible/near-infrared remote sensing period through benchmark MODIS sensor on board Aqua and Terra satellites to Suomi NPP VIIRS sensor. Indian remote sensing satellites such as Resourcesat having sensors LISS III and AWIFS consisting of shortwave infrared band (1550–1770 nm) provide a high degree of detectability and mapping potential over the globe. Harnessing strength of SWIR, MIR regions coupled with thermal regions at about 3900, 11,000 and 12,000 nanometres offer clear scope of smouldering and incandescent fires, which can detect forest fires of different nature from a geostationary and polar orbiting platforms.

An exhaustive review of optical remote sensing sensors employed in forest fire mapping along with traditional and neural network-based computing procedures to assist early fire warning systems provides an extensive survey on both flame and smoke detection algorithms employed, encompassing data retrieved between 1990 and October 2020 from Web of Science (Barmpoutis et al. 2020). Data streams from terrestrial, airborne and space-borne-based systems with various models aiming to detect fire occurrences with high accuracy in challenging environments have been assessed. Massive peaking of works during 2019 is observed (Fig. 15), which may correspond to works coinciding with peak activity of fire vulnerability and events.

Sun synchronous satellites have been deployed since earliest days to monitor various land cover characteristics using multispectral imaging of which AVHRR, MODIS and VIIRS form key sensors, which have been of great use in mapping forest fires. AVHRR (advanced very-high-resolution radiometer) which has

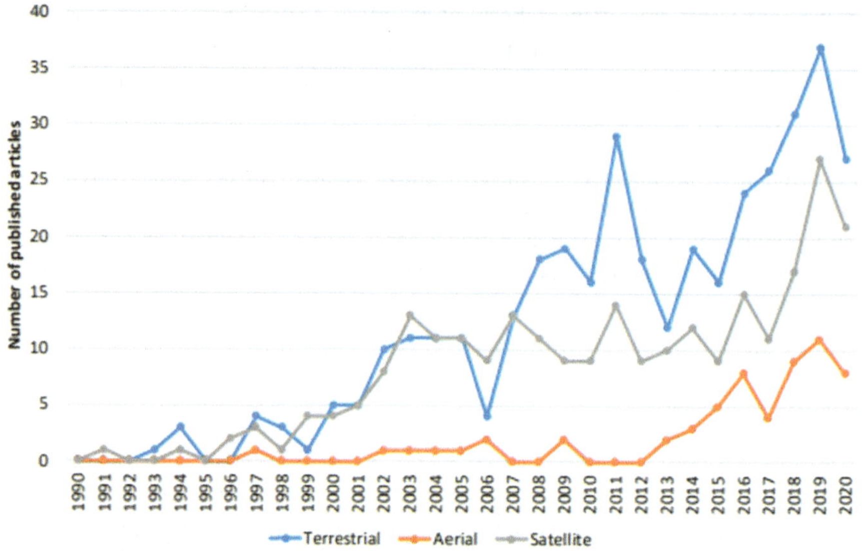

Fig. 15 Trend in studies published on forest fire from 1990 through three streams of remote sensing data, using Web of Science source. (Shukla and Pal 2009)

six channels, three in the visible/near-infrared region and three thermal infrared channels, with 1 km spatial resolution. MODIS (Moderate Resolution Imaging Spectroradiometer) on board Aqua and Terra with 1–2 days revisit time, image data in 36 spectral bands ranging in wavelengths from 0.4 to 14.4 μm and at varying spatial resolutions (2 bands at 250 m, 5 bands at 500 m and 29 bands at 1 km). This is followed by VIIRS sensor (Visible Infrared Imaging Radiometer Suite) on board Suomi NPP satellites which provides 22 different spectral bands, i.e. 16 moderate-resolution bands (M-bands, 750 m), 5 imaging resolution bands (I-bands, 375 m) and 1 day/night panchromatic band (750 m). Himawari is also used to detect the forest fire using its Advanced Himawari Imager. Apart from this, geostationary satellites with sensors such as GOES Baseline Imager and ESA-SEVIRI (Spinning Enhanced Visible and InfraRed Imager) have also been used for forest fire detection. Differentiation of smoke and clouds has been a key factor in achieving the clear identification of forest fires since clouds can produce similar signatures. Multiband thresholding technique for discriminating between smoke plumes and clouds suggested that it was able to isolate smoke pixels in the presence of other scene types, such as clouds. However, this approach performed better in identifying fresh dense smoke as compared to highly diffused smoke (Shukla and Pal 2009). Global Forest Watch, an online web GIS platform, provides latest updated content on ground fires prevalent across the globe using VIIRS 375 m pixel-based information and is meant to be an open data source for empowering participatory forest management.

5 Earth Observation to Support Governance of Climate Change

Argument about the climate change scenario is that contract of science and society has been broken (Glavovic et al. 2021). In spite the understanding that climate change is anthropogenic and science is settled in the issue of providing clear proofs, climate change is being reversed. The indicators of change are rising alarmingly at an unacceptable scale. The 'uncertainty' factors that are naturally part of scientific conclusions have been exploited by interest groups promoting financial viability of investments. Today, solutions to reverse climate change seem to be oriented about society and economy, rather than science alone. Activism is being repeatedly argued as a truly plausible alternative since policymaking has hardly kept pace with true needs of natural resources handling, to the limit that the next IIPC assessment report (AR7) need not be compiled at all. Severity of climate change impacts in Europe causing forest blazes, floods, dry weather and even melting of tarmacs has opened unprecedented perspectives coupled with compulsions of war disrupting all the energy production patterns. Though countries need to pledge for reduction of fossil fuel usage, war has pushed many countries to look for newer projects exploiting fossil-based energy unwillingly. Current plans to address climate change need to ideally result in cuts to greenhouse gas emissions by around 7.5% by 2030. Yet, to meet the 1.5 °C target, the world would require a cut of 55% by 2030 (Moody 2022).

Existing governance frameworks are poorly equipped with skills and procedures to handle the unprecedented uncertainty in climate change research results making it the biggest impediment. Intra-agency information sharing alone can bring in accountability of the actions taken. The essence of new governance needs to inculcate 'adaptive governance' framework that demands a systematic monitoring and adaptation approaches in decisions and programmes (Camacho 2009).

An approach suggested to bring in adaptive governance to climate change by managing uncertainty through a learninginfrastructure focuses on fourfold framework. It aims to build case studies to illustrate valuable lessons of challenges to create effective natural resource management, followed by deliberating the specific implications of climate change considering the interagency information sharing and adaptive governance. Effort needs to be done to also engage the growing theoretical literature in adaptive management and federalism. Insight about how agencies manage uncertainty that has far-reaching implications for other areas of administrative regulation is also important. The fragmented regulatory patchwork prevalent in handling natural resources has been the biggest hurdle in addressing the concerns. In spite of having sound investments on managing natural resources such as large waterbodies and their basins, multitude of institutions with their discrete approaches not communicating to each other will not create any true resilient or adaptive mechanism, instead leaving the system possibly intensely vulnerable. Recent spurt in flooding in major cities of India points to such a phenomenon of not being able to cater to the extremes and discovering the intentional misappropriation of land parcels for development damaging the economy itself. In spite of bringing in policy-level adjustments to foster collaboration between interagency activity, fear is expressed about less than expected level of collaborative innovations.

Such shortcomings or lack of learning outcomes from instances of natural resource handling across stakeholder, in principle, need to harness the capacity of Earth observation along with advances in android technology followed by open-source technologies in serving geospatial data. Geospatial data streaming from various satellite-based or non-orbital sensor into an information system can be integrated as a spatial or process model to provide analysis of the status and dynamics of the climate change-related indicator or its value addition into existing national-level governance service mechanism. Capacity building of the functionaries to handle such information either as consuming the satellite image-based thematic content followed by training and handholding of the devices to carry out spatial inventory is central to scale up applications of governance-related projects.

Urban systems are the biggest consumers of energy on Earth and regulating their patterns of energy usage lies at the heart of mitigating the change. Wide-ranging experiments have been recorded across the globe which are initiated as either local governance driven or sociotechnical or strategic to arrive at reduced carbon footprints of urban activities. Bulkeley and Broto (2013) analysed 627 such experiments across 100 cities of the globe to understand the overall trend, which points towards creation of new forms of political space within the city, as public and private authorities merge. Such actions are primarily enacted through forms of technical intervention in infrastructure networks, drawing attention to the importance of such sites in urban climate politics. The diverging arguments arising out of multiple actors demonstrating solutions also cast a doubt about the overall efficacy of attaining low carbon economy and climate-resilient urban future, since actors with state authority and actors in voluntary sectors may operate cross purposes. Role of Earth observation systems in such contexts of conflict handling can be quite relevant since geospatial systems along with mapping and monitoring ability of remote sensing offer spatio-temporal capability to resolve the jurisdictions of each of the experiments and understand the hotspots of effects of each treatment. At least empirically low carbon patterns induced may be catalogued to assess the contribution of each intervention, so as to bring in the synergy of state- and citizen-induced changes.

Advent of night-time higher-resolution imaging by Suomi NPP VIIRS can provide insight into spatial pattern of energy usage and hence enable a direct insight into carbon consumption. A system for using VIIRS data for remote carbon estimates, which provides monthly independent, unbiased estimates of per-capita carbon emissions, has been implemented (Jasmin et al. 2015). Full spatial resolution (750 meter) mosaics of VIIRS were regressed against census data of US followed by emission estimates from the US Department of Energy. Due consideration in of fraction of energy use from renewables was also taken care to provide estimates at country level. Opening of up opportunistic remote sensing using volunteered passenger aircraft night-time imaging offers further scope of strengthening up

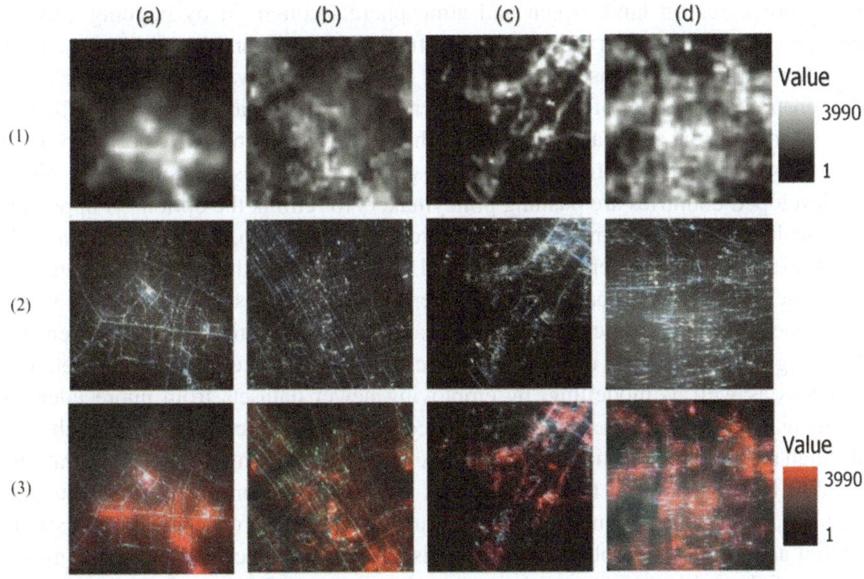

Fig. 16 Comparative images of satellite-derived and aerial camera night-time lights. Brightness contrast between VIIRS-DNB and VPAN-RS images at different scales. (1) VIIRS-DNB. (2) VPAN-RS. (3) Overlapped image of VIIRS-DNB and VPAN-RS images. (**a**) Puning, Guangdong. (**b**) Part of Shanghai. (**c**) Edge of Wuhan, Hubei. (**d**) Downtown area of Changsha, Hunan. (Liu et al. 2021)

carbon accounting of urban systems. In a study covering 16 cities of China or one from Japan (Liu et al. 2021), a reliable approach has been developed (Fig. 16) using VIIRS imagery as reference to align night-time passenger aircraft-derived data so as to obtain high resolution (up to even 1 mtr) at 5–10 pixel error. This method provides good frequency of observation and can have huge implications in monitoring second- and third-tier cities which are accumulating investment in current growth scenarios, especially in India.

6 Conclusion

Addressing the aspects of remote sensing for climate change is a challenging task in terms of representing the prevalent technologies and findings that lead to the understanding of climate change along with its management. Alarm raised at COP27 about the extremity of events in developed world, and need to assist majority of global population to combat the impacts as well as support affected livelihoods requires formulation of mechanisms to fund, implement and monitor the mitigative and adaptation measures. Geospatial observation coupled with remote sensing systems offer reliable and at times irreplaceable approaches to watch the

Earth processes on land, ocean and atmosphere, brought in by ongoing climate change. Long-term observations derived from earlier ground-based sensors, later coupled with satellite-based observations with sufficient validation, have clearly demonstrated the effect of anthropogenic emissions onto changing climate. Though uncertainty factors associated with any of the scientific methods have been exploited by undesirable market forces, heavily invested in fossil fuels, disasters at the doors of developed countries are pushing policymakers to rethink the options of approach drastically. It sounds at times that terms of reference at Kyoto, Bali and Paris have all been relegated by commercial and national interests, while the vulnerable countries continue to suffer the impact of unabated emissions by industrial establishments.

Ironically, increasing climate change has fuelled innovations in remote sensing focusing on atmospheric chemistry, physics, geophysics oceanography, vegetation sciences as well as modelling, by employing newer datasets from unprecedented domains. Though science has demonstrated sufficient evidence of climate change through all these observations and experiments, it is important now that observations need to be dovetailed to the requirements of citizenry in bracing the impacts and being alerted about the impending losses and threats. Earth observations in various spatial and temporal resolutions help in visualizing the processes and their impact ahead of the critical events such as drought, cyclone, floods and storms at each administrative levels through well-developed web-enabled geographic information systems coupled with information on smart phone applications served by governments and organizations handling stewardship in climate change such as IPCC and UNFCCC (IPCC 2022).

Insight into possibilities and capabilities prevalent at global and national level has potential to assimilate the information hierarchically as per the scale of observation, so as to match measures to combat the impacts using state-of-the-art data analytics involving machine learning. Though many of the events are obvious as the scale of impact is global, boundary phenomenon having high degree of fuzziness as well as temporal precision in predicting the movements of atmospheric systems require huge computation power and innovation. Two- and three-dimensional awareness of climate systems and their interaction with Earth-based land cover along with terrain complexities is key to give forecasts and nowcasts so as to keep citizens aware and alert of the impending dangers especially in areas of historically well-known vulnerabilities. Our country being a curious mixture of high resourcefulness in terms of funds and hardware at one end of the spectrum coupled with totally unaware remote and hapless societal proportions requires perspectives from various angles. Exploration of existing ITK (indigenous and traditional knowledge) and their degree of relevance in new climate normal should be a priority and it needs to be catalogued, analysed and given a technologically value-added outlook so as to render the readiness against hazards in sufficient manner. Translation of satellite-derived content through tools of artificial intelligence and machine learning into a communication understandable by highly vulnerable sections of society can be one among the important tasks towards adaptation.

Acknowledgement The authors would like to express gratitude for the support and insights provided for this task by Shri Bishwadip Gharai, Former Head of the Atmospheric Sciences Group. The kind support provided by Dr. Prakash Chauhan, Director of NRSC, Dr. K. Vinod Kumar, Deputy Director of RSA and Dr. Rajashree V. Bothale, Deputy Director of ECSA, in developing this review is greatly acknowledged. Suggestions on addressing critical impacts aspects by Dr. K. Vinod Kumar has improved the scope of the work immensely. We would also like to express our deep gratitude to the support extended by Dr. M V Ramana, Group Head of AS and LSP CSG, ECSA. The authors also thank Mr. Mahesh P. (ASD, ECSA) for his resourceful help in compilation of this document. Timely and valuable inputs by Dr. Rajadeep Roy, scientist at RRSC-E, Kolkata, have added relevant strength in this review. We also thank all the support and guidance provided by Dr. K M Reddy, Head of Rural Development Division of RSA. The authors also thank the effort of Dr. Suneetha Manne, Professor and Head of IT Department at Siddhartha Engineering College, Vijayawada, and his team in supporting the initial compilation of information using customized tools.

References

Alahacoon N, Edirisinghe M (2022) A comprehensive assessment of remote sensing and traditional based drought monitoring indices at global and regional scale. Geomat Nat Haz Risk 13(1):762–799

Arora M, De K, Chaudhury NR, Nanajkar M, Chauhan P, Pateriya B (2022) Climate change induced thermal stress caused recurrent coral bleaching over gulf of Kachchh and Malvan marine sanctuary, west coast of India. Climate Change in Asia and Africa: Examining the Biophysical and Social Consequences, and Society's Responses 35

Azarderakhsh M, Prakash S, Zhao Y, AghaKouchak A (2020) Satellite-based analysis of extreme land surface temperatures and diurnal variability across the hottest place on earth. IEEE Geosci Remote Sens Lett 17(12):2025–2029

Banger K, Tian H, Tao B, Lu C, Ren W, Yang J (2015) Magnitude, spatiotemporal patterns, and controls for soil organic carbon stocks in India during 1901–2010. Soil Sci Soc Am J 79(3):864–875

Barmpoutis P, Papaioannou P, Dimitropoulos K, Grammalidis NA (2020) Review on early Forest fire detection systems using optical remote sensing. Sensors 20(22):6442. https://doi.org/10.3390/s20226442

Bell JR, Schultz LA, Molthan AL, Meyer FJ (2018) Investigations of Hail Damage Swaths using Various Satellite Remote Sensing Platforms. In Annual Meeting of the National Weather Association (No. MSFC-E-DAA-TN60390))

Bhattacharyya T, Pal DK, Easter M, Batjes NH, Milne E, Gajbhiye KS et al (2007) Modelled soil organic carbon stocks and changes in the Indo-Gangetic Plains, India from 1980 to 2030. Agric Ecosyst Environ 122(1):84–94

Bolch T, Kulkarni A, Kaab A, Huggel C, Paul F, Cogley JG et al (2012) The state and fate of Himalayan glaciers. Science 336(6079):310–314

Brown ME, De Beurs KM, Marshall M (2012) Global phenological response to climate change in crop areas using satellite remote sensing of vegetation, humidity and temperature over 26 years. Remote Sens Environ 126:174–183

Bulkeley H, Castan Broto V (2013) Government by experiment? Global cities and the governing of climate change. Trans Inst Br Geogr 38(3):361–375

Camacho AE (2009) Adapting governance to climate change: managing uncertainty through a learning infrastructure. Emory LJ 59:1

Canadian Space Agency (2022). https://letstalkscience.ca/educational-resources/backgrounders/7-ways-satellites-help-fight-climate-change

Candanosa RM (2022) NASA finds 2022 Arctic Winter Sea ice maximum extent 10th-lowest on record. scitechdaily.Com

Chakraborty A, Seshasai MVR, Dadhwal VK (2014) Geo-spatial analysis of the temporal trends of kharif crop phenology metrics over India and its relationships with rainfall parameters. Environ Monit Assess 186(7):4531–4542

Chakraborty A, Seshasai MVR, Rao SVC, Dadhwal VK (2017) Geo-spatial analysis of temporal trends of temperature and its extremes over India using daily gridded (1° × 1°) temperature data of 1969–2005. Theor Appl Climatol 130(1):133–149

Chakraborty A, Seshasai MVR, Reddy CS, Dadhwal VK (2018) Persistent negative changes in seasonal greenness over different forest types of India using MODIS time series NDVI data (2001–2014). Ecol Indic 85:887–903

Chaturvedi RK, Kattumuri R, Ravindranath D (2014) Mainstreaming adaptation to climate change in Indian policy planning. Int J Appl Econ Econ 22(1):23–56

Chen R, Jones WL (2018, March) Creating a consistent multi-decadal oceanic TRMM-GPM brightness temperature data record. In: 2018 IEEE 15th specialist meeting on microwave radiometry and remote sensing of the environment (MicroRad). IEEE, pp 1–6

Cleland EE, Chuine I, Menzel A (2007) Shifting plant phenology in response to global change. Trends Ecol Evol 22(7):357–365

Climateinformation (n.d.) What do different RCPs mean?. https://climateinformation.org/wp-content/uploads/sites/6/2019/06/rcps.png. Accessed on 13 Dec 2022

Cracknell AP, Cracknell AP (2001) Remote sensing and climate change: role of earth observation. Springer, Berlin

D'Mello JR, Prasanna KS (2018) Processes controlling the accelerated warming of the Arabian Sea. Int J Climatol 38(2):1074–1086

Das PK, Chakraborty A, Seshasai MV (2014) Spatial analysis of temporal trend of rainfall and rainy days during the Indian Summer Monsoon season using daily gridded (0.5° × 0.5°) rainfall data for the period of 1971–2005. Meteorol Appl 21(3):481–493

Dubash NK, Raghunandan D, Sant G, Sreenivas A (2013) Indian climate change policy: exploring a co-benefits based approach. Econ Polit Wkly:47–61

ESA (n.d.) Climate change. https://www.esa.int/Applications/Observing_the_Earth/Space_for_our_climate/Climate_change

Falloon P, Jones CD, Cerri CE, Al-Adamat R, Kamoni P, Bhattacharyya T et al (2007) Climate change and its impact on soil and vegetation carbon storage in Kenya, Jordan, India and Brazil. Agric Ecosyst Environ 122(1):114–124

Fei S, Morin RS, Oswalt CM, Liebhold AM (2019) Biomass losses resulting from insect and disease invasions in US forests. Proc Natl Acad Sci 116(35):17371–17376

Forzieri G, Dakos V, McDowell NG et al (2022) Emerging signals of declining forest resilience under climate change. Nature 608:534–539. https://doi.org/10.1038/s41586-022-04959-9

Gaddam VK, Boddapati R, Kumar T, Kulkarni AV, Bjornsson H (2022) Application of ("OTSU") an image segmentation method for differentiation of snow and ice regions of glaciers and assessment of mass budget in Chandra basin, Western Himalaya using Remote Sensing and GIS techniques. Environ Monit Assess 194(5):1–18

GCOS (2021) The status of the Global Climate Observing System 2021: The GCOS status report (GCOS-240). World Meteorological Organization Rep. 384p. https://gcos.wmo.int/en/publications/gcos-status-report-2021

Glavovic BC, Smith TF, White I (2021) The tragedy of climate change science. Clim Dev:1–5

Gouveia CM, Martins JP, Russo A, Durão R, Trigo IF (2022) Monitoring heat extremes across Central Europe using land surface temperature data records from SEVIRI/MSG. Remote Sens 14(14):3470

Gupta S (2015) Simulating climate change impact on soil erosion & soil carbon sequestration. M. Tech RS&GIS) thesis, Andhra University, Visakhapatnam

Ha T, Shen Y, Duddu H, Johnson E, Shirtliffe SJ (2022) Quantifying hail damage in crops using Sentinel-2 imagery. Remote Sens 14(4):951

Halsnæs K, Larsen MAD, Kaspersen PS (2018) Climate change risks for severe storms in developing countries in the context of poverty and inequality in Cambodia. Nat Hazards 94:261–278. [CrossRef]

Hartmann H, Bastos A, Das AJ, Esquivel-Muelbert A, Hammond WM, Martínez-Vilalta J et al (2022) Climate change risks to global forest health: emergence of unexpected events of elevated tree mortality worldwide. Annu Rev Plant Biol 73:673–702

Hartmann H, Moura CF, Anderegg WR, Ruehr NK, Salmon Y, Allen CD et al (2018) Research frontiers for improving our understanding of drought-induced tree and forest mortality. New Phytol 218(1):15–28

Hooker J, Duveiller G, Cescatti A (2018) A global dataset of air temperature derived from satellite remote sensing and weather stations. Scientific Data 5(1):1–11

Hulley GC, Malakar NK, Islam T, Freepartner RJ (2017) NASA's MODIS and VIIRS land surface temperature and emissivity products: a long-term and consistent earth system data record. IEEE J Select Topics Appl Earth Observ Remote Sens 11(2):522–535

IPCC (2022) Climate change 2022: mitigation of climate change. In: Shukla PR, Skea J, Slade R, Al Khourdajie A, van Diemen R, McCollum D, Pathak M, Some S, Vyas P, Fradera R, Belkacemi M, Hasija A, Lisboa G, Luz S, Malley J (eds) Contribution of working group III to the sixth assessment report of the intergovernmental panel on climate change. Cambridge University Press, Cambridge, UK and New York, NY. https://doi.org/10.1017/9781009157926

Jain J, Mitran T (2020) A geospatial approach to assess climate change impact on soil organic carbon in a semi-arid region. Trop Ecol 61(3):412–428

Jasmin T, Desai AR, Pierce RB (2015) Estimating global per-capita carbon emissions with VIIRS nighttime lights satellite data. In: AGU fall meeting abstracts, vol. 2015, pp GC11B–G1029

Jones PD, Moberg A (2003) Hemispherical and large scale surface air temperature variations; an extensive revision and an update to 2001. J Clim 16:206–223

Keen RM, Voelker SL, Wang SYS, Bentz BJ, Goulden ML, Dangerfield CR et al (2022) Changes in tree drought sensitivity provided early warning signals to the California drought and forest mortality event. Glob Chang Biol 28(3):1119–1132

Kopp RE, Kemp AC, Bittermann K, Horton BP, Donnelly JP, Gehrels WR et al (2016) Temperature-driven global sea-level variability in the common era. Proc Natl Acad Sci 113(11):E1434–E1441

Kumar R, Manzoor S, Vishwakarma DK, Al-Ansari N, Kushwaha NL, Elbeltagi A et al (2022) Assessment of climate change impact on snowmelt runoff in Himalayan region. Sustainability 14(3):1150

Li T, Zhang Y, Chang C-P, Wang B (2001) On the relationship between Indian Ocean sea surface temperature and Asian summer monsoon. Geophys Res Lett 28(14):2843–2846

Lima CB, Prijith SS, Sesha Sai MV, Rao PV, Niranjan K, Ramana MV (2019) Retrieval and validation of cloud top temperature from the geostationary satellite INSAT-3D. Remote Sens 11(23):2811

Lindsey R, Dahlman L (2020) Climate Change: Ocean Heat Content). NOAA Climate.gov

Liu C, Tang Q, Xu Y, Wang C, Wang S, Wang H et al (2021) High-spatial-resolution nighttime light dataset acquisition based on volunteered passenger aircraft remote sensing. IEEE Trans Geosci Remote Sens 60:1–17

Mazzoglio P, Laio F, Balbo S, Boccardo P, Disabato F (2019) Improving an extreme rainfall detection system with GPM IMERG data. Remote Sens 11(6):677

McCarl BA, Thayer AW, Jones JPH (2016) The challenge of climate change adaptation for agriculture: an economically oriented review. J Agric Appl Econ 48(4):321–344

McSweeney R (2015) Analysis: the most 'cited' climate change papers. https://www.carbonbrief.org

Menzel A, Estrella N, Fabian P (2001) Spatial and temporal variability of the phenological seasons in Germany from 1951 to 1996. Glob Chang Biol 7(6):657–666

Merchant CJ, EmMcbury O, Bulgin CE et al (2019) Satellite-based time-series of sea-surface temperature since 1981 for climate applications. Sci Data 6:223. https://doi.org/10.1038/s41597-019-0236-x

Midha N, Mathur PK (2010) Assessment of forest fragmentation in the conservation priority Dudhwa landscape India using FRAGSTATS computed class level metrics. J Indian Soc Remote Sens 38:487–500
Mitran T, Lal R, Mishra U, Meena RS, Ravisankar T, Sreenivas K (2018) 12 climate change impact on soil carbon stocks in India. Soil and Climate
Mondal A, Khare D, Kundu S, Meena PK, Mishra PK, Shukla R (2015) Impact of climate change on future soil erosion in different slope, land use, and soil-type conditions in a part of the Narmada River basin, India. J Hydrol Eng 20(6):C5014003
Moody J (2022) Europe caught napping on climate change as heatwaves spread by Jessica Moody | 29 Jul 2022 | Climate change, Environment, Europe, News Decoder
Minnett PJ, Alvera-Azcárateb A, Chin TM, Corlett GK, Gentemann CL, Karagali I, Li X, Marsouin A, Marullo S, Maturi E, Santoleri R, Picart SS, Steele M, Vazquez-Cuervo J (2019) Half a century of satellite remote sensing of sea-surface temperature, 111366. Remote Sens Environ 233. https://doi.org/10.1016/j.rse.2019.111366. ISSN 0034-4257,
Nagler T, Rott H, Malcher P, Müller F (2008) Assimilation of meteorological and remote sensing data for snowmelt runoff forecasting. Remote Sens Environ 112(4):1408–1420
NASA (n.d.) Taking a global perspective on earth's climate. https://climate.nasa.gov/nasa_science/history/
Nela BR, Singh G, Kulkarni AV (2023) Ice thickness distribution of Himalayan glaciers inferred from DInSAR-based glacier surface velocity. Environ Monit Assess 195(1):1–20
Nes EH et al (2016) What do you mean, 'tipping point'? Trends Ecol Evol 31:902–904
New M, Todd M, Hulme M, Jones P (2001) Precipitation measurements and trends in the twentieth century. Int J Climatol 21:189–1922
Newburger (2020) Climate change is driving widespread forest death and creating shorter, younger trees. www.cnbc.com/2020/05/28/climate-change-is-driving-widespread-forest-death-creating-shorter-trees.html
NRSC (2022) National Information system for climate and environment Studies. https://www.nrsc.gov.in/sites/default/files/doc_to_html/NICES_brochure_final_July_2022.pdf
O'Carroll AG, Armstrong EM, Beggs HM, Bouali M, Casey KS, Corlett GK et al (2019) Observational needs of sea surface temperature. Front Mar Sci 6:420
Pandey R, Alatalo JM, Thapliyal K, Chauhan S, Archie KM, Gupta AK et al (2018) Climate change vulnerability in urban slum communities: investigating household adaptation and decision-making capacity in the Indian Himalaya. Ecol Indic 90:379–391
Parida BR, Pandey AC, Patel NR (2020) Greening and browning trends of vegetation in India and their responses to climatic and non-climatic drivers. Climate 8(8):92
Penuelas J, Fiella I (2001) Response to a warming world. Science 294(5543):793–794
Prijith SS, Rao PVN, Mohan M, Sesha Sai MVR, Ramana MV (2018) Trends of absorption, scattering and total aerosol optical depths over India and surrounding oceanic regions from satellite observations: role of local production, transport and atmospheric dynamics. Environ Sci Pollut Res 25:18147–18160. https://doi.org/10.1007/s11356-018-2032-0
Rahman MS, Di L (2020) A systematic review on case studies of remote-sensing-based flood crop loss assessment. Agriculture 10(4):131
Ramanathan V, Ramana M, Roberts G et al (2007) Warming trends in Asia amplified by brown cloud solar absorption. Nature 448:575–578. https://doi.org/10.1038/nature06019
Rao KG, Goswami BN (1988) Interannual variations of sea surface temperature over the Arabian Sea and the Indian monsoon: a new perspective. Mon Weather Rev 116(3):558–568
Rao RR, Girish Kumar MS, Ravichandran M, Rao AR, Gopalakrishna VV, Thadathil P (2010) Interannual variability of Kelvin wave propagation in the wave guides of the equatorial Indian Ocean, the coastal Bay of Bengal and the Southeastern Arabian Sea during 1993–2006. Deep-Sea Res I Oceanogr Res Pap 57(1):1–13
Remya SN, Syed TH, Kulkarni AV, Anand R (2022) Manifestation of topography and climate variations on long-term glacier changes in the Alaknanda Basin of Central Himalaya, India. Geocarto Int:1–20

Reynolds RW, Rayner NA, Smith TM, Stokes DC, Wang W (2002) An improved in situ and satellite SST analysis for climate. J Climate 15(13):1609–1625
Roxy MK, Ritika K, Terray P, Murtugudde R, Ashok K, Goswami BN (2015) Drying of Indian subcontinent by rapid Indian ocean warming and a weakening land-sea thermal gradient. Nat Commun 6(1):1–10
Roxy MK, Gnanaseelan C, Parekh A, Chowdary JS, Singh S, Modi A, Kakatkar R, Mohapatra S, Dhara C, Shenoi SC, Rajeevan M (2020) Indian ocean warming assessment of climate change over the Indian region: a report of the Ministry of Earth Sciences (MoES), Government of India ed R Krishnan et al. Springer, Singapore, pp 191–206
Scheffer M, Carpenter S, Foley JA, Folke C, Walker B (2001) Catastrophic shifts in ecosystems. Nature 413:591–596
Senf C, Buras A, Zang CS, Rammig A, Seidl R (2020) Excess forest mortality is consistently linked to drought across Europe. Nat Commun 11(1):1–8
Shetye SS, Kurian S, Gauns M, Vidya PJ (2019) 2015-16 ENSO contributed reduction in oil sardines along the Kerala coast, south-west India. Mar Ecol 40(6) https://onlinelibrary.wiley.com/doi/abs/10.1111/maec.12568
Shukla BP, Pal PK (2009) Automatic smoke detection using satellite imagery: preparatory to smoke detection from Insat-3D. Int J Remote Sens 2009, 30:9–22. [CrossRef]
Singh R, Kumar R, Latief SU, Kumar R, Shekhar M (2022) Recession of Gaglu Glacier, Chandra Basin, Western Indian Himalaya. In: Rani S, Kumar R (eds) Climate change. Springer Climate. Springer, Cham. https://doi.org/10.1007/978-3-030-92782-0_5
Sosa L, Justel A, Molina ´i (2021) Detection of crop hail damage with a machine learning algorithm using time series of remote sensing data. Agronomy 11(10):2078
Sreenivas G, Mahesh P, Mahalakshmi DV, Kanchana AL, Chandra N, Patra PK, Dadhwal VK (2022) Seasonal and annual variations of CO_2 and CH_4 at Shadnagar, a semi-urban site. Sci Total Environ 819:153114
Subrahmanyam B, Robinson IS (2000) Sea surface height variability in the Indian Ocean from TOPEX/POSEIDON altimetry and model simulations. Mar Geod 23(3):167–195
Taori A, Suryavanshi A, Pawar S et al (2022) Establishment of lightning detection sensors network in India: generation of essential climate variable and characterization of cloud-to-ground lightning occurrences. Nat Hazards 111:19–32
Taori A, Suryavanshi A, Bothale RV (2023) Cloud-to-ground lightning occurrences over India: seasonal and diurnal characteristics deduced with ground-based lightning detection sensor network (LDSN). Nat Hazards 116:4037–4049
Thapliyal A, Kimothi S, Taloor AK, Bisht MPS, Mehta P, Kothyari GC (2023) Glacier retreat analysis in the context of climate change impact over the Satopanth (SPG) and Bhagirathi-Kharak (BKG) g Newberger, 2020laciers in the Mana basin of the Central Himalaya, India: a geospatial approach. Geosystems and Geoenvironment 2(1):100128
Verron J, Bonnefond P, Aouf L, Birol F, Bhowmick SA, Calmant S, Conchy T, Crétaux J-F, Dibarboure G, Dubey AK, Faugère Y, Guerreiro K, Gupta PK, Hamon M, Jebri F, Kumar R, Morrow R, Pascual A, Pujol M-I, Rémy E, Rémy F, Smith WHF, Tournadre J, Vergara O (2018) The benefits of the Ka-band as evidenced from the SARAL/AltiKa altimetric mission: scientific applications. Remote Sens 10:163. https://doi.org/10.3390/rs10020163
Vibhute A, Halder S, Singh P, Parekh A, Chowdary JS, Gnanaseelan C (2020) Decadal variability of tropical Indian Ocean Sea surface temperature and its impact on the Indian summer monsoon. Theor Appl Climatol 141(1):551–566
Vignudelli S, Birol F, Benveniste J, Fu LL, Picot N, Raynal M, Roinard H (2019) Satellite altimetry measurements of sea level in the coastal zone. Surv Geophys 40(6):1319–1349
Vineetha G, Karati KK, Raveendran TV, Idrees Babu KK, Riyas C, Muhsin MI, Shihab BK, Simson C, Anil P (2018) Responses of the zooplankton community to peak and waning periods of El Niño 2015-2016 in Kavaratti reef ecosystem, northern Indian Ocean. Environ Monit Assess 190(8):1–22. https://doi.org/10.1007/s10661-018-6842-9

Wang J, Sun R, Zhang H, Xiao Z, Zhu A, Wang M et al (2021) New global MuSyQ GPP/NPP remote sensing products from 1981 to 2018. IEEE J Select Topics Appl Earth Observ Remote Sens 14:5596–5612

White MA, Thornton PE, Running SW (1997) A continental phenology model for monitoring vegetation response to interannual climatic variability. Glob Biogeochem Cycles 11(3):217–234

Satellite-Based Remote Sensing Approaches for Estimating Evapotranspiration from Agricultural Systems

Abhilash Chandel

Abstract Quantifying the actual amount of water used by agricultural cropping systems is deemed essential for irrigation scheduling and management. Conventionally, this has been either not pursued or done using single-point measurement or estimation tools. Since crop water uses may vary spatially within the field and over time, understanding its spatiotemporal dynamics has emerged as a critical need for sustainable agricultural production. To address this, satellite-based remote sensing (RS) has evolved as a rapid and high-throughput tool for mapping geospatial evapotranspiration (ET) from agricultural production systems, globally. Such data is either used through various biophysical models or empirical data-run approaches towards improving the accuracy of ET estimates which are deemed to serve as decision support for precision irrigation and water management. This chapter discusses fundamentals of computing ET through various energy balance and empirical models that have evolved or refined over time. The chapter also summarizes up-to-date case studies with identified accuracies of water use estimations using satellite-based RS approaches. Such approaches demonstrate potentials to be coupled with automated irrigation systems for envisioned precision irrigation scheduling and management at spatiotemporal scales.

Keywords Evapotranspiration · Satellite-based remote sensing · Energy balance models · Empirical models · Agricultural cropping systems · Precision irrigation

1 Introduction

The rate of exchange of water packets from land surface to the atmosphere is termed as evapotranspiration (ET). The rate of water exchanged from soil surface is termed as evaporation, while that exchanged from vegetation surfaces is termed

---

A. Chandel (✉)
Department of Biological Systems Engineering, Virginia Tech Tidewater AREC, Suffolk, VA, USA
e-mail: abhilashchandel@vt.edu

P. M. Priyadarshan et al. (eds.), *Digital Agriculture*,
https://doi.org/10.1007/978-3-031-43548-5_9

as transpiration (Fig. 1). The water is exchanged in the form of energies that in turn require energy sources to carry out this process. Direct solar radiation and ambient temperatures are the primary variables that drive exchange of water from soil or vegetation surfaces to the atmosphere depending upon the deficit in water vapor pressure between the evaporating surface and the surrounding atmosphere. The evaporation rate subsides as it proceeds and surrounding air enriches with water in it. Eventually, the evaporation may stop if the surrounding atmosphere becomes completely saturated. The saturation is also affected by the wind speed which is responsible for the replacement of saturated air by the dry air. Therefore, in a nutshell, solar radiation, air temperature, relative humidity, and wind speed are the major climatological variables which drive water exchange from land surface to atmosphere. Soil evaporation is also affected by the amount of vegetation shading on the soil and the amount of water available within (due to rainfall or irrigation) to fulfill the evaporative demand. If the soil surface can fulfill the evaporative demand

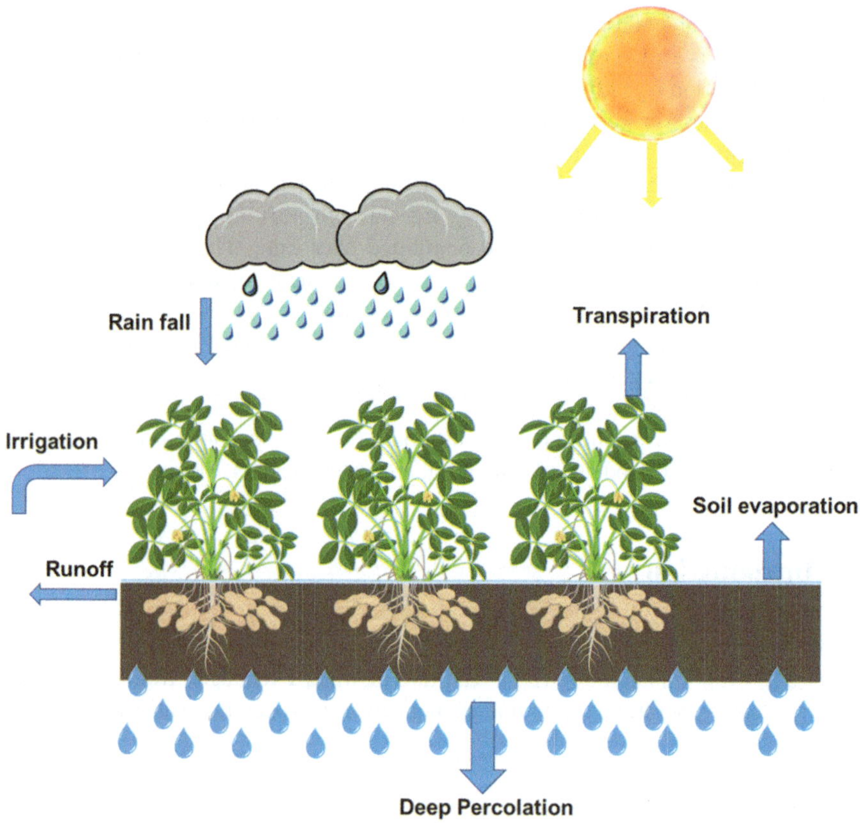

Fig. 1 The water cycle representing evapotranspiration in an agricultural cropping system

easily, the evaporation can be determined solely by the meteorological variables. However, when the water inputs (irrigation or rainfall) to the soil occur at very large intervals, the transport of water from deep soil to pear evaporating surface slows down and the soil surface dries out. This drying slows down the soil evaporation and in case of no soil water availability, the soil evaporation almost ceases in some (Borrelli et al. 1998).

Transpiration occurs from stomata, the small openings on the leaf surface through which water vapor and gases are exchanged with the surrounding atmosphere. This water is taken up by the roots from soil and transported throughout the plant. Almost all the water taken up by the plant is lost as transpiration and a minute fraction of it is used in overall plant development. Alike soil evaporation, transpiration is driven by excitation energy, vapor pressure deficit, and wind which stem from solar radiation, air temperature, relative humidity, and wind speed variables. The soil water availability and soil's ability to allow water pumping by the roots influence transpiration rates. Additionally, transpiration rate is affected by the crop type, vigor and physiology, ambient weather, and land management practices such as tillage.

The transpiration and soil evaporation, together termed as ET, occur simultaneously. When the vegetation or crop is small, the ET is predominantly accounted by soil evaporation and as the crop develops, the evaporation slows down and transpiration increases. Eventually when the soil is completely covered by the crop/vegetation, ET almost gets completely dominated by transpiration. It is estimated that at crop sowing 100% of ET comes from soil evaporation, and at full crop cover over 90% of ET comes from transpiration. ET is expressed as the water depth lost per unit time (hour, day, month, etc.).

Regional ET is generally more than half of the total precipitation and tends to be almost equal to precipitation in semiarid regions. Therefore, an in-depth understanding of ET is critical for evaluating hydrological cycle, water resource management, environmental sustentation, hydrometeorological predictions, global climate, and water cycle shift simulations and most importantly for agricultural water use mitigation and management (Jiang et al. 2009; Mcshane et al. 2017). Majority of agricultural production regions of the world today are irrigation-dependent, especially the arid and semiarid ones which are prone to drought risks. To mitigate such risks and ensure healthy crop production, about 70% of the fresh water is utilized for irrigation. As the food demands, climate change impacts continue to grow, the freshwater utilization rates are further expected to multiply, leaving reduced water sources to meet growing household consumption demands (Misra 2014; Mancosu et al. 2015). Utilizing ET to determine actual crop water requirements can help precise optimization of freshwater irrigation without compromising the crop yield and quality (Adeyemi et al. 2017). Numerous methods of ET estimation have evolved over time which range from point to regional scales as discussed in the following sections.

2 ET Estimation: Small-Scale Methods

2.1 Point-Scale Methods

Point-scale approaches can be categorized into invasive and noninvasive ones. Common invasive point-scale approaches include canopy water content retrieval from leaf sample weights before and after oven-drying, sap flow measurements, and measurement of leaf photosynthesis or stomatal conductance rates in the field, while common non-canopy-invasive point-scale approaches include soil moisture depletion measurements in the root zone and water budget rate change measurements using lysimeters. Lysimeters are the tank-like structures installed in the soil bed which isolate the crop root zone, and water loss (ET) from that tank is measured in terms of change of mass in case of weighing lysimeters. In non-weighing-type lysimeters, ET for a given time duration is calculated by subtracting the total water collected at the tank bottom (discharge) from the total input water (irrigation and/or rainfall). Lysimeters and leaf photosynthesis measurement systems are expensive and may not be readily affordable by the researchers or end users. At point scale, ET can also be calculated by measuring the water flux components within the crop root zone (Fig. 1, Eq. 1, Allen et al. 1998).

$$\mathrm{ET} = \mathrm{I} + \mathrm{P} - \mathrm{RO} - \mathrm{DP} + \mathrm{CR} \pm \Delta\ \mathrm{SF} \pm \Delta\ \mathrm{SW} \tag{1}$$

Where, irrigation (I) and rainfall (P) are the water inputs to crop root zone, RO is the surface runoff and is the part of I and P, which does not stay within the soil. DP is the deep percolation, also a part of I and P that will eventually recharge the water table. CR is the capillary rise, which is the water that may transport upward, and Δ SF is the rate of horizontal movement of water from shallow depths towards the root zone. In non-slopy land surfaces, Δ SF is assumed to be negligible, while the CR is assumed negligible within short time periods. Soil evaporation and crop transpiration deplete water from the root zone. Once all the water inputs and outputs of a crop root zone are known, ET can be calculated from the change in soil water content (Δ SW) over time. The soil water balance approach can be better applied when estimating ET over long time periods ranging at least from week to months. Some of the other indirect but point-scale methods of estimating ET are measurement of stem and leaf water potentials and leaf/canopy level thermometry. Although point-scale approaches are accurate at leaf/point level, all such approaches are constrained due to limited sampling and inaccuracies of spatial variability assessments when intended for implementing precision irrigation management at field scales. Mobile point-scale approaches of ET estimation such as photosynthesis, leaf conductance, manual thermometry, and leaf/stem water potential measurements, among others, require extensive human effort and time and therefore expenses to gather sufficient measurement samples to assess spatial variations in crop ET at field-scales. This is one of the reasons for limited adaptation of point-scale approaches for ET estimation and irrigation management by the crop producers.

2.2 *Gross-Scale Method: Eddy Covariance*

The eddy covariance (EC) technique measures and calculates turbulent energy fluxes within atmospheric boundary layers by analyzing frequencies and magnitudes of wind, energies, and various gases including H_2O, CO_2, CH_4, and N_2O above an area of land surface. These measurements are ultimately assessed in the form of gas emission and consumption rates, sensible heat (H), and latent heat fluxes (LE). The latent heat fluxes are precisely estimated based on the covariance between vertical wind velocities and specific humidity over land areas of various sizes ranging from square-hundreds to square-meters (Burba 2013; Denager et al. 2020). EC measuring towers (Fig. 2) are typically equipped with 3D sonic anemometers installed at 12 m above ground, an open-path gas analyzer, net radiometer, air temperature and humidity sensor, and a data logging system. EC has been a widely used method in micrometeorology for over 30 years now (Monteith and Unsworth 2008; Baldocchi 2013). The integrity of EC estimated LE is evaluated from the energy balance (Eq. 2) considering the fact that an equal amount of energy enters (net radiation: R_n) and exits (G: soil heat flux, H, and LE) the earth system over a given time period (Li et al. 2009). It must be noted that the energy inputs and outputs of the

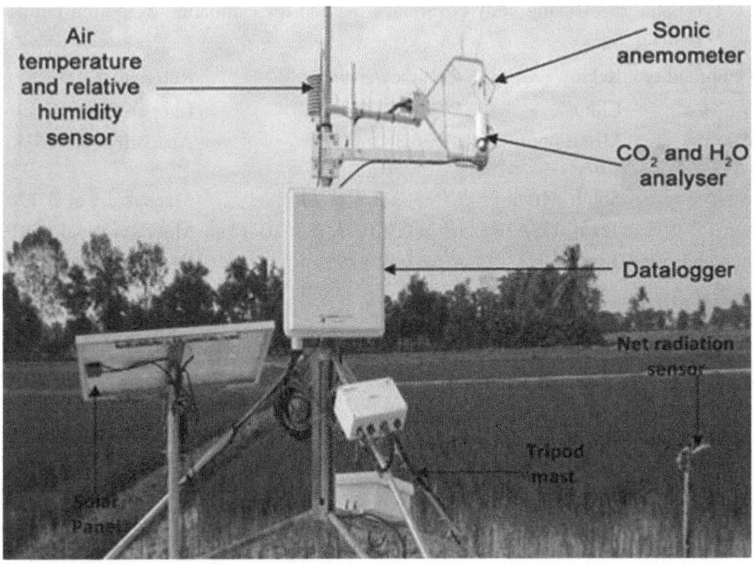

Fig. 2 A close view of eddy covariance flux measurement tower used for gas exchange measurement from paddy field. (From Bhattacharyya et al. 2013).

earth system can reverse their directions based on time of the day and agroclimatic conditions. Several studies have documented with evidence that the land surface energy balance is always incomplete when using energy fluxes out of EC method (Foken et al. 2011; Leuning et al. 2012). This is mostly due to uncertainties or underestimation of turbulent fluxes that compute the difference of G and R_n energy components to be larger than the sum of H and LE energy components (Twine et al. 2000; Foken 2008; Foken et al. 2011). Therefore, an energy balance closure ratio (sum of H and LE fluxes to the difference of R_n and G fluxes) is mostly used to represent the uncertainty in the energy balance computed by the EC method. This ratio mostly ranges between 0.7 and 0.9 depending on the surface (forests, orchards, short vegetation or crops, or bare soil). As a result, it has been reported that the EC method often underestimates actual ET. EC approach is a single point-in-time (or gross) estimator of ET and does not account for spatial variations for a given region or crop field; as a result, temporal precision irrigation could be scheduled but not the site-specific irrigation. Furthermore, EC flux towers can be expensive to install and operate and incoming data would need further processing to convert it to decisions. Therefore, this approach often lacks adaptation by crop producers for irrigation management. Some of the case studies where EC method have been used for ET estimation in agricultural cropping systems are summarized in Table 1.

$$R_n = \mathrm{LE} + \mathrm{H} + \mathrm{G} \tag{2}$$

Table 1 Case studies of using eddy covariance method for estimating evapotranspiration from cropping systems

Crop/ commodity	Region	Accuracy/errors	References
Corn	China	$R^2 = 0.84$, $E = 6\%$	Li et al. (2008)
Soybean	Mississippi, USA	$E = 6.8$–18%	Anapalli et al. (2018)
Potato	South Africa	$R^2 = 0.92$, $E = 4\%$	Machakaire et al. (2021)
Sorghum	Texas, USA	$R^2 = 0.9$–0.93, $E = 10$–15%	Moorhead et al. (2019)
Sorghum and corn	Texas, USA	$R^2 = 0.91$–0.94, $E = 0.51$–1.34 mm	Dhungel et al. (2021)
Grapevines	Spain	$R^2 = 0.69$–0.75, $E = 0.5$ mm	Sánchez et al. (2019)
Wheat	China	$R^2 = 0.96$–0.98, $E_{season} = 6$–25%	Wang et al. (2020)
Corn	China	$R^2 = 0.94$–1.0, $E_{season} = 9$–27%	Wang et al. (2020)

E: error, r: correlation coefficient, R^2: coefficient of determination. Unless otherwise specified, E refers to error in daily ET estimates

2.3 Gross-Scale Method: Standard-Crop Coefficients

The other commonly used method for ET estimation at a scale similar to EC is the standard-crop coefficient approach. In this method, a factor determinant of crop's physiological status termed as crop coefficient is multiplied to the reference ET (ET_r) computed from meteorological data. ET_r is the estimate of atmospheric water demand from the land surface based on weather conditions for a given time duration. Hypothetically, ET_r is equal to the actual ET from a well-watered short grass or alfalfa crop surface. The term ET_o is used when using short grass as reference crop and ET_r is used when using alfalfa as reference crop. A large number of empirical or semiempirical equations have been developed for computing ET_r from a single or few weather variables. However, most of these equations are valid under very specific climatic and agronomic conditions and are not globally applicable. In an expert consultation meeting held in May 1990, the UN Food and Agricultural Organization (FAO) recommended the use of the Penman–Monteith (PM) equation as standard for ET_r computation. The PM equation (Eq. 3) includes weather inputs of rainfall, wind speed, air temperature, relative humidity, net radiation, and inputs pertaining to hypothetical crop surface (short grass or alfalfa).

$$\lambda \text{ET} = \frac{\Delta \times (R_n - G) + \rho_a \times C_p \times \frac{(e_s - e_a)}{r_{\text{ah}}}}{\Delta + \gamma \times \left(1 + \frac{r_s}{r_{\text{ah}}}\right)} \tag{3}$$

where $(e_s - e_a)$ is the vapor pressure deficit of the air, ρ_a is the mean air density at constant pressure, c_p is the specific heat of the air, Δ is the slope of the saturated vapor pressure–temperature relationship, γ is the psychrometric constant (0.066 kPa °C^{-1}), and r_s and r_{ah} are the (bulk) surface and aerodynamic resistances to the water exchange. The PM method includes all parameters governing energy exchange and LE flux (or ET) from uniform vegetation surface. All the parameters can be measured or calculated from standard weather data. The PM equation can be directly utilized for ET calculation for any crop, given that the surface and aerodynamic resistances (r_{ah}) are crop specific (Eq. 4).

$$r_{\text{ah}} = \frac{\ln\left[\frac{z_m - d}{z_{\text{om}}}\right] \times \ln\left[\frac{z_h - d}{z_{\text{oh}}}\right]}{k^2 \times u_z} \tag{4}$$

where z_m is the height of wind speed measurement (m), z_h is the height of humidity measurement (m), d is the zero-plane displacement height (m), z_{om} is the roughness length governing momentum transfer (m), z_{oh} is the roughness length governing transfer of heat and vapor (m), k is the von Karman constant (0.41), and u_z is the wind speed measured at height z (m s^{-1}). This equation is restricted for neutral stability, i.e., adiabatic conditions where no heat exchange occurs. The variables d,

z_{om}, and z_{oh} are considered when the land surface is covered by vegetation as they depend on the crop height and architecture. For a standard hypothetical and well-irrigated reference crop (short grass, height: 0.12 m; r_s: 70 s m^{-1}; and albedo: 0.23), Eq. 3 translates to Eq. 5.

$$ET_0 = \frac{0.408 \times \Delta \times (R_n - G) + \gamma \times \left(\frac{900}{T_a}\right) \times u_2 \times \frac{(e_s - e_a)}{r_{ah}}}{\Delta + \gamma \times (1 + 0.34 \times u_2)} \tag{5}$$

where T_a and u_2 are the air temperature (Kelvin) and wind speed (m s^{-1}) measured at 2 m height above ground. ET_r (or ET_o) is multiplied by a crop-specific factor termed as crop coefficient (K_c) to estimate actual ET from a specific crop field. K_c is determined experimentally as the ratio of actual ET from lysimeter/soil water balance to the ET_r. However, since lysimeters cannot be affordable for all, standard K_c values have been determined which can be adjusted as per local agroclimatic conditions (Allen et al. 1998). K_c values represent integrated effects of canopy leaf area, height, crop health and vigor, development rate, management practices, canopy resistance to evaporative losses, and soil and climate conditions. K_c varies over the season with phenological growth stage and represents different rates of ET or crop water use as the season progresses. Typically for standard usage, K_c values have been documented for three growth stages (initial, mid, and late) in the FAO irrigation and drainage paper 56 (Allen et al. 1998). There are two types of crop coefficients generally used for estimating crop water use: (1) single crop coefficient and (2) dual crop coefficient. The single crop coefficient approach combines both transpiration and evaporation components, while in dual crop coefficient approach the two components are split. The single crop coefficient approach is mostly used for irrigation management, while dual crop coefficient approach is used for detailed understanding of soil evaporation. As the name suggests, dual crop coefficient is formed by a basal crop coefficient (K_{cb}) and soil evaporation coefficient (K_e). All these coefficients are represented in Fig. 3. As the experimented values of crop coefficients may not be readily generated across different agroclimatic conditions due to resource constraints, standard equation (Eq. 6) as in the FAO irrigation and drainage paper 56 (Allen et al. 1998) is used to adjust default crop/basal crop coefficients for given agroclimatic conditions.

$$K_c = K_{c,tab} + [0.04 \times (u_2 - 2) - 0.004 \times (RH_{min} - 45)] \times \left(\frac{h}{3}\right)^{0.3} \tag{6}$$

where $K_{c,tab}$ is the generalized standard crop coefficient for a given crop as tabulated in the FAO irrigation and drainage paper 56, RH_{min} is the mean daily minimum relative humidity (%), and h is the mean crop height at the given growth stage. This equation is applicable for both single crop coefficient and basal crop coefficient. It must be noted that this equation is designed with short grass as the reference crop. A ratio adjustment to this equation should be done when computing coefficients with alfalfa as the reference crop (Allen et al. 1998). Relative to EC method,

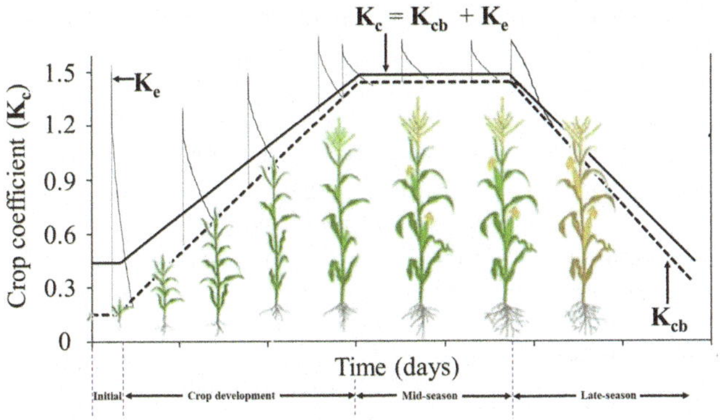

Fig. 3 A generalized conceptual view of single and dual crop coefficients with basal crop and soil evaporation coefficients progressing during the crop growing season.

crop coefficient-based approaches are nearly inexpensive to utilize as there are free weather datasets available for actual ET computation. However, alike EC method, this approach is also a single point-in-time (or gross) estimator of ET and does not account for spatial heterogeneities within a crop field towards site-specific irrigation scheduling. Furthermore, crop coefficient approaches are relatively inaccurate in ET estimation compared to EC or other point-scale methods. This is due to its empirical nature when adjusting crop coefficients and not accounting for all possible soil, crop, or weather variations (Chandel et al. 2020, 2021). Nonetheless, due to its simplistic nature of ET computation, it is fairly used by the crop producers for blanket irrigation scheduling. Some of the case studies where crop coefficient approaches have been used for ET estimation in agricultural cropping systems are summarized in Table 2.

3 ET Estimation: Regional-Scale Methods

3.1 Remote Sensing

Remote sensing (RS) is the method of obtaining information about objects, surfaces, or phenomenon without getting into physical contact. RS data is typically acquired from spectral sensors such as spectroradiometers or spectral cameras at various spatial resolutions (μm to m to km). The spatial resolution is the characteristic of the sensors being used and the platform on which such sensors are mounted ranging from handheld frames to fixed poles to mobile ground vehicles to occupied/unoccupied aircrafts to satellites (Sankaran et al. 2015; Ranjan et al. 2019; Sinha et al. 2021). As the demand for effortless and high-throughput RS has increased significantly over time, satellite-based RS has gained tremendous attention and adaptation for land surface monitoring especially agricultural and

Table 2 Case studies of using crop coefficient approaches for estimating evapotranspiration from cropping systems

Crop coefficient approach	Crop/commodity	Region	Accuracy and error	Reference
Dual	Peach orchard	Portugal	$R^2 = 0.9$, $E = 0.32$ mm	Paço et al. (2012)
Dual	Wheat and corn	China	$E = 0.5$ mm	Zhang et al. (2013)
Dual	Phacelia, hairy vetch, rye, mustard	Austria	$E_{dry} = 6.7\%$ $E_{wet} = 1.4\%$	Bodner et al. (2007)
Dual	Potato, lima bean, dolichos	Kenya	$R^2 = 0.77–0.92$ $E = 0.03–0.09$ mm	Nyawade et al. (2021)
Dual	Maize	China	$R^2 = 0.73–0.80$ $E_{season} = 4.6–12.6$ mm	Li and Ma (2019)
Single	Tomato	China	$R^2 = 0.78–0.95$ $E = 0.25–0.43$ mm	Gong et al. (2020)
Single	Sugarcane	India	$E = 4–25.5\%$	Dingre and Gorantiwar (2020)

E: error, r: correlation coefficient, R^2: coefficient of determination. Unless otherwise specified, E refers to error in daily ET estimates.

forestry systems. Major benefits of satellite-based RS include nondestructive data retrieval for wider geographical regions (hundreds of kilometers), data acquisition over remote locations, plethora of spatial resolution options ranging at least from 0.1 m to 1 km, temporal data acquisition, negligible sampling biases, and free of cost/less expensive, among others (Liaghat and Balasundram 2010; Zhang et al. 2020). Some of the most commonly used satellite-based sensors for crop health diagnostics provide data in visible range (RGB), infrared ranges (near infrared, shortwave infrared, mediumwave infrared, longwave infrared [thermal]), and radar ranges.

3.2 *Satellite RS for ET Estimation*

Satellite-based RS data has been extensively used for ET estimation. The main advantage is water use mapping at high spatiotemporal resolutions and from field to regional scales. We have grouped RS-based approaches into three categories: single-source energy balance, dual-source energy balance, and empirical models, where spectral imagery typically in the visible and near-infrared and thermal infrared ranges are used as inputs along with the weather data or ET_r to estimate actual ET at hourly, daily, weekly, or seasonal intervals. The following sections describe the most widely used satellite RS-based ET estimation approaches in detail along with latest use cases specific to agricultural cropping systems.

3.2.1 Surface Energy Balance Concept

Sun is the primary energy source of the Earth. This energy reaches earth surface in radiation form and raises the atmospheric and land surface temperatures. The difference between these two temperatures aided with other meteorological variations forms the source of exciting energy transfer from the earth surface to the atmosphere to maintain energy equilibrium. Given the law of energy conservation, the energy can neither be created nor destroyed but can be transferred from one form to other. Based on this fact, energy enters the earth system in the form of radiation of which a portion is reflected back and the remaining comes in contact with the land surface, termed as net radiation (R_n). A portion of R_n is lost to the ground as soil/ground heat flux (G), one portion is lost to the atmosphere as sensible heat flux (H) due to temperature gradient between surface and the overlying air, and the remaining portion is lost as the latent heat flux (LE) to the atmosphere due to evaporation or condensation at the surface (Fig. 4). The proportions of all these fluxes vary over time and geographical locations but maintain a balance within the earth system. Equations 7, 8, 9, 10 and 11 present general equations utilized to compute all the energy balance flux components of Eq. 2.

$$R_n = R_{s\downarrow} + R_{s\uparrow} + R_{l\downarrow} - R_{l\uparrow} \tag{7}$$

where $R_{s\downarrow}$ and $R_{s\uparrow}$ are the incoming and outgoing shortwave radiations (W m^{-2}) and their net sum is calculated using Eq. 8. $R_{l\downarrow}$ and $R_{l\uparrow}$ are the incoming and outgoing longwave radiations (W m^{-2}) and can be calculated from Eqs. 9 and 10.

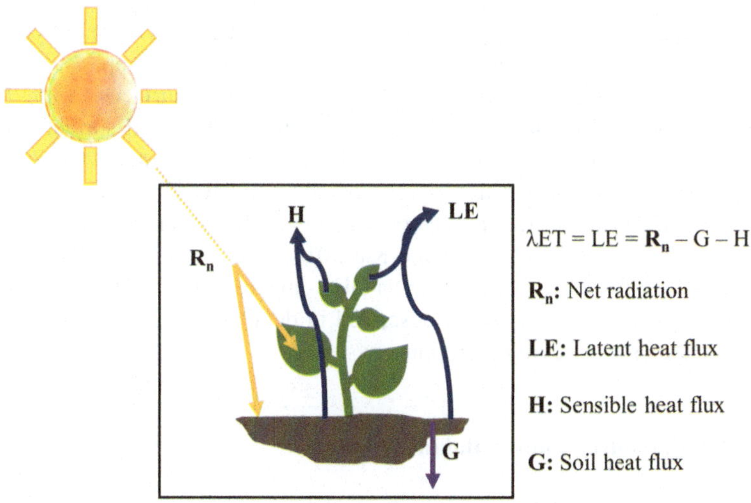

Fig. 4 The energy balance concept for actual evapotranspiration. λ is the latent heat of water vaporization

$$\sum R_s = (1 - \alpha)\, R_{s\downarrow} = (1 - \alpha) \times \left(S_c \times \cos\beta \times \mathrm{d}_y \times z_j\right) \tag{8}$$

$$R_{l\downarrow} = \varepsilon_a \times \sigma \times T_a^4 \tag{9}$$

$$R_{l\uparrow} = \varepsilon_0 \times \sigma \times T_s^4 \tag{10}$$

where α is the surface albedo, S_c is the solar constant (1367 W m^{-2}), β is solar incidence angle, d_y is relative distance between the Earth and Sun, z_j is the atmospheric transmissivity, ε_a is the atmospheric emissivity, σ is the Stefan–Boltzmann constant (W m^{-2} K^{-4}), T_a is the absolute air temperature (K), ε_0 is surface emissivity, and T_s is the absolute surface temperature (K). The H energy component is calculated from Eq. 11.

$$H = \frac{\rho \times C_P \times \mathrm{d}T}{r_{\mathrm{ah}}} \tag{11}$$

where ρ is the air density (kg m^{-2}), C_p is the specific heat of air (1004 J kg^{-1} K^{-1}), r_{ah} is the aerodynamic resistance, and dT is the gradient between air temperature and aerodynamic temperature near the surface. Over the past two decades, research studies have been applying surface energy balance (SEB) to satellite RS inputs for actual ET estimation. This is because SEB takes an analytical approach using physics-driven models of varied complexities where combinations of ground-based weather data and RS data inputs are required and effectively analyzed. As mentioned earlier, SEBs are primarily of two types, single-source energy balance (SSEB) where vegetation and soil energy budgets are combinedly analyzed and two-source energy balance (TSEB) where the two budgets are analyzed separately. SSEBs are best used for estimating transpiration from vegetation surfaces, while TSEBs are used for better estimation of evaporation from bare land/soil surfaces. TSEBs require larger number of input data variables and parametrization but do to not improve ET estimations significantly compared to SSEBs. The primary RS input to SEBs is the thermal infrared imagery which is useful for land's biophysical and ecological process modeling (Liou and Kar 2014). The output of SEBs is the map with ET values for each pixel of the satellite images. In the following sections, variants of SSEBs and TSEBs are described in detail with some of the latest use cases evaluated for agricultural cropping systems.

3.2.2 Single-Source Energy Balance Models

Some of the most widely used SSEB models include surface energy balance algorithm for land (SEBAL; Bastiaanssen et al. 1998a, b), Modified SEBAL (M-SEBAL; Long and Singh 2012), simplified surface energy balance index (S-SEBI;

Roerink et al. 2000), surface energy balance system (SEBS; Su 2002), simplified surface energy balance (SSEB, Senay et al. 2007), mapping evapotranspiration at high resolution with internalized calibration (METRIC; Allen et al. 2007a, 2007b), and operational simplified surface energy balance (SSEBop; Senay et al. 2007, 2013).

3.2.2.1 SEBAL and M-SEBAL

SEBAL was developed in 1998 which computes actual and potential ET exchanges between land and atmosphere as a residual of SEB (Eq. 2). SEBAL uses an integration of a few empirical relationships and physical parametrizations to estimate ET at local and regional scales. The major inputs to SEBAL are surface radiances in visible, NIR, and thermal infrared spectral ranges which are capable of providing surface temperature (T_S), normalized difference vegetation index (NDVI), and surface albedo (α). Such all-in-one dataset are available from sensors on board Landsat series satellites (4/5/7/8/9; spatial resolution: 30–120 m/pixel; temporal resolution: ~16 days) and moderate resolution imaging spectroradiometer (MODIS)/ visible infrared imaging radiometer suite (VIIRS) sensors on board Terra and Aqua satellites (spatial resolution: 0.250–1 km; temporal resolution: ~2 days). Net radiation is calculated using Eqs. 7, 8, 9, and 10 from the balance of shortwave and longwave radiations. Soil heat flux can be calculated using an empirical relationship for all vegetation and soil types (Eq. 12) that have been validated in over 30 countries worldwide with accuracies of 85–95%. In the next step, sensible heat flux (H) is computed using air temperatures measured at two reference points (one closer to the surface and other at an upper height). In SEBAL, the temperature gradient in Eq. 11 is assumed to have a linear relationship with the surface temperature (Kelvin, Eq. 13) under homogenous meteorological and surface conditions.

$$G = (T_s - 273.15) \times [0.0032 + 0.0062 \times \alpha] \times \left[1 - 0.98 \times \mathrm{NDVI}^4\right] \times R_n \tag{12}$$

$$\mathrm{d}T = k + a \times T_s \tag{13}$$

where k and a are the empirical coefficients obtained from the anchor "hot" and "cold" pixels in a given satellite image. In SEBAL, the "hot" pixel is the point where evaporation is almost zero and is typically a bare soil surface at high temperature, while at the "cold" pixel point, the surface transpires at its full capacity, is at low temperature, and is typically a free water surface. For the "cold" pixel, dT values and sensible heat flux are assumed to be zero as most of the energy is consumed by evaporation. For the "hot" pixel, dT is calculated from Eq. 14 where H_{hot} is the sensible heat at the "hot" pixel and is equal to R_n-G for that pixel. Once coefficients k and a are computed, dT can be computed from Eq. 13. Next, H is computed iteratively with $r_{ah,hot}$ corrected for stability through extrapolation of wind speed (u_z) between ground level to a height of 100–200 m above ground.

$$dT_{hot} = \frac{H_{hot} \times r_{ah,hot}}{\rho \times C_P} \tag{14}$$

Major noted advantages of SEBAL for ET estimation include (1) minimum ground data requirement, (2) automatic internal correction of surface temperature from atmospheric interferences, and (3) internal calibration of each image using "hot" and "cold" anchor pixels which reduces bias in surface roughness and aerodynamic stability corrections. Some major limitations of using SEBAL are (1) subjective specification of "hot" and "cold" pixels for internal calibration which induce uncertainties in H and ET estimates, (2) unaccountability of surface temperature and wind speed variations and lapses in mountains or variable terrains, (3) uncertainties in H and ET estimates due to non-corrected surface air temperature gradients or surface temperature measurements, and (4) unaccounted variations from sensor-viewing angles that can vary surface temperature estimates by several degrees. SEBAL works on rectangular contextual relationship between surface temperature and vegetation fraction due to subjectivity of anchor pixels. This can distort spatial distribution of latent heat flux (or ET) by several degrees. To avoid this, a modified version of SEBAL (M-SEBAL) was developed where a trapezoidal contextual relationship has been defined between surface temperature and vegetation fractions on ground. Further details on SEBAL can be availed from literature by Bastiaanssen et al. (1998a, b, 2005) and Ahmad et al. (2006) and for M-SEBAL from literature by Long and Singh (2012). Table 3 mentions some of the use cases where SEBAL or its improved versions were used for geospatial ET estimation in agricultural–forestry systems.

3.2.2.2 METRIC

METRIC is an advanced version of SEBAL developed in 2007 and well-applied in almost all agroclimatic conditions (Allen et al. 2007a,b). METRIC was especially developed for Landsat satellite RS data (~30 m/pixel) for specific areas smaller than hundreds of square kilometers. METRIC modifies over SEBAL by not choosing "cold" pixel subjectively, rather utilizing weather data-based ET_r to perform energy balance at that "cold" pixel thereby serving as a check on actual ET estimates especially in agricultural areas. This weather-based ET_r for alfalfa as reference crop is utilized to refine and obtain automated internal calibration of energy balance for entire satellite-imagery. This refined calibration further minimizes the biases in estimating aerodynamic stability correction and surface roughness. METRIC provides several advantages over other satellite RS-based energy balance model first due to its refined internal calibration using weather data, extrapolation of ET estimates to daily, weekly, and seasonal estimates, thereby compensating for regional advection effects, applicable where ET can exceed net radiation especially

Table 3 Application use cases of using SEBAL and its versions for geospatial evapotranspiration mapping of agricultural/forestry systems

SEBAL version	Satellite/RS sensor	Crop/commodity	Region	Accuracy and error	References
SEBAL	MODIS	Land use and cover	China	$R^2 = 0.92$ $E = 20\%$ (daily), 9% (seasonal)	Du et al. (2013)
SEBAL	Landsat 7 ETM+	Wheat	India	$R^2 = 0.84$ $E = 0.583$ mm	Bala et al. (2017)
SEBAL	Landsat 8	Multiple crops	China	$R^2 = 0.73$–0.99 $E = 1.5\%$	Tan et al. (2021)
SEBAL	Landsat TM, MODIS, AVHRR	Agro–pastures	China	$E_{\text{Landsat}} = 12\%$ $E_{\text{MODIS}} = 33\%$ $E_{\text{AVHRR}} = 3.6\%$	Li et al. (2021)
geeSEBAL	Landsat 5/7/8	Multiple agro–forestry systems	Brazil	$R^2 = 0.2$–0.77 $E = 15$–23%	Laipelt et al. (2021)
SEBALIGEE	Landsat 7/8, MODIS	Corn soybean, winter wheat	Contiguous USA	$R^2 = 0.78$–0.83 $E_{\text{Monthly}} = 14\%$	Mhawej et al. (2022)

E: error, r: correlation coefficient, R^2: coefficient of determination. Unless otherwise specified, E refers to error in daily ET estimates

in arid and semiarid regions. METRIC unlike other models or traditional approaches has very limited dependency on crop or surface specifics where phenological stage, cultivar, or other information is not needed.

In METRIC, R_n is computed from shortwave and longwave radiation components computed from satellite-measured narrow-band reflectance, surface temperature, and digital elevation model (DEM) and ground weather data being collected near target area (Eq. 15), unlike SEBAL. Next, G is computed from R_n using the refined version of Eq. 12 (Eq. 19, Bastiaanssen 2000). H is computed from surface temperature, surface roughness, and wind speed through buoyancy corrections. METRIC leverages SEBAL's concept of estimating dT as a linear function of surface temperature to eliminate the need for measurement of accurate aerodynamic surface temperature and air temperature to estimate sensible heat flux. METRIC also modifies the computation of surface albedo (α) for global applicability, following the procedure laid out by Tasumi et al. (2008). Further details on computing intermediate parameters of METRIC can be obtained from Allen et al. (2007a).

$$R_n = R_{s\downarrow} + R_{s\uparrow} + R_{l\downarrow} - R_{l\uparrow} - (1 - \varepsilon_0) \times R_{l\downarrow} \tag{15}$$

where component (1-ε_0) $\times$ $R_{1\downarrow}$ is the fraction of incoming longwave radiation reflected back from the surface. This component was not accounted for in other methods of R_n computations. $R_{s\downarrow}$ in METRIC is computed as in Eq. 16. The surface emissivity (ε_0) in METRIC for computing outgoing longwave radiation ($R_{1\uparrow}$) is calculated using a linear empirical Eq. 17, as a function of leaf area index (LAI).

When LAI exceeds 3 $m^2\ m^{-2}$, ε_0 is fixed equal to 0.98. The atmospheric emissivity (e) used for computing incoming longwave radiation ($R_{l\downarrow}$) is also calculated using an empirical Eq. 18. The constants in Eq. 18 were derived for alfalfa crop grown in Idaho, USA, and can be experimentally derived for a given agroclimatic condition.

$$R_{s\downarrow} = \frac{S_c \times \cos\beta \times z_j}{{d_y}^2} \tag{16}$$

$$\varepsilon_0 = 0.95 + 0.01 \times \text{LAI} \qquad \text{when LAI} \leq 3 \tag{17}$$

$$e = 0.85 \times \left(-\ln z_j\right)^{0.09} \tag{18}$$

$$G = (T_s - 273.15) \times [0.0038 + 0.0074 \times \alpha] \times \left[1 - 0.98 \times \text{NDVI}^4\right] \times R_n \tag{19}$$

METRIC follows a different approach than SEBAL for computing sensible heat flux (H) through calibration of H function. For calculation of coefficients in Eq. 13, dT_{hot} is calculated from Eq. 14 same as SEBAL, following same assumptions, but also including a provision for supplying nonzero value of ET (or LE) for "hot" pixel in case there is a residual evaporation from bare soil (Eq. 20). For the "cold" pixel, METRIC defines H_{cold} in Eq. 21. In the agricultural setting, the coldest pixel has LAI over 4 $m^2\ m^{-2}$, lower temperature, high NDVI, and typically actual ET about 5% more than ET_r for alfalfa crop (Tasumi 2003, Tasumi et al. 2005). Therefore, LE_{cold} in Eq. 21 can be replaced by $1.05 \times ET_r \times \lambda$ following which dT_{cold} and coefficients a and k of Eq. 13 are computed from Eq. 23.

$$H_{\text{hot}} = (R_n - G)_{\text{hot}} + \text{LE}_{\text{hot}} \tag{20}$$

$$H_{\text{cold}} = (R_n - G)_{\text{cold}} + \text{LE}_{\text{cold}} \tag{21}$$

$$dT_{\text{cold}} = \frac{H_{\text{cold}} \times r_{ah,\text{cold}}}{\rho \times C_P} \tag{22}$$

$$k = \frac{dT_{\text{hot}} - dT_{\text{cold}}}{T_{s,\text{hot}} - T_{s,\text{cold}}}, \qquad a = \frac{dT_{\text{hot}} - k}{T_{s,\text{hot}}} \tag{23}$$

Finally, instantaneous ET (mm h^{-1}) is calculated using Eq. 2 and latent heat of water vaporization. Instantaneous ET is converted to daily water use (or ET) by computing reference ET fraction (ET_rF) for each pixel as the ratio of instantaneous actual ET and ET_r for that hour instance. ET_rF is then multiplied to daily or 24-h ET_r to obtain daily or 24-h actual ET from one satellite image. It must also be noted that ET_rF can be used as a surrogate of crop coefficient (K_c). The cumulative ET_rF and actual ET for any given period (month or year or further) with limited number of satellite imageries can be computed from Eqs. 24 and 25. With these equations, seasonal actual ET can be estimated even from one satellite imagery obtained per month. However, during rapid crop development stages, using multiple satellite imageries over a month is recommended for accurate estimations.

$$\mathrm{ET}_r F_{\mathrm{Period}} = \frac{\sum_{i=m}^{n} [(\mathrm{ET}_r F_i) \times (\mathrm{ET}_{r24i})]}{\sum_{i=m}^{n} ET_{r24i}} \tag{24}$$

$$\mathrm{ET}_{\mathrm{Period}} = \sum_{i=m}^{n} [(\mathrm{ET}_r F_i) \times (\mathrm{ET}_{r24i})] \tag{25}$$

Since development, METRIC model has been well validated across various agroclimatic regions with estimation errors not exceeding 17%. Some of the use cases of estimating geospatial ET for agroecosystems with METRIC energy balance model are presented in Table 4.

3.2.2.3 SEBI, S-SEBI, and SEBS

SEBI is a crop water stress index (CWSI, Jackson et al. 1981) based energy balance model where an index of evaporative fraction (EF, Eq. 26) is calculated using surface temperature satellite imagery and minimum (T_{hot}) and maximum

Table 4 Application use cases of using METRIC for geospatial evapotranspiration mapping of agricultural cropping systems

Crop/commodity	Region	Accuracy and errors	References
Sugar beet and meadow	Idaho, USA	E: 1–4%	Allen et al. (2007b)
Corn and cotton	Texas, USA	E: 5–13%	Gowda et al. (2008)
Corn and soybean	Iowa, USA	R^2: 0.9, E: 0.25 mm	Gonzalez-Dugo et al. (2009)
Grapevine	Chile	E: 11%	Ortega-Farıás et al. (2016)
Alfalfa	Kingdom of Saudi Arabia	R^2: 0.8, E: 1.7 mm	Elkatoury et al. (2020)
Tall and short crops under dry and irrigated regimes	Texas, US	R^2: 0.62–0.65, E: 0.14–0.16 mm	Hashem et al. (2020)
Olive orchard	Chile	E: 5%	Ortega-Salazar et al. (2021)
Almond and pistachio	California, USA	E: 0.30–0.33 mm	He et al. (2022)

E: error, r: correlation coefficient, R^2: coefficient of determination. Unless otherwise specified, E refers to error in daily ET estimates

temperatures (T_{cold}) computed for wet and dry surfaces, respectively, using external meteorological data sources (Menenti and Choudhury 1993).

$$EF = \frac{T_{hot} - T_s}{T_{hot} - T_{cold}} \tag{26}$$

Due to complexities in calculating surface temperature from satellite RS, biases are often incurred in ET estimation. To minimize this bias, reflectance-based identification of minimum and maximum temperature pixels within satellite imagery is available with simplified-SEBI (S-SEBI) version (Roerink et al. 2000). At low reflectance ranges, surface temperature remains pretty much constant with reflectance increase, for example, in case of well-irrigated lands or free water surfaces where all the available energy is consumed in evaporation. While at higher reflectance ranges, the surface temperature increases with the increase in reflectance up to a certain threshold point. Increase in temperature till this threshold is termed as "evaporation-controlled" because the change in temperature then is a result of decreased evaporation due to decreased soil moisture availability. Beyond this threshold point of reflectance, the surface temperature decreases with increasing reflectance. This is because of the soil moisture level extent that no more aids evaporation and all available energy just heats up the surface. In this situation, the available energy reduces as the result of reduced net radiation because of higher reflection fraction. This temperature decrease with increased reflectance is termed as "radiation-controlled." S-SEBI simplifies the EF computation by directly using extreme pixels (wet/cold or dry/hot) if present within the surface temperature satellite image under consistent atmospheric conditions. If consistency condition is not met, extreme temperatures are computed as in SEBI, i.e., from external meteorological data sources. When using SEBI or S-SEBI, net radiation can be computed from Eq. 5, soil heat flux from Eq. 27, sensible heat flux from Eq. 28, and latent heat (or ET) from Eq. 29.

$$G = (T_s - 273.15) \times [0.32 + 0.62 \times \alpha] \times \left[1 - 0.98 \times \text{NDVI}^4\right] \times R_n \tag{27}$$

$$H = (1 - \text{EF}) \times (R_n - G) \tag{28}$$

$$\lambda \text{ET} = \text{EF} \times (R_n - G) \tag{29}$$

SEBS is a modified version of SEBI (Su 2002) where H and LE components are computed from satellite imagery and weather data. Energy balance is computed through an instantaneous evaporative fraction (Λ, Eq. 30) using dry and wet regions in the imagery. SEBM fixes LE to be zero in Eq. 20 so that H satisfies to be maximum for dry (or hot) surfaces. Cold surface is assumed to be fully covered by vegetation and therefore G_{cold} is substituted as $0.05 \times R_n$ and hot surface is assumed to be bare soil for which G_{hot} is substituted as $0.315\times R_n$. For the wet (or cold) surface, H_{cold} is calculated from Eq. 30. Next, LE_{cold} can be calculated from Eq. 21, and sensible heat flux component (H) is calculated using the Monin–Obukhov similarity functions (Brutsaert 1999). Instantaneous EF is calculated using Eq. 32 and is assumed to stay constant throughout the 24-h period; it is multiplied to the daily reference ET (ET_o or ET_r) to compute 24-h actual ET.

$$\Lambda_r = \frac{H_{\text{hot}} - H}{H_{\text{hot}} - H_{\text{cold}}} \tag{30}$$

$$H_{\text{cold}} = \frac{\gamma \times (R_n - G)_{\text{hot}}}{\gamma + \Delta} - \frac{\rho \times C_P \times (e_s - e_a)_{\text{cold}}}{r_{\text{ah,cold}} \ (\gamma + \Delta)} \tag{31}$$

$$\Lambda = \frac{\text{LE}}{R_n - G} = \frac{\Lambda_r \times \text{LE}_{\text{cold}}}{R_n - G} \tag{32}$$

SEBS reduces uncertainty in ET estimation stemming from uncertainty in satellite-based surface temperature imagery, by defining "hot" (or wet) and "cold" (wet) surface conditions for calibrating energy balance. However, SEBS is demanding large number of data inputs which if not available will limit ET estimation using SEBS. Some of the case studies that utilized SEBI, S-SEBI, and SEBS models for ET estimation in agricultural setting are listed in Table 5.

3.2.2.4 SSEB, SSEBelvi, and SSEBop

SSEB (Senay et al. 2013) operates similar to METRIC in that ET is calculated from satellite-based thermal imagery and meteorological data only by identifying and utilizing temperatures of "hot" and "cold" anchor pixels within satellite imagery. SSEB runs a partial energy balance as it does not compute sensible energy flux like other energy balance models. SSEB assumes that actual ET can be inferred from the ratio of surface temperature gradient with hot pixel to the temperature gradient of "hot" and "cold" pixels. Actual ET at "hot" pixel is assumed to be zero and equal to ET_r for "cold" pixel. "Hot" pixel is identified as the one with high temperature and low NDVI and "cold" pixel as the one with low temperature and high NDVI. An evaporative fraction (Eq. 26) is then calculated and multiplied with the ET_r to obtain daily actual ET map. SSEB ignores albedo and soil heat flux and therefore underestimates ET for surfaces with low albedo and overestimates for surfaces with high albedo and soil heat flux such as for bare soil. The assumption of linearity between temperature and ET also does not stand true for the landscapes or vegetation surfaces different than reference crops (alfalfa or short grass) as well as under complex terrains (Senay et al. 2011a). For these limitations, an improved version SSEBelvi includes elevation variation factor for refining the computation of land surface temperature ($T_{s,\text{elv}}$, Eq. 33) and also includes empirical vegetation parameterization through NDVI for computing adjusted EF (EF_{elvi}, Eq. 34).

Table 5 Application use cases of using S-SEB, and SEBS models for geospatial evapotranspiration mapping of agricultural cropping systems

Crop/commodity	Model and satellite imagery	Region	Accuracy and errors	Reference
Cotton	S-SEBI and Landsat 5	Brazil	R^2: 0.84–0.87, E: 4–16%	Santos et al. (2010)
Paddy	S-SEBI and Landsat 8	India	R^2: 0.71–0.77, E: 0.52 mm	Kumar et al. (2020)
Grass, wheat, barley, and grapevines	S-SEBI and Landsat 8	Spain	R^2: 0.8, E: 0.63–1.71 mm	Sobrino et al. (2021)
Barley	S-SEBI and Landsat 8	Switzerland and Italy	R^2: 0.7–0.9, E: 0.9 mm	Santos et al. (2022)
Fennel, maize, ryegrass	SEBS and Landsat 7	Italy	R^2: 0.6, E: 0.7 mm	Nisa et al. (2021)
Sugar beet, dry bean, barley	SEBS and Landsat 7/8	Wyoming, USA	R^2: 0.87, E: 12.3%	Acharya et al. (2021)
	S-SEBI & Landsat 7/8		R^2: 0.76, E: 3.9%	
Rapeseed, wheat, barley, green peas, rye	SEBS and MODIS	Spain	R^2: 0.7, E: 0.34–0.45 W m^{-2}	Pardo et al. (2014)

E: error, r: correlation coefficient, R^2: coefficient of determination. Unless otherwise specified, E refers to error in daily ET estimates

$$T_{s,\mathrm{elv}} = T_s + K_L \times \mathrm{DEM} \tag{33}$$

$$\mathrm{EF}_{\mathrm{elvi}} = \left(0.35 \times \frac{\mathrm{NDVI}}{0.7} + 0.65\right) \times \mathrm{EF}_{\mathrm{elv}} \tag{34}$$

where K_{L} is the standard temperature lapse rate of 0.0065 K m^{-1}, DEM is the land surface elevation (m) obtained from digital elevation model, and $\mathrm{EF}_{\mathrm{elv}}$ is the EF obtained from Eq. 25 for surface temperature corrected for elevation ($T_{s,\mathrm{elv}}$). SSEBelvi assumes that if NDVI is greater than 0.7, then the surface is well-vegetated and will have ET higher than the reference crop if the available water is limited. This assumption is similar to the assumption of "cold" pixel ET in METRIC. For using SSEBelvi, $\mathrm{EF}_{\mathrm{elvi}}$ is multiplied with ET_r for computing daily ET. SSEB and SSEBelvi can induce biases during "hot" and "cold" pixel selection due to human subjectivity. To eliminate this, a further improved version SSEBop is used for ET computations which determines the "hot" and "cold" boundary conditions for each image pixel similar to SEBS (Senay et al. 2013). The only required data inputs for SSEBop are T_s (from satellite imagery), T_a, and ET_r from weather data. SSEBop is based on the argument that R_n drives surface energy balance and under clear skies, the "hot" and "cold" boundary conditions do not vary for years and their difference is constant for a given location and day of year. Under clear sky conditions, T_{cold} and T_{hot} can be calculated using weather data from Eq. 35, and ultimately actual ET can be calculated from Eq. 36, where a calibration coefficient (k_f) is introduced to scale ET_o to maximum ET for a lesser aerodynamic crop. Standard value of 1.2 is used for k_f but can be obtained from calibration of soil water balance or energy balance.

$$T_{\text{cold}} = c \times T_a, \quad T_{\text{hot}} = T_{\text{cold}} + dT \tag{35}$$

$$\text{ET} = \text{EF} \times k_f \times \text{ET}_o \tag{36}$$

where c is the coefficient relating T_s to T_a for a well-irrigated vegetation at maximum ET. Some of the case studies that utilized SSEB, SSEBelvi, and SSEBop models for ET estimation in agricultural settings are listed in Table 6.

3.2.3 Two-Source Energy Balance Models

As discussed earlier, the two-source energy balance (TSEB) model evaluates vegetation and soil energy budgets separately and are mostly used for accurate estimation of soil evaporation fluxes. Surface temperatures of soil and vegetation are used as inputs to estimate evaporation and transpiration components. Since the temperature measurements from RS sensor is a single layer combination of soil and vegetation, TSEBs deploy approaches to decompose soil and vegetation temperatures into two different input layers for energy computations. Most widely used TSEB models are the Priestley–Taylor two-source energy balance model (TSEB-PT), Penman–Monteith two-source energy balance model (TSEB-PM), two-source energy balance atmosphere land exchange inverse (TSEB ALEXI), and enhanced two-source evapotranspiration model for land (ETEML). These models are discussed in detail in the following sections along with their use cases in agricultural settings.

3.2.3.1 Priestley-Taylor and Penman–Monteith Two-Source Energy Balance Models

TSEB-PT was developed by Norman et al. (1995) through which soil and vegetation energy fluxes are partitioned based on their respective surface temperatures sensed remotely (Eqs. 37 and 38). The temperature partitioning considers the sensor viewing angle (ϕ) and fraction of vegetation cover calculated from LAI (Eq. 39). Along these lines, H is computed as the sum of sensible heat fluxes for vegetation canopy (H_c) and soil (H_s).

Table 6 Application use cases of using SSEB, SSEBelvi, and SSEBop for geospatial evapotranspiration mapping of agricultural cropping systems

Crop/commodity	Model and Satellite imagery	Region	Accuracy and errors	Reference
Corn and sorghum	SSEB- Landsat 7/8	Texas, USA	E: 16%	Gowda et al. (2009)
Agricultural subbasins	SSEB-MODIS	Conterminous USA	R^2: >0.9, E: 11%	Senay et al. (2011b)
Corn and soybean	SSEB-Landsat 7/8	South Dakota, USA	R^2: >0.9 with METRIC-ET	Senay et al. (2007)
Agricultural fields	SSEBelvi-Landsat 7/8	Idahoa, USA	R^2: 0.95 with METRIC-ET	Senay et al. (2011a)
Diverse agroecosystems	SSEB-op-Landsat 7/8	Contiguous USA	R^2: 0.64, $E_{Monthly}$: 27 mm	Senay et al. (2013)
Diverse croplands	SSEBop-Landsat 7/8	Contiguous USA	R^2: 0.92, $E_{Monthly}$: 13 mm	Chen et al. (2016)
Maize and soybean	SSEBop-Landsat	Midwest, USA	R^2: 0.92, $E_{Seaonal}$: 84 mm	Singh and Senay (2015)
Diverse agroecosystems	SSEBop-Landsat 8	Colorado, USA	R^2: 0.78–0.95, E: 10%	Singh et al. (2013)
Diverse agroecosystems	SSEBop-Landsat 8 and MODIS	Colorado, USA	R^2_{Daily}: >0.82, E_{Daily}: 0.6 mm, R^2_{Annual}: >0.78, E_{Annual}: 77 mm,	Singh et al. (2014)
Wheat	SSEBop-Landsat 7/8	Brazil	R^2: 0.82, E: 13.6%	Lopes et al. (2019)
Cropland	SSEBop-Landsat 8/MODIS	China	R^2: 0.93, E: 13.3%	Zhuang et al. (2022)
Beans	SSEBop-Landsat 7/8	Brazil	R^2: 0.82, E: 0.62 mm	Paula et al. (2019)

E: rror, r: correlation coefficient, R^2: coefficient of determination. Unless otherwise specified, E refers to error in daily ET estimates

$$T_b(\phi) = \left[\varepsilon(\phi) \times \left(T_{\text{Rad}}(\phi)\right)^4 + (1 - \varepsilon(\phi)) \times T_{\text{sky}}^4\right)\right]^{\frac{1}{4}} \tag{37}$$

$$T_{\text{Rad}}(\phi) = \left[f(\phi) \times T_{\text{cs}}^4 + (1 - f(\phi))\, T_{\text{ss}}^4\right]^{\frac{1}{4}} \tag{38}$$

where T_b is the brightness temperature, $\varepsilon(\phi)$ is the surface emissivity, ϕ is the view zenith angle of the sensor, and $f(\phi)$ is the fraction of field or view of the sensor that is occupied by the canopy and calculated using Eq. 39 assuming random canopy with spherical leaf angle distribution. T_{Rad} is the representation of combined soil (T_{ss}) and canopy (T_{ss}) surface temperatures.

$$f(\phi) = 1 - \exp\left(\frac{-0.5 \times \mathrm{LAI}}{\cos\phi}\right) \tag{39}$$

The net radiation component (R_n) is first calculated from Eq. 7 or 15 and then partitioned into canopy (R_{nc}) and soil net radiation (R_{ns}) fluxes (Eq. 40), following which soil and canopy sensible heat fluxes (H_s and H_c) are calculated using soil and canopy temperatures (Eq. 40) and respective latent heat fluxes (evaporation: LE_s; transpiration: LE_c) from energy balance (Eq. 40).

$$R_{ns} = R_n \times \exp[0.9 \times \ln(1 - f_c)] \quad \text{and} \quad R_{nc} = R_n - R_{ns} \tag{40}$$

$$H = H_c + H_s = \rho \times C_p \times \left(\frac{T_{cs} - T_a}{r_{ah}} + \frac{T_{ss} - T_a}{r_{ah} + r_{ss}}\right) \tag{41}$$

$$\mathrm{LE} = \mathrm{LE}_c + \mathrm{LE}_s = (R_{nc} - H_c) + (R_{ns} - H_s - G) \tag{42}$$

where f_c is the fraction canopy cover and the empirical constant of 0.9 is approximated for random leaf with spherical distribution and R_n extinction coefficient of 0.45. Flux G is approximated as $0.35 \times R_{ns}$ (Choudhury et al. 1987). r_{ss} is the resistance to heat flow above soil surface, and T_{cs} and T_{ss} are obtained by solving Eqs. 37 and 38 by having measurements of T_b and T_{Rad} at two view angles (e.g., $\phi_1 = 50°$, $\phi_2 = 25°$). Satellites with thermal remote sensing and capable of making measurements at two view angles (e.g., along-track scanning radiometer (ATSR) satellite) can help in the calculation of T_{cs} and T_{ss}. Quite frequently, thermal data inputs from satellites are not available at two viewing angles; therefore, T_c can be obtained by partitioning R_{nc} into H_c and LE_c using an iterative Priestly–Taylor approximation (Priestly and Taylor 1972) for the green vegetation fraction (f_g, Eqs. 43 and 44). This approximation is based on experimental and theoretical assumption that R_{nc} almost directly relates to the intercepted photosynthetically active radiation to cause photosynthesis and thereby the transpiration. However, as the vapor pressure deficit with the surrounding atmosphere varies, the transpiration rate will vary despite constant R_{nc}. Initial canopy temperature (T_{csi}) in the iteration is calculated from Eq. 43, following which H_c, T_{ss}, and H_s are calculated and recalculated until the energy balance is achieved. Along those lines, for a negative LE_s, it is replaced by a zero to recalculate correct H_s, T_{ss}, T_{cs}, and H_c. Furthermore, if H_c is greater than R_{nc} which is not possible, then LE_c is made zero, to recalculate H_c, T_{cs}, T_{ss}, and H_s. This iterative and check-based approach ensures correctness and accuracy of soil and canopy energy balances.

$$T_{csi} = T_a + \frac{R_{nc}}{\rho \times C_p} \times \left[1 - 1.3 \times f_g \times \left(\frac{\Delta}{\Delta + \gamma}\right)\right] \tag{43}$$

$$\mathrm{LE}_c = 1.3 \times f_g \times R_{\mathrm{nc}} \times \left(\frac{\Delta}{\Delta + \gamma}\right) \quad \text{and}$$
$$\mathrm{LE}_s = R_{\mathrm{nc}} \times \left[1 - 1.3 \times f_g \times \left(\frac{\Delta}{\Delta + \gamma}\right)\right] \tag{44}$$

The advantage of TSEB-PT is that unlike single-source models, it does not require calibration of momentum roughness length for heterogeneous vegetation surfaces. Furthermore, TSEB-PT distinguishes between composite surface temperature and aerodynamic temperature and therefore does not require surface-specific calibration like in the single-source models. Pertaining to transpiration component, a further accurate method to estimate the fraction of senesced vegetation (1-f_c) is needed that does not contribute to LE_c but H_c. This can be done by experimental evaluation of vegetation index LAI during the crop growing cycle. The same experiment can be used to refine empirical calculation of G (0.35 R_{ns}). These steps will further minimize empirical dependencies within the model. Further TSEB-PT model details can be availed from literature by Norman et al. (1995) and Kustas and Norman (2000).

The TSEB-PM version was developed in 2012 (Colaizzi et al. 2014, 2012) as the revised version of TSEB-PT where the Penman–Monteith equation is used as a replacement to compute canopy transpiration (or LE_{c}), parameterized also for semiarid climate. Here, unlike TSEB-PT, G is calculated from a phase-difference approach (Eq. 45, Santanello Jr. and Friedl 2003) based on the finding that during daytime G has a strong phase difference with R_{ns}. The nighttime G is calculated as a constant fraction of R_{ns} and as then there is no phase difference.

$$G_{\mathrm{Day}} = R_{\mathrm{ns}} \times \left\{A.\cos\left[\frac{2\Pi}{D}(t + C)\right]\right\} \quad \text{and} \quad G_{\mathrm{Night}} = \quad w \times R_{\mathrm{ns}} \tag{45}$$

where A is the amplitude, D is the period length, C is the shift, and t is the solar time angle. All the constants A, D, C, t, and w can be derived by calibrating calculated G and that measured for the surface using heat flux plates (Evett et al. 2012). TSEB-PM also modifies the computation of sensible heat fluxes (Eq. 46) by (1) considering all resistances to heat flows from canopy and soil surfaces to be in series unlike TSEB-PT, (2) including resistance to heat flow near canopy in the boundary layer (r_{c}), and (3) including aerodynamic temperature (T_{ac}). The iterative computation of T_{cs} and T_{ss} in TSEB-PM is modified by using a Penman–Monteith approach (Eqs. 47 and 48) as a replacement to Eq. 43 that used a Priestley–Taylor parameter of 1.3. The TSEB-PM approach assumes non-water-stressed conditions but contains provisions to increase canopy resistance if the stress increases unlike the TSEB-PT approach where a similar parameter decreased with the canopy stress. Some case studies that deployed TSEB-PT and TSEB-PM with satellite RS for water use estimation of agricultural systems are presented in Table 7.

Table 7 Application use cases of using TSEB-PT and TSEB-PM with satellite RS for geospatial evapotranspiration mapping of agricultural cropping systems

Crop/commodity	Model and satellite imagery	Region	Accuracy and errors	Reference
Diverse vegetation	TSEB-PT-ASTER	China	E: 8%	Yang et al. (2018)
Diverse vegetation	TSEB-PM-ASTER	China	E: 8.2%	Yang et al. (2018)
Diverse vegetation	TSEB-PT-ASTER	China	E: 11.7%	Song et al. (2016)
Multiple crops	TSEB-PT-MODIS	Morocco	r: 0.70, E: 30 W m^{-2}	Diarra et al. (2022)
Multiple crops	TSEB-PT-Landsat 7/8 and MODIS	Australia and Europe	r: 0.83, E: 10%	Jaafer et al. (2022)
Cotton	TSEB-PM-Landsat 7/8 and MODIS	Texas, USA	R^2: 0.83, E: 12%	Colaizzi et al. (2014)
Cotton	TSEB-PT-Landsat 7/8 and MODIS	Texas, USA	R^2: 0.66, E: 13%	Colaizzi et al. (2014)
Tree-grass ecosystems	TSEB-PT- MODIS	Spain	r: 0.78, BE: 34 W m^{-2}	Burchard-Levine et al. (2019)
Multiple agroecosystems	TSEB-PT-Landsat 8	Ghana	R^2: 0.85–0.96, E: 7%	Alhassan and Jin (2020)
Olive grove	TSEB-PT-Landsat 5	Portugal	R^2: 0.5, E: 5%	Häusler et al. (2018)

E: error, r: correlation coefficient, R^2: coefficient of determination. Unless otherwise specified, E refers to error in daily ET estimates

$$H = H_c + H_s = \rho \times C_p \times \left(\frac{T_{\text{cs}} - T_{\text{ac}}}{r_c} + \frac{T_{\text{ss}} - T_{\text{ac}}}{r_{\text{ss}}} \right) = \rho \times C_p \times \left(\frac{T_{\text{ac}} - T_a}{r_{\text{ah}}} \right) \tag{46}$$

$$T_{\text{csi}} = T_a + \frac{R_{\text{nc}} \times r_{ah} \times \gamma \times \left(1 + \frac{r_c}{r_{\text{ah}}}\right)}{\rho c_p \left(\Delta + \gamma \times \left(1 + \frac{r_c}{r_{\text{ah}}}\right)\right)} - \frac{e_s - e_a}{\Delta + \gamma \times \left(1 + \frac{r_c}{r_{\text{ah}}}\right)} \tag{47}$$

$$\text{LE}_c = \left[\frac{\Delta \times R_{\text{nc}}}{\Delta + \gamma \times \left(1 + \frac{r_c}{r_{\text{ah}}}\right)} + \frac{\rho \times c_P \times (e_s - e_a)}{r_{\text{ah}} \times \left(\Delta + \gamma \times \left(1 + \frac{r_c}{r_{\text{ah}}}\right)\right)} \right] \tag{48}$$

3.2.3.2 TSEB Atmosphere Land Exchange Inverse and Spatially Disaggregated ALEXI

Two-source energy balance atmosphere land exchange inverse (TSEB ALEXI) (Anderson 1997) originally known as two-source time-integrated model (TSTIM) is the extended version of TSEB-PT model. ALEXI was developed to resolve a limitation of TSEB-PT or TSEB-PM, i.e., for their regional-scale application, meteorological boundary conditions in air temperature are needed at the spatial resolution similar to thermal imagery which otherwise cannot be interpolated accurately due to continuous and localized cross-feedbacking between the land surface and atmosphere. ALEXI resolves this by coupling existing TSEB-PM with atmospheric boundary layer model to internally simulate land-atmosphere feedbacks and their effects on local air temperature. TSEB uses instantaneous surface temperature measurements, while TSEB ALEXI uses brightness temperatures acquired twice in the day from Geostationary Operational Environmental Satellite (GOES, spatial resolution: 5–10 km/pixel) 1.5 hours after the sunrise and next after 4 h to evaluate temporal energetics coupled with the atmospheric boundary layer model. This reduces sensor bias, eliminates dependency on air temperature measurements, and provides continuous flux estimates at continental scales. ALEXI has been improved over time and now also estimates moisture stress and soil water availability from ET estimates. ALEXI also includes techniques to estimate ET on cloudy days when surface thermal measurements are unavailable as well as extrapolate instantaneous ET for larger durations.

ALEXI uses the series arrangement of resistances instead of parallel in TSEB-PT and approximates composite surface temperature (T_{Rad}) as the linear aggregation of soil and canopy surface temperatures (Eq. 49). Energy balance closure between the thermal data acquisition instances relates the air temperature rise in the mixed atmospheric layer to the time-integrated influx of sensible heat from land surface. ALEXI therefore relies only on time-differential temperature signals, thereby minimizing errors due to sensor calibration and atmospheric and spatial aberrations. Specific inputs to ALEXI are surface temperatures at two instances (source: GEOS), LAI (MODIS), landcover (UMD Global), wind speed (ASOS/AWOS), lapse rates (Radiosonde), atmospheric corrections (Radiosonde), cloud amount (GEOS), hourly net radiation (GEOS), and soil texture (STATSGO, for cloudy days). The instantaneous TSEB and sensible heat fluxes at two instances are computed using Eqs. 41–44, 46, and 50. A linear form of H(t) is assumed to compute time-integrated heat flux during the morning interval. The time-integrated sensible heating is related to the height rise and potential temperature of the mixed layer through conservation Eqs. 51 and 52 (given by McNaughton and Spriggs (1986)). Potential evaporation and transpiration components are calculated from Priestley–Taylor approximation (Eq. 44). The hourly and daily values of ET and H are computed from evaporative fraction obtained at ALEXI modeling time (i.e., $\Lambda_{\mathrm{ALEXI,\,t2}}$) and hourly values of R_n and G measured by GOES (Eqs. 53, 54 and 55). Studies have shown evaporative fractions obtained at midday underestimate actual ET by 5–10%; therefore, $\Lambda_{\mathrm{ALEXI,\,t2}}$ is approximated as 1.1 times the evaporative fraction obtained at t_2.

$$T_{\mathrm{Rad}}\,(\phi) \approx [f\,(\phi)\,T_{\mathrm{cs}} + (1 - f\,(\phi)\,T_{\mathrm{ss}})] \tag{49}$$

$$\int_{t1}^{t2} H(t)\,dt = \frac{1}{2}\,[H_2 t_2 - H_1 t_1] \tag{50}$$

$$\int_{t1}^{t2} H(t)\,\mathrm{dt} = \rho \times C_p \left[(z_{m2} T_{m2} - z_{m2} T_{m2}) - \int_{z_{m1}}^{z_{m2}} T_{ss}\,(z_m)\,dz_m \right] \tag{51}$$

$$T_m = T_a \times \left(\frac{100}{p}\right)^{0.286}, \text{ p : atmospheric pressure (kPa)} \tag{52}$$

$$\Lambda_{\mathrm{ALEXI}} = 1.1 \times \frac{\mathrm{LE}_2}{R_{n2} - G_2} \tag{53}$$

$$\text{For soil } \left(\mathrm{i^{th}\ hour}\right): \quad \Lambda_{s,\ \mathrm{ALEXI}} = 1.1 \times \frac{\mathrm{LE}_{s2}}{R_{\mathrm{ns}2} - G_{s2}}, \quad \mathrm{LE}_{si} = \Lambda_{s,\ \mathrm{ALEXI}} \times (R_{\mathrm{nsi}} - G_{\mathrm{si}}) \tag{54}$$

$$\text{For canopy } \left(\mathrm{i^{th}\ hour}\right): \quad \mathrm{LE}_{\mathrm{ci}} = \mathrm{LE}_i - \mathrm{LE}_{\mathrm{si}},\ H_{\mathrm{ci}} = H_i - H_{\mathrm{si}} \tag{55}$$

ALEXI is advantageous as it moves the TSEB's upper boundary condition of temperature from near-surface to atmospheric boundary layer where more spatial uniformity is achieved at continental scales. However, this is also a limitation that forces ALEXI to be applicable only at spatial resolutions of 5–10 km specific to GEOS. This limitation was alleviated by arriving at a modified version of ALEXI called as disaggregated ALEXI (DisALEXI, Norman et al. 2003). DisALEXI generates energy fluxes at much higher spatial resolutions of 0.3–1 km by combining ALEXI outputs with the high-resolution satellite (Landsat 4/5/7/8, ASTER, MODIS) or aerial imagery. DisALEXI relies on the concept of blending height (100–200 m above ground) where the wind speeds become constant over the land surface. The first step follows same as ALEXI of computing air temperature from GEOS satellite (5 km/pixel), while in the next step the TSEB approach is applied on higher-resolution imagery of vegetation fraction (computed using NDVI) and radiometric surface temperature instead of low resolution as in ALEXI. At this step, TSEB is applied on total "N" number of pixels contained in one GEOS pixel (5 km) for which the air temperature is held constant. This hybrid approach uses atmospheric boundary layer component of ALEXI at large scale, while the surface component at finer scales.

DisALEXI operates in the following steps. Air temperature, wind speed, downwelling short- and longwave radiation, and blending height wind speed variables are held constant over 5 km disaggregation scale. High-resolution land surface inputs of radiometric surface temperature (T_{Rad}) and vegetation fraction (f_c) are developed on a grid with each pixel designated an id "i" for total "N" pixels made to coincide exactly with the 5 km grid from GEOS. This surface temperature is acquired around the second GEOS observation time and corrected for atmospheric and surface emissivity differences. Next, using the soil and canopy temperatures from ALEXI, the surface temperature at 5 km acquired at view angle ϕ is adjusted to the average angle at which the high-resolution surface temperature was acquired. Alternatively, a rigorous correction can also be applied by iteratively adjusting each pixel of high-resolution image to the GOES angle and then comparing the average adjusted surface temperature to the unadjusted GOES surface temperature. However, the two approaches have been reported to produce non-different adjusted surface temperatures at high resolution. Vegetation height is then scaled between season maximum and minimum based on f_c for each pixel and used to estimate aerodynamic roughness and resistance for each pixel. For each pixel, net radiation is computed using equation from ALEXI but with high-resolution surface temperature, surface albedo, and surface emissivity (from f_c). Rest all the flux computation steps are similar to ALEXI. An important step in DisALEXI is compensating for the differences in sensor calibration, atmospheric correction, and view angles between surface temperature estimates from GOES and high-resolution satellites. If not addressed, significant biases may occur in computations of sensible and latent heat fluxes. Recent case studies that have used ALEXI or DisALEXI for ET estimation in agricultural systems are mentioned in Table 8. A major limitation with ALEXI or DisALEXI is requirement of large number of data inputs. A simpler version of ALEXI is also available as the dual temperature difference (DTD) model that utilizes two radiometric surface temperatures like ALEXI but operates on a simpler structure with a smaller number of other inputs. This makes DTD more applicable than ALEXI or DisALEXI. The surface temperature partitioning for soil and vegetation cover is based on the fraction of radiometric view (ϕ) following which the sensible heat flux is calculated from Eq. 56 and soil and canopy LE fluxes from Priestley–Taylor expression (Eq. 44 and 55).

$$
\begin{aligned}
H_i = \rho C_p & \left[\frac{\left(T_{\text{Rad},i} - T_{\text{Rad},0}\right) - \left(T_{a,i} - T_{a,0}\right)}{(1 - f(\phi)) \times \left(r_{\text{ah},i} + r_{\text{ss},i}\right)} \right] \\
& + H_{c,i} \left[1 - \frac{f(\phi) \times r_{\text{ah},i}}{(1 - f(\phi)) \times \left(r_{\text{ah},i} + r_{\text{ss},i}\right)} \right] \\
& + H_{c,i} \left[\frac{f(\phi) \times r_{\text{ah},0}}{(1 - f(\phi)) \times \left(r_{\text{ah}} + r_{\text{ss},i}\right)} \right]
\end{aligned} \tag{56}
$$

where subscript i refers to any hour of the day when fluxes are computed and subscript 0 refers to the initial flux computation time, i.e., 1 h after sunrise.

Table 8 Application use cases of using ALEXI/DisALEXI with satellite RS for geospatial evapotranspiration mapping of agricultural cropping systems

Crop/commodity	Model and satellite imagery	Region	Accuracy and errors	References
Grapevines	DisALEXI and Landsat 8 and Sentinel-2	California	R^2: 0.9, E: 1.36 mm	Knipper et al. (2023)
Almonds			R^2: 0.86, E: 1.61 mm	
Multiple crops	DisALEXI and Landsat 8	Continental USA	R^2: 0.8, E: 0.81 mm	Cawse-Nicholson et al. (2021)
Corn, soybean, wheat	DisALEXI and Landsat 7	Maryland, USA	R^2: 0.8, E: 9%	Sun et al. (2017)
Forests and olive groves	DisALEXI and Landsat 7/8 and MODIS	Spain	E: 0.67 mm	Carpintero et al. (2021)
Grapevines, alfalfa, paddy, and wetlands	DisALEXI and Landsat 7/8 and MODIS	California, USA	R^2: 0.8, E: 0.91–0.95 mm	Anderson et al. (2019)

E: error, r: correlation coefficient, R^2: coefficient of determination. Unless otherwise specified, E refers to error in daily ET estimates

3.2.3.3 TSEB-Trapezoidal Framework Model

Primary TSEB models estimate ET using surface temperature measurements and vegetation features independently acquired at single-viewing or dual-viewing RS angles. These models can be categorized as non-space-based models where the interrelationships between the input parameters are not utilized for computing energy fluxes. As a result, such models require a large number of independent modelling inputs of vegetation, meteorology, and soil factors. To overcome this challenge, a version of TSEB was developed that utilizes soil wetness isopleths and trapezoidal space formed by the scatter plot relationship between surface temperature and vegetation index-derived fraction canopy cover to compute energy fluxes. Inclusion of soil isolines in trapezoidal space represents soil temperature (Long and Singh 2012) and overall aids acquisition of sufficient surface temperature measurements from a single RS observation. The trapezoidal space encompasses four boundary points (Eq. 57): (1) driest bare surface at highest temperature ($f_c = 0$, $T_{Rad} = T_{ss,max}$); (2) wettest surface at lowest temperature ($f_c = 0$, $T_{Rad} = T_{ss,min}$); (3) full vegetation surface at lowest temperature ($f_c = 1$, $T_{Rad} = T_{cs,min}$); and (3) full vegetation surface at highest temperature ($f_c = 1$, $T_{Rad} = T_{cs,max}$). The warm regime joining extreme points 1 and 4 experience largest water stress and therefore no LE flux. Cold edge joining extreme points 2 and 3 represents surfaces with no water stress and equilibrium in ET (i.e., EF = 1). These warm and cold edges are the boundary conditions for TSEB-TFM that are met theoretically under given meteorological and surface conditions. Trapezoidal space also helps improve estimation of water scarcity, plant water stress and transpiration, and aerodynamic resistances. Unlike the previously discussed non-space-based TSEB models, space-based TSEB-TFM does not require parametrization of surface (soil and canopy) resistance network and this approach avoids the need for excessive data inputs of crop physiology and meteorology. Net radiation fluxes are computed from Eq. 58. Following which, the EFs and net LE flux are estimated using patch approach that weights soil and canopy LE using fraction canopy cover, f_c (Eqs. 59 and 60).

$$T_{\mathrm{ss,max}} = T_a + \frac{R_{\mathrm{ns},0}}{4\varepsilon_s\sigma T_a{}^3 + \frac{\rho C_p}{0.65\times r_{ss}}} \quad , \quad T_{\mathrm{cs,max}} = T_a + \frac{R_{\mathrm{nc},0}}{4\varepsilon_c\sigma T_a{}^3 + \frac{\rho C_p}{r_{ah}}} \tag{57}$$

$$R_{\mathrm{ns}} = (1-\alpha_s)\,R_{s\downarrow} + \varepsilon_s\varepsilon_a\sigma T_a{}^4 - \varepsilon_s\sigma T_a{}^4 - 4\varepsilon_s\sigma T_a{}^3\,(T_{\mathrm{ss}} - T_a) \text{ and}$$

$$R_{\mathrm{nc}} = (1-\alpha_c)\,R_{s\downarrow} + \varepsilon_c\varepsilon_a\sigma T_a{}^4 - \varepsilon_c\sigma T_c{}^4 \tag{58}$$

For the bare soil hottest surface, $T_{\mathrm{ss,max}}$ is equal to T_a, and $R_{\mathrm{ns}} = R_{\mathrm{ns},0}$. Similarly, for the hottest vegetation, $T_{\mathrm{cs,max}}$ is equal to T_a, and $R_{\mathrm{nc}} = R_{\mathrm{nc},0}$. The EF$s$ for soil and canopy can be calculated from Eq. 59.

$$\begin{gathered} \mathrm{EF}_s = \frac{T_{\mathrm{ss,max}} - T_{ss}}{T_{\mathrm{ss,max}} - T_a} \times \frac{R_{\mathrm{ns},0}}{R_{\mathrm{ns}}}, \\ EF_c = \frac{T_{\mathrm{cs,max}} - T_{cs}}{T_{\mathrm{cs,max}} - T_a} \times \frac{R_{\mathrm{nc},0}}{R_{\mathrm{nc}}}, \text{ and} \\ \mathrm{EF} = f_c\mathrm{EF}_c + (1-f_c)\,\mathrm{EF}_s \end{gathered} \tag{59}$$

$$\mathrm{LE}_c = R_{\mathrm{nc}}\mathrm{EF}_c,\ LE_s = (R_{\mathrm{ns}} - G)\,\mathrm{EF}_s, \text{ and } \mathrm{LE} = f_c\mathrm{LE}_c + (1-f_c)\,\mathrm{LE}_s \tag{60}$$

It is worth noting that the TFM approach computes EF without parametrizing for the heterogeneity of aerodynamic and surface roughness and assumes those to be equal and uniform over the entire atmosphere–land surface nexus. It therefore essentially ignores the atmospheric stability within flux computations and gets restricted only to surfaces that are sufficiently uniform but may fail under heterogeneous conditions such as terrains, forests, and orchard crops, among others. A hybridized version of TSEB-TFM is therefore available (HTFM) where the resistance network parametrization as in fundamental TSEB approach is integrated with the patch approach of splitting radiometric temperature into soil and canopy temperature as in conventional TFM (Yang and Shang 2013). In HTFM, the fundamental TSEB is used to split R_{n} into R_{nc} and R_{ns} (Eq. 40) and TFM is used to breakdown those net radiation components into sensible, soil, and latent energy fluxes through weighted surface temperature decomposition for soil and canopy. The soil heat flux is calculated using Eq. 19, sensible heat fluxes using Eq. 41, and latent heat fluxes as residue of surface energy balance (Eq. 61).

$$\mathrm{LE}_c = \frac{R_{\mathrm{nc}}}{f_c} - H_c \quad \text{and} \quad \mathrm{LE}_s = \frac{R_{\mathrm{ns}-G}}{1 - f_c} - H_s \tag{61}$$

Although VI surface temperature, space-based models such as TFM and HTFM are advantageous for their relative independence over site-specific calibration of energy fluxes, such models are constrained due to four key limitations: (1) requirements of large areas in satellite images with sufficient vegetation and soil pixels under both dry/hot and cold/wet conditions, since such conditions cannot be readily identified in heterogeneous land surfaces especially with low-resolution satellites (e.g., MODIS, ASTER, and AVHRR, among others); (2) correct and unbiased selection of hot/dry and cold/wet pixels is not well-validated and documented; (3) non-validated isoline assumptions of soil moisture status; and (4) difficult hybridization of space-based models with other energy balance models due to difference in anchor pixel (hot/cold) selection. An enhanced two-source evapotranspiration model for land (ETEML) was developed to overcome such limitations by theoretically defining the criteria for VI surface temperature space isolines. Similar to TFM and HTFM, the patch approach is applied to obtain latent heat fluxes for soil and canopy but through calibration of potential latent heat fluxes (PLE_s, PLE_c) using soil water-deficit index (SWDI, Moran et al. 1994) and crop water stress index (CWSI, Jackson et al. 1988) for soil and canopy, respectively (Eqs. 62, 63 and 64).

$$\mathrm{LE}_s = (1 - \mathrm{SWDI}) \times \mathrm{PLE}_s \text{ and } LE_c = (1 - \mathrm{CWSI}) \times \mathrm{PLE}_c \tag{62}$$

$$\begin{aligned} \mathrm{CWSI} &= \frac{(T_{\mathrm{cs}} - T_a) - (T_{\mathrm{cs}} - T_a)_{\min}}{(T_{\mathrm{cs}} - T_a)_{\max} - (T_{\mathrm{cs}} - T_a)_{\min}} \\ SWDI &= \frac{(T_{\mathrm{ss}} - T_a) - (T_{\mathrm{ss}} - T_a)_{\min}}{(T_{\mathrm{ss}} - T_a)_{\max} - (T_{\mathrm{ss}} - T_a)_{\min}} \end{aligned} \tag{63}$$

$$\mathrm{PLE_s} = 1.26\left(\frac{\Delta}{\Delta + \gamma}\right) \times (R_{ns} - G) \qquad \mathrm{PLE}_c = \left(\frac{\Delta}{\Delta + \gamma}\right) \times (R_{\mathrm{nc}} + E_a) \tag{64}$$

where $(T_{\mathrm{cs}}\text{-}T_a)_{\min}$ and $(T_{\mathrm{cs}}\text{-}T_a)_{\max}$ and $(T_{\mathrm{ss}}\text{-}T_{\mathrm{a}})_{\min}$ and $(T_{\mathrm{ss}}\text{-}T_a)_{\max}$ are the lower and upper temperature gradient limits of crop water and soil water stresses. E_a is the drying power of the air calculated using Brutsaert (1982) equation. PLE_s and PLE_c are calculated by first determining a theoretical VI surface temperature space for each pixel where the surface air temperature difference is used as a surrogate instead of absolute soil and canopy surface temperatures. Using these identifications, four boundary points (dry bare soil, saturated bare soil, well-watered full vegetation, and water-stressed full vegetation) around each pixel of satellite image (represented by a point M within trapezoidal space) are defined as in TFM. Solving for Eqs. 65, 66, 67, 68 and 69 will determine canopy and soil temperatures, following which net radiation flux, soil heat flux, and latent heat fluxes can be calculated from Eqs. 62, 63 and 64.

$$T_{cs} - T_a = S_s(1 - f_c) + (T_s - T_a) \quad \text{and} \quad T_{ss} - T_a = (T_s - T_a) - S_s f_c \tag{65}$$

$$S_s = S_{s,\text{dry}} + \frac{q}{p+q} \times \left[S_{s,\text{wet}} - S_{s,\text{dry}}\right] \tag{66}$$

$$S_{s,\text{dry}} = (T_{cs} - T_a)_{\max} - (T_{\text{ss}} - T_a)_{\max} \qquad S_{s,\text{wet}} = (T_{\text{cs}} - T_a)_{\min} - (T_{ss} - T_a)_{\min} \tag{67}$$

$$p = (T_s - T_a) - S_{s,\text{wet}}(1 - f_c) - (T_{cs} - T_a)_{\min} \tag{68}$$

$$q = (1 - f_c)(T_{\text{ss}} - T_a)_{\max} + f_c(T_{cs} - T_a)_{\max} - (T_s - T_a) \tag{69}$$

where S_s is the slope of isoline that passes through a point M within the trapezoidal space, q is the temperature gradient between point M and the warm edge (joining hot/dry soil and hot/dry vegetation), and p is the temperature gradient between point M and the cold edge of trapezoidal space (joining saturated/wet soil and wet/cold vegetation). Some of the case studies where TSEB-TFM and HTFM models have been used for ET estimation in agricultural cropping systems are summarized in Table 9.

Table 9 Application use cases of using TSEB-TFM and HTFM models with satellite RS for geospatial evapotranspiration mapping of agricultural cropping systems

Crop/commodity	Model and satellite imagery	Region	Accuracy and errors	References
Vegetables and paddy	TSEB-TFM and MODIS	China	R^2: 0.52–0.61, E: 36.3 W m^2	Chen et al. (2020)
Forests and farmland	TSEB-TFM and MODIS	China	R^2: 0.74, E: 9–16%	Chen et al. (2022)
Corn and sunflower	HTFM and HJ-1A/1B	China	E: 11.9%	Yu and Shang (2020)
Corn and sunflower	HTFM and MODIS	China	R^2: 0.7–0.74, E: 9.6–12.7%	Yu et al. (2019)

E: error, r: correlation coefficient, R^2: coefficient of determination. Unless otherwise specified, E refers to error in daily ET estimates.

3.2.4 Empirical Crop Coefficient Vegetation Index Approaches

The aforementioned energy balance models more or less require large number of data inputs to compute actual ET. As an alternative, the crop coefficient approaches (Sect. 2.3) are still among the most preferred ones. However, their appropriateness is dependent on localized calibration and adjustments at spatiotemporal scales on how well the dynamics of crop growth and phenological developments are represented. Therefore, empirical approaches of deriving crop coefficients (single and basal) from remote sensing data are also used to estimate actual ET or crop water use. It is to be highly considered that since the spatiotemporal dynamics of crop development is dependent on localized crop, soil, and meteorological conditions, site-specific calibrations become critical. The rate of water use is influenced by the active vegetation cover typically represented by LAI. However, LAI varies between crops, their cultivars, and localized agroclimatic conditions. Therefore, using a single LAI approximation to determine actual ET for different cropping systems becomes impractical (Campos et al. 2017). To alleviate this concern, RS data in the form of VIs have been widely used as an integrated surrogate for crop physiology, phenology, growth stage, effective vegetation cover, biomass, and LAI parameters, among others (Fig. 5). Relationships between VIs and single or basal crop coefficients are established and multiplied with the reference ET (ET_r or ET_o) to estimate actual geospatial ET. The K_c-VI relationships have been derived for a large range of agricultural crops including barley, wheat, cotton, corn, sugar beet, alfalfa, garlic, peach, grapevines, apples, olives, and others. Some of the widely used VIs to derive such relationships include NDVI (normalized difference

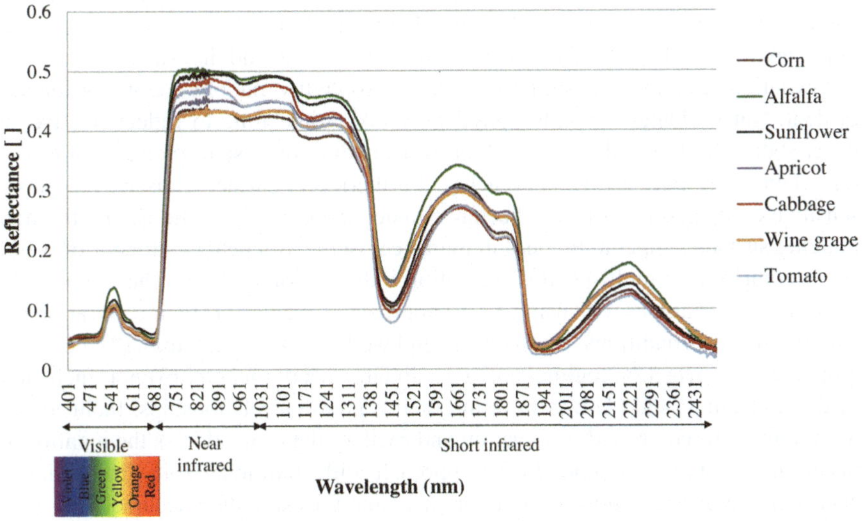

Fig. 5 Canopy spectral reflectance curves for different crops within visible, near-infrared, and shortwave-infrared ranges. (From Hosgood et al. 1993)

VI), GNDVI (green NDVI), SAVI (soil-adjusted VI), EVI (enhanced VI), NDRE (normalized difference red-edge index), optimized SAVI, transformed SAVI, and many others. Some of the fundamental steps to empirically obtain crop coefficients (K_c or K_{cb}) include (1) determining crop coefficients as the ratio of measured ET (from lysimeters, eddy covariance fluxes, or other methods) to reference ET (ET_r or ET_o), and the basal crop coefficient is determined as the ratio of transpiration to reference ET; (2) modelling the determined crop coefficients as the function of VIs computed from RS data for the same location where actual ET measurements were carried out; (3) repeating step 2 over space and time for enhanced accuracy of crop coefficient and VI relationship (linear, polynomial, exponential, etc.); and (4) multiplying the relationship from step 3 with the reference ET to obtain geospatial ET estimates. Some case studies that developed empirical relationships between satellite RS and crop coefficients (K_c or K_{cb}) are detailed in Table 10 with evaluation results for different agricultural cropping systems.

4 Practical Implications of Satellite-Based Remote Sensing for ET Estimation

Crop coefficient-PM (including K_c-VI) approaches are purely empirical or semiempirical in nature that exclude surface temperature and heat/energy flow resistances pertaining to actual surfaces. These factors are the primary and practical drivers of surface water exchange with the atmosphere. Exclusion of these factors results in accuracy and robustness of actual ET estimates to be tremendously compromised. Nonetheless, crop coefficient-based approaches require much lesser data inputs compared to energy balance models and therefore are handy for ET estimation of larger areas. K_c-VI approaches are crop and site-specific and the water stress/water use variations can only be estimated when chlorophyll variations are proportionally evident. Since changes in chlorophyll may take some time to reflect relative to water status variations, there is a risk of over-, under-, or misinterpretation of actual ET when using pure K_c-VI approaches. Furthermore, surface emissivity is more sensitive compared to surface reflectance under crop water variations, and thermal imaging/surface temperature measurements can serve as a better estimator of water stress compared to multispectral vegetation indices. Some of the other commonly identified uncertainties with using K_c-VI approaches include (1) their calibration for extreme surface conditions, i.e., bare soil and well-watered vegetation, (2) their site, crop, cultivar, irrigation regime, and agroclimate-specific nature, (3) non-inclusion of ET_r dynamics, and (4) bias in fitting their regression relationships. Inclusion of minimum required ground observation and axillary data can address the mentioned uncertainties, while coupling K_c-VI approach with traditional soil water balance and/or energy balance models can improve actual ET estimation accuracies.

Effective application of surface energy balance models depends on correct land surface temperature imagery. The correctness is very often impacted by the cloud

Table 10 Use cases of empirically deriving crop coefficients and actual evapotranspiration using visible and near-infrared remote sensing data

Crop	Crop coefficient-VI relationship	Region	Accuracy	References
Corn	$K_{cb} = 1.092\text{NDVI} - 0.053$	Utah, USA	E: 2.6%	Neale et al. (1989)
	$K_{cb} = 1.181\text{NDVI} - 0.026$		E: 4.7%	
	$K_{cb} = 1.416\text{SAVI} + 0.017$	Colorado, USA	E: 6%	Bausch (1993)
	$K_{cb} = 1.414\text{SAVI} - 0.02$	Nebraska, USA	R^2: 0.84, E: 10%	Campos et al. (2017)
	$K_c = 1.317\text{NDVI} + 0.023$		R^2: 0.86, E: 14%	Singh et al. (2009)
Wheat	$K_{cb} = 1.46\text{NDVI} - 0.26$	Arizona, USA	R^2: 0.80, E: 25%	Choudhary et al. (1994)
	$K_{cb} = 1.69\text{SAVI} - 0.16$		R^2: 0.88, E: 12%	
	$K_{cb} = 1.54\text{TSAVI} + 0.03$		R^2: 0.86, E: 21%	
	$K_{cb} = 1.63\text{NDVI}_n - 2.57\text{NDVI}_n^2 + 1.93\text{NDVI}_n^3 + 0.18$		R^2: 0.90, E: 4%	Hunsaker et al. (2005)
	$K_{cb} = 1.64\text{NDVI} - 0.14$	Morocco	R^2: 0.80, E: 15%	Duchemin et al. (2006)
	$K_c = 1.5141\text{SAVI} + 0.4077$	India	R^2: 0.90	Gontia and Tiwari (2010)
	$K_c = 2.711\text{NDVI} + 0.424$		R^2: 0.80	
Wheat	$K_{c,\ 1\text{st stage}} = 654.943\text{NDVI} - 437.75\text{SAVI} + 0.1099$ $K_{c,\ 2\text{nd stage}} = 18.405\text{SAVI} - 28\text{NDVI} + 1.877$ $K_{c,\ 3\text{rd stage}} = 12.067\text{SAVI} - 17.90\text{NDVI} + 0.745$	Egypt	1st stage: R^2: 0.81, E: 0.0091 2nd stage: R^2: 0.90, E: 0.0014 3rd stage: R^2: 0.98, E: 0.0007	Farg et al. (2012)
Soybean	$K_{cb} = 1.258\text{SAVI} - 0.006$	Nebraska, USA	R^2: 0.84, E: 10%	Campos et al. (2017)
	$K_c = 1.217\text{NDVI} + 0.93$		R^2: 0.93	Singh et al. (2009)
Sorghum	$K_c = 1.453\text{NDVI} - 0.112$		R^2: 0.93	
Grapevine	$K_c = 1.44\text{NDVI} - 0.10$	Spain	R^2: 0.84, E: 5%	Campos et al. (2010)
	$K_c = 0.181e^{1.314\text{NDVI}}$	Mexico	R^2: 0.77, E: 18%	Er-Raki et al. (2013)
Apple	$K_{cb} = 1.82\text{SAVI} - 0.07$	Chile	R^2: 0.95, E: 25%	Odi-Lara et al. (2016)
Potato	$K_{cb} = 1.044\text{NDVI} + 0.4159$	Canada	R^2: 0.96	Ali (2022)
Bell pepper	$K_{cb} = 1.451\text{NDVI} - 0.124\text{NDVI}^2 - 0.063$	California, USA	E: 9.5%	Johnson and Trout (2012)
Broccoli	$K_{cb} = 2.636\text{NDVI} - 0.823\text{NDVI}^2 - 0.165$		E: 6.1%	
Garlic	$K_{cb} = 2.663\text{NDVI} - 1.564\text{NDVI}^2 - 0.077$		E: 5.9%	
Lettuce	$K_{cb} = 1.393\text{NDVI} - 0.111\text{NDVI}^2 + 0.012$		E: 9.8%	

E: error, r: correlation coefficient, R^2: coefficient of determination. Unless otherwise specified, E refers to error in daily ET estimates

covers and atmospheric aberrations (water vapor, surface emissivity, air quality, etc.) which are not very straightforward to resolve. On top of that, the restricted temporal resolution of the satellites carrying a thermal-infrared imaging sensor with desent spatial resolution (e.g., Landsat series, spatial resolution: 30 m, temporal resolution: 16 days) further leads to missing opportunity of actual ET estimation. Although, interpolation techniques can be applied to estimate the missing estimates, the accuracy is still questionable for temporal agroclimatic conditions. Ongoing efforts to obtain thermal imaging data at high spatial as well as temporal resolution can help alleviate data quality concerns with precision (Chandel et al. 2020, 2021). Another concern with satellite-based thermal imaging is ignoring the difference between surface temperature and aerodynamic temperature. Since the aerodynamic temperature is unmeasurable and its relationship is still not clear with the surface temperature due to thermodynamic properties of vegetation and soil surface, errors in precision ET estimation are quite possible. This is although a bigger concern for single-source energy balance models that relatively consider lower surface roughness and resistance parametrization compared to the two-source models. Considering absolute surface temperature estimates can induce significant uncertainties in sensible heat flux estimations when applying surface energy modeling to partially vegetated surfaces with varying architectures and geometries. Instead, utilizing the aerodynamic surface temperature gradient as in METRIC and further decomposing the surface temperature to soil and canopy vegetation through resistance network can minimize errors in estimating sensible heat fluxes.

Studies have reported underestimation of actual latent heat flux compared to available energy, causing non-closure of energy balance. This non-closure is mostly due to sensors (e.g., wind speed, air temperature, solar radiation, soil heat flux, net radiation), measurement and computation errors, and errors due to ignoring other possible energy sources. Parametrization of unmeasurable resistance network is also critical for computing surface fluxes. The network is affected by atmospheric stability, wind speed profiles, and vegetation distributions on ground. Under unstable conditions, iterative methods of energy flux computations can be used.

5 Conclusion

Remote sensing data when coupled with meteorological or empirical models can be very promising to estimate actual exchange of water between the land surface and atmosphere nexus. Selection of appropriate ET estimation model and input parameters are equally critical and must be paid due attention based on resource availability and accuracy limits. Specific to agricultural ecosystems, estimating actual ET is estimating the actual amount of water plants utilized to maintain their water-healthy status. This estimation is highly critical from precision irrigation management perspective amid the concerns of freshwater reserve shortages, population growth, and food demands. Estimation of ET using RS data can help develop site-/zone-specific precision irrigation management prescriptions. Currently, small unmanned aerial systems are also under heavy exploration to estimate actual crop ET at much finer spatiotemporal resolutions. Modeling RS-based ET estimates through cloud computing tools and web and smartphone applications can come handy, labor-efficient, and cost-effective for deriving and implementing actual irrigation management decisions on ground by the crop growers.

References

Acharya B, Sharma V (2021) Comparison of satellite driven surface energy balance models in estimating crop evapotranspiration in semi-arid to arid inter-mountain region. Remote Sens 13(9):1822

Adeyemi O, Grove I, Peets S, Norton T (2017) Advanced monitoring and management systems for improving sustainability in precision irrigation. Sustainability 9:353–372

Ahmad M-D, Biggs T, Turral H, Scott CA (2006) Application of SEBAL approach and MODIS time-series to map vegetation water use patterns in the data scarce Krishna river basin of India. Water Sci Technol 53:83–90. https://doi.org/10.2166/wst.2006.301

Alhassan A, Jin M (2020) Evapotranspiration in the Tono reservoir catchment in upper east region of Ghana estimated by a novel TSEB approach from ASTER imagery. Remote Sens 12:569. https://doi.org/10.3390/rs12030569

Allen RG, Pereira L, Raes D, Smith M (1998) Crop evapotranspiration—guidelines for computing crop water requirements. Food and Agriculture Organization of the United Nations Irrigation and Drainage Paper 56

Allen RG, Tasumi M, Morse A, Trezza R, Wright JL, Bastiaanssen W, Kramber W, Lorite I, Robison CW (2007a) Satellite-based energy balance for mapping evapotranspiration with internalized calibration (METRIC)—applications. J Irrig Drain Eng 133:395–406. https://doi.org/10.1061/(ASCE)0733-9437(2007)133:4(395)

Allen RG, Tasumi M, Trezza R (2007b) Satellite-based energy balance for mapping evapotranspiration with internalized calibration (METRIC)—model. J Irrig Drain Eng 133:380–394. https://doi.org/10.1061/(ASCE)0733-9437(2007)133:4(380)

Ali U (2022) Application of satellite-based indices for precise agricultural land management and estimation of evapotranspiration. Masters Thesis. Dalhousie University Halifax, Nova Scotia

Anapalli SS, Fisher DK, Reddy KN, Wagle P, Gowda PH, Sui R (2018) Quantifying soybean evapotranspiration using an eddy covariance approach. Agric Water Manag 209:228–239

Anderson M (1997) A two-source time-integrated model for estimating surface fluxes using thermal infrared remote sensing. Remote Sens Environ 60:195–216. https://doi.org/10.1016/S0034-4257(96)00215-5

Anderson M, Diak G, Gao F, Knipper K, Hain C, Eichelmann E, Hemes KS, Baldocchi D, Kustas W, Yang Y (2019) Impact of insolation data source on remote sensing retrievals of evapotranspiration over the California Delta. Remote Sens 11(3):216

Bala A, Pawar PS, Misra AK, Rawat KS (2017) Estimation and validation of actual evapotranspiration for wheat crop using SEBAL model over Hisar District, Haryana, India. Curr Sci 113:134. https://doi.org/10.18520/cs/v113/i01/134-141

Baldocchi D (2013) A brief history on Eddy covariance flux measurements: a personal perspective. FluxLetter 5(2):1–8

Bastiaanssen WGM, Menenti M, Feddes RA, Holtslag AAM (1998a) A remote sensing surface energy balance algorithm for land (SEBAL). 1. Formulation J Hydrol 212–213:198–212. https://doi.org/10.1016/S0022-1694(98)00253-4

Bastiaanssen WGM, Pelgrum H, Wang J, Ma Y, Moreno JF, Roerink GJ, van der Wal T (1998b) A remote sensing surface energy balance algorithm for land (SEBAL). J Hydrol 212–213:213–229. https://doi.org/10.1016/S0022-1694(98)00254-6

Bastiaanssen WG (2000) SEBAL-based sensible and latent heat fluxes in the irrigated Gediz Basin, Turkey. J Hydrol 229(1–2):87–100

Bastiaanssen WGM, Noordman EJM, Pelgrum H, Davids G, Thoreson BP, Allen RG (2005) SEBAL model with remotely sensed data to improve water-resources management under actual field conditions. J Irrig Drain Eng 131:85–93. https://doi.org/10.1061/(ASCE)0733-9437(2005)131:1(85)

Bausch WC (1993) Soil background effects on reflectance-based crop coefficients for corn. Remote Sens Environ 46:213–222. https://doi.org/10.1016/0034-4257(93)90096-G

Bhattacharyya P, Neogi S, Roy KS, Rao KS (2013) Gross primary production, ecosystem respiration and net ecosystem exchange in Asian rice paddy: an eddy covariance-based approach. Curr Sci:67–75

Bodner G, Loiskandl W, Kaul HP (2007) Cover crop evapotranspiration under semi-arid conditions using FAO dual crop coefficient method with water stress compensation. Agric Water Manag 93(3):85–98

Brutsaert W (1999) Aspects of bulk atmospheric boundary layer similarity under free-convective conditions. Rev Geophys 37(4):439–451

Burba G (2013) Eddy covariance method for scientific, industrial, agricultural and regulatory applications: a field book on measuring ecosystem gas exchange and areal emission rates. LI-Cor Biosciences

Burchard-Levine V, Nieto H, Riaño D, Migliavacca M, El-Madany TS, Perez-Priego O, Carrara A, Martín MP (2019) Adapting the thermal-based two-source energy balance model to estimate energy fluxes in a complex tree-grass ecosystem. Hydrol Earth Syst Sci Discuss:1–37

Borrelli J, Fedler CB, Gregory JM (1998) Mean crop consumptive use and free-water evaporation for Texas. Texas Water Development Board Grant No. 95-483-137

Campos I, Neale CMU, Calera A, Balbontín C, González-Piqueras J (2010) Assessing satellite-based basal crop coefficients for irrigated grapes (Vitis vinifera L.). Agric Water Manag 98:45–54. https://doi.org/10.1016/j.agwat.2010.07.011

Campos I, Neale CM, Suyker AE, Arkebauer TJ, Gonçalves IZ (2017) Reflectance-based crop coefficients REDUX: for operational evapotranspiration estimates in the age of high producing hybrid varieties. Agric Water Manag 187:140–153

Carpintero E, Anderson MC, Andreu A, Hain C, Gao F, Kustas WP, González-Dugo MP (2021) Estimating evapotranspiration of Mediterranean Oak Savanna at multiple temporal and spatial resolutions. Implications for water resources management. Remote Sens 13(18):3701

Cawse-Nicholson K, Anderson MC, Yang Y, Yang Y, Hook SJ, Fisher JB, Halverson G, Hulley GC, Hain C, Baldocchi DD, Brunsell NA (2021) Evaluation of a CONUS-wide ECOSTRESS DisALEXI evapotranspiration product. IEEE J Sel Top Appl Earth Observ Remote Sens 14:10117–10133

Chandel AK, Molaei B, Khot LR, Peters RT, Stöckle CO (2020) High resolution geospatial evapotranspiration mapping of irrigated field crops using multispectral and thermal infrared imagery with metric energy balance model. Drones 4(3):52

Chandel AK, Khot LR, Molaei B, Peters RT, Stöckle CO, Jacoby PW (2021) High-resolution spatiotemporal water use mapping of surface and direct-root-zone drip-irrigated grapevines using uas-based thermal and multispectral remote sensing. Remote Sens 13(5):954

Chen M, Senay GB, Singh RK, Verdin JP (2016) Uncertainty analysis of the Operational Simplified Surface Energy Balance (SSEBop) model at multiple flux tower sites. J Hydrol 536:384–399

Chen H, Huang JJ, Berg A, McBean E (2020) Development of a trapezoidal framework-based model (PCALEP) for partition of land evapotranspiration. J Hydrol 589:124994

Chen H, Huang JJ, Dash SS, McBean E, Li H, Zhang J, Lan Z, Gao J, Zhou Z (2022) Assessing the effects of end-members determination on regional latent heat flux simulation in trapezoidal framework based model. Agric For Meteorol 312:108734

Choudhury BJ, Idso SB, Reginato RJ (1987) Analysis of an empirical model for soil heat flux under a growing wheat crop for estimating evaporation by an infrared-temperature based energy balance equation. Agric For Meteorol 39(4):283–297

Colaizzi PD, Agam N, Tolk JA, Evett SR, Howell TA, Gowda PH, Shaughnessy SAO, Kustas WP, Anderson MC (2014) Two-source energy balance model to calculate E, T, and ET: comparison of Priestley-Taylor and Penman-Monteith formulations and two time scaling methods. Trans ASABE:479–498. https://doi.org/10.13031/trans.57.10423

Colaizzi PD, Kustas WP, Anderson MC, Agam N, Tolk JA, Evett SR, Howell TA, Gowda PH, O'Shaughnessy SA (2012) Two-source energy balance model estimates of evapotranspiration using component and composite surface temperatures. Adv Water Resour 50:134–151. https://doi.org/10.1016/j.advwatres.2012.06.004

Denager T, Looms MC, Sonnenborg TO, Jensen KH (2020) Comparison of evapotranspiration estimates using the water balance and the eddy covariance methods. Vadose Zone J 19(1):e20032

Dhungel R, Aiken R, Evett SR, Colaizzi PD, Marek G, Moorhead JE, Baumhardt RL, Brauer D, Kutikoff S, Lin X (2021) Energy imbalance and evapotranspiration hysteresis under an advective environment: evidence from lysimeter, eddy covariance, and energy balance modeling. Geophys Res Lett 48(1):e2020GL091203

Diarra A, Jarlan L, Khabba S, Le Page M, Er-Raki S, Balaghi R, Charafi S, Chehbouni A, El Alami R (2022) Medium-resolution mapping of evapotranspiration at the catchment scale based on thermal infrared MODIS data and ERA-interim reanalysis over North Africa. Remote Sens 14:5071

Dingre SK, Gorantiwar SD (2020) Determination of the water requirement and crop coefficient values of sugarcane by field water balance method in semiarid region. Agric Water Manag 232:106042

Du J, Song K, Wang Z, Zhang B, Liu D (2013) Evapotranspiration estimation based on MODIS products and surface energy balance algorithms for land (SEBAL) model in Sanjiang Plain, Northeast China. Chin Geogr Sci 23:73–91. https://doi.org/10.1007/s11769-013-0587-8

Duchemin B, Hadria R, Erraki S, Boulet G, Maisongrande P, Chehbouni A, Escadafal R, Ezzahar J, Hoedjes JCB, Kharrou MH, Khabba S, Mougenot B, Olioso A, Rodriguez JC, Simonneaux V (2006) Monitoring wheat phenology and irrigation in Central Morocco: on the use of relationships between evapotranspiration, crops coefficients, leaf area index and remotely-sensed vegetation indices. Agric Water Manag 79:1–27. https://doi.org/10.1016/j.agwat.2005.02.013

Elkatoury A, Alazba AA, Mossad A (2020) Estimating evapotranspiration using coupled remote sensing and three SEB models in an arid region. Environ Proc 7(1):109–133

Er-Raki S, Rodriguez JC, Garatuza-Payan J, Watts CJ, Chehbouni A (2013) Determination of crop evapotranspiration of table grapes in a semi-arid region of Northwest Mexico using multi-spectral vegetation index. Agric Water Manag 122:12–19

Evett SR, Agam N, Kustas WP, Colaizzi PD, Schwartz RC (2012) Soil profile method for soil thermal diffusivity, conductivity and heat flux: comparison to soil heat flux plates. Adv Water Resour 50:41–54

Farg E, Arafat SM, Abd El-Wahed MS, El-Gindy AM (2012) Estimation of evapotranspiration ETc and crop coefficient Kc of wheat, in South Nile Delta of Egypt using integrated FAO-56 approach and remote sensing data. Egypt J Remote Sens Space Sci 15(1):83–89

Foken T (2008) The energy balance closure problem: an overview. Ecol Appl 18(6):1351–1367

Foken, T., Aubinet, M., Finnigan, J.J., Leclerc, M.Y., Mauder, M. and Paw U, K.T., 2011. Results of a panel discussion about the energy balance closure correction for trace gases. Bull Am Meteorol Soc, 92(4), ES13-ES18

Gong X, Qiu R, Sun J, Ge J, Li Y, Wang S (2020) Evapotranspiration and crop coefficient of tomato grown in a solar greenhouse under full and deficit irrigation. Agric Water Manag 235:106154

Gontia NK, Tiwari KN (2010) Estimation of crop coefficient and evapotranspiration of wheat (Triticum aestivum) in an irrigation command using remote sensing and GIS. Water Resour Manag 24:1399–1414

Gonzalez-Dugo MP, Neale CMU, Mateos L, Kustas WP, Prueger JH, Anderson MC, Li F (2009) A comparison of operational remote sensing-based models for estimating crop evapotranspiration. Agric For Meteorol 149:1843–1853

Gowda PH, Chávez JL, Howell TA, Marek TH, New LL (2008) Surface energy balance based evapotranspiration mapping in the Texas high plains. Sensors 8(8):5186–5201

Häusler M, Conceição N, Tezza L, Sánchez JM, Campagnolo ML, Häusler AJ, Silva JMN, Warneke T, Heygster G, Ferreira MI (2018) Estimation and partitioning of actual daily evapotranspiration at an intensive olive grove using the STSEB model based on remote sensing. Agric Water Manag 201:188–198. https://doi.org/10.1016/j.agwat.2018.01.027

He R, Jin Y, Jiang J, Xu M, Jia S (2022) Sensitivity of METRIC-based tree crop evapotranspiration estimation to meteorology, land surface parameters and domain size. Agric Water Manag 271:107789. https://doi.org/10.1016/j.agwat.2022.107789

Hosgood B, Jacquemound S, Andreeoli G, Verdebout J, Pedrini A, Schmuck G (1993) Leaf Optical properties experiment database (LOPEX93). Available on-line [http://ecosis.org] from the Ecological Spectral Information System (EcoSIS). Accessed on October 2022

Hunsaker DJ, Pinter PJ Jr, Kimball B (2005) Wheat basal crop coefficients determined by normalized difference vegetation index. Irrigation Sci 24:1–14. https://doi.org/10.1007/s00271-005-0001-0

Jackson RD, Kustas WP, Choudhury BJ (1988) A reexamination of the crop water stress index. Irrig Sci 9:309–317

Jiang L, Islam S, Guo W, Jutla AS, Senarath SU, Ramsay BH, Eltahir E (2009) A satellite-based daily actual evapotranspiration estimation algorithm over South Florida. Glob Planet Chang 67(1–2):62–77

Johnson LF, Trout TJ (2012) Satellite NDVI assisted monitoring of vegetable crop evapotranspiration in California's San Joaquin Valley. Remote Sens-Basel 4:439–455. https://doi.org/10.3390/rs4020439

Knipper K, Anderson M, Bambach N, Kustas W, Gao F, Zahn E, Hain C, McElrone A, Belfiore OR, Castro S, Alsina MM (2023) Evaluation of partitioned evaporation and transpiration estimates within the DisALEXI Modeling framework over irrigated crops in California. Remote Sens 15(1):68

Kumar U, Sahoo B, Chatterjee C, Raghuwanshi NS (2020) Evaluation of simplified surface energy balance index (S-SEBI) method for estimating actual evapotranspiration in Kangsabati reservoir command using Landsat 8 imagery. J Indian Soc Remote Sens 48:1421–1432. https://doi.org/10.1007/s12524-020-01166-9

Kustas WP, Norman JM (2000) A two-source energy balance approach using directional radiometric temperature observations for sparse canopy covered surfaces. Agron J 92:847–854. https://doi.org/10.2134/agronj2000.925847x

Liaghat S, Balasundram SK (2010) A review: the role of remote sensing in precision agriculture. Am J Agric Biol Sci 5(1):50–55

Laipelt L, Henrique Bloedow Kayser R, Santos Fleischmann A, Ruhoff A, Bastiaanssen W, Erickson TA, Melton F (2021) Long-term monitoring of evapotranspiration using the SEBAL algorithm and Google Earth Engine cloud computing. ISPRS J Photogramm Remote Sens 178:81–96. https://doi.org/10.1016/j.isprsjprs.2021.05.018

Leuning R, van Gorsel E, Massman WJ, Isaac PR (2012) Reflections on the surface energy imbalance problem. Agric For Meteorol 156:65–74

Li F, Ma Y (2019) Evaluation of the dual crop coefficient approach in estimating evapotranspiration of drip-irrigated summer maize in Xinjiang, China. Water 11(5):1053

Li S, Kang S, Li F, Zhang L (2008) Evapotranspiration and crop coefficient of spring maize with plastic mulch using eddy covariance in Northwest China. Agric Water Manag 95(11):1214–1222

Li X, Xu X, Wang X, Xu S, Tian W, Tian J, He C (2021) Assessing the effects of spatial scales on regional evapotranspiration estimation by the SEBAL model and multiple satellite datasets: a case study in the agro-pastoral ecotone. Northwestern China Remote Sens 13:1524. https://doi.org/10.3390/rs13081524

Li Z-L, Tang R, Wan Z, Bi Y, Zhou C, Tang B, Yan G, Zhang X (2009) A review of current methodologies for regional evapotranspiration estimation from remotely sensed data. Sensors 9:3801–3853. https://doi.org/10.3390/s90503801

Liou YA, Kar SK (2014) Evapotranspiration estimation with remote sensing and various surface energy balance algorithms—a review. Energies 7(5):2821–2849

Long D, Singh VP (2012) A modified surface energy balance algorithm for land (M-SEBAL) based on a trapezoidal framework: a MODIFIED SEBAL. Water Resour Res 48. https://doi.org/10.1029/2011WR010607

Dias Lopes J, Neiva Rodrigues L, Acioli Imbuzeiro HM, Falco Pruski F (2019) Performance of SSEBop model for estimating wheat actual evapotranspiration in the Brazilian Savannah region. Int J Remote Sens 40(18):6930–6947

Machakaire ATB, Steyn JM, Franke AC (2021) Assessing evapotranspiration and crop coefficients of potato in a semi-arid climate using Eddy Covariance techniques. Agric Water Manag 255:107029

Mancosu N, Snyder RL, Kyriakakis G, Spano D (2015) Water scarcity and future challenges for food production. Water 7:975–992

McNaughton KG, Spriggs TW (1986) A mixed-layer model for regional evaporation. Bound-Layer Meteorol 34(3):243–262

McShane RR, Driscoll KP, Sando R (2017) A review of surface energy balance models for estimating actual evapotranspiration with remote sensing at high spatiotemporal resolution over large extents. Scientific Investigations Report. Department of the Interior and U.S. Geological Survey: Reston, VA, USA, 2017

Menenti M, Choudhury BJ (1993) Parameterization of land surface evaporation by means of location dependent potential evaporation and surface temperature range. IAHS Publ

Mhawej M, Gao X, Reilly J, Abunnasr Y (2022) SEBALIGEE v2: global evapotranspiration estimation replacing hot/cold pixels with machine learning (preprint). Agric Vet Food Sci. https://doi.org/10.1002/essoar.10512468.1

Misra AK (2014) Climate change and challenges of water and food security. Int J Sustain Built Environ 3(1):153–165

Monteith JL, Unsworth MH (2008) Principles of environmental. Physics xxi:418

Moorhead JE, Marek GW, Gowda PH, Lin X, Colaizzi PD, Evett SR, Kutikoff S (2019) Evaluation of evapotranspiration from eddy covariance using large weighing lysimeters. Agronomy 9(2):99

Moran MS, Clarke TR, Inoue Y, Vidal A (1994) Estimating crop water deficit using the relation between surface-air temperature and spectral vegetation index. Remote Sens Environ 49(3):246–263

Neale CMU, Bausch WC, Heermann DF (1989) Development of reflectance-based crop coefficients for corn. Trans ASAE 32:1891–1899

Nisa Z, Khan MS, Govind A, Marchetti M, Lasserre B, Magliulo E, Manco A (2021) Evaluation of SEBS, METRIC-EEFlux, and QWaterModel actual evapotranspiration for a Mediterranean cropping system in southern Italy. Agronomy 11(2):345

Norman JM, Anderson MC, Kustas WP, French AN, Mecikalski J, Torn R, Diak GR, Schmugge TJ, Tanner BCW (2003) Remote sensing of surface energy fluxes at 10^{1} -m pixel resolutions: REMOTE SENSING OF SURFACE ENERGY FLUXES. Water Resour Res 39. https://doi.org/10.1029/2002WR001775

Norman JM, Kustas WP, Humes KS (1995) Source approach for estimating soil and vegetation energy fluxes in observations of directional radiometric surface temperature. Agric For Meteorol 77:263–293. https://doi.org/10.1016/0168-1923(95)02265-Y

Nyawade SO, Gitari HI, Karanja NN, Gachene CK, Schulte-Geldermann E, Parker ML (2021) Yield and evapotranspiration characteristics of potato-legume intercropping simulated using a dual coefficient approach in a tropical highland. Field Crop Res 274:108327

Odi-Lara M, Campos I, Neale CM, Ortega-Farías S, Poblete-Echeverría C, Balbontín C, Calera A (2016) Estimating evapotranspiration of an apple orchard using a remote sensing-based soil water balance. Remote Sens 8(3):253

Ortega-Farías S, Fonseca D, De La Fuente D, Kilic A, Ortega-Salazar S, Allen R, Carrasco-Benavides M (2016) Remote sensing model to evaluate the spatial variability of vineyard water requirements. In: X international symposium on grapevine physiology and biotechnology, vol 1188, pp 235–242

Ortega-Salazar S, Ortega-Farías S, Kilic A, Allen R (2021) Performance of the METRIC model for mapping energy balance components and actual evapotranspiration over a superintensive drip-irrigated olive orchard. Agric Water Manag 251:106861

Paço TA, Ferreira MI, Rosa RD, Paredes P, Rodrigues GC, Conceição N, Pacheco CA, Pereira LS (2012) The dual crop coefficient approach using a density factor to simulate the evapotranspiration of a peach orchard: SIMDualKc model versus eddy covariance measurements. Irrig Sci 30:115–126

Pardo N, Sánchez ML, Timmermans J, Su Z, Pérez IA, García MA (2014) SEBS validation in a Spanish rotating crop. Agric For Meteorol 195:132–142

Paula ACPD, Silva CLD, Rodrigues LN, Scherer-Warren M (2019) Performance of the SSEBop model in the estimation of the actual evapotranspiration of soybean and bean crops. Pesq Agrop Brasileira 54

Priestley CHB, Taylor RJ (1972) On the assessment of surface heat flux and evaporation using large-scale parameters. Mon Weather Rev 100(2):81–92

Ranjan R, Chandel AK, Khot LR, Bahlol HY, Zhou J, Boydston RA, Miklas PN (2019) Irrigated pinto bean crop stress and yield assessment using ground based low altitude remote sensing technology. Inf Process Agric 6(4):502–514

Roerink GJ, Su Z, Menenti M (2000) S-SEBI: a simple remote sensing algorithm to estimate the surface energy balance. Phys. Chem Earth Part B Hydrol Oceans Atmosphere 25:147–157. https://doi.org/10.1016/S1464-1909(99)00128-8

Sanchez JM, López-Urrea R, Valentín F, Caselles V, Galve JM (2019) Lysimeter assessment of the Simplified Two-Source Energy Balance model and eddy covariance system to estimate vineyard evapotranspiration. Agric For Meteorol 274:172–183

Sankaran S, Khot LR, Espinoza CZ, Jarolmasjed S, Sathuvalli VR, Vandemark GJ, Miklas PN, Carter AH, Pumphrey MO, Knowles NR, Pavek MJ (2015) Low-altitude, high-resolution aerial imaging systems for row and field crop phenotyping: a review. Eur J Agron 70:112–123

Santos CACD, Bezerra BG, Silva BBD, Rao TVR (2010) Assessment of daily actual evapotranspiration with SEBAL and S-SEBI algorithms in cotton crop. Rev Bras Meteorol 25:383–392

Santanello JA Jr, Friedl MA (2003) Diurnal covariation in soil heat flux and net radiation. J Appl Meteorol 42:851–862

Senay GB, Budde M, Verdin JP, Melesse AM (2007) A coupled remote sensing and simplified surface energy balance approach to estimate actual evapotranspiration from irrigated fields. Sensors 7(6):979–1000

Senay GB, Budde ME, Verdin JP (2011a) Enhancing the Simplified Surface Energy Balance (SSEB) approach for estimating landscape ET: validation with the METRIC model. Agric Water Manag 98(4):606–618

Senay GB, Leake S, Nagler PL, Artan G, Dickinson J, Cordova JT, Glenn EP (2011b) Estimating basin scale evapotranspiration (ET) by water balance and remote sensing methods. Hydrol Process 25(26):4037–4049

Senay GB, Bohms S, Singh RK, Gowda PH, Velpuri NM, Alemu H, Verdin JP (2013) Operational evapotranspiration mapping using remote sensing and weather datasets: a new parameterization for the SSEB approach. JAWRA J Am Water Resour Assoc 49(3):577–591

Singh R, Irmak A (2009) Estimation of crop coefficients using satellite remote sensing. J Irrig Drain Eng 135:597–608. https://doi.org/10.1061/(ASCE)IR.1943-4774.0000052

Singh RK, Senay GB, Velpuri NM, Bohms S, Scott RL, Verdin JP (2013) Actual evapotranspiration (water use) assessment of the Colorado River Basin at the Landsat resolution using the operational simplified surface energy balance model. Remote Sens 6(1):233–256

Singh RK, Senay GB, Velpuri NM, Bohms S, Verdin JP (2014) On the downscaling of actual evapotranspiration maps based on combination of MODIS and Landsat-based actual evapotranspiration estimates. Remote Sens 6(11):10483–10509

Singh RK, Senay GB (2015) Comparison of four different energy balance models for estimating evapotranspiration in the Midwestern United States. Water 8(1):9

Sinha R, Khot LR, Gao Z, Chandel AK (2021) Sensors III: spectral sensing and data analysis. In: Fundamentals of agricultural and field robotics. Springer, Cham, pp 79–110

Sobrino JA, Souza da Rocha N, Skoković D, Suélen Käfer P, López-Urrea R, Jiménez-Muñoz JC, Alves Rolim SB (2021) Evapotranspiration estimation with the S-SEBI method from Landsat 8 data against Lysimeter measurements at the Barrax site, Spain. Remote Sensing 13(18):3686

Song L, Liu S, Kustas WP, Zhou J, Xu Z, Xia T, Li M (2016) Application of remote sensing-based two-source energy balance model for mapping field surface fluxes with composite and component surface temperatures. Agric For Meteorol 230:8–19

Su Z (2002) The surface energy balance system (SEBS) for estimation of turbulent heat fluxes. Hydrol Earth Syst Sci 6:85–100. https://doi.org/10.5194/hess-6-85-2002

Sun L, Anderson MC, Gao F, Hain C, Alfieri JG, Sharifi A, McCarty GW, Yang Y, Yang Y, Kustas WP, McKee L (2017) Investigating water use over the C hoptank R iver W atershed using a multisatellite data fusion approach. Water Resour Res 53(7):5298–5319

Tan L, Zheng K, Zhao Q, Wu Y (2021) Evapotranspiration estimation using remote sensing technology based on a SEBAL model in the upper reaches of the Huaihe River basin. Atmos 12:1599. https://doi.org/10.3390/atmos12121599

Tasumi M (2003) Progress in operational estimation of regional evapotranspiration using satellite imagery. PhD thesis. University of Idaho

Tasumi M, Trezza R, Allen RG, Wright JL (2005) Operational aspects of satellite-based energy balance models for irrigated crops in the semi-arid U.S. Irrig Drain Syst 19:355–376. https://doi.org/10.1007/s10795-005-8138-9

Tasumi M, Allen RG, Trezza R (2008) At-surface reflectance and albedo from satellite for operational calculation of land surface energy balance. J Hydrol Eng 13(2):51–63

Twine TE, Kustas WP, Norman JM, Cook DR, Houser P, Meyers TP, Prueger JH, Starks PJ, Wesely ML (2000) Correcting eddy-covariance flux underestimates over a grassland. Agric For Meteorol 103(3):279–300

Wang F, Ma D, Zhao W, Lu Y, Zhou D, Zhang J, Chen L, Huang P (2020) A validation of eddy covariance technique for measuring crop evapotranspiration on different time scales in the North China plain. Can J Soil Sci 101(1):134–146

Yang Y, Qiu J, Zhang R, Huang S, Chen S, Wang H, Luo J, Fan Y (2018) Intercomparison of three two-source energy balance models for partitioning evaporation and transpiration in semiarid climates. Remote Sens 10(7):1149

Yang Y, Shang S (2013) A hybrid dual-source scheme and trapezoid framework–based evapotranspiration model (HTEM) using satellite images: algorithm and model test. J Geophys Res Atmos 118(5):2284–2300

Yu B, Shang S (2020) Estimating growing season evapotranspiration and transpiration of major crops over a large irrigation district from HJ-1A/1B data using a remote sensing-based dual source evapotranspiration model. Remote Sens 12(5):865

Yu B, Shang S, Zhu W, Gentine P, Cheng Y (2019) Mapping daily evapotranspiration over a large irrigation district from MODIS data using a novel hybrid dual-source coupling model. Agric For Meteorol 276:107612

Zhang C, Marzougui A, Sankaran S (2020) High-resolution satellite imagery applications in crop phenotyping: an overview. Comput Electron Agric 175:105584

Zhang B, Liu Y, Xu D, Zhao N, Lei B, Rosa RD, Paredes P, Paço TA, Pereira LS (2013) The dual crop coefficient approach to estimate and partitioning evapotranspiration of the winter wheat–summer maize crop sequence in North China plain. Irrig Sci 31:1303–1316

Zhuang Q, Shao H, Guan D (2022) Operational daily evapotranspiration mapping at field scale based on SSEBop model and spatiotemporal fusion of multi-source remote sensing data. PLoS One 17(2):e0264133

Satellite Imagery in Precision Agriculture

Joel Segarra

Abstract Many national and international institutions recognize that novel agriculture paradigms are needed to address the current challenges of adaptation to and mitigation of climate change. In this sense, digital agriculture and, specifically, satellite images in precision farming allow efficient monitoring of crops to ameliorate the management impacts to the environment. These data allow estimating yields or fertilization requirements, as well as water-related aspects, such as evapotranspiration and crops hydric status. In this chapter, I aim to describe satellite imagery applications in precision agriculture, and I present the nature of remotely sensed data, the types of satellites, and data access, management, and processing in the case of precision farming applications. Landsat 9, Sentinel-2, as well as other commercial satellites orbiting the Earth are described as feature relevant characteristics for agriculture monitoring, especially regarding the visible, near-infrared, and red-edge parts of the spectrum, which can be related to biomass, canopy vigor, or chlorophyll content and subsequently be matched with agronomic features. Regarding data access and coverage, openly accessible datasets and commercial satellites are discussed. Moreover, data management and processing have also been presented in regard to the limitations that processing and analyzing such large amounts of data (i.e., images from vast agricultural regions on a daily basis) has and the potential of cloud computing and processing. I conclude that in industrial agriculture settings, openly accessible satellite imagery can contribute significantly to an overview the status of crops, guide specific and timely actions, and reduce production losses and the impacts on the environment. Satellite imagery has a spatial dimension that can be used at the field to regional level. The assessment of agricultural performance can also be matched to several agroecological and environmental levels; however, satellite imagery in precision farming has several limitations and knowledge gaps in its application in heterogenous and agricultural landscapes with small-scale fields.

J. Segarra (✉)
Plant Physiology Section, Department of Evolutionary Biology, Ecology and Environmental Sciences, Faculty of Biology, University of Barcelona, Barcelona, Catalonia
e-mail: joel.segarra@ub.edu

P. M. Priyadarshan et al. (eds.), *Digital Agriculture*,
https://doi.org/10.1007/978-3-031-43548-5_10

Keywords Remote sensing · Precision farming · Landsat · Sentinel-2 · Multispectral

1 Introduction

Since agriculture started thousands of years ago, humans have shaped the Earth by producing agricultural landscapes in most of the habitats that we live in today (Blondel 2006). Rice terraces in Asia, high-yielding maize hydroponic islands in Mexico (Chinampas), or Mediterranean agroforestry systems have traditionally articulated local agricultural landscapes. The technique and the geographic milieu simultaneously change with human's actions (Santos 2000), and so happens currently with a technique that has gone beyond using local resources and species for farming and has become a global highly technological and productive activity, such as many others in society. Since the mid-twentieth century, the "Green Revolution" transformed traditional agriculture into an industrial system by providing high yielding genotypes, fertilizers, and other chemically derived products, as well as improved machinery for sowing and harvesting. This turning point changed the agricultural paradigm in most areas of the world by improving agricultural yields and reducing the human labor force needed. Yet, the current industrial production system is recognized as a major source of global pollution, and its sustainability is discussed.

The current industrial farming practices, characterized by a generalized use of chemical fertilizers and pesticides together with fossil-fueled machinery, have caused a tremendous negative impact to the environment. Agriculture is responsible for 21.2% of global anthropogenic greenhouse gas emissions when including land-use changes (Tubiello et al. 2015). Hence, a significant contribution to climate change and temperature increases is related with agriculture, and many national and international institutions recognize that novel agriculture paradigms are needed to address the current challenges of adaptation to and mitigation of climate change (Rhodes 2016). In this sense, remote sensing data used in precision framing, such as that obtained with satellite technologies, allow an efficient monitoring of crops by acquiring satellite data. These data allow monitoring yields (Segarra et al. 2020b; Wolanin et al. 2019) or fertilization needs (Nutini et al. 2018), as well as water-related aspects, such as evapotranspiration and irrigation needs (Rozenstein et al. 2018) to ameliorate the agricultural management impacts to the environment. Many satellites orbiting the Earth have relevant characteristics for agriculture monitoring (Segarra et al. 2020a), especially regarding the visible, near-infrared, and red-edge parts of the spectrum, which can be related to biomass, canopy vigor, or chlorophyll content (Gitelson and Merzlyak 1996). These plants' physiological features can be related to agronomic characteristics of crops and drive management decisions. This information is central for crops production, irrigation planification, and yield stability. The role of remote sensing and spatial analysis in adaptation to and mitigation of climate change is certainly relevant (Yang et al. 2013); however, there are still scale and knowledge gaps which need to advance to find adequate management strategies when following these remotely sensed data in precision farming.

Precision farming is an agricultural management paradigm which is based on observing, measuring, and responding to crops' temporal and spatial variability with the aim to improve agricultural production sustainability (Cisternas et al. 2020; Zhang et al. 2002). Hence, this paradigm goes beyond the classical "Green Revolution" framework, and the use of agricultural inputs is optimized regarding crop demands and its impacts to the agro-environment. The multispectral sensors mounted on board of satellites have resolution features (Adams and Gillespie 2006) which can be used to determine crops' physiological and agronomic characteristics. The temporal and spatial variability of crops is central for its monitoring in precision farming, and it suits satellites resolution features. Satellites have a temporal resolution in the sense that an orbiting satellite has a specific period in which it returns to the same geographic area after orbiting the Earth, and the revisit time is central to follow the emergence of crops or features related to phenology and the evolution of the crop. Moreover, the spatial resolution of satellites is central to precision farming as agriculture fields and landscapes are generally heterogenous and meet singularities within the field that can be observed if the spatial resolution of the satellite is enough to differentiate certain objects. The spectral resolution refers to the number of bands and the width, namely, the parts of the reflected spectrum that satellite sensors can capture and the spectral resolution allows monitoring different physiological characteristics of the plant. Finally, the radiometric resolution refers to how much information the satellite sensors can capture. All these sensor features are central for understanding the corresponding plants' agronomic and physiological characteristics.

This chapter presents satellite imagery use in precision farming, namely, it is focused on understanding the nature of satellite data and match it with farming. Moreover, we present how data can be accessed and different data management approaches. Finally, we discuss the advantages and limitations of using satellite data for agriculture and the implications it has regarding global sustainability and planetary boundaries. This chapter mainly focuses on passive remote sensing, comprehending optical and thermal spheres. Active remote sensing, those sensors using radar and other active technologies are presented but do not occupy a central part of this section, reviews on the use of active sensors in agriculture can be found elsewhere (McNairn and Shang 2016). The novelty of remote sensing applications in agriculture at multiple levels (Weiss et al. 2020) or for specific satellites such as Sentinel-2 (Segarra et al. 2020a) have been addressed in the last years with increasing interest. In this chapter, I present a general overview of satellite data applications in agriculture and how these data can be accessed and managed to finally discuss the advantages and limitations of satellites application in agriculture regarding the multiscale framework of digital agriculture for a sustainable food production.

2 Satellite Data and Farming

Satellite remote sensing is a technological field which senses Earth surfaces using multispectral, hyperspectral, and other instruments mounted on satellites orbiting the Earth. These Earth observation systems are available as a diverse array of sensors and, in regard of the source of illumination used on the sensed objects, can be divided between passive and active sensors. Passive sensors, such as optical and thermal systems, rely on reflected sunlight or emitted thermal energy. Passive multispectral sensors can acquire data beyond the visible wavelengths (i.e., infrared and thermal wavelengths) across the electromagnetic spectrum (Lechner et al. 2020). Earth surfaces reflect and absorb sunlight at different wavelengths; these differences in spectral reflectance properties (i.e., spectral signatures) work as distinct fingerprints to differentiate surface types (Shaw and Burke 2003) which allow, for instance, identifying different crop types. Active sensors, meanwhile, emit a pulse and measure the backscatter reflecting to the sensor. Such sensors can penetrate clouds and operate at night. Active sensors such as SAR (synthetic aperture radar) on board of Sentinel-1 can differentiate crops features according to their surface roughness and the three-dimensional structure of the targets (d'Andrimont et al. 2021; Ndikumana et al. 2018). Other active sensors such as LiDAR (light detection and ranging or laser imaging, detection, and ranging) systems emit a pulse from lasers and measure distance to a target and the reflected light; differences in laser return times and wavelengths can then be used for making digital three-dimensional representations of the target (Lechner et al. 2020). Satellites from the European Space Agency (ESA), such as ADM-Aeolus, has a LiDAR mounted on board, although its use is not focused on agriculture. Meanwhile, NASA (National Aeronautics and Space Administration) has several satellites with LiDAR such as ICESat-2 (Ice, Cloud, and Land Elevation Satellite-2), which is used to monitor vegetation across the globe and determine vegetation structure, also with some potential applications for farming (Brown et al. 2023); in 2019, ICESat-2 data was made available (Martino et al. 2019).

Remote sensing satellite sensors feature a trade-off between the spatial, temporal, and spectral resolutions (Shen et al. 2016). Spatial resolution refers to the pixel size, which is very relevant as the spatial dimension allows differentiating objects within the Earth surface. The temporal resolution is the frequency with which satellite images of the same area are taken, that is, the time it takes for the sensor to revisit the same location on Earth. This depends on the features of the satellite and the mission itself; while some satellites are single devices, others are a constellation of them, such as Sentinel-2 A+B which is a constellation of two twin satellites (A and B) and therefore synchronically orbit the Earth increasing temporal resolution with relevant applications in precision farming (Segarra et al. 2020a). Spectral resolution is also relevant, and optical sensors vary in terms of the number of bands (and the widths of those bands) from which data are captured. The spectral resolution allows to extract more accurate information on the sensed surface as several parts of the

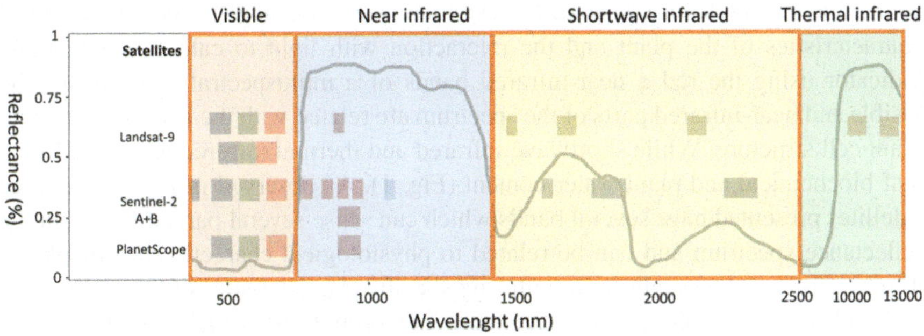

Target trait	Spectral Information	Applications
Field vegetation cover and greenness	Visible and near infrared	Stress detection Canopy cover Leaf area index Growth dynamics Senescence Crop greenness Agronomic and yield traits Plant emergence Phenology Leaf nitrogen content
Chlorophyll content	Visible, near infrared, and red edge	Crop greenness Stress detection Chlorophyll content Leaf nitrogen content Grain nitrogen content
Photosynthesis and yield	Visible and near infrared	Photosynthetic status Biomass and yield Senescence Chlorophyll content Stress detection
Water content	Shortwave infrared and thermal infrared	Water status Evapotranspiration Stress detection

Fig. 1 Crop reflectance signature with the bands available for the satellite sensors (Landsat 9, Sentinel-2, and PlanetScope). The spectral regions are indicated, and the agronomic and physiological traits described in relation with its application in agriculture and plant monitoring

reflected spectrum can be detected. As an example, Sentinel-1 SAR has a six-day revisit period at a high spatial resolution of about 20 m.

In Fig. 1, the spectral signature of a crop is shown together with satellites Sentinel-2, Landsat 9, and PlanetScope. The physiological characteristics of the crop vegetation cover are appreciated in the reflectance spectrum in the sense that in the visible green parts of the spectrum the electromagnetic radiation is reflected, while in the blue and red areas the reflectance is inferior as the absorption of sunlight by the chlorophylls to carry out the photosynthetic activity happens in these regions of the spectrum. Moreover, in the area between red and near infrared, the so-called red edge, the reflectance increases greatly as in wavelengths over 700 nm the energy of the photon is not sufficient to synthesize organic molecules (Taiz and Zeiger 2015), and it is hence highly reflected. The differences between photosynthetic active regions (between 400 and 700 nm) and the near infrared allows understanding the status of the vegetation cover, the biomass, and the photosynthetic activity. It is the case of the widely used vegetation index NDVI (normalized difference vegetation index) (Rouse Jr. et al. 1974) which takes advantage of the physiological characteristics of the plant and the interaction with light to calculate a biomass indicator using the red a near-infrared bands of a multispectral instrument. The visible and near-infrared parts of the spectrum are related with the leaf pigments and plant cell structure. While shortwave infrared and thermal infrared are related with leaf biochemical and plant water content (Fig. 1). As observed in Fig. 1, the three satellites presented have several bands which can sense several parts throughout the reflectance spectrum and can be related to physiological characteristics of plants. The spectral resolution allows monitoring specific characteristics of the crops.

Field vegetation cover and greenness are crop traits which can be sensed with visible and near infrared spectral information obtained from multispectral instruments (Gracia-Romero et al. 2017). In this sense, this data can be used to detect plant stress (both biotic and abiotic), to assess the canopy cover as well as to understand growth dynamics or phenology. Moreover, the chlorophyll content can be monitored with mainly green and red-edge bands, specially the red-edge band is very relevant to monitor chlorophyll content which can also be used as a proxy for the nitrogen status of the plant (Segarra et al. 2022b). The photosynthesis activity of the crop is directly linked to the yield as it is the source of organic matter for the plant (Sanchez-Bragado et al. 2014); hence, understanding this activity through satellite imagery allows developing yield estimation models which can be useful for both prediction of final yield and guiding management action to stabilize the potential final yield. The shortwave infrared and the thermal infrared are especially relevant regarding the water status of the plant (Guan et al. 2017). The water that the plant needs to grow can be monitored with the evapotranspiration which is a balance of the water transpired through the plants' stomata during photosynthesis plus the evaporation of the water in the plant and soil surfaces within the agricultural fields in this case.

Generally, as shown in Table 1, the spatial resolution of thermal bands obtained from satellites is coarse. Sentinel-3 provides 1 km spatial resolution thermal data, while Landsat 9 provides 100 m resolution thermal bands. These resolutions do not provide sufficient precise information to understand at the field level, for instance, the water status of a crop and drive the management decision of the farmer, namely, applying the precision farming framework in the case of irrigation. However, the combination of other satellite spectral information such as higher-resolution 10 to 20 m Sentinel-2 bands allows fine-scaling some thermal remotely sensed data and obtain higher-resolution evapotranspiration products such as Sen-ET (https://www.esa-sen4et.org/) which resamples 1 km pixels into 20 m by combining it with Sentinel-2 higher-resolution images. A few decades ago, estimates of crop water demand from Landsat satellite data (Allen et al. 2005) were already addressed, however, for a regional level. The combination of thermal and multispectral visible and near-infrared satellite-based imagery to empirically solve surface energy balance equations and provide estimates of crop actual evapotranspiration from fractional vegetation cover and latent heat flux is almost operational currently for precision agriculture with the 20 m evapotranspiration grids available through the Sen-ET plugin from the ESA.

Table 1 Satellite missions, data type, and characteristics

Sensor type	Satellite name	Data type	Revisit capability and spatial resolution
Passive	Sentinel-2 A+B	Visible and multispectral	Every five days, pixels of 10 to 20 m size (archives since 2015)
	Sentinel-3	Thermal	1 km of spatial resolution
	Landsat 8 and 9	Visible, thermal, and multispectral	Every 15 days pixels of 15 to 30 m size, thermal 100 m (archives since 2013)
	Landsat 1, 2, 3, 4, 5, 6, 7, 8 and 9	Visible, thermal, and multispectral	Archives available since 1972, ongoing active missions Landsat 8 and 9
	PlanetScope	Visible and multispectral	Scenes taken daily, high-resolution images below 5 m
	WorldView-3	Visible and multispectral	Scenes taken daily, spatial resolution of 0.34 to 4.1 m
	Pléiades 1A/1B	Visible and near infrared	Scenes taken daily, spatial resolution of 0.5 m
	Amazônia-1	Visible and near infra-red	Every five days, pixels of 60 m
	Cartosat	Visible	Every five days, archives available since 2005 with Cartosat-1, current Cartosat-3 has a 0.25 m spatial resolution
Active	Sentinel-1	SAR (radar)	Every six days, 20 m of spatial resolution
	ICESat-2	LiDAR	1 km spatial resolution and 90-day revisit time

For the case of grain yield or nitrogen status monitoring, the applicability of satellite data in precision farming is more advanced. In this sense, some studies have addressed the use of Sentinel-2 images to map grain yield within field variability at 10 to 20 m resolution (Cavalaris et al. 2021; Hunt et al. 2019; Segarra et al. 2022a). These products take advantage of the several elements used in the digital agriculture paradigm such as geolocated combine harvesters, which allow obtaining the harvested grain, for instance, with a geolocated reference. In contrast with obtaining single field values on the yield, or carrying out crop cuts by researchers, the combination of these technological advances allows creating high-resolution within-field performance maps. Moreover, beyond the performance maps themselves, the spectral data obtained from satellites and its relationship with physiological characteristics of the plant can determine the logic behind higher and lower yielding areas within a field. Whether the water status, the emergence of the crop, or the nitrogen status, just to mention some, are the reasons behind having lower-yielding areas in a field, they can be understood by linking the reflectance characteristics with the actual understanding of the plant physiology, either by using vegetation indices as proxies or biophysical variables obtained from more complex radiative transfer models such as those developed by (Weiss and Baret 2016) in the case of Sentinel-2 images.

Hence, the capacity of satellite data is not limited to generate within field actual agronomic grain yield or nitrogen status maps but can moreover be used to assess the physiological features of the plant hindering the specific limitation of the crop to subsequently guide specific management decisions. The vast dimension and potentialities for the use of satellite imagery for precision agriculture could be totally deployed when hyperspectral openly available satellite data can be accessed for agricultural monitoring. At UAV (unmanned aerial vehicle) and aircraft level, such hyperspectral data have been used in biotic stress monitoring in olive groves (Poblete et al. 2021) or grain nitrogen status monitoring in wheat (Zhao et al. 2019). However, these demonstrations of the potentialities of remote sensing do not represent the operationality of precision farming. Mainly due to the lack of general availability of data and the intrinsic cost of many of these devices which make it operationally unlikely for most farmers to use them. However, on the scientific basis and future application, these pathways are of pivotal interest.

Beside hyperspectral instruments, which are in the frontier regarding agricultural applications, high-resolution multispectral instruments on board of commercial satellites capture images with potential applications in precision farming. As shown in Table 1, PlanetScope, WorldView-3, or Pléiades 1A/1B provide daily high-resolution images. Pléiades 1A/1B are a constellation of two satellites which have very-high optical resolution (0.5 m resolution); the satellites have four bands: the red–green–blue (RGB) visible bands and near infrared. Meanwhile, Worldview-3 multispectral instrument collects images at 0.31 m panchromatic (RGB) and 1.24 m in the eight near infrared bands, 3.7 m in the eight shortwave near infrared bands, and a 30 m resolution in the clouds, aerosols, vapors, ice, and snow bands. WorldView-3 has bands for enhanced multispectral analysis (coastal blue, yellow, and red edge) designed to improve segmentation and classification of land features, such as agricultural production. In this sense, several authors have used WorldView high-resolution images in agriculture monitoring, such as for segmenting olive tree crowns (Solano et al. 2019) or macadamia trees (Johansen et al. 2020), which needs a resolution that Sentinel-2 and Landsat 9 do not have. PlanetScope multispectral instruments, on board of the three orbiting satellites that constitute the constellation, operate currently in eight bands: red edge, red, green (2), yellow, blue, coastal blue, and near infrared. A PlanetScope RGB scene is shown in Fig. 2 together with a Sentinel-2 RGB scene; as observed, the delineation of the agricultural fields has a higher resolution in the PlanetScope image (below 5 m spatial resolution) than in the Sentinel-2 images (10 m spatial resolution). In both cases, nonetheless, the agricultural fields can be clearly observed. Even within-field variability can be visually assessed in the case of the RGB scene, other parts of the spectrum, in the case of these two satellites shown, cannot be sensed with the current sensors or the resolution is too coarse, as discussed, for example, in the case of thermal bands before. Regarding the findings of (Skakun et al. 2021), by comparing several spatial resolution of satellite imagery, they observed that spatial resolution of below 3 m is critical to explaining 100% of the within-field yield variability for corn and soybean.

Fig. 2 RGB composites of Sentinel-2 and PlanetsScope scenes at different spatial resolution (10 m and 5 m, respectively), coordinates of the scene (N 10.215075 E -71.943003; decimal degree WGS84)

The results also showed that moving to coarser resolution data of 10 m, 20 m, and 30 m reduced the explained variability to 86%, 72%, and 59%, respectively. I continue by analyzing data accessibility, management, and processing for the case of satellite imagery for precision farming.

3 Data Access, Management, and Processing

The access to satellite data is an important aspect to consider. Some missions such as those from NASA and ESA provide accessibility to archives when logging in with a user, as well as other mission with limited satellite data availability and coverage such as Brazilian and Indian space missions. In Table 2, the accessibility to several satellites is presented. The Copernicus mission archives can be accessed through the Copernicus Open Access Hub (https://scihub.copernicus.eu/); this provides complete, free, and open access to Sentinel-1, Sentinel-2, and Sentinel-3 products. On the Copernicus Open Access Hub, a user-friendly platform allows defining the regions of interests and downloading the satellite imagery directly from the previous year and on demand from previous years as data need to be restored from the archives. Regarding NASA, on the US Geological Survey site (https://earthexplorer.usgs.gov/), Landsat archives are available since 1972 until the current ongoing active missions Landsat 8 and 9. Moreover, other satellites such as MODIS (Moderate Resolution Imaging Spectroradiometer) are available but their spatial resolution is not suitable for the case of precision farming, and it is rather used in ecosystems monitoring. Other missions from NASA such as LiDAR ICESat-2 can be accessed elsewhere (https://openaltimetry.org/data/icesat2/), albeit its processing needs more complex transformations and its use is not specifically intended to agriculture as its spatial resolution of 1 km is limiting. Commercial satellites have their own platforms where scenes can be purchased and downloaded. In Table 2, the access of several satellites is described. The European Union provides

Table 2 Access to satellite data

Satellite name	Access and coverage	References
Sentinel-2 A+B	Publicly accessible (European Commission and European Space Agency), global	https://scihub.copernicus.eu/
Sentinel-3		
Sentinel-1		
Landsat 8 and 9	Publicly accessible (NASA, USA), global	https://earthexplorer.usgs.gov/
Landsat 1, 2, 3, 4, 5, 6, 7, 8 and 9		
PlanetScope	Private, global on-demand	https://www.planet.com/nicfi/
WorldView-2,3,4	Private, global on-demand	https://www.maxar.com/worldview-legion
Amazônia-1	Publicly accessible (Brazilian Space Agency), global theoretically but on the catalog only scenes in South America are available	http://www2.dgi.inpe.br/catalogo/explore
Cartosat	Publicly accessible (Indian Space Research Organization), scenes only cover India subcontinent	https://bhuvan-app3.nrsc.gov.in/data/download/index.php

access to some scenes already purchased for European programs and to archives of PlanetScope and WorldView; however, it is only intended for research institutions and innovative projects, which need to be justified (https://earth.esa.int/eogateway/catalog/worldview-3-full-archive-and-tasking).

The accessibility to these data is central for precision farming. In this sense, besides some exceptions made in research or conservation initiatives, private satellites such as WorldView-3 or PlanetScope offer expensive services that capture high-resolution images on demand. For most farmers on Earth, cooperatives, and even small to middle companies, these data are far from their economic capacities. Hence, by understanding that precision farming involves observing, measuring, and responding to crops' temporal and spatial variability and sustainability, one recognizes the importance of open accessibility to satellite data in this agriculture paradigm. Moreover, as most research institutions cannot afford these images and the research carried out with these data is not always reproducible (due to copyrights on the data and paywalls to access it), the potentialities of high-resolution satellite imagery in precision farming cannot be fully deployed.

Meanwhile, publicly accessible satellite data, such as Sentinel-2, features many studies and applications due to the open access nature of the data. Although the spatial resolution of 10 m cannot explain all the variability within an agricultural field, the resolution of the satellite makes it almost fully operational for its use in precision farming (especially in regions with standardized agricultural managements). Regarding its access, there are several ways to freely and openly download Sentinel-

2 imagery; one of them is the direct download of the imagery from ESA's website, through Copernicus (https://scihub.copernicus.eu/dhus/#/home), as mentioned. Furthermore, third-party tools for downloading the imagery are available. For instance, there is the US Geological Survey (USGS) (https://earthexplorer.usgs.gov/) which allows comprehensive searching and downloading of full-resolution Sentinel-2 images as well as Landsat archives. On the open-source software QGIS, there are various plug-ins that take advantage of the ESA's Application Programing Interface at the Copernicus Open Access Hub (https://scihub.copernicus.eu/apihub/). Furthermore, Google Earth Engine has daily updated copies of all the available Sentinel-2, Landsat, MODIS, and other accessible satellite data and provides both access to this data repository along with high processing capacity using their image processing servers. Many other similar tools and services exist on other software applications and web portals and are being developed continuously.

National and international agencies such as ESA or NASA provide specific access tools, algorithms, and software in support of the use and processing of their satellites, such as the Sentinel-2 Toolbox within the Sentinel Application Platform (SNAP, https://sentinel.esa.int/web/sentinel/toolboxes/sentinel-2), which can be used for agriculture monitoring. Besides vegetation indices, more deterministic biophysical parameters, such as LAI (leaf area index) or FAPAR (fraction of absorbed photosynthetically active radiation), can be calculated on SNAP with Sentinel-2 data following the algorithms developed by (Weiss and Baret 2016). Most of these algorithms ready for the user are developed with thousands of training and validation points and follow complex inverse radiative transfer models. The availability of these models already developed improved the capacities to take most from satellite information.

Another key point is data processing. For most research teams, farmers, or local agricultural companies, the computing capacity to operate large datasets is limited, especially when requiring visual interpretation of imagery and heavy processing. In the next few years, data accessibility will likely be widespread, including images from high-resolution satellites with increasing processing capacity demands. Currently, data acquisition is no longer a major challenge with Landsat and Sentinel-2 archives; instead, it is the capacity to process and analyze such large amounts of data (i.e., images from vast agricultural regions on a daily basis) that is becoming the bottleneck. In this sense, besides the features of satellites and the data that can be obtained, the large amount of data and its potential use has generated commercial analytically oriented initiatives such as Google Earth Engine (Gorelick et al. 2017) or EarthServer (Baumann et al. 2016) that process these data on high-capacity cloud servers. In this sense, RUS (Research and User Support) virtual machines from the ESA also offer high storage and processing capacities on cloud servers (unfortunately only accessible to European-based institutions).

Besides the technological advancements in satellite remote sensing, a central aspect when working with its applications in agriculture is modelling crops development and forecasting agronomic variables (i.e., yield, quality traits). Advanced models regarding artificial intelligence and machine learning, as a general frame, have shown considerable promise in agricultural remote sensing applications

(Chlingaryan et al. 2018). These computer algorithms are particularly useful for studying complex biological systems, as they can capture complex interactions among variables and find generalizable predictive patterns (Bzdok et al. 2018), which can eventually be useful in guiding agricultural management decisions and can take the most of the data obtained from the agricultural fields. The use of machine learning to retrieve crop performance has been considered one of the most important areas to develop associated with remote sensing and agriculture (Weiss et al. 2020).

4 Advantages and Limitations of Satellites Use in Precision Farming

The advantages of using publicly available satellite data for precision farming are multiple: having up-to-date crop-related data, having an overview of the status of crops, guiding specific and timely actions, reducing production losses and the impacts to the environment, and achieving a sustainable production. An example of satellite data potentialities being deployed in precision farming is the Belgian WatchITgrow platform (Curnel 2017). This platform uses Sentinel-2 data and algorithms developed by national research institutions in partnership with other administrations and farming enterprises to monitor potato production in Belgium at the field level. The farmers using the platform can access the information collected from the satellites and the products generated (potential yield maps, nitrogen status, etc.). Moreover, the data is secured for each user, and it is intended to improve the management of fields and is not sold to other enterprises. Namely, the data of the user always remain property of the user. The public agricultural institutions of the country and research institutes together with farmers and other agricultural enterprises can lead the creation of accessible platforms to guide specific precision farming management decision, such as in Belgium and the platform WatchITgrow to monitor potato production with Sentinel-2 data.

Satellite data applications in precision farming present significant potentialities in standardized and relatively homogenous agricultural settings, such as those common in industrial agriculture. However, most farming activities in the Earth are carried out in relatively heterogenous agricultural landscapes, with polycultures, trees, and herbaceous crops being simultaneously grown within the field, and relatively small fields (Altieri and Nicholls 2017). In such agricultural settings, in contrast with middle-resolution satellites, high-resolution satellites can best capture the variability within the fields and give a significant overview of crops status to guide the management. It is true, however, that Sentinel-2 imagery has been used in assessing heterogenous and diverse agricultural landscapes, such as in the case of Mali (Lambert et al. 2018), with relatively promising results. Nonetheless, higher-resolution WorldView scenes were also used to map trees within the field and clear pixels for an improved assessment of field's main crop.

The current availability of satellite data, which has significant restrictions to high-resolution scenes, is an important limitation for heterogenous agricultural landscapes. Such landscapes are often located in low-income countries in which securing yield and optimizing the use of inputs is pivotal. Moreover, in these regions, the complexity of interactions in agricultural production makes it difficult to monitor many variables. For instance, in the case of monitoring biotic stresses at regional level and guiding specific field-level management approaches, Buchaillot et al. (2022) observed that PlanetScope high-resolution images offered greater benefits in contrast with Sentinel-2 imagery. However, the complexity on using these scenes is also presented, especially due to the polycultures and diverse management in the fields, as well as the heterogenous agricultural mosaic present in Southern and Eastern Africa. Therefore, in heterogeneous agricultural landscapes, precision farming has several limitations regarding the sensing of the actual crops in the field, the sizes of the fields, as well as the management approaches, and the resources that farmers have in order to address them. In this sense the multiscale approach of digital agriculture together with the understanding of agroecological dynamics (many retrievable with remote sensing data) together with high-resolution satellite data can support the application of precision farming in such agro-environments.

5 Conclusion

In summary, in this section, I have presented satellite imagery use in precision agriculture. After defining the nature of satellite multispectral data, I have linked it with plants physiological and agronomic characteristics for its application in precision farming. Moreover, several satellites have been presented, regarding data access and coverage, as well as their resolution features, which are pivotal to understanding the data needed for its application in precision farming. Data management and processing have also been presented in regard with the limitations that processing and analyzing such large amounts of data (i.e., images from vast agricultural regions on a daily basis) have. In this sense, high-processing capacity cloud servers, such as Google Earth Engine, have been introduced.

I conclude that paywalls to high-resolution satellite data limit the application of precision farming in heterogenous agricultural landscapes, which are especially present in low-income countries. In contrast with the potentialities that satellite imagery uses in precision farming have in standardized agricultural settings, in heterogenous agro-environments, the variability of crops within field level is not easily retrievable with current publicly accessible data. Meanwhile, I conclude that in industrial agriculture settings, openly accessible satellite imagery can contribute significantly to overview the status of crops, guide specific and timely actions, and reduce production losses and the impacts to the environment. Satellite imagery has a spatial dimension which covers field to regional levels; moreover, the assessment of agricultural performance can be matched with several agroecological and environmental levels. Hence, satellite imagery plays a pivotal role in the multiscale framework of digital agriculture for a sustainable food production.

Acknowledgments I acknowledge the support of the project PID2019-106650RB-C21 from the Ministerio de Ciencia e Innovacion, Spain. I am a recipient of a FPI fellowship from the same institution (Grant PRE2020- 091907).

References

Adams JB, Gillespie AR (2006) Remote sensing of landscapes with spectral images: a physical modeling approach. In: Remote sensing of landscapes with spectral images: a physical modeling approach, pp 1–362. https://doi.org/10.1017/CBO9780511617195

Allen RG, Tasumi M, Morse A et al (2005) A landsat-based energy balance and evapotranspiration model in Western US water rights regulation and planning. Irrig Drain Syst 19(3–4). https://doi.org/10.1007/s10795-005-5187-z

Altieri MA, Nicholls CI (2017) The adaptation and mitigation potential of traditional agriculture in a changing climate. Clim Chang 140(1). https://doi.org/10.1007/s10584-013-0909-y

Baumann P, Mazzetti P, Ungar J et al (2016) Big data analytics for earth sciences: the EarthServer approach. Int J Digit Earth 9(1):3–29. https://doi.org/10.1080/17538947.2014.1003106

Blondel J (2006) The 'design' of Mediterranean landscapes: a millennial story of humans and ecological systems during the historic period. Hum Ecol 34(5):713–729. https://doi.org/10.1007/s10745-006-9030-4

Brown ME, Arias SD, Chesnes M (2023) Review of ICESat and ICESat-2 literature to enhance applications discovery. Remote Sens Appl Soc Environ 29:100874. Elsevier. https://doi.org/10.1016/J.RSASE.2022.100874

Buchaillot ML, Cairns J, Hamadziripi E et al (2022) Regional monitoring of fall armyworm (FAW) using early warning systems. Remote Sens 14(19):5003. https://doi.org/10.3390/rs14195003

Bzdok D, Altman N, Krzywinski M (2018) Points of significance: statistics versus machine learning. Nat Methods 15(4):233–234. Nature Publishing Group. https://doi.org/10.1038/nmeth.4642

Cavalaris C, Megoudi S, Maxouri M et al (2021) Modeling of durum wheat yield based on sentinel-2 imagery. Agronomy 11(8). https://doi.org/10.3390/agronomy11081486

Chlingaryan A, Sukkarieh S, Whelan B (2018) Machine learning approaches for crop yield prediction and nitrogen status estimation in precision agriculture: a review. Comput Electron Agric 151(May):61–69. Elsevier. https://doi.org/10.1016/j.compag.2018.05.012

Cisternas I, Velásquez I, Caro A et al (2020) Systematic literature review of implementations of precision agriculture. Comput Electron Agric 176:105626. Elsevier. https://doi.org/10.1016/J.COMPAG.2020.105626

Curnel Y (2017) Watch it grow, an innovative platform for a sustainable growth of the Belgian potato production. FACCE MACSUR reports

d'Andrimont R, Verhegghen A, Lemoine G et al (2021) From parcel to continental scale – a first European crop type map based on Sentinel-1 and LUCAS Copernicus in-situ observations. Remote Sens Environ 266(October). https://doi.org/10.1016/j.rse.2021.112708

Gitelson AA, Merzlyak MN (1996) Signature analysis of leaf reflectance spectra: algorithm development for remote sensing of chlorophyll. J Plant Physiol 148(3–4):494–500. https://doi.org/10.1016/S0176-1617(96)80284-7

Gorelick N, Hancher M, Dixon M et al (2017) Google earth engine: planetary-scale geospatial analysis for everyone. Remote Sens Environ 202:18–27. Elsevier. https://doi.org/10.1016/J.RSE.2017.06.031

Gracia-Romero A, Kefauver SC, Vergara-Díaz O et al (2017) Comparative performance of ground vs. aerially assessed rgb and multispectral indices for early-growth evaluation of maize performance under phosphorus fertilization. Front Plant Sci 8(November):1–13. https://doi.org/10.3389/fpls.2017.02004

Guan K, Wu J, Kimball JS et al (2017) The shared and unique values of optical, fluorescence, thermal and microwave satellite data for estimating large-scale crop yields. Remote Sens Environ 199:333–349. https://doi.org/10.1016/j.rse.2017.06.043

Hunt ML, Blackburn GA, Carrasco L et al (2019) High resolution wheat yield mapping using Sentinel-2. Remote Sens Environ 233(December 2018):111410. Elsevier. https://doi.org/10.1016/j.rse.2019.111410

Johansen K, Duan Q, Tu Y-H et al (2020) Mapping the condition of macadamia tree crops using multi-spectral UAV and WorldView-3 imagery. ISPRS J Photogramm Remote Sens 165:28–40. https://doi.org/10.1016/j.isprsjprs.2020.04.017

Lambert MJ, Traoré PCS, Blaes X et al (2018) Estimating smallholder crops production at village level from Sentinel-2 time series in Mali's cotton belt. Remote Sens Environ 216(June):647–657. Elsevier. https://doi.org/10.1016/j.rse.2018.06.036

Lechner AM, Foody GM, Boyd DS (2020) Applications in remote sensing to forest ecology and management. One Earth 2(5):405–412. Cell Press. https://doi.org/10.1016/J.ONEEAR.2020.05.001

Martino AJ, Neumann TA, Kurtz NT et al (2019) ICESat-2 mission overview and early performance. In: Neeck SP, Kimura T, Martimort P (eds) Sensors, systems, and next-generation satellites XXIII. SPIE, p 11. https://doi.org/10.1117/12.2534938

McNairn H, Shang J (2016) A review of multitemporal synthetic aperture radar (SAR) for crop monitoring. Remote Sens Digit Image Process. https://doi.org/10.1007/978-3-319-47037-5_15

Ndikumana E, Ho Tong Minh D, Baghdadi N et al (2018) Deep recurrent neural network for agricultural classification using multitemporal SAR Sentinel-1 for Camargue. France Remote Sens 10(8):1217. https://doi.org/10.3390/rs10081217

Nutini F, Confalonieri R, Crema A et al (2018) An operational workflow to assess rice nutritional status based on satellite imagery and smartphone apps. Comput Electron Agric 154(February):80–92. Elsevier. https://doi.org/10.1016/j.compag.2018.08.008

Poblete T, Navas-Cortes JA, Camino C et al (2021) Discriminating Xylella fastidiosa from Verticillium dahliae infections in olive trees using thermal- and hyperspectral-based plant traits. ISPRS J Photogramm Remote Sens 179. https://doi.org/10.1016/j.isprsjprs.2021.07.014

Rhodes CJ (2016) The 2015 Paris climate change conference: COP21. Sci Prog 99(1):97–104. https://doi.org/10.3184/003685016X14528569315192

Rouse JW Jr, Haas R, Schell J et al (1974) Monitoring vegetation systems in the great plains with ERTS, vol 351(309). NASA Special Publication

Rozenstein O, Haymann N, Kaplan G et al (2018) Estimating cotton water consumption using a time series of Sentinel-2 imagery. Agric Water Manag 207(June):44–52. https://doi.org/10.1016/j.agwat.2018.05.017

Sanchez-Bragado R, Molero G, Reynolds MP et al (2014) Relative contribution of shoot and ear photosynthesis to grain filling in wheat under good agronomical conditions assessed by differential organ δ13C. J Exp Bot 65(18):5401–5413. https://doi.org/10.1093/jxb/eru298

Santos M (2000) La naturaleza del espacio. Ariel geografía

Segarra J, Buchaillot ML, Araus JL et al (2020a) Remote sensing for precision agriculture: sentinel-2 improved features and applications. Agronomy (Precision Agriculture):1–18

Segarra J, González-Torralba J, Aranjuelo ´i et al (2020b) Estimating wheat grain yield using Sentinel-2 imagery and exploring topographic features and rainfall effects on wheat performance in Navarre. Spain Remote Sens 12(14):1–24. https://doi.org/10.3390/rs12142278

Segarra J, Araus JL, Kefauver SC (2022a) Farming and earth observation: Sentinel-2 data to estimate within-field wheat grain yield. Int J Appl Earth Obs Geoinf 107. https://doi.org/10.1016/j.jag.2022.102697

Segarra J, Rezzouk FZ, Aparicio N et al (2022b) Multiscale assessment of ground, aerial and satellite spectral data for monitoring wheat grain nitrogen content. Inf Process Agric. Elsevier. https://doi.org/10.1016/J.INPA.2022.05.004

Shaw GA, Burke HK (2003) Spectral imaging for remote sensing. Lincoln Lab J

Shen H, Meng X, Zhang L (2016) An integrated framework for the spatio-temporal-spectral fusion of remote sensing images. IEEE Trans Geosci Remote Sens 54(12):7135–7148. Institute of Electrical and Electronics Engineers Inc. https://doi.org/10.1109/TGRS.2016.2596290

Skakun S, Kalecinski NI, Brown MGL et al (2021) Assessing within-field corn and soybean yield variability from WorldView-3, planet, Sentinel-2, and Landsat 8 satellite imagery. Remote Sens 13(5):872. https://doi.org/10.3390/rs13050872

Solano F, Di Fazio S, Modica G (2019) A methodology based on GEOBIA and WorldView-3 imagery to derive vegetation indices at tree crown detail in olive orchards. Int J Appl Earth Obs Geoinf 83:101912. https://doi.org/10.1016/j.jag.2019.101912

Taiz L, Zeiger E (2015) Plant physiology, 6th edn

Tubiello FN, Salvatore M, Ferrara AF et al (2015) The contribution of agriculture, forestry and other land use activities to global warming, 1990–2012. Glob Chang Biol 21(7):2655–2660. https://doi.org/10.1111/gcb.12865

Weiss M, Baret F (2016) S2toolbox level 2 products: Lai, Fapar, Fcover. Avignon

Weiss M, Jacob F, Duveiller G (2020) Remote sensing for agricultural applications: a meta-review. Remote Sens Environ 236:111402. Elsevier. https://doi.org/10.1016/J.RSE.2019.111402

Wolanin A, Camps-Valls G, Gómez-Chova L et al (2019) Estimating crop primary productivity with Sentinel-2 and Landsat 8 using machine learning methods trained with radiative transfer simulations. Remote Sens Environ 225(March):441–457. Elsevier. https://doi.org/10.1016/j.rse.2019.03.002

Yang J, Gong P, Fu R et al (2013) The role of satellite remote sensing in climate change studies. Nat Clim Chang 3(10):875–883. https://doi.org/10.1038/nclimate1908

Zhang N, Wang M, Wang N (2002) Precision agriculture – a worldwide overview. Comput Electron Agric 36(2–3):113–132. Elsevier. https://doi.org/10.1016/S0168-1699(02)00096-0

Zhao H, Song X, Yang G et al (2019) Monitoring of nitrogen and grain protein content in winter wheat based on Sentinel-2A data. Remote Sens 11(14):1–25. https://doi.org/10.3390/rs11141724

Applications of UAVs: Image-Based Plant Phenotyping

Suchitra M. Patil, Sunita Choudhary, Jana Kholova, Magesh Chandramouli, and Adinarayana Jagarlapudi

Abstract Plant phenotyping plays an important role in the qualitative and quantitative assessment of plant growth in its growth environment. Traditional data-collection process used in plant breeding applications is mostly manual, time-consuming, labor-intensive, and highly subjective. Recent advancements in imaging sensors and platforms have significantly enhanced the speed and precision of image-based automated high-throughput plant phenotyping (HTPP). Current automated HTPP is mostly done in controlled environment where the plants are moved to phenotyping platforms. Such technologies are not feasible for open-field phenotyping. Satellite-based remote sensing has been used from decades but is not much effective in small-scale field phenotyping. Nowadays, satellite imagery with good resolution of up to few centimeters (~10–50 cm) is available, but due to fixed revisit time its temporal resolution is still limited. For crops' trait estimation, high spatial, spectral, and temporal resolutions are mandatory. On the other hand, unmanned aerial vehicle (UAV)-drone-assisted image-based HTPP is current state-of-the-art for open-field phenotyping, and is known for providing data with high spatiotemporal resolution, with wide coverage in shorter duration. UAV (drone)-assisted HTPP is used in quantitative phenotyping for traits like plant height, biomass, and leaf area index,

S. M. Patil (✉) · A. Jagarlapudi
Centre of Studies in Resources Engineering (CSRE), Indian Institute of Technology Bombay, Mumbai, India
e-mail: suchitrapatil@somaiya.edu; adi@iitb.ac.in

S. Choudhary
Crop Physiology & Modelling, International Crops Research Institute for Semi-Arid Tropics, Hyderabad, India

J. Kholova
Crop Physiology & Modelling, International Crops Research Institute for Semi-Arid Tropics, Hyderabad, India

Department of Information Technologies, Faculty of Economics and Management, Czech University of Life Sciences Prague, Prague, Czech Republic

M. Chandramouli
Computer Graphics Technology, Purdue University Northwest, Hammond, IN, USA

P. M. Priyadarshan et al. (eds.), *Digital Agriculture*,
https://doi.org/10.1007/978-3-031-43548-5_11

and qualitative phenotyping for traits like leaf nitrogen content; it is also used in biotic and abiotic stress quantification in plants. As of now the UAVs (drones) are popular for scouting and pesticide spraying in open field. The use of UAVs (drones) for phenotyping is a newer research area that is not matured enough till date. The main objective of this chapter is to explore the use of UAVs (drones) with different types of sensors mounted on it, for lean field phenotyping so that it will be used to assist the breeders in speeding up the selective breeding process using image-based HTPP with high precision and accuracy.

Keywords High-throughput plant phenotyping (HTPP) · Unmanned ariel vehicle (UAV) · Image-based phenotyping

Abbreviations

AGBM	Above ground biomass
ANN	Artificial neural network
CCC	Canopy chlorophyll content
CGM	Crop growth model
CHM	Crop height model
CI1	Red-edge chlorophyll index 1
CI2	Red-edge chlorophyll index 2
CIGR	Chlorophyll index green
CIRE	Chlorophyll index red edge
CIVE	Color index vegetation index
CNN	Convolutional neural network
CSM	Crop surface model
CWSI	Crop water stress index
DCNN	Deep convolutional neural network
DEM	Digital elevation model
DSM	Digital surface model
DVI	Difference vegetation index
EVI	Enhanced vegetation index
EVI2	Enhanced vegetation index 2
ExG	Excessive green index
ExG-R	Excess green index minus red
FAPAR	Fraction of absorbed photosynthetically active radiation
FCOVER	Fractional vegetation cover
GCC	Green chromaticity coordinate
GI	Green index
GLI	Green leaf index
GLM	General linear model
GNDVI	Green NDVI
GNYLI	Named by the developers GNYP and LI
GRVI	Green red vegetation index
GWAS	Genome-wide association study (two step mean adjusted model)
GY	Grain yield
HS	Hyperspectral
iPLS	interval partial least square
KNN	K nearest neighbour
LAI	Leaf area index

LCC	Leaf chlorophyll content
LDM	Leaf dry matter
LM	Linear model
LNA	Leaf N accumulation
LR	Linear regression
LSWI	Land surface water index
MCARI	Modified chlorophyll absorption ratio index
MCARI2	Modified chlorophyll absorption ratio index 2
MLM	Mixed linear model
MLR	Multiple linear regression
MS	Multispectral
MSAVI	Modified soil-adjusted vegetation index
MSI	Moisture stress index
MSR	Modified simple ratio
MTVI	Modified triangular vegetation index
MTVI2	Modified triangular vegetation index 2
NDI	Normalized difference index
NDRE	Normalized difference red edge
NDVI	Normalized difference vegetation index
NDWI	Normalized difference water index
NGRDI	Normalized green red difference index
NIR	Near infrared
NLM	Non-linear model
NLR	Non-linear regression
NNI	Nitrogen nutrition index
NPLR	Non-parametric linear regression
NPRFM	Non-parametric random forest model
NRMSE	Normalized root-mean-square error
OBIA	Object based image analysis
OSAVI	Optimized soil-adjusted vegetation index
PDM	Plant dry matter
PGLM	Parametric generalized linear model
PLR	Parametric linear regression
PLS	Partial least squares
PLSR	Partial least squares regression
PNA	Plant nitrogen accumulation
PRI	Photochemically refractive index
R	Pearson's correlation coefficient
RDVI	Renormalized difference vegetation index
REIP	Red edge inflection point
RERVI	Red edge ratio vegetation index
RF	Random forest
RFR	Random forest regression
RGBVI	Red–green–blue vegetation index
RTM	Radiative transfer model
RTVI	Red edge simple ratio
RVI	Ratio vegetation index
SAR	Synthetic aperture radar
SAVI	Soil-adjusted vegetation index
SIPI	Structure-insensitive pigment index
SM	Surface model
SR	Simple ratio
SVM	Support vector
SVR	Support vector regression

SWIR	Short-wave infrared
TCARI	Transformed chlorophyll absorption reflectance index
TVI	Traiangular vegetation index
VI	Vegetation index
VNIR	Visible and near-infrared
WDRVI	Wide dynamic range VI

1 Introduction

The population of the world is about to reach 11 billion by 2100. Increasing demand for food, changes in climate, and limited arable land pose immense challenges for sustainable agriculture. For example, as an effect of global warming, more frequent, severe flooding and drought are experienced, which destroys the crops. To face this food security challenge, breeders and crop scientists are working together on the genetic improvement of crops and crop management practices. One particular effort focuses on crop cultivars improvement programs for breeding a new ideotype of crop that can sustain the adverse climate conditions (excessive temperature, saline soil or diseases, pest attack, etc.) (Donald 1968) and still provide higher yields with good quality. Crop productivity deteriorates if crops fail to adapt to the variability in climate conditions. To handle such scenarios, the agricultural community needs an in-depth understanding of the relationship among genotype, environment, and phenotype (physical and biochemical characteristics) in the selective breeding programs. On the other hand, improving crop management practices is also equally important as it explores the adaption of advanced farming concepts like precision farming. In precision farming, right input is provided in the right quantity at right time to maximize productivity and quality where an agricultural production system with technological innovations like sensing technologies, automation, and data science techniques is followed (Karunathilake et al., 2023).

Recent advancements in plant phenotyping platforms and imaging sensors are game changers in high-throughput plant phenotyping (HTPP) with high-precision phenotyping. The popular HTPP platforms use potted plants grown in a controlled environment (like a greenhouse) that are often taken to the phenotyping platforms, which are mostly the conveyer belt (Demidchik et al. 2020). Potted plants are placed on the conveyer belt and the belt moves at a fixed speed. Various imaging sensors, as per applications requirement, are installed around the belt, as per predefined arrangement (imaging angle, height, etc.). These sensors image the plants moving on conveyer belt from a particular angle. Another HTPP platform is stationary platform, where, unlike the conveyer belt-based platform, the plant is fixed at one position (on table) and the sensor moves to the plant with the help of robotic system to capture the images (Demidchik et al. 2020). This process generates thousands of images per second. Such huge image databases can be used in data-hungry advanced machine learning (ML) and deep learning (DL) algorithms for automatic phenotyping of traits like leaf count, leaf angle, leaf length, stem height, and stem diameter. (Demidchik et al. 2020).

In a selective breeding program, thousands of crop lines are crossed to get new cultivars and timely quantification of several traits in resultant cultivars is needed. It is not feasible to use greenhouses to accommodate such a huge number of crop lines. Hence, generally such experimentation is carried out in the open field wherein the technologies that use a fixed platform for image-based HTPP are not suitable for such high-scale breeding programs (White et al. 2012). Field phenotyping can be intensive with highly equipped phenotyping platforms with sensors, weather station, shelters, etc. installed in field itself, or it can be lean (Pieruschka and Schurr 2019), where minimal equipment, such as a drone (type of unmanned aerial vehicle [UAV], which is not piloted remotely by human)-mounted imaging sensor, is needed (Morisse et al. 2022). Hereafter, drones are referred to as UAVs in this chapter. Breeders can track genotype performance in field plots using a UAV, which is a mobile image-based HTPP platform on which various sensors can be attached and farmers can use cutting-edge technology for precise crop management (Pieruschka and Schurr 2019). This chapter provides an overview of image-based HTPP and types of UAVs, sensors, and photogrammetry software suitable for agriculture applications, especially for morphological, physiological, and stress phenotyping (Danzi et al. 2019; Pasala and Pandey 2020). UAVs are flown at low altitude with different types of imaging sensors mounted on it. Generally, even if UAV is flown at high altitude, it generates imagery with higher spatial resolution compared to satellite (Anderson and Gaston 2013). With different advanced sensors mounted on UAV imagery, it is possible to capture an imagery with much higher spectral resolution (Colomina and Molina 2014). Unlike satellites, revisit time of UAVs is not fixed, so user can plan the frequency of the flying as required and hence higher temporal resolution can also be obtained. Using such high spectral and spatiotemporal resolution imagery, it is possible to perform image-based structural and physiological traits phenotyping as well as quantification of both biotic and abiotic stresses in plant with high precision.

The chapter focuses on the feasibility of UAV as image-based HTPP platforms for lean field phenotyping. Breeders can accelerate the process of selection of lines in the selective breeding process, where multiple lines are crossed to get new cultivars, for which timely quantification of traits on large scale is needed. The chapter starts with Sect. 1 introducing the concept of image-based plant phenotyping, along with its pros and cons. Furthermore, in Sect. 2, an introduction to HTPP platforms is covered. In Sect. 3, UAV as HTPP platform for lean field phenotyping is discussed, where different types of UAVs, various imaging sensors compatible with UAVs, software for UAV mission planning, and software for UAV imagery processing are covered. In Sect. 4, the general framework used in UAV-assisted image-based phenotyping is discussed. In Sect. 5, 43 latest research papers covering different crop species, where UAV-assisted image-based phenotyping is used, are summarized. In Sect. 6, different real-time challenges faced in UAV-based lean field phenotyping are discussed and the chapter ends with a summary section.

2 Image-Based Plant Phenotyping

Plant phenotype refers to the set of visible biophysical characteristics of a plant, and it is determined by the interaction between genotype (the genetic makeup of the plant) and environmental conditions in which the plant actually grows (Hickey et al. 2019). This interaction is shown in Fig. 1. A quantitative assessment of a plant's morphological, physiological, and component traits at the cell, tissue, organ, plant, canopy, or population level is known as phenotyping (Demidchik et al. 2020). Phenotypic traits can be as simple as morphological parameters such as leaf surface area, stem diameter, leaf angle, width, the height of the plant, tiller count, etc., which are related to the architecture of the plant or can be as complex as the physiological principles, such as the evapotranspiration rate, which controls the plant functions. Measuring morphological traits (leaf length, width, area, stem diameter, etc.) is easier compared to quantification of functional or physiological traits (water content, chlorophyll contents of plant, temperature of individual leaf) (Das Choudhury et al. 2020). In selective breeding applications, breeders monitor and analyze different traits controlling plant growth on large scale and on regular basis. This can be achieved collectively with effective image processing for feature extraction and machine learning (ML) for data analysis. Recent ML literature has a deep learning (DL)-based state-of-the-art explored for various image-processing tasks like object detection, localization, image semantic segmentation, and classification. These techniques can help the plant science community to close the genotype-to-phenotype gap by accelerating research in plant

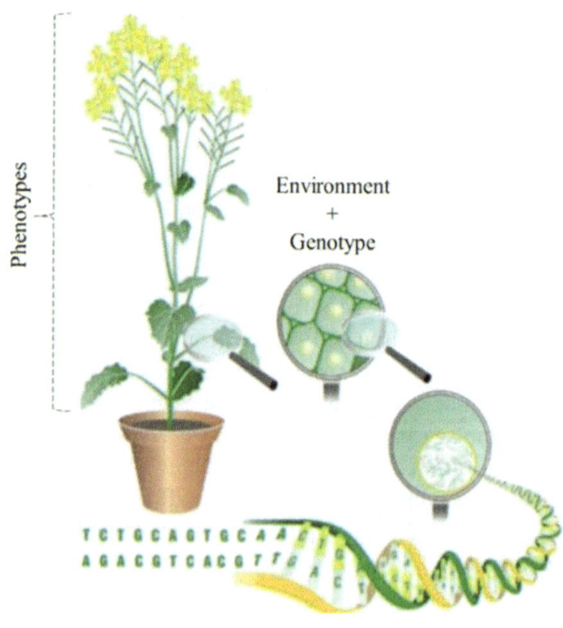

Fig. 1 Phenotype is function of environment and genotype. (Source: Furbank and Tester 2011)

phenotyping. Some DL techniques such as deep convolution networks combine feature extraction and regression/classification into a single pipeline, which is trained end to end simultaneously. These techniques do not rely on hand-engineered features that are based on human expertise and hence prone to error sometimes. Also, ML/DL-based algorithms have the potential to improve the robustness of image-based phenotyping and support automatic extraction of more complex and abstract phenotypic features required for the genotype–phenotype association.

Phenotyping using traditional tools and techniques is subjective and generally relies on manual measurements of selected traits from small number of samples of plants; hence, it is called a low- or limited-throughput technique. In field-based phenotyping platforms, the throughput is in terms of plant population (White et al. 2012). Manual measurement of morphological traits does not require any special skill as such, but obtaining frequent and robust measurements of multiple plant traits across many cultivars becomes a labor-intensive, error-prone, and imprecise task. Also, such techniques cannot be extended to quantify functional traits related to dynamic and complex plant processes (Pasala and Pandey 2020). For example, phenotypes such as early stress quantification cannot be performed directly in many cases. This is a phenotyping bottleneck. Hence, as an improvement in phenotyping efficiency and throughput, simultaneous measurement and analysis of environmental conditions is needed. To address this bottleneck it is required to develop a fully automatic, high-resolution, high-throughput system for quantitative measurement of plant structure- and function-related traits with a capacity to perform a comprehensive analysis of phenotypes. Image-based phenotyping offers a viable solution to overcome these limitations, wherein plant traits are automatically extracted from images with the help of image analysis algorithms.

Images needed for image-based plant phenotyping are generally acquired by HTPP platforms, in which robotic arms or/and conveyer belt are used to image the plants grown in controlled environments like greenhouses. Both, ground-based and aerial imaging platforms are used in fields to take pictures of plants and crops, matching the scale and throughput of image-based phenotyping (Mochida et al. 2018). However, a key requirement for such image-based phenotyping tools is the ability to automatically convert plant images into accurate and reliable phenotypic measurements across a wider range of phenotypes for different scientific applications. With the advancements in imaging techniques, multiple images of plants from different angles are captured and further temporal dimension can also be added to same. Similarly, the use of thermal infrared imaging for water stress detection, quantitative analysis of photosynthesis by chlorophyll fluorescence imaging, near-infrared spectroscopy for identifying nutrient deficiencies related to changes in plant organs, visible–near-infrared hyperspectral imaging for shoot biomass estimation, and short-wave infrared hyperspectral imaging for water absorbance allows getting complex phenotype information (Demidchik et al. 2020). All these images obtained from the sensor can be further analyzed using advanced image-processing algorithms, for instance, image segmentation can be used for automatic and pixel-wise phenotype data extraction for various phenotypic traits, such as leaf length, width, area, plant height (PH), and stem diameter, and functional traits such as chlorophyll content, etc. (Lobet et al. 2013; Gibbs et al. 2017). Image-based phenotyping techniques have been used for the quantification of a wide range of properties of roots, shoots, leaves, etc. by constructing a 3D model of the plant using the images captured from different view angles (Gibbs et al. 2017). A simple analysis of the color of plant images can be important while studying abiotic and biotic stresses experienced by the plant. As discussed above, in high-throughput image-based plant phenotyping, the major bottleneck is an assessment of different phenotypes for dense population. Hence, here the main focus is on building high-throughput automatic pipeline for processing and assessment of the plant phenotypes on large scale.

2.1 Advantages and Disadvantages of Image-Based Plant Phenotyping

Advantages

- This methodology is non-destructive.
- The same plant can be imaged to obtain a sequence in its entire life cycle.
- Imaging the entire plant or even the entire field in one single image is possible.
- Analysis of a single image enables quantification of various characteristics.
- Infrared data facilitates obtaining and assessing the data that cannot be seen by human eyes.

Disadvantages

- High cost of imaging setup consisting of hyperspectral cameras, drones, and controlled environment chambers.
- Data analytics software must be capable of handling complex and huge data with superfast computing.

2.2 High-Throughput Plant Phenotyping Platforms

Of late, both environmentally controlled plants and field phenotyping platforms have witnessed significant advancements. Generally, phenotyping in environmentally controlled conditions (greenhouses) is used in educational and some research institutes that conduct research of small potted plants such as *Arabidopsis* and some other crops/plants (Demidchik et al. 2020). Crop breeders can also use them in a selective breeding program when a limited number of lines are used. In such platforms, in-depth measurement of plants is taken with the help of imaging sensors mounted on robotic arm, or fixed around conveyer belt, and image analysis tools integrated as a system. There are many limitations for plant growth in a controlled environment. For instance, factors such as limited soil and small spaces generally affect flower and seed production in such plants compared to plants in open fields. However, certain obstacles like wind speed, light, evaporation, pest, and diseases attack are comparatively lesser than in an open-air environment. Measurement of abiotic stress is restricted in controlled environments (Gilliham et al. 2017), so breeders and crop researchers have now focused on field-level enhancements in yield productivity or abiotic stress resistance quantification using field-based phenotyping. In field-based phenotyping platforms, the throughput is in terms of plant population (White et al. 2012). Furthermore, field-based platforms such as ground-based platforms (modified vehicles) and aerial platform with remote sensing sensors mounted on it have great potential, because it covers wider area, real-time data are acquired by sensing plant conditions, and it has some useful instruments, such as remote sensing tools, a global positioning system (GPS), and geographic information system (GIS) for exploring spatial changeability (Kang et al. 2019). Pheno-towers is one such advanced platform, which has eight sets of sensors, and two 3D flight cameras, an RGB (red–green–blue) camera, three laser distance sensors, a spectral imaging camera, and two light curtain imaging structures. This platform is capable of collecting data on the height of plants, fresh weight density, moisture content, growth stage, and tiller density nitrogen content of all plots by screening almost 250 plots per hour (Busemeyer et al. 2013; Li et al. 2014).

Plant phenomics deals with the registration, accumulation, and mathematical analysis of arrays of data on the phenotype of plant organisms. This field is undergoing rapid development and has opened up completely new possibilities for fundamental research on genotype–phenotype relationship, which is critical for the transition to high-tech agriculture and forestry (Demidchik et al. 2020). Like genomics, where the entire sequenced genome is completely characterized, in phenomics, the entire phenome cannot be characterized due to its highly dynamic nature and multidimensional properties. Nevertheless, one can go for high-throughput phenotyping on a set of certain traits; here throughput refers to the number of individual units at certain structural levels in plants. Dimensionality addresses various phenotypic characteristics, such as plant compositions, physiology, and performance in different spatial and temporal systems, along with the number of genotypes and the different environmental conditions considered at the time of phenotyping (Dhondt et al. 2013). ML, especially DL–based techniques, can provide robust measurement, and this can be extended to handle complex phenotype too, considering the above issues of invariant features and dimensionality (Ubbens et al. 2018).

Aerial phenotyping platform is a good alternative to overcome the difficulties faced in ground-based phenotyping platforms. Aerial phenotyping is a quick and non-invasive method of quantification of plants and plots of a plant (large population). The primary aerial phenotyping platforms use UAVs, which can be small-size aeroplanes or multicopters with considerable payloads (4–10 kg) needed for this task. Multicopters are more popular for lean field phenotyping. More on UAVs is discussed in the next section.

3 UAV for Image-Based Plant Phenotyping

As far as lean field phenotyping is concerned, the use of UAV-based high-resolution imagery for phenotyping is current state-of-the-art (Colomina and Molina 2014). In the USA and many Asian countries, for crop breeding, the steep use of UAV-based imaging for data collection is noticed. It is due to its reliability, cost-effectiveness, and resolution. Current applications of UAV imagery are seen in weed detection, pathogen detection, drought stress assessment, nutrient status, growth assessment, and in yield prediction (Maes and Steppe 2019).

3.1 Types of Drones Used in Phenotyping

Remote sensing using UAVs is proving to be a game changer in precision agriculture. UAVs have sorted out different problems, such as cloud cover effect, temporal and spatial resolution, faced in satellite-based remote sensing. Researchers use unmanned aerial systems (UAS) as an umbrella term to refer to the entire system comprising of UAVs, all other equipment, and the software (for flight mission planning as well as data analysis) related to it. This can include global positioning system (GPS), communications equipment, and sensors. Rotary and fixed wings are two popular types of UAV described below.

Rotary UAVs are generally recognized by the number of rotors (propellers) they have, like a quadcopter has four rotors, a hex-copter has six rotors, and so on. These UAVs have the helicopter-like appearance and are more suitable for agricultural applications like field scouting (Khot n.d.). It is possible to hover rotary UAV over the specific research plots or some important area in the experiment, to image that specific area for more detailed analysis. They have vertical landing and take-off ability and hence need minimum landing space. The speed limit for such UAVs is 10–20 mph and is mostly unstable with wind speeds of more than 15 mph. Rotary UAVs are better for small-size fields due to their shorter battery life (most of the battery power is consumed for operating the multiple rotors).

Single-rotor UAV has a larger rotor on top and a smaller rotor at the tail. These UAVs can fly for a longer time compared to multirotors as these are operated by gas engines. These UAVs have heavier payload capability than fixed wings, but it is challenging to fly such UAVs, which comes with operational dangers due to their larger rotor (Khot n.d.; Kakarla and Ampatzidis 2021).

Fixed-wings UAVs are used for longer flights of 1–2 h duration. They have longer battery life as they use the aerodynamic lift provided by their structure to stay floating in the air. These UAVs can operate at the maximum speed of 70 miles/h, but it is not recommended, as such high speed will induce blur in images. It requires a more prominent and uniform area for landing and take-off. This UAV cannot be hovered over a specific location and need a lot of training to fly it. Fixed wings are costlier than rotary UAVs (Kakarla and Ampatzidis 2021).

3.2 Types of Sensors

To capture the images, different imaging sensors like RGB, multispectral, hyperspectral, thermal, and light detection and ranging (LiDAR) can be mounted on UAVs. RGB camera is used for visible spectrum (450–750 nm; number of bands ≤3) imaging to cover larger areas and capture more images to get higher spatial resolution. The multispectral camera is an advanced version of RGB camera with imaging capability till the infrared region (450–1000 nm; 4–10 bands) whereas the hyperspectral camera can capture the higher spectral resolution (450–1000 nm) with 100–200 very narrow bands. LiDAR sensors can be single as well as multiband. Researchers convert this spectral information into measurable quantities using band arithmetic-based vegetation indices (VIs). Later, these VIs can be correlated with various traits of crops, for instance, the use of hyperspectral images with ML for the detection, identification, and differentiation of plant diseases having the same visual characteristics (Hariharan et al. 2019; Abdulridha et al. 2020). Similarly, thermal sensors measure the thermal energy emitted by an object at the wavelength matching its surface temperature, and hence the thermal cameras can be utilized in measuring parameters such as canopy temperature and based on it to determine the canopy waters stress required in precision irrigations (Zhou et al. 2020) and in leaf moisture detection (Swarup et al. 2020). In agricultural applications, LiDAR sensors are used to get high-resolution DEM (digital elevation model) and DSM (digital surface model) from which crop height, crop density, etc. is computed (García et al. 2018). Comparisons of different sensors discussed above are represented in Table 1 and discussed in Guo et al. (2021).

Before purchasing the sensors, the application for which it is used and its compatibility with the UAV on which it will be mounted need to be checked

Table 1 Comparative analysis of various imaging sensors compatible with UAVs (Guo et al. 2021)

Sensor	No. of bands (commonly available)	Spectral range covered	Cost	Weight	Resolution (megapixels)	Ease of use
RGB	3	450–750 nm	Low	Low-medium	Low-high	Easy
Multispectral	3–10	450–1000 nm	Medium	Low-medium	Medium	Medium
Hyperspectral	>10	450–1000 nm	High	High	Low	Difficult
Thermal	1	3500–7500 nm	Medium	Low	Low	Medium
LiDAR	1[a]	905 nm	Medium-high	Medium-high	Medium-high[b]	Difficult

[a]There are some multiband LiDARs
[b]LiDAR resolution is given as point cloud density

3.3 Flight Mission Planning and Photogrammetry Software

As mentioned earlier, UAS is the entire system comprising of UAVs, sensors, and different software required for operating UAV and processing UAV imagery.

Flight Mission Planning Before taking flight using a UAV, it is very important to plan the flight mission carefully. In this mission planning, the region over the flight is to be taken is marked in satellite map and different parameters like flight speed, flying altitude, front and side overlap, camera aperture and shutter speed, and imaging angle etc., can be set. There are several software and mobile applications available for flight mission planning, for instance, DroneDeploy (www.dronedeploy.com n.d.) is paid software and few free software include Pix4Dcapture (Pix4D n.d.), DJI GS Pro (DJI Official n.d.), Precision Flight (flightsims-dev.10web.site n.d.), etc.; more on such software is discussed in Kakarla and Ampatzidis (2021). Furthermore, this imagery is processed using photogrammetry software.

Photogrammetry Software Photogrammetry is the key technique used in UAV-based agricultural applications. Photogrammetric techniques also have played significant role in 3D reconstruction employed in geographic information systems and various other areas with considerable success (Chandramouli et al. 2016). In this study, geometric information is extracted in three dimensions from the high-resolution imagery (Kakarla and Ampatzidis 2021) taken with 70–90% front and side overlap, using structure from motion algorithm. First, the common tie points from selected pairs of images are obtained, and then a dense 3D point cloud is generated. Further, this point cloud is rasterized to get the digital elevation model (DEM), and using DEM a single orthomosaic from these multiple images is obtained. Various photogrammetry software that can be used for this purpose are Pix4D (Pix4D 2011) and Agisoft Metashape (Agisoft 2019), which are proprietary, and OpenDroneMapper (WebODM) (OpenDroneMap 2018), a web-based open-source software. Furthermore, with use of various software, such as QGIS (QGIS 2017) and ArcGIS (Esri 2019), this orthomosaic is analyzed and processed to extract the valuable information (traits) based on vegetation indices that the crop growers and researchers are interested in.

4 Framework for UAV-Based Phenotyping

As discussed in the previous sections, the raw imagery is captured by on-board application-specific sensor. Photogrammetry software Agisoft Metashape provides a facility to check the quality of these raw images based on contrast. Good-quality image (which has good contrast and no blur) is further processed with the help of photogrammetric software. In this process, the imagery is first aligned. In the process of alignment, the common tie points are selected from the pair of overlapping images using shift invariant feature transform (SIFT) technique (Zhao et al. 2016), to get sparse point cloud. Furthermore, using structure from motion technique the dense point cloud is generated from sparse point cloud. This dense point cloud is rasterized into 2D digital surface model (DSM) also known as DEM in Agisoft. Finally, the orthomosaic is generated from the DEM. DEM is processed further to get crop height information. Orthomosaic is processed by some other software like ArcGIS and QGIS to extract the individual plot image from the entire field and from that plot level information like representative plant height (PH), nitrogen content (N), and leaf area index (LAI) of the plot is extracted. Furthermore, the extracted information is correlated with the observed ground truth data using statistical methods like linear regression. For ground truth generation, manual destructive techniques are used. Plant height readings are measured with rulers for selected plant samples from plot. For leaf area, the Licor 3100 (www.licor.com n.d.) instrument is used and all the leaves of these sample plants are scanned; nitrogen content is obtained using SPAD-502 (www.specmeters.com n.d.), such as spectroradiometers. Advanced ML and DL algorithms extract these features directly from the images and these are further correlated with ground truth using CNN-based predictive modeling. The information extracted from the image correlates with many traits, such as 2D canopy cover, leaf area, crop height, canopy nitrogen content, canopy temperature, and canopy water content. Based on these traits further assessment of stress can be done. For instance, using canopy water content, water stress quantification is done, and using canopy cover and crop height information, crop's biomass prediction is done. The general framework for UAV-assisted image-based HTPP is shown in the Fig. 2.

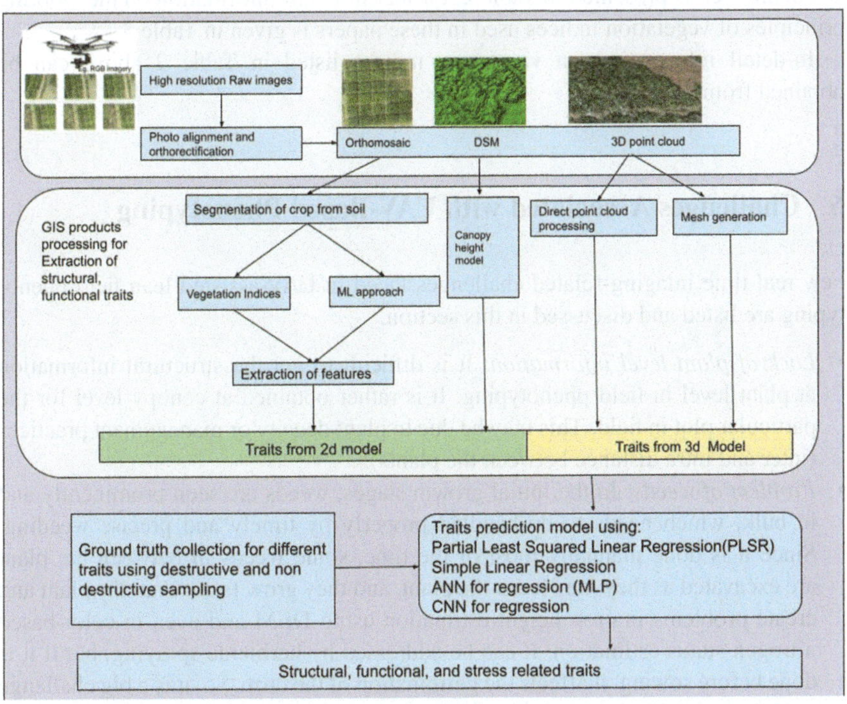

Fig. 2 General framework for UAV-assisted image-based high-throughput plant phenotyping

5 Applications of UAV in Plant Phenotyping

UAV drones are known as movable plant phenotyping platform suitable for lean field phenotyping that assists in image-based HTPP. The UAV-captured high-resolution imagery can be processed to extract the canopy structure-related information, which is used for quantification of morphological traits such as plant height using DSM (Bendig et al. 2015), 2D canopy cover, or 2D leaf area (Makanza et al. 2018). The high spectral resolution can be used to extract the spectral information from the crops in various narrower bands, and further, using the vegetation indices as metric for spectral transformation, the presence and state of vegetation can be quantified. During spectral transformation, the spectral information in images is transformed into some quantity that can be used to extract the functional traits like canopy nitrogen content (Jiang et al. 2020) and chlorophyll concentration (Zhu et al. 2020). Using these, vegetation indices like excessive green index (ExG), green chromaticity coordinate index (GCC), red–green–blue vegetation index (RGBVI), normalized difference vegetation index (NDVI), normalized difference red edge (NDRE), and normalized difference water index (NDWI) are used for estimating the nitrogen, greenness, canopy cover, leaf area index, and water stress quantification in the crop. Review of 43 latest papers from 2010 to 2022, where UAV-assisted plant phenotyping is explored for structural and functional phenotyping in different crops like maize, wheat, barley, potato, pea, cotton, soybean, dry beans, grapevine, beet, and sunflower, is presented in Table 2. Further in-detail information on the working principles of vegetation indices used in these papers is given in Table 3.

In-detail information on vegetation indices listed in Table 2 above can be obtained from Table 3.

6 Challenges Associated with UAV-Based Phenotyping

Few real-time imaging-related challenges faced in UAV-assisted lean field phenotyping are listed and discussed in this section.

- *Lack of plant-level information*: It is difficult to get the structural information at plant level in field phenotyping. It is rather obtained at canopy level for the particular plot in field. This may be due to plant density or management practices (inter and intra distance between the plants).
- *Problem of weeds*: In the initial growth stages, weeds are seen prominently and in bulk, which needs to be handled properly by timely and precise weeding. Since it is done manually most of the time, some weeds in between the plant are excavated as they may harm the plant, and they grow faster than the plant and create problems in crop height estimation using DEM and even in color-based nitrogen status estimation. It can be addressed by herbicide spraying, but if it is done before sowing, it affects the germination of the crop. So, it is a big challenge in UAV-based phenotyping.

Table 2 Summary of various crops and their phenotypes estimation using UAV-imagery

S. no.	Crop	Sensor	UAV type	Estimated variable	Type of plant traits/stress	Plant phenotype (traits)	Plant trait model	Altitude	Image resolution (unit/pixel)	Citation
1	Wheat	MS	Pheno-copter	NDVI	Agronomical functional	Biomass, leaf greenness	LR	30–50 m	1.8 cm	Duan et al. (2017)
2	Barley	RGB, HS	Multirotor	NDVI, SAVI, MSAVI, OSAVI, GNYLI	Agronomical morphological	Biomass, plant height	MLR, NLR, CSM	Low-altitude	1 cm	Bendig et al. (2015)
3	Corn, soybean	MS	–	NDVI, SR, GNDVI, MTVI2 LSWI, MSI, RTVI	Agronomical	LAI, biomass	MLR, NLR	Low-altitude	–	Kross et al. (2015)
4	Rye grass	MS	Balloon system	NDVI	Agronomical functional	Biomass, nitrogen	LR, MLR	Low-altitude	15 cm	Kawamura (2011)
5	Pea	RGB, MS	Quadrotor	CIGR, CIRE, EVI2, GNDVI, MCARI2, MTVI2, NDRE, NDVI, NDWI, OSAVI, RDVI, RGBVI	Agronomical morphological	Biomass (AGBM), canopy volume	LR, DSM, 3D reconstruction model	10–20 m	0.21–1.36 cm	Sangjan et al. (2022)
6	Wheat	RGB, MS	Fixed-wings	RedEdge, NIR and RGB VIs	Morphological agronomical	Plant height, LAI, LCC, CCC, plant yield	PLR, RFR, SVR	Low-altitude	5 cm	Ganeva et al. (2022)
7	Wheat	RGB, MS, HS	–	RERVI, CIRE, DVI	Agronomical functional	LAI, nitrogen content	LR	0.5–1 m	–	Jiang et al. (2020)

(continued)

Table 2 (continued)

S. no.	Crop	Sensor	UAV type	Estimated variable	Type of plant traits/stress	Plant phenotype (traits)	Plant trait model	Altitude	Image resolution (unit/pixel)	Citation
8	Potato	RGB, HS	Quadcopters	NDVI, MSR, MSAVI, OSAVI, MCARI, MCARI2, TCARI, NDI, CI1, CI2, SIPI, ExG	Agronomical morphological	Biomass, crop yield, plant height, crop yield	RFR, PLSR, LR	30 m	0.5, 2.2, 3.1 cm	Li et al. (2020)
9	Maize	HS	Air borne	VNIR, SWIR	Functional agronomical	Nitrogen, grain yield	RTM, PLSR	500 m	–	Wang et al. (2021)
10	Maize	MS + RGB	Multicopter	NDVI, GNDVI, SR, SAVI	Biotic stress agronomical	MSV (maize streak virus) grain yield	MLR, LR	–	–	Chivasa et al. (2021)
11	Maize	RGB (Panasonic camera)	Multirotor	NDVI	Morphological	Crop cover	Broad-sense heritability and genetic correlations	80 m	1.5, 1 cm	Makanza et al. (2018)
12	Corn	RGB	Multirotor	ExG, OBIA	Biotic stress	Weed detection	CNN, RF, SVM	3 m above ground	–	Bah et al. (2018)
13	Potato	RGB + HS	Quadcopters	NDVI, MSR,	Agronomical	Biomass crop yield	RF, LR	30 m	–	Li et al. (2020)
14	Barley	RGB	–	NDVI, PRI, SAVI	Functional	Nitrogen use efficiency	MLR	50 m	–	Kefauver et al. (2017)
15	Soybean	RGB	Multirotor	Color, gradient, texture, shape	Biotic stress	Foliar disease	SVM, KNN	1, 2, 4, 8, 16	–	Castelao Tetila et al. (2017)

16	Cotton	RGB	Multirotor	NDVI	Morphological	Height	LR	50, 29, 13	–	Xu et al. (2019)
17	Cotton	RGB	Multirotor	Crop yield	Biotic stress	Cotton boll detection	LR, OBIA	13	–	Yeom et al. (2018)
18	Peanut	HS, MS	Multirotor	GRVI, NDRE	Biotic stress	Spot wilt	VI	–	–	Patrick et al. (2017)
19	Beet	HS	Multirotor	NDVI, NDWI	Abiotic stress	Beet cyst nematode	Decision tree	80 m	–	Joalland et al. (2018)
20	Tomato	RGB, MS	Multirotor	NDVI	Abiotic stress	Salinity stress plant area	OBIA	13 m	–	Johansen et al. (2019)
21	Drybean	MS	Multirotor	GNDVI	Agronomical abiotic stress	Seed yield, biomass, flowering, drought	Pearson *r*	50, 120	7.2, 3 cm	Sankaran et al. (2018)
22	Citrus tree	Multispectral	Multirotor	GNDVI	Morphological	Counting trees	DCNN	–	5 cm	Ampatzidis and Partel (2019)
23	Sunflower	RGB, MS + NIR	Multirotor	ExG, NDVI	Biotic stress	Weed	OBIA	–	0.5, 1.12 cm	López-Granados et al. (2015)
24	Vineyard	Multispectral	Multirotor	NDVI, NDRE	Functional	Stem water potential, water stress	ANN	30	~2.2 cm, 1.11	Romero et al. (2018)

(continued)

Table 2 (continued)

S. no.	Crop	Sensor	UAV type	Estimated variable	Type of plant traits/stress	Plant phenotype (traits)	Plant trait model	Altitude	Image resolution (unit/pixel)	Citation
25	Wheat	RGB, multi-spectral	Multirotor	Lodging index	Functional	Lodging	Heritability, correlation, GWAS	25m	–	Singh et al. (2019)
26	Corn	3D sensor	–	Points cloud	Morphological	Height, weight, length, area	3D model	–	–	Li and Tang (2017)
27	Grapevine	HS	Fabry-Pérot interferom-eter	Narrow-band VI	Agronomical functional	LAI, plant and leaf nitrogen content, transpiration	PLS, iPLS	32 m	2 cm	Matese et al. (2022)
28	Barley	HS	Multirotor	NDVI, GNDVI, ExG, ExG-R, CIVE, GRVI	Agronomical morphological	Biomass, plant height	3D model	70 m	0.01 m, 0.04 m	Di Gennaro et al. (2017)
29	Wheat	RGB	Rotary-wings	NDV, GI, GLI	Biotic stress	Foliage disease severity	3D model	25 m	0.7 cm	Bhandari et al. (2020)
30	Barley	RGB, MS	Quadcopter	Various MS and RGB VIs	Morphological agronomical	Canopy height, vegetation cover, growth yield	PGLM, NPRFM	30 m, 50 m	0.82 cm, 1.3 cm	Herzig et al. (2021)
31	Winter wheat	RGB	ARF Mikrokopter Okto XL	NDVI, REIP, GRVI	Morphological	Canopy cover, plant height	CSM, DSM	50 m, 75 m, 30 m	0.10 m, 0.06 m, 0.04 m	Roth and Streit (2017)

32	Winter wheat jointing, flagging, and flowering periods	HS	Multicopter	NDVI, OSAVI, many other HS and RGB spectrum VIs	Agronomical morphological	LAI, biomass, crop height	RFR, PLSR	50 m	–	Yue et al. (2018)
33	Winter barley	HS	Rotary-wings UAV	NDVI	Agronomical morphological	Biomass, plant height	LR or PLSR analysis	80 m	–	Roth and Streit (2017)
34	Wheat	HS	Microcopter	NDVI, MCARI	Agronomical morphological	Biomass, nitrogen content	LM, NLM	–	–	Pölönen et al. (2013)
35	Rapeseed	MS	Fixed-wings	NDVI, NDRE, GNDVI	Agronomical	Crop yield	SM	–	10 cm, 5 cm, 2.5 cm	Nebiker et al. (2016)
36	Barley	MS	Fixed-wings	NDVI, NDRE, GNDVI	Agronomical	Crop yield	SM	–	10 cm, 5 cm, 2.5 cm	Nebiker et al. (2016)
37	Onion	MS	Fixed-wings	NDVI	Biotic stress	Disease detection	SM	–	10 cm, 5 cm, 2.5 cm	Nebiker et al. (2016)

(continued)

Table 2 (continued)

S. no.	Crop	Sensor	UAV type	Estimated variable	Type of plant traits/stress	Plant phenotype (traits)	Plant trait model	Altitude	Image resolution (unit/pixel)	Citation
38	Potato	MS	Fixed-wings	NDVI	Biotic stress	Disease detection (potato blight infestation sites)	SM	–	10 cm, 5 cm	Nebiker et al. (2016)
39	Winter wheat	HS	Multicopter	NDVI, GI, RVI, WDRVI, EVI, OSAVI, MSAVI	Agronomical morphological	Biomass, crop height	PLSR	36 m	1 cm	Yue et al. (2017)
40	Olive	HS	Fixed-wings	NDVI, RDVI, OSAVI, TVI, MTVI, SR, MSR, CWSI	Biotic stress	Detection of verticillium	GLM, RTM	550 m	53 cm*42 cm	Calderón et al. (2013)
41	Wheat	MS	–	NDVI, NDRE, NGRDI	Physiological	Grain yield	MLM	30 m, 40 m	2.5 cm, 3 cm	Hassan et al. (2019)
42	Maize	LiDAR	Quadcopter	NRMSE, RMSE	Agronomical	Change in LAI	NRMSE model	15 m	–	Lei et al. (2019)
43	Maize	LiDAR	Multirotor	–	Morphological	Plant height	DSM	15 m	5 cm	Zhou et al. (2020)

Table 3 More information on various vegetation indices listed in Table 2

Vegetation index	References/Citations
CIGR	Sangjan et al. (2022)
CIRE	Jiang et al. (2020)
CIVE	Di Gennaro et al. (2017)
CWSI	Calderón et al. (2013)
DVI	Jiang et al. (2020)
ExG	Di Gennaro et al. (2017)
ExG-R	Di Gennaro et al. (2017)
GI	Bhandari et al. (2020)
GLI	Bhandari et al. (2020)
GNDVI	Nebiker et al. (2016)
GNYLI	Bendig et al. (2015)
GRVI	Patrick et al. (2017)
LAI	Wang et al. (2021)
LSWI	Kross et al. (2015)
MCARI2	Li et al. (2020)
MSAVI	Bendig et al. (2015)
MSI	Kross et al. (2015)
MSR	Calderón et al. (2013)
MTVI	Calderón et al. (2013)
NDI	Li et al. (2020)
NDRE	Sangjan et al. (2022)
NDVI	Duan et al. (2017)
NDWI	Sangjan et al. (2022)
NGRDI	Hassan et al. (2019)
OSAVI	Bendig et al. (2015)
OSAVI	Yue et al. (2018)
PLSR	Li et al. (2020)
PNA	Jiang et al. (2020)
PRI	Kefauver et al. (2017)
REIP	Roth and Streit (2017)
RERVI	Jiang et al. (2020)
RTM	Wang et al. (2021)
RTVI	Kross et al. (2015)
SAVI	Bendig et al. (2015)
SR	Chivasa et al. (2021)
TVI	Calderón et al. (2013)
WDRVI	Yue et al. (2017)

- *The problem of wind*: Wind speed of <12 mph is ideal for flying a UAV; if the wind is more than this, then the crop canopy and UAV both are unstable. During high wind speed if the flight is taken then the structural information of crop canopy (crop height, canopy cover, etc.) is affected in images and further agronomically essential traits like yield and biomass for which canopy information serves as a component trait is affected too. Also, though it is possible to fly UAV in wind speed of >12 mph, in order to stay floating and stable in such wind resistance, the batteries of UAV are consumed faster.

- *Ground truth and time of flight*: Though plants do not grow too fast, it is a recommended practice to obtain ground truth and flight data simultaneously. This is not always feasible because for a few traits, such as 3D leaf area, fresh and dry biomass, and nitrogen content, destructive sampling has to be done to record the observed values. It is a laborious and time-consuming process and is difficult to complete at the same time when plants are toward maturity. In all the growth stages, the number of samples, genotypes, treatment combinations, and further processing of it is required to be checked. So, a maximum duration of one to two days can be considered between ground truth and flight if it is not done on the same day.
- *Solar radiation*: Flight data must be collected under a clear sky, in an environment that is as evenly distributed as is practical. Due to cloudy skies or uneven sunlight during flight, the quality of images, especially the color and contrast, is hampered. In case RGB sensor is used to phenotype color-based traits such as chlorophyll concentration, then solar radiation is the biggest challenge.
- *Uneven climate conditions*: Due to uneven climate, drought-related experiments are affected a lot. The sudden raining spoils the drought experimentation carried out in the field and creates a problem. Unfortunately, in case of lean field phenotyping such inevitable conditions cannot be avoided where as in intense field phenotyping rainout shelters can handle this to some extent.

7 Summary

The increasing population and unfavorable climate conditions are exaggerated and it will lead to the worsening of the pre-existing food security challenge, by 2100. Breeders and crop scientists are working to handle these by working on the genetic improvement of crops and crop management practices. Breeders use selective breeding where many crop lines are used, and the introgression is explored, and timely monitoring and quantification of many traits on large scale is needed. Such experimentation is generally carried out in the open field rather than the controlled environment, which has space limitations for the huge number of lines that can be accommodated in it. Traditional methods used for phenotyping are manual, laborious, time-consuming, and more prone to human error. So, an automatic phenotyping solution for plant phenotyping in time-efficient manner is required. With the advancements in HTPP platform and image-based high precision, plant phenotyping is possible. UAVs are the most suitable high-throughput imaging platform suitable for lean field plant phenotyping. This chapter provided an overview of the use of UAVs as a high-throughput plant phenotyping platform. Different types of sensors can be used for different applications. Generally, for morphological traits, CSM and DEM/DTM-based approaches are more popular. Further, the different vegetation indices were generated based on the spectral range captured by the sensor during imaging. There is a trade-off in the spatial and spectral resolutions of various sensors and based on the need for application appropriate sensor as per their compatibility with UAV is selected. Thus UAV-based HTPP gives higher precision but it comes with many real-time challenges as discussed.

References

Abdulridha J, Ampatzidis Y, Qureshi J, Roberts P (2020) Laboratory and UAV-based identification and classification of tomato yellow leaf curl, bacterial spot, and target spot diseases in tomato utilizing hyperspectral imaging and machine learning. Remote Sens 12(17):2732. https://doi.org/10.3390/rs12172732

Agisoft (2019) Agisoft Metashape. [online] Agisoft.com. Available at: https://www.agisoft.com/. Accessed 15 Nov 2022

Ampatzidis Y, Partel V (2019) UAV-based high throughput phenotyping in citrus utilizing multispectral imaging and artificial intelligence. Remote Sens 11(4):410. https://doi.org/10.3390/rs11040410

Anderson K, Gaston KJ (2013) Lightweight unmanned aerial vehicles will revolutionize spatial ecology. Front Ecol Environ 11(3):138–146. https://doi.org/10.1890/120150

Bah M, Hafiane A, Canals R (2018) Deep learning with unsupervised data labeling for weed detection in line crops in UAV images. Remote Sens 10(11):1690. https://doi.org/10.3390/rs10111690

Bendig J, Yu K, Aasen H, Bolten A, Bennertz S, Broscheit J, Gnyp ML, Bareth G (2015) Combining UAV-based plant height from crop surface models, visible, and near infrared vegetation indices for biomass monitoring in barley. Int J Appl Earth Obs Geoinf 39:79–87. https://doi.org/10.1016/j.jag.2015.02.012

Bhandari M, Ibrahim AMH, Xue Q, Jung J, Chang A, Rudd JC, Maeda M, Rajan N, Neely H, Landivar J (2020) Assessing winter wheat foliage disease severity using aerial imagery acquired from small Unmanned Aerial Vehicle (UAV). Comput Electron Agric 176:105665. https://doi.org/10.1016/j.compag.2020.105665

Busemeyer L, Mentrup D, Möller K, Wunder E, Alheit K, Hahn V, Maurer H, Reif J, Würschum T, Müller J, Rahe F, Ruckelshausen A (2013) BreedVision – a multi-sensor platform for non-destructive field-based phenotyping in plant breeding. Sensors 13(3):2830–2847. https://doi.org/10.3390/s130302830

Calderón R, Navas-Cortés JA, Lucena C, Zarco-Tejada PJ (2013) High-resolution airborne hyperspectral and thermal imagery for early detection of Verticillium wilt of olive using fluorescence, temperature and narrow-band spectral indices. Remote Sens Environ 139:231–245. https://doi.org/10.1016/j.rse.2013.07.031

Castelao Tetila E, Brandoli Machado B, de Belete NA, Guimaraes DA, Pistori H (2017) Identification of soybean foliar diseases using unmanned aerial vehicle images. IEEE Geosci Remote Sens Lett 14(12):2190–2194. https://doi.org/10.1109/lgrs.2017.2743715

Chandramouli M, Tsu-Chiang Lei PE, Chou T-Y, Huang Y-M, Kolanuvada SR (2016) Design and implementation of GIS virtual learning environments for interactive visualization using desktop virtual reality (VR) & iSpace. In: In 2016 ASEE proceedings. The 123rd ASEE annual conference & exposition

Chivasa W, Mutanga O, Burgueño J (2021) UAV-based high-throughput phenotyping to increase prediction and selection accuracy in maize varieties under artificial MSV inoculation. Comput Electron Agric 184:106128. https://doi.org/10.1016/j.compag.2021.106128

Colomina I, Molina P (2014) Unmanned aerial systems for photogrammetry and remote sensing: a review. ISPRS J Photogramm Remote Sens 92:79–97. https://doi.org/10.1016/j.isprsjprs.2014.02.013
Danzi D, Briglia N, Petrozza A, Summerer S, Povero G, Stivaletta A, Cellini F, Pignone D, De Paola D, Janni M (2019) Can high throughput phenotyping help food security in the Mediterranean area? Front Plant Sci 10. https://doi.org/10.3389/fpls.2019.00015
Das Choudhury S, Maturu S, Samal A, Stoerger V, Awada T (2020) Leveraging image analysis to compute 3D plant phenotypes based on voxel-grid plant reconstruction. Front Plant Sci 11. https://doi.org/10.3389/fpls.2020.521431
Demidchik VV, Shashko AY, Bandarenka UY, Smolikova GN, Przhevalskaya DA, Charnysh MA, Pozhvanov GA, Barkosvkyi AV, Smolich II, Sokolik AI, Yu M, Medvedev SS (2020) Plant phenomics: fundamental bases, software and hardware platforms, and machine learning. Russ J Plant Physiol 67(3):397–412. https://doi.org/10.1134/s1021443720030061
Dhondt S, Wuyts N, Inzé D (2013) Cell to whole-plant phenotyping: the best is yet to come. Trends Plant Sci 18(8):428–439. https://doi.org/10.1016/j.tplants.2013.04.008
Di Gennaro SF, Rizza F, Badeck FW, Berton A, Delbono S, Gioli B, Toscano P, Zaldei A, Matese A (2017) UAV-based high-throughput phenotyping to discriminate barley vigour with visible and near-infrared vegetation indices. Int J Remote Sens 39(15–16):5330–5344. https://doi.org/10.1080/01431161.2017.1395974
DJI Official (n.d.) DJI GS Pro – DJI. [online]. Available at: https://www.dji.com/ground-station-pro. Accessed 30 Jan 2023
Donald CM (1968) The breeding of crop ideotypes. Euphytica 17(3):385–403. https://doi.org/10.1007/bf00056241
Duan T, Chapman SC, Guo Y, Zheng B (2017) Dynamic monitoring of NDVI in wheat agronomy and breeding trials using an unmanned aerial vehicle. Field Crop Res 210:71–80. https://doi.org/10.1016/j.fcr.2017.05.025
Esri (2019) arcgis. [online] Arcgis.com. Available at: https://www.arcgis.com/index.html Accessed 29 Jan 2023
Furbank, R.T. and Tester, M. (2011). Phenomics - technologies to relieve the phenotyping bottleneck. Trends in Plant Science, [online] 16(12), p635–644. https://doi.org/10.1016/j.tplants.2011.09.005.
flightsims-dev.10web.site (n.d.) Precision flight controls | FAA approved AATD flight simulators. [online]. Available at: https://flightsims-dev.10web.site/. Accessed 30 Jan 2023
Ganeva D, Roumenina E, Dimitrov P, Gikov A, Jelev G, Dragov R, Bozhanova V, Taneva K (2022) Phenotypic traits estimation and preliminary yield assessment in different phenophases of wheat breeding experiment based on UAV multispectral images. Remote Sens 14(4):1019. https://doi.org/10.3390/rs14041019
García M, Saatchi S, Ustin S, Balzter H (2018) Modelling forest canopy height by integrating airborne LiDAR samples with satellite radar and multispectral imagery. Int J Appl Earth Obs Geoinf 66:159–173. https://doi.org/10.1016/j.jag.2017.11.017
Gibbs JA, Pound M, French AP, Wells DM, Murchie E, Pridmore T (2017) Approaches to three-dimensional reconstruction of plant shoot topology and geometry. Funct Plant Biol 44(1):62. https://doi.org/10.1071/fp16167
Gilliham M, Able JA, Roy SJ (2017) Translating knowledge about abiotic stress tolerance to breeding programmes. Plant J 90(5):898–917. https://doi.org/10.1111/tpj.13456
Guo W, Carroll ME, Singh A, Swetnam TL, Merchant N, Sarkar S, Singh AK, Ganapathysubramanian B (2021) UAS-based plant phenotyping for research and breeding applications. Plant Phenomics 2021:1–21. https://doi.org/10.34133/2021/9840192
Hariharan J, Fuller J, Ampatzidis Y, Abdulridha J, Lerwill A (2019) Finite difference analysis and bivariate correlation of hyperspectral data for detecting laurel wilt disease and nutritional deficiency in avocado. Remote Sens 11(15):1748. https://doi.org/10.3390/rs11151748
Hassan M, Yang M, Rasheed A, Yang G, Reynolds M, Xia X, Xiao Y, He Z (2019) A rapid monitoring of NDVI across the wheat growth cycle for grain yield prediction using a multi-spectral UAV platform. Plant Sci 282:95–103. https://doi.org/10.1016/j.plantsci.2018.10.022

Herzig P, Borrmann P, Knauer U, Klück H-C, Kilias D, Seiffert U, Pillen K, Maurer A (2021) Evaluation of RGB and multispectral unmanned aerial vehicle (UAV) imagery for high-throughput phenotyping and yield prediction in barley breeding. Remote Sens 13(14):2670. https://doi.org/10.3390/rs13142670

Hickey LT, Hafeez N, Robinson H, Jackson SA, Leal-Bertioli SCM, Tester M, Gao C, Godwin ID, Hayes BJ, Wulff BBH (2019) Breeding crops to feed 10 billion. Nat Biotechnol [online] 37(7):744–754. https://doi.org/10.1038/s41587-019-0152-9

Jiang J, Zhang Z, Cao Q, Liang Y, Krienke B, Tian Y, Zhu Y, Cao W, Liu X (2020) Use of an active canopy sensor mounted on an unmanned aerial vehicle to monitor the growth and nitrogen status of winter wheat. Remote Sens 12(22):3684. https://doi.org/10.3390/rs12223684

Joalland S, Screpanti C, Varella H, Reuther M, Schwind M, Lang C, Walter A, Liebisch F (2018) Aerial and ground based sensing of tolerance to beet cyst nematode in sugar beet. Remote Sens 10(5):787. https://doi.org/10.3390/rs10050787

Johansen K, Morton MJL, Malbeteau YM, Aragon B, Al-Mashharawi SK, Ziliani MG, Angel Y, Fiene GM, Negrão SSC, Mousa MAA, Tester MA, McCabe MF (2019) Unmanned aerial vehicle-based phenotyping using morphometric and spectral analysis can quantify responses of wild tomato plants to salinity stress. Front Plant Sci 10. https://doi.org/10.3389/fpls.2019.00370

Kakarla SC, Ampatzidis Y (2021) Types of unmanned aerial vehicles (UAVs), sensing technologies, and software for agricultural applications EDIS. 2021(5). https://doi.org/10.32473/edis-ae565-2021

Karunathilake, E.M.B.M., Le, A.T., Heo, S., Chung, Y.S. and Mansoor, S. (2023). The Path to Smart Farming: Innovations and Opportunities in Precision Agriculture. Agriculture, [online] 13(8), p.1593. https://doi.org/10.3390/agriculture13081593

Kang J, Sui L, Yang X, Wang Z, Huang C, Wang J (2019) Spatial pattern consistency among different remote-sensing land cover datasets: a case study in Northern Laos. ISPRS Int J Geo Inf 8(5):201. https://doi.org/10.3390/ijgi8050201

Kawamura K (2011) Mapping herbage biomass and nitrogen status in an Italian ryegrass (Lolium multiflorum L.) field using a digital video camera with balloon system. J Appl Remote Sens 5(1):053562. https://doi.org/10.1117/1.3659893

Kefauver SC, Vicente R, Vergara-Díaz O, Fernandez-Gallego JA, Kerfal S, Lopez A, Melichar JPE, Serret Molins MD, Araus JL (2017) Comparative UAV and field phenotyping to assess yield and nitrogen use efficiency in hybrid and conventional barley. Front Plant Sci 8. https://doi.org/10.3389/fpls.2017.01733

Khot L (n.d.) Unmanned aerial systems in agriculture: Part 1 (systems) UAS in AG series unmanned aerial systems in agriculture: Part 1 (systems)

Kross A, McNairn H, Lapen D, Sunohara M, Champagne C (2015) Assessment of RapidEye vegetation indices for estimation of leaf area index and biomass in corn and soybean crops. Int J Appl Earth Obs Geoinf 34:235–248. https://doi.org/10.1016/j.jag.2014.08.002

Lei L, Qiu C, Li Z, Han D, Han L, Zhu Y, Wu J, Xu B, Feng H, Yang H, Yang G (2019) Effect of leaf occlusion on leaf area index inversion of maize using UAV–LiDAR data. Remote Sens [online] 11(9):1067. https://doi.org/10.3390/rs11091067

Li J, Tang L (2017) Developing a low-cost 3D plant morphological traits characterization system. Comput Electron Agric 143:1–13. https://doi.org/10.1016/j.compag.2017.09.025

Li L, Zhang Q, Huang D (2014) A review of imaging techniques for plant phenotyping. Sensors 14(11):20078–20111. https://doi.org/10.3390/s141120078

Li B, Xu X, Zhang L, Han J, Bian C, Li G, Liu J, Jin L (2020) Above-ground biomass estimation and yield prediction in potato by using UAV-based RGB and hyperspectral imaging. ISPRS J Photogramm Remote Sens 162:161–172. https://doi.org/10.1016/j.isprsjprs.2020.02.013

Lobet G, Draye X, Périlleux C (2013) An online database for plant image analysis software tools. Plant Methods 9(1):38. https://doi.org/10.1186/1746-4811-9-38

López-Granados F, Torres-Sánchez J, Serrano-Pérez A, de Castro AI, Mesas-Carrascosa F-J, Peña J-M (2015) Early season weed mapping in sunflower using UAV technology: variability of herbicide treatment maps against weed thresholds. Precis Agric 17(2):183–199. https://doi.org/10.1007/s11119-015-9415-8

Maes WH, Steppe K (2019) Perspectives for remote sensing with unmanned aerial vehicles in precision agriculture. Trends Plant Sci 24(2):152–164. https://doi.org/10.1016/j.tplants.2018.11.007

Makanza R, Zaman-Allah M, Cairns J, Magorokosho C, Tarekegne A, Olsen M, Prasanna B (2018) High-throughput phenotyping of canopy cover and senescence in maize field trials using aerial digital canopy imaging. Remote Sens 10(2):330. https://doi.org/10.3390/rs10020330

Matese A, Di Gennaro SF, Orlandi G, Gatti M, Poni S (2022) Assessing grapevine biophysical parameters from unmanned aerial vehicles hyperspectral imagery. Front Plant Sci 13. https://doi.org/10.3389/fpls.2022.898722

Mochida K, Koda S, Inoue K, Hirayama T, Tanaka S, Nishii R, Melgani F (2018) Computer vision-based phenotyping for improvement of plant productivity: a machine learning perspective. GigaScience [online] 8(1):153. https://doi.org/10.1093/gigascience/giy153

Morisse M, Wells DM, Millet EJ, Lillemo M, Fahrner S, Cellini F, Lootens P, Muller O, Herrera JM, Bentley AR, Janni M (2022) A European perspective on opportunities and demands for field-based crop phenotyping. Field Crop Res 276:108371. https://doi.org/10.1016/j.fcr.2021.108371

Nebiker S, Lack N, Abächerli M, Läderach S (2016) Light-weight multispectral UAV sensors and their capabilities for predicting grain yield and detecting plant diseases. Int Arch Photogramm Remote Sens Spat Inf Sci XLI-B1:963–970. https://doi.org/10.5194/isprs-archives-xli-b1-963-2016

OpenDroneMap (2018) WebODM drone software – OpenDroneMap. [online]. Available at: https://www.opendronemap.org/webodm/. Accessed 29 Jan 2023

Pasala R, Pandey BB (2020) Plant phenomics: high-throughput technology for accelerating genomics. J Biosci 45(1). https://doi.org/10.1007/s12038-020-00083-w

Patrick A, Pelham S, Culbreath A, Holbrook C, José De Godoy I, Li C (2017) IEEE instrumentation & measurement magazine high throughput phenotyping of tomato spot wilt disease in peanuts using unmanned aerial systems and multispectral imaging

Pieruschka R, Schurr U (2019) Plant phenotyping: past, present, and future. Plant Phenomics 2019:1–6. https://doi.org/10.34133/2019/7507131

Pix4D (2011) Pix4D. [online]. Available at: https://www.pix4d.com/. Accessed 15 Nov 2022

Pix4D (n.d.) PIX4Dcapture: free drone flight planning mobile app. [online]. Available at: https://www.pix4d.com/product/pix4dcapture. Accessed 30 Jan 2023

Pölönen I, Saari H, Kaivosoja J, Honkavaara E, Pesonen L (2013) Hyperspectral imaging based biomass and nitrogen content estimations from light-weight UAV. Remote Sens Agric Ecosyst Hydrol XV. https://doi.org/10.1117/12.2028624

QGIS (2017) Welcome to the QGIS project! [online] Qgis.org. Available at: https://www.qgis.org/en/site/. Accessed 29 Jan 2023

Romero M, Luo Y, Su B, Fuentes S (2018) Vineyard water status estimation using multispectral imagery from an UAV platform and machine learning algorithms for irrigation scheduling management. Comput Electron Agric 147:109–117. https://doi.org/10.1016/j.compag.2018.02.013

Roth L, Streit B (2017) Predicting cover crop biomass by lightweight UAS-based RGB and NIR photography: an applied photogrammetric approach. Precis Agric 19(1):93–114. https://doi.org/10.1007/s11119-017-9501-1

Sangjan W, McGee RJ, Sankaran S (2022) Optimization of UAV-based imaging and image processing orthomosaic and point cloud approaches for estimating biomass in a forage crop. Remote Sens 14(10):2396. https://doi.org/10.3390/rs14102396

Sankaran S, Zhou J, Khot LR, Trapp JJ, Mndolwa E, Miklas PN (2018) High-throughput field phenotyping in dry bean using small unmanned aerial vehicle based multispectral imagery. Comput Electron Agric 151:84–92. https://doi.org/10.1016/j.compag.2018.05.034

Singh D, Wang X, Kumar U, Gao L, Noor M, Imtiaz M, Singh RP, Poland J (2019) High-throughput phenotyping enabled genetic dissection of crop lodging in wheat. Front Plant Sci 10. https://doi.org/10.3389/fpls.2019.00394

Swarup A, Lee WS, Peres N, Fraisse C (2020) Strawberry plant wetness detection using color and thermal imaging. J Biosyst Eng 45(4):409–421. https://doi.org/10.1007/s42853-020-00080-9

Ubbens J, Cieslak M, Prusinkiewicz P, Stavness I (2018) The use of plant models in deep learning: an application to leaf counting in rosette plants. Plant Methods 14(1). https://doi.org/10.1186/s13007-018-0273-z

Wang S, Guan K, Wang Z, Ainsworth EA, Zheng T, Townsend PA, Liu N, Nafziger E, Masters MD, Li K, Wu G, Jiang C (2021) Airborne hyperspectral imaging of nitrogen deficiency on crop traits and yield of maize by machine learning and radiative transfer modeling. Int J Appl Earth Obs Geoinf 105:102617. https://doi.org/10.1016/j.jag.2021.102617

White JW, Andrade-Sanchez P, Gore MA, Bronson KF, Coffelt TA, Conley MM, Feldmann KA, French AN, Heun JT, Hunsaker DJ, Jenks MA, Kimball BA, Roth RL, Strand RJ, Thorp KR, Wall GW, Wang G (2012) Field-based phenomics for plant genetics research. Field Crop Res 133:101–112. https://doi.org/10.1016/j.fcr.2012.04.003

www.dronedeploy.com (n.d.) Drone mapping software | drone mapping app | UAV mapping | surveying software | DroneDeploy. [online]. Available at: https://www.dronedeploy.com. Accessed 30 Jan 2023

www.licor.com (n.d.) LI-COR biosciences – impacting lives through science. [online]. Available at: https://www.licor.com/env/products/leaf_area/LI-3100C/. Accessed 29 Jan 2023

www.specmeters.com (n.d.) SPAD 502 plus chlorophyll meter | spectrum technologies. [online]. Available at: https://www.specmeters.com/nutrient-management/chlorophyll-meters/chlorophyll/spad502p/

Xu R, Li C, Paterson AH (2019) Multispectral imaging and unmanned aerial systems for cotton plant phenotyping. PLoS One 14(2):e0205083. https://doi.org/10.1371/journal.pone.0205083

Yeom, J., Jung, J., Chang, A., Maeda, M. and Landivar, J. (2018). Automated Open Cotton Boll Detection for Yield Estimation Using Unmanned Aircraft Vehicle (UAV) Data. Remote Sensing, 10(12), p.1895. https://doi.org/10.3390/rs10121895

Yue J, Yang G, Li C, Li Z, Wang Y, Feng H, Xu B (2017) Estimation of winter wheat above-ground biomass using unmanned aerial vehicle-based snapshot hyperspectral sensor and crop height improved models. Remote Sens 9(7):708. https://doi.org/10.3390/rs9070708

Yue J, Feng H, Jin X, Yuan H, Li Z, Zhou C, Yang G, Tian Q (2018) A comparison of crop parameters estimation using images from UAV-mounted snapshot hyperspectral sensor and high-definition digital camera. Remote Sens 10(7):1138. https://doi.org/10.3390/rs10071138

Zhao Y, Zhai Y, Dubois E, Wang S (2016) Image matching algorithm based on SIFT using color and exposure information. J Syst Eng Electron [online] 27(3):691–699. https://doi.org/10.1109/JSEE.2016.00072

Zhou L, Gu X, Cheng S, Yang G, Shu M, Sun Q (2020) Analysis of plant height changes of lodged maize using UAV-LiDAR data. Agriculture 10(5):146. https://doi.org/10.3390/agriculture10050146

Zhu W, Sun Z, Yang T, Li J, Peng J, Zhu K, Li S, Gong H, Lyu Y, Li B, Liao X (2020) Estimating leaf chlorophyll content of crops via optimal unmanned aerial vehicle hyper-spectral data at multi-scales. Comput Electron Agric 178:105786. https://doi.org/10.1016/j.compag.2020.105786

Digital Yield Predictions

Tarmo Lipping and Petteri Ranta

Abstract Yield prediction is a vast area of study involving different fields of science such as agriculture, plant physiology, informatics, and machine learning. Numerous review papers have been published on various aspects of yield prediction. Instead of focusing on certain types of models, data sources, or crops, we provide a general overview of the methods used for forecasting crop yield. We first consider various sources of data used in yield prediction efforts as well as the various measures to assess prediction accuracy. We then give a brief overview on plant physiology-based yield simulation models. Although the main aim of these models is usually not to forecast crop yield as accurately as possible, they describe the phenomena of plant growth that ultimately underlie all efforts related to yield prediction. After that, a more comprehensive overview is given on the various types of machine learning methods applied to yield prediction in exponentially increasing number of studies. We first describe the conventional feature-based machine learning techniques after which the use of several deep learning methods for yield prediction is considered.

Keywords Yield prediction · Prediction accuracy · Crop growth modelling · Machine learning · Deep neural networks

1 Introduction

Yield prediction is a general term describing an attempt to forecast crop yield in a forthcoming harvest season. Virtually all the efforts aiming at yield prediction

T. Lipping (✉)
Faculty of Information Technology and Communication Sciences, Tampere University, Pori, Finland
e-mail: tarmo.lipping@tuni.fi

P. Ranta
Geoloc Oy, Petäjävesi, Finland
e-mail: petteri.ranta@geoloc.fi

P. M. Priyadarshan et al. (eds.), *Digital Agriculture*,
https://doi.org/10.1007/978-3-031-43548-5_12

involve data and some kind of an underlying model. A variety of different efforts can be considered when talking about yield prediction, depending on the aim of the prediction, the type of the model used (from gut feeling to complex plant physiology-based models to rigorous deep learning algorithms), the type of crop under consideration, the time horizon of the prediction, the scale of yield assessment (intra-field vs. regional), etc.

Yield prediction can be done at a global or national scale in which case the aim of the prediction is often economical (to predict the market prices, for example) or related to the security of supply. As market prices and food supply are complex issues depending, in addition to yield, on geopolitical situation, logistics, financial issues, etc., yield prediction driven by these aims is not considered in this chapter. Also, the data sources used to predict the yield are different in these efforts including official statistics, questionnaires to the farmers, etc. Instead, in this chapter we focus on yield prediction performed at the scale of a single crop field (or even at the subfield scale) over which the soil properties and growth conditions can be considered more or less constant. Indeed, this kind of effort will produce information for yield prediction at the global or national scale, but the immediate goal is different.

The aim of the yield prediction effort depends also on the time horizon of the prediction. One can develop a general model using the data from multiple years (together with general knowledge on plant physiology) to inform the farmers which crops and varieties would have higher potential in their particular environment or when to sow the crops. Slightly different aims can be considered when the predictions are made during the actual growth season the yield of which is predicted. Having yield prediction maps at the subfield level in the budding phase of plant growth could still inform the farmer on possible actions to be taken before harvest. Also, yield prediction can be performed using either a 'snapshot' of data at a certain time or a time series of data (weekly acquired weather data or remote sensing data, for example).

In this chapter we mainly focus on yield prediction based on the data acquired from the growth environment during the particular (and possibly also previous) growth season. While some data remain relatively constant (such as soil properties or climate conditions), other kinds of data can change radically from year to year (weather conditions). The main goals of this kind of yield prediction are to better understand the relationship between the environmental parameters and the yield and to provide as accurate prediction as possible. The practical value of these kinds of efforts is to provide the involved stakeholders (farmers, food industry, and regulators) information to support their decision-making (what crops to cultivate and how to improve the growth conditions by means of, for example, drainage, irrigation, soil shaping, or fertilization). The primary type of crops considered is cereals, although the models can be applied also to growing vegetables (but not so well to growing fruits or to the subject area of horticulture).

We will first have a look at the various data sources used in yield prediction and the measures used to assess the goodness of the prediction result. We will then consider two kinds of models—those based on plant physiology and those based on data alone—in separate sections of the chapter. The models based on plant physiology are presented only briefly and mostly for reference purposes, the main emphasis being on data-driven yield prediction models using machine learning.

2 Prerequisites of Yield Prediction

Whatever the method or model of yield prediction, it is always based on data. In the case of physics-based models, the data is used to calibrate or tune the model while in the case of data-driven models, such as those based on deep learning, the data are used for training the model parameters. In this section we first describe some common sources of data used in calibration or training yield prediction models. We then discuss various measures used in the evaluation of the accuracy of yield prediction models.

2.1 Source Data

Soil Data Soil and its properties (composition and structure) play a major role in how plants grow and produce yield. A wide variety of variables can be derived such as:

- Soil acidity (pH)
- Cation exchange capacity (CEC)
- Soil type
- Soil chemical content (potassium and magnesium, for example)
- Soil structure (clay or sand content, soil texture, etc.)
- Water-related properties such as water holding capacity or water permeability.

Different ways can be used to acquire subsets of these variables. A chemical or structural analysis of soil samples in a laboratory setting is the most direct way to estimate soil properties. However, the collected data is sparse and it is often difficult to decide, where in the field the samples should be taken, especially in areas where soil properties change abruptly. More efficient sampling of soil chemical content can be done using portable hand-held X-ray fluorescence (XRF) devices (Weindorf and Chakraborty 2020). These devices enable determination of the amount of chemical elements in soil at a certain location in the field without the need for preprocessing the sample. However, the data acquisition is still manual and requires a license to operate the device. Other ways of acquiring soil property data are scanning for Electrical Conductivity (EC) (Stadler et al. 2015) or using the Ground Penetrating Radar (GPR) (Linna et al. 2022).

Features such as soil moisture or soil temperature can also be used in yield prediction models. These features are closely related to weather or climate data and can be directly measured using soil sensors. A variety of solutions are available from miniature wireless underground sensor systems (Tiusanen 2013) to tubes equipped with sensors at various depths providing stratigraphic data (Shah et al. 2012) of soil moisture and temperature. On the other hand, soil properties can also be obtained as target variables when applying machine learning methods to remote sensing data (Tantalaki et al. 2019). In this case soil properties, estimated from satellite or drone data, can be fed into the yield prediction models.

Remote Sensing Remote sensing settings can be divided according to the host platform of the sensor. During recent years Unmanned Aerial Vehicles (UAVs), more commonly called drones, have become popular in remote sensing of agricultural land. Initial expectations of autonomous data acquisition with drones have not fully realized as in most countries restrictions on using UAVs are in place. In Finland, for example, there has to be constant visual contact between the operator and the drone and the operator has to have a license.

The Unmanned Aerial Systems (UASs) used in the context of smart farming usually contain a separate sensor mounted to the UAV platform, while in consumer systems, an RGB camera is often integrated to the drone. The most common sensor types used with UAVs include RGB cameras, multispectral cameras, hyperspectral cameras, thermal sensors, and lidar devices (Messina and Modica 2020; Tsouros et al. 2019). While RGB cameras use three wavelength bands in the visual range of 400–700 nm, multispectral cameras typically add one or more additional bands at the Near-InfraRed (NIR) region. In agricultural applications the main role of these additional bands is to cover the red edge in the spectrum caused by chlorophyll in plants. Hyperspectral cameras differ from the multispectral ones in that they cover a certain spectral range (usually either up to about 1100 nm or about 2500 nm) in consecutive wavelength bands. Thermal sensors (wavelength of 3–8 μm) measure the surface temperature of the foliage and are mainly used for monitoring plant water stress and detecting plant diseases (Messina and Modica 2020). Lidar is the only active measurement technique in the above list as it measures the reflection of an emitted light beam from the surface. Using lidar techniques, the elevation map of a crop field can be produced. In addition, by analyzing the waveform of the reflected pulse, the structure of the targets can be characterized.

Remote sensing data from high-altitude satellite systems form another important data source in smart farming and yield prediction applications. Data from satellites were available long before UAVs became available, however, after the launches of higher resolution systems such as Landsat 8 in 2013 and Sentinel 2 in 2015, and after several operators have started to offer their data for free over a well-defined interface, the number of studies and services based on these kinds of data has increased significantly. A comprehensive list of satellite missions can be found in the Satellite Missions Catalogue,[1] the most common platforms employed in agricultural applications being Landsat 7&8, Sentinel 2, WorldView 2&3, and Geofen 1&2. All

[1] https://www.eoportal.org/satellite-missions

these missions provide remote sensing data in the optical range of the spectrum starting from 400 nm. The spatial resolution of the data varies from 0.31 m/pixel for commercial WorldView satellites to 10...60 m/pixel for open access Landsat and Sentinel missions. In addition to the optical range, satellite data from Synthetic Aperture Radar (SAR) missions such as Sentinel 1 or TerraSAR-X have been used in crop yield prediction (Alebele et al. 2021). As mentioned above, remote sensing data can either be used directly for developing yield prediction models or they can be used to derive features such as soil or plant moisture, soil temperature, nitrogen level, etc. to be further used in plant physiology-based yield prediction models. Even if satellite data are freely available, processing and interpretation of the data requires expert knowledge and the farmers usually rely on either public or commercial service providers.

Weather Data Weather data are probably the most common data source when decisions are made on immediate actions in agricultural production. In contrast to other data sources considered, weather data are often freely available from publicly maintained weather stations. Various derived parameters such as growing degree days may also be available. However, if more accurate and location-specific weather data is required, a private weather station can be installed. More advanced weather stations can provide data on a wide variety of environmental factors such as air temperature, wind speed and direction, atmospheric pressure, light intensity, solar radiation, and precipitation. Indeed, weather data are related to soil temperature and moisture, and due to easy access and interpretation, weather data provide valuable additional information for yield prediction models. Physical models for yield prediction usually involve weather-related parameters directly, whereas in data-driven models they can be used as additional data features. Some studies have even built deep learning models solely on weather data to predict crop growth stages (Yue et al. 2020).

Yield Maps To validate yield prediction models, reference data on actual yield is required. The traditional approach to measuring crop yield is to weigh the harvested grain and calculate the average in a field by field basis. This kind of yield data can be used if the scope of yield prediction is county level, for example (Wang et al. 2020). To obtain data on intra-field variability of crop yield, yield monitoring devices can be mounted to harvesters. These devices may be based on optical measurement or on kinetic mass flow sensors. Also, accurate logging of the location of the harvester is required using satellite navigation systems. While harvester-mounted yield monitors are becoming more common among farmers, the skills required to extract and preprocess the data often hinder their use locally. Different vendors use different data formats and the data need to be corrected for several factors such as the properties of the grain (moisture level, for example) or incomplete swathes of harvesting. Also, point data obtained from the yield monitors need to be aggregated and rasterized. Other methods have also been proposed for intra-field yield assessment such as manual yield assessment within a standard frame at several locations of the field (Narra et al. 2022). The yield map can then be formed by coarse interpolation of the sampled yield values.

2.2 Assessment of Prediction Accuracy

In validating crop yield models for either parameter calibration of physical models or training of machine learning models, some kind of metrics is needed to estimate prediction error. Let us denote the predicted and true yield at location i by $\hat{y}_i$ and y_i, respectively. The most common error metrics include

$$\text{Mean Absolute Error:}\;\; \text{MAE} = \frac{1}{N}\sum_{i=1}^{N}|\hat{y}_i - y_i| \tag{1}$$

or

$$\text{Root Mean Squared Error:}\;\; \text{RMSE} = \sqrt{\frac{1}{N}\sum_{i=1}^{N}(\hat{y}_i - y_i)^2}, \tag{2}$$

where N is the number of individual units of yield measurement. If the units are of different size (as in the case of yield prediction on a field-by-field basis), the yield values should be normalized by the area of the corresponding field. In the case of intra-field yield assessment, usually yield in equal-sized units (say, 10×10 m) is considered.

The MAE and RMSE error metrics are useful if prediction errors obtained for the same crop in similar growing conditions are compared. Otherwise, it would be more useful to calculate relative error metrics such as

$$\text{Mean Absolute Percentage Error:}\;\; \text{MAPE} = \frac{1}{N}\sum_{i=1}^{N}\left|\frac{\hat{y}_i - y_i}{y_i}\right| \cdot 100\%, \tag{3}$$

or

$$\text{Relative Root Mean Square Error:}\;\; \text{RRMSE} = \sqrt{\frac{\frac{1}{N}\sum_{i=1}^{N}(\hat{y}_i - y_i)^2}{\sum_{i=1}^{N}\hat{y}_i^{\,2}}} \cdot 100\%. \tag{4}$$

Another popular performance metric of crop yield prediction models is the coefficient of determination R^2. R^2 evaluates how well the true versus predicted yield values follow the linear regression line and can be calculated as

$$R^2 = 1 - \frac{\sum_{i=1}^{N}(\hat{y}_i - y_i)^2}{\sum_{i=1}^{N}(y_i - \mu_y)^2}, \tag{5}$$

where μ_y is the average over the true yield values. In their review on crop yield prediction using machine learning, van Klompenburg et al. have found that in 50 selected studies RMSE was used 29, R^2 19, and MAE 8 times as the metric of the prediction error (van Klompenburg et al. 2020).

Other metrics for model efficiency used in the context of yield prediction include

$$\text{Coefficient of Residual Mass: CRM} = \frac{\sum_{i=1}^{N} y_i - \sum_{i=1}^{N} \hat{y}_i}{\sum_{i=1}^{N} y_i}, \tag{6}$$

used in the assessment of crop yield prediction models in Chipanshi et al. (2015), for example, or

$$\text{Lin's Concordance Correlation Coefficient: LCCC} = \frac{2r_{y\hat{y}}\sigma_y\sigma_{\hat{y}}}{\sigma_y^2 + \sigma_{\hat{y}}^2 + (\mu_y - \mu_{\hat{y}})^2}, \tag{7}$$

where σ_y^2 and $\sigma_{\hat{y}}^2$ are the variances of the true and predicted yield, respectively, and r is the correlation coefficient between the two variables. LCCC measures the goodness of linear regression between predicted and true yield and is used, for example, in Filippi et al. (2019). Still other metrics, more suitable for usage in the context of physics-based models, include the Skill Score (SS) (Johnson et al. 2016) and the ecological distance measure (Tian et al. 2020). It is common to use several error metrics in a single study to better characterize the model behavior.

The above list of yield prediction accuracy assessment measures is not comprehensive, and in individual studies several other metrics have been used. The selection of appropriate metrics should take into account the type of prediction model as well as the usage of the metrics (i.e., for what comparison is the metrics used for).

3 Physics-Based Models for Crop Yield Prediction

There are many plant physiology-based crop growth models available. EU Joint Research Center (JRC) launched the Monitoring Agricultural ResourceS (MARS) initiative in 1988 to acquire information on crop production using remote sensing technology (van der Velde et al. 2019). The crop monitoring and yield forecasting are currently performed by the Food Security Unit of the European Commission's Joint Research Center using the MARS Crop Yield Forecasting System (MCYFS). Part of this system is the crop simulation module relying on crop models. The main crop growth model used within the MCYFS is the WOFOST (acronym for WOrld FOod STudies) model (de Wit et al. 2019), introduced already in 1989 (van Diepen et al. 1989) and updated continuously since. WOFOST explains crop growth based on the underlying processes such as photosynthesis and respiration. The effects of environmental conditions on these processes are considered when monitoring and forecasting crop growth and yield. WOFOST is open source and numerous

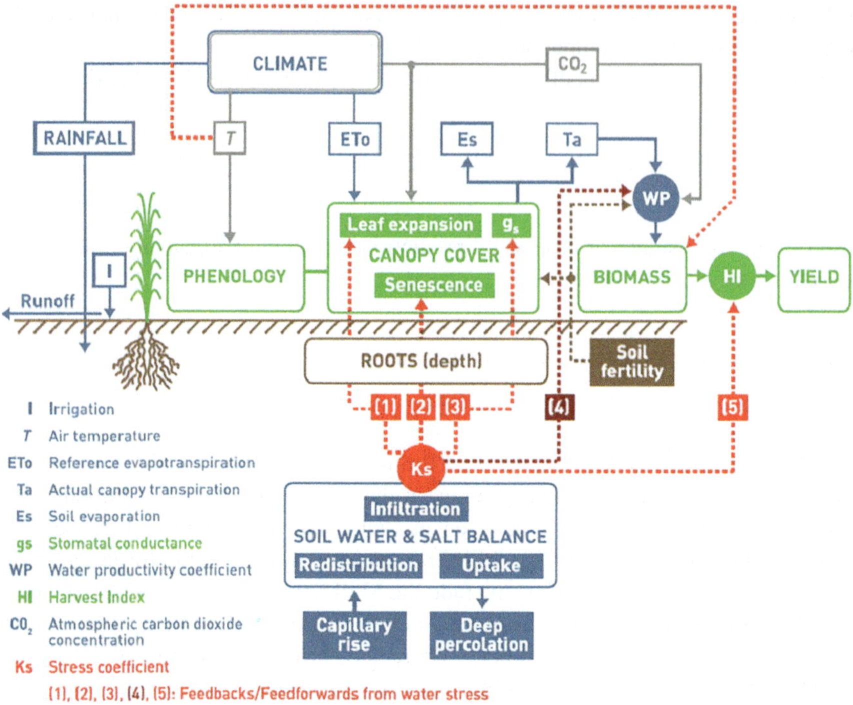

Fig. 1 Schematic of the Aquacrop model (https://www.fao.org/3/i6321e/i6321e.pdf)

implementations of its conceptual framework exist. The model has been used for modeling a wide variety of crops such as wheat, barley, maize, potato, sunflower, and rice in different growing conditions from Europe to China.[2] Other more limited models used in the MCYFS context include:

- WARM: a simplified user-friendly growth model for paddy rice crops
- CropSyst: a multi-layer multi-crop model designed to study the effect of cropping systems management on productivity
- CANERGO: sugarcane growth model based on daily weather data, soil properties, and data on management.

At the global level, the Food and Agriculture Organization (FAO) of the United Nations has developed the Aquacrop model, widely used to simulate the dependence of crop growth on water and nutrient availability (Steduto et al. 2009). The model is based on converting transpiration into biomass through water productivity. Biomass is connected to yield via the Harvest Index (HI) parameter (see Fig. 1). Similarly to WOFOST, numerous open source implementations of the Aquacrop model exist.

[2] https://marswiki.jrc.ec.europa.eu/agri4castwiki/index.php/Crop_Simulation

In Todorovic et al. (2009) the Aquacrop model is compared with the WOFOST and Cropsyst models in the simulation of sunflower growth under different water regimes. The authors note that whereas Aquacrop is water-driven, Cropsyst can be considered both water- and radiation-driven and the WOFOST model is carbon-driven. It is found that the performance of the three models is similar in simulating biomass and yield, while Aquacrop requires less input parameters. In Mkhabela and Bullock (2012) the performance of the Aquacrop model in simulating yield and soil moisture for wheat is assessed. The model appears to model soil moisture better than yield (R^2 of 0.90 vs 0.66, respectively). Aquacrop is compared to the WOFOST model for potato crop in Quintero and Díaz (2020). Both models gave correlation over 0.99 between the true and simulated harvestable biomass.

A major challenge in applying physics-based crop growth models for yield simulation and forecasting is model calibration. For example, the Aquacrop model has more than 50 input variables or model parameters that should be determined to run the model. Modeling can be performed at a field scale with more precise parameter values or at a regional scale with different calibration for different crops and their varieties as well as different climatic conditions. In Silvestro et al. (2017) the sensitivity of the Aquacrop model to its parameters has been studied using the Morris and EFAST (Extended Fourier Amplitude Sensitivity Test) techniques. In the study, Aquacrop is compared to a more simple SAFYE (Simple Algorithm For Yield expanded with the evapotranspiration component) model (Duchemin et al. 2008) in complexity and plasticity for wet and dry conditions. SAFYE was found to be less complex but of less plasticity.

The main aim of the plant physiology-based crop models is usually not to estimate the crop yield as accurately as possible but rather to understand the factors affecting crop growth, biomass generation, and yield production. The target variables in these models can be other than yield (biomass or leaf area index, for example). The brief presentation of these models here is meant to underline the importance of relating the data-driven yield prediction models to the physiology of plant growth. Performing yield prediction using remote sensing or environmental data is, in fact, an indirect way to assess the factors of crop growth and yield production.

4 Data-Driven Yield Prediction Using Machine Learning

In this section yield prediction methods relying completely on the underlying data are discussed, i.e., no physical model of plant growth or growth environment is considered. Although the prediction algorithm can be called a *model* also in this case, the model is purely computational and its parameters are determined based on the data by some learning algorithm. If the learning algorithm involves training data (i.e., data for which the true yield value is known), it is called supervised, otherwise unsupervised learning or clustering is in question. The amount of training data required to train a supervised learning algorithm depends on the complexity of the computational model, its structure, and the number of parameters.

While recently more attention has been paid on deep learning models, the so-called conventional classification or regression models are still intensively used. Although it is difficult to draw a strict line between the two types of models from the application point of view, the main difference is that the conventional methods are usually based on precalculated features or properties of the data while deep learning models work on raw data. The number of parameters is usually much higher in deep learning models, and therefore, more data need to be used in their training. Deep learning models are usually not as sensitive to occasional errors in data as the conventional methods; on the other hand, they are only as good as their training data and biased training data will produce biased predictions. Deep learning models can comprehend a large amount of available data of different modalities being capable of combining virtually all the data sources available for a particular task (see Sect. 2.1). The results obtained with deep learning models are difficult to track or interpret, and although methods exist to pinpoint the features in the source data that affect the prediction results most, it is still difficult to relate the performance of the model to certain phenomena.

In the following a brief overview of the conventional machine learning methods and their usage in yield prediction is given. After that, three main types of deep learning models, Convolutional Neural Networks (CNNs), Recurrent Neural Networks (RNNs), and Transformer Neural Networks (TNNs) are discussed. The following is not a comprehensive literature review of the usage of these models; the reader is directed to the numerous review papers on the subject. The aim is to provide a general overview on the various methods with examples of their use for yield prediction.

4.1 Feature-Based Methods in Yield Prediction

The conventional classification models can be roughly divided into three categories: regression analysis, Bayesian models, and decision trees.

4.1.1 Regression Analysis

The main idea behind these models is to divide the feature space into subareas based on what is known about the true yield in the form of the training data. For example, the training data samples can be projected to the feature space formed by two or more wavelength bands of a remotely sensed data set and a discrimination curve can be defined to optimally separate the data points according to the true yield values. Probably the most common method in this category is the Support Vector Machine (SVM). In its basic form the SVM works in a two-dimensional feature space producing a linear separation line between two classes (in our case, the data points corresponding to yield higher or lower with respect to a certain threshold). Using modifications of the SVM such as kernel functions, SVMs can fit nonlinear discrimination functions and be used in higher dimensional feature space with more than two classes.

SVMs have been widely used for the classification of remotely sensed data acquired from crop fields, especially when evaluating the performance of more advanced deep learning (DL) methods in their early applications (Kim and Lee 2016; Ji et al. 2018). They are still commonly used in agricultural applications, including yield prediction (Kuradusenge et al. 2023). A common usage of SVMs is in combination with the CNNs (see Sect. 4.2) as a classification layer working on the features provided by the convolutional layers of the CNN (Tao and Wei 2022).

4.1.2 Bayesian Methods in Yield Prediction

In its simplest form Bayesian yield prediction models are based on the probabilities:

$$p(Y_k|\mathbf{x}) = \frac{p(Y_k)p(\mathbf{x}|Y_k)}{p(\mathbf{x})}, \tag{8}$$

where $p(Y_k|\mathbf{x})$ is the probability of certain yield range k given feature vector $\mathbf{x}$ (posterior probability), $p(Y_k)$ is the prior probability of having yield in the range k, $p(\mathbf{x}|Y_k)$ is the likelihood that if yield values are in the range k, certain feature vector $\mathbf{x}$ has occurred, and $p(\mathbf{x})$ is the probability of having a certain feature vector $\mathbf{x}$ in the first place (i.e., evidence). Thus, to determine the model one should have the knowledge on how the probability of observing certain source data values (wavelength band values in remote sensing or temperature/precipitation sums, for example) relates to the probability of having yield in certain ranges. Once the probabilities have been determined using the training data, the model can be used for obtaining the posterior probability of future yield values given the input feature vector.

The Bayesian method has the additional advantage of obtaining the uncertainty of the predicted yield values. Also, information about the sensitivity of the model output to changes in the input variables is inherently present in the model, while in the case of DL models, Monte Carlo analysis should be performed to assess the sensitivity of the model to its input. Bayesian inference is also widely used with physics-based model. The probabilities in Eq. 8 can be based on physical models and the knowledge on the underlying phenomena instead of using the training data.

An example of maize yield prediction based on temperature and precipitation using Bayesian inference is presented in Shirley et al. (2020). In Bazrafshan et al. (2022) Bayesian analysis is used to quantify the uncertainty of the parameters and input variables of yield prediction models that rely on other techniques such as multi-layer perceptrons or neuro-fuzzy models.

4.1.3 Decision Trees in Yield Prediction

The basic idea behind decision trees is to use expert knowledge in classifying the input feature vector by comparing the values of the features to predetermined thresholds in a step-by-step manner. As the models described in Sect. 4.1.1, decision trees also divide the feature space into subareas, however, the resulting subareas are rectangles bordered by threshold values used in the tree. From their basic form, decision tree models have developed into ensemble structures where a large number of individual decision trees are applied and their outputs are aggregated according to some rules. These methods are commonly referred to as Random Forest (RF) classifiers. In the context of machine learning, the thresholds used at the tree nodes are determined based on the training data. Also, the structure of the trees can be optimized (referred to as tree pruning). A deep learning approach to decision trees is provided by the eXtreme Gradient Boosting (XGBoost) software library including algorithms for penalization of trees, tree pruning, randomization (to avoid overfitting), and automatic feature selection.

Several studies applying SVMs to perform yield prediction also use RF classifiers in comparison (Kim and Lee 2016; Jhajharia et al. 2023). In Jhajharia et al. (2023) the RF classifier outperformed several other methods including SVM and LSTM (see Sect. 4.3). This indicates that the conventional prediction models have still their advantages despite the shift in the main focus of machine learning-based yield prediction toward DL models. In Huber et al. (2022) the XGBoost model is compared to DL models in soybean yield prediction with the advantage of more transparent prediction process. The authors encourage further experiments with the XGBoost model for other crops and geographical areas.

4.2 Convolutional Neural Network Models

Convolutional neural networks are probably the most widely used deep learning neural network architecture so far. The introduction of the pioneering 7-level LeNet-5 architecture meant the beginning of a new area in image analysis (Lecun et al. 1998). The main component of the model is the convolution operation, where a set of trainable kernels is applied to the input image, resulting in a set of features describing the data. The model learns basic features in the first layers and composite features in further layers. A fully connected (FC) network layer is then used after the convolutional layers to perform the classification. Structures where the FC layer is replaced by other classifiers such as the SVM have also been widely used.

In addition to the feature-extracting convolutional layers, several other properties of the CNNs have contributed to their popularity. The Rectified Linear Unit (ReLU) activation function used after the convolution operator, the batch normalization and pooling layers, as well as using regularization in the loss function used in error backpropagation to avoid overfitting constitute some of the properties behind the success of CNNs. As the most common application area of CNNs is image analysis, they are especially suitable for yield prediction based on remote sensing imagery. However, the kernel filters of CNNs can also be applied to one-dimensional input such as time series. On the other hand, using three-dimensional kernels (3D CNN), sequences of images (or other type of input data) can be used for yield prediction (Nevavuori et al. 2020).

The use of CNNs has been extensively studied in the context of smart farming and agriculture and several comprehensive reviews have been published on the subject. In a review published in 2018, the use of CNNs in agriculture has been considered in a set of 23 papers published between 2014 and 2017 (Kamilaris and Prenafeta-Boldú 2018). It was found that the most popular application areas of CNNs were fruit counting, plant recognition, land cover classification, weed identification, and disease detection, with one paper considering maize yield estimation (Kuwata and Shibasaki 2015). In a later review on using machine learning techniques specifically for crop yield prediction, 50 papers were considered (van Klompenburg et al. 2020). Of these, 30 papers applied deep learning models, CNN being the most popular with 15 cases. In some cases CNNs were combined with LSTMs (see Sect. 4.3) or some modification of the basic CNN architecture (such as Region-based CNN, R-CNN) was used.

4.3 Recurrent Neural Network Models

Recurrent Neural Networks (RNNs) form a subclass of deep learning architectures designed to analyze sequential data. As the term *recurrent* implies, the output of a network node can be used as an input to the same node at the next step of the sequence, forming loops. Another way to look at the network structure is having multiple nodes operating on consecutive elements of the sequence (feature vectors corresponding to consecutive sampling instances of data sources, for example). In addition to the input values, a state variable from the previous node is fed to each network node. The output of the network can be taken from all nodes forming an output sequence or just from the last node (if, for example, a single crop yield value is to be obtained based on a sequence of input feature vectors). Also, CNN layers can be applied to the input data before feeding them to the RNN nodes to automatically extract features, or FC layers can be applied to the RNN outputs for classification.

A node of an RNN structure is more complex compared to what is usually considered a node in a conventional neural network or in CNN architecture, containing several trainable parameter matrices and gates. Several modifications of RNN nodes have been introduced. The most popular RNN subclass in agricultural applications seems to be that of Long Short-Term Memory (LSTM). The main idea behind LSTM node architecture is to avoid vanishing or exploding gradients when training the network using backpropagation. There are two general concepts in the LSTM that help it learn temporal features from data. The first is the concept of memory, introduced as the cell state. The other one is the concept of gates, effectively trainable FC layers, manipulating the cell state in response to new inputs from the data and past outputs of the model. To handle the sequence of data, the model loops over the sequence, altering its cell (C) and hidden (H) states in the process using a combination of learned parameters and nonlinear activation functions.

In the review by van Klompenburg et al. (2020) 8 papers were found applying either LSTMs or hybrid methods including LSTMs to yield prediction. In a more recent review on deep learning methods for crop yield prediction using remote sensing data, 44 papers were considered. It was found that since 2018 the number of papers on the subject has been increasing exponentially and that LSTMs are gaining popularity with 30% of the studies applying this model (Muruganantham et al. 2022). Also, various hybrid architectures and subclasses of CNNs and LSTMs have been applied. In Nevavuori et al. (2020) we tested four different models (pretrained CNN, CNN-LSTM, convolutional LSTM, and 3D CNN) for the prediction of wheat, barley, and oats yield based on a sequence of UAV-based RGB data and found that the least prediction error was obtained with the 3D CNN model, while the CNN-LSTM model performed in a more stable manner (i.e., did not produce ill-fitted predictions for individual inputs).

4.4 Transformer Networks

Recently, a new deep learning architecture, generally called transformer network, has been presented. The basic transformer architecture was first introduced in Vaswani et al. (2017) for natural language processing applications such as translation. Transformer networks are based on the encoder-decoder architecture with a connection between the two. As RNNs, transformer networks are designed for the analysis of sequences of data, however, instead of sequential data processing by network nodes, joint information between all pairs of the elements of the sequence is considered by a set of computations in the multi-head attention block. A desired output sequence is fed to the decoder part for training the model and is processed by the masked multi-head attention block, the output of which is combined with the information coming from the encoder and fed to another attention block. FC layers are also used in both encoder and decoder.

Transformer networks have outperformed other deep learning structures in language models and in linguistic Artificial Intelligence. However, they have recently been successfully applied also to image analysis (using the Vision Transformer (ViT) architecture (Dosovitskiy et al. 2021)) as well as to other forms of source data. In this case, the image blocks are considered as the elements of a sequence. The blocks are encoded together with the information about the position of the block within the image before feeding to the multi-head attention. When considering yield prediction based on remote sensing data, transformer networks have the advantage of making better use of long-range and multi-level dependencies across the regions within the image (spatial dependencies) as well as long-term time dependencies in a sequence of images.

As of writing this chapter, only a few studies could be found applying transformer networks for crop yield prediction. In Liu et al. (2022) a modified version of the transformer network, called Informer (Zhou et al. 2021), was used for rice yield prediction across the Indian Indo-Gangetic Plains by combining time-series satellite data and environmental variables. The Informer model was found to give higher R^2 and lower RMSE and MAPE than the other tested models (the Least Absolute Shrinkage and Selection Operator (LASSO), RF, XGBoost, and a modification of the LSTM) almost consistently. In Bi et al. (2022) two transformer networks, the ViT for image analysis and another transformer module for time series analysis, were used for the prediction of soybean yield. The authors claim a reduction of 40% in the prediction error compared to the baseline models of CNN combined with Linear Regression and CNN-LSTM. In other studies the transformer networks have been used for crop disease detection (Jubair et al. 2021) and crop classification (Weilandt et al. 2023). These early studies are promising, and given the success of the transformer models in other application areas we can expect rapid growth in their application for crop yield prediction as well.

5 Discussion and Conclusions

This chapter is an attempt to give a brief overview on the techniques and technologies used for the prediction of crop yield. The number of studies dealing with the task has increased exponentially during recent years. One reason might be that smart farming and precision agriculture have gained a lot of attention and the amount and variety of available data to develop methods for yield prediction has also increased, especially in the area of remote sensing. Satellite data have become freely available from various sources and drones are now in the reach of all the interested users. The role of data in agriculture has been intensively discussed and rules are being developed to determine the ownership and value of data. This gives incentives to develop algorithms and tools that would make use of the data and provide additional value for stakeholders. The largest increase in studies concerning yield prediction is related to applying novel deep learning methods to the task. Yield prediction is a favorable task to test and apply these methods as the reference data is relatively easy to obtain using yield monitors, for example.

We have included a brief overview of yield monitoring models based on plant physiology in this chapter. This is usually considered as a separate subject compared to machine learning-based yield prediction. This can be justified as the aims of the two types of models are different and obtaining an accurate yield forecast is not the primary goal of physics-based models. However, we suggest that combining these two branches of research would be worth paying more attention. From the point of view of yield forecasting, the machine learning models can be considered as metamodels for physics-based crop growth models. Also, machine learning can be used within physics-based models to assist in determining model parameters and in the calibration of the model.

There is virtually an infinite set of possibilities to test and evaluate various models for crop yield prediction. The models vary according to the crops and their varieties, climatic conditions, model structures, soil types, crop management, etc. For the model to be used in practical decision-making, the use cases and limitations of the models should be well defined. Linking the deep learning models to physical properties of the growth conditions and plant physiology makes the models more reliable and encourages their use.

References

Alebele Y, Wang W, Yu W, Zhang X, Yao X, Tian Y, Zhu Y, Cao W, Cheng T (2021) Estimation of crop yield from combined optical and SAR imagery using Gaussian Kernel regression. IEEE J Sel Top Appl Earth Obs Remote Sens 14:10520–10534. https://doi.org/10.1109/JSTARS.2021.3118707. Conference Name: IEEE Journal of Selected Topics in Applied Earth Observations and Remote Sensing

Bazrafshan O, Ehteram M, Dashti Latif S, Feng Huang Y, Yenn Teo F, Najah Ahmed A, El-Shafie A (2022) Predicting crop yields using a new robust Bayesian averaging model based on multiple hybrid ANFIS and MLP models. Ain Shams Eng J 13(5):101724. https://doi.org/10.1016/j.asej.2022.101724

Bi L, Hu G, Wally O, Tenuta AU, Kandel YR, Mueller DS (2022) A transformer-based approach for early prediction of soybean yield using time-series images. In: Deep learning approaches for yield prediction and crop disease recognition, *Doctoral dissertation*, vol. 2022. Iowa State University. https://dr.lib.iastate.edu/server/api/core/bitstreams/5464feed-3fd3-4a17-abb3-7221aa41c228/content

Chipanshi A, Zhang Y, Kouadio L, Newlands N, Davidson A, Hill H, Warren R, Qian B, Daneshfar B, Bedard F, Reichert G (2015) Evaluation of the integrated Canadian crop yield forecaster (ICCYF) model for in-season prediction of crop yield across the Canadian agricultural landscape. Agric For Meteorol 206:137–150. https://doi.org/10.1016/j.agrformet.2015.03.007

de Wit A, Boogaard H, Fumagalli D, Janssen S, Knapen R, van Kraalingen D, Supit I, van der Wijngaart R, van Diepen K (2019) 25 years of the WOFOST cropping systems model. Agric Syst 168:154–167. https://doi.org/10.1016/j.agsy.2018.06.018

Dosovitskiy A, Beyer L, Kolesnikov A, Weissenborn D, Zhai X, Unterthiner T, Dehghani M, Minderer M, Heigold G, Gelly S, Uszkoreit J, Houlsby N (2021) An image is worth 16x16 words: transformers for image recognition at scale. https://doi.org/10.48550/arXiv.2010.11929. http://arxiv.org/abs/2010.11929. ArXiv:2010.11929 [cs]

Duchemin B, Maisongrande P, Boulet G, Benhadj I (2008) A simple algorithm for yield estimates: Evaluation for semi-arid irrigated winter wheat monitored with green leaf area index. Environ Model Softw 23(7):876–892. https://doi.org/10.1016/j.envsoft.2007.10.003

Filippi P, Jones EJ, Wimalathunge NS, Somarathna PDSN, Pozza LE, Ugbaje SU, Jephcott TG, Paterson SE, Whelan BM, Bishop TFA (2019) An approach to forecast grain crop yield using multi-layered, multi-farm data sets and machine learning. Precis Agric 20(5):1015–1029. https://doi.org/10.1007/s11119-018-09628-4

Huber F, Yushchenko A, Stratmann B, Steinhage V (2022) Extreme gradient boosting for yield estimation compared with deep learning approaches. Comput Electron Agric 202:107346. https://doi.org/10.1016/j.compag.2022.107346

Jhajharia K, Mathur P, Jain S, Nijhawan S (2023) Crop yield prediction using machine learning and deep learning techniques. Procedia Comput Sci 218:406–417. https://doi.org/10.1016/j.procs.2023.01.023

Ji S, Zhang C, Xu A, Shi Y, Duan Y (2018) 3D convolutional neural networks for crop classification with multi-temporal remote sensing images. Remote Sens 10(1):75. https://doi.org/10.3390/rs10010075

Johnson MD, Hsieh WW, Cannon AJ, Davidson A, Bédard F (2016) Crop yield forecasting on the Canadian Prairies by remotely sensed vegetation indices and machine learning methods. Agric For Meteorol 218–219:74–84. https://doi.org/10.1016/j.agrformet.2015.11.003

Jubair S, Tucker JR, Henderson N, Hiebert CW, Badea A, Domaratzki M, Fernando WGD (2021) GPTransformer: a transformer-based deep learning method for predicting Fusarium related traits in barley. Front Plant Sci 12. https://doi.org/10.3389/fpls.2021.761402

Kamilaris A, Prenafeta-Boldú FX (2018) A review of the use of convolutional neural networks in agriculture. J Agric Sci 156(3):312–322. https://doi.org/10.1017/S0021859618000436 (Publisher: Cambridge University Press)

Kim N, Lee YW (2016) Machine learning approaches to corn yield estimation using satellite images and climate data: a case of Iowa state. J Korean Soc Surv Geod Photogramm Cartogr 34(4):383–390. https://doi.org/10.7848/ksgpc.2016.34.4.383 (Publisher: Korean Society of Surveying, Geodesy, Photogrammetry and Cartography)

Kuradusenge M, Hitimana E, Hanyurwimfura D, Rukundo P, Mtonga K, Mukasine A, Uwitonze C, Ngabonziza J, Uwamahoro (2023) A crop yield prediction using machine learning Models: case of Irish potato and maize. Agriculture 13(1):225. https://doi.org/10.3390/agriculture13010225 (Number: 1 Publisher: Multidisciplinary Digital Publishing Institute)

Kuwata K, Shibasaki R (2015) Estimating crop yields with deep learning and remotely sensed data. In: 2015 IEEE international geoscience and remote sensing symposium (IGARSS), pp 858–861. https://doi.org/10.1109/IGARSS.2015.7325900. ISSN: 2153-7003

Lecun Y, Bottou L, Bengio Y, Haffner P (1998) Gradient-based learning applied to document recognition. Proc IEEE 86(11):2278–2324. https://doi.org/10.1109/5.726791 (Conference Name: Proceedings of the IEEE)

Linna P, Halla A, Narra N (2022) Ground-penetrating radar-mounted drones in agriculture. In: Lipping T, Linna P, Narra N (eds) New developments and environmental applications of drones. Springer, Cham, pp 139–156

Liu Y, Wang S, Chen J, Chen B, Wang X, Hao D, Sun L (2022) Rice yield prediction and model interpretation based on satellite and climatic indicators using a transformer method. Remote Sens 14(19):5045. https://doi.org/10.3390/rs14195045 (Number: 19 Publisher: Multidisciplinary Digital Publishing Institute)

Messina G, Modica G (2020) Applications of UAV thermal imagery in precision agriculture: state of the art and future research outlook. Remote Sens 12(9):1491. https://doi.org/10.3390/rs12091491 (Number: 9 Publisher: Multidisciplinary Digital Publishing Institute)

Mkhabela MS, Bullock PR (2012) Performance of the FAO AquaCrop model for wheat grain yield and soil moisture simulation in Western Canada. Agric Water Manag 110:16–24. https://doi.org/10.1016/j.agwat.2012.03.009

Muruganantham P, Wibowo S, Grandhi S, Samrat NH, Islam N (2022) A systematic literature review on crop yield prediction with deep learning and remote sensing. Remote Sens 14(9):1990. https://doi.org/10.3390/rs14091990 (Number: 9 Publisher: Multidisciplinary Digital Publishing Institute)

Narra N, Halla A, Linna P, Lipping T (2022) A minimalist approach to yield mapping of standing wheat crop with unmanned aerial vehicles. In: Lipping T, Linna P, Narra N (eds) New developments and environmental applications of drones. Springer, Cham, pp 157–171. https://doi.org/10.1007/978-3-030-77860-6_9

Nevavuori P, Narra N, Linna P, Lipping T (2020) Crop yield prediction using multitemporal UAV data and spatio-temporal deep learning models. Remote Sens 12(23):4000. https://doi.org/10.3390/rs12234000

Quintero D, Díaz E (2020) A comparison of two open-source crop simulation models for a potato crop. Agron Colomb 38(3):382–387. https://doi.org/10.15446/agron.colomb.v38n3.82525

Shah N, Ross M, Trout K (2012) Using soil moisture data to estimate evapotranspiration and development of a physically based root water uptake model. https://doi.org/10.5772/18040

Shirley R, Pope E, Bartlett M, Oliver S, Quadrianto N, Hurley P, Duivenvoorden S, Rooney P, Barrett AB, Kent C, Bacon J (2020) An empirical, Bayesian approach to modelling crop yield: maize in USA. Environmental Research Communications 2(2):025002. https://doi.org/10.1088/2515-7620/ab67f0 (Publisher: IOP Publishing)

Silvestro PC, Pignatti S, Yang H, Yang G, Pascucci S, Castaldi F, Casa R (2017) Sensitivity analysis of the Aquacrop and SAFYE crop models for the assessment of water limited winter wheat yield in regional scale applications. PLOS ONE 12(11):e0187485. https://doi.org/10.1371/journal.pone.0187485

Stadler A, Rudolph S, Kupisch M, Langensiepen M, van der Kruk J, Ewert F (2015) Quantifying the effects of soil variability on crop growth using apparent soil electrical conductivity measurements. Eur J Agron 64:8–20. https://doi.org/10.1016/j.eja.2014.12.004

Steduto P, Hsiao TC, Raes D, Fereres E (2009) AquaCrop–the FAO crop model to simulate yield response to water: I. Concepts and underlying principles. Agron J 101(3), 426–437. https://doi.org/10.2134/agronj2008.0139s. _eprint: https://onlinelibrary.wiley.com/doi/pdf/10.2134/agronj2008.0139s

Tantalaki N, Souravlas S, Roumeliotis M (2019) Data-driven decision making in precision agriculture: the rise of big data in agricultural systems. J Agric Food Inf 20(4):344–380. https://doi.org/10.1080/10496505.2019.1638264

Tao T, Wei X (2022) A hybrid CNN–SVM classifier for weed recognition in winter rape field. Plant Methods 18(1):29. https://doi.org/10.1186/s13007-022-00869-z

Tian L, Wang C, Li H, Sun H (2020) Yield prediction model of rice and wheat crops based on ecological distance algorithm. Environ Technol Innov 20:101132. https://doi.org/10.1016/j.eti.2020.101132

Tiusanen J (2013) Soil scouts: description and performance of single hop wireless underground sensor nodes. Ad Hoc Netw 11(5):1610–1618. https://doi.org/10.1016/j.adhoc.2013.02.002

Todorovic M, Albrizio R, Zivotic L, Saab MA, Stöckle C, Steduto P (2009) Assessment of AquaCrop, CropSyst, and WOFOST models in the simulation of sunflower growth under different water regimes. Agron J 101(3):509–521. https://doi.org/10.2134/agronj2008.0166s

Tsouros DC, Bibi S, Sarigiannidis PG (2019) A review on UAV-based applications for precision agriculture. Information 10(11):349. https://doi.org/10.3390/info10110349 (Number: 11 Publisher: Multidisciplinary Digital Publishing Institute)

van der Velde M, van Diepen CA, Baruth B (2019) The European crop monitoring and yield forecasting system: celebrating 25 years of JRC MARS Bulletins. Agric Syst 168:56–57. https://doi.org/10.1016/j.agsy.2018.10.003

van Diepen C, Wolf J, van Keulen H, Rappoldt C (1989) WOFOST: a simulation model of crop production. Soil Use and Manag 5(1):16–24. https://doi.org/10.1111/j.1475-2743.1989.tb00755.x. _eprint: https://onlinelibrary.wiley.com/doi/pdf/10.1111/j.1475-2743.1989.tb00755.x

van Klompenburg T, Kassahun A, Catal C (2020) Crop yield prediction using machine learning: a systematic literature review. Comput Electron Agric 177:105709. https://doi.org/10.1016/j.compag.2020.105709

Vaswani A, Shazeer N, Parmar N, Uszkoreit J, Jones L, Gomez AN, Kaiser L, Polosukhin I (2017) Attention is all you need. https://doi.org/10.48550/arXiv.1706.03762. ArXiv:1706.03762 [cs]

Wang X, Huang J, Feng Q, Yin D (2020) Winter wheat yield prediction at county level and uncertainty analysis in main wheat-producing regions of China with deep learning approaches. Remote Sens 12(11):1744 . https://doi.org/10.3390/rs12111744 (Number: 11 Publisher: Multidisciplinary Digital Publishing Institute)

Weilandt F, Behling R, Goncalves R, Madadi A, Richter L, Sanona T, Spengler D, Welsch J (2023) Early crop classification via multi-modal satellite data fusion and temporal attention. Remote Sens 15(3):799. https://doi.org/10.3390/rs15030799 (Number: 3 Publisher: Multidisciplinary Digital Publishing Institute)

Weindorf DC, Chakraborty S (2020) Portable X-ray fluorescence spectrometry analysis of soils. Soil Sci Soc Am J 84(5):1384–1392. https://doi.org/10.1002/saj2.20151

Yue Y, Li JH, Fan LF, Zhang LL, Zhao PF, Zhou Q, Wang N, Wang ZY, Huang L, Dong XH (2020) Prediction of maize growth stages based on deep learning. Comput Electron Agric 172:105351. https://doi.org/10.1016/j.compag.2020.105351

Zhou H, Zhang S, Peng J, Zhang S, Li J, Xiong H, Zhang W (2021) Informer: beyond efficient transformer for long sequence time-series forecasting. Proc AAAI Conf Artif Intell 35(12):11106–11115. https://doi.org/10.1609/aaai.v35i12.17325 (Number: 12)

Part III
Digital Agriculture Roles in Genetic Conservation, Speed Breeding/Fast Forward Breeding

Crop Phenomics and High-Throughput Phenotyping

Priya Kumari, Himanshi Gangwar, Vishal Kumar, Vandana Jaiswal, and Vijay Gahlaut

Abstract Abiotic stresses, like drought, salinity, and high temperature pose significant challenges to global agriculture, jeopardizing crop yields and food security. Traditional breeding methods struggle to efficiently develop stress-tolerant crop varieties due to the complex genetic basis of stress responses. Phenomics, the comprehensive study of plant traits, has emerged as a valuable approach to accelerate abiotic stress breeding. This chapter reviews recent advances in phenomics techniques applied to abiotic stress research, highlighting their potential to enhance stress tolerance in crops. We discuss cutting-edge technologies, including high-throughput phenotyping and imaging systems, which enable the rapid and accurate assessment of stress-induced morphological and physiological changes. Moreover, we explore how multidimensional data generated by these techniques can be harnessed through data analytics and machine learning to uncover key stress-responsive traits and genes. Through this synthesis, we emphasize the transformative impact of phenomics on breeding programs and its pivotal role in developing stress-resilient crop varieties, ensuring sustainable agricultural productivity in the face of changing environmental conditions.

Keywords Abiotic stress · Data management · High-throughput phenotyping · Image analysis · Root · Shoot

P. Kumari · H. Gangwar · V. Kumar · V. Jaiswal
Biotechnology Division, CSIR-Institute of Himalayan Bioresource Technology, Palampur, Himachal Pradesh, India

V. Gahlaut (✉)
Biotechnology Division, CSIR-Institute of Himalayan Bioresource Technology, Palampur, Himachal Pradesh, India

Department of Biotechnology, University Center for Research and Development, Chandigarh University, Mohali, Punjab, India

P. M. Priyadarshan et al. (eds.), *Digital Agriculture*,
https://doi.org/10.1007/978-3-031-43548-5_13

1 Introduction

Plant phenomics is a comprehensive evaluation of plant traits including plant growth, development, architecture, physiology, and yield (Gaudin et al. 2013). Plant phenotyping is important for the development of elite plant varieties. Forward phenomics employs phenotyping technologies to identify the most favorable genotypes with the most desirable attributes within a vast collection. Reverse phenomics is the in-depth analysis of qualities that have been demonstrated to be useful in revealing underlying concepts and enabling the exploitation of a mechanism in novel techniques (Furbank and Tester 2011). Additionally, high-throughput phenotyping (HTP) for crop development in response to climate scenarios is made possible by the core science of phenomics (Yang et al. 2013). Crop plant phenomics is based on vast amounts of plant phenotyping data, including morphological qualities, physiological variables, and biochemical traits, that have been collected using high-throughput systems (Fig. 1) (Rahaman et al. 2015).

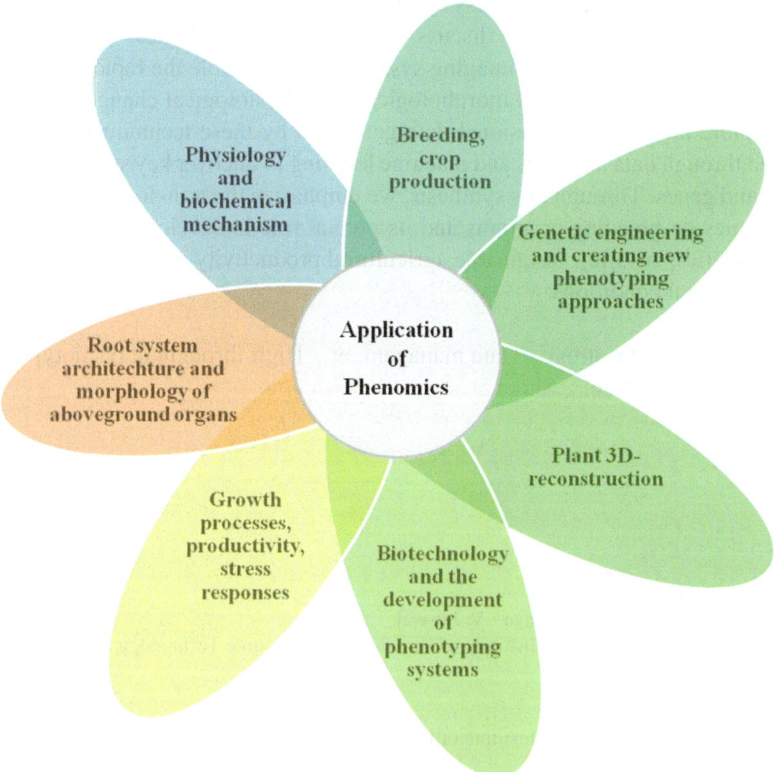

Fig. 1 Application of phenomics in fundamental and applied plant sciences

Phenomics implements multidisciplinary technologies including software and hardware components (Lin 2015). Many countries have made investments in phenomics, including the United States, Australia, Belgium, England, France, Germany, Japan, China, India, and Korea. These countries set up high-throughput plant phenotyping (HTPP) facilities using visible (VIS), near-infrared (NIR), infrared (IR), and hyperspectral images. The major purpose of the facilities was to analyze rice, wheat, maize, vegetables, and fruits. There are many phenotypic trait analyses, including biomass, root shape, yield-related attributes, leaf characteristics, and stress responses (Arvidsson et al. 2011; Balachandran et al. 1997; Duan et al. 2011; Golzarian et al. 2011; Kumar et al. 2014).

Data management is the process of organizing, storing, and disseminating research data (Brown et al. 2014). It can be difficult, particularly whenever studies involve multiple researchers and are done in complicated situations. Throughout the study phase, the way data are handled is determined by the types of data involved, how they are gathered and kept, and how data are used. The success of research is influenced by the manner in which data are maintained. Data management assists researchers in organizing research files and data for better access and analysis (Li et al. 2013). It contributes to the overall excellence of the research as well as validates the published results in terms of ongoing data analysis accountability. The gathering of large-scale plant phenotypic data is rising at an exponential rate, and it must be effectively handled prior to, throughout, and following the research period. The massive amounts of phenotypic data received from various phenomics platforms, both raw and metadata, are put into analytic workflows (Yang et al. 2013), where adequate data management is required for optimal applicability. Plant phenomics data management is a vital procedure for crop development programs. Therefore, this book chapter tries to describe the workflow of data management in plant phenomics programs.

One of the major problems in current plant breeding is the genotype-to-phenotype gap (Houle et al. 2010). Although studies in genomics have revealed plenty of information about the genetic structure of many plant species, sequencing techniques and the data generated by them much exceed our present abilities for plant phenotyping (Yang et al. 2014). Conventional plant phenotyping technologies, particularly depending on the laborious measurement of selected attributes from a small sample of plants, have very low throughput and thus impede complete analysis of traits within and across cultivars. This referred to as phenotyping bottleneck inhibits the chance to figure out how expressed phenotypes correspond with underlying genetic variables and environmental conditions, and it has hampered progress on critical breeding challenges like drought resistance (Furbank and Tester 2011).

Image-based approaches have the potential to significantly expand the area of study and efficiency of plant phenotyping efforts. In the last eight years, the ability to capture images of plants and crops has grown considerably due to the incorporation of new imaging technologies, robotic and conveyer belt systems in greenhouses, and ground-based and aerial imaging platforms in fields (Fahlgren et al. 2015). It has been suggested that future advancements in image-based plant phenotyping will necessitate a collaborative effort in the areas of image processing based on obtaining features and machine learning for analyzing the data (Tsaftaris et al. 2016). Deep learning methods now dominate the present state of advancement in many picture-based tasks, such as object recognition, among the localization of operations segmentation based on semantics, classification of images, and others, according to the current machine learning literature. However, limited applications for deep learning are currently presented in the field of plant phenotyping literature, and few general-purpose tools have been made available for the plant phenotyping group in order to encourage such techniques.

Our aim is to help the plant phenotyping community with the opportunity to utilize cutting-edge deep learning techniques in machine vision with the objective to accelerate plant phenotyping research and contribute to closing the genotype-to-phenotype gap. The majority of HTPPs, including those operated through large transnational seed manufacturers as well as those managed by the world's leading open plant-related organizations, including the Australian Plant Phenomics Facility, the European Plant Phenotyping Network, and the USDA (United States Department of Agriculture), have fully automated amenities in greenhouses or growth chambers that use robotics, accurate monitoring of the environment, along with remote sensing methods to analyze plant development and yield. In this book chapter, we discuss phenomics for abiotic stresses, high-throughput plant phenotyping techniques, data management, and the role of HTPs in crop breeding programs.

2 Phenomics for Abiotic Stresses

Crop growth and development are severely hampered by abiotic factors such as drought, salt, water logging, extremely high temperatures, and heavy metals, which reduce yield (Bray et al. 2000). Non-stress agricultural area accounts for just 10% of total arable land worldwide (Dita et al. 2006). According to Furbank and Tester (2011), there is a significant knowledge gap between genotype and phenotype, or, to put it another way, the relationship between the two is mostly illusory. In fact, a new obstacle in plant breeding and stress biology has been identified: high-throughput phenotyping (Yang et al. 2013). As proteins play a role in the stress response of plants, it is crucial to examine proteome changes under diverse stress situations. The cellular processes of stress sensing and signaling are the initial mechanisms by which plants react to stressful situations. When plants are under stress, it is important to comprehend how proteins are modified during post-translation. Proteomics research offers a wealth of knowledge regarding the fine-tuning of cellular pathways involved in stress mitigation in the past.

The leaf surface temperature, which represents the plant's transpiration rate, is a crucial characteristic in drought research. Feher-Juhasz et al. (2014) employed a combination of shoot digital imaging, infrared imaging, and automated weighing and watering to explore WUE (Water Use Efficiency). To choose drought-tolerant transgenic wheat plants, these authors used a platform based on a self-built greenhouse. The platform enables the assessment of the leaf surface temperature by side-view thermal camera recording the variations in temperatures of plant shoots, as well as the monitoring of the growth of mature cereal plants by multiple-view RGB (red-green-blue) imaging (Feher-Juhasz et al. 2014). Additionally, various vegetation indices and unmanned aerial vehicle (UAV) platform are used for HTP to assess many aspects of plant development and wellness, including soil properties, water content, and nutrient levels (Tayade et al. 2022). The assessment of drought tolerance in barley was conducted using the same platform and phenotyping experimental methodology. The system offers an in-depth investigation of plant physiology and development, but its application to large-scale analysis is constrained by a semi-automated regime that necessitates manually loading the plants into the system (Cseri et al. 2013).

It is possible to research plants' resistance to both drought and high temperatures using the same methods since physiological responses to both stressors are closely related. A review by Gupta et al. (2012) provides information on the usage of high-throughput phenotyping for high-temperature tolerance as well as a description of the required sensors. The effects of the high temperature on the *Arabidopsis* plants have been investigated (Vasseur et al. 2014). In order to show different adaptation responses to the pressures of high temperature and drought, the authors employed a commercial prototype platform that allowed top-view RGB photography and WUE analysis, followed by a highly sophisticated statistical method (Vasseur et al. 2014).

The other process that is frequently linked to stress from drought and high temperatures is the salinization of soil. In a study, an example of a procedure for salt-stress research on different cereals, including wheat, was provided that combined RGB imaging with destructive leaf sampling to determine Na^+ concentration (Berger et al. 2012). Using digital RGB imaging in a commercial system situated in a greenhouse, the effects of salt stress were studied by Rajendran et al. (2009). This study gave a thorough understanding of the physiological mechanisms linked to salt in wheat. For the purpose of quantifying the senescent region, the authors estimated a digital area of the shoot and visualized changes in leaf color using multiple-view RGB imaging. Through non-invasive plant phenotyping and examination of Na^+ concentration in fourth leaf, the authors projected a plant salinity tolerance index that correlated well with the findings of standard salt-tolerant assays (Rajendran et al. 2009). The physiological study in wheat and barley used conventional RGB imaging (Harris et al. 2010). A similar method was also used by Schilling et al. to choose a transgenic barley line that can tolerate salt (Schilling et al. 2014). In order to choose rice cultivars that are resistant to salt, digital RGB imaging and SLCFIM (Strasbourg Laser-induced Chlorophyll Fluorescence Imaging) were combined (Hairmansis et al. 2014). These salt-stress tolerance investigations were carried out on the same commercial platform that included the SLCFIM sensor. This form of chlorophyll fluorescence imaging (CFIM) only gives an estimation of a senescent region, which may be acquired using an earlier method of estimation based on color detection using RGB imaging, as mentioned in Sect. 5.3, "Chlorophyll Fluorescence Imaging" (CFIM). In order to quantify the quantum yield of photochemistry and other competitive processes, the application of KCFIM (Karlsruhe Chlorophyll Fluorescence Imaging) is thus required in order to enhance the usefulness of the physiological evaluation (Lazar et al. 2015).

The pioneering study of Chaerle et al. (2006), who studied the effects of moderate mottle virus infection on tobacco and bean plants, employed a combination of RGB imaging, thermal imaging, and TLCFIM (Terrestial Laser-induced Chlorophyll Fluorescence Imaging). The concept of the approach based on RGB imaging of leaf growth was explained by Moreau et al. (2009). In a study, a comprehensive investigation on the phenotypic impacts of nitrogen and phosphorus nutritional statuses was conducted in *Brachypodium*, using RGB imaging to assess growth rate (Poire et al. 2014). In a study conducted by Neilson et al. (2015), a similar methodology was used to assess the effects of nitrogen deficiency and drought using RGB imaging, NIR imaging, and automated weighing. In addition, the authors also developed software that retrieved additive characteristics from the images, including predicted plant height and the height to the ligule of the youngest fully grown leaf. These traits had excellent correlates with standard manually observed agronomical parameters (Neilson et al. 2015). Chaerle et al. (2007a, b) employed RGB imaging, thermal imaging, and TLCFIM to assess the phenotypes associated with magnesium shortage and biotic stress in beans when they earliest described the multiple-sensor technique. Chlorophyll fluorescence (ChlF) analysis is a common non-invasive technique used to study the effects of cold stress on plant growth and physiology; however, fluorescence sensors incorporated into sophisticated growth-analyzing platforms are rarely used (Mishra et al. 2011). In another study by Humplik et al. (2015), an automated screening method based on RGB imaging and KCFIM analysis was developed for the selection of pea cultivars with various levels of cold sensitivity. The described study was designed for investigations of plant cold-response strategies generally, not just the selection of cold-sensitive/tolerant types. The presented approach should potentially be used for shoot assessments of different plant species since the CFIM analysis is not restricted to plant shape and the image analysis was sensitive enough to detect small tendrils (Humplik et al. 2015).

3 Phenomics Techniques for Plant Shoots and Canopies

In 1729, the French astronomer Jean Jacques Ortous de Mairan discovered the existence of circadian rhythms in plants after observing the daily leaf motions of the heliotrope plant (Mimosa), which continued throughout continuous darkness (de Mairan 1729). It is generally known that one output of a plant's circadian clock is the daily rhythmic movements of its leaves (Engelmann et al. 1992); this is known as TRiP, or Tracking Rhythms in Plants. Using this rhythmic movement, it is possible to calculate the internal clock's period. Using time-lapse photography and consistent lighting, images are taken every 10–20 min over a span of 5–10 days to track the timing of leaf movement. Large image series are generated, which are evaluated for rhythmicity by monitoring the precise position of the leaves in every photograph. A number of techniques for performing this analysis have been devised; however, each of them requires user input at various stages of the process (Bours et al. 2012). One popular technique, for instance, uses the Biological Rhythms Analysis Software System (BRASS) and MetaMorph® software to analyze each cotyledon's movement and fit period, phase, and amplitude data using the Fast Fourier Transform Nonlinear Least Squares (FFT-NLLS) algorithm (Edwards and Millar 2007). The process of creating the input data for BRASS in MetaMorph® or a comparable image analysis program is a significant bottleneck for the analysis. The region tool in MetaMorph® is used to choose the area around specific leaves. This region needs to be drawn with enough space around it to encompass the leaf throughout the image stack as it develops and moves during the time series. The study of a huge population is extremely labor-intensive and time-consuming due to the requirement to process each plant individually. Another disadvantage of utilizing a single cotyledon is that its movement is dependent on the petiole's active growth, and that when growth stops, the movement substantially slows down, making period detection inaccurate (Engelmann et al. 1992).

Unmanned aerial vehicles (UAVs) were utilized in research to produce 3D reconstructions of winter wheat from several photos in order to determine crop height (Khanna et al. 2015) and to create 3D digital surface models of barley from hyperspectral data (Aasen et al. 2015). In another research study, a laser scanner placed on a UAV was used to measure maize crop height. The so-called 3D digitizer, which creates 3D images of individual plants or plant sections as a whole using ultrasonic or electromagnetic sensors, is an extremely precise and intriguing technology. To construct light models in plant canopies in rice (Zheng et al. 2008) and cucumber (Wiechers et al. 2011) canopies, the plant architecture of various crops was assessed using 3D digitizers. Due to the requirement to physically point the digitizing pen at significant plant landmarks in order to record the architecture of the plant in 3D, 3D digitizing requires a lot of labor and time. As a result, this technology cannot be utilized as a high-throughput, automated phenotyping system.

Plant phenotyping under laboratory or field circumstances is a significant study area in which terrestrial laser scanning (TLS) is used. Numerous morphological plant factors have been studied, including canopy height (Tilly et al. 2014) and leaf area (Gebbers et al. 2011). Besides morphological parameters, structural and functional information has also been studied (Sirault et al. 2013). After height growth, biomass is likely the second-most crucial factor (Lumme et al. 2008). In another method, the identification of single maize plants has been carried out to enhance crop management strategies or plant growth models (Hofle 2013). TLS measurements were mainly performed on single plants in pots; therefore, it was difficult to generalize the results to crops, which limited their use for field studies (Paulus et al. 2014). Additionally, these observations were frequently performed in controlled and entirely artificial environments, such as greenhouses or climate chambers (Kjaer and Ottosen 2015). In the field, if TLS measurements were

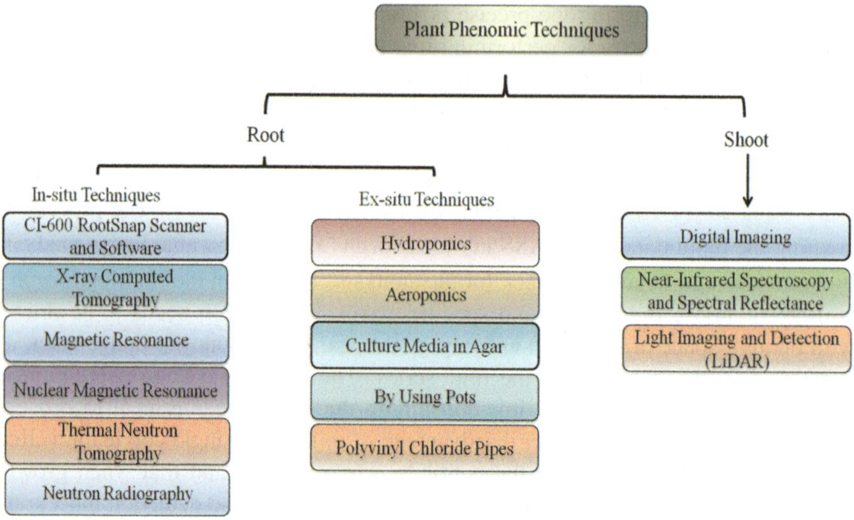

Fig. 2 High-throughput plant phenotyping techniques used for root system architecture (RSA; includes in situ and ex situ approaches) and shoot phenotyping for a comprehensive study of plant growth and development

made, they were typically made on extremely small regions or with low resolution (Hoffmeister et al. 2013).

The most popular technique for determining shoot biomass in phenotyping platforms is to take digital pictures of the plants after putting them at a specific position toward a camera under specific lighting conditions (Fig. 2). Using color and brightness evaluation, digital image processing after capture makes it possible to isolate plant features from the background of the image (Fiorani et al. 2012). The primary drawbacks of assessing biomass using imaging techniques like color imaging in 2D spatial dimensions are: (1) overlapped leaves and stems cause the shoot area to be underestimated and frequently limit this application to a specific plant size or developmental stage; and (2) image segmentation necessitates quite complicated processing pipelines (Paproki et al. 2012). A phenotyping technique known as a light curtain array (LC) has been utilized to measure canopy height in the field with success (Busemeyer et al. 2013). Depending on whether plants are transferred to the sensor or vice versa, phenotyping platforms can be managed utilizing plant-to-sensor or sensor-to-plant measurement protocols. There are several restrictions when using imaging techniques as a sensor-to-plant strategy, with small rosette plants being the exception (Arvidsson et al. 2011). Multiple cameras would need to be positioned above the plant in a specific direction, and the lighting during imaging would need to be tightly regulated. This requirement partially explains the implementation of a plant-to-sensor strategy, in which plants are transported to specific imaging stations, using existing platforms for the phenotypic evaluation of plants of various sizes (Golzarian et al. 2011). Table 1 describes the image analysis software available for high-throughput phenotyping.

Table 1 Selected image analysis software available for high-throughput phenotyping

Plant tissue	Software	Phenotypic trait/parameter	References
Shoot and leaves	TraitMill	Platform for monitoring different agronomic traits	Reuzeau et al. (2006)
	PHENOPSIS	Automated evaluation of characteristics associated to water deficit	Granier et al. (2006)
	LeafAnalyser	Leaf shape variation analysis	Weight et al. (2008)
	LAMINA	Rapid measurement of leaf size and shape	Bylesjo et al. (2008)
	HYPOTrace	Hypocotyl growth and shape	Wang et al. (2009)
	LEAFPROCESSOR	Analysis of leaf shape	Backhaus et al. (2010)
	Lamina2Shape	Leaf length and shape: width ratio and leaf area	Dornbusch and Andrieu (2010)
	Easy leaf area	Total leaf area and non-invasive canopy area estimation	Easlon and Bloom (2014)
	LeafByte	Leaf dimensions, herbivory extent	Getman-Pickering et al. (2020)
	LI-3000C	Leaf dimensions, leaf area	–
	WinDIAS	Leaf area, length, width, perimeter, proportion of diseased area	–
	WinFOLIA	Morphological measurements of broad leaves, herbivory extent, disease extent, and color profiles	http://www.regent.qc.ca/products/folia/WinFOLIA.html
Root	KineRoot	Root diameter and growth	Basu et al. (2007)
	EZ-Rhizo	Root length, insertion-angles, and branches	Armengaud et al. (2009)
	PlaRoM	Lateral root formation, extension, and root hair development	Yazdanbakhsh and Fisahn (2009)
	DART	Root system architecture	Le Bot et al. (2010)
	RootTrace	Root length and curvature	Naeem et al. (2011)
	Root reader 3D	2D analysis of root length, depth, convex-hull, and volume	Clark et al. (2011)
	SmartRoot	Measurements of growth and architecture	Lobet et al. (2011)
	GiA roots	2D analysis of total root length, area, and volume	Galkovskyi et al. (2012)
	GROWSCREEN-Rhizo	Root and shoot growth	Nagel et al. (2012)

(continued)

Table 1 (continued)

Plant tissue	Software	Phenotypic trait/parameter	References
	DIRT	Root tissue angle in soil, root dispersion and density, root network size, shape, and depth	Das et al. (2015)
	archiDART v3.0	Topology of root system	Delory et al. (2018)
	RootNav 2.0	Primary and lateral root count, lengths, insertion angles	Yasrab et al. (2019)
	saRIA	Total root length, area, volume, and diameter, all aspects of root morphology, both globally and locally	Narisetti et al. (2019)
	SegRoot	Root length	Wang et al. (2019)
	4DRoot	Root architecture	Herrero-Huerta et al. (2022)
	WinRHIZO	Morphological characteristics including root area, volume, length, surface, and root color	http://www.regent.qc.ca/products/rhizo/RHIZOTron.html
Seed	SHAPE	Quantitative evaluation of shape parameters	Iwata and Ukai (2002)
	ImageJ	Seed area, size, and shape	Herridge et al. (2011), http://rsb.info.nih.gov/ij/
	SmartGrain	Seed shape, size, dimensions, and count	Tanabata et al. 2012
	GrainScan	Seed size, color, dimensions, and seed count	Whan et al. 2014
	WinSEEDLE	Measurements of the volume and surface area of seeds and needles	http://www.regent.qc.ca/products/needle/WinSEEDLE.html
	SeedCount	Seed size, dimensions, color, seed count	–

3.1 *Digital Imaging*

Digital image evaluation offers a quick and affordable method for accurately assessing plant traits that would otherwise take a lot of effort. The measuring of canopy characteristics is a good example (Fiorani et al. 2012). Digital photographs have several benefits over conventional ways of estimating light interception, including the ability to immediately process photos by computer. Video image analysis enables a rapid, low-cost, and non-destructive evaluation of canopy characteristics and crop development (Elsayed et al. 2011). Digital imaging is also useful for monitoring root features in research, which are sometimes limited by a lack of acceptable methods for continuous, non-destructive observations (Blouin et al. 2007). Furthermore, the analysis of digital images (Armengaud et al. 2009) enables accurate evaluation at higher resolution scales, which is required to explore the kinetics of the mechanisms that regulate root growth. In this regard, Chavarria-Krauser et al. (2008) used a non-invasive technique based on digital imaging to quantify highly resolved spatiotemporal dynamics within the root development zone of *Arabidopsis*.

3.2 Near-Infrared Spectroscopy and Spectral Reflectance

High-throughput phenotyping platforms may benefit from remote sensing using near-infrared spectroscopy and the spectral reflectance of plant canopies (Montes et al. 2007) and offer intriguing chances to gather integrative features with great temporal resolution (Gutierrez et al. 2010). Sensors attached on tractors gather spectral reflectance in the visible and near-infrared wavelengths of the electromagnetic spectrum from the crop canopy (Montes et al. 2007) or by employing handheld gadgets with digital cameras attached on them (Casadesus et al. 2007). With the help of remote sensing, we now understand how species, leaf thickness, canopy structure, leaf age, nutritional status, and, most crucially, water status affect changes in leaf reflectance and emittance. On the basis of these data, numerous vegetative indices for crop canopies have been developed to quantify agronomic characteristics. The usage of calibration models for the phenotypic value prediction is necessary to get relevant information from the plot spectra. Under carefully controlled experimental circumstances, spectral reflectance used to track plant photosynthetic pigment composition, evaluate the state of the water, and identify abiotic stress in its early stages (Babar et al. 2006; Gray et al. 2010).

3.3 Light Imaging and Detection

The introduction of light detection and ranging (LiDAR) technology has given rise to a new goal for 3D plant phenotype analysis (Lefsky et al. 2002). Short-wavelength lasers like ultraviolet to near-infrared light are used in this innovative remote sensing approach to calculate the distance between sensor and target object using laser beam speed and flight time captured by a timer (Lin 2015; Shan and Toth 2018). The angle encoder captures laser emitting angles and converts the distance into 3D structure information. This technique gathers information about the canopy and leaves' numerous characteristics, including the vegetation's height, structure, and leaf area index as well as the nitrogen level (Lin 2015; Madec et al. 2017; Zhang and Grift 2012). LiDAR provides a number of benefits, including high-throughput phenotyping, excellent repeatability, high spatial resolution, and independence from light; this makes the technique suitable for field use (Llorens et al. 2011; Madec et al. 2017). The short-wavelength laser can characterize the interior structures of plants by penetrating the plant canopy, making up for the shortcomings of the other optical image techniques (Berk et al. 2016).

4 Phenomics Techniques for Roots/RSA (Root System Architecture)

Plant roots play important roles in water and mineral absorption, anchoring, development as well as growth, storing food, and as interaction areas for many biological communities (Urfan et al. 2022). Several biochemical processes occur in plant roots to adapt to stress, which can be rapid or long term. As a result, numerous breeders realize that the secret to producing genotypes resistant to a variety of abiotic stresses is located beneath the soil surface (DoVale and Fritsche-Neto 2015). The structure and arrangement of a root system in a particular environmental state is termed as root system architecture (RSA) (Urfan et al. 2022). Plant root phenotyping is a difficult undertaking that represents a significant gap in plant root study. Because of this, the genetic and physiological bases of roots are less developed than those of above-ground phenes. In order to overcome this "phenotyping gap," classical phenotyping has given way to image-based phenotyping, which allows for comparatively high throughput while retaining root measurement precision (Kumar et al. 2022). In the beginning of the 2000s, certain devices and programs were developed that allowed the assessment of various root statistics such as volume, length, area of surface, and projected area, among others (Danilevicz et al. 2021; DoVale and Fritsche-Neto 2015). It is important to examine the stages of root phenotyping along the traditional, single-trait and developing, multi-trait paths to accelerate the adaptation of root phenotypes to field environments. Root phenotyping generally starts in controlled environments and advances to field validation for early, quick success due to the high-throughput performing approach that prevents challenging conditions (Watt et al. 2020). The two main categories of root phenotyping methods are ex situ (where the whole system of roots has been collected and examined outside of living conditions) and in situ (where the entire root system is obtained and evaluated in natural conditions) (Mir et al. 2019), as shown in Fig. 2.

4.1 Ex Situ Techniques

Since 1727, traditional root phenotyping techniques have been damaging, low-resolution, and time-consuming such as the pinboard method, excavation methods, and the trench profile technique (Shi et al. 2022). A variety of soil-free techniques—such as aeroponics, where the target plant's roots are suspended in the air and treated with a fine spray of nutrient solutions (Gangopadhyay et al. 2021), and hydroponics, static hydroponics that involves growing plants in containers and water-circulating hydroponics that uses PVC (Polyvinyl Chloride) pipes—are used for RSA studies (Bonato et al. 2022). The paper roll technique is also commonly employed in the study and monitoring of early RSA characteristics under various environmental stimuli in the seedlings of wheat (Alemu et al. 2021). It is possible to thoroughly explore their implications on RSA.

Interestingly, RSA in *Arabidopsis* and other related plants can be studied using rhizoponics, a system that combines hydroponics and rhizotron (Mathieu et al. 2015; Urfan et al. 2022). The rhizoponics technique allows for the exact measurement of plant root growth. Using the rhizoslide method, a plant is grown inside a layer of large, two-dimensional (2D) plates. In this, germination sheets that offer substrate, water, and nutrients for the growing embryo are covered on both sides of the central glass shelter, which stabilizes the root system (Urfan et al. 2022; Yang et al. 2020). For a high-throughput system study of RSA in wheat seedlings, transparent pots have been employed successfully (Richard et al. 2015).

Ex situ root phenotyping requires less expense, does not call for specialized knowledge, and does not require access to expensive equipment. However, the ex situ method of examination has a number of drawbacks, such as: (i) only 2D data of root development; (ii) it is difficult to perform phenotyping on a large scale; and (iii) because small roots break during the washing process during phenotyping in PVC pipes, the contribution of these roots is underestimating (DoVale and Fritsche-Neto 2015). So digital scanning paired with computerized image processing offers a quicker high-throughput technique for analyzing root morphological features, including the length of the root, width, structure, and branching (Mir et al. 2019).

4.2 *In Situ Techniques*

There are now a number of advanced non-destructive approaches for root assessment that have been effectively applied in in situ root phenotype analysis.

4.2.1 X-Ray Computed Tomography

The inner 3D volume can be seen using X-ray computed tomography (CT), a three-dimensional (3D) structural visualization application, based on the changes in the X-ray attenuation of various materials, such as soil and roots (DoVale and Fritsche-Neto 2015). A beneficial version of the CT scanner, developed by Hounsfield in 1976, rapidly followed the invention of the original CT scanner, for which its creators were awarded the 1979 Nobel Prize in Physiology and Medicine (Yang et al. 2020). Detailed root phenotyping, superior spatial accuracy, and three-dimensional interaction of root hairs in soil—where root architectural aspects are crucial for water and nutrient uptake—are all possible with X-ray CT (Mir et al.

2019). Comparing this technology to other methods, it is deemed to be quicker, more consistent, and more adaptable, especially when used to evaluate field samples collected in situ, as was previously done in wheat, canola, and barley. Additionally, it is important to remember the following problems with CT scanning of roots: (i) low dose of radiation (Zappala et al. 2013); and (ii) the use of CT imaging frequently includes trade-offs; for example, bigger pots with greater resolution will restrict the number of samples or rate of data capture because the scanning volume and resolution both lengthen the scanning process (Mairhofer et al. 2012).

4.2.2 CI-600 RootSnap Scanner and Software

This phenotyping method is intended for extensive field research on plants that are still growing, allowing for several assessments of each plant at various stages of its life cycle. This technique helps to examine how roots grow, develop, and work to adapt to a certain environment. The CI-600 scanner takes high-resolution digital photos without causing any damage. As the plants begin to build their root "networks" around the tube, images of the roots can be collected using the scanner and viewed with the CI-690 RootSnap program (DoVale and Fritsche-Neto 2015). WinRHIZO, RootReader, SmartRoot software, ImageJ: IJ-Rhizo, DynamicRoots software, and Digital Imaging of Root Traits (DIRT) are some other software that help to study root architecture. Uses of WinRHIZO in wheat allowed for the examination of numerous features and processes, such as root length density modeling (Zuo et al. 2004). In wheat research, SmartRoot was successfully used to identify germplasm (Roselló et al. 2019), analyze interactions among plants (Finch et al. 2017), and investigate prospective breeding targets for root architectural features (Cane et al. 2014). RootNav's capacity to analyze lateral roots in complex networks of roots has offered it various applications in wheat research, i.e., research relating seedling traits to yield components (Xie et al. 2017) and intake of nitrogen (Kenobi et al. 2017), as well as research into the genetic aspects of root architecture (Griffiths et al. 2019). DIRT holds potential for wheat research because of its unique capacity to measure excised root systems without the additional technical skills required for installing and operating independent software. This is already used to investigate the response of wheat toward phosphorus deprivation (Nguyen et al. 2019).

4.2.3 Magnetic Resonance Imaging

Magnetic resonance imaging (MRI) is classified based on the magnetic field strength functional range, with high-field MRI (HF-MRI) commonly working in the 1–10 T (Tesla) range and low-field MRI (LF-MRI) operating below one Tesla (1 T) (Bagnall et al. 2020). MRI is used to observe and measure root development in 3D in opaque soil (Mir et al. 2019). MRI can be performed separately or in conjunction with various methods to determine root shape, length, and volume (Urfan et al. 2022).

But this approach is very susceptible to sample moisture content. Moisture level exceeding 80% and dense soil can significantly impede the identification of lateral roots and, to a lesser extent, seminal roots in the soil (DoVale and Fritsche-Neto 2015).

4.2.4 Neutron Radiography

Beginning in 1985, neutron radiography was put forward and used to acquire roots' growing images (Moradi et al. 2009). However, because of the long-term impact of radiation on the growth of roots, as well as the cost and discomfort of the equipment, neutron radiography is not applicable for in situ root phenotyping (Shi et al. 2022).

5 Recent Advances in High-Throughput Phenomics

High-throughput plant phenomics enables rapid phenotyping of several plant populations at each plant level (Pasala and Pandey 2020). Majority of high-throughput phenomics approaches are thermal-infrared imaging, fluorescence imaging, visible-light scanning, spectroscopy imaging, and tomographic imaging including computed tomography, magnetic resonance imaging, and positron emission tomography (PET), as given in Fig. 3 (Sozzani et al. 2014; Yang et al. 2020). These advanced software systems and imaging-based, automated, high-throughput plant phenotyping technologies have become essential tools for plant biology (Paproki et al. 2012). Plants must adapt to an environment that is always changing, including stressful situations that are unfavorable to plant growth and development. These unfavorable conditions include biotic and abiotic stresses (such as heat, drought, cold, and salinity). Plant stress phenotyping provides chances for early intervention to stop the spread of illnesses/damages and to aid in the selection of elite lines to direct plant breeding efforts (Li et al. 2020).

5.1 Visible-Light Imaging

Visible-light imaging devices are basic tools for assessing the characteristics of plants, such as their color, size, texture, leaf biomass, leaf physiology, imbibition, and germination rates, panicle traits, yield traits, seed morphology, and root architecture (Arvidsson et al. 2011; Duan et al. 2011; Kumar et al. 2015; Li et al. 2014; Zhang and Zhang 2019). The estimated leaf area in plants, including *Arabidopsis thaliana* and maize, is provided through commercial systems that are based on visual imaging in one example for shoot biomass in a controlled environment (Li et al. 2014). Visible-light imaging has been widely utilized in plant science due to its low cost and clarity (Rahaman et al. 2015). Typically, this imaging approach

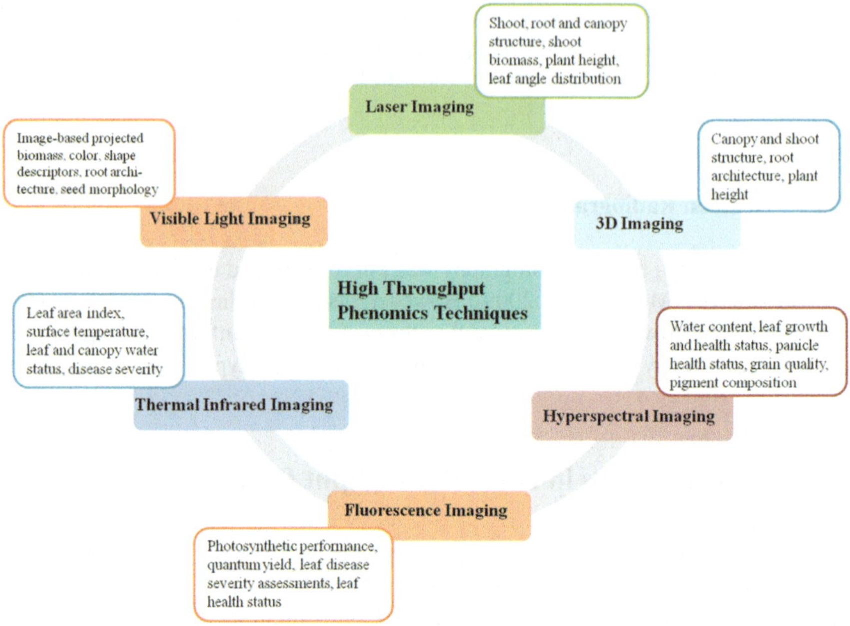

Fig. 3 High-throughput plant phenomics techniques and their phenotypic parameters used to capture the desired plant trait and gain insight into plant adaptation, performance, and responses to different environmental conditions

is carried out by using conventional color cameras with electromagnetic spectrum wavelengths between 400 and 750 nm, and two-dimensional (2D) images can be used to monitor changes in plant biomass and analyze various phenotypic traits (Bylesjo et al. 2008; Golzarian et al. 2011). In order to do this, three-dimensional (3D) imaging was developed to produce more precise detail on complex phenotypes. The 2D and 3D imaging technologies have been combined to improve phenotyping accuracy (Rahaman et al. 2015). Traditional digital or RCB/CIR (Red-Green-Blue/Color Infrared) cameras are typically employed in visible imaging due to their rapid detection (Jangra et al. 2021). Previously, Golzarian et al. (2011) precisely evaluated the dry weight of shoots for assessing wheat seedlings for salt stress using LemnaTec 3D Scanalyzer. Bowman et al. (2015) analyzed canopy spectral reflectance to assess wheat grain yield during an extended drought. Phenotyping with RGB has been applied for various abiotic stressors in various crops utilizing many different platforms; for example, PlantScreen and GROWSCREEN have been used for chilling tolerance in *Arabidopsis* and peas, respectively (Jansen et al. 2009; Humplik et al. 2015). In addition, PHENOPSIS and WIWAM have been used for drought stress in *Arabidopsis*, LemnaTec for drought stress in barley and maize and for salt stress in rice, wheat, and barley (Ge et al. 2016; Granier et al. 2006; Hairmansis et al. 2014; Honsdorf et al. 2014; Humplik et al. 2015, Meng et al. 2017).

5.2 Infrared- and Thermal-Based Imaging

Infrared imaging technologies are being utilized to examine objects for interior molecular motions that generate infrared light (Kastberger and Stachl 2003). The most often utilized wavelengths for thermal imaging are 3–5 μm or 7–14 μm (Kaplan 2007). This imaging technique offers better insight into plant wellness under various stress conditions both in the field and in greenhouse conditions. It measures the leaf surface temperature of the plant's response to changes in water status and transpiration rate as well as differences in the plant's stomatal conductance for the adoption of abiotic stress (Yang et al. 2013). Reduced rates of photosynthesis and transpiration are frequently caused by biotic or abiotic challenges, and thermal imaging remote sensing of leaf temperature can be an effective way for recognizing changes in the physiological status of plants in response to various biotic or abiotic stresses (Chaerle and van der Straeten 2000). In a study, 92 distinct maize genotypes were screened for their ability to tolerate drought using thermal-infrared imaging (Romano et al. 2013). Additionally, it can assess the stomatal conductance, leaf area, and relative chlorophyll content in wheat response to water deficit (Munns et al. 2010). Stomatal behavior under various stress situations has been measured using infrared thermal imaging systems, for instance, to monitor salt tolerance in wheat genotypes (Bayoumi et al. 2014). This imaging technique is also used for the estimation of rust disease severity, stored fungal infection detection, and estimation of crop canopy leaf area index in various moisture stress conditions (Banerjee et al. 2018; Chelladurai et al. 2010; Singh et al. 2022).

5.3 Fluorescence Imaging

Fluorescence is the process of a substance absorbing some shorter-wavelength light and then emitting low-wavelength light (Li et al. 2014). Fluorescence imaging illuminates the plants with flashes of blue light (500 nm or less), and the plants themselves emit fluorescence light in the red spectrum between 600 and 750 nm (Singh et al. 2018). The chlorophyll complex is the portion of the plant that normally fluoresces. Chlorophyll fluorescence is typically employed in phenomics to identify the impact of various environmental events and the capacity of plants to continue photosynthesis under these conditions, because abiotic pressures largely affect chlorophyll concentration (Weirman 2010). Photosynthetic performance, stomatal mobility, an association between spatial and temporal fluctuations of photosynthesis, detection of genetic disease resistance, identifying growth-related QTLs (Quantitative Trait Loci), and plants with improved or delayed metabolism and growth under stress can be rapidly recognized using fluorescence imaging (Cardon et al. 1994; Chen et al. 2014; El-Lithy et al. 2004; Fiorani and Schurr 2013; Li et al. 2014; Walter et al. 2004). Swarbrick et al. (2006) investigated the resistance response of barley leaves infected with *Blumeria graminis* using quantitative imaging of chlorophyll fluorescence. Chaerle et al. (2006) screened sugar beet susceptible and resistant lines infected with *Cercospora beticola* and estimated fluorescence intensity using fluorescence imaging. Burling et al. (2010) examined variations in the degree of wheat cultivar resistance to *Puccina triticina.* Under salt stress, *Arabidopsis thaliana* and rice morphology, growth, and photosynthetic performance were measured using RGB and chlorophyll fluorescence (ChlF) imaging (Awlia et al. 2016; Hairmansis et al. 2014). In wheat, the combined effects of heat and drought were examined using fluorescence imaging phenotyping (Abdelhakim et al. 2021). Additionally, wheat seedling leaves were observed under salt and osmotic stresses for physiological and chloroplast proteome analyses (Zhu et al. 2021). Through this technique, wheat plants were investigated under heat stress to confirm that the PS II (Photosystem II) system protects through the methyl jasmonate pathway (Fatma et al. 2021; Kim et al. 2021).

5.4 Spectroscopy Imaging

Spectroscopy imaging is to accomplish vegetation remote sensing and it measures the effect of solar radiation on plants. This imaging technique displays immense potential for plant phenotyping (Kokaly et al. 2009). Multispectral or hyperspectral sensors, which can periodically scan wavelengths of interest, can be used to gather spectral measurements of the electromagnetic spectrum (Fiorani and Schurr 2013). Spectral imaging employed for plant phenotyping includes rapid, non-destructive measures of green biomass, pigment content, canopy chlorophyll content, leaf, canopy senescence, water status, and yield in many crop species. For large-scale phenotyping and dynamic assessments of the biomass, greenness, nitrogen content, pigment composition, photosynthetic state, and canopy water content, a number of indices have been established in both field research and breeding programs (Cheng et al. 2011; Claudio et al. 2006; Din et al. 2017; Mistele and Schmidhalter 2008; Penuelas and Filella 1998; Schlemmer et al. 2005; Ullah et al. 2013). Near-infrared spectroscopy was used to precisely predict genotypic changes in the nitrogen and leaf ash content and in the kernel of maize grown under various water treatments (Cabrera-Bosquet et al. 2011). The severity of the disease in wheat leaves under stress from powdery mildew infection was monitored using hyperspectral imaging (Jiang et al. 2010). Early stress symptoms are detected through hyperspectral imaging by combing with other analysis tools. Simple volume maximization (SiVM) is one of the popular tools for the early detection of drought stress in plants (Thurau et al. 2010). Moshou et al. (2014) analyzed drought stress in wheat through spectral reflectance and fluorescence imaging.

5.5 *Integrated Imaging Techniques*

Functional imaging and optical 3D structural tomography are recent technological developments that have shifted more and more toward live visualization in all directions, high accuracy, and low noise level in plants (Zhao et al. 2019). These imaging techniques are used for plant phenotyping including canopy, shoot, and root architecture as well as for plant height. Functional imaging focuses on physiological changes to assess photosynthetic performance under stress, such as ChlF imaging and positron emission tomography (Baker 2008). Magnetic resonance imaging (MRI), another cutting-edge imaging method, is utilized to image physiological processes occurring in vivo (Borisjuk et al. 2013). A unique functional and structural imaging technique involves screening the dynamic changes in plant functions and structures using MRI and PET (Jahnke et al. 2009). Fluorescence resonance energy transfer (FRET) is another innovative non-invasive technique for high-resolution molecular phenotyping of tiny molecules in live tissue (Jones et al. 2014). Another high-throughput imaging technique is 3D imaging. The devices for 3D imaging include laser scanners, time-of-flight cameras, stereo vision, and light detection and ranging sensors (Omasa et al. 2007). These imaging tools are used for plant height, leaf area, and leaf shape (Takizawa et al. 2005). Wheat plants were phenotyped using the PlantEye, a high-resolution 3D laser scanner, while they were developing in a controlled environment under both salt stress and control conditions. The system uses a data cloud that is created when PlantEye scans plants from above to calculate attributes like 3D leaf area, plant height, and leaf number (Maphosa et al. 2017).

6 Data Management and Their Tool Assembly for Plant Phenomics

Through the use of phenotyping tools, phenomics produces a large number of images and metadata, hence efficient data processing and management are required (Kim et al. 2017). It can be challenging, especially when numerous researchers are involved in the study or when it is carried out in a complex environment. Throughout the research life cycle, the types of data involved, the methods used to gather and store them, and the uses for which they are intended all affect how data are handled, and hence how well the data are managed impacts the conclusion of the study (Kim et al. 2017). According to Yang et al. (2013), Brown et al. (2014), and Klukas et al. (2014), data management is the process of organizing, storing, and sharing research data. Data management assists researchers in organizing study files and data for easy access and analysis (Li et al. 2013). It contributes to the long-term accountability of data analysis by ensuring the quality of the study. If the data are adequately maintained, researchers may simply find information that will aid them in producing the intended outcomes. Large-scale plant phenotypic data collection

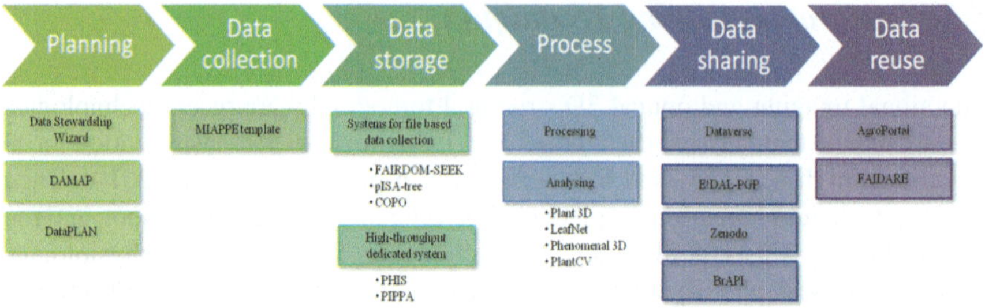

Fig. 4 Data management and tool assembly are used to handle and organize data on large scale during phenotyping, which advances our understanding of plant biology and improving crop productivity

is expanding extensively, and it is essential to handle it efficiently before, during, and after the research period. An automated HTP device, such as the largest robotic field scanner in the world, TERRA-REF (Transportation Energy Resources from Renewable Agriculture Phenotyping Reference Platform), may collect phenotypic data and can output up to 10 TB of data every day, with an estimated 10 PB over the course of three years (Kim 2017). The huge amounts of phenotypic data, both raw and metadata, that are received from a range of phenomics platforms are transferred into analytic pipelines (Yang et al. 2013), where they require adequate data management for the best use. Generally, an Integrated Analysis Platform supports a broad set of functionalities, including both data management and data processing. Data management describes the planning of data, how data are collected, stored, and analyzed, as well as how we share and reuse the data throughout the research cycle, as described in Fig. 4.

Considerations for data management in plant phenomics include the following:

1. *Planning*: It includes different elements that describe description, documentation, process, and archive of data. There are many different tools available for data management planning, which includes Data Stewardship Wizard (DSW) (Pergl et al. 2019), DAMAP (https://damap.org/), DataPLAN (https://plan.nfdi4plants.org/), DMPonline (https://dmponline.dcc.ac.uk/), DMPTool (https://dmptool.org/), EasyDMP (https://easydmp.no/), and many more.
2. *Data collection*: Data should be arranged to make effective data administration and analysis possible. To maintain uniformity, this might involve creating a data dictionary and adopting standardized folders and file names. To fill up the information and description of your experiments, MIAPPE (Minimum Information About a Plant Phenotyping Experiment) template should be used. There is a readme file that covers each field, as well as its kind and optional or necessary status. This will further enable the processing and validation of data using specific tools (Krajewski et al. 2015).
3. *Data storage*: It is crucial to decide how the data will be organized and kept. Scalable and affordable choices for data storage may be offered through cloud-

based storage systems. The output files are saved on the file server after being processed by scripts, and copies may be downloaded by each partner (Billiau et al. 2012). Many public databases maintain phenotypic data to organize and gather the information (Cobb et al. 2013), such as Soybase (http://soybase.org; Grant et al. 2010), MaizeGDB (http://www.maizegdb.org; Schaeffer et al. 2011), PHENOPSIS DB (http://bioweb.supagro.inra.fr/phenopsis; Fabre et al. 2011), and T3 Triticeae toolbox (http://triticeaetoolbox.org; Blake et al. 2016).

4. *Data management process*: It is an essential aspect of the research process. It can be difficult, especially when studies include numerous researchers and/or are done from multiple locations. Throughout the study period, data management is determined by the types of data involved, and how they are collected, kept, and used. Processing offers the methods and tools needed to convert raw primary data, including imaging or observational data, into a suitable quality and processable state. Then, the analysis focuses on obtaining information from the data that have been processed in order to aid in the acquisition of knowledge. Some analytic tools like Plant 3D (Ziamtsov and Navlakha 2020), LeafNet (Li et al. 2022), PlantCV (Gehan et al. 2017), and Phenomenal 3D (Artzet et al. 2019) are registered in bio.tools and dedicated to plant phenotyping experiments.
5. *Data sharing*: It must be considered to think about if and how the data will be shared with the larger scientific community. Data sharing may be facilitated by standardized data formats and metadata, and data repositories can be used to grant restricted access to data. However, data sharing must be done responsibly to ensure data privacy and intellectual property protection. Some data repository software are Dataverse (https://dataverse.org/), e!DAL-PGP (https://edal-pgp.ipk-gatersleben.de/), Zenodo (https://zenodo.org/), and BrAPI (https://www.brapi.org/).
6. *Data reuse*: Plant phenomics researchers should contribute to ensuring that their data are as helpful as possible for increasing scientific knowledge and avoiding redundancy in data collecting by encouraging data sharing, correct recording, standardization, analysis, and citation. Many data repositories are available like AgroPortal (Jonquet et al. 2018), and FAIDARE for reusing plant phenotyping data that follows MIAPPE specifications.

7 Role of HTP Phenomics in Accelerating Plant Breeding

The scope of phenomics data has expanded, which has made it easier to determine whether analytical and quantitative genetic approaches for the study of plant breeding in multi-environment trials have become more apparent. It is critical for plant breeding and the advancement of plant genomics to connect phenotypes and genotypes in order to find genetic structures that regulate significant features. Genome-wide association studies (GWAS) have been used widely in recent years to decipher the connections between genes and phenotypes (Xiao et al. 2022). Bai et al.

(2016) developed a multi-sensor system for high-throughput phenotyping platform suitable for evaluating canopy data in wheat. This particular system includes five sensors for plant phenotyping, including (i) ultrasonic distance sensor, (ii) thermal-infrared radiometer, (iii) NDVI (Normalized Difference Vegetation Index) sensor, (iv) portable spectrometer, and (v) RGB web camera. Besides, two sensors (solar radiation sensor and temperature/relative humidity sensor) are also included to record the environmental data. LabVIEW program enabled the synchronization of multiple sensors and data storage. Number of platforms are available to hold the sensors for phenotyping but each has its own limitations and advantages. For instance, platforms like self-propelled tractors cause more mechanical disturbance and soil compactness. Unmanual aerial vehicles are limiting while the payload is a concern. For a limited area (few acres) of phenotyping, manually operated platform was found best to carry the multiple sensors and to ensure the minimum mechanical disturbance. A strong correlation between manual phenotyping and sensor-based phenotyping suggested the applicability of this system for HTP in wheat. Although the primary purpose of phenotyping in plant breeding is to discover plants with enhanced characteristics, phenotyping is currently primarily used to monitor crops for pest identification and fertilizer requirements in crop management. The advancement of phenotyping techniques and processes for both proximal and remote sensing speeds up germplasm screening and selection, which increases the genetic variety in breeding material since germplasm with such qualities can then be kept in breeding programs (Zheng et al. 2021). Some HTPP techniques for plant breeding includes satellite imaging, UAVs (Unmanned Aerial Vehicles), and proximal phenotyping (Chawade et al. 2019; Pinto et al. 2023; Zhang et al. 2020).

In order to ensure precise phenotyping, particularly under drought/heat stress, it is important to develop a large-scale platform that controls the environmental condition along with the HTP. One such platform called "PhénoField®" is created at Ouzouer-le-Marché/Beaucela Romaine in 2013 (Beauchene et al. 2019). It occupies the area of 7.5 hectares. This platform has three major parts: (i) automated rainout shelters, which protect the crop from rainfall and reduces cofounding in phenotyping, (ii) high-throughput field phenotyping sensors, and (iii) data processing and storage unit. The platform is also equipped with automated irrigation systems and fertilization management systems. Micrometeorological measurement systems, available in the system, access the environmental data and also record the responses of different genotypes under such conditions. These features are good to study genotype–environment interactions. The availability of various automated systems, like rainout shelters, irrigation systems, fertilization management systems, and micrometeorological measurement system, makes "PhénoField®" the best suitable system for HTP under various abiotic stresses. The suitability of "PhénoField®" for wheat production under abiotic stress is also demonstrated (Beauchene et al. 2019). The "PhénoField®" is used in BREEDWHEAT (an initiative to strengthen the French wheat breeding sector), and in field trial conducted in 2017. The large-scale precision phenotyping applicability of the system was well demonstrated, particularly under water deficit and nitrogen stress conditions. Varietal differences in growth and wheat development under stress conditions were well characterized by "PhénoField®." Drought stress during stem elongation severely impacted the wheat yield. Similarly, yield goes down under nitrogen deficiency as compared to optimum nitrogen conditions (Beauchene et al. 2019). To enhance traditional breeding methods and enhance genetic gain, HTP approaches are essential. Incorporating these novel technologies into conventional breeding pipelines will help cultivars with resilient yields be delivered in the face of the anticipated unfavorable future climatic circumstances brought on by climate change and as a result of a rise in the biotic and abiotic stressors.

8 Conclusion

In recent years, HTP has advanced significantly, and it now plays a crucial role in the analysis and comprehensive measurement of various observable traits, including morphological, physiological, biochemical, and behavioral traits, in order to identify the complex relationship between phenotype and genotype in different crops. It is feasible to do crop planning on agricultural lands, determine the factors that influence crop production, and determine the processes by which plants work against abiotic stresses by using phenomics techniques. Phenomics technologies are being used in different areas such as physiology and biochemical mechanism, the morphology of above-ground organs and RSA, growth processes, productivity, identification of biotic and abiotic stresses, genetic engineering, breeding, selection of lines and varieties, yield forecasting, etc., which makes it possible to monitor the environment more effectively and find solutions to major issues in the area of food and environmental safety. In the upcoming decade, crop phenomics will also need to overcome hurdles such as the creation of novel artificial intelligence (AI)-based approaches and techniques to advance image-based phenotyping, which efficiently evaluates and precisely interprets large digital image-based phenotypes, and identifies useful quantitative traits for functional analysis of crop plants. We also need to work in the area of intelligent data-mining of multi-omics data, which offers a potent tool to explore biological mechanisms regulating plant growth and development and assist in plant breeding for the development of high-yielding and climate-resilient crops.

Acknowledgments Vijay Gahlaut acknowledges the DST-INSPIRE Faculty Awards received from the Department of Science and Technology, Ministry of Science and Technology, Government of India.

Conflict of Interest The authors have no competing interest to be declared.

References

Aasen H, Burkart A, Bolten A, Bareth G (2015) Generating 3D hyperspectral information with lightweight UAV snapshot cameras for vegetation monitoring: from camera calibration to quality assurance. ISPRS J of Photogramm. Remote Sens 108:245–259

Abdelhakim LO, Rosenqvist E, Wollenweber B, Spyroglou I, Ottosen CO, Panzarová K (2021) Investigating combined drought-and heat stress effects in wheat under controlled conditions by dynamic image-based phenotyping. Agron 11:364

Alemu A, Feyissa T, Maccaferri M, Sciara G, Tuberosa R, Ammar K, Badebo A, Acevedo M, Letta T, Abeyo B (2021) Genome-wide association analysis unveils novel QTLs for seminal root system architecture traits in Ethiopian durum wheat. BMC Genomics 22:1–6

Armengaud P, Zambaux K, Hills A, Sulpice R, Pattison RJ, Blatt MR, Amtmann A (2009) EZ-Rhizo: integrated software for the fast and accurate measurement of root system architecture. Plant J 57:945–956

Artzet S, Chen TW, Chopard J, Brichet N, Mielewczik M, Cohen-Boulakia S, Cabrera-Bosquet L, Tardieu F, Fournier C, Pradal C (2019) Phenomenal: an automatic open source library for 3D shoot architecture reconstruction and analysis for image-based plant phenotyping. BioRxiv 21:805739

Arvidsson S, Pérez-Rodríguez P, Mueller-Roeber B (2011) A growth phenotyping pipeline for Arabidopsis thaliana integrating image analysis and rosette area modeling for robust quantification of genotype effects. New Phytol 191:895–907

Awlia M, Nigro A, Fajkus J, Schmoeckel SM, Negrão S, Santelia D, Trtìlek M, Tester M, Julkowska MM, Panzarová K (2016) High-throughput non-destructive phenotyping of traits that contribute to salinity tolerance in *Arabidopsis thaliana*. Front Plant Sci 7:1414

Babar MA, Van Ginkel M, Klatt AR, Prasad B, Reynolds MP (2006) The potential of using spectral reflectance indices to estimate yield in wheat grown under reduced irrigation. Euphytica 150:155–172

Backhaus A, Kuwabara A, Bauch M, Monk N, Sanguinetti G, Fleming A (2010) LEAFPROCESSOR: a new leaf phenotyping tool using contour bending energy and shape cluster analysis. New Phytol 187:251–261

Bagnall GC, Koonjoo N, Altobelli SA, Conradi MS, Fukushima E, Kuethe DO, Mullet JE, Neely H, Rooney WL, Stupic KF, Weers B (2020) Low-field magnetic resonance imaging of roots in intact clayey and silty soils. Geoderma 370:114356

Bai G, Ge Y, Hussain W, Baenziger PS, Graef G (2016) A multi-sensor system for high throughput field phenotyping in soybean and wheat breeding. Comput Electron Agric 128:181–192

Baker NR (2008) Chlorophyll fluorescence: a probe of photosynthesis in vivo. Annu Rev Plant Biol 59:89–113

Balachandran S, Hurry VM, Kelley SE, Osmond CB, Robinson SA, Rohozinski J, Seaton GG, Sims DA (1997) Concepts of plant biotic stress. Some insights into the stress physiology of virus-infected plants, from the perspective of photosynthesis. Physiol Plant 100:203–213

Banerjee K, Krishnan P, Mridha N (2018) Application of thermal imaging of wheat crop canopy to estimate leaf area index under different moisture stress conditions. Biosyst Eng 166:13–27

Basu P, Pal A, Lynch JP, Brown KM (2007) A novel image-analysis technique for kinematic study of growth and curvature. Plant Physiol 145:305–316

Bayoumi TY, El-Hendawy S, Yousef MS, Emam MA, Okasha SA (2014) Application of infrared thermal imagery for monitoring salt tolerant of wheat genotypes. J Am Sci 10:227–234

Berger B, de Regt B, Tester M (2012) Trait dissection of salinity tolerance with plant phenomics. In: Plant salt tolerance: methods and protocols. Humana Press, pp 399–413

Berk P, Hocevar M, Stajnko D, Belsak A (2016) Development of alternative plant protection product application techniques in orchards, based on measurement sensing systems: a review. Comput Electron Agric 124:273–288

Billiau K, Sprenger H, Schudoma C, Walther D, Köhl KI (2012) Data management pipeline for plant phenotyping in a multisite project. Funct Plant Biol 39:948–957

Blake VC, Birkett C, Matthews DE, Hane DL, Bradbury P, Jannink JL (2016) The Triticeae toolbox: combining phenotype and genotype data to advance small-grains breeding. The Plant Genome 9:2014–2112

Blouin M, Barot S, Roumet C (2007) A quick method to determine root biomass distribution in diameter classes. Plant Soil 290:371–381

Bours R, Muthuraman M, Bouwmeester H, van der Krol A (2012) OSCILLATOR: a system for analysis of diurnal leaf growth using infrared photography combined with wavelet transformation. Plant Methods 8:1–2

Bowman BC, Chen J, Zhang J, Wheeler J, Wang Y, Zhao W, Nayak S, Heslot N, Bockelman H, Bonman JM (2015) Evaluating grain yield in spring wheat with canopy spectral reflectance. Crop Sci 55:1881–1890

Bray EA, Bailey-Serres J, Weretilnyk E (2000) Biochemistry and molecular biology of plants. American Society of Plant Physiologists, Rockville

Brown TB, Cheng R, Sirault XR, Rungrat T, Murray KD, Trtilek M, Furbank RT, Badger M, Pogson BJ, Borevitz JO (2014) TraitCapture: genomic and environment modelling of plant phenomic data. Curr Opin Plant Biol 18:73–79

Burling K, Hunsche M, Noga G (2010) Quantum yield of non-regulated energy dissipation in PSII (Y (NO)) for early detection of leaf rust (*Puccinia triticina*) infection in susceptible and resistant wheat (*Triticum aestivum* L.) cultivars. Precis Agric 11:703–716

Busemeyer L, Mentrup D, Möller K, Wunder E, Alheit K, Hahn V, Maurer HP, Reif JC, Würschum T, Müller J, Rahe F (2013) BreedVision—A multi-sensor platform for non-destructive field-based phenotyping in plant breeding. Sensors 13:2830–2847

Bylesjo M, Segura V, Soolanayakanahally RY, Rae AM, Trygg J, Gustafsson P, Jansson S, Street NR (2008) LAMINA: a tool for rapid quantification of leaf size and shape parameters. BMC Plant Biol 8:1–9

Beauchene K, Leroy F, Fournier A, Huet C, Bonnefoy M, Lorgeou J, De Solan B, Piquemal B, Thomas S, Cohan JP (2019) Management and characterization of abiotic stress via PhénoField®, a high-throughput field phenotyping platform. Front Plant Sci 10:904

Borisjuk L, Rolletschek H, Neuberger T (2013) Nuclear magnetic resonance imaging of lipid in living plants. Prog Lipid Res 52(4):465–487

Cabrera-Bosquet L, Sanchez C, Rosales A, Palacios-Rojas N, Araus JL (2011) Near-infrared reflectance spectroscopy (NIRS) assessment of δ18O and nitrogen and ash contents for improved yield potential and drought adaptation in maize. J Agric Food Chem 59:467–474

Cane MA, Maccaferri M, Nazemi G, Salvi S, Francia R, Colalongo C, Tuberosa R (2014) Association mapping for root architectural traits in durum wheat seedlings as related to agronomic performance. Mol Breed 34:1629–1645

Cardon ZG, Mott KA, Berry JA (1994) Dynamics of patchy stomatal movements, and their contribution to steady-state and oscillating stomatal conductance calculated using gas-exchange techniques. Plant Cell Environ 17:995–1007

Casadesús J, Kaya Y, Bort J, Nachit MM, Araus JL, Amor S, Ferrazzano G, Maalouf F, Maccaferri M, Martos V, Ouabbou H (2007) Using vegetation indices derived from conventional digital cameras as selection criteria for wheat breeding in water-limited environments. Ann Appl Biol 150:227–236

Chaerle L, Van Der Straeten D (2000) Imaging techniques and the early detection of plant stress. Trends Plant Sci 5:495–501

Chaerle L, Pineda M, Romero-Aranda R, Van Der Straeten D, Baron M (2006) Robotized thermal and chlorophyll fluorescence imaging of pepper mild mottle virus infection in Nicotiana benthamiana. Plant Cell Physiol 47:1323–1336

Chaerle L, Hagenbeek D, De Bruyne E, Van Der Straeten D (2007a) Chlorophyll fluorescence imaging for disease-resistance screening of sugar beet. Plant Cell Tissue Organ Cult 91:97–106

Chaerle L, Hagenbeek D, Vanrobaeys X, Van Der Straeten D (2007b) Early detection of nutrient and biotic stress in Phaseolus vulgaris. Int J Remote Sens 28:3479–3492

Chavarría-Krauser A, Nagel KA, Palme K, Schurr U, Walter A, Scharr H (2008) Spatio-temporal quantification of differential growth processes in root growth zones based on a novel combination of image sequence processing and refined concepts describing curvature production. New Phytol 177:811–821

Chawade A, van Ham J, Blomquist H, Bagge O, Alexandersson E, Ortiz R (2019) High-throughput field-phenotyping tools for plant breeding and precision agriculture. Agronomy 9:258

Chelladurai V, Jayas DS, White ND (2010) Thermal imaging for detecting fungal infection in stored wheat. J Stored Prod Res 46:174–179

Cheng T, Rivard B, Sanchez-Azofeifa A (2011) Spectroscopic determination of leaf water content using continuous wavelet analysis. Remote Sens Environ 115:659–670

Clark RT, MacCurdy RB, Jung JK, Shaff JE, McCouch SR, Aneshansley DJ, Kochian LV (2011) Three-dimensional root phenotyping with a novel imaging and software platform. Plant Physiol 156:455–465

Claudio HC, Cheng Y, Fuentes DA, Gamon JA, Luo H, Oechel W, Qiu HL, Rahman AF, Sims DA (2006) Monitoring drought effects on vegetation water content and fluxes in chaparral with the 970 nm water band index. Remote Sens Environ 103:304–311

Cobb JN, DeClerck G, Greenberg A, Clark R, McCouch S (2013) Next-generation phenotyping: requirements and strategies for enhancing our understanding of genotype–phenotype relationships and its relevance to crop improvement. Theor Appl Genet 126:867–887

Cseri A, Sass L, Torjek O, Pauk J, Vass I, Dudits D (2013) Monitoring drought responses of barley genotypes with semi-robotic phenotyping platform and association analysis between recorded traits and allelic variants of some stress genes. Aust J Crop Sci 7:1560–1570

Danilevicz MF, Bayer PE, Nestor BJ, Bennamoun M, Edwards D (2021) Resources for image-based high-throughput phenotyping in crops and data sharing challenges. Plant Physiol 187:699–715

Das A, Schneider H, Burridge J, Ascanio AK, Wojciechowski T, Topp CN, Lynch JP, Weitz JS, Bucksch A (2015) Digital imaging of root traits (DIRT): a high-throughput computing and collaboration platform for field-based root phenomics. Plant Method 11:1–2

Delory BM, Li M, Topp CN, Lobet G (2018) archiDART v3. 0: a new data analysis pipeline allowing the topological analysis of plant root systems. F1000research 7

Din M, Zheng W, Rashid M, Wang S, Shi Z (2017) Evaluating hyperspectral vegetation indices for leaf area index estimation of *Oryza sativa L.* at diverse phenological stages. Front Plant Sci 8:820

Dita MA, Rispail N, Prats E, Rubiales D, Singh KB (2006) Biotechnology approaches to overcome biotic and abiotic stress constraints in legumes. Euphytica 147:1–24

Dornbusch T, Andrieu B (2010) Lamina2Shape—an image processing tool for an explicit description of lamina shape tested on winter wheat (Triticum aestivum L.). Comput Electron Agric 70:217–224

DoVale JC, Fritsche-Neto R (2015) Root phenomics. In: Phenomics: how next-generation phenotyping is revolutionizing plant breeding. Springer, Heidelberg, pp 49–66

Duan L, Yang W, Huang C, Liu Q (2011) A novel machine-vision-based facility for the automatic evaluation of yield-related traits in rice. Plant Methods 7:1–3

Easlon HM, Bloom AJ (2014) Easy leaf area: automated digital image analysis for rapid and accurate measurement of leaf area. Appl Plant Sci 2:1400033

Edwards KD, Millar AJ (2007) Analysis of circadian leaf movement rhythms. In: Arabidopsis thaliana. Circadian rhythms: methods and protocols. Humana Press, pp 103–113

El-Lithy ME, Clerkx EJ, Ruys GJ, Koornneef M, Vreugdenhil D (2004) Quantitative trait locus analysis of growth-related traits in a new Arabidopsis recombinant inbred population. Plant Physiol 135:444–458

Elsayed S, Mistele B, Schmidhalter U (2011) Can changes in leaf water potential be assessed spectrally? Funct Plant Biol 38:523–533

Engelmann W, Simon K, Phen CJ (1992) Leaf movement rhythm in Arabidopsis thaliana. Z Naturforsch C 47:925–938

Fabre J, Dauzat M, Nègre V, Wuyts N, Tireau A, Gennari E, Neveu P, Tisné S, Massonnet C, Hummel I, Granier C (2011) PHENOPSIS DB: an information system for Arabidopsis thalianaphenotypic data in an environmental context. BMC Plant Biol 11:1–7

Fahlgren N, Gehan MA, Baxter I (2015) Lights, camera, action: high-throughput plant phenotyping is ready for a close-up. Curr Opin Plant Biol 24:93–99

Fatma M, Iqbal N, Sehar Z, Alyemeni MN, Kaushik P, Khan NA, Ahmad P (2021) Methyl jasmonate protects the PS II system by maintaining the stability of chloroplast D1 protein and accelerating enzymatic antioxidants in heat-stressed wheat plants. Antioxidants 10:1216

Fehér-Juhász E, Majer P, Sass L, Lantos C, Csiszár J, Turóczy Z, Mihály R, Mai A, Horváth GV, Vass I, Dudits D (2014) Phenotyping shows improved physiological traits and seed yield of transgenic wheat plants expressing the alfalfa aldose reductase under permanent drought stress. Acta Physiol Plant 36:663–673

Finch JA, Guillaume G, French SA, Colaco RD, Davies JM, Swarbreck SM (2017) Wheat root length and not branching is altered in the presence of neighbours, including blackgrass. PLoS One 12:e0178176

Fiorani F, Schurr U (2013) Future scenarios for plant phenotyping. Ann Rev Plant Biol 64:267–291

Fiorani F, Rascher U, Jahnke S, Schurr U (2012) Imaging plants dynamics in heterogenic environments. Curr Opin Biotechnol 23:227–235

Furbank RT, Tester M (2011) Phenomics–technologies to relieve the phenotyping bottleneck. Trends Plant Sci 16:635–644

Gaudin AC, Henry A, Sparks AH, Slamet-Loedin IH (2013) Taking transgenic rice drought screening to the field. J Exp Bot 64(1):109–117

Galkovskyi T, Mileyko Y, Bucksch A, Moore B, Symonova O, Price CA, Topp CN, Iyer-Pascuzzi AS, Zurek PR, Fang S, Harer J (2012) GiA roots: software for the high throughput analysis of plant root system architecture. BMC Plant Biol 12:1–2

Gangopadhyay M, Das AK, Sahu R, Saha A, Dey S, Bandyopadhyay S, Mitra A (2021) Evaluation of growth response for mass production and accumulation of 2-hydroxy-4-methoxybenzaldehyde in endangered Hemidesmus indicus by an aeroponic system. Ind Crop Prod 172:114072

Ge Y, Bai G, Stoerger V, Schnable JC (2016) Temporal dynamics of maize plant growth, water use, and leaf water content using automated high throughput RGB and hyperspectral imaging. Comput Electron Agric 127:625–632

Gebbers R, Ehlert D, Adamek R (2011) Rapid mapping of the leaf area index in agricultural crops. J Agron 103:1532–1541

Gehan MA, Fahlgren N, Abbasi A, Berry JC, Callen ST, Chavez L, Sax T (2017) PlantCV v2: image analysis software for high-throughput plant phenotyping. PeerJ 5:e4088

Getman-Pickering ZL, Campbell A, Aflitto N, Grele A, Davis JK, Ugine TA (2020) LeafByte: a mobile application that measures leaf area and herbivory quickly and accurately. Methods Ecol Evol 11:215–221

Golzarian MR, Frick RA, Rajendran K, Berger B, Roy S, Tester M, Lun DS (2011) Accurate inference of shoot biomass from high-throughput images of cereal plants. Plant Methods 7:1–11

Granier C, Aguirrezabal L, Chenu K, Cookson SJ, Dauzat M, Hamard P, Thioux JJ, Rolland G, Bouchier-Combaud S, Lebaudy A, Muller B (2006) PHENOPSIS, an automated platform for reproducible phenotyping of plant responses to soil water deficit in Arabidopsis thaliana permitted the identification of an accession with low sensitivity to soil water deficit. New Phytol 169:623–635

Grant D, Nelson RT, Cannon SB, Shoemaker RC (2010) SoyBase, the USDA-ARS soybean genetics and genomics database. Nucleic Acids Res 38:843–846

Gray SB, Dermody O, DeLucia EH (2010) Spectral reflectance from a soybean canopy exposed to elevated CO2 and O3. J Exp Bot 61:4413–4422

Griffiths M, Atkinson JA, Gardiner LJ, Swarup R, Pound MP, Wilson MH, Well DM (2019) Identification of nitrogen-dependent QTL and underlying genes for root system architecture in hexaploid wheat. BioRxiv, p 591529

Gupta PK, Balyan HS, Gahlaut V, Kulwal PL (2012) Phenotyping, genetic dissection, and breeding for drought and heat tolerance in common wheat: status and prospects. Plant Breed Rev 36:85–168

Gutierrez M, Reynolds MP, Klatt AR (2010) Association of water spectral indices with plant and soil water relations in contrasting wheat genotypes. J Exp Bot 61:3291–3303

Hairmansis A, Berger B, Tester M, Roy SJ (2014) Image-based phenotyping for non-destructive screening of different salinity tolerance traits in rice. Rice 7:1–10

Harris BN, Sadras VO, Tester M (2010) A water-centred framework to assess the effects of salinity on the growth and yield of wheat and barley. Plant Soil 336:377–389

Herrero-Huerta M, Raumonen P, Gonzalez-Aguilera D (2022) 4DRoot: root phenotyping software for temporal 3D scans by X-ray computed tomography. Front Plant Sci 13:986856

Herridge RP, Day RC, Baldwin S, Macknight RC (2011) Rapid analysis of seed size in Arabidopsis for mutant and QTL discovery. Plant Methods 7:1–11

Hoffmeister D, Waldhoff G, Curdt C, Tilly N, Bendig J, Bareth G (2013) Spatial variability detection of crop height in a single field by terrestrial laser scanning. Precis Agric 13:267–274

Hofle B (2013) Radiometric correction of terrestrial LiDAR point cloud data for individual maize plant detection. IEEE Geosci Remote Sens Lett 11:94–98

Honsdorf N, March TJ, Berger B, Tester M, Pillen K (2014) High-throughput phenotyping to detect drought tolerance QTL in wild barley introgression lines. PLoS One 9:e97047

Houle D, Govindaraju DR, Omholt S (2010) Phenomics: the next challenge. Nat Rev Genet 11:855–866

Humplík JF, Lazár D, Fürst T, Husičková A, Hýbl M, Spíchal L (2015) Automated integrative high-throughput phenotyping of plant shoots: a case study of the cold-tolerance of pea (*Pisum sativum* L.). Plant Methods 11:1–11

Iwata H, Ukai Y (2002) SHAPE: a computer program package for quantitative evaluation of biological shapes based on elliptic Fourier descriptors. J Hered 93:384–385

Jangra S, Chaudhary V, Yadav RC, Yadav NR (2021) High-throughput phenotyping: a platform to accelerate crop improvement. Phenomics 1:31–53

Jansen M, Gilmer F, Biskup B, Nagel KA, Rascher U, Fischbach A, Briem S, Dreissen G, Tittmann S, Braun S, De Jaeger I (2009) Simultaneous phenotyping of leaf growth and chlorophyll fluorescence via GROWSCREEN FLUORO allows detection of stress tolerance in Arabidopsis thaliana and other rosette plants. Funct Plant Biol 36:902–914

Jiang JB, Huang WJ, Chen YH (2010) Using canopy hyperspectral ratio index to retrieve relative water content of wheat under yellow rust stress. Guang Pu Xue Yu Guang Pu Fen Xi/Spectrosc. Spectr Anal 30:1939–1943

Jonquet C, Toulet A, Arnaud E, Aubin S, Yeumo ED, Emonet V, Graybeal J, Laporte MA, Musen MA, Pesce V, Larmande P (2018) AgroPortal: a vocabulary and ontology repository for agronomy. Comput Electron Agric 144:126–143

Jahnke S, Menzel MI, Van Dusschoten D, Roeb GW, Bühler J, Minwuyelet S, Blümler P, Temperton VM, Hombach T, Streun M, Beer S (2009) Combined MRI-PET dissects dynamic changes in plant structures and functions. The Plant J 59(4):634–644

Jones AM, Danielson J, ManojKumar SN, Lanquar V, Grossmann G, Frommer WB (2014) Abscisic acid dynamics in roots detected with genetically encoded FRET sensors. Elife 3:e01741

Kokaly RF, Asner GP, Ollinger SV, Martin ME, Wessman CA (2009) Characterizing canopy biochemistry from imaging spectroscopy and its application to ecosystem studies. Remote Sens Environ 113:78–91

Kaplan H (2007) Practical applications of infrared thermal sensing and imaging equipment. SPIE Press, p 75

Kastberger G, Stachl R (2003) Infrared imaging technology and biological applications. Behav Res Methods Instrum Comput 35:429–439

Kenobi K, Atkinson JA, Wells DM, Gaju O, De Silva JG, Foulkes MJ, Dryden IL, Wood AT, Bennett MJ (2017) Linear discriminant analysis reveals differences in root architecture in wheat seedlings related to nitrogen uptake efficiency. J Exp Bot 68:4969–4981

Khanna R, Möller M, Pfeifer J, Liebisch F, Walter A, Siegwart R (2015) Beyond point clouds-3d mapping and field parameter measurements using UAVs. 20th Conference on Emerging Technologies & Factory Automation (ETFA), pp 1–4

Kim JY (2020) Roadmap to high throughput phenotyping for plant breeding. J Biosyst Eng

Kim SL, Solehati N, Choi IC, Kim KH, Kwon TR (2017) Data management for plant phenomics. J Plant Biol 60:285–297

Kim M, Lee C, Hong S, Kim SL, Baek JH, Kim KH (2021) High-throughput phenotyping methods for breeding drought-tolerant crops. Int J Mol Sci 22:8266

Kjaer KH, Ottosen CO (2015) 3D laser triangulation for plant phenotyping in challenging environments. Sensors 15:13533–13547

Klukas C, Chen D, Pape J-M (2014) Integrated analysis platform: an open-source information system for high throughput plant phenotyping. Plant Physiol 165:506–518

Krajewski P, Chen D, Ćwiek H, van Dijk AD, Fiorani F, Kersey P, Klukas C, Lange M, Markiewicz A, Nap JP, van Oeveren J (2015) Towards recommendations for metadata and data handling in plant phenotyping. J Exp Bot 66:5417–5427

Kumar A, Kaushik P (2022) High throughput phenotyping approach. Retrieved from Preprints.org. https://doi.org/10.20944/preprints202204.0228.v

Kumar P, Huang C, Cai J, Miklavcic SJ (2014) Root phenotyping by root tip detection and classification through statistical learning. Plant Soil 380:193–209

Kumar J, Pratap A, Kumar S (2015) Phenomics in crop plants: trends, options and limitations. Springer International, New Delhi

Lazár D (2015) Parameters of photosynthetic energy partitioning. J Plant Physiol 175:131–147

Le Bot J, Serra V, Fabre J, Draye X, Adamowicz S, Pagès L (2010) DART: a software to analyse root system architecture and development from captured images. Plant Soil 326:261–273

Lefsky MA, Cohen WB, Parker GG, Harding DJ (2002) Lidar remote sensing for ecosystem studies: Lidar, an emerging remote sensing technology that directly measures the three-dimensional distribution of plant canopies, can accurately estimate vegetation structural attributes and should be of particular interest to forest, landscape, and global ecologists. Biosci 52:19–30

Li YF, Kennedy G, Ngoran F, Wu P, Hunter J (2013) An ontology-centric architecture for extensible scientific data management systems. Future Gener Comp Syst 29:641–653

Li L, Zhang Q, Huang D (2014) A review of imaging techniques for plant phenotyping. Sensors 14:20078–20111

Li Z, Guo R, Li M, Chen Y, Li G (2020) A review of computer vision technologies for plant phenotyping. Comput Electron Agric 176:105672

Li S, Li L, Fan W, Ma S, Zhang C, Kim JC, Wang K, Russinova E, Zhu Y, Zhou Y (2022) LeafNet: a tool for segmenting and quantifying stomata and pavement cells. Plant Cell 34:1171–1188

Lin Y (2015) LiDAR: an important tool for next-generation phenotyping technology of high potential for plant phenomics? Comput Electron Agric 119:61–73

Llorens J, Gil E, Llop J, Escolà A (2011) Ultrasonic and LIDAR sensors for electronic canopy characterization in vineyards: advances to improve pesticide application methods. Sensors 11:2177–2194

Lobet G, Pagès L, Draye X (2011) A novel image-analysis toolbox enabling quantitative analysis of root system architecture. Plant Physiol 157:29–39

Lumme J, Karjalainen M, Kaartinen H, Kukko A, Hyyppä J, Hyyppä H, Jaakkola A, Kleemola J (2008) Terrestrial laser scanning of agricultural crops. Int Arch Photogramm Remote Sens Spat Inf Sci 37:563–566

Madec S, Baret F, De Solan B, Thomas S, Dutartre D, Jezequel S, Hemmerlé M, Colombeau G, Comar A (2017) High-throughput phenotyping of plant height: comparing unmanned aerial vehicles and ground LiDAR estimates. Front Plant Sci 8:2002

Mairhofer S, Zappala S, Tracy SR, Sturrock C, Bennett M, Mooney SJ, Pridmore T (2012) RooTrak: automated recovery of three-dimensional plant root architecture in soil from X-ray microcomputed tomography images using visual tracking. Plant Physiol 158:561–569

Maphosa L, Thoday-Kennedy E, Vakani J, Phelan A, Badenhorst P, Slater A, Spangenberg G, Kant S (2017) Phenotyping wheat under salt stress conditions using a 3D laser scanner. Isr J Plant Sci 64:55–62

Mathieu L, Lobet G, Tocquin P, Périlleux C (2015) “Rhizoponics”: a novel hydroponic rhizotron for root system analyses on mature Arabidopsis thaliana plants. Plant Methods 11:1–8

Meng R, Saade S, Kurtek S, Berger B, Brien C, Pillen K, Tester M, Sun Y (2017) Growth curve registration for evaluating salinity tolerance in barley. Plant Methods 13:1–9

Mir RR, Reynolds M, Pinto F, Khan MA, Bhat MA (2019) High-throughput phenotyping for crop improvement in the genomics era. Plant Sci 282:60–72

Mishra A, Mishra KB, Höermiller II, Heyer AG, Nedbal L (2011) Chlorophyll fluorescence emission as a reporter on cold tolerance in Arabidopsis thaliana accessions. Plant Signal Behavior 6:301–310

Mistele B, Schmidhalter U (2008) Spectral measurements of the total aerial N and biomass dry weight in maize using a quadrilateral-view optic. Field Crop Res 106:94–103

Montes JM, Melchinger AE, Reif JC (2007) Novel throughput phenotyping platforms in plant genetic studies. Trends Plant Sci 12:433–436

Moradi AB, Conesa HM, Robinson B, Lehmann E, Kuehne G, Kaestner A, Oswald S, Schulin R (2009) Neutron radiography as a tool for revealing root development in soil: capabilities and limitations. Plant Soil 318:243–255

Moreau D, Schneider C, Huguet T, Salon C, Munier-Jolain N (2009) Can differences of nitrogen nutrition level among medicago truncatula genotypes be assessed non-destructively? Probing with a recombinant inbred lines population: probing with a recombinant inbred lines population. Plant Signal Behav 4:30–32

Moshou D, Pantazi XE, Kateris D, Gravalos I (2014) Water stress detection based on optical multisensor fusion with a least squares support vector machine classifier. Biosyst Eng 117:15–22

Munns R, James RA, Sirault XR, Furbank RT, Jones HG (2010) New phenotyping methods for screening wheat and barley for beneficial responses to water deficit. J Exp Bot 61:3499–3507

Naeem A, French AP, Wells DM, Pridmore TP (2011) High-throughput feature counting and measurement of roots. Bioinformatics 27:1337–1338

Nagel KA, Putz A, Gilmer F, Heinz K, Fischbach A, Pfeifer J, Faget M, Blossfeld S, Ernst M, Dimaki C, Kastenholz B (2012) GROWSCREEN-Rhizo is a novel phenotyping robot enabling simultaneous measurements of root and shoot growth for plants grown in soil-filled rhizotrons. Funct Plant Biol 39:891–904

Narisetti N, Henke M, Seiler C, Shi R, Junker A, Altmann T, Gladilin E (2019) Semi-automated root image analysis (saRIA). Sci Rep 9:1–10

Neilson EH, Edwards AM, Blomstedt CK, Berger B, Møller BL, Gleadow RM (2015) Utilization of a high-throughput shoot imaging system to examine the dynamic phenotypic responses of a C4 cereal crop plant to nitrogen and water deficiency over time. J Exp Bot 66:1817–1832

Nguyen VL, Stangoulis J (2019) Variation in root system architecture and morphology of two wheat genotypes is a predictor of their tolerance to phosphorus deficiency. Acta Physio Plant 41:1–13

Omasa K, Hosoi F, Konishi A (2007) 3D lidar imaging for detecting and understanding plant responses and canopy structure. J Exp Bot 58:881–898

Paproki A, Sirault X, Berry S, Furbank R, Fripp J (2012) A novel mesh processing based technique for 3D plant analysis. BMC Plant Boil 12:1–13

Pasala R, Pandey BB (2020) Plant phenomics: high-throughput technology for accelerating genomics. J Biosci 45:111

Paulus S, Dupuis J, Riedel S, Kuhlmann H (2014) Automated analysis of barley organs using 3D laser scanning: an approach for high throughput phenotyping. Sensors 14:12670–12686

Pergl R, Hooft R, Suchánek M, Knaisl V, Slifka J (2019) Data stewardship wizard: a tool bringing together researchers, data stewards, and data experts around data management planning. Data Sci J 18:1

Pinto F, Zaman-Allah M, Reynolds M, Schulthess U (2023) Satellite imagery for high-throughput phenotyping in breeding plots. Front Plant Sci 14:1114670

Poiré R, Chochois V, Sirault XR, Vogel JP, Watt M, Furbank RT (2014) Digital imaging approaches for phenotyping whole plant nitrogen and phosphorus response in *Brachypodium distachyon*. J Integr Plant Biol 56:781–796

Penuelas J, Filella I (1998) Visible and near-infrared reflectance techniques for diagnosing plant physiological status. Trends Plant Sci 3(4):151–156

Rahaman MM, Chen D, Gillani Z, Klukas C, Chen M (2015) Advanced phenotyping and phenotype data analysis for the study of plant growth and development. Front Plant Sci 6:619

Rajendran K, Tester M, Roy SJ (2009) Quantifying the three main components of salinity tolerance in cereals. Plant Cell Environ 32:237–249

Romano G, Zia S, Spreer W, Cairns J, Araus JL, Müller J (2013) Rapid phenotyping of different maize varieties under drought stress by using thermal images

Roselló M, Royo C, Sanchez-Garcia M, Soriano JM (2019) Genetic dissection of the seminal root system architecture in Mediterranean durum wheat landraces by genome-wide association study. Agronomy 9:364

Schaeffer ML, Harper LC, Gardiner JM, Andorf CM, Campbell DA, Cannon EK, Sen TZ, Lawrence CJ (2011) MaizeGDB: curation and outreach go hand-in-hand. Database 2011

Schilling RK, Marschner P, Shavrukov Y, Berger B, Tester M, Roy SJ, Plett DC (2014) Expression of the A rabidopsis vacuolar H+-pyrophosphatase gene (AVP 1) improves the shoot biomass of transgenic barley and increases grain yield in a saline field. Plant Biotechnol J 12:378–386

Schlemmer MR, Francis DD, Shanahan JF, Schepers JS (2005) Remotely measuring chlorophyll content in corn leaves with differing nitrogen levels and relative water content. J Agron 97:106–112

Shan J, Toth CK (eds) (2018) Topographic laser ranging and scanning: principles and processing. CRC Press

Shi X, Li J, Mukherjee S, Datta S, Rathod V, Wang X, Lu W, Udpa L, Deng Y (2022) Ultra-wideband microwave imaging system for root phenotyping. Sensors 22:2031

Singh B, Mishra S, Bohra A, Joshi R, Siddique KH (2018) Crop phenomics for abiotic stress tolerance in crop plants. In: Biochemical, physiological and molecular avenues for combating abiotic stress tolerance in plants. Academic Press, pp 277–296

Singh RN, Krishnan P, Singh VK, Banerjee K (2022) Application of thermal and visible imaging to estimate stripe rust disease severity in wheat using supervised image classification methods. Ecol Inform 71:101774

Sirault XR, Fripp J, Paproki A, Kuffner P, Nguyen C, Li R, Daily H, Guo J, Furbank R (2013) PlantScan: a three-dimensional phenotyping platform for capturing the structural dynamic of plant development and growth. In Proceedings of the 7th International Conference on Functional-Structural Plant Models, Saariselkä, pp 45–48

Sozzani R, Busch W, Spalding EP, Benfey PN (2014) Advanced imaging techniques for the study of plant growth and development. Trends Plant Sci 19:304–310

Swarbrick PJ, Schulze-Lefert PA, Scholes JD (2006) Metabolic consequences of susceptibility and resistance (race-specific and broad-spectrum) in barley leaves challenged with powdery mildew. Plant Cell Environ 29:1061–1076

Takizawa H, Ezaki N, Mizuno S, Yamamoto S (2005) Plant recognition by integrating color and range data obtained through stereo vision. JACIII 9:630–636

Tanabata T, Shibaya T, Hori K, Ebana K, Yano M (2012) Smart grain: high-throughput phenotyping software for measuring seed shape through image analysis. Plant Physiol 160:1871–1880

Tayade R, Yoon J, Lay L, Khan AL, Yoon Y, Kim Y (2022) Utilization of spectral indices for high-throughput phenotyping. Plan Theory 11:1712

Thurau C, Kersting K, Bauckhage C (2010) Yes we can: simplex volume maximization for descriptive web-scale matrix factorization. Proceedings of the 19th ACM International Conference on Information and Knowledge Management, pp 1785–1788

Tilly N, Hoffmeister D, Cao Q, Huang S, Lenz-Wiedemann V, Miao Y, Bareth G (2014) Multitemporal crop surface models: accurate plant height measurement and biomass estimation with terrestrial laser scanning in paddy rice. J Appl Remote Sens 8:083671
Tsaftaris SA, Minervini M, Scharr H (2016) Machine learning for plant phenotyping needs image processing. Trends Plant Sci 21:989–991
Ullah S, Skidmore AK, Groen TA, Schlerf M (2013) Evaluation of three proposed indices for the retrieval of leaf water content from the mid-wave infrared (2–6 μm) spectra. Agric For Meteorol 171:65–71
Urfan M, Sharma S, Hakla HR, Rajput P, Andotra S, Lehana PK, Bhardwaj R, Khan MS, Das R, Kumar S, Pal S (2022) Recent trends in root phenomics of plant systems with available methods-discrepancies and consonances. Physiol Mol Biol Plants 28:1311–1321
Vasseur F, Bontpart T, Dauzat M, Granier C, Vile D (2014) Multivariate genetic analysis of plant responses to water deficit and high temperature revealed contrasting adaptive strategies. J Exp Bot 65:6457–6469
Walter A, Rascher U, Osmond B (2004) Transitions in photosynthetic parameters of midvein and interveinal regions of leaves and their importance during leaf growth and development. Plant Biol 6:184–191
Wang L, Uilecan IV, Assadi AH, Kozmik CA, Spalding EP (2009) HYPOTrace: image analysis software for measuring hypocotyl growth and shape demonstrated on Arabidopsis seedlings undergoing photomorphogenesis. Plant Physiol 149:1632–1637
Wang T, Rostamza M, Song Z, Wang L, McNickle G, Iyer-Pascuzzi AS, Qiu Z, Jin J (2019) SegRoot: a high throughput segmentation method for root image analysis. Comput Electron Agric 162:845–854
Watt M, Fiorani F, Usadel B, Rascher U, Muller O, Schurr U (2020) Phenotyping: new windows into the plant for breeders. Annu Rev Plant Biol 71:689–712
Weight C, Parnham D, Waites R (2008) Technical advance: LeafAnalyser: a computational method for rapid and large-scale analyses of leaf shape variation. The Plant J 53:578–586
Weirman A (2010) Plant phenomics teacher resource. www.plantphenomics.org.au/files/teacher/FinalPhenomicsforwordwith_image.doc
Whan AP, Smith AB, Cavanagh CR, Ral JP, Shaw LM, Howitt CA, Bischof L (2014) GrainScan: a low cost, fast method for grain size and colour measurements. Plant Methods 10:1–10
Wiechers D, Kahlen K, Stützel H (2011) Evaluation of a radiosity based light model for greenhouse cucumber canopies. Agric For Meteorol 151:906–915
Xiao Q, Bai X, Zhang C, He Y (2022) Advanced high-throughput plant phenotyping techniques for genome-wide association studies: a review. J Adv Res 35:215–230
Xie Q, Fernando KM, Mayes S, Sparkes DL (2017) Identifying seedling root architectural traits associated with yield and yield components in wheat. Ann Bot 119:1115–1129
Yang W, Duan L, Chen G, Xiong L, Liu Q (2013) Plant phenomics and high-throughput phenotyping: accelerating rice functional genomics using multidisciplinary technologies. Curr Opin Plant Biol 16:180–187
Yang W, Guo Z, Huang C, Duan L, Chen G, Jiang N, Fang W, Feng H, Xie W, Lian X, Wang G (2014) Combining high-throughput phenotyping and genome-wide association studies to reveal natural genetic variation in rice. Nat Commun 5:5087
Yang W, Feng H, Zhang X, Zhang J, Doonan JH, Batchelor WD, Xiong L, Yan J (2020) Crop phenomics and high-throughput phenotyping: past decades, current challenges, and future perspectives. Mol Plant 13:187–214
Yasrab R, Atkinson JA, Wells DM, French AP, Pridmore TP, Pound MP (2019) RootNav 2.0: deep learning for automatic navigation of complex plant root architectures. GigaScience 8:giz123
Yazdanbakhsh N, Fisahn J (2009) High throughput phenotyping of root growth dynamics, lateral root formation, root architecture and root hair development enabled by PlaRoM. Funct Plant Biol 36:938–946
Zappala S, Helliwell JR, Tracy SR, Mairhofer S, Sturrock CJ, Pridmore T, Bennett M, Mooney SJ (2013) Effects of X-ray dose on rhizosphere studies using X-ray computed tomography. PloS One 8:e67250

Zhang L, Grift TE (2012) A LIDAR-based crop height measurement system for Miscanthus giganteus. Comput Electron Agric 85:70–76

Zhang Y, Zhang N (2019) Imaging technologies for plant high-throughput phenotyping: a review. Front Agric Sci Eng 5:406–419

Zhang C, Marzougui A, Sankaran S (2020) High-resolution satellite imagery applications in crop phenotyping: an overview. Comput Electron Agric 175:105584

Zhao C, Zhang Y, Du J, Guo X, Wen W, Gu S, Wang J, Fan J (2019) Crop phenomics: current status and perspectives. Front Plant Sci 10:714

Zheng B, Shi L, Ma Y, Deng Q, Li B, Guo Y (2008) Comparison of architecture among different cultivars of hybrid rice using a spatial light model based on 3-D digitising. Funct Plant Biol 35:900–910

Zheng C, Abd-Elrahman A, Whitaker V (2021) Remote sensing and machine learning in crop phenotyping and management, with an emphasis on applications in strawberry farming. Remote Sens 13:531

Zhu D, Luo F, Zou R, Liu J, Yan Y (2021) Integrated physiological and chloroplast proteome analysis of wheat seedling leaves under salt and osmotic stresses. J Proteome 234:104097

Ziamtsov I, Navlakha S (2020) Plant 3D (P3D): a plant phenotyping toolkit for 3D point clouds. Bioinformatics 36:3949–3950

Zuo Q, Jie F, Zhang R, Meng L (2004) A generalized function of wheat's root length density distributions. Vadose Zone J 3:271–277

Speed Breeding to Accelerate Crop Improvement

Sobhan B. Sajja, Anurag Mathew, Janila Pasupuleti, and Radhakrishnan T.

Abstract Speed breeding through controlled environments as a new technique on the block offers advantages over conventional field-based generation advancement methods. Physiological parameters, especially light, is altered to induce early flowering to reduce generation time. Germinating immature seeds will reduce the generation time further. Several experiments were conducted in the past and are being conducted to develop speed breeding protocols for many crops. Speed breeding protocols were standardized for some crops, for example, chickpea, that allow six to seven generations per year as opposed to two to three earlier. Besides being faster, speed breeding enables savings on resources as advancing generations is cheaper through speed breeding as compared to field experiments. Rapid generation advancement through speed breeding integrated with the advanced techniques of genomic tools, gene editing, early- and high-throughput phenotyping, rapid population development, etc. would boost the genetic analysis and increase the rate of genetic gain in the cultivar development of the crop plants. In the backdrop of increasing food and nutrition demands, gains in crop improvement need to be increased. Speed breeding offers one of the feasible ways to achieve this. The costs involved may pose an obstacle to many enthusiasts, but cheaper alternatives can be explored. Integrating artificial intelligence with speed breeding makes it more valuable in crop improvement programs.

S. B. Sajja · J. Pasupuleti (✉)
International Crops Research Institute for the Semi-Arid Tropics (ICRISAT), Hyderabad, Telangana, India
e-mail: P.Janila@cgiar.org; janila.pasupuleti@icrisat.org

A. Mathew
International Crops Research Institute for the Semi-Arid Tropics (ICRISAT), Hyderabad, Telangana, India

Acharya N G Ranga Agricultural University (ANGRAU), Guntur, Andhra Pradesh, India

Radhakrishnan T.
ICAR-Directorate of Groundnut Research, Junagadh, Gujarat, India

P. M. Priyadarshan et al. (eds.), *Digital Agriculture*,
https://doi.org/10.1007/978-3-031-43548-5_14

Keywords Rapid generation advancement · Speed breeding · Genetic gain · Crop breeding · Photoperiod · Controlled environment

1 The Necessity

From one billion to eight billion, human population witnessed a big jump in a short period of 200 years considering that the growth was relatively slow up till 1950 (Fig. 1). Post 1950, the growth was phenomenal. If observed, the growth in population was exponential in Asia, as compared to other continents. Africa was the next to record such growth in the last few decades. Out of the eight billion population, six billion are in Asia (4.69 billion) and Africa (1.39 billion). During the early 60s, barring few developed countries, majority of the countries in Africa and Asia were struggling with food security issues, and several countries continue to be food insecure to date. Global hunger map (Fig. 2) shows that more than 800 million people do not get enough to eat (https://reliefweb.int/map/world/hunger-map-2018).

The global population has increased 2.7 times since 1960, while the production of primary cereals during the same period has increased by 3.5 times (Fig. 3). The growth of food grains can be largely attributed to the improvements made in crop genetics, agronomy, and plant protection through new high-yielding cultivars, improved cultivation practices, and pest and disease resistance/tolerance,

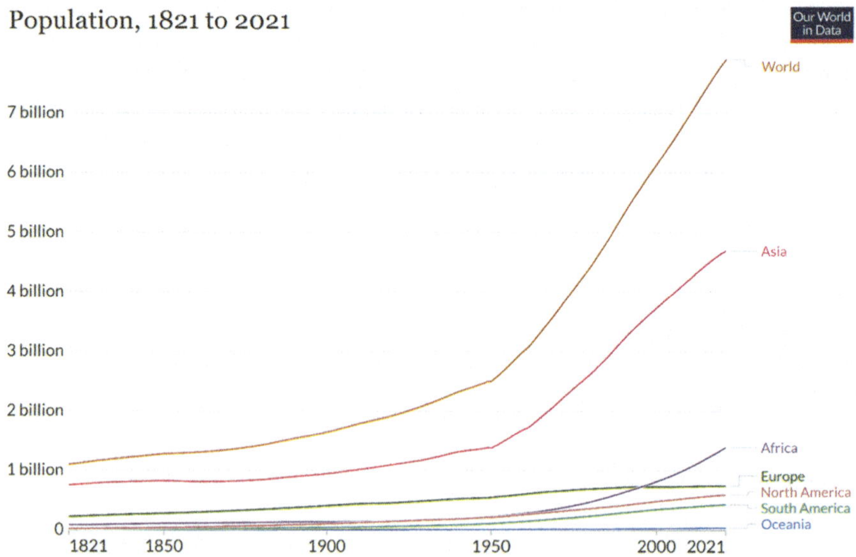

Fig. 1 Global growth in population over the last 200 years. (Source:https://ourworldindata.org/grapher/population)

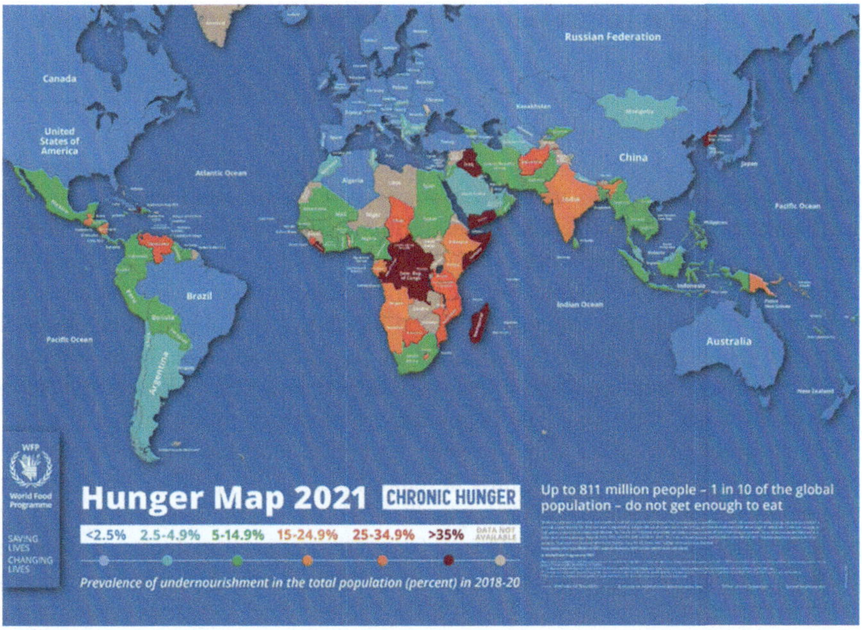

Fig. 2 Global hunger map (Source: https://reliefweb.int/map/world/hunger-map-2018)

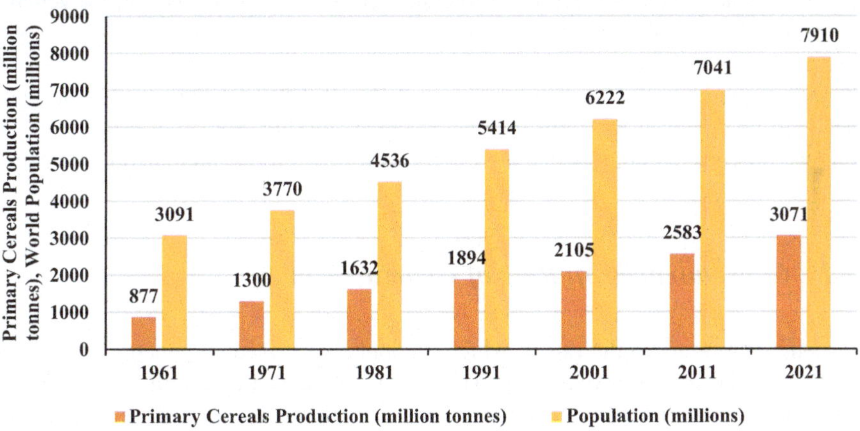

Fig. 3 Global production of primary cereals and population growth since the 1960s. (Source: https://www.fao.org/faostat/en/#data/QCL;https://www.worldometers.info/world-population/world-population-by-year/)

respectively (Spanne 2021). Plant breeding efforts produced new high-yielding crop cultivars with resistance to diseases and drought, contributing to the food grain production.

2 Role of Crop Improvement in the Phenomenal Food Grain Production

Crop improvement has been instrumental in increasing food grain production and addressing food insecurity. As the global population continues to grow, the development of new crop varieties with improved yield potential, pest and disease resistance, abiotic stress tolerance, and nutritional value will remain a critical tool in meeting the needs of the world's population. Traditional breeding techniques like introduction, selection, and interspecific hybridization have been successful over the years in developing crop varieties. Mutation breeding has also played an essential role in crop improvement and food production. By 2000, techniques of recombinant DNA technology were available to plant breeders to make further improvement in crop yields. Genetically modified plants were the last step to improve crop plants when all other techniques failed to solve perennial problems of stresses, biotic and abiotic. Some of the successfully developed genetically modified plants are Bt cotton, golden rice, roundup ready soybean, etc. However, the concerns associated with the safety of GM crops paved way for another new technology called "gene editing," which, on the other hand, involves precise modifications of a plant's DNA using technologies such as CRISPR/Cas9 to introduce targeted changes in the genetic code. The gene editing technique has been used in several crops like tomato to delay ripening (Tiwari et al. 2023), rice for bacterial blight resistance (Zaidi et al. 2016), and canola for herbicide tolerance (Li et al. 2013). The rate of genetic gain in major cereals saw an impressive growth from 1970 onwards. To meet the projected demand for food in future for the growing population, the rate of genetic improvement must double across (Voss-Fels et al. 2019). Enhanced genetic gain needs to be achieved not only for major staples, but also for other crops such as millets, oilseeds, and pulses that are important for both food and nutrition security.

2.1 *Consequences on Rate of Genetic Gain*

Genetic gain refers to the rate at which a population's average genetic value for a particular trait increases over time. This rate is influenced by a number of factors, such as selection intensity, genetic variance of the trait, accuracy of selection, and breeding cycle time. It is important to consider all of these factors when attempting to increase the rate of genetic gain for a particular trait or population. The following is the equation to measure genetic gain per year (Begna 2022):

$$\Delta G_{\text{year}} = \frac{i \; r_{AI} \; \sigma_A}{L}$$

i = Selection intensity

r_{AI} = Accuracy
σ_A = Genetic variance
L = Generation interval

Traditionally, genetic improvement programs have focused on increasing the numerator or the additive genetic variance of a trait. This involves selecting individuals with superior genetics and using them as parents for the next generation. But the impressive gains made in the last few decades are difficult to achieve currently, as yields are plateaued in many crops. However, recent research has shown that the denominator, or the generation interval, also plays an important role in genetic improvement. Reducing the breeding cycle time can have a significant impact on the rate of genetic gain and help to improve crop yields, especially in crops that have already plateaued in terms of yield improvement.

3 Reducing the Breeding Cycle Time

Development of new cultivars requires production of homozygous breeding lines following self-fertilization for four to five generations. Homozygous breeding lines thus developed constitute the candidates for selection. Further, these candidates will be subjected to selection for morphological traits, biotic and abiotic stress tolerance/resistance, and other parameters to select approximately 20% of lines (selected candidates) that will be advanced to next selection cycle or yield evaluation in multilocation trials. Production of homozygous lines in the field depends on the number of seasons a crop can be grown at a location. Generally, crops are grown once or twice in a year, depending on the local conditions. If only one season is feasible at a location, it takes 6–7 years to produce homozygous lines from hybridization to production of F_5 or F_6 lines, and about 3–3.5 years if two seasons of a crop can be taken. Therefore, considerable time is invested in the production of homozygous lines in the crop breeding programs.

To accelerate breeding cycles, there are different methods adopted by the crop breeding programs, such as shuttle breeding, doubled haploidy, anther culture, and rapid generation advancement (RGA) under controlled conditions to take two or more generations in a year. The controlled conditions required for each crop can vary depending on the geographical location where RGA will be taken up, the resources available, and the targeted breeding cycles per year.

3.1 Shuttle Breeding

Shuttle breeding is a method used for testing genetic material during the off-season, where plants are grown in different environments to grow two plant generations within a year. This technique has been successful in reducing the time taken to complete a breeding cycle by half (Alahmad et al. 2022). The process involves screening and selecting populations that are segregating, while simultaneously advancing the generations. However, this has a limitation of finding a suitable place and involves cost on logistics. Many crop improvement programs follow this route of shuttle breeding to reduce the time to achieve homozygosity in breeding lines.

The wheat shuttle breeding at CIMMYT (International Maize and Wheat Improvement Centre) that was first developed by Norman Borlaug to speed up breeding cycles resulted in development of varieties faster for Mexican wheat farmers (Alahmad et al. 2018). Mallik et al. (2002) have developed several rice varieties suitable for rain-fed and irrigated ecosystems in India. DBW 14, an early maturing wheat variety, suitable for late sown and irrigated conditions, was developed using the shuttle breeding approach and suitable for rice–wheat system of India (CVRC 2003). ICRISAT had successfully sped up the process of developing new chickpea varieties better adapted to different environments, with higher yields and resistance to pests and diseases using the shuttle breeding approach. They employed two approaches to produce three generations per year in a short-season environment in southern India. The first approach involved growing one crop in the field during the regular crop season and two additional crops in a glasshouse during the off-season. The second approach featured three crop cycles per year, with the initial crop in the field, the second crop in the field under late-sown conditions accompanied by irrigation, and the third crop in a nursery located off-season at Hiriyur in Karnataka, India. These approaches mainly targeted short- and medium-duration crosses (Samineni et al. 2020).

3.2 *Anther Culture*

Anther culture is used to generate haploid plants, which then be used to create homozygous lines through chromosome doubling, resulting in a pure breeding line that is genetically identical to the parent plant. This process can save time and resources compared to the traditional breeding methods. Anther culture can also be used to introduce genetic variability into a plant population, after subjecting anthers to mutagenesis or genetic modification; thereby, scientists can generate new traits that can be selected for subsequent generations (Maluszynski et al. 2003). This technique has been used to create new crop varieties that are resistant to pests, diseases, or stresses, and help to increase crop yields and improve food security.

3.3 *Doubled Haploidy*

Doubled haploidy (DH) is the quickest method to achieve homozygosity. This technique involves inducing a plant to produce haploid cells, which have only one set of chromosomes instead of the usual two sets. Haploid cells are then stimulated to double their chromosome number through a process called "doubling." This results in the production of plants that are genetically identical to the parent plant and have a complete set of chromosomes. Doubled haploid production is a valuable tool in plant breeding because it can significantly reduce the time and effort required to produce new crop varieties with desirable traits. Doubled haploidy has been used to develop improved varieties in several crops like wheat, maize, tobacco, rapeseed (also known as canola), and barley for yield, oil content, improved resistance to disease, and stress (Maluszynski et al. 2003).

3.4 Speed Breeding

Speed breeding is a novel plant breeding technique that involves manipulating the plant growth conditions to accelerate the breeding process and reduce the time required for crop development. This technique allows breeders to develop new breeding lines in a short time, which is particularly beneficial for developing crops that are resistant to disease, drought, and other environmental stresses (Watson et al. 2018). Indirectly, this will also contribute to increase in genetic gains for the crop. By altering the photoperiod, light quality and intensity (using artificial LED lights) under controlled conditions, it is possible to induce early flowering, thus reducing the time for completing the life cycle of crop plants. Speed breeding has also gained attention in recent years as a promising tool for improving global food security and addressing the challenges of climate change. Several research institutions and companies have been working on refining this technique and applying it to a wide range of crops, including wheat, barley, chickpea, and potato (Hickey et al. 2019).

3.4.1 Origin

The speed breeding method has its origins from the experiments conducted on photoperiod and photoperiodism by Garner and Allard (Thomas 2003). Since then, scientists have been studying the plants from the perspective of their light requirements. This forms the basis for all future speed breeding experiments dealing with the plants. Experiments can be broadly divided into two categories—one with long hours of light and the other with long hours of dark. Based on the requirements of the plants to enter the flowering stage, photoperiod can be adjusted to induce flowering. For example, long-day plants need additional hours of light beyond daylight hours. This can be done through artificial lighting during evening hours. Similarly, for short-day plants, long dark hours are needed to enter the flowering stage. In such situations, the area where plants are grown needs to be covered to avoid daylight or any other form of light.

3.4.2 Requirements

Speed breeding procedures essentially require controlled environment, which can be achieved through either dedicated structures like growth chambers or glass/polyhouses. The choice of structure will depend on factors such as the size of the breeding program, the type of plant being bred, and the available resources. Such structures are equipped with artificial lighting with precise control on the wavelength and intensity, shade nets in glass/polyhouses to regulate daylight intensity, cooling/heating systems to maintain precision temperature, and a system to keep the humidity at control (humidifier/de-humidifiers). There can be an automated system to maintain the soil moisture level when large glass/polyhouses are used. The entire parameters are to be monitored by sensors, which are then integrated to a programmable control system, which can be monitored in real time and controlled remotely over networks. There can be dedicated mobile apps or web portals for the purpose. Integration with the artificial intelligence system can provide instant and real-time corrections in the set parameters, as a useful warning system and to generate meaningful data.

Various studies have explored the use of controlled growth structures to develop rapid generation advancement protocols for different crops. For example, Ochatt et al. (2002) studied the use of controlled glasshouse conditions to develop a rapid generation advancement protocol for Bambara groundnuts (*Vigna subterranea*) and peas (*Pisum sativum*). Ohnishi et al. (2011) developed a speed breeding protocol for rice using the biotron breeding system in combination with tiller removal and embryo rescue, which allowed for the control of photoperiod, temperature, and CO_2 levels in an environmental chamber. Similarly, Tanaka et al. (2016) reduced the duration of the rice crop by 3 months without additional interventions like embryo rescue or tiller removal, in the rice variety Nipponbare. Baier et al. (2012) reported early flowering in transgenic chestnut (*Castanea dentata*) trees grown under artificial conditions in a chamber. Collard et al. (2017) evaluated the use of rapid generation advancement (RGA) in rice breeding and found that it had better cost-effectiveness compared to conventional methods and was effective in both glasshouse and field conditions. Watson et al. (2018) used growth chambers to enhance generation turnover and phenotype for pod shattering in canola cultivars. Nagatoshi and Fujita (2019) demonstrated a speed breeding method for soybean using compact growth chambers. Rana et al. (2019) modified the biotron system and used it for introgressing the *hst1* gene to the salinity-susceptible rice cultivar Yukinko-mai, with the help of a SNP-based marker system. Edet and Ishii (2022) reported a rapid generation advancement system for cowpea using breeding lines IT86D-1010, IT97K-499-35, and the Japanese cultivar Sasaque grown in growth chambers.

Artificial lighting can be provided by installing electric bulbs or lights. Several options are available in the market, i.e., incandescent bulbs, high-pressure sodium lamps, fluorescent lights, and LED (light emitting diode) lights. Incandescent bulbs are cheap and do not provide any option to adjust the light spectrum. Intensity can be adjusted by altering the height of lights from the crop canopy. Fluorescent lights provide options to switch to broad-spectrum light, in addition to adjust light intensity. LED lights are commonly used in RGA or speed breeding, as they are energy-efficient and also customizable to provide specific wavelengths of light that are optimal for plant growth (Watson et al. 2018). Lights should offer 100–1000 μmol/m^2/s intensity, based on the requirements of different plant species at different growth stages, as light is one of the essential factors for photosynthesis and plant growth and also should offer adjustable spectrum control, height adjustment mechanism, and active cooling options.

O'Connor et al. (2013) reported peanut growth under continuous light (24 h) using photosynthetically active radiation (PAR) lamps and specific temperature and humidity conditions. Jahne et al. (2020) used far-red-deprived and blue-light-enriched LED spectrum systems that allowed *Glycine max, Oryza sativa*, and *Amaranthus* spp. plants to advance five generations/year by adjusting the photoperiod to 10 h. Harrison et al. (2021) developed a speed breeding protocol for US soybean genotypes using red and blue LED lights in combination with photothermal conditions, reducing the growing time to 83–81 days against 120 days in field. Mobini et al. (2016) studied two cultivars of lentil viz. CDC Greenland and CDC Maxim, grown under T5 fluorescent bulbs (R:FR = 5.6); T5 supplemented with near far-red bulbs (R:FR = 3.1) and LEDs (R:FR = 3.09). Though both the cultivars responded similarly to different R:FR ratios, plants grown under the R:FR = 3.1 or less could flower 10–11 days earlier than the ones grown under the R:FR = 5.6. Watson et al. (2018) used LED-supplemented glasshouses for accelerating the generations of crops to four to six generations/year in wheat, barley, oat pea, chickpea, and various *Brassica* plants. Cazzola et al. (2020) reported advancement of pea plants, five generations/year, using 22 h of photoperiod through T5 fluorescent tubes under controlled conditions.

Speed breeding can be useful for long-day plants. Most of the legume crops, which fall under this category, can be the ideal crops of interest to develop protocols for faster generation advancement. These crops can respond to extended photoperiod that can be achieved through artificial lighting. On the other hand, short-day plants need less duration of photoperiod, which can be achieved in closed structures with environmental control like growth chambers or glasshouses using shade net to reduce the number of hours of natural light. Appropriate modification of the quality of the light spectrum in such cases was found to be useful in inducing early flowering of short-day plants (Jähne et al. 2020).

Durum wheat (*Triticum durum*), spring bread wheat (*Triticum aestivum*), barley (*Hordeum vulgare*), and model grass (*Brachypodium distachyon*) were grown under environment-controlled growth chambers, glasshouses with temperature control (22 °C), and low-cost homemade growth rooms and found that up to six generations could be achieved under these speed breeding conditions (Watson et al. 2018). Pazos-Navarro et al. (2017) reported generation acceleration in subterranean clover (*Trifolium subterraneum* L.), a forage crop, using an in vitro-assisted single seed descent (SSD) protocol by which 2.7–6.1 generations per year could be generated. The protocol involved the use of controlled conditions of temperature (25 °C) and photoperiod to minimize the duration for flowering and truncation of seed filling period. Saxena et al. (2017) developed an RGA protocol for pigeon pea (*Cajanus cajan* L.) by integrating SSD with the germination of immature seeds. Thirty-five-day-old seeds could be germinated with 100% success at 32 °C, which have three to four successive generations per year.

In pulses, long generation time is one of the major impediments to cultivar development. Further, pulses are also not easily amenable to in vitro manipulations (Ochatt et al. 2002). The in vitro approach of speed breeding in pulses involves regenerating and growing plants till flowering and seed development. The immature seeds thus produced can be used further for generation advancement. Such protocols have been reported in chickpea (*Cicer arietinum* L.); pea (*Pisum sativum* L.) by Espósito et al. (2012); common bean (*Phaseolus vulgaris* L.); faba bean (*Vicia faba* L.) by Mobini et al. (2015); and lentil (*Lens culinaris Medikus*) by Mobini et al. (2015). In pulses, the combination of in vitro and in vivo techniques (controlled environment) has been reported to be more useful in reducing generation time. Bhattarai et al. (2009) reported an in vitro protocol to rescue tomato embryos and reduce the generation advancement time. This approach could produce five generations per year, while the conventional methods could produce only two. Gebologlu et al. (2011) reported in vitro germination of immature tomato seeds in Murashige and Skoog (MS) medium supplemented with different growth regulators and observed that 28- to 32-day-old embryos gave better germination and the protocol could shorten the flowering time by 53–36 days. In most of the fruit crops, the juvenile phase is relatively long and it may take even more than 20 years to flower in some cases (Van Nocker et al. 2014). Hence, speed breeding techniques in such crop species primarily aim at reducing the juvenile phase by inducing vigorous vegetative growth and early flowering.

Several crop species, i.e., Amaranth, *Arabidopsis thaliana,* barley, canola, chickpea, faba bean, groundnut, lentil, pea, pigeon pea, rice, sorghum, soybean and wheat, have undergone rapid generation advancement through the implementation of various techniques aimed at reducing their time to flowering and increasing the number of generations per year (Wanga et al. 2021).

3.4.3 Costs and Consequences

The cost of building a structure is onetime investment. The maintenance takes centerstage afterwards. So, it is important to consider long-term sustainability. Let us investigate the options available and costs associated with them.

A growth chamber is the best choice to develop a speed breeding protocol. Precision is guaranteed and operations are automated. However, apart from initial investment (starting price around US $40,000), the maintenance costs are huge. For example, a growth chamber of 1.5 m^2 consumes around 200 units of electricity per day. The energy cost alone is US $1000 per month. Thus, it becomes a hindrance to conduct experiments in it. Even developing a protocol will be a huge cost. The growth chamber does offer precise conditions. When a speed breeding protocol is developed using a growth chamber, it needs to be scaled up for practical use in breeding programs. The scaling up is usually done in polyhouse type of construction, where the control over light, humidity, and temperature is not precise, as in a growth chamber. That means the protocol needs to be tweaked to make use of the conditions that exist in a polyhouse. This is the reason why a different solution is needed to bring the protocol to scaling up facility. Conducting the protocol development experiment on the lines of a polyhouse in a similar but smaller structure would be a better idea. Then, the protocol can directly be applied to scaling up experiments.

3.4.4 Low-Cost Alternatives

Small chambers fitted with cooling pads and LED lights on top of work benches will be an ideal option to save costs. These chambers can be further subdivided into ultra-small units by arranging separators in the form of polycarbonate sheets of sufficient thickness and lined up with reflective material on inner side. A thick cloth sheet with Velcro tapes for closing can be used to cover these ultra-small chambers to contain leakage of light. LED lights available in the local market can be used. Care should be taken to choose right capacity and spectral control. Lights that allow control via the internet cost a bit higher as compared to manually controlled lights. Alternatively, some species do not need variable spectrum. In this case, normal cool LED lights of sufficient capacity can be used.

For temperature control, AC units or cooling pads can be used. For a small area, the cost will be under control. Dehumidifier is a must to check the humidity, inside the chamber.

The automated irrigation system with a nutrient supply system is ideal, but not mandatory. Manual irrigation and nutrient supply will a cheaper solution. Accessories such as a rose can and small handheld sprayers of 1–2 liters of capacity can also serve the purpose.

Finally, the space can be optimized, if a second and third layer of planting is possible. This needs structural interventions backed up by strong light interventions. Suitable methods need to be identified to undertake operations in the upper tiers. For example, instead of work benches, the first tier can be on the floor. The second tier can be accommodated on a two-level workbench bottom layer and a third on the top. This way, volume of plants in the same area goes up and reduces the cost. Utilizing cost-effective methods such as benchtop cabinets and LED-supplemented glasshouses, Ghosh et al. (2018) achieved successful early generation advancements in crops including pea (*Pisum sativum*) and wheat (*Triticum aestivum*), while Watson et al. (2018) reported similar achievements in crops such as barley (*H. vulgare* cvs Gus and Baudin), oat (*Avena sativa* cv. Swan), and triticale (*Tritico secale* cvs Jackie and Coorong).

3.4.5 Speed Breeding for Accelerating Basic and Applied Research

In addition to the advantage of reducing the generation time of crops, speed breeding (SB) has tremendous potential, when used in combination with other tools of crop improvement and basic research. Speed breeding integrated with the phenotyping of complex traits will make trait-based selection easy and fast (Christopher et al. 2015). Since SB is done under controlled environments, this approach can readily be adapted to study the adaptation of crop species under changing environments and other stress conditions like water deficit, in addition to the accelerated generation advancement (Wang et al. 2019).

More directed approaches to genetic improvement like transgenic and gene editing techniques, like CRIPSPR-CAS9, have been a component of modern plant breeding to enhance genetic diversity (Richardson et al. 2014; Doudna et al. 2014; Wolter et al. 2019). This integration facilitates the breeder to snip out undesirable traits from the crop and, with the rapid generation advancement, cultivar development in crops with complex genomics like polyploids.

Speed breeding can also be profitably utilized in the artificial domestication of wild plants through early habitations, selective breeding, and reduced duration and generation advancement (O'Connor et al. 2013; Hickey et al. 2019). For rapid development of breeding or mapping populations like recombinant inbred lines (RILs) for marker discovery, the rapid generation advancement system is potentially relevant. Speed breeding can be used in the development of complex mapping populations, accelerated backcrossing and pyramiding of genes/quantitative trait locus (QTL) (Varshney et al. 2021), and marker-assisted and genomic selection (Gosal et al. 2020; Croser et al. 2021; Dadu et al. 2021). Marker-assisted selection was employed in soybean to identify hybrid progenies with specific E1–E4 alleles associated with growth period, rather than relying on phenotypic performance. This approach facilitated the identification of stable genotypes, leading to improved selection efficiency and a significant reduction in the time required for generation advancement in the soybean speed breeding program (Fang et al. 2021). Hickey et al. (2017) used a revamped backcross method with speed breeding to develop 87 Scarlett introgression lines (ILs) in BC1F3:4 within 2 years. There will be cost saving also on the space requirements to grow a large number of inbred lines as they could be planted in high density (Yao et al. 2017). Another technique is genomic selection, which enables the selection of plants with higher genomic estimated breeding values (GEBVs) and advancing them to the next generation. This technique not only contributes to greater genetic gain, but also expedites the overall process of generation advancement (Gorjanc et al. 2018). Advancement of segregating populations using SSD, combined with speed breeding, is cost-effective and less time-consuming and gives better turnover than the conventional pedigree method or shuttle breeding (Ortiz et al. 2007; Jähne et al. 2020; Sinha et al. 2021).

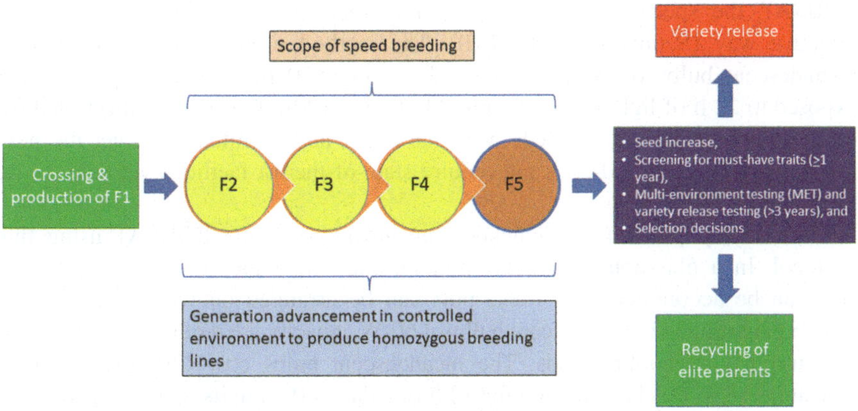

Fig. 4 General breeding cycle to develop homozygous breeding lines

3.4.6 Speed Breeding for Crop Breeding and Cultivar Development

Fig. 4 depicts a general breeding cycle to develop a homozygous breeding line. It takes 3–6 years depending on the number of generations possible in the given geographical area. Slight changes are possible, if shuttle breeding is practiced. Availability of DH technology rapidly reduces this duration. However, it is not available in several crops of interest. At this juncture, speed breeding can be handy to save time attaining homozygosity rapidly. Depending on the need, either three or four generations can be rapidly advanced followed by screening and initial evaluation for two seasons. If markers are available, the population can be screened in F_2 or subsequent generations to save resources. Only the desirable genotypes can be advanced further. The varietal release follows certain norms for evaluation of performance. This duration is nonnegotiable, usually taking 3–4 years. Thus, a new variety can be released potentially in 6–10 years.

4 Case Studies

Speed breeding has successfully reduced the generation interval in various long-day and short-day plants, resulting in an accelerated plant improvement process (Samantara et al. 2022). A few selected crops are discussed hereunder.

4.1 Chickpea

The concept was proven in many crops, such as wheat, barley, canola, and pea (Wanga et al. 2021). However, many of these experiments involved huge energy costs besides infrastructure facilities. One of the exceptions was chickpea. At

ICRISAT, the speed breeding protocol for chickpea was developed in a cost-effective way (Samineni et al. 2020). The method used a regular glasshouse, incandescent bulbs of 60-W capacity delivering 870 lm. Chickpea plants were exposed to 22 h of light (10 h of natural light and 12 h of artificial light) to induce early flowering. Immature seeds were harvested and germinated to get the next generation. This method does not require state-of-the-art facilities yet delivers on the target.

Currently, populations of chickpea are advanced at the ICRISAT using this protocol. In a glasshouse bay of 6-m long and 9-m wide, 1368 pots of 5-inch size can be accommodated. These pots can be arranged on work benches. For artificial lighting, 32 incandescent bulbs of 60-W capacity are required. Temperature is controlled by cooling pads. The incandescent bulbs cost ₹20 apiece. Power consumption of this bulb if used for 12 h in a day is 0.72 units. Cooling pads (two numbers in one glasshouse bay) consume 48 units of power in a day (assuming on-and-off action to maintain temperature). So, a generation advancement experiment of 60 days would cost 2923 units of power. By far, this is the cheapest to advance chickpea population at the ICRISAT.

4.2 *Lentil*

Idrissi et al. (2019) applied extended photoperiod for accelerating lentil growth and development. Plants were exposed to an extended period of 22 h of light and 2 h of dark under growth chamber conditions, using far-red-enriched LED and blue LED lights. The extended photoperiod results in shorter cycles of development, such as fewer number of days to first flower, first pod, physiological maturity, and also shorter period for pod filling. A reduction of 46% was recorded for early physiological maturity due to photoperiod extension. Harvested seeds (F_3) were sown and advanced three generations per year under growth chamber conditions instead of one generation in the field.

4.3 *Wheat*

Watson et al. (2018) and Ghosh et al. (2018) facilitated speed breeding protocols for shortening wheat generation time using controlled environment chamber and glasshouse conditions. The process involves the exposure of plants to a longer period of 22 h of light and 2 h of dark at 22 °C and 17 °C, respectively, using LED lights, which cover the photosynthetically active radiation (PAR) of 400–700 nm focusing on blue, red, and far-red regions of spectrum. They discovered that extended photoperiod reduced the total number of days of flowering to half compared to the wheat genotypes grown with 12/12-h light/dark.

4.4 Groundnut

A rapid generation turnover (RGT) for peanut under greenhouse conditions with continuous light (24 h) using 450-W PAR lamps and temperature 32 °C/22 °C maximum/minimum with 65% relative humidity (RH) was reported by O'Connor et al. (2013). The peanut lines, Farnsfield and D147-p3-115, generally mature in 140–145 days in the field, while the F_2 and F_3 generations from the cross of the two lines matured in 113 and 89 days, respectively, under the speed breeding conditions. The final plant recovery ranged from 68% to 74% in F_2 to F_4 generations. A combination of speed breeding with single seed chipping (a rapid protocol for sampling for DNA extraction using a small portion of the cotyledon without losing the viability of the seed) and high-throughput genotyping in peanut has been reported recently, which also could save 6–8 months of crop duration (Parmar et al. 2021).

4.5 Pigeon Pea

Saxena et al. (2017) reported an RGA technology combining the germination of immature seeds with the single seed descent (SSD) technique in pigeon pea, a legume crop that is widely grown in many parts of the world. They reported that, using speed breeding, they were able to significantly reduce the time required for pigeon pea plants to reach maturity and produce seeds. The outcome demonstrated that 35 days old immature seeds could be used to successfully produce a new generation of pigeon pea with 100% seed germination, at 32 °C and 60% relative humidity RGA technique allowed them to grow three generations of pigeon pea in a single year, as opposed to the usual one generation per year under normal growing conditions.

4.6 Cowpea

Cowpea is an important dryland crop in sub-Saharan Africa that has potential to improve food security. Edet and Ishii (2022) developed and validated an efficient speed breeding protocol for cowpea that accommodates seven to eight breeding generations per year for three cowpea genotypes. The protocol involves using controlled growth conditions in two different chamber types and cultivating new plant generations from seeds of oven-dried immature pods, thereby reducing 62% time from pollination to sowing of the next plant generation.

5 Conclusion and Prospectives

Plant breeding has played a crucial role in ensuring the food security of the world's population over the past century through high-yielding varieties in major crops. Most of this effort was through traditional breeding techniques. To meet the demands of food and nutritional security, modern techniques need to be adopted. Speed breeding has just begun and picking up globally by scientists involved in crop improvement as it saves time and resources. Protocols are being developed for many crops that respond to the techniques of speed breeding.

Currently, the focus is on reducing the generation time of a crop. Going forward, speed breeding can be complemented with various advanced techniques to accelerate the breeding programs. The integration of techniques such as backcrossing, gene pyramiding, gene editing, and genomic selection with speed breeding can help to reduce project costs and to accelerate crop improvement.

In recent times, artificial intelligence (AI) has emerged as a valuable tool for exploring the biological and molecular processes that govern plant functions in response to environmental factors (Rai 2022). AI will have a significant impact on plant breeding offering various benefits. AI-assisted breeding systems will play a pivotal role in different aspects of research, evaluation, selection, development of breeding procedures, and field management. These AI-driven systems possess remarkable capabilities in designing and predicting outcomes through model simulation and optimization. Moreover, robots equipped with AI are now involved in data collection, storage, and analysis, thereby significantly enhancing breeding information systems (Xu et al. 2022). A list of successful implementation events of AI models in plant breeding studies of various crops like *Glycine max* L., *Phaseolus vulgaris* L., *Zea mays* L., *Brassica rapa* L., and *Triticum aestivum* L. were reported by Rai (2022). Further, AI has the capacity to integrate and incorporate OMICS datasets, which are essential requirements for effectively implementing speed breeding protocols. As a result, speed breeding becomes not only more compelling but also more cost-effective.

References

Alahmad S, Dinglasan E, Leung KM, Riaz A, Derbal N, Voss-Fels KP, Able JA, Bassi FM, Christopher J, Hickey LT (2018) Speed breeding for multiple quantitative traits in durum wheat. Plant Methods 14:36. https://doi.org/10.1186/s13007-018-0302-y

Alahmad S, Rambla C, Voss-Fels KP, Hickey LT (2022) Accelerating breeding cycles. In: Wheat improvement: food security in a changing climate. Springer, Cham, pp 557–571. https://doi.org/10.1007/978-3-030-90673-3_30

Baier K, Maynard C, Powell WA (2012) Early flowering in chestnut species induced under high intensity, high dose light in growth chambers. J Am Chestnut Foundation 26:8–10

Begna T (2022) Speed breeding to accelerate crop improvement. Int J Agric Sci Food Technol 8(2):178–186. https://doi.org/10.17352/2455-815X.000161

Bhattarai SP, de la Pena RC, Midmore DJ, Palchamy K (2009) In vitro culture of immature seed for rapid generation advancement in tomato. Euphytica 167(23–30):70. https://doi.org/10.1007/s10681-008-9855-6

Cazzola F, Bermejo CJ, Guindon MF, Cointry E (2020) Speed breeding in pea (*Pisum sativum* L.), an efficient and simple system to accelerate breeding programs. Euphytica 216(11):178. https://doi.org/10.1007/s10681-020-02715-6

Christopher J, Richard C, Chenu K, Christopher M, Borrell A, Hickey L (2015) Integrating rapid phenotyping and speed breeding to improve stay-green and root adaptation of wheat in changing, water limited, Australian environments. Procedia environmental sciences. Proceedings of the Environ Sci 29:175–176. https://doi.org/10.1016/j.proenv.2015.07.246

Collard BC, Beredo JC, Lenaerts B, Mendoza R, Santelices R, Lopena V, Verdeprado H, Raghavan C, Gregorio GB, Vial L, Demont M (2017) Revisiting rice breeding methods–evaluating the use of rapid generation advance (RGA) for routine rice breeding. Plant Prod Sci 20(4):337–352. https://doi.org/10.1080/1343943X.2017.1391705

Croser J, Mao D, Dron N, Michelmore S, McMurray L, Preston C, Bruce D, Ogbonnaya FC, Ribalta FM, Hayes J, Lichtenzveig J (2021) Evidence for the application of emerging technologies to accelerate crop improvement–A collaborative pipeline to Introgress herbicide tolerance into chickpea. Front Plant Sci 2839. https://doi.org/10.3389/fpls.2021.779122

CVRC, Central varietal release committee (2003) Developing high yielding and climate resilient wheat varieties: breeding wheat genotypes for North Western Plains of India. ICAR-Indian Institute of Wheat and Barley Research, Karnal

Dadu RHR, Bar I, Ford R, Sambasivam P, Croser J, Ribalta F, Kaur S, Sudheesh S, Gupta D (2021) Lens orientalis contributes quantitative trait loci and candidate genes associated with Ascochyta Blight resistance in lentil. Front Plant Sci 12:703283. https://doi.org/10.3389/fpls.2021.703283

Doudna JAC, Charpentier E (2014) Genome editing. The new frontier of genome engineering with CRISPR-Cas9. Science 346(6213):1258096. https://doi.org/10.1126/science.1258096

Edet OU, Ishii T (2022) Cowpea speed breeding using regulated growth chamber conditions and seeds of oven-dried immature pods potentially accommodates eight generations per year. Plant Methods 18(1):106. https://doi.org/10.1186/s13007-022-00938-3

Espósito MA, Almirón P, Gatti I, Cravero VP, Anido FSL, Cointry EL (2012) A rapid method to increase the number of F1 plants in pea (Pisum sativum) breeding programs. Genet Mol Res 11(3):2729–2732. https://doi.org/10.4238/2012.June.18.1

Fang Y, Wang L, Sapey E, Fu S, Wu T, Zeng H, Sun X, Qian S, Khan MAA, Yuan S, Wu C (2021) Speed-breeding system in soybean: integrating off-site generation advancement, fresh seeding, and marker-assisted selection. Front Plant Sci 12:717077. https://doi.org/10.3389/fpls.2021.717077

Gebologlu N, Bozmaz S, Aydin M, Cakmak P (2011) The role of growth regulators, embryo age and genotypes on immature embryo germination and rapid generation advancement in tomato (*Lycopersicon esculentum* mill.). Afr J Biotechnol 10:4895–4900. https://doi.org/10.5897/AJB10.2275

Ghosh S, Watson A, Gonzalez-Navarro OE, Ramirez-Gonzalez RH, Yanes L, Mendoza-Suárez M, Simmonds J, Wells R, Rayner T, Green P, Hafeez A (2018) Speed breeding in growth chambers and glasshouses for crop breeding and model plant research. Nat Protoc 13(12):2944–2963. https://doi.org/10.1038/s41596-018-0072-z

Gorjanc G, Gaynor RC, Hickey JM (2018) Optimal cross selection for long-term genetic gain in two-part programs with rapid recurrent genomic selection. Theor Appl Genet 131:1953–1966. https://doi.org/10.1007/s00122-018-3125-3

Gosal SS, Pathak D, Wani SH, Vij S, Pathak M (2020) Accelerated breeding of plants: methods and applications. In: Gosal SS, Wani SH (eds) Accelerated plant breeding, 1. Springer, Cham, pp 1–29. ISBN 978-3-030-41865-6

Harrison D, Da Silva M, Wu C, De Oliveira M, Ravelombola F, Florez-Palacios L, Acuña A, Mozzoni L (2021) Effect of light wavelength on soybean growth and development in a context of speed breeding. Crop Sci 61(2):917–928. https://doi.org/10.1002/csc2.20327

Hickey LT, Germán SE, Pereyra SA, Diaz JE, Ziems LA, Fowler RA, Platz GJ, Franckowiak JD, Dieters MJ (2017) Speed breeding for multiple disease resistance in barley. Euphytica 213(3):64–78. https://doi.org/10.1007/s10681-016-1803-2

Hickey LT, Hafeez AN, Robinson H, Jackson SA, Leal-Bertioli SC, Tester M, Gao C, Godwin ID, Hayes BJ, Wulff BB (2019) Breeding crops to feed 10 billion. Nat Biotechnol 37(7):744–754. https://doi.org/10.1038/s41587-019-0152-9

Idrissi O, Sahri A, Udupa S, Kumar S (2019) Single seed descent under extended photoperiod as a simple, rapid and efficient breeding method for accelerated genetic gain in lentil. In: Proceedings of the 3rd international legume society conference, Poznan, Poland, pp 21–24

Jähne F, Hahn V, Würschum T, Leiser WL (2020). Speed breeding short-day crops by LED-controlled light schemes. Theor Appl Genet, 133(8), 2335–2342. https://doi.org/10.1007/s00122-020-03601-4

Jahne F, Hahn V, Würschum T, Leiser WL (2020) Speed breeding short-day crops by LED-controlled light schemes. TAG. Theor Appl Genet Theoretische und Angewandte Genetik 133(8):2335–2342. https://doi.org/10.1007/s00122-020-03601-4

Li JF, Norville JE, Aach J, McCormack M, Zhang D, Bush J, Church GM, Sheen J (2013) Multiplex and homologous recombination–mediated genome editing in *Arabidopsis* and *Nicotiana benthamiana* using guide RNA and Cas9. Nat Biotechnol 31(8):688–691. https://doi.org/10.1038/nbt.2654

Mallik S, Mandal BK, Sen SN, Sarkarung S (2002) Shuttle-breeding: an effective tool for rice varietal improvement in rainfed lowland ecosystem in eastern India. Curr Sci 83(9):1097–1102

Maluszynski M, Kasha K, Forster BP, Szarejko I (2003) Doubled haploid production in crop plants: a manual. Springer. https://doi.org/10.1007/978-94-017-1293-4

Mobini SH, Lulsdorf M, Warkentin TD, Vandenberg A (2015) Plant growth regulators improve in vitro flowering and rapid generation advancement in lentil and faba bean. In Vitro Cell Dev Biol Plant 51(1):71–79. https://doi.org/10.1007/s11627-014-9647-8

Mobini SH, Lulsdorf M, Warkentin TD, Vandenberg A (2016) Low red: far-red light ratio causes faster in vitro flowering in lentil. Can J Plant Sci 96(5):908–918. https://doi.org/10.1139/cjps-2015-0282

Nagatoshi Y, Fujita Y (2019) Accelerating soybean breeding in a COT supplemented growth chamber. Plant Cell Physiol 60(1):77–84. https://doi.org/10.1093/pcp/pcy189

O'Connor DJ, Wright GC, Dieters MJ, George DL, Hunter MN, Tatnell JR, Fleischfresser DB (2013) Development and application of speed breeding technologies in a commercial peanut breeding program. Peanut Sci 40(2):107–114. https://doi.org/10.3146/PS12-12.1

Ochatt SJ, Sangwan RS, Marget P, Assoumou Ndong Y, Rancillac M, Perney P (2002) New approaches towards the shortening of generation cycles for faster breeding of protein legumes. Plant Breed 121(5):436–440. https://doi.org/10.1046/j.1439-0523.2002.746803.x

Ohnishi T, Yoshino M, Yamakawa H, Kinoshita T (2011) The biotron breeding system: A rapid and reliable procedure for genetic studies and breeding in rice. Plant Cell Physiol 52(7):1249–1257. https://doi.org/10.1093/pcp/pcr066

Ortiz R, Trethowan R, Ferrara GO, Iwanaga M, Dodds JH, Crouch JH, Crossa J, Braun HJ (2007) High yield potential, shuttle breeding, genetic diversity, and a new international wheat improvement strategy. Euphytica 157(3):365–384. https://doi.org/10.1007/s10681-007-9375-9

Parmar S, Deshmukh DB, Kumar R, Manohar SS, Joshi P, Sharma V, Chaudhari S, Variath MT, Gangurde SS, Bohar R, Singam P (2021) Single seed-based high-throughput genotyping and rapid generation advancement for accelerated groundnut genetics and breeding research. Agronomy 11(6):1226. https://doi.org/10.3390/agronomy11061226

Pazos-Navarro M, Castello M, Bennett RG, Nichols P, Croser J (2017) In vitro-assisted single-seed descent for breeding-cycle compression in subterranean clover (Trifolium subterraneum L.). crop and pasture. Science 68(11):958. https://doi.org/10.1071/CP17067

Rai KK (2022) Integrating speed breeding with artificial intelligence for developing climate-smart crops. Mol Biol Rep 49(12):11385–11402. https://doi.org/10.1007/s11033-022-07769-4

Rana MM, Takamatsu T, Baslam M, Kaneko K, Itoh K, Harada N, Sugiyama T, Ohnishi T, Kinoshita T, Takagi H, Mitsui T (2019) Salt tolerance improvement in rice through efficient SNP marker-assisted selection coupled with speed-breeding. Int J Mol Sci 20(10):2585. https://doi.org/10.3390%2Fijms20102585

Richardson T, Thistleton J, Higgins TJ, Howitt C, Ayliffe M (2014) EfficientAgro bacterium transformation of elite wheat germplasm without selection. Plant Cell Tissue Organ Cult 119(3):647–659. https://doi.org/10.1007/s11240-014-0564-7

Samantara K, Bohra A, Mohapatra SR, Prihatini R, Asibe F, Singh L, Reyes VP, Tiwari A, Maurya AK, Croser JS, Wani SH (2022) Breeding more crops in less time: A perspective on speed breeding. Biology 11(2):275. https://doi.org/10.3390/biology11020275

Samineni S, Sen M, Sajja SB, Gaur PM (2020) Rapid generation advance (RGA) in chickpea to produce up to seven generations per year and enable speed breeding. The Crop J 8(1):164–169. https://doi.org/10.1016/j.cj.2019.08.003

Saxena K, Saxena RK, Varshney RK (2017) Use of immature seed germination and single seed descent for rapid genetic gains in pigeonpea. Plant Breed 136(6):954–957. https://doi.org/10.1111/pbr.12538

Sinha P, Singh VK, Bohra A, Kumar A, Reif JC, Varshney RK (2021) Genomics and breeding innovations for enhancing genetic gain for climate resilience and nutrition traits. TAG Theoretical and Applied Genetics Theoretische und Angewandte Genetik 134(6):1829–1843. https://doi.org/10.1007/s00122-021-03847-6

Spanne A (2021) Green revolution: history, technologies, and impact. https://www.treehugger.com/green-revolution-history-technologies-and-impact-5189596

Tanaka J, Hayashi T, Iwata H (2016) A practical, rapid generation-advancement system for rice breeding using simplified biotron breeding system. Breed Sci 66(4):542–551. https://doi.org/10.1270/jsbbs.15038

Thomas B (2003) REGULATORS OF GROWTH | Photoperiodism. In: Thomas B (ed) Encyclopedia of applied plant sciences, vol 2003. Elsevier, pp 1077–1084., ISBN 9780122270505. https://doi.org/10.1016/B0-12-227050-9/00245-3

Tiwari JK, Singh AK, Behera TK (2023) CRISPR/Cas genome editing in tomato improvement: advances and applications. Front Plant Sci 14. https://doi.org/10.3389/fpls.2023.1121209

Van Nocker S, Gardiner SE (2014) Breeding better cultivars, faster: applications of new technologies for the rapid deployment of superior horticultural tree crops. Hortic Res 1:14022. https://doi.org/10.1038/hortres.2014.22

Varshney RK, Bohra A, Yu J, Graner A, Zhang Q, Sorrells ME (2021) Designing future crops: genomics-assisted breeding comes of age. Trends Plant Sci 26(6):631–649. https://doi.org/10.1016/j.tplants.2021.03.010

Voss-Fels KP, Stahl A, Hickey LT (2019) Q&A: modern crop breeding for future food security. BMC Biol 17:18. https://doi.org/10.1186/s12915-019-0638-4

Wang X, Xuan H, Evers B, Shrestha S, Pless R, Poland J (2019) High-throughput phenotyping with deep learning gives insight into the genetic architecture of flowering time in wheat. GigaScience 8(11):giz120. https://doi.org/10.1093/gigascience/giz120

Wanga MA, Shimelis H, Mashilo J, Laing MD (2021) Opportunities and challenges of speed breeding: A review. Plant Breed 140(2):185–194. https://doi.org/10.1111/pbr.12909

Watson A, Ghosh S, Williams MJ, Cuddy WS, Simmonds J, Rey MD, Asyraf Md Hatta M, Hinchliffe A, Steed A, Reynolds D, Adamski NM (2018) Speed breeding is a powerful tool to accelerate crop research and breeding. Nat Plants 4(1):23–29. https://doi.org/10.1038/s41477-017-0083-8

Wolter F, Schindele P, Puchta H (2019) Plant breeding at the speed of light: the power of CRISPR/Cas to generate directed genetic diversity at multiple sites. BMC Plant Biol 19(1):176. https://doi.org/10.1186/s12870-019-1775-1

Xu Y, Zhang X, Li H, Zheng H, Zhang J, Olsen MS, Varshney RK, Prasanna BM, Qian Q (2022) Smart breeding driven by big data, artificial intelligence and integrated genomic-enviromic prediction. Mol Plant. https://doi.org/10.1016/j.molp.2022.09.001

Yao Y, Zhang P, Liu H, Lu Z, Yan G (2017) A fully in vitro protocol towards large scale production of recombinant inbred lines in wheat (Triticum aestivum L.). Plant Cell Tissue Organ Cult 128(3):655–661. https://doi.org/10.1007/s11240-016-1145-8

Zaidi SSEA, Tashkandi M, Mansoor S, Mahfouz MM (2016) Engineering plant immunity: using CRISPR/Cas9 to generate virus resistance. Front Plant Sci 7:1673. https://doi.org/10.3389/Ffpls.2016.01673

Digital Agriculture for Enhancing Yield, Nutrition, and Biological Stress Resistance

Mohamed A. M. El-Tabakh, Mariam A. Abd El-Samea, Yasmeen A. Roby, Yomna A. Mohamed, Hossam E. Harb, Ahmed M. Saleh, Mahmoud H. Abd El-Aziz, and Sozan E. El-Abeid

Abstract The application of artificial intelligence (AI) has fundamentally altered the agriculture industry by introducing novel approaches that boost crop yields while lowering the cost of production. The monitoring and estimation of preyield requirements, including water and energy for agriculture, disease prediction, and pest and weed control, have been made by the development of methodologies and applications based on AI. In addition, computer vision is a branch of artificial intelligence that focuses on enabling computers to interpret visual information from images or videos. It uses technologies such as object recognition, image segmentation, and motion estimation to identify objects in an image or video frame and track them over time. Computer vision has been used for applications such as imaging analysis of plants and prediction of plant health status. Also, artificial intelligence-based smart irrigation systems are being employed to improve crop yields while reducing water consumption. In addition, artificial intelligence-based crop yield forecasting algorithms have been developed in order to reliably anticipate crop yields. This enables governments to plan crops in an appropriate manner. Additionally, strategies based on AI have been applied to improve transportation

---

M. A. M. El-Tabakh
Zoology Department, Faculty of Science, Al-Azhar University, Cairo, Egypt

M. A. Abd El-Samea · Y. A. Roby · Y. A. Mohamed · H. E. Harb
Biotechnology Department - Faculty of science, Cairo University, Cairo, Egypt

A. M. Saleh
Department of Pharmaceutical Chemistry, Faculty of Pharmacy, Horus University, Dumyat, Egypt
e-mail: asaleh@horus.edu.eg

M. H. Abd El-Aziz
Plant Pathology Research Institute, Agriculture Research Center, Giza, Egypt

S. E. El-Abeid (✉)
Nanotechnology & Advanced Nano-Materials Laboratory (NANML), Plant Pathology Research Institute, Agricultural Research Center, Giza, Egypt

Mycology and Disease Survey Research Department, Plant Pathology Research Institute, Agricultural Research Center, Giza, Egypt

P. M. Priyadarshan et al. (eds.), *Digital Agriculture*,
https://doi.org/10.1007/978-3-031-43548-5_15

demands, purchasing processes, storage facilities, and the agricultural sector's economy as a whole. It is therefore possible to draw the conclusion that artificial intelligence has completely transformed the agriculture industry by delivering effective solutions that simultaneously raise levels of productivity and cut the expenses of production.

Keywords Monitoring · Artificial intelligence · Nanosensor · Sensor · Smart irrigation

1 Introduction

The use of AI in agricultural and weather forecasting has become more widespread in recent years. This has sparked a surge in study in these areas, which has improved AI technology. Given the need to optimize productivity and efficiency and the fact that 40% of the world's population is employed in agriculture (Ramankutty et al. 2018), research on using artificial intelligence to increase crop yields with little waste was initiated as a result of the world's waste of essential resources. Manual labors were used in traditional farming for years (Al-Arif et al. 2012). Inaccurate calculations and a lack of comprehension resulted in damaged crops and decreased soil productivity (Patil and Thorat 2016). Consequently, objectives of the early stages of artificial intelligence research in an effort to adapt it to agricultural requirements and alleviate farmers' problems are as follows:

1. Determine contemporary agriculture farmer difficulties.
2. Evaluate AI technology for agricultural use.
3. Create a prototype of an AI system to support farmers with daily duties.
4. Test and improve the prototype depending on farmer feedback.
5. Use the AI system in agriculture and track its success.
6. Collect data from the deployed AI system and use it to improve its performance further.
7. Provide farmers a simple way to use the AI system.
8. Teach farmers how to gain from AI. This coincided with industrial revolution inventions and mechanism advances (Kait et al. 2007). The collection of information on agricultural practices like farming, irrigation, crop forecasting, planning, and organizing the global supply and demand for agricultural products requires a combination of artificial intelligence techniques and the Internet (Vinayak and Malavade 2016). It also allows the collection of information on crops, their growth rate, and the amount of water they need for optimal growth. This chapter describes artificial intelligence's agricultural and crop monitoring technologies. In this chapter, we have highlighted the various tools and contributions of artificial intelligence in agriculture and crop monitoring.

2 Some Effective Technologies

Number of effective technologies used in climate change monitoring is summarized in Fig. 1:

1. Climate modeling: This technology uses mathematical models to simulate the climate system and predict future climate change.
2. Remote sensing: This technology uses satellites, aircraft, and unmanned aerial vehicles to collect data on land cover, land use, and other environmental parameters such as changes in the Earth's atmosphere, land, and humidity.
3. Geographic information system (GIS): GIS is a computer-based system that combines spatial data with other information to analyze and visualize relationships between different elements of the environment.
4. Soil moisture sensors: These sensors measure the amount of water in the soil and can be used to monitor irrigation needs or detect water stress in plants.
5. Weather stations: Weather stations measure temperature, humidity, wind speed, rainfall, and other meteorological variables that can be used to monitor climate change and predict weather patterns.
6. Drones: Unmanned aerial vehicles (UAVs) are increasingly being used for environmental monitoring applications such as mapping land cover changes or monitoring air quality.
7. Plant health monitoring systems: These systems use sensors to monitor plant health parameters such as leaf temperature, leaf wetness, and nutrient levels in order to detect early signs of pest infestations or disease outbreaks.

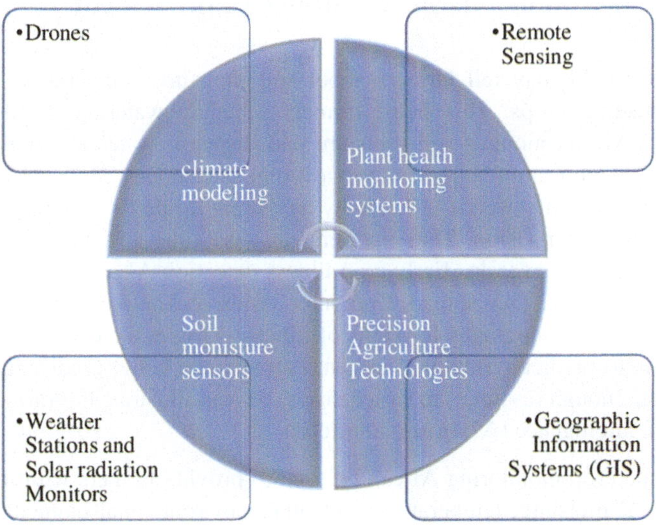

Fig. 1 Number of effective technologies used in climate change monitoring

8. Precision agriculture technologies: Precision agriculture technologies such as yield monitors and variable rate application systems allow farmers to optimize their inputs for maximum efficiency while minimizing their environmental impact.
9. Solar radiation monitors: These instruments measure the amount of solar radiation reaching the Earth's surface from the Sun.

They have helped to monitor climate change, frequent weather, nutrition and moisture efficiency of plants, pest and weed control, and plant health monitoring, which leads researchers in developing subtechnology such as GPS satellites (Gondchawar and Kawitkar 2016), sensors, and satellite imagery (images of satellite imagery), and to monitor areas that influence factors such as temperature, moisture, soil pH, and other factors. This other factors that can be monitored using sensors and satellite imagery include vegetation health, land use, water quality, air quality, and topography. Additionally, satellite imagery can be used to monitor changes in land cover and land use over time. All these huge data are collected through proximity, remote, and temperature/humidity sensors (Jha et al. 2019). These advanced Internet means enable data to be disseminated to different channels, such as Internet cloud technology (Kodali and Sahu 2016). With the increase in the technical ability in data processing, the computational power and accuracy have increased its effectiveness in dealing with agricultural challenges and setting agricultural forecasts useful, accurate, and reliable (Roopaei et al. 2017) (Fig. 2).

3 Digital Agriculture for Nutrient Security and Environmental Sustainability Improvement

Crop tracking AI may tell farmers about soil moisture, temperature, and other factors affecting crops. This helps farmers schedule watering, fertilizing, and harvesting. AI can monitor soil moisture and dispense water in smart irrigation systems. This innovation helps farmers save time and money while caring for their crops. AI can identify weeds in fields by analyzing drone or camera photographs and using computer vision. Thus, farmers can save time and money by focusing on the most important tasks. Early crop illness detection AI can help farmers spot crop illnesses before they spread. Computer vision technology is used to analyze drone or camera photographs for signs of illness. AI uses meteorological and soil quality data to predict agricultural output. Farmers can better predict future needs and allocate enough resources to varied crops. We can illustrate the various abilities of artificial intelligence (AI) in agriculture as:

- Real-time crop monitoring AI can be used to provide farmers with information about soil moisture, temperature, and other environmental elements affecting their crops. This is helpful for farmers because it allows them to better plan when to water, fertilize, and harvest their crops.

Fig. 2 A graph showing the different abilities of potential AI roles in agriculture

- Smart irrigation systems can be easily achieved with the assistance of AI by monitoring soil moisture levels and dispensing water accordingly. Farmers may save both time and money while still giving their crops the care they need, thanks to this innovation.
- Artificial intelligence (AI) can be used to spot weeds in fields by analyzing photographs captured by drones or other cameras and utilizing computer vision technologies to identify the plants. As a result, farmers can save both time and money by focusing their efforts where they will have the greatest impact.
- Early detection of crop illnesses. Artificial intelligence can be used to help farmers identify crop diseases before they cause widespread damage. Doing so involves the use of computer vision technologies to analyze images captured by drones or other cameras in order to spot telltale symptoms of sickness.
- Predicting crop yields with AI requires analyzing past data and current conditions, such as weather and soil quality. As a result, farmers are better able to anticipate future needs and dedicate sufficient resources to the cultivation of various crops.

Now, it is crucial that we make an effort to discover the important features that best explain this technology and the significant role it plays in agriculture.

3.1 Information Production

Agricultural informatization is essential to achieving agricultural modernization. In China, agricultural information refers to the widespread use of information technologies in agricultural production and management, which encompasses the transformation of information technology into means of production and the industrialization of agricultural information products or services (Liu and Gao 2016).

As part of a study effort, information technology (IT) supports the appropriate service procedure was built to optimize the precision farming process. In the information production system (IPS) project, a software system that facilitates efficient and transparent service provision was developed using service engineering methodologies (Information Production System for Precision Farming). Today, site-specific farming, which allows for efficient field production, is adopted in response to rising global food demand. However, owing to the lack of arable land and the requirement of employing farming practices, it is crucial that agricultural output be optimized in a way that takes into consideration more than just self-serving interests. Over the past three decades, significant technics have been made in agricultural field management (Mulla 2013). A prerequisite is efficient and detailed data collecting. Sensors are offered for various precision farming (PF) applications (e.g., nitrogen fertilization and plant protection). Although preliminary techniques for sensor-based analysis for particular nutrients exist, no solution for complete and reliable sensor-based detection is predicted in the medium future (Kim et al. 2009). The creation of IPS demonstrates how adopting a service engineering methodology in agriculture leads to service quality and effectiveness gains. It gives farmers greater traceability and transparency when using the cloud-based AgriPort. Using IPS as an example, numerous effects of digitalizing cooperative business operations were demonstrated. Weaknesses have been discovered based on business process modeling. The entire service became more efficient by standardizing data formats, reducing the number of tools used, and predefining error routines, benefiting all stakeholders (Friedrich et al. 2016). Agriculture includes crop production and animal husbandry for food, wool, and other items. Agriculture sustains every economy. Long-term economic growth and structural transformation depend on it. Previously, agriculture only produced food and crops. Agriculture now includes animal processing, production, marketing, and distribution. Nowadays, agriculture generates most revenue, reduces unemployment, supplies raw materials to other businesses, and boosts economic growth. Agriculture made sedentary human civilization possible by producing food surpluses. Sheep, goats, pigs, and cattle were domesticated 10,000 years ago. Plants were grown individually in 11 locations. Throughout the twentieth century, industrial agriculture based on large-scale monoculture dominated agricultural output despite 2 billion people still practicing subsistence agriculture (2002). Intelligent or e-agricultural farming is digital agriculture. These agricultural technologies capture, store, analyze, and distribute digital data. The UN Food and Agriculture Organization calls digitalizing agriculture the "digital agricultural revolution" (Klerkx et al. 2019). Digital technologies may change farming. Agricultural revolutions boost productivity and technology. The five agricultural revolutions were the First, Arab, British, Scottish, and Green. Despite increased production, prior agricultural revolutions ignored some issues. The Green Revolution caused inequity and environmental devastation. The Green Revolution first favored large farmers who could afford new technologies, worsening interfarm and interregional inequities. Second, its detractors say its policies promoted excessive input use and pesticide use, causing soil deterioration and chemical runoff. Digital agriculture may mitigate Green Revolution effects (Shepherd et al. 2020). To feed 9 billion people by 2050, the FAO predicts 56% greater food production than in 2010. Hunger, climate change, food waste, and dietary changes are global challenges. Food production, greenhouse gas emissions, and agricultural land protection must all be done at once to establish a "sustainable food future." Digital agriculture can improve the agricultural value chain's efficacy, egalitarianism, and sustainability (Alexandratos 1995).

3.2 Artificial Intelligence (AI) Role in Agriculture

Throughout the nineteenth century, during the time of the industrial revolution, machines were frequently utilized in place of or to reduce the need for human labor. The gradual replacement of human labor by AI is a fact that cannot be ignored in the present day and age (Dharmaraj and Vijayanand 2018). AI in agriculture increases agricultural production and sustainability. For agricultural decision-making, AI can be used for crop monitoring and yield prediction, precision irrigation, weed and pest identification, and automated machine learning. Farmers use AI to detect areas that need irrigation, fertilizer, pesticides, or production enhancements. Agronomists research with AI. AI has been used to predict tomato ripening, monitor soil moisture, control agricultural robots, do predictive analytics, profile livestock pig throw emotions, automate greenhouses, detect illnesses and pests, and save water (Talaviya et al. 2020).

AI's adaptability, high performance, precision, and affordability are key to agriculture. AI in agriculture will be achievable because of big data analytics, robotics, the Internet of Things, inexpensive sensors and cameras, drone technology, and extensive Internet connectivity on geographically scattered fields. By analyzing soil management data sources like temperature, weather, soil analysis, moisture, and historical crop performance, AI systems will be able to predict which crop to plant in a given year and when the optimal sowing and harvesting dates are in a specific region, increasing crop yields and reducing water, fertilizer, and pesticide use. Workers' safety and natural ecosystems may benefit from AI technology. In turn, this will keep food costs low and guarantee food production stays pace with population growth (Eli-Chukwu 2019).

In the context of a wide variety of agri-food applications and supply chain phases, comprehending a model's identification, service creation, and decision-making approaches can considerably benefit from AI methodologies. AI technologies deliver algorithms for evaluating performance, classifying patterns, and predicting unanticipated problems or phenomena to solve comprehension issues in the agricultural field, as well as for the identification of pests and their appropriate treatment methods, as well as for the management of irrigation, using remote sensing and sensors to assess abiotic and biotic factors in order to optimize agriculture and animal management. Other applications of AI tools include the management of irrigation, as well as the identification of pests and their appropriate treatment methods. In addition, the implementation and use of AI provide enormous benefits that have the potential to transform the agri-food sector and associated businesses. Initially, AI provides more efficient techniques for growing, harvesting, and marketing agricultural products, emphasizing analyzing problematic crops and increasing the possibility of healthy crop yield. AI is also used in applications such as the automated adjustment of machines for weather forecasting and the 98% accurate identification of plant diseases or pests (Eli-Chukwu 2019).

The initial phase in the agricultural supply chain is the preproduction cluster. It focuses mostly on crop production forecasts, soil parameters, and irrigation needs. Numerous experts stress the importance of agricultural yield generation for enhancing plant support management. Indeed, by employing input data, precision agriculture solutions strive to make stakeholders and farmers more aware of their demands (such as water, nutrients, and fertilizers) by projecting effective models based on machine learning (ML) algorithms. Additionally, they assist with optimum crop output forecasting judgments and enhance prudent agricultural practices. Various machine learning (ML) methods, including deep learning, Bayesian networks, regression, decision trees, clustering, and artificial neural networks (ANNs), have recently been used in agricultural production prediction. There is currently a dearth of expertise with sophisticated machine learning systems on farms throughout the globe, despite AI providing vast prospects for agricultural applications. AI systems need substantial data for machine training to produce correct predictions. The future of agriculture will rely greatly on adaptable cognitive solutions. Extensive research has resulted in the availability of several applications, even though the agriculture market remains underserved. AI farming is still in its infancy, despite using AI decision-making systems and predictive solutions to address genuine farmer difficulties and expectations (Dharmaraj and Vijayanand 2018).

3.3 *Weather Forecasting*

It may be finished day by day or in advance over multiple days. Weather forecasting is a crucial procedure that influences many people's everyday lives and may save countless lives by preventing accidents in industries such as agriculture, irrigation, and maritime commerce. Weather forecasting has several uses; it impacts numerous facets of life, including industry, transportation, disaster management, and energy management (Fathi et al. 2021).

Weather forecasting uses science and technology to forecast the atmospheric conditions for a certain location and time. Weather predictions have been done informally for millennia and professionally since the eighteenth century. Meteorologists make weather forecasts by gathering quantitative data on the current state of the atmosphere, land, and water and then using meteorology to determine how the atmosphere will change at a certain location. Globally, agricultural communities must adjust to climate change. Simultaneously, they must increase the food supply for a rising population while guaranteeing the sustainable use of natural resources. In this dynamic, Agriculture is one of the factors contributing to increased greenhouse gas emissions and a victim of extreme weather events and rising temperatures. Agriculture is also an ally in climate change mitigation and adaptation because ad hoc crop types and animals may lower farming's environmental effects, and improved management techniques can encourage soil conservation. The influence of agriculture on climate change adaptation and mitigation will be defined by the adopted rural policies, as well as a knowledge of the links between weather, climate, and farming (Parolini 2022).

As a crucial and necessary aspect of people's everyday life, weather forecasting examines changes in the existing atmosphere condition. Big data analytics analyzes enormous volumes of data to reveal hidden patterns and pertinent information that might lead to improved outcomes. Multiple sectors of society, including the meteorological institution, are now interested in big data. Consequently, big data analytics will provide more accurate weather forecasts and aid forecasters in making more precise weather predictions. Several big data approaches and technologies have been developed to handle and analyze a large amount of weather data from a variety of sources in order to accomplish this objective and to provide effective remedies (Fathi et al. 2021). Previously, weather forecasting was computed manually based mostly on changes in barometric pressure, present weather conditions, and sky conditions or cloud cover. Weather forecasting depends on computer-based models that account for several atmospheric elements.

The selection of the optimal prediction model still involves human input, which necessitates pattern identification abilities, teleconnections, model performance information, and model bias awareness. Inaccurate forecasting results from the chaotic character of the atmosphere, the huge processing capacity necessary to solve the equations describing the atmosphere, land, and ocean, measurement error, and a lack of knowledge of atmospheric and associated processes. As the time gap between now and the period for which the prediction is being produced (the forecast range) rises, the accuracy of forecasts decreases. Using ensembles and model consensus aids in reducing error and providing prediction certainty (Fathi et al. 2021).

There are several applications for weather forecasting. Important weather forecasts are weather warnings because they protect people and property. Agriculture and, by extension, commodity traders need forecasts based on temperature and precipitation. The demand forecasts of utility companies are based on temperature estimates. Numerous folks depend on weather forecasts to choose how to dress each day. Because severe rain, snow, and wind chill greatly limit outdoor activities, forecasts may be used to plan around these occurrences and prepare for their occurrence. In 2009, the United States spent around $5,1 billion on weather forecasting, with expected benefits six times higher. The ability to anticipate the weather is essential in agriculture since it enables farmers to produce healthy and abundant crops. Estimated precipitation and temperatures and historical data are the most important meteorological variables for agriculture to organize field activities from planting through harvesting, with fertilizer and herbicide applications in between. Accurate climate projections 3–6 months in advance may enable farmers and others in the agricultural industry to take steps to minimize negative consequences or capitalize on anticipated pleasant weather. Due to several physical, biological, economic, social, and political considerations, however, the potential advantages of climate projections vary considerably (Lazo et al. 2009; Jones et al. 2000).

The practice of weather forecasting dates back to the nineteenth century. Weather forecasting assesses atmospheric variables such as temperature, radiation, air pressure, wind speed, wind direction, humidity, and precipitation. A massive amount of data must be collected or developed to predict the weather. In addition, the information is unorganized. Consequently, utilizing meteorological data to forecast the weather is difficult with excessive changeable components. These variables fluctuate as a result of rapidly changing meteorological conditions. To propose a weather forecasting algorithm, we must consider its unique properties, such as continuity, data density, and multidimensional and chaotic behavior. Forecasting the weather has transitioned from a labor-intensive activity to a computer one that requires high-tech equipment. Numerous variables may alter the accuracy of forecasts. Effective criteria include the season, geographical location, input data accuracy, weather classifications, lead time, and validity (Fathi et al. 2021).

In contrast, climate change and severe weather events provide new problems for agricultural meteorology study and application in industrialized nations. Over the last century, agricultural meteorology has seen many changes, but the focus on aiding farmers wherever needed has remained constant, although with less-than-expected outcomes (Parolini 2022).

In the nineteenth century, the practice of weather forecasting began. Assessing atmospheric data, including temperature, radiation, air pressure, wind speed, wind direction, humidity, and precipitation, defines weather forecasting. Collecting or generating a vast amount of data is necessary to predict the weather.

3.4 Smart Irrigation

Global food consumption is rising due to the fast population increase, placing extra pressure on water resources (Ochoa-García and Rist 2018). Irrigation accounts for over 70% of worldwide water withdrawals, and agriculture is the largest consumer of water (Simionesei et al. 2020). Water shortage is one of agriculture's greatest challenges in arid and semiarid areas (Nazari et al. 2018). This issue emerges as a result of farmers regularly irrigating all portions of an agricultural field without addressing the crop's water needs. Some portions of a farm may be over- or underirrigated, a downside of the irrigation mentioned above since it may cause unpleasant water stress on the crops (Abioye et al. 2020). Therefore, good irrigation water management is necessary to ensure global water security (Nazari et al. 2018). In agriculture, the disparity between water demand and availability is considered an issue that should be solved by optimizing irrigation water utilization using cutting-edge technology (Pereira et al. 2022).

Remote sensing (RS), soil moisture sensor (SMS), evapotranspiration (ET)-based controllers, and optical sensors are generally the four smart irrigation systems available on the market to reduce water waste during field watering. This section focuses on the effect of these techniques on water conservation and crop quality. It was discovered that the Internet of Things-based automated agricultural field watering system with soil, temperature, and humidity sensors was 92% more successful than conventional human approaches. Smart irrigation requires little to no human input and only uses water where it is required. Additionally, it has high cost-effectiveness since less water is utilized and the process is more precise, resulting in lower prices and expenditures overall (Jain and Vani 2018).

Additionally, the technique considerably decreases energy consumption since fewer hours are spent operating the equipment, and regulated pauses are performed during the process to minimize total energy usage. Moreover, since resources are limited and enterprises must control expenditures to some level, it is essential to reduce expenses and save materials. Smart irrigation takes into account cost, allowing related tasks to be completed successfully while spending less money. Lastly, improved irrigation efficiency and water management ensure that crops and plants receive only the necessary amount of water, reducing crop loss due to inadequate or excessive watering (Jain and Vani 2018).

Advantages of using smart irrigation are to (1) reduce water waste to save money, (2) improve your landscape's health, (3) prepare you for water's future, and (4) avoid paying fines. (5) There are numerous advantages to using Internet of Things (IoT) systems in irrigation, featuring decreased total water usage, high cost-effectiveness, high-performance efficiency, decreased energy consumption, and decreased crop waste. Reduced water use is one of the major advantages of IoT irrigation systems (Touil et al. 2022).

The principle of intelligent irrigation uses sensors that assess soil moisture, expected precipitation, and external temperatures to gather crucial data. The intelligent controller may utilize this data to guide the sprinkler system's actuators to switch on and off in response to user requests. The user merely has to define a moisture and temperature range for the sensors to capture this information. When this threshold is met, an action is triggered (such as starting or stopping smart irrigation). The intelligent irrigation controller can detect water flow and if pipes are blocked or leaking (massive water savings here). Typically, a smartphone app will enable you to operate and monitor your intelligent irrigation system. In contrast, your smart controller/irrigation system will be set up in an atomized manner, determining the optimal watering timings, durations, and quantities for your garden based on weather and sensor data (García et al. 2020).

A sensor is a device capable of detecting measured data and turning it into an electrical signal or other kinds of information output depending on a set of rules to meet the requirements of information transmission, processing, storage, display, recording, and control. As the Internet of Things technology advances, sensors are used in various industries, including agriculture and industry. Agriculture sensors such as air temperature and humidity, soil moisture, pH, light intensity, and carbon dioxide are often used to collect data on crop production's seedling, growth, and harvest stages. Using agricultural conductivity and pH sensors, water and fertilizers are monitored. The integrated monitoring system monitors the liquid mixture of fertilizer sensors of temperature. The temperature and humidity sensor is a piece of electronic equipment that detects and transmits the temperature and humidity levels in the shed. It is also a type of sensor with many applications in smart agriculture. There are two kinds of sensors in agricultural greenhouses: air temperature and humidity sensors and soil temperature and humidity sensors. During installation, the crop's various root depths determine the sensor's depth. It detects soil temperature, moisture content, and changes during crop growth and development, allowing for timely and appropriate watering. Here is a more in-depth look at the two temperature and humidity sensors. Using temperature and humidity sensors in greenhouses may help plant development, and employing such sensors for monitoring and management can successfully avert significant droughts and floods. We can also calculate the appropriate amounts of temperature and humidity for crops. When the temperature and humidity data exceed the standard, the sensor sends a signal to connect the heating/dehumidification equipment in parallel to effectively control the humidity and temperature in the shed, allowing the crop to grow normally. It determines whether crops lack water based on soil temperature and humidity. If crops are water-stressed, intelligent irrigation should be implemented promptly to meet crop water demands. When the temperature and humidity sensor detects enough water in the soil, it signals the irrigation system to turn off the water supply, completing the irrigation system's automation. It meets crop needs without wasting water resources (Aniley et al. 2017).

The World Meteorological Organization (WMO) characterizes soil temperature as a physical quantity that indicates the average random movement of molecules in a physical body (IAEA 2008). The temperature may also determine whether an object is warm or cold. It pertains to the random thermal mobility of a material's molecules. It measures a substance's average translational kinetic energy (Fahrenheit and Kelvin n.d.). From 0 to 40 °C is the most changeable soil temperature range (Liu et al. 2011). Between 20 and 30 °Celsius is the best range of typical soil temperature for plant growth.

The Importance of Soil Temperature Measurement Temperature influences several processes inside the soil and its environment. Consequently, soil temperature measurements are required (Valente et al. 2006). The soil temperature influences photosynthesis, respiration, transpiration, the soil's water potential, soil translocation, and microbiological activity.

The Value of Measuring Soil Temperature Several processes inside the soil and its ecosystem are affected by temperature. Therefore, soil temperature measurements are essential (Valente et al. 2006). The temperature of the soil impacts photosynthesis, respiration, transpiration, the water potential of the soil, soil translocation, and microbial activity (Lehnert 2014).

3.5 Sensors of Soil Moisture

Soil moisture sensors are also referred to by the moniker "soil moisture meters." The principal uses for this technology are agricultural irrigation, forest protection, monitoring the volumetric water content of the soil, and measuring the soil's moisture content. FDR and TDR, which stand for frequency and temporal domains, respectively, are the two types of soil moisture sensors that are now available. The most common types of soil moisture sensors include tensiometers, capacitance, dielectric methods, volumetric sensors, and neutron analyzers. Gypsum blocks are another common type of soil moisture sensor. When inserted into the ground, these sensors determine the soil tension as well as the volumetric water content. The soil moisture sensor is a device that measures the level of moisture that is currently present in the soil. The incorporation of sensors into agricultural irrigation systems helps to enhance the effectiveness of water distribution. These meters help detect whether the amount of watering should be decreased or increased to promote optimal plant growth (Aniley et al. 2017).

3.6 The TDR Approach and the FDR Method

The time domain reflectometer method, also referred to as the TDR method, is a technique for determining the moisture content of the soil by measuring the dielectric constant using a time domain reflectometer. As soil moisture content increases, so does the value of the dielectric constant, and the speed of propagation of electromagnetic waves in the medium is proportional to the dielectric constant. Because the dielectric constant of water in the soil is significantly greater than that of solid particles and air in the soil, the value of the dielectric constant increases as soil moisture content increases. Because the square root has an inverse proportionality, the waveguide rod causes an increase in the amount of time it takes for electromagnetic waves to travel down the rod. It is possible to quantify the moisture content of the soil if one takes into account the speed at which high-frequency electromagnetic pulses travel through the soil along the waveguide rod and measures the speed at which they do so (Babaeian et al. 2019).

The term "FDR" refers to a method known as the frequency domain soil moisture sensor. It uses the electromagnetic pulse principle to compute the soil volumetric water content (v) by measuring the apparent permittivity of the soil in relation to the frequency of propagation of electromagnetic waves through the medium. This is done so in order to determine how much water is contained in the soil volumetrically. After the soil has been calibrated, the measurement accuracy is very high, the probe form is unrestricted, and it is possible to detect multiple depths at the same time, which makes data gathering much easier. The current method of monitoring soil moisture in agricultural settings is handled by a small number of sensors. The application of these sensors is severely limited due to their low sample volume, high cost, requirement for close contact between the soil and the sensor, and poor performance in salty, vertical, and rocky soils. This study was carried out to investigate a wide range of new and cutting-edge soil moisture sensors and to assess the range of possible applications those sensors could have in agriculture (Babaeian et al. 2019).

The collection and analysis of soil moisture data is essential for a number of ecosystem services, including agricultural production, watershed hydrology, flood forecasting, and landslide prediction. Agricultural is the industry that uses the most water globally, accounting for around 70% of the overall water use. The rising demand for diminishing water supplies all over the world has rekindled interest in the research and development of proximal soil moisture sensors for the purpose of improved irrigation and better control of soil moisture in agricultural settings. A handful of "trusted" technologies are currently in control of the monitoring of soil moisture in agricultural settings. These technologies include frequency domain reflectometry (FDR) or capacitance, gypsum block sensors, time domain reflectometry (TDR), neutron moisture meters (NMMs), and amplitude domain reflectometry (ADR) (Babaeian et al. 2019). Because soil moisture is dynamic in space and time, it must be regularly monitored. There are several ways to assess the state of soil moisture. Depending on the needs and objectives of the project, each of these techniques has benefits and downsides and should be used with care. Compared to the total quantity of water on the globe, soil moisture is the temporary storage of water in a thin layer of the Earth's surface. It is crucial to agronomic, hydrological, and meteorological processes at all geographical scales. It is essential for identifying water stress and controlling irrigation. In addition to predicting natural catastrophes like drought and floods, soil moisture data may also be used to forecast environmental changes like dust storms and erosion. However, a reliable estimate of soil moisture using in situ measurement is prohibitively costly owing to the need for replication sampling to assess the periodic change in soil moisture. Various approaches have been used to assess soil moisture, including spot measurements and remote sensing. Gravimetric, neutron probe, time domain reflectometry, capacitance, FDR, tensiometer techniques, and hygrometric techniques are examples of traditional methods or point measurement techniques. The typical gravimetric approach is precise and economical, but it is damaging, sluggish, and time-consuming, resulting in restricted coverage. At an Relative accuracy of the permittivity of ±4%, dielectric probe techniques are the most reliable means of measuring surface. Land surface parameters such as land use, land cover, soil moisture, surface water area, surface temperature, and snow cover have been collected using optical remote sensing. As a result, numerous researchers have investigated the relationship between soil moisture and reflectance. This method is well-suited for automated irrigation system monitoring and control and needs little maintenance. Hydrometers are part of a vast, sophisticated, and costly system, making their usage impracticable is one of its major drawbacks (Pavan et al. 2018).

Likewise, methods including remote sensing provide more accurate measurements of near-surface soil moisture over a vast region with a spatial and temporal variation. Although the soil moisture at the surface looks small, this thin layer controls all agricultural activities. In this work, we examined all remote sensing techniques. Due to their inability to penetrate weather, optical and thermal remote sensing techniques are inapplicable for estimating soil moisture under plant cover. However, microwave remote sensing methods are suitable for assessing soil moisture under vegetation cover. Because point measurement techniques give point estimates, they cannot be used on a vast area with high precision, as determined by our review. Active microwave remote sensing technologies can estimate the spatial distribution of soil moisture over vast agricultural regions with high spatial resolution, but the temporal distribution of soil moisture over a broad area cannot be established. Passive microwave remote sensing can assess soil moisture with greater spatial and temporal precision across a broader observation region. Sensors for moisture humidity is the concentration of water vapors in the atmosphere. Generally speaking, water vapor, or water in its gaseous condition, is not visible to the human eye. The humidity level predicts the presence of precipitation, dew, or fog. Temperature and pressure influence the humidity of the system under study. Since it contains the same water vapors, cool air has greater relative humidity than warm air. The dew point is an associated variable. As the temperature rises, the quantity of water vapor necessary for saturation increases. As the temperature of an air parcel decreases, it reaches saturation without gaining or losing water mass. The quantity of water vapors in a given air volume may vary considerably (Brun et al. 2022).

A parcel of near-saturated air may contain 28 g of water per cubic meter of air at 30 °C (86 °F) but only 8 g at 8 °C (46 °F) (Brun et al. 2022). Humidity is measured by humidity sensors, which turn the data into an electrical signal that may be utilized for several reasons. You may have heard humidity sensors referred to as hygrometers. They are sometimes referred to as hygrometers, while "humidity sensor" is the more prevalent name.(Brun et al. 2022).

Humidity sensors are often constructed from ceramic, semiconductor, and polymeric materials. When exposed to a humid environment, these sensors detect changes in the conductivity or dielectric permittivity of hygroscopic sensing materials induced by water vapor absorption and desorption. As an essential environmental characteristic, humidity (often represented as relative humidity (RH)) significantly impacts the economy, agriculture, and human life. Consequently, humidity sensors serve a crucial role in detecting ambient humidity, and several kinds of humidity sensors have been created and extensively used in various situations. It directly and indirectly affects fruit crop water relations, leaf development, photosynthesis, disease incidence, and economic output. It influences the rate of evapotranspiration as well as the water needs of fruit crops. Because of the high turgor pressure and the ease with which fungal spores germinate on plant leaves, high humidity promotes leaf enlargement—humidity too low, humidity too high, and wilting soft growth (Aniley et al. 2017).

Digital Agriculture for Biotic Stress

There are three primary applications of AI in plant health and biostress management—disease profile, symptom fingerprinting, and the need to collect ever-increasing numbers of symptom developments in plant hosts—all play a role in disease detection and prognosis. Second, fungicides or other methods are used to control this disease quickly and directly to prevent the development of the pathogen, including the quantities and times at which they should be added. The collection of all weather data facilitates the growth of pathogens and results in crop losses, in order to make timely recommendations about who should use chemical control, plant protection, and disease tolerance. Due to the widespread use by potato farmers and producers, there was an urgent need for disease prediction and speeding up control when the warning is made that conditions are conducive to the spread of the disease. This is because the devastating effects of the disease occurred after suitable weather for infection in Ireland, leading to the loss of the potato crop (Anderson et al. 2004). The programs' conclusions depend on taking individual readings of symptoms, environmental factors (such as temperature and humidity), and experimental conditions (such as the number of hours of daylight) Fig. 3.

3.7 Crop Disease Prediction and Health Monitoring

3.7.1 Disease Detection

The most popular mobile program used in AI plant disease diagnosis is Plantix. Other programs in the market that offer similar services include Agrivi, Plant Village, and Farm Logs. Also using artificial intelligence, Albattah et al. (2022) established a new way to identify plant disease. As a novel deep learning method for plant disease detection and classification, these programs pass throw steps of AI to help in plant disease diagnosis Fig. 4.

1. Data collection: Collect data related to plant disease, such as images of diseased plants, symptoms, and environmental conditions.

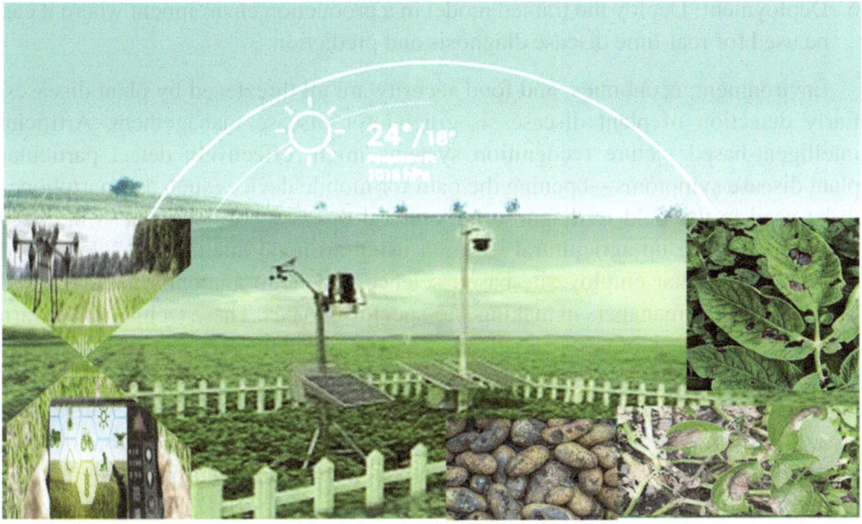

Fig. 3 Role of programs of forecasting in protecting potato plants against late blight disease

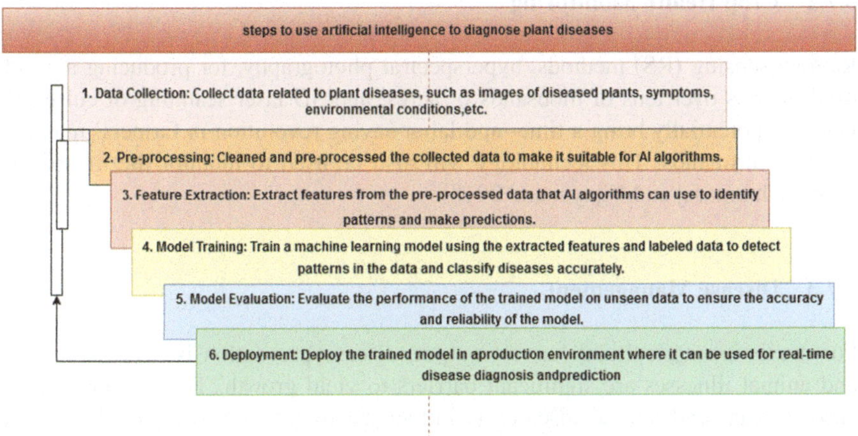

Fig. 4 Step progression to embed artificial intelligence into plant disease diagnoses

2. Preprocessing: Clean and preprocess the collected data to make it suitable for AI algorithms.
3. Feature extraction: Extract features from the preprocessed data that AI algorithms can use to identify patterns and make predictions.
4. Model training: Train a machine learning model using the extracted features and labeled data to detect patterns in the data and classify diseases accurately.
5. Model evaluation: Evaluate the performance of the trained model on unseen data to ensure the accuracy and reliability of the model.

6. Deployment: Deploy the trained model in a production environment where it can be used for real-time disease diagnosis and prediction.

Environment, economics, and food security are all threatened by plant diseases. Early detection of plant diseases is crucial for disease management. Artificial intelligent-based picture recognition systems might effectively detect particular plant disease symptoms—opening the path for mobile devices such as smartphones to be used in the field to diagnose agricultural diseases. Bestelmeyer et al. 2020 reported "Scaling up agricultural research using artificial intelligence," developed AI-based tools that employ site-based science and large amounts of data to aid farmers and land managers in making site-specific choices. These technologies warn farmers the outbreak of pests and diseases and facilitate the selection of sustainable agricultural management strategies. Image sensing and analysis guarantee that plant leaf pictures are split into surface regions such as the leaf's backdrop, sick area, and healthy area. The sick or infected region is removed and submitted for further analysis. This also helps with insect identification and nutritional deficiency diagnosis (Dharmaraj and Vijayanand 2018).

3.7.2 Crop Health Monitoring

Remote sensing (RS) methods, hyperspectral photography, for producing agricultural metrics over tens of thousands of acres, and 3D laser scanning of cultivable land can potentially bring a time- and labor-saving revolution in farmers' management of farmlands. This technology will also be used to monitor the lifespan of crops (Dharmaraj and Vijayanand 2018).

3.7.3 Disease Management

For maximum agricultural harvest output, disease prevention is necessary. Plant and animal illnesses are significant barriers to yield growth. The development of various plant and animal diseases is influenced by genetic, soil, weather, wind, and temperature variables, among others. Due to these characteristics and the fickle nature of certain illnesses' causal effects, managing the consequences in large-scale agriculture is especially difficult. A farmer should use an integrated disease control and management model incorporating physical, chemical, and biological techniques to successfully control illnesses and reduce losses. This is time-consuming and inefficient, underscoring the need for an AI-based disease management and control approach. The explanation block (EB) gives a transparent picture of the logic followed by the kernel of an expert system. A unique fuzzy logic-based technique to rule promotion is used in the system for intelligent agricultural disease management conclusions. A text-to-speech (TTS) converter enables a text-to-speaking user interface. It provides a very effective interactive online user interface for real-time communications. A rule-based and forward-chaining inference engine was employed to construct the system, which assists in illness identification and therapy suggestion (Sladojevic et al. 2016).

Both direct and indirect approaches were capable of detecting and identifying plant diseases. Typically, indirect methods are used to assess plant pathogens (bacteria, oomycetes, fungi, and viruses) or biomolecular markers (nucleic acids, proteins, and carbohydrates) isolated from diseased plant tissues. By detecting changes in physiological or histological markers, such as leaf surface temperature or humidity, spectroscopic features of plant tissues, shape, growth rate, and volatile organic compound emissions, indirect diagnostics identifies plant disease (VOCs). Numerous spectroscopic, electrochemical, and molecular technologies might be used as direct or indirect detection approaches. Nanodiagnostic instruments for agricultural biotechnology (AgBio) research are still in their infancy of development. While several nanosensing technologies have been created and shown for monitoring human health, the application of nanosensors in agriculture started in 2009 (Valdés et al. 2009; Li et al. 2020), a relatively new development. Other innovative sensing technologies, such as nanopore sequencing and array-based nanosensors, have just lately entered the AgBio industry, and the development of plant wearable and microneedle instruments is even more recent. In recent years, the acceptance and implementation of innovative agriculture and plant science technologies have increased. For instance, CRISPR (clustered regularly interspaced short palindromic repeats) technology has already found several uses in agriculture and the food business (Hanson et al. 2017).

3.7.4 Weed/Pest Management

Major pests damage crop production, which include fruit-piercing moths, fruit borers, leaf rollers, leaf-feeding caterpillars, and beetle borers. It needs more than one method to control pests and weeds. Control methods include biological, cultural, chemical, and physical. Weed control aims to prevent or reduce the growth of weeds, especially noxious weeds, to reduce competition with desirable flora and fauna, including domesticated plants and livestock, and to prevent nonnative species from competing with native species in natural settings. Weed management is important in agriculture. Hand cultivation with hoes, power cultivation with cultivators, mulch smothering, deadly wilting with high heat, burning, and herbicide chemical control are all examples of cultivation techniques for weeding in the field (weed killers) (Klingman 1961).

Pest control refers to regulating or managing any animal, plant, or fungus that is detrimental to human activity or the environment. Depending on the magnitude of the harm, the human reaction will range from tolerance to deterrent and control to efforts to eliminate the pest. A plan for integrated pest management may involve pest control methods (Dent and Binks 2020). Rodents, birds, insects, and other species that share human environments and feed on or damage property are considered pests in houses and towns. Exclusion or quarantine, repulsiveness, physical removal, and chemical agents are utilized to manage these pests. Alternative biological control strategies include sterilization campaigns (Flint and Van den Bosch 2012).

Weed continually affects the anticipated profit and productivity of farmers. Research indicated that weed infestations diminish the output of dry beans and maize by 50%. Approximately 48% of wheat output is lost due to weed competition. These losses might even surpass 60%. Gene technologies such as gene silencing (e.g., RNA interference) and gene drive offer a tremendous capacity for population control. It is anticipated that the deployment of gene technologies as weed control techniques would be intrinsically complicated, with several variables impacting their implementation. RNA molecules' environmental stability and distribution into plants affect exogenous gene silencing applications. Drive efficiency, resistance alleles, and plant ecological complexity influence gene drive. Despite the harmful consequences on the environment and human health, farmers rely on agrochemicals for plant disease, insect, and weed management, and they stick to conventional crop protection tactics (which utilize a huge number of chemicals). For example, herbicides are sprayed on more than 90% of cropland in the United States (Gianessi and Reigner 2007). Herbicides have decreased the need for physical labor in weeding fields. Using herbicides has reduced production costs and increased agricultural yields in the United States. An estimated $26 billion is spent annually on herbicides in the United States. Herbicides account for about 65% of the overall expenditures of American farmers (Gianessi and Reigner 2006).

Pests reduce up to 40% of the worldwide potential crop production. This number might be quadrupled without application of agrochemicals (Deutsch et al. 2018; Oerke 2006). In 2015, global pesticide consumption was projected at 3.5 billion kg, or $45 billion (Pretty and Bharucha 2015).

In addition to the advantages of employing agrochemicals for pest and weed management, there are disadvantages, most notably the restrictions of conventional spraying methods. Reducing the detrimental effects of agrochemicals is a big worldwide socioeconomic and human health concern (especially spraying technologies); 72% of people regard agrochemical residues as one of their top three food-related concerns. The European Food Safety Authority (EFSA) issued a report about screening food products within 2013 warning of potential dangers to the food supply (European Food Safety Authority 2015). According to the EFSA, 99. % of food products have some trace of pesticides (with 1.5% of them over the legal limits). The threat to agricultural output in many countries is posed by plant agrochemicals (such as herbicides) (European Food Safety Authority 2015). In recent decades, there has been a rise in interest in pest and disease identification, as well as weed spraying automation (Abdulridha et al. 2018; Cruz et al. 2017, 2019; Fernández-Quintanilla et al. 2018). Moller (2010) concluded that computer vision technology in agricultural operations minimizes operator stress. A smart sprayer system must be able to recognize weeds in real time and only apply the required chemical where it is necessary. Etienne and Saraswat (2019) investigated a variety of weed identification sensors and methodologies, including machine vision, spectrum analysis, remote sensing, and thermal pictures. Machine vision has been used for many years to differentiate a plant from dirt backgrounds using picture segmentation techniques based on color differences (McCarthy et al. 2010).

Weed control was achieved by utilizing several strategies, such as mechanical weed control, crop rotation, and pesticides. Various biochemical weed and pest control methods are available on the market; however, they all result in a decline in crop output. Regular and consistent pesticide use in a field has a negative effect on crop output. AI provides a clever answer to this challenge (McAllister et al. 2019). Furthermore, they have developed "Agbots," an autonomous, AI-based robot, that can do weed control in the field in an intriguing manner. AI-based weed management aims to automatically detect weeds (using pictures captured by autonomous robots' cameras) and take the necessary remedial measures (such as mechanical weed removal or pesticide spraying). Other AI-based solutions, such as Blue River Technologies' See and Spray, utilize AI to recognize and spray individual plants in milliseconds (Partel et al. 2019; Allmendinger et al. 2022). Actually, we can depend on drones, also known as unmanned aircraft systems (UAS), remotely piloted aircraft, and unmanned aerial vehicles, to offer farm management reaches accurate pixel sizes, coverage on demand, and rapid delivery of information they have been requesting from the remote sensing platform for a long time.

3.7.5 Crop Management System (CMS)

Determine the Maturity of the Crop Images of different crops were taken under white and ultraviolet A (UVA) light to determine the degree of ripeness of green fruits. Based on this study, producers might establish several degrees of preparedness for each fruit or crop, and then arrange them into various stacks before transporting them to the market.

Agricultural Readiness Identification A system based on artificial intelligence gathers photographs of a crop and analyzes them to determine its readiness for harvest in a specific region. Before shipping the crops to market, they might be classified according to their maturity and other quality factors. Utilizing different pattern clustering algorithms, such as K-means, fuzzy C-means (FCM), expectation maximization (EM), and hierarchical clustering, is crucial for classification. As part of the Internet of Things (IoT) platform layer, an online crop management system (OCMS) was designed. The system attempted to provide measurements of the farm's environmental parameters, including temperature, water level, pH value, and dissolved oxygen, from an arable farm where food and horticultural crops were cultivated. The OCMS will aid farmers in optimizing farm operation and management, data management, file sharing, farm environmental data analysis, and analytical reporting administration. The graphical user interface of the online crop management system (OCMS) displays temperature (°C), pH or water acidity in moles per liter, dissolved oxygen in milligrams per liter (mg/L), and water level data (ft) (Aggarwal et al. 2022).

4 Nanotechnology and Nanosensor Agricultural Management Tools

In recent years, nanotechnology has emerged as one of the most important and reliable technologies in the agricultural sector. Agricultural production can be controlled by nanotechnology through the regulation of the use of nutrients through the use of nanofertilizers; the rationalization of the use of agrochemicals, the detection and treatment of disease-host interactions; molecular interactions with nanocarriers (nanobiosensors) for plant disease diagnosis; the removal of contaminants from water and soil; farm management; and current trends including the treatment of salinity soil. In addition to detecting pathogens, fertilizers, moisture, and pH levels in the soil, nanosensors can help enhance the use of plant protection products, cut down on nutrient loss, and boost crop output through more efficient nutrient management (Kaushal and Wani 2017).

Sensors can identify analytes in samples. The bioreceptor layer's connection to the transducer determines the biosensor's success. The goal is to establish a strong biological–sensory relationship (converter). Nanomaterials increase system sensitivity.

A nanosensor uses at least one nanostructure to detect gases, chemicals, biological agents, electric fields, light, heat, etc. The right system helps biosensors attach analytics and accurately detect biological elements (such as antibodies, enzymes, and DNA strands). Nanomaterials bridge the nanoscale gap between the converter and the bioreceptor in these systems (Kaushal and Wani 2017). The biomaterial detection mechanism determines whether electrochemical biosensors are catalytic or propulsive. Electrochemical sensors use potentiometric, chronometric, voltametric, impedance measurement, and field effect transistors. Nanostructures and electrochemical technologies have produced sensors with high sensitivity and decomposition power. Nanoparticles, nanotubes, nanowires, nanopores, self-adhesive monolayers, and nanocomposites can improve sensor performance.

Nanomaterials are more active as pesticide agents or transporters. It can boost crop productivity and economic and social equity in agriculture (Prasad et al. 2014). Biodegradability, solubility, permeability, and thermal stability make nanomaterials useful for sensors and crop protection in agriculture. Their surface regions also attract the target organism. Nanopores, nanoencapsulates, nanocaps, and nanoemulsions are used as transporters, pesticides, and crop disease control. Controlled release mechanisms from nanoparticles (NPs) reduce pesticide and fertilizer use while growing crops. Nanoparticles are ideal for electrochemical and biosensors due to their unique features (Peng and Miller 2011). Nanosensors can detect waterborne pollutants, pathogens, and mercury (Selid et al. 2009). Nanosensors were constructed using nanomaterials for hormone regulation, agricultural pests, viruses, soil nutrient levels, and stress variables. Auxin and oxygen nanosensors have been produced (Koren et al. 2015). Fabiyi et al. (2020) reported that nanoparticle biomarkers could detect bacteria, viruses, fungi, and nematodes of economic value in agriculture. Nanosensors may also detect chemicals in plants that evoke disease symptoms. Nanosensors detect pesticide residue accurately. Nanotechnology involves the creation and utilization of nanoparticles as nanosensors like soil sensors.

5 Digital Agriculture for Yield Development

5.1 Remote Sensing-Based Yield Prediction

Through machine learning and statistical methods, technological advances have made yield prediction more accessible and accurate. Random forests, linear regression, and ensemble methodologies have historically been used to estimate agricultural productivity. However, deep learning techniques have recently dominated crop yield estimates. Recent research has combined observed phenotypic data with environmental data to forecast wheat, corn, and strawberry yield. In addition, a rapidly developing corpus of research combines convolutional neural networks with unmanned aerial vehicle (UAV) data to provide a prediction. Recent research demonstrates a definite movement toward employing deep neural networks for agricultural production forecasting, irrespective of the data collection technique. Regarding our issue, remote sensing technologies have been used globally to estimate agricultural yields at different sizes (field, county, and state), Fig. 5. Several research has been conducted to predict the yield of maize, wheat, grapes, rice, corn, and soybeans utilizing random forests, neural networks, multiple linear regression, partial least squares regression, and crop models based on several vegetation indices. In the past, standard machine learning and statistical methods dominated remote sensing yield predictions, such as tabular yield data and UAV pictures (Kogan 1990; Khaki et al. 2021). Standard machine learning and statistical methods have been used to predict crop yields from remote sensing data for decades. These methods include linear regression, logistic regression, decision trees, and support vector machines. Tabular yield data such as soil type, crop type, and weather conditions are used to train the models. UAV pictures are also used to create high-resolution images of the crops, which can be used to identify areas of stress or disease in the crops. The models are then used to predict the yield of a given crop based on these inputs. Additionally, more recent developments in machine learning such as deep learning have been applied to remote sensing data for yield prediction (Khaki et al. 2021). These models use convolutional neural networks (CNNs) to extract features from the imagery and then use these features to make predictions about crop yields.

Nonetheless, the current trend in crop production prediction is to combine convolutional neural networks with satellite images. These studies demonstrate that deep learning and remote sensing have the capacity, power, and precision to estimate yields on a massive scale. Remote sensing is the study of acquiring information about an object or phenomenon without having direct contact with it and, thus, without being intrusive. Electromagnetic radiation (EMR) emissions, such

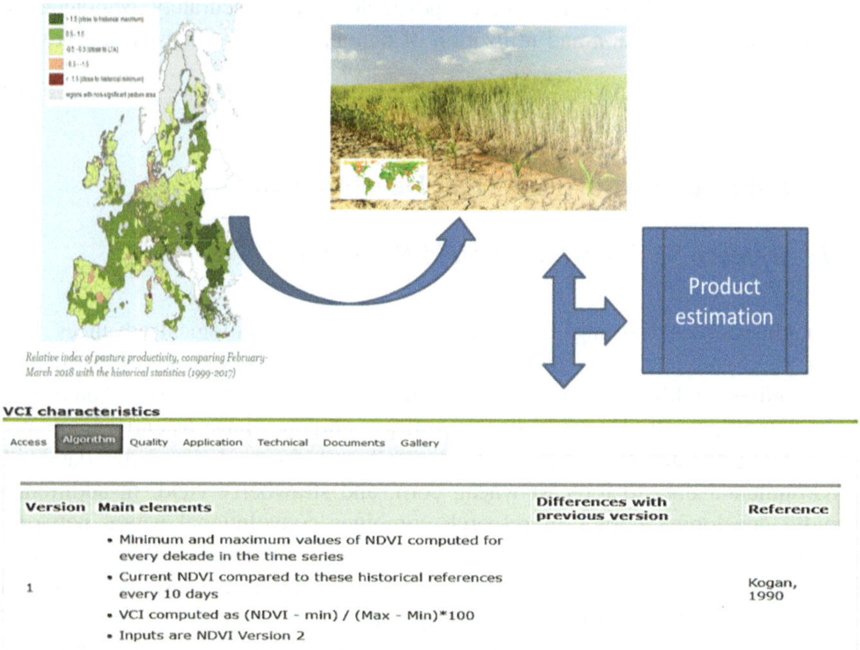

Fig. 5 yield data and UAV pictures globally to estimate agricultural yields. Neural networks, multiple linear regression, partial least squares regression, and crop models based on several vegetation indices. https://land.copernicus.eu/global/products/vci

as ultraviolet, radio waves, infrared light, visible light, and microwaves, are used for remote sensing. Agriculture uses agricultural remote sensing to analyze crop growth conditions over time. In addition, it provides information for drought monitoring, recognizing excessive soil moisture, assessing meteorological effects on plants, evaluating vegetation health and productivity, and forecasting agricultural output. The availability of remote sensing data, which permits continuous monitoring of crops throughout their development cycle, enables the calculation of agricultural production on a wide scale. This information enables stakeholders and farmers to maximize crop potential in real time. In addition, satellite images may be used to determine many environmental factors, such as surface temperature and precipitation. In addition to monitoring crop growth and production, remote sensing data may be used to identify soil moisture and salt levels, assess insect infestation levels, and monitor environmental pollution levels. Commonly derived descriptive indices from remote sensing include the normalized difference vegetation index (NDVI), the temperature condition index, the enhanced vegetation index, and the leaf area index (Khaki et al. 2021).

Several research based on the parallel development of crop models and remote sensing has combined agricultural models and remote sensing methods for regional or global crop yield estimates. Numerous remote sensing methods and equipment, including satellite images, aerial photography, radar systems, and lidar, are utilized to collect this data (Jin et al. 2018; Cai et al. 2018; Yue et al. 2018). There are three types of remote sensing: ground-based, aircraft, and satellite. First, ground-based remote sensing is applied in Schneider et al. (2012)'s research on winter wheat, where ground-based spectra are used to anticipate yield at the start of the shooting stage. For example, the highest accurate yield projections of winter oilseed rape were obtained when spectral measurements were taken during the crop's complete budding phase (Wójtowicz et al. 2005).

However, Piekarczyk et al. (2011) found that the highest connection between spectral data and winter rape output was obtained during the early blooming stage, but wheat yields were most accurately predicted when the plants were in the vegetative stage. According to Piekarczyk et al. (2011), indices based on reflectance in green and near-infrared (NIR) wavelengths exhibited the strongest relationship with yield before oilseed rape blooming (550 and 775 nm, respectively). During rape flowering, indices based on reflectance at NIR wavelengths and their logarithmic adjustment beat nontransformed spectral data for yield predictions (Piekarczyk 2011).

Second, airborne remote sensing—even though satellite data are often accessible in various spatial, temporal, and spectral resolutions—may be used to predict agricultural yield over a broad range of geographical locations and scales. The availability of Earth observation (EO) data has created new opportunities for agricultural mapping on a wide scale. EO data enable a novel method for gathering agricultural information over vast regions and regularly updating it to generate crop production and yield maps. Due to the high spatial resolution necessary for precise yield estimations, unmanned aerial vehicles (UAVs) have been promoted for data collection (Wójtowicz et al. 2016).

Although UAV platforms have exhibited greater picture capture capabilities, it is not viable to precisely measure the yield over wide areas without a huge crew. Piloted aircraft are predominantly used for airborne remote sensing; in recent years, they have been superseded by unmanned aerial vehicles (UAVs), which are aircraft remotely flown from a ground station. UAVs are often low-cost, lightweight, and slow-moving aircraft well suited for remote sensing data collection (Wójtowicz et al. 2016).

There are presently two basic platforms for unmanned aerial vehicles: "fixed wing" versions that can fly at high speeds for extended durations with less aerodynamic characteristics and "rotary wing" variants that can take off and land vertically while hovering above a target. However, their flying range is restricted due to mechanical complexity and diminished battery capacity. UAVs provide several benefits, including the capacity to deploy rapidly and often, to be adaptable in terms of flight altitude and mission schedule, and to produce images with very high quality. Platforms for unmanned aerial vehicles might provide the high-resolution data necessary for site-specific crop management. Very high-resolution UAVs might also be exploited in agronomic research, specialized crop management, and within-field variation studies (Wójtowicz et al. 2016).

Third, satellite imagery has been used to map crop varieties, evaluate crop conditions, and calculate agricultural acreage. Due to the low spatial resolution of sensors, these applications were often deployed across expansive regions. However, the better resolutions of more modern satellite sensors enable in-field evaluations of issues such as drought stress, floods, and hail damage. Even if there are more satellite remote sensing uses, this technology has limitations. Stafford (2000) stressed that changing weather conditions may influence satellite images. According to Lamb and Brown (2001), low-resolution satellite images only apply to large-scale research and may not be suitable for small-scale farms.

5.2 *Data Capturing*

In agriculture, remote sensing utilizes information gathered by many pieces of equipment over time without personal touch with the object. The information collected can then be used to investigate various aspects of the crop and production. This research is utilized to make crop changes to maximize yield. The method could be used to conduct numerous tests and apply suitable measures. Remote sensing in agriculture can help farmers detect frequent crop threats early on, such as pest infestation and weeds, and alert farmers to take the appropriate countermeasures to ensure crop health. Sensors of various types are used to collect data from various regions of the ground. As previously said, sensors are classified into three logistical types: satellite sensors, airborne sensors, and ground sensors (Jin et al. 2018).

Ground-based remote sensing is further classified into many types of ground sensors. Portable ground sensors, vehicle-mounted ground sensors, such as those connected to tractors, and free-standing ground sensors, which are generally affixed to posts and larger trees, are the three kinds of ground sensors. Small-scale operational field monitoring of biotic and abiotic stress variables, including nutrient levels, soil moisture content, and weather, is facilitated using ground-based remote sensing technology. Numerous adjustments may be made to the use of fertilizers and irrigation to maintain a good yield. This technology provides better temporal, spectral, and geographical resolutions than airborne and satellite remote sensing, but it cannot compete with the efficiency of airborne and satellite remote sensing when collecting data across larger regions simultaneously. Utilizing field spectrometer capabilities is most often used in agricultural research for forecasting crop output, determining plant nutrient needs, detecting insect damage, determining water requirements, and managing weeds (Schneider et al. 2012). Aerial sensors are the sort of sensor that follows. Due to the availability of drones, aerial sensors have become extraordinarily affordable and accessible. While floating at low elevations above the crops for lengthy periods, these airborne sensors may acquire high-resolution photographs of the land and gather other sensor data. The information gathered by these sensors may assist in weed identification, yield estimate, and other extensive research, such as detecting soil salinity and chlorophyll content. While they have become more accessible and affordable, they are still ineffective in heavy winds and dismal weather, which is a drawback to their usage. Data from aerial remote sensing may significantly enhance agricultural production and forecast models. Launay and Guerif (2005) developed a model that integrates growth-season photograph data. The root means square error (RMSE) was decreased from 20% to about 10 %, resulting in better yield projections. The quantity and timing of images, which influenced the number and kind of plant biophysical parameters that could be assessed, impacted the robustness of the model. If late-season remote sensing data were used, the model's forecasts improved (RMSE decreased from 21% to 15%) when yield estimates were created for areas with poorly defined soil. According to the experts, the crop model was much less accurate during severe drought conditions. It is possible to generate yield estimates using data from an aerial photography platform, such as an unmanned aerial vehicle (UAV). For instance, Swain et al. (2010) used an uncrewed aircraft to collect multispectral images to estimate rice (*Oryza sativa* L.) output. Swain and Zaman (2012) used a linear estimate of the rice (*Oryza sativa* L.) yield using an unmanned aircraft to collect multispectral images.

Scientists generally divide remote sensing systems into three components: the space foundation system, the ground infrastructure system, and the data storage system for distant sensing. Moreover, we will focus on the same three components for UAV remote sensing systems. Although satellite remote sensing is the most established technology in developing RS technology, since it is often a national-level technology, it may not be practical for other businesses that need remote sensor technology to assist in their growth. The cost is expensive since satellites have a time cycle when gathering data such as photographs. In addition, the picture quality is inadequate, making it difficult to catch fine details. This resulted in development of several novel delivery platforms, such as unmanned aerial vehicles (UAVs) and unmanned airships. UAVs offer various benefits over satellite sensors, including cheap flying costs, a broad range of flying heights, high flexibility, high work efficiency, no significant demand for take-off and landing sites, lower risk, less environmental impact, and the possibility to conduct secondary measurements at any time. UAV remote sensing data collecting and preprocessing disadvantages include tiny volume, restricted power, and limited payload. Nevertheless, some advanced remote sensors, such as multispectrum, hyperspectrum, and high-resolution image sensors, can compensate for these limitations (Yue et al. 2018).

Because the movement of the UAV relies entirely on its control system, it is essential for successful remote sensing that the control system performs admirably. To address UAVs' ever-increasingly complex remote sensing needs, linear control methods are rapidly being superseded by nonlinear control approaches. Although some research has proven that these nonlinear control techniques tackle a certain classical control method problem, they will have additional shortcomings. Then, several methods integrating expert systems and neural networks appeared to strengthen the nonlinear control approach even more. Because remote sensing information contains a huge quantity of data, fast data transfer speed and antijamming capabilities are essential to assuring data integrity. In other words, UAV remote sensing must choose several highly efficient and dependable data transmission routes to ensure the continuous transfer of data in real time. The preprocessing and storage of data must also be addressed to guarantee accurate and complete data (Wójtowicz et al. 2016; Yue et al. 2018).

Image sensor devices that are the most distant are satellite sensors. Utilizing satellite imagery in agriculture helps cover a vast land region and may improve crop monitoring. After a natural catastrophe, it may be beneficial for assessing loss and calculating agricultural output. There are various advantages to using satellite sensor data but also some disadvantages. For starters, it is costly, and even if the expense is considered, the imaging must be ordered months in advance. This may be for naught if cloud cover is present at the desired place and time. Governments throughout the globe have started to make satellite pictures accessible to the public, which might make the practice far simpler in the future. Regional crop production estimate was accomplished using vegetation indicators derived from National Oceanic and Atmospheric Administration Advanced Very-High-Resolution Radiometer (AVHRR/NOAA) satellite image data (Prasad et al. 2006). The authors' model, which characterized the connections between satellite spectral data and crop production in Iowa, produced high R^2 values for corn (0.78) and soybean (0.80, 0.78, and 0.86). Using AVHRR/NOAA images, Dabrowska-Zielinska et al. (2008, 2016) tracked grain growth and yield in Poland using this approach. Using the leaf area index (LAI) and evapotranspiration indices derived from AVHRR images, the scientists developed a model to predict wheat yield (with a 13% margin of error; RMSE). Due to satellite, airplane, and ground-based remote sensing developments, reflectance data are increasingly being utilized in agriculture. In the information carrier in remote sensing, electromagnetic radiation travels at the speed of light in a vacuum as waves of variable lengths. Visible light (VIS), near-infrared (NIR), shortwave infrared (SWIR), thermal infrared (TIR), and microwave bands are the most helpful wavelengths in remote sensing. Active sensors generate radiation interacting with the researched target and returning to the measuring device. Data preparation is a crucial aspect of deep learning projects, including a substantial amount of the analytical pipeline. Data preparation includes cleansing, normalization, transformation, feature extraction, and selection (Wójtowicz et al. 2016; Altalak et al. 2022).

5.3 Data Interpretation

After data capture and storage using different IoT devices, data curation and storage techniques, as well as statistical methodology and programming models to extract usable information, are necessary. Data interpretation is analyzing data and drawing relevant conclusions using different analytic techniques. Consequently, AI operates on external information collected from IoT and other big data sources, utilizes knowledge-based rules (provided by developers), or finds the underlying rules and patterns using artificial intelligence technology to push systems toward specified objectives. Utilizing sophisticated algorithms and assessing the system's performance in relation to the intended objective enables a system to make autonomous, localized decisions and take necessary action. This degree of autonomy in sensing, decision-making, and action defines an "intelligent" IoT system. A fully intelligent system can learn, generalize (if such a capability exists), amass information, establish goals and priorities, and reduce decision-making risks. Such algorithms outperform merely static programming instructions by generating data-driven predictions/decisions (Misra et al. 2020).

A scientific paper uses remote sensing, crop modeling, and machine learning to overcome current limitations in estimating maize crop production. MODIS NDVI, MODIS Land Surface Temperature (LST), and SMOS (Soil Moisture and Ocean Salinity) Surface Soil Moisture (SSM) were used to compute phenological metrics and drought and heat stress indices. In addition, in order to calibrate remote sensing-based models, the SARRA-O crop model (Baron et al. 2005) was used to simulate AGB-F (aboveground biomass at flowering), Cstr (water stress coefficient), and final maize yields, and ground-based data were used to confirm our findings. To get field data continuously, following the gathering of data, which includes a variety of factors or variables such as phytosanitary treatment, fertilization, climate (which includes five factors—temperature, precipitation, humidity, wind speed, and solar radiation), as well as information on irrigation and parcels (le Roux et al. 2019).

Several factors are associated. For instance, we may see that climate data and phytosanitary treatment parameters are connected. These data suggest that certain pathogens target trees at particular periods of the year (for example, *Ceratitis capitata*, which attacks citrus fruits in winter). In addition, the link between fertilization and climate is proven by the fact that the quantity of fertilizers needed throughout the year is often dependent on high temperatures or water stress, both of which are influenced by temperature. This reliance is crucial for the performance of prediction algorithms during training, and it may help us in feature engineering (Moussaid et al. 2022).

5.4 Yield Prediction

After the interpretation of the collected data, the next step is yield prediction. Crop production is important to the economy because it provides food, raw materials, and employment. Crop production now includes marketing, processing, distribution, and after-sales service. Yield prediction and agriculture monitoring are required for lower human intervention in practice and to assist in labor reduction and productivity increase. Food consumption is increasing daily, and it will be difficult to meet this demand unless contemporary agricultural methods are implemented. Artificial intelligence has been used in crop selection to assist farmers in picking harvestable crops and fertilizers that promote the highest growth. Moreover, artificial intelligence is considered the most important issue in precision agriculture because of its significant role in yield mapping, yield estimation, matching crop supply and demand, and crop management to increase production and reduce labor (Liakos et al. 2018; Chlingaryan et al. 2018).

Agriculture has new prospects for more accurate crop forecasts due to machine learning and simulation crop modeling advances. Using larger datasets affects algorithm performance and results in greater accuracy or a lower RMSE value. Each of these technologies has contributed distinctive capabilities and significant accomplishments, with the understanding that the use of larger datasets affects algorithm performance and results in greater accuracy or a lower RMSE value (Shah et al. 2018).

The prediction of the crop method predicts the optimal crop by analyzing multiple soil characteristics and meteorological conditions. Parameters to consider include soil type, depth, pH, organic carbon, phosphate, potassium, nitrogen, manganese, magnesium, sulfur, copper, iron, calcium, temperature, precipitation, and humidity (Eli-Chukwu 2019).

Here are some examples of ML applications currently under development: an effective, inexpensive, and nondestructive method for mechanically counting coffee fruits on a branch. The research categorized coffee fruits into three categories: harvestable, unharvestable, and those whose maturity stage was neglected. In addition, the method computed the weight and maturity percentage of the coffee fruit. This project aimed to provide coffee growers with information that would enable them to maximize economic benefits and coordinate agricultural operations. The outcomes of this research will motivate the development of a new generation of instruments for use by coffee producers. It is an effective, nondestructive, and cost-effective method that gives critical data for planning agricultural activities and gaining economic benefits from resource management (Martins et al. 2019).

Another study proposed a method for tomato identification based on electromagnetic (EM) and remotely sensed red–green–blue (RGB) images obtained by an unmanned aerial vehicle (UAV). The authors used ANNs and multitemporal remote sensing data in a separate study to build a model for predicting grassland biomass (kg dry matter/ha/day). Another generalized approach for estimating agricultural yields based on an ENN application to long-period agronomical data (1997–2014) has been developed. The research focuses on regional predictions (especially in Taiwan) to aid farmers in avoiding market supply and demand mismatches caused or accelerated by harvest crop quality (yue 2018; Weerakkody and Mawalagedera 2020).

In addition, the scientists constructed a method for forecasting the phases of rice growth based on support vector machine (SVM) and basic geographic data collected from meteorological stations. Incorporating field-specific rainfall data and meteorological parameters for each region is another use of machine learning in rice production prediction. Adjusting ANNs affects the accuracy of rice yield forecasts.

Another effort is to build a system that reduces the amount of physical labor necessary for harvesting and handling. As part of their yield prediction system, they designed a machine vision system to shake and catch cherries during harvest automatically. The algorithm segments and identifies even when densely leafed cherry branches are hidden. In another research, the scientists developed a strategy for detecting immature green citrus in an outdoor citrus orchard using early yield mapping. As with other prior studies of this kind, the objective of this investigation was to provide farmers with yield-specific data to assist them in maximizing their orchards in terms of profit and productivity. Another research focused on yield prediction, specifically wheat yield prediction, built a system that utilizes satellite imagery to give integrated crop development characteristics and soil data for more accurate predictions (Liakos et al. 2018).

There is also a groundnut yield prediction study in which weather data play a key role in determining agricultural crop yield statistics. On 8 years of groundnut data, researchers used the K-nearest neighbor (KNN) method to evaluate the results of multiple linear regression, regression tree, K-nearest neighbor, and artificial neural Network. They made their predictions based on soil, environmental, and abiotic parameters (Shah et al. 2018).

5.5 Decision-Making

Combining agriculture with AI has shown a great opportunity with the potential to tackle a number of agricultural production efficiency concerns.

Artificial intelligence (AI) studies theory and computer systems capable of performing tasks that normally require human intellect, such as sensory perception and decision-making. Conventional agricultural operations have been replaced by automation, giving the end user real-time data analysis for more precise and fast decisions.

Moreover, compared to conventional agricultural process management systems, AI techniques simplify data processing and give the end user increased decision-making skills. Digital agriculture, which encompasses agri-technology and precision agriculture, is the consequence of the use of data-driven technologies in agriculture, which increases agricultural output while lowering environmental impacts (Liakos et al. 2018).

AI, or cognitive-based technology, is the most disruptive and powerful advanced analytics tool enterprises may utilize to make supply chain decisions (Liakos et al. 2018).

Zhai et al. (2020) assessed 13 decision support systems (DSSs) based on recognized criteria for accessibility, scalability, interoperability, and other variables, highlighting mainly graphical user interface (GUI) and system operations improvement opportunities.

Kanter et al. (2015) developed a framework that included key aspects of agricultural processes, involved stakeholders and the supply chain, and focused on the sociotechnical and socioeconomic drivers and challenges of big data and AI, which they perceived to exist in two contexts: developed countries with a severe labor shortage and developing countries with a nascent technology infrastructure.

Huang et al. (2018) examine the possibility of automated identification and interpretation of satellite images of the Earth using big data analysis and AI. Jha et al. (2019) investigated the current status of automation and agricultural applications of artificial neural networks, machine learning, and IoT. Using machine learning, they also presented an IoT-based system for flower and leaf identification (Zhai et al. 2020).

A decision support system (DSS) is a computerized system that gathers, analyzes, and synthesizes data to provide integrated data reports. In agriculture, the DSS is an interactive visualization system that collects and processes multiple inputs (raw data from farm-deployed sensors, compound data, agronomic data, and microbe data) to provide the farm manager with a solution to a deriving problem, such as disease pressure, ice, and snow, or the need to irrigate. A DSS is not a computer that makes choices but helps managers make better judgments (Kaur et al. 2022).

As the name implies, the DSS is a system that supports the end user in achieving the final goals from various sources. It collects and analyzes data from various sources before generating relevant results. Decision support systems (DSSs) in agriculture collect and analyze data from many sources to offer end users visibility into their critical decision-making responsibilities (Kaur et al. 2022).

These technologies assist farmers in addressing complex crop yield concerns in the agricultural arena. In this regard, DSSs are critical components of contemporary agriculture. Better decision-making, faster problem-solving, and higher efficiency in dealing with difficulties, operations, planning, and management are all aided by decision support systems (Kaur et al. 2022).

6 Role of the Decision Support System in Agriculture

Big Data–Based Yield Prediction

A strategy for estimating agricultural production using machine learning techniques has been developed under the paradigm of big data computing. Therefore, yield forecast accuracy is a crucial problem that must be addressed. Farmers worldwide are usually perplexed when it comes to making sound decisions, but big data helps to maximize production and boost the economic sector to improve productivity. In extreme circumstances, the forecast will also assist farmers in making choices, such as selecting alternative crops or abandoning a crop at an early stage. Consequently, crop yield must be simulated and anticipated prior to cultivation for effective crop management and desired outcomes. By preserving the experience of farmers, weather conditions, and other influencing elements in a massive database, early prediction is possible. Typical input factors include precipitation, temperature, humidity, solar radiation, crop population density, fertilizer application, irrigation, tillage, soil type, depth, farm capacity, and soil organic matter. Recently, projecting crop yield at the field level has become increasingly prevalent. Weather has the biggest influence on agricultural yield. If weather prediction gets more precise, farmers can be warned far in advance, saving significant losses and helping economic growth. Due to the nonlinear relationship between agricultural production and its influencing variables, machine learning techniques may be useful for yield prediction (Palanivel and Surianarayanan 2019).

Cutting-edge technologies like blockchain, IoT, big data, and machine learning (ML) are required to provide sustainable agricultural output. The majority of the world's food problems could be solved if farms had access to massive amounts of data and used it to guide agricultural decisions. If farmers had access to datasets or maps for various environmental parameters around the farm, they might implement practices like smart farming, precision farming, and vertical farming. Scientific evidence shows that data-driven farming improves crop yields, cuts expenses, and guarantees the industry's long-term viability (Bhat and Huang 2021).

Big data can be categorized into three types: geospatial data, metadata, and telematics data. Geospatial data refers to information about items, events, or phenomena that have a physical location on the planet's surface. Metadata includes information such as application dates for pesticides, herbicides, and fertilizers; cultivar selection; and the depth at which agricultural seed was sowed. Telematics data includes collected agricultural equipment (sensors installed on combines, tractors, sprayers, etc.) and sensors indicating the fuel and maintenance needed by machinery to accomplish a certain task. It also includes how much time a piece of equipment has spent doing a job (Bhat and Huang 2021).

Big data is the most advanced stage of development of technology in the agricultural sector; it encompasses ideas, tools, and procedures used in various settings, from fieldwork to crop planting. According to these same authors, informatization, intelligence, and precision could be the key to overcoming the problems plaguing conventional farming methods. However, the study of agricultural big data is just getting started, so there is a lot of room for exploration (Cravero et al. 2022).

The most crucial enabling technologies included software-based decision support systems (which collect and analyze data to address system dynamics and optimization issues), sensors (which collect data on the functioning of agricultural equipment and resources), and digital communication tools. Some believe that the future of agriculture depends heavily on technologies that can be used in the field to collect data in real time, such as geographic information and geo-locating systems. In reference to the various approaches to analyzing big data, sensors attached to farm machinery could listen to the vibrations of the ground and detect any abnormalities in the machinery's operation through sound waves. Second, using historical weather data, it may be possible to foretell the weather in the future using predictive analytics. Third, social media analytics are becoming increasingly important as social media's role in everyday life expands. The data collected through these channels may be used as a supplementary resource for decision-making. It is possible to collect data from these platforms and use "big data" methods to analyze it. In order to find useful information, AI algorithms sift through mountains of text. Video analytics: cameras on machinery use artificial intelligence to recognize patterns (Lassoued et al. 2021).

Fan et al. (2015) developed a big data analytics-based method for enhancing the accuracy of agricultural production forecasts. The "Big data prediction of agricultural output" study describes a reliable infrastructure for handling big data in agriculture. It consists of three key components. A MapReduce weather data processing component computes enormous datasets on a computer cluster. The second step involves selecting comparable years using the closest neighbors, and the last step involves developing an autoregressive moving average (ARMA) model based on the most recent year and generating the forecast number. The experimental assessment of the previous paper reveals that the nearest-neighbor strategy is effective, suggesting that crop yield is strongly tied to weather patterns.

AI is increasingly being used in agriculture to increase production and effectiveness. AI applications in agriculture include yield prediction algorithms, image recognition algorithms for plant pest and disease diagnosis, and agricultural harvesting robots. For instance, autonomous AI robots can pick up food at a higher capacity and quicker pace than human labor. Yield prediction algorithms utilize meteorological and historical yield data to accurately predict crop yields. Image recognition algorithms can detect pests and diseases in plants quickly and accurately, allowing for timely interventions that can save crops from destruction. Agricultural harvesting robots are able to pick up food at a higher capacity and quicker pace than human labor, increasing efficiency in the harvesting process (Al-Turjman 2019).

While AI applications in agriculture are expanding, our survey anticipates that AI deployment will contribute 35%, significantly (33%), or substantially (23%) to the sector as a whole. AI will assist agricultural equipment and logistics, according to 90% of respondents; market information, according to 89%; plant breeding, according to 81%; and risk management, according to 71%.

According to experts, applying digital decision-support systems, big data analytics, and artificial intelligence may increase agricultural productivity in several ways, including climate forecasting, yield prediction, crop selection, and disease/pest control. For the foreseeable future, however, the advantages of big data technologies will likely be used by developed countries that make substantial upfront expenditures. According to experts, big data and AI will enhance global food security by lowering crop losses and increasing crop yields (Lassoued et al. 2021).

7 Conclusion

Successful farming requires monitoring crops, predicting yields, and smart irrigation systems that rely on artificial intelligence. It has been proven through excellent prototypes and solutions to match the requirements of the current scenario after analyzing many of the existing systems. In forecasting agricultural productivity and irrigation requirements, AI-based methods, along with additional hardware components such as Raspberry Pi, soil moisture sensors, temperature, and moisture content, have been shown to play an important role in predicting productivity and proper intelligent irrigation and flour. This plays a major role in reducing the total agricultural costs and the amount of wasted resources such as water and energy during the agricultural process, which improves economic growth, and reduces the wastage of major resources. These methods are excellent in reducing human efforts and speeding up the planning of agricultural practices. However, reducing the cost of implementing the system and training unskilled and technically illiterate farmers remains challenging.

References

Abioye EA, Abidin MSZ, Mahmud MSA, Buyamin S, Ishak MHI, Abd Rahman MKI et al (2020) A review on monitoring and advanced control strategies for precision irrigation. Comput Electron Agric 173:105441

Abdulridha J, Ampatzidis Y, Ehsani R, de Castro AI (2018) Evaluating the performance of spectral features and multivariate analysis tools to detect laurel wilt disease and nutritional deficiency in avocado. Comput Electron Agric 155:203–211

Aggarwal A, Kumar S, Bhatt A, Shah MA (2022) Solving user priority in cloud computing using enhanced optimization algorithm in workflow scheduling. Comput Intell Neurosci: 2022

Allmendinger A, Spaeth M, Saile M, Peteinatos GG, Gerhards R (2022) Precision chemical weed management strategies: a review and a design of a new CNN-based modular spot sprayer. Agronomy 12(7):1620

Al-Arif SMMR, Ferdous AHMI, Nijami SH (2012) Comparative study of different path planning algorithms: a water based rescue system. Int J Comput Appl 39

Albattah W, Nawaz M, Javed A, Masood M, Albahli S (2022) A novel deep learning method for detection and classification of plant diseases. Complex Intell Syst:1–18

Alexandratos N (1995) The outlook for world food and agriculture to the year 2010. Islamochristiana 1995:25–48

Altalak M, Ammad uddin M, Alajmi A, Rizg A (2022) Smart agriculture applications using deep learning technologies: a survey. Appl Sci 12(12):5919

Al-Turjman F (2019) The road towards plant phenotyping via WSNs: an overview. Comput Electron Agric 161:4–13

Anderson PK, Cunningham AA, Patel NG, Morales FJ, Epstein PR, Daszak P (2004) Emerging infectious diseases of plants: pathogen pollution, climate change and agrotechnology drivers. Trends Ecol Evol 19(10):535–544

Aniley AA, Kumar N, Kumar A (2017) Soil temperature sensors in agriculture and the role of nanomaterials in temperature sensors preparation. Int J Eng Manuf Sci 7(2):2249–3115

Babaeian E, Sadeghi M, Jones SB, Montzka C, Vereecken H, Tuller M (2019) Ground, proximal, and satellite remote sensing of soil moisture. Rev Geophys 57(2):530–616

Baron C, Sultan B, Balme M, Sarr B, Traore S, Lebel T, Dingkuhn M (2005) From GCM grid cell to agricultural plot: scale issues affecting modelling of climate impact. Philos Trans R Soc B Biol Sci 360(1463):2095–2108

Bestelmeyer BT, Marcillo G, McCord SE, Mirsky S, Moglen G, Neven LG et al (2020) Scaling up agricultural research with artificial intelligence. IT Prof 22(3):33–38

Bhat SA, Huang NF (2021) Big data and ai revolution in precision agriculture: survey and challenges. IEEE Access 9:110209–110222

Brun P, Zimmermann NE, Hari C, Pellissier L, Karger DN (2022) Global climate-related predictors at kilometer resolution for the past and future. Earth Syst Sci Data 14(12):5573–5603

Cai Q, He B, Kogel KH, Jin H (2018) Cross-kingdom RNA trafficking and environmental RNAi—nature's blueprint for modern crop protection strategies. Curr Opin Microbiol 46:58–64

Chlingaryan A, Sukkarieh S, Whelan B (2018) Machine learning approaches for crop yield prediction and nitrogen status estimation in precision agriculture: a review. Comput Electron Agric 151:61–69

Cravero A, Pardo S, Sepúlveda S, Muñoz L (2022) Challenges to use machine learning in agricultural big data: a systematic literature review. Agronomy 12(3):748

Cruz AC, Luvisi A, De Bellis L, Ampatzidis Y (2017) X-FIDO: An effective application for detecting olive quick decline syndrome with deep learning and data fusion. Front Plant Sci 8:1741

Cruz A, Ampatzidis Y, Pierro R, Materazzi A, Panattoni A, De Bellis L, Luvisi A (2019) Detection of grapevine yellows symptoms in Vitis vinifera L. with artificial intelligence. Comput Electron Agric 157:63–76

Dąbrowska-Zielińska K, Ciołkosz A, Budzyńska M, Kowalik W (2008) Monitorowanie wzrostu i plonowania zbóż metodami teledetekcji. Problemy Inżynierii Rolniczej 16:45–54

Dabrowska-Zielinska K, Budzynska M, Tomaszewska M, Malinska A, Gatkowska M, Bartold M, Malek I (2016) Assessment of carbon flux and soil moisture in wetlands applying Sentinel-1 data. Remote Sens 8(9):756

Dent D, Binks RH (2020) Insect pest management. CabinetMaker

Deutsch CA, Tewksbury JJ, Tigchelaar M, Battisti DS, Merrill SC, Huey RB, Naylor RL (2018) Increase in crop losses to insect pests in a warming climate. Science 361(6405):916–919

Dharmaraj V, Vijayanand C (2018) Artificial intelligence (AI) in agriculture. Int J Curr Microbiol App Sci 7(12):2122–2128

Eli-Chukwu NC (2019) Applications of artificial intelligence in agriculture: a review. Eng Technol Appl Sci Res 9(4):4377–4383

Etienne A, Saraswat D (2019, May) Machine learning approaches to automate weed detection by UAV based sensors. In: Autonomous air and ground sensing systems for agricultural optimization and phenotyping IV, vol. 11008. SPIE, pp 202–215). SPIE

European Food Safety Authority (2015) The 2013 European Union report on pesticide residues in food. EFSA J 13(3):4038

Fabiyi OA, Alabi RO, Ansari RA (2020) Nanoparticles' synthesis and their application in the management of phytonematodes: an overview. In: Management of phytonematodes: recent advances and future challenges, pp 125–140

Fan S, Lau RY, Zhao JL (2015) Demystifying big data analytics for business intelligence through the lens of marketing mix. Big Data Res 2(1):28–32

Fathi M, Haghi Kashani M, Jameii SM, Mahdipour E (2021) Big data analytics in weather forecasting: a systematic review. Arch Comput Methods Eng:1–29

Fernández-Quintanilla C, Peña JM, Andújar D, Dorado J, Ribeiro A, López-Granados F (2018) Is the current state of the art of weed monitoring suitable for site-specific weed management in arable crops?. Weed Res 58(4):259–272

Flint ML, Van den Bosch R (2012) Introduction to integrated pest management. Springer

Friedrich J, Klingner S, Becker M, Schneider M (2016) IPS–an information production system for precision farming. landtechnik 71(2):35–43

García L, Parra L, Jimenez JM, Lloret J, Lorenz P (2020) IoT-based smart irrigation systems: an overview on the recent trends on sensors and IoT systems for irrigation in precision agriculture. Sensors 20(4):1042

Gianessi L, Reigner N (2006) The importance of fungicides in US crop production. Outlooks Pest Manag 17(5):209

Gianessi LP, Reigner NP (2007) The value of herbicides in US crop production. Weed Technol 21(2):559–566

Gondchawar N, Kawitkar RS (2016) IoT based smart agriculture. Int J Adv Res Comput Commun Eng 5(6):838–842

Hanson AMGJ, Joel MG, Joy A, Francis J (2017) Plant leaf disease detection using deep learning and convolutional neural network. Int J Eng Sci 5324:2–4

Huang X, Liu MY, Belongie S, Kautz J (2018) Multimodal unsupervised image-to-image translation. In: Proceedings of the European conference on computer vision (ECCV), pp 172–189

Huang X, Zhu Y, Kianfar E (2021) Nano biosensors: properties, applications and electrochemical techniques. J Mater Res Technol 12:1649–1672

IAEA (2008) https://www.iaea.org/sites/default/files/publications/reports/2008/anrep2008_full.pdf
Jain S, Vani KS (2018) A survey of the automated irrigation systems and the proposal to make the irrigation system intelligent. Int J Comput Sci Eng 6:357–360
Jha K, Doshi A, Patel P, Shah M (2019) A comprehensive review on automation in agriculture using artificial intelligence. Artif Intell Agric 2:1–12
Jin X, Kumar L, Li Z, Feng H, Xu X, Yang G, Wang J (2018) A review of data assimilation of remote sensing and crop models. Eur J Agron 92:141–152
Jones JW, Hansen JW, Royce FS, Messina CD (2000) Potential benefits of climate forecasting to agriculture. Agric Ecosyst Environ 82(1–3):169–184
Kait LK, Kai CZ, Khoshdelniat R, Lim SM, Tat EH (2007) Paddy growth monitoring with wireless sensor networks. In: 2007 international conference on intelligent and advanced systems. IEEE, pp 966–970
Kanter R, Walls HL, Tak M, Roberts F, Waage J (2015) A conceptual framework for understanding the impacts of agriculture and food system policies on nutrition and health. Food Secur 7(4):767–777
Kaur P, Joshi JC, Aggarwal P (2022) A multi-model decision support system (MM-DSS) for avalanche hazard prediction over north-west Himalaya. Nat Hazards 110(1):563–585
Kaushal M, Wani SP (2017) Nanosensors: frontiers in precision agriculture. In: Nanotechnology. Springer, Singapore, pp 279–291
Khaki S, Pham H, Wang L (2021) Simultaneous corn and soybean yield prediction from remote sensing data using deep transfer learning. Sci Rep 11(1):1–14
Kim HJ, Sudduth KA, Hummel JW (2009) Soil macronutrient sensing for precision agriculture. J Environ Monit 11(10):1810–1824
Klerkx L, Jakku E, Labarthe P (2019) A review of social science on digital agriculture, smart farming and agriculture 4.0: new contributions and a future research agenda. NJAS-Wageningen J Life Sci 90:100315
Klingman GC (1961) Weed control: as a science. In: Weed control: as a science. Wiley, New York
Kodali RK, Sahu A (2016) An IoT based soil moisture monitoring on Losant platform. In: 2016 2nd international conference on contemporary computing and informatics (IC3I). IEEE, pp 764–768
Kogan FNF (1990) Remote sensing of weather impacts on vegetation in non-homogeneous areas. Int J Remote Sens 11:1405–1419
Koren K, Brodersen KE, Jakobsen SL, Kühl M (2015) Optical sensor nanoparticles in artificial sediments–a new tool to visualize O2 dynamics around the rhizome and roots of seagrasses. Environ Sci Technol 49(4):2286–2292
Lamb DW, Brown RB (2001) Pa—precision agriculture: remote-sensing and mapping of weeds in crops. J Agric Eng Res 78(2):117–125
Lassoued R, Macall DM, Smyth SJ, Phillips PW, Hesseln H (2021) Expert insights on the impacts of, and potential for, agricultural big data. Sustainability 13(5):2521
Launay M, Guerif M (2005) Assimilating remote sensing data into a crop model to improve predictive performance for spatial applications. Agric Ecosyst Environ 111(1–4):321–339
Lazo JK, Morss RE, Demuth JL (2009) 300 billion served: sources, perceptions, uses, and values of weather forecasts. Bull Am Meteorol Soc 90(6):785–798
le Roux L, Juba R, Mbanjwa W (2019) Exploring current and emerging irrigation and drainage management to reduce the impact of extreme events and mitigate droughts and floods. WRC Report, (2861/1/18)
Lehnert M (2014) Factors affecting soil temperature as limits of spatial interpretation and simulation of soil temperature. Acta Universitatis Palackianae Olomucensis–Geographica 45(1):5–21
Li Z, Yu T, Paul R, Fan J, Yang Y, Wei Q (2020) Agricultural nanodiagnostics for plant diseases: recent advances and challenges. Nanoscale Adv 2(8):3083–3094
Liakos KG, Busato P, Moshou D, Pearson S, Bochtis D (2018) Machine learning in agriculture: a review. Sensors 18(8):2674

Liu YC, Gao HM (2016) Development and applications of mobile farming information system for food traceability in health management. In: Applied computing in medicine and health. Morgan Kaufmann, pp 244–268

Liu JY, Chen CH, Lin CH, Tsai HF, Chen CH, Kamogawa M (2011) Ionospheric disturbances triggered by the 11 March 2011 M9. 0 Tohoku earthquake. J Geophys Res Space Physics 116(A6)

Martins PMM, Ribeiro LS, Miguel MGDCP, Evangelista SR, Schwan RF (2019) Production of coffee (Coffea arabica) inoculated with yeasts: impact on quality. J Sci Food Agric 99(13):5638–5645

McAllister W, Osipychev D, Davis A, Chowdhary G (2019) Agbots: weeding a field with a team of autonomous robots. Comput Electron Agric 163:104827

McCarthy C, Rees S, Baillie C (2010, January) Machine vision-based weed spot spraying: a review and where next for sugarcane?. In: Proceedings of the 32nd Annual Conference of the Australian Society of Sugar Cane Technologists (ASSCT 2010), vol 32, pp. 424–432)

Möller J (2010) Computer vision-a versatile technology in automation of agricultural machinery. J Agric Eng 47(4):28–36

Misra NN, Dixit Y, Al-Mallahi A, Bhullar MS, Upadhyay R, Martynenko A (2020) IoT, big data and artificial intelligence in agriculture and food industry. IEEE Internet Things J

Moussaid D, Khallouk K, El Khalfaouy R, Tagnaouti Moumnani F, Kherbeche A, Barakat A (2022) Solution combustion synthesis of β-Cu2V2O7 nanoparticles: photocatalytic degradation of crystal violet under UV and visible light illumination. React Kinet Mech Catal 135(5):2797–2812

Mulla DJ (2013) Twenty five years of remote sensing in precision agriculture: key advances and remaining knowledge gaps. Biosyst Eng 114(4):358–371

Nazari M, Oroojlooy A, Snyder L, Takác M (2018) Reinforcement learning for solving the vehicle routing problem. Adv Neural Inf Proces Syst 31

Ochoa-García H, Rist S (2018) Water justice and integrated water resources management: constitutionality processes favoring sustainable water governance in Mexico. Hum Ecol 46(1):51–64

Oerke EC (2006) Crop losses to pests. J Agric Sci 144(1):31–43

Palanivel K, Surianarayanan C (2019) An approach for prediction of crop yield using machine learning and big data techniques. Int J Comput Eng Technol 10(3):110–118

Parolini G (2022) Weather, climate, and agriculture: historical contributions and perspectives from agricultural meteorology. Wiley Interdiscip Rev Clim Chang 13(3):e766

Patil SS, Thorat SA (2016) Early detection of grapes diseases using machine learning and IoT. In: 2016 second international conference on cognitive computing and information processing (CCIP). IEEE, pp 1–5

Partel V, Nunes L, Stansly P, Ampatzidis Y (2019) Automated vision-based system for monitoring Asian citrus psyllid in orchards utilizing artificial intelligence. Comput Electron Agric 162:328–336

Peng HI, Miller BL (2011) Recent advancements in optical DNA biosensors: exploiting the plasmonic effects of metal nanoparticles. Analyst 136(3):436–447

Pereira LB, Gambarini VMDO, de Menezes AB, Ottoboni LM, Vicentini R (2022) Influence of sugarcane variety on rhizosphere microbiota under irrigated and water-limiting conditions. Curr Microbiol 79(9):1–10

Piekarczyk J (2011) Szacowanie plonów roślin uprawnych na podstawie naziemnych pomiarów spektralnych. Teledetekcja Środowiska 46:23–28

Piekarczyk J, Sulewska H, Szymańska G (2011) Winter oilseed-rape yield estimates from hyperspectral radiometer measurements. Quaest Geogr 30:77–84

Prasad AK, Chai L, Singh RP, Kafatos M (2006) Crop yield estimation model for Iowa using remote sensing and surface parameters. Int J Appl Earth Obs Geoinf 8(1):26–33

Prasad R, Kumar V, Prasad KS (2014) Nanotechnology in sustainable agriculture: present concerns and future aspects. Afr J Biotechnol 13(6):705–713

Pretty J, Pervez Bharucha Z (2015) Integrated pest management for sustainable intensification of agriculture in Asia and Africa. Insects 6(1):152–182

Ramankutty N, Mehrabi Z, Waha K, Jarvis L, Kremen C, Herrero M, Rieseberg LH (2018) Trends in global agricultural land use: implications for environmental health and food security. Annu Rev Plant Biol 69(1):789–815
Roopaei M, Rad P, Choo KKR (2017) Cloud of things in smart agriculture: intelligent irrigation monitoring by thermal imaging. IEEE Cloud Computing 4(1):10–15
Schneider M, Barthlott S, Hase F, González Y, Yoshimura K, García OE et al (2012) Ground-based remote sensing of tropospheric water vapour isotopologues within the project MUSICA. Atmos Meas Tech 5(12):3007–3027
Selid PD, Xu H, Collins EM, Striped Face-Collins M, Zhao JX (2009) Sensing mercury for biomedical and environmental monitoring. Sensors 9(7):5446–5459
Shah S, Dey D, Lovett C, Kapoor A (2018) Airsim: high-fidelity visual and physical simulation for autonomous vehicles. In: Field and service robotics. Springer, Cham, pp 621–635
Shepherd M, Turner JA, Small B, Wheeler D (2020) Priorities for science to overcome hurdles thwarting the full promise of the 'digital agriculture' revolution. J Sci Food Agric 100(14):5083–5092
Simionesei L, Ramos TB, Palma J, Oliveira AR, Neves R (2020) IrrigaSys: a web-based irrigation decision support system based on open source data and technology. Comput Electron Agric 178:105822
Sladojevic S, Arsenovic M, Anderla A, Culibrk D, Stefanovic D (2016) Deep neural networks based recognition of plant diseases by leaf image classification. Comput Intell Neurosci:2016
Stafford JV (2000) Implementing precision agriculture in the 21st century. J Agric Eng Res 76(3):267–275
Swain KC, Thomson SJ, Jayasuriya HP (2010) Adoption of an unmanned helicopter for low-altitude remote sensing to estimate yield and total biomass of a rice crop. Trans ASABE 53(1):21–27
Swain KC, Zaman QU (2012) Rice crop monitoring with unmanned helicopter remote sensing images. In: Remote sensing of biomass-principles and applications, pp 253–272
Talaviya T, Shah D, Patel N, Yagnik H, Shah M (2020) Implementation of artificial intelligence in agriculture for optimisation of irrigation and application of pesticides and herbicides. Artif Intell Agric 4:58–73
Touil S, Richa A, Fizir M, Argente García JE, Skarmeta Gomez AF (2022) A review on smart irrigation management strategies and their effect on water savings and crop yield. Irrig Drain
Valdés MG, Valdés González AC, García Calzón JA, Díaz-García ME (2009) Analytical nanotechnology for food analysis. Microchim Acta 166:1–19
Valente A, Morais R, Tuli A, Hopmans JW, Kluitenberg GJ (2006) Multi-functional probe for small-scale simultaneous measurements of soil thermal properties, water content, and electrical conductivity. Sensors Actuators A Phys 132(1):70–77
Vinayak N, Malavade PK (2016) Akulwar, role of IoT in agriculture. In: IOSR Journal of Computer Engineering, National Conference on "Changing Technology and Rural Development
Weerakkody WAP, Mawalagedera SMMR (2020) Recent developments in vegetable production technologies in Sri Lanka. In: Agricultural research for sustainable food systems in Sri Lanka. Springer, Singapore, pp 189–214
Wójtowicz A, Wójtowicz M, Piekarczyk J (2005) Zastosowanie teledetekcji do monitorowania i oceny produktywnosci plantacji rzepaku. Rośliny Oleiste-Oilseed Crops 26(1):269–276
Wójtowicz M, Wójtowicz A, Piekarczyk J (2016) Application of remote sensing methods in agriculture. Commun Biometry Crop Sci 11(1):31–50
Yue J, Feng H, Jin X, Yuan H, Li Z, Zhou C, Tian Q (2018) A comparison of crop parameters estimation using images from UAV-mounted snapshot hyperspectral sensor and high-definition digital camera. Remote Sens 10(7):1138
Zhai Z, Martínez JF, Beltran V, Martínez NL (2020) Decision support systems for agriculture 4.0: survey and challenges. Comput Electron Agric 170:105256

Plant-Based Electrical Impedance Spectroscopy for Plant Health Monitoring

Lee Bar-On, Umberto Garlando, Harpreet Singh, Adi Avni, Nir Sade, Yarden Zeron, Danilo Demarchi, and Yosi Shacham-Diamand

Abstract Food security and increasing agricultural yields have become one of this century's most essential and challenging topics. The global population is projected to reach 9.7 billion by 2050, so food production must increase significantly to meet the growing demand. Increasing agricultural yields is one of the ways to address the issue of food security. This can be achieved through various means, such as improving crop varieties, using better agricultural practices, and adopting advanced technologies such as precision agriculture and genetically modified crops. One of the ways to promote this is to improve understanding and activity within plants using electrical methods. This was the objective of the presented research. In this research, a hypothesis for signal conduction through the plant medium is suggested, modeled, and characterized. The results show that this approach could be included where the plant is used as the actual sensor, and changes in its internal activity indicate changes in the environment and the plant's needs. It hereby allows the detection of water stress, different daylight conditions, and possibly future pathogenic attacks. Another new theoretical representation and approach were also presented and supported with various experimental methods showing that the plant's physiological response and status can be derived from its electrical characteristics, similar to methods used in plant physiology studies. It paves the path for designing and applying new sensing technologies to promote plant monitoring and serve as an additional method in precision agriculture.

Keywords Plant impedance · Spectroscopy · Plant sensor system · Plant electrical model · Plant equivalent circuit · Precision agriculture

---

L. Bar-On · H. Singh · A. Avni · N. Sade · Y. Zeron
Tel Aviv University, Ramat Aviv, Israel

U. Garlando · D. Demarchi
Politecnico di Torino, Torino, Italy

Y. Shacham-Diamand (✉)
The Scojen Institute for Synthetic Biology Center, Reichman University, Herzliya, Israel
e-mail: yosef.shacham@runi.ac.il

P. M. Priyadarshan et al. (eds.), *Digital Agriculture*,
https://doi.org/10.1007/978-3-031-43548-5_16

1 Introduction

One of today's most significant challenges is to ensure food security for a rapidly increasing global population while resources, like land and water, are limited. The United Nations Food and Agriculture Organization's (UNFAO) latest report (Garlando et al. 2020) estimated that, by 2050, the world population will have reached over 10 billion. Simultaneously, trends like urbanization, economic changes, and migration will increase nutrition dependence on fruits and vegetables rather than cereal produce. Climate change threatens the limited natural resources needed for agricultural growth and production (Cervantes-Godoy et al. 2020).

The need for holistic solutions that involve a technological system approach is vital, according to the latest UNFAO report: "High-input, resource-intensive farming systems, which have caused massive deforestation, water scarcities, soil depletion and high levels of greenhouse gas emissions, cannot deliver sustainable food and agricultural production. Needed are innovative systems that protect and enhance the natural resource base while increasing productivity. Needed is a transformative process towards 'holistic' approaches, such as agroecology, agroforestry, climate-smart agriculture, and conservation agriculture, which also build upon indigenous and traditional knowledge. Technological improvements and drastic cuts in economy-wide and agricultural fossil fuel use would help address climate change and the intensification of natural hazards, which affect all ecosystems and every aspect of human life. Greater international collaboration is needed to prevent emerging threats to transboundary agriculture and food system, such as pests and diseases" (Garlando et al. 2020).

Providing food security, based on the availability of agricultural produce worldwide, has become of the highest significance. A few issues where closer crop monitoring will have an impact are as follows:

1. The amount of available food or agricultural produce
2. The nutritional level and quality of crops grown
3. The management of available produce: maximizing the number of crops that reach the population while minimizing the amount of produce that goes to waste

One of the ways to increase agricultural yield, improve crop quality, and meet these increasing demands for agricultural produce, or "farm to fork," is the incorporation of new technologies and monitoring methods into agriculture. This field has been named "precision agriculture." Here, the combination of new accurate monitoring methods, i.e., technology-driven solutions, will allow data collection from areas, land, and when correctly used, leading to improved data analysis, decision-making, and crop management. These tasks depend on the quality and accuracy of the monitoring systems' data. Technological solutions need to be researched and developed by combining knowledge and studies in agriculture, plant biology, and different engineering fields. One aspect of monitoring plants in agriculture is sensor technology. Here, the development of new methods to sense changes in plant well-being is needed. This research attempts to improve the understanding of electrical behavior in a plant by incorporating electrical sensing methods for monitoring. The development of such devices and approaches relies directly on the electronic and physical interpretation of the biological and chemical processes involved. Applying an interdisciplinary approach to developing new monitoring tools for precision agriculture will promote and improve monitoring methods, advancing food security throughout communities and crops.

We describe here a novel approach for a study of the plant in electronic terms, the plant network methods, sensor technologies, and an example of the electrical methodology that can be applied.

1.1 Internet of Things and Precision Agriculture

The Internet of Things describes a network in which different devices can be connected and exchange information. It can be used for monitoring and controlling an environment, machines, or a set of devices. Examples of available systems today are smart homes, intelligent cities, and smart cars. The ability to monitor and exchange information in the Internet of Things depends on sensing changes and transmitting this information into a network. Measurement of such changes relies both on network abilities and communication, but before that, on the ability to sense the device or specimen to be monitored.

This suggests that the existing sensing technology and ability to collect signals that can indicate a change and interpret and understand these signals is crucial for monitoring any specimen and creating an Internet of Things.

The increasing demand for agricultural produce corresponds with the growth in the worldwide population. Consequently, the ability to monitor, forecast, and affect plant well-being and health is incredibly significant (Bar-On et al. 2019a). Precision agriculture refers to incorporating technology into agriculture to improve all aspects, including crop monitoring and quality, development, use of resources such as water. Here, the ability to collect information from crops in the field and adapt their care quickly depends on the available sensor technologies and data interpretation for improved decision-making.

Combining these concepts suggests creating a Plant-Internet-of-Things. Plants (green plants, trees, fields, etc.) will be the monitored specimen in the network and will provide information on their status and receive information from the network. The concept is to address aspects in the development of an intelligent sensor system that will be interconnected with an external sensing system, similar to the methods of body channel communication, where the body serves as both the sensing interface and a medium for the signal transmission within the network (Lodi et al. 2020). The plant and its monitored responses will act as the sensing interface and a medium for signal transmission within the network created. The system will then be interfaced with the cloud via a propriety network. Such a network requires the use of a range of sensor technologies.

The sensor connected to the plant could combine a range of existing sensors with novel electronic sensors using concepts for both sensing and signal transmission within the plant. Today, various electrical measurement techniques exist for assessing plant well-being. Despite spanning a range of measurement methods and technologies, very few techniques offer either a direct assessment of the plant status or the overall status of the plant. For a comprehensive sensor system and network, an assessment indicating the status of the entire plant is needed. This status should be measured directly from the plant rather than a local measurement, showing a localized change in the plant or its surroundings. Therefore, this work aims to create a new technological aspect for direct plant monitoring using electrical measurements. A review of existing direct sensing methods is shown in Fig. 1.

1.2 Sensors in Agriculture

Many sensors are employed for monitoring in agriculture (Luvisi 2016). They span across many technologies alongside different parameters that are measured to determine crop status, treatment, health, and resource provisions. Here, we refer to different sensing areas: (a) indirect plant sensors and (b) direct plant sensors. While indirect sensors refer to different devices that collect data from the plant surrounding environment, and according to those readings, plant treatment is adapted. Usually, such sensors include temperature sensing, humidity, and soil moisture. These sensors are positioned in physical proximity to the plant to reflect environmental information closest to the plant experience. Direct plant sensors, also called functional sensors, collect a signal directly from the plant and use it to indicate its status. The novelty is that the plant itself is used as the sensor, while the electronic device "reads" its change in signaling due to the change in its status. A review of existing technologies, new developments, and gaps in the field is brought here, focusing on electronic measurement methods employed in the area and newer upcoming technologies. Data acquisition and analysis methods are currently emerging. These include direct plant measurements and methods for monitoring the plant environment (Mogili and Deepak 2018; Walter et al. 2017). This work aims to focus on the approach of sensing changes within the plant in a direct manner as a measure of plant physiological status and well-being. Different direct monitoring methods have been reported. Among these are sophisticated imaging and radar technologies used to monitor visual changes in crop status (Luvisi 2016; Zhang and Willison 1991). Others focus on root behavior, soil quality, and trunk health for tree stability assessments, using rather costly and not field-deployable tools (Yongzong et al. 2016; Sambuelli et al. 2003). Plant leaf changes are another monitored area. Here, temperature detection using thermocouples or capacitance measurements, imaging technology, and various electrical measurements are translated for plant treatment adaption (Repo et al. 2000; Volkov 2000; Volkov et al. 2016; Zhao et al. 2013). Sap flow monitoring has also been reported. It is considered as it represents

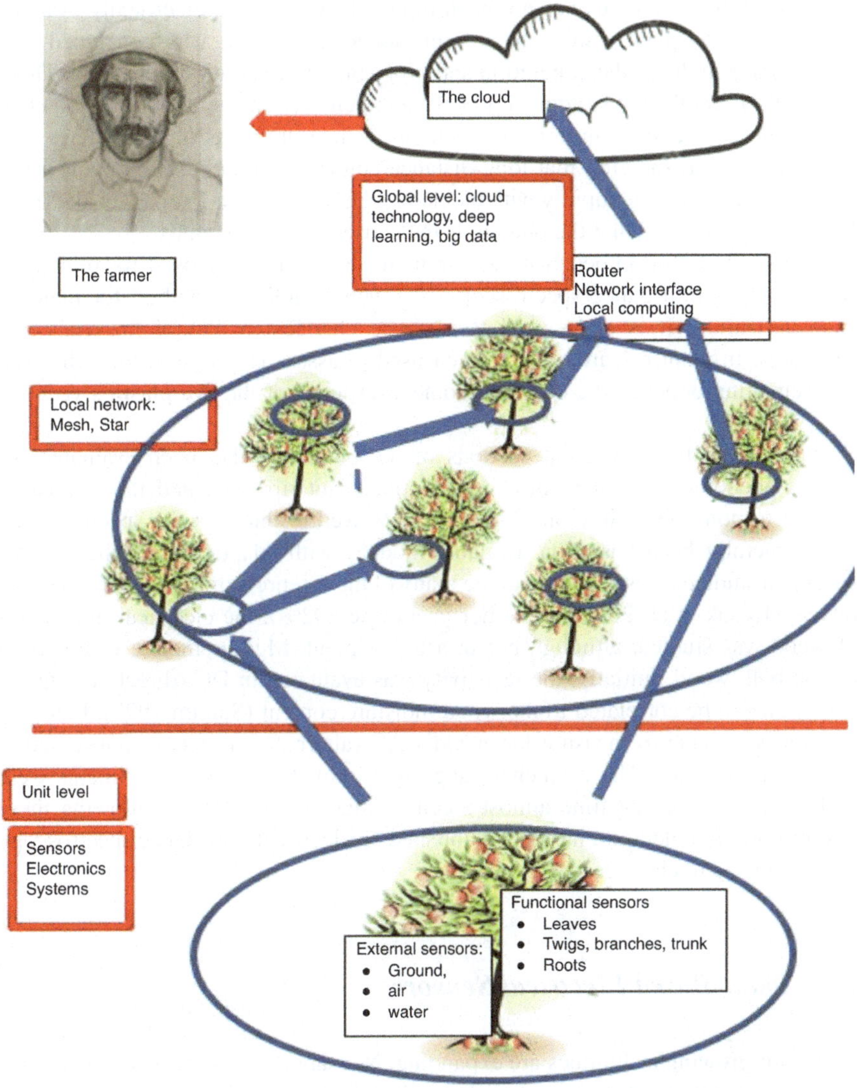

Fig. 1 Schematic description of "Plant-Internet-of-Things" (by Prof. Yosi Shacham, 2019)

the flow of nutrients toward the plant roots. However, it is plant-type-specific and seasonal.

In plant research, several electrical evaluations have been undertaken. The assessment of internal plant signaling among cells has been studied, showing action potentials and longer traveling signals called variation potentials (Volkov 2000; Volkov et al. 2016; Brown and Volkov 2006). These signals, however, indicate chemically induced changes or a response to local physical changes. Information on the overall plant status has not been reported. Attempts to electrically evaluate the response to plant trauma have been suggested, and it was concluded that local damage induces the generation and propagation of variation potentials. These potentials affect the physiological processes in plants (Repo et al. 2000; Zhao et al. 2013). Electrochemical and bioelectrochemical measurements of plants have also been carried out, showing that long-distance communication between plant tissue and cells can propagate rapidly with bioelectrochemical signals (Volkov et al. 2016). It has been suggested that the phloem is the carrier for these signals.

Sensors based on impedance measurements are commonly used for biological specimens, as impedance spectroscopy is a well-established method for material characterization. It is often used to evaluate changes in natural materials and structures. In plant research, it has been used to assess response at the cell level, differentiating between the cell membrane's response or the cytoplasm's vacuole, etc.

Plant tissue from different sections of the plant has also been evaluated for several phenomena, such as disease detection, fruit ripening, and the evaluation of frost response (Jócsák et al. 2019). Models used to interpret the spectrum data have generally been based on available models, with adaptations to the specific specimen studied. Jócsák et al. include many of these reports in their comprehensive review (Jócsák et al. 2019). In earlier years (the 1920s), the electrical impedance of wood was studied, although not in a living plant. Measurements were carried out on bulk wood. Initially, the resistivity was evaluated in DC (direct current). It was shown to be correlated to the wood moisture content (Stamm 1927). Later, an attempt was made to measure the wood's AC (alternate current) response across various frequencies. The frequency range was limited, as these experiments were undertaken for the first time almost a century ago (Luyet 1932). Following these findings, limited literature has been published on electrical impedance spectroscopy measurements to characterize live plants.

1.3 Plant-Based Electrical Sensors

New crop-growing techniques are expanding. New approaches are available among growing greenhouse farms, including vertical farming, lighting technologies for indoor growth, hydroponics, and many more. Although these environments are well monitored and allow almost complete control of the growing conditions, they still rely on plant well-being measures derived from the environment, prior research, or product feedback (once we taste the tomato, we decide whether growing conditions should be adapted). Yet, there need to be more technology-orientated methods that monitor plant health directly from a growing plant to adjust its nutrient and conditions. This implies that early detection of plant physiological change, alongside the ability to adapt treatment before the crop is produced and distributed, may significantly improve overall agriculture yield. Furthermore, across outdoor farmlands, where it is impossible to control the environment and predict climate

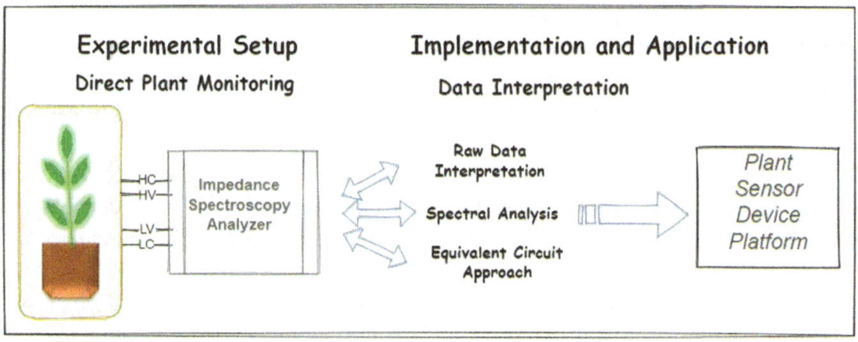

Fig. 2 Monitoring methods that read a direct plant signal have been presented

fluctuations, early detection of plant response, resulting in the ability to adapt irrigation and fertilization, would even more affect yield for the better (Fig. 2).

In this research, we suggest incorporating a sensing method that follows the behavior in the plant vascular cambium, mounted onto the plant stem, while observing the readings continuously and examining changes in induced external stress factors. The approach suggests exploring the plant's basic anatomy and physiology and applying physics and engineering modeling methods and measurement techniques adapted to the plant structure. These applications are expected to allow a more rigorous quantitative assessment of plant physiological status, which can be adapted for a complete electronic system of sensors in the future.

The work aims to improve our ability to describe the plant structure and internal changes that indicate a physiological change in electronic terms. Once these terms are defined, they could be applied to any species of plants in the future, ranging from small shrubs to large trees, as the primary plant anatomical structure across the species is similar (Taiz and Zeiger 2010).

2 Impedance Spectroscopy for Plant Monitoring

Impedance is defined as the ability of a material to resist the flow of electric current through it. It can be measured across frequencies, yielding a response of the material or system to different excitations or using singular frequency measurements. Intrinsically, the impedance at different frequencies will estimate the ability of dipoles in the material to respond to the excitation. The magnitude of change and the response time will be incorporated into the term of impedance values. The response of the system or material across frequencies is named electrical impedance spectroscopy. Impedance spectroscopy is often used to characterize a material or a design and asses its frequency response. The measured frequency values are determined according to the studied specimen, expected physical affects, and measurement setup considerations. Impedance measurements are commonly used to assess change in a material due to an external change, shock, wear over time, and more. It is also often applied for sensing applications, which consider biological tissue or material changes. These detectable changes can often be used to assess a physiological phenomenon. As impedance spectroscopy has been found helpful for sensor applications in biology, it has also been applied in plant studies. A recent review (Jócsák et al. 2019) shows the different applications of impedance spectroscopy to plant studies. She mentions the different experimental work carried out and explains the use of frequency range, electrical circuit modeling, and parameter extraction. Additional work for plant monitoring based on impedance spectroscopy has also become available. Here is an attempt to present a few challenges to overcome for field implementation of impedance spectroscopy measurements in precision agriculture.

2.1 Coupling to the Plant and Electrodes

Electronically two approaches to coupling to the specimen are defined: galvanic and capacitive (or faradic). The galvanic coupling means the electrode is in direct contact with the measured material, creating a resistive junction between the electrode conducting material and the specimen. A capacitive coupling means that a junction layer that behaves capacitively is formed between the electrode-bearing material and the sample measured. An example can be taken from electrocardiogram (ECG) measurements, where an adhesive is used to couple the electrode to the patient's skin. When the measurement results are extracted, the effect of the adhesive material and skin contact need to be extracted. This requires prior information regarding the geometry and dimensions of the contact and the material properties. This knowledge allows for extracting the actual signal being studied and obtaining data about a patient. The advantage of such coupling often means that direct contact may not be necessary or that invasive measures can be avoided. The electrode coupling and contact formation are significant for determining the signal measured and understanding the signal-to-noise ratios to determine whether an effect can be measured using a particular electrode contact configuration. When examining changes in a signal, as in sensor applications, the contributions from the surrounding setup are crucial to determining whether the change results from the electrode/contact interface or arises from an actual change in the measured specimen. The electrode configuration should determine the interface being measured. As in the literature, two-, three-, or four-point setups are suitable for different types of measurements. However, in each setup, the electrode specimen interface contributes to the overall measurement values. Therefore, the electrode interface and configuration need to be carefully tailored and adapted to the specimen monitored. An example can be taken from semiconductor interfaces used for microelectronics fabrication. Here, ohmic contacts are often formed between metals and semiconductors. Once the contact is characterized as ohmic, it behaves linearly, and its resistance can be estimated for different working regimes. In other cases, where a non-ohmic contact exists, the behavior must be considered part of the device functionality. In plant measurements, different electrode setups have been demonstrated. A four-probe measurement using capacitive coupling to the leaf as shown using graphene-based square electrodes mounted onto the leaf. Results showed that detection of induced water stress could be detected (Zheng et al. 2015). Another example is the electrochemical sensing setup, using a three-electrode setup, where the sensor device is mounted close to the leaf and can detect a change in gas composition near the stomatal opening. Another form of electrochemical sensing has been demonstrated within the plant stem, where the device and electrodes have been inserted into the plant stem and show the ability to measure sugar changes due to plant transpiration activity (Cervantes-Godoy et al. 2020). Other forms of electrodes have been used to measure fruit ripening; an example is the use of medically prevalent sticky electrode patches. These were attached to the fruit in a four-point configuration, and different ripening stages were detected. Frost detection in fruits has also been studied using four-point probe impedance measurements. The electrodes also seem to have been inserted into the specimen under test. More recent technology for flexible microneedle fabrication has also been demonstrated to measure impedance in plants, showing results similar to other electrode configurations. The coupling to the specimen under test is highly significant for electrical impedance spectroscopy measurement. The method relies on the specimen's response to a signal applied across a frequency range. As the reaction is composed of resistive, capacitive, and possibly inductive parts, the actual electrode's proximity, location, and interface can affect the results of the measurements. With advances in nanotechnology and material sciences, several electrode options are available to improve the electrode material and mounting to the plant. However, a better and more in-depth study of the electrode interface with the plant and the internal effects on a living plant is needed. In addition, geometrical considerations and mechanical stability need to be considered—also durability in harsher outdoor conditions—possibly feeling the plant anatomy in the electrode design and optimization.

2.2 *System Design: Supporting Electronic Requirements*

Continuous monitoring and easily mounted field devices are needed in precision agriculture. For such applications, embedded electronic circuits and devices need to be designed. While for continuous long-term measurements, additional considerations are required, such as, signal integrity over time and nondestructive connectivity to the plant being measured. The need for a compact system that allows continuous data collection and transfer to the cloud is clear.

Furthermore, it should consist of low-power electronics and be robust for outdoor conditions. A requirement would also be that it would be easy for untrained farming workers to support multiple plants or measurements simultaneously. Lab systems have been suggested to show prolonged continuous monitoring of tobacco plants

(Garlando et al. 2022). The proposed method was multiplexed to measure various plants and combined with designated environmental sensors. However, a complete outdoor system is yet to be developed. Some of the challenges posed have been addressed for different biomedical devices, which with adaptions, may also offer solutions that could be applied to plants in precision agriculture. Yet for agricultural monitoring, the low cost of the devices is also a priority, as many of the food security issues span across the poorest countries, with low income and resources for technological development.

2.3 Data

Collection and interpretation of the signals acquired using impedance spectroscopy measurements are also needed. The specificity of the plant response to different environmental changes of disease is unknown and needs to be quantified. Furthermore, the sensitivity of the measured data also requires study in plant physiological terms, not only electronically. The question of how often measurements should be collected and according to what amount of data improved decision-making will occur is also unknown. Beyond these research questions also lies the technological aspect of collecting, analyzing, and interpreting the data using extensive data algorithmic methods alongside engineering and physiological understanding.

2.4 Biological Background

All seed plants have similar basic body plans. However, diversity is apparent. The vegetative body comprises three organs: leaf, stem, and root (see Fig. 3). The primary function of the leaf is photosynthesis, that of the stem, support and root, anchorage, and absorption of water and minerals. Leaves are attached to the stem at nodes, and the region of the stem between two nodes is termed the internode. An evolutionary difference exists between flowering and nonflowering plants. We do not take this into account in this work.

The stem consists of a vascular cambium, structured as a cylinder, and consists of supportive fiber cells and conduction vessels called the xylem and the phloem. These vessels conduct water and nutrients across the plant. The xylem provides a low-resistance pathway for water movement, thus reducing the pressure gradients needed to transport water from the soil to the leaves, while the phloem allows for nutrient flow from the leaves downwards. The stem growth is depicted in Fig. 4. Plant growth occurs from the center of the stem outwards, increasing the vascular cambium portion in the stem as the plant ages (Taiz and Zeiger 2010).

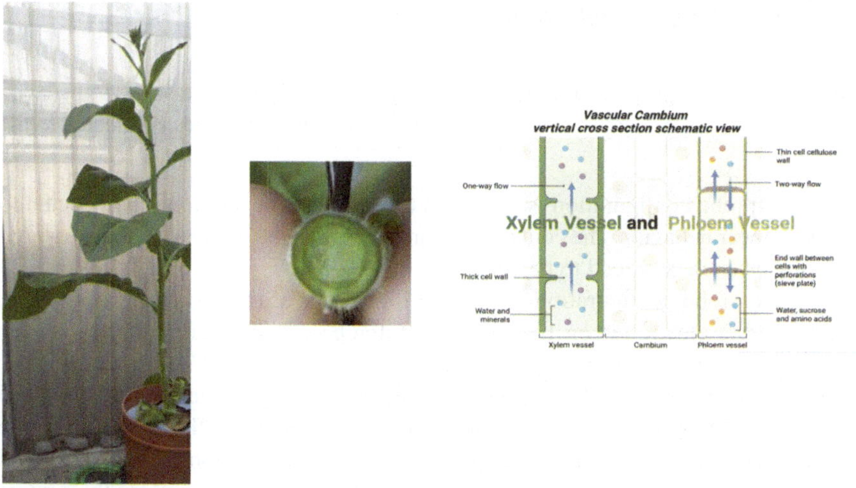

Fig. 3 A tobacco plant displaying general plant structure showing (from left to right) the stem and leaves, a cross section of a the stem showing the different areas and the vascular cambium and a detailed schematic description of the vascular cambium

Fig. 4 Schematic of a manual impedance measurement setup

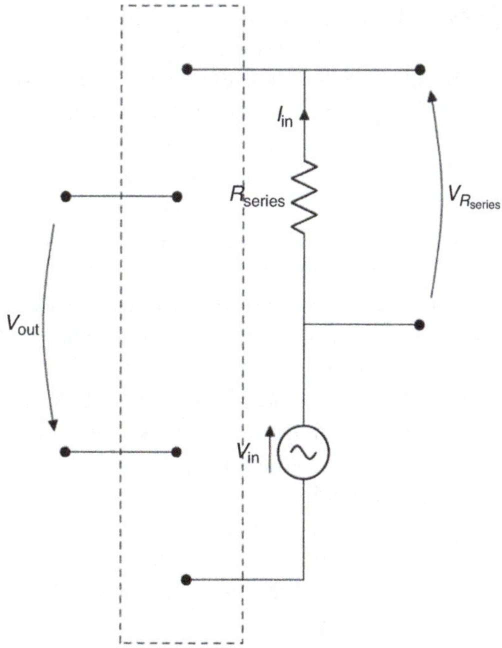

2.5 Electrical-Based Model

We chose a four-point configuration based on the following reasons. Impedance spectroscopy can be carried out using a different number of probes connected to the device:

- A two-probe configuration is typical in cases where the specimen is expected to have linear behavior. The contact resistance between the probes and the measured sample can be assumed to be ohmic.
- A three-probe configuration is suitable for interface interaction measurements, where one electrode can be used as a reference. This is commonly used for electrochemical measurements.
- A four-point probe configuration allows the decoupling of the contact resistance of the behavior of the probe from the actual measurement of the device under the test itself. It is often used in cases where the contact interaction with the specimen is unknown or may change over time. It is also helpful in cases where the contact resistance may introduce high resistance values compared to the device under test.

The advantage of this configuration is that the voltage is sampled using different probes where the current is applied. Since the voltage is measured using a very high impedance operational amplifier, the effect of the contact impedance between the metal electrodes and the device under test is reduced. This is important since the exact characteristics of those contacts are often unknown. Four-point probe measurements are well known in the semiconductor industry for the characterization of different materials, from thin film dielectric layers to bulk metal lines and so on. The four-point probe configuration, both in direct and alternating currents, uses two probes to induce the current and the other two to measure the voltage drop across the specimen. The probes are often placed in a line, where the two outer probes are used for the current and the two inner probes for the voltage. In this manner, the current is forced across the specimen under test through a single set of leads (called "force"), while the voltage is measured through the second set (called "sense"). The voltage drop across the sense leads will be negligible, so the measured voltage is essentially the voltage across the specimen under test. This has several advantages over a two-probe configuration. The leads and connections resistances can be almost eliminated, allowing assessment of the actual contact resistances while providing more accurate and less noise-sensitive measurements.

There are two ways to connect to the device under test using a four-point measurement configuration. One is using four physical contacts to the device under test, or using only two.

The four physical contacts separate the "force" and "sense" connections. While with two physical contacts, each connected to a force and a sensor measurement. Therefore, a two-contact measurement requires specialized equipment and calibration due to current leakage. In our case, a four-point configuration was used to avoid calibration uncertainty and separate the current source and voltmeter.

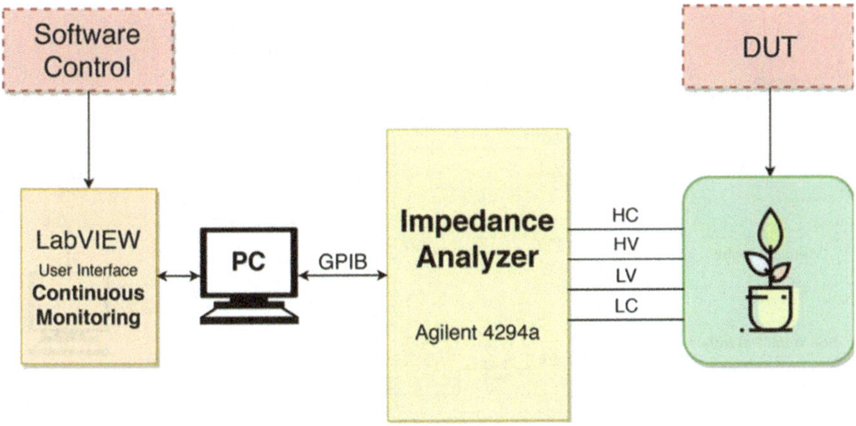

Fig. 5 Impedance measurement setup schematic diagram including the plant connected and the control system designed (Garlando et al. 2020)

In addition, this also removes the uncertainty of current leakage via the sense leads (Keithley 2016). During this work, different equipment was used to collect impedance data. Initially, a manual impedance setup was devised. Here, a signal generator was used as the source for signal excitation at different frequencies (for example, Keysight waveform generator 33500B), the input was measured across a resistor, and the output was registered using an oscilloscope (Agilent MSOX2012A (Keithley 2016)). The response of the plant stem was calculated according to the difference in voltage drop across the output about the input. A schematic illustration of the setup is shown in Fig. 5.

2.6 Continuous Monitoring Systems

Continuous monitoring setups were devised during the research to acquire multiple ongoing readings of the plant across different intervals. The setups rely on an impedance analyzer that was controlled using dedicated software, which was developed during this work. Environment sensors and monitoring tools complemented these setups to study the plant status and response to different conditions comprehensively.

2.6.1 Setup 1—Lab Indoor Plant Monitoring System

An impedance analyzer (Keysight Technologies model 4294A (Keithley 2016)) was connected directly to the plant stem in a four-terminal setup. The analyzer was programmed and run continuously using a designated LabView© software

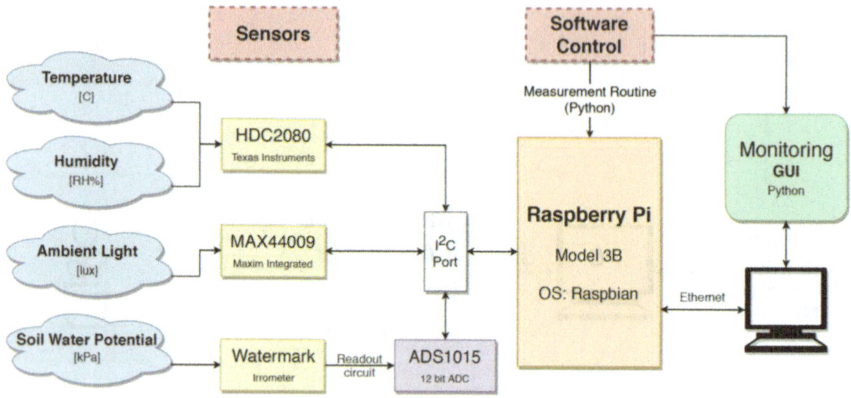

Fig. 6 Sensor system block diagram (Bar-On et al. 2019a)

interface. Measurements were conducted every 15 minutes, and data were logged and collected using an additional LabView© software. A diagram of the system can be seen in Fig. 6 (Bar-On et al. 2019a).

The system is described in detail in Bar-On et al. (2019a). In addition to the impedance analyzer setup, an electronic system with controlling software was devised. This system included electronic circuitry to support multiplexing of the impedance analyzer to measure two channels without affecting the accuracy of measurements. This enabled measuring two plants under similar environmental conditions. In addition to the multiplexer option, a set of external sensors were configured onto an embedded board, designed as a system for environmental monitoring. This setup included a range of commercial sensors, including a temperature and humidity sensor, a soil moisture gauge, and a light sensor. The software was developed to allow simultaneous data collection and synchronized readings from the impedance analyzer and the sensor system (Bar-On et al. 2019a).

The details for the system of sensors to collect information on the environmental status are described below.

The system allows continuous data collection of the surrounding plant environment and samples every 15 minutes. It is controlled and programmed using a Python interface, and the hardware is based on the Raspberry Pi© platform. It includes three generic sensors:

1. HDC2080 (Texas Instruments Ltd)—temperature and relative air humidity sensor (Texas Instruments 2018)
2. MAX44009 (Maxim Integrated Ltd)—ambient light sensor monitoring (Maxim Integrated 2011)
3. 200SS WATERMARK Sensor (Irrometer Ltd)—soil moisture monitoring (Irrometer 1978)

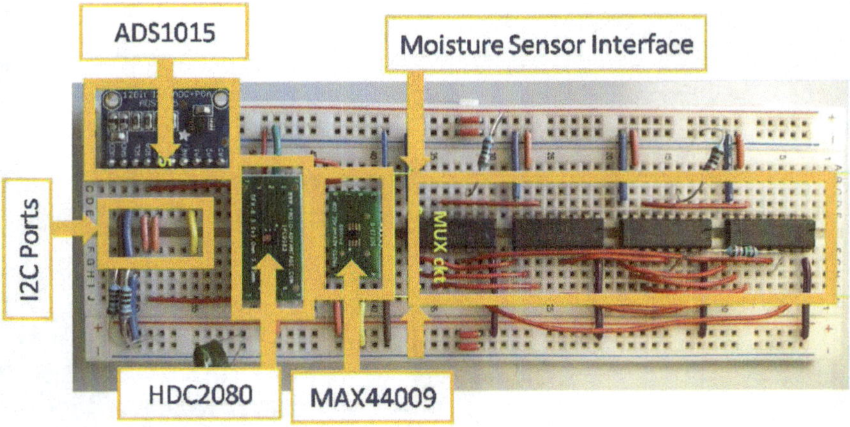

Fig. 7 The designed environment sensor board including the direct plant electronics interface (see Bar-On et al. (2019a))

The sensors are all connected to the GPIO (General-Purpose Input Output) port of the Raspberry Pi© using the I2C (Inter-integrated Circuit) protocol (port #1). At the same time, the moisture sensor requires an analog-to-digital converter and a readout circuit. The circuit is constituted by the schematics and components recommended by the manufacturer, also using the I2C protocol and sharing the port with the previously mentioned sensors. The complete supporting circuit allows to control and collect data in a synchronized manner while having the readout from each sensor collected serially every 15 minutes and saved to a data file. A block diagram of the system is shown in Fig. 5. The circuit with the sensors connected, indicating the ports that are then connected to the Raspberry Pi, can be seen in Fig. 7.

A graphic user interface that included plotting abilities and calibration of all the connected sensors and impedance measurements were also established (Fig. 8).

2.6.2 Field Outdoor System

An outdoor field system for continuous impedance monitoring was also devised. The plan was set up in an outdoor greenhouse in Tel Aviv, Israel. The setting allowed exposure to natural light and outdoor growing conditions while the temperature in the greenhouse was controlled. All other parameters were closely monitored as in standard commercial greenhouse growing facilities.

For the plant impedance monitoring, a Hioki IM3570 (Hioki Ltd n.d.) impedance analyzer was used in an analyzer mode. Measurements were taken at 500-mV RMS, while each measurement was averaged with a factor of 4. The frequency was swept logarithmically across 50 Hz to 4 MHz, collecting 801 points per sweep. Calibration, including cables, was completed before measurements. The electrodes were coupled galvanically to the plant by direct insertion into the plant

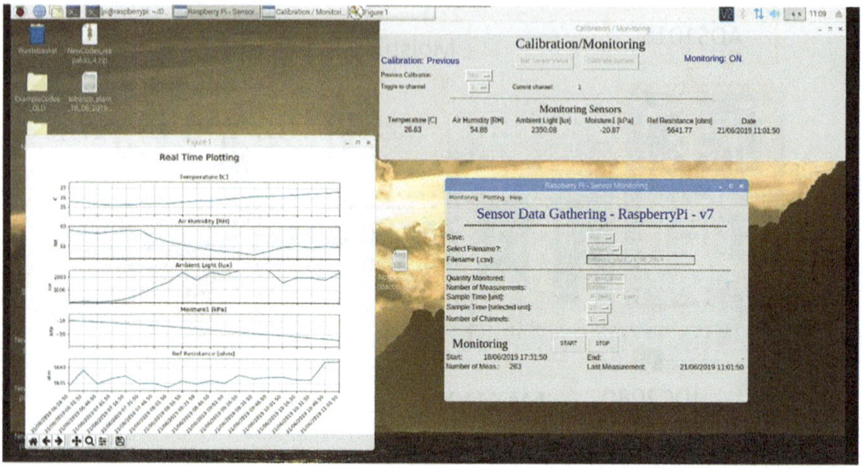

Fig. 8 User interface, calibration, and plotting utility

stem at a distance of 5 cm (as described in Bar-on et al. 2019b). Au electrodes (0.5-mm diameter, approx. 4 cm in length) were inserted into the stem, ensuring direct contact with the vascular tissues inside the stem. Measurements were carried out continuously over a few days, and the results were logged and analyzed. Experiments were repeated across multiple plants, while they were each prolonged across 30–60 days at a time, and data were collected every 9–24 minutes, yielding experiment sets of over 10,000 measurements each.

2.6.3 Plant Choice

This study tested a few different plants, including tobacco, tomato, and the cherry tree. However, we chose to focus the research on tobacco plants. The plant type for this purpose was selected regarding the knowledge of the vascular structure, robustness, and the study of its genetic makeup. *Nicotiana tabacum* L. cv. Samsun-NN (tobacco) plants were a suitable match for this study. Young plants grown for 3–4 months were used, having a stem diameter of 0.7–1.1 cm (Fig. 9).

Plants were grown at different locations, providing growth conditions tailored per experiment.

2.6.4 Comparative Study

An assessment of the ability of impedance measurement to detect physiological change was completed in a comparative study manner.

A comparative study can be carried out by comparing the same sample to itself under the same conditions with controlled change across time. Such an experiment

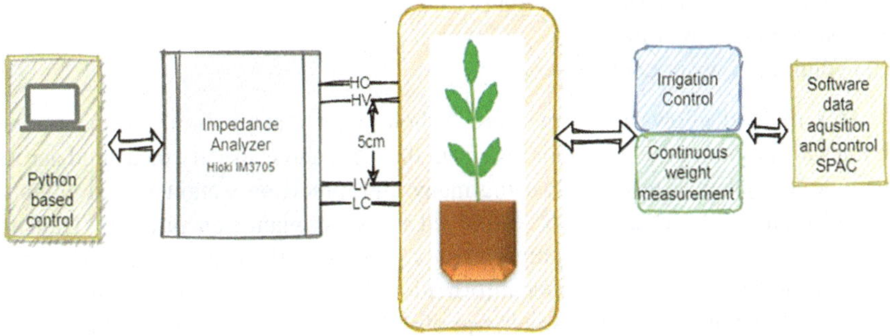

Fig. 9 Outdoor impedance monitoring setup schematics

would be carried out multiple times across multiple specimens to have a correct experimental procedure. Another way is to compare two similar specimens in the same environment, under a different condition, keeping one in a control condition and the other in the test. Both these approaches were used during the research and experiments.

For example, this research completed a series of experiments to evaluate changes in impedance measurement values due to water stress. Initially, the plant was compared to its characteristics under normal hydration conditions, and once water stress causing dehydration was induced, values and changes in measurements were reevaluated. Next, a comparative study was done by multiplexing the measurement system and measuring two plants consecutively (with a difference of a few minutes between measurements so that this is almost simultaneous) in the same conditions.

Once this was achieved, the newly suggested measurement method was compared to well-known techniques commonly used in plant science to evaluate plant well-being and physiological state. The following section describes the different plant physiological measurements used.

3 Plant Physiological Monitoring

Plant monitoring in plant science and biology research is a well-established field. Only a few nondestructive measurement methods that estimate whole-plant status are available. One includes measuring plant water usage based on a continuous weight measurement technique. This method is called gravimetry, where weight changes are measured and attributed to changes in the monitored specimen. The estimated weight in gravimetry includes the overall plant weight, roots, stem and leaves, pot, and ground. These are counted multiple times, and the weight change across time can indicate plant water usage.

The system allows assessing a few terms that are used to determine the plant's physiological behavior, which we define below:

(a) Plant transpiration (transpiration rate)
(b) Daily and across time
(c) Volumetric water content

This method allows us to measure "whole-plant transpiration." Transpiration defines the amount of water vapor loss from the plant and is usually measured across a single leaf or area; nowadays, using many high-precision weight measurements, overall plant water usage can be estimated to assess plant transpiration. Transpiration is a fundamental and highly complex process in the plant system. Ongoing gravimetric data collection provide information regarding overall plant transpiration activity, allowing us to study water use efficiency, improve the understanding of the physiological activity, and give a comprehensive view of the plant status. Therefore, this method was chosen as a comparative baseline for inspecting the electrical impedance spectroscopy data acquired during this research. To study changes in impedance values measured, plants were also monitored using this method simultaneously. This way, modifications could be compared to well-known effects and values and attributed to different physiological phenomena.

3.1 System Description—Gravimetric Measurements Using Plant Array

Whole-plant physiological performance was monitored with the functional phenotyping system. Plant array platform is described in detail. The system includes a weight sensory system with an accuracy of a few milligrams. In addition, it consists of a soil moisture sensor placed inside the pot as an indicator of changes in the soil water content. Next, the system is connected to a controlled irrigation system, where the plant can be hydrated according to experiment requirements. In addition to the local monitoring of the plant, the greenhouse environment is continuously monitored and maintained. This includes measurements of temperature, light condition, and air humidity. The soil moisture is continuously monitored as well using soil probes. The data collected by the system are logged into a soil–plant–atmosphere control software tool.

3.2 Physiological Studies Alongside Impedance Data

To establish whether the impedance data collected might indicate the plant status, a range of different conditions were tested experimentally during the research for the electrical plant response.

In plant physiology, the plant response to different situations, which can be induced externally, is known and can be expected. A simple example is water stress, i.e., once the plant lacks water resources, its leaves will wilt. Other stress conditions, such as lighting changes or disease, can also be induced.

3.2.1 Water Stress/Drought

Experiments were conducted to study the impedance measurements' response to water stress situations. To accomplish this, plants were regularly watered, showing repetitive behavior across the days (watering saturation). Once These stable conditions were achieved study of different water stress cycles were completed by skipping specific watering time slots.

3.2.2 Light

Plant growth, development, and physiological regulation depend strongly on light. Therefore, exposure to changes in lighting and daylight hour fluctuations is monitored in plant research and taken into account for physiological change studies. While daylight fluctuations are present across the different seasons and light exposure of a plant can depend on its location and external factors, sometimes growth chambers are used. Here, a light that imitates the sun spectra is used, and daylight hours can be controlled. Throughout the experiments presented in this work, the systems included continuous light monitoring within each system (each sensor is described in the system setup description).

3.2.3 Soil

The ground where plants grow varies in different locations and areas worldwide. Due to these variations, different types of land have been studied and adapted for growing agricultural crops around the world. Every kind of ground has different attributes and is better suited for specific terrain, climate, and crop. In our case, two types of land were used: standard soil and coarse sand. While comparing the behavior of these two, and is known to have lower water retention than soil, this can be utilized while examining drought or weight changes.

3.3 Experimental Procedure Using the Gravimetric System

Wild-type *Nicotiana tabacum* seeds were sown in growing soil (Avin Ari Ltd., Israel). The plants were grown in fully controlled growth chamber equipment with "ELIXIA" lights of "Heliospectra" (ref) set on a long day (18/6 hours) with 500 PAR, and the temperature was controlled with AC (26/19). After 3 weeks, the plants were gently washed from the soil and transferred to the designated Plant Array 3.9 pot full of coarse sand of "Negev Industrial Minerals Ltd.," Israel. Coarse sand is a highly homogeneous, inert medium; using this medium minimizes the noise caused by the soil absorption of water and nutrients (ref). An "EVA" foam sheet covered the pot's top to minimize the evapotranspiration from the soil. The 3.9-L pot with the seedlings was transferred to the Plants Array room in the semicontrol greenhouse at the Institute for Cereal Crops Improvement, Tel Aviv University, for acclimation. AC controlled the greenhouse temperature (28–19), and natural light conditions were used; the weather station recorded the humidity, temperature, and light during the experiments. The Plant Array system is supplied with a fully controlled irrigation system. During the acclimation period (from 3 weeks to the start of the measurements), the plants were watered throughout the day to stabilize and form their root system. When the plants showed full acclimation determined by active growth and transpired more than 300 mL per day, we started irrigating only at night (21-02) at list 2 L to reach full saturation. After 2–3 days of night irrigation period, the experiment started. "Shaphir Nitrate Solutions" fertilization (4:2:6) of "Deshen Gat" was added to the irrigation to ensure healthy plant growth. When plants were in high turgor in the morning, four probes were galvanically coupled to one plant's stem while the other was used as a control. The probes were made of noble metals to avoid corrosion effects. The two pairs of probes were placed at a distance of 5 cm from each other. The impedance measurement and the Plant Array systems were checked and synchronized to collect data continuously. To collect sufficient data, multiple rounds of experiments were run, each match including measurement under no-stress and stress water deficiency (drought) conditions. To measure the plant at control conditions (nonstressed), plants were irrigated to saturation (at list 2 L) and complete drain of the remaining water during the night (21-02). To measure the plants under stress conditions, the plant connected to the impedance was not irrigated overnight. After the plants showed stress phenotypically (low transpiration and withered leaves) and the soil probe showed low moisture, the plant was given recovery irrigation overnight for at least 3 days before the repetition of another stress cycle. Each experiment was run for approximately 3–4 weeks. A more detailed description of our model has been published in "Frontiers in Electronics" (Bar-On et al. 2021).

3.3.1 Setup 1—in Lab Monitoring—Turin, Italy

The results in the section have been published in IEEE over the research period (Bar-On et al. 2019a; Garlando et al. 2021). The indoor lab setup includes an impedance system alongside a set of environment sensors, which was established for this work. It allowed data collection across long periods, with different sampling rates. In addition to the impedance data collected, environmental data were also collected. A range of experiments were completed using two types of plants: tomato and tobacco (Fig. 10).

Nicotiana tabacum (tobacco) plants were grown in labs from seeds provided by the lab at the Faculty for Plant Science at Tel Aviv University (TAU); the monitored plants were grown for 3–4 months, reaching a stem diameter of 0.5–1 cm. In addition, as tomatoes are quickly grown in Italy, tomato plants bought for edible tomato growing were tested. Here, plants were small (about 50 cm in height), and

Fig. 10 Plants used in the Turin lab setup, visualizing the connected stem to an electrical design. Right: tobacco, left: tomato

the stem diameter was chosen to be similar to the used tobacco plants, yielding a measurement of 0.5–1 cm in diameter.

Initially, the system was set up to allow continuous measurements of a single plant. Later, the electronics allowing multiplexing the system to measure two plants consecutively was established. The system included (as described in Sect. 2.6) an impedance analyzer controlled using a dedicated LabView© software and a set of environmental sensors controlled using reliable electronics and controlled with Raspberry Pi© to run in a synchronized manner and log the data collected. Once continuous measurements of a single plant over time were completed and a sampling rate allowed the system to provide minimal noise during sampling, a multiplexer was added to collect data from two plants at a time. Here, a measurement was taken from each plant consecutively and continuously without affecting the quality of the data collected (i.e., parasitic impedances, signal averaging time on the impedance analyzer). Figure 11 shows this system setup with two plants connected for measurements.

Impedance measurements were carried out across time, and data were collected. The impedance analyzer was calibrated according to the manufacturer's manual, allowing the establishment of impedance spectra of the plant stem. The measurements were set to measure across a frequency range of 40 Hz to 1 MHz. At the same time, in this domain, the impedance analyzer was fully calibrated with the additional multiplexing circuit and cables and connectors to the plant. Calibration included all wires except the actual probes in contact with the plant, as they were assumed to be perfect conductors (Au wires of approximately 4 cm in length and

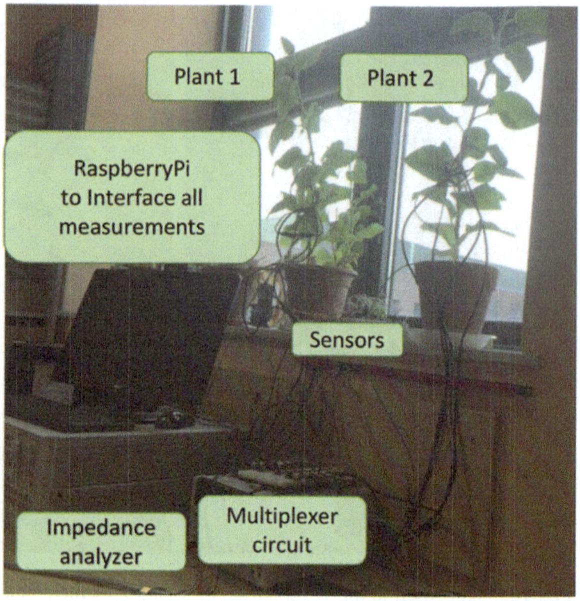

Fig. 11 Continuous monitoring system depicting two plants monitored simultaneously in Turin, Italy

0.6 mm in diameter were used). Voltage was set to 500 mV VRMS while averaged across 4 for each sweep.

This setup allowed data collection across a wide range of frequencies and time, so examining the results can be done in various ways. During this work, we presented the impedance data by looking at the impedance magnitude and impedance phase separately, as done in Bode plots (Agarwal and Lang 2005; Sze and Ng 2006); we found this useful for our study (other representations are available and presented briefly in this study).

Impedance spectra were initially collected, showing that they coincided with previous results. In addition, the orders of magnitude measured agree with our initial results (published in Bar-on et al. 2019b, using a manual setup for impedance measurements). The continuous manner of data collection allowed us to distinguish that a change in the measured impedance values of the plant occurs during the day. By dividing the day into three regions: morning 8–12 AM, afternoon: 12–16, evening: and averaging the measurements taken every 4 minutes across these hours. We could distinguish that a shift in the measured curve exists, indicating the sensitivity of the size to the time of day (possibly corresponding to the plant's daily cycle, daylight hours). An increase in impedance can be seen as the day proceeds across frequencies, while in phase, a shift is present, with decreasing absolute values of the angle.

Next, the behavior of the plant across several consecutive days was examined (shown in Fig. 13). Here, to examine the variation across frequencies and time, we examined the behavior at the centroid frequency. This data, for both impedance magnitude and phase, show that the measurement is sensitive to a daily trend in the plant. However, plant behavior under similar conditions across days is repetitive. These would require further studies but indicate that the collected data may be a significant monitor for the plant.

Next, using the multiplexer electronics, two plants were monitored simultaneously. This setup allowed us to carry out case studies in a comparative manner. Several experiments were carried out across 2–3 weeks each time (Fig. 12).

In the following example, the experiment was carried out across multiple days (up to 3 weeks at a time). Two tomato plants were monitored continuously, collecting impedance data at 15-minute intervals. Here, we present an example of measurements taken across 6 days, examining the response of the two plants: one to complete dehydration over time, which was started on day 6; the second plant, which was kept well hydrated with regular watering every day. The result can be seen in Fig. 22. In this experimental setup, it is to be noted that the plant's watering was completed manually, using approximately 300 mL of tap water, at 17:00 each day. The daily trend is visible in both the impedance magnitude and phase (as in Fig. 21). A change due to hydration/dehydration is also present in both measures. It should be noted that irrigation was done manually and exposed to variations.

In addition to the impedance data, the sensor system collected environmental data simultaneously. The data acquired here were used comparatively. Presented below are impedance compared with data from a soil moisture sensor inserted into the pot plant soil and a light sensor.

Measurement of soil moisture is a well-known method for irrigation planning. Here, the soil moisture sensor was calibrated and inserted into the plant pot. For convenience, we inspect the time derivative of the soil moisture sensor reading to see a "spike" each time an irrigation event occurs (an example can be seen in Fig. 19). These values are shown alongside the impedance magnitude and phase across

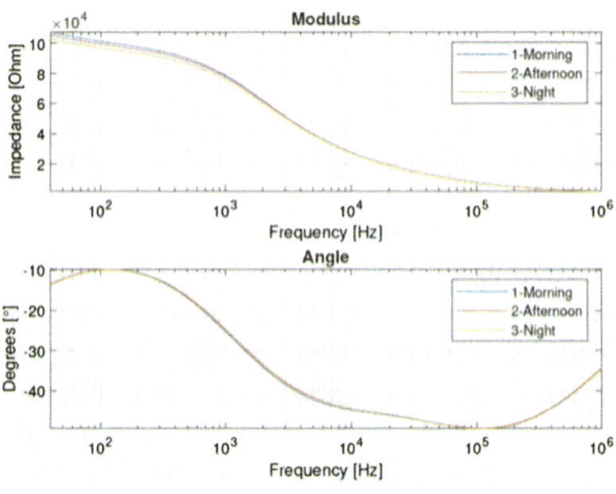

IMPEDANCE MEAN VALUES OF TOBACCO PLANT

Frequency [Hz]	Modulus [kΩ]	Angle [deg]
100	94755 ±6320	-9±0.68
1000	74388± 3882	-24.3±0.93
10000	27320± 1024	-44±0.51
100000	7855±153	-49.5±0.98

Fig. 12 Impedance measurements across a day, showing the differences between different times (classified as morning, afternoon, and evening). Right: Impedance magnitude and phase values across different frequencies. (Similar results are presented in our publication (Bar-On et al. 2019a))

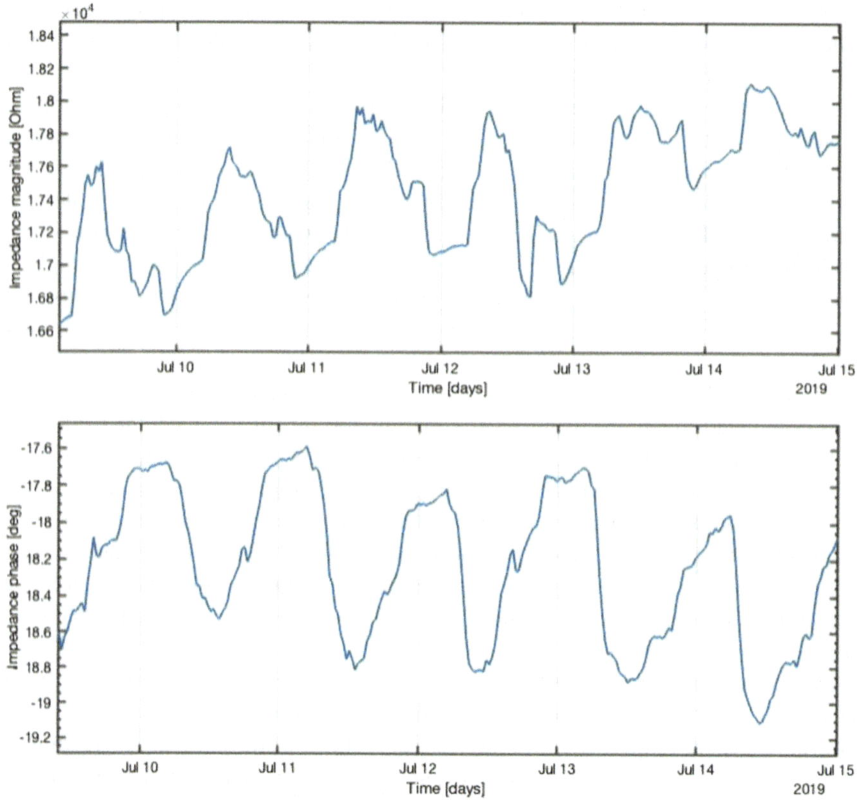

Fig. 13 Impedance measurements shown at the centroid frequency across 6 days. The plant was manually watered once a day. The impedance magnitude (top) and phase (bottom) are shown separately across time. Both measures show a repetitive behavior across different days, indicating that the impedance data may follow the plant's daily cycle trend

time, clearly showing a response in the plant impedance to these irrigation events (Fig. 13).

Examination of impedance data alongside external illumination conditions was also completed. A light sensor sensitive to visible light was connected in the system to provide indication to daylight hours and lighting fluctuations. Here, the light sensor was set up in the lab near the plants and a window exposing the outside lighting conditions. Here, we could see that the impedance values fluctuate qualitatively, with a similar trend to the daylight cycle. This indicates that the impedance measurement follows activity within the plant vascular cambium, which is known to respond directly to light exercise. This is shown in Figs. 14 and 15.

The results shown on the system established in the indoor lab conditions present the possibilities and sensitivities of the suggested impedance measurement of

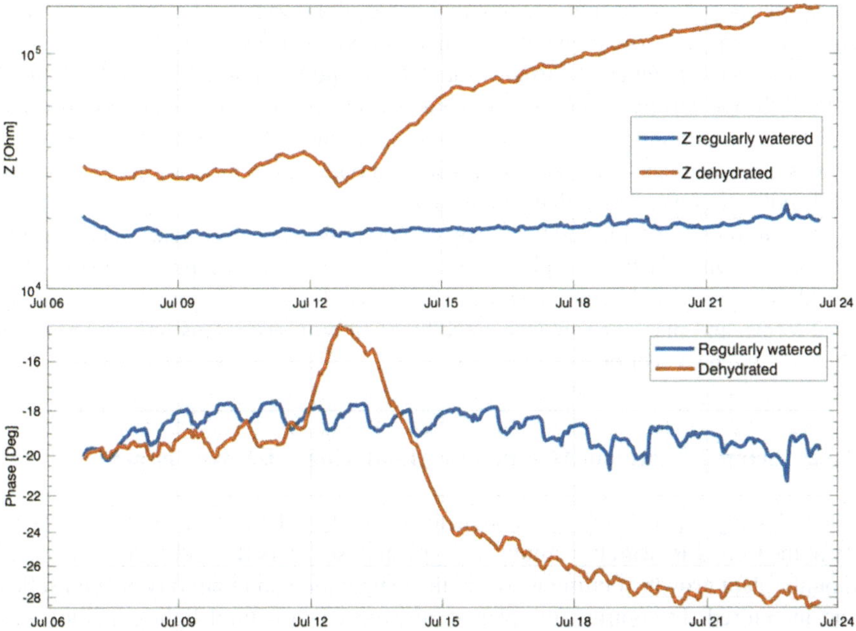

Fig. 14 Impedance measured across time of two tomato plants, the one regularly watered and the other dehydrated—both impedance magnitude (top) and phase (bottom) present the response to regular hydration versus dehydration

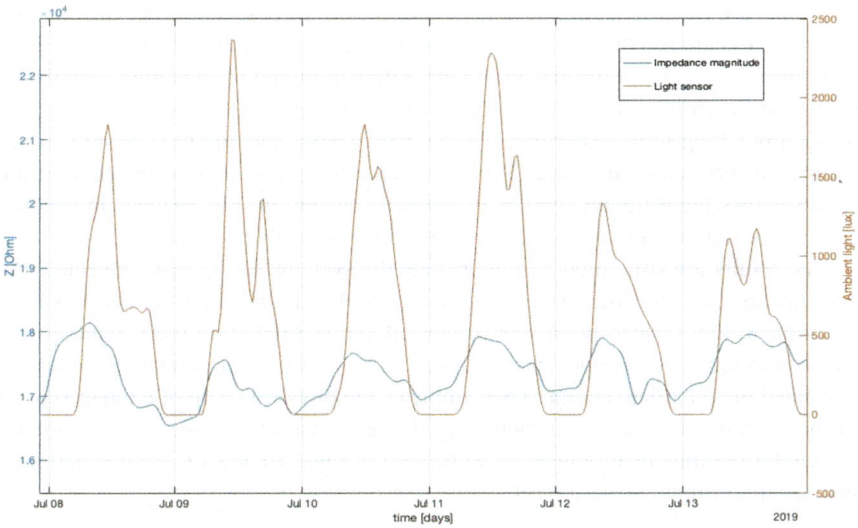

Fig. 15 Tobacco plant signal across time

living plants. These results suggest that close monitoring and combination with environmental sensors can yield useful data for precision agriculture.

As the Department of Electronics and Telecommunications (DET) at the Politecnico di Torino lab has expertise in embedded electronics for sensor systems, the collaboration continued with a focus on developing additional multiplexing electronics, improvements, and adding sensors to monitor the surrounding environment while collecting data from plants in this lab.

In addition, we established that comparative studies must be carried out alongside well-known and established plant-monitoring methods in the plant physiological world. Here, we could better understand the data collected, compare it to commonly used forms, and plan experiments suited by plant biologists. This was set up and is shown in the next section.

3.3.2 Setup 2—Outdoor Greenhouse Monitoring—Tel-Aviv, Israel

The following experiments were conducted in Tel Aviv, Israel, in a greenhouse facility at the Cereal Research Institute. The greenhouse offers the ability to control and monitor plant growth conditions so that the temperature and humidity are controlled and monitored. In contrast, the plants are exposed to natural light conditions, as these greenhouses are situated outside. To carry out our impedance experiments, the impedance monitoring system was connected inside the greenhouse to collect data continuously across time. A significant advantage of this setup was that the greenhouse included a fully automated irrigation control system that allowed one to determine the exact irrigation conditions, such as timing, speed, water quality, and nutrients. To enhance the research, the chosen greenhouse included a state-of-the-art plant physiological monitoring system that offers high-end plant weight monitoring alongside electronic sensors (PlantArray©, DiTech Ltd.).

Multiple long-term experiments of a few weeks and up to 2 months were completed throughout the year. This provides information and consistency of the results across the different seasons of the year, accounting for changes in weather conditions, light temperatures, etc., as well as for multiple plant trials.

The results are structured to demonstrate the systems set up and connected, the specifications of the different experiments completed, and their effects. For each experiment, the tobacco plant was connected and stabilized on the weight system and connected to the impedance analyzer. The exact experimental procedure is explained in an earlier section. An example of the physical setup of a tobacco plant in the greenhouse attached to both impedance measurement and the gravimetric system for continuous monitoring can be seen in Figs. 16 and 17. Each experiment was set up to run for several weeks.

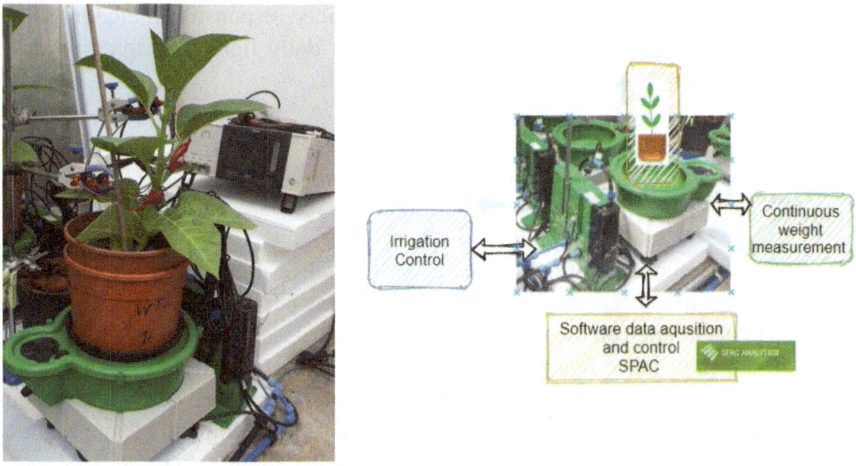

Fig. 16 Continuous monitoring of a tobacco plant in the TAU greenhouse; Left: Actual plant monitored during experiment, right: schematic description of the PlantArray System (DiTech Ltd.)

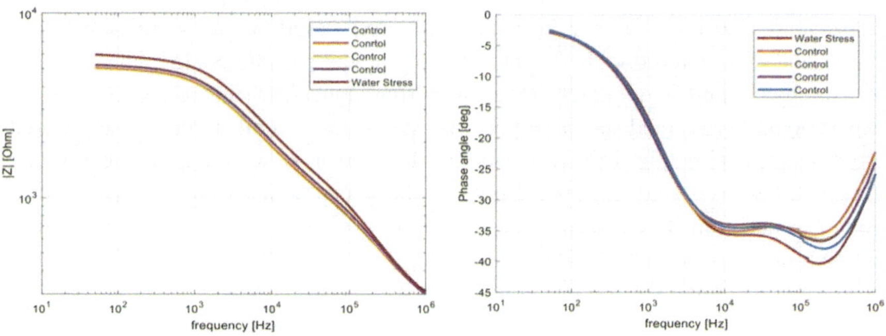

Fig. 17 Impedance spectra of a tobacco plant at different points in time. Right: impedance magnitude; left: impedance phase

3.4 *Impedance and Plant Physiological Response—Light, Daily Cycle, Water Stress*

It is well known that activity in the plant, including transpiration and photosynthesis, depends on light conditions. The plant's diurnal cycle depends on daylight hours and light exposure. To examine the response of the plant impedance spectra to a controlled light environment, the plant was kept fully hydrated, and constant intensity lighting was induced for a steady number of hours each day. The results show that, across the different frequencies, the plant impedance responds directly to the light changes.

Normal daily behavior—shown with the impedance response to muted response to different hydration amounts graphs to prepare: daily light, Transpiration Rate (TR), weight, etc.

4 Data Analysis

The suggested model was fitted to the experimental results, and an evaluation of the model's accuracy across the frequency range was performed. The spectra were provided to the lumped element model using a standard least square fitting algorithm based on the impedance magnitude. An example of the fit and an experimentally obtained spectra are shown in Fig. 20. The fitting algorithm was derived from the mathematical representation of each component in the lumped element model presented. It was run using the Matlab® curve fitting toolbox. The measurement data were imported and organized into the program and then run sequentially using an appropriate algorithm that considers least-squares fitting. The fitting provides an estimate for each model coefficient and can be presented parametrically. Initial values and lower and upper bounds were used based on measured impedance values and expected model behavior, allowing model parameters to converge across measurements and frequencies. Thus, the fitting algorithm yielded a good fit for all experimentally obtained sets of impedance spectra, with a mean relative appropriate error of approximately 1.06% ± 0.12%. This error is well within the limits of experimental error and the error limits of the instruments used. The error across time, i.e., across different days measured, is shown in Figs. 16, 17, 18, 19, and 20, where minor deviations are visible (Fig. 21).

During data analysis, the following analysis has been done:

1. Analysis of the impedance at prechosen representative frequencies.
2. Spectral analysis using the dominant pole parameters as indicators.
3. Fitting the spectrum, at each time point, to the physically based lumped element circuit model, using the model parameters as indicators. This analysis was also published by Bar-On et al. (2019a) and appeared in the section below.

Three analysis methods are presented, followed by a comparison to gravimetry. The results were analyzed from the plant's electrical response each day. Across those days, a series of hydration–dehydration cycles were performed daily. A more extended dehydration period (about 24 hours) was applied every few days, followed by a return to the daily hydration/dehydration sequence. The more extended dehydration period was used to study the plant's longer-term characterlike, such as post-dehydration recovery time. The results depended on the dehydration/hydration cycles and the ambient plant conditions, i.e., temperature, time of the day, humidity, and lighting. Therefore, the effect of dehydration/hydration cycles on the electrical measured parameters was studied at the same time daily where the temperature and lighting provide similar comparative conditions.

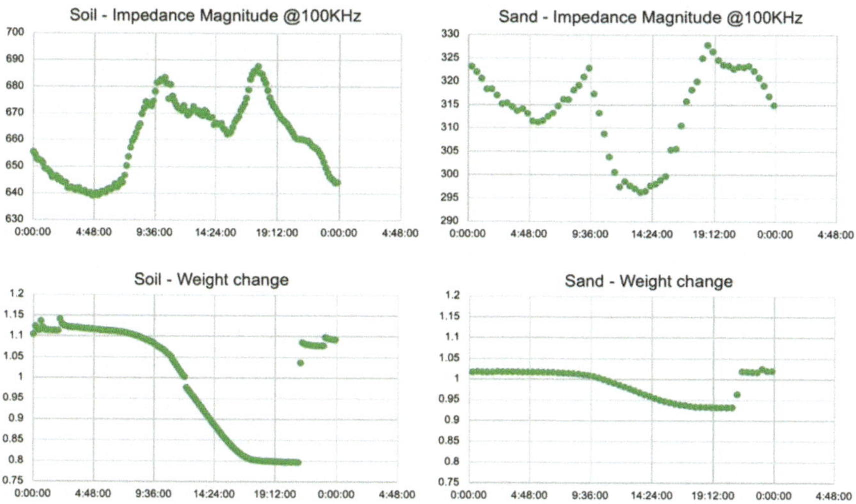

Fig. 18 Soil versus sand behavior

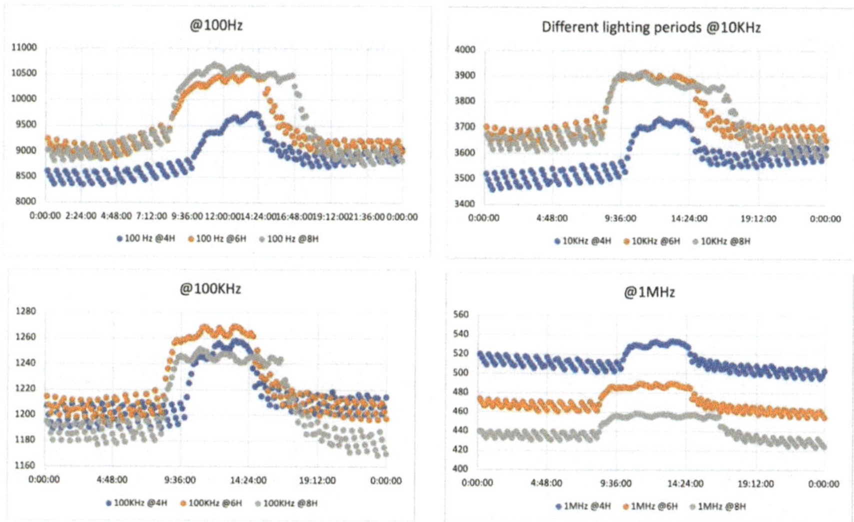

Fig. 19 Water stress across time

The lumped element modeling approach is commonly used to analyze a biologically based system using electronics. An equivalent circuit using the lumped element modeling has been suggested for this measurement setup (see Bar-On et al. (2019a)). A lumped element circuit approach attempts to consider the different physical components in the device under test and represent each of their contributions across the collection of frequencies used. The model parameter graphs across time

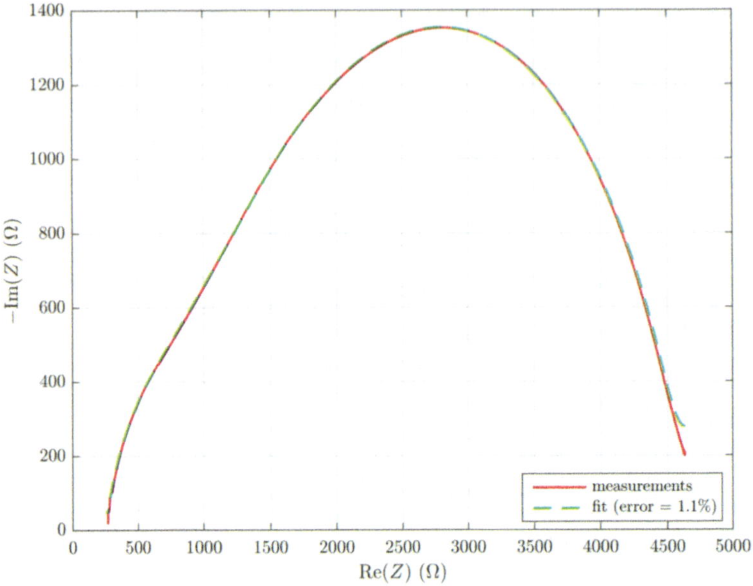

Fig. 20 An example of the model fitting for both impedance magnitude and phase shown for a representative experimental result

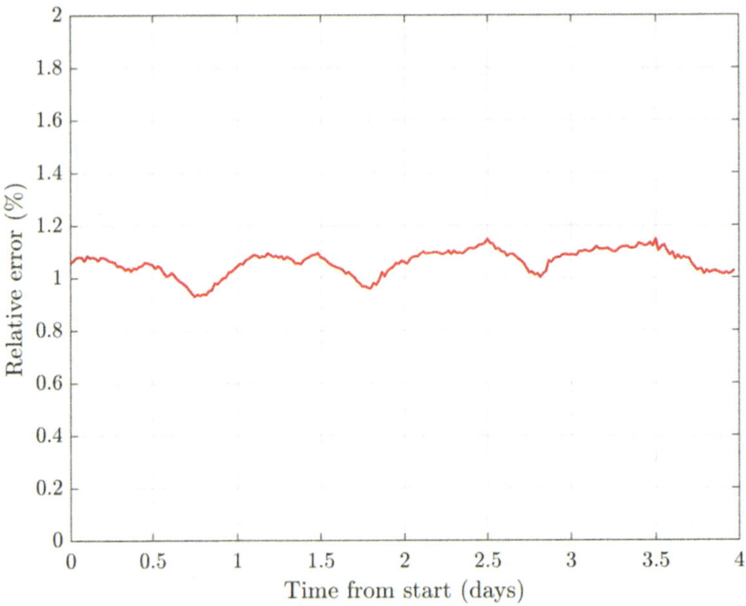

Fig. 21 Model error relative error calculated and shown across time

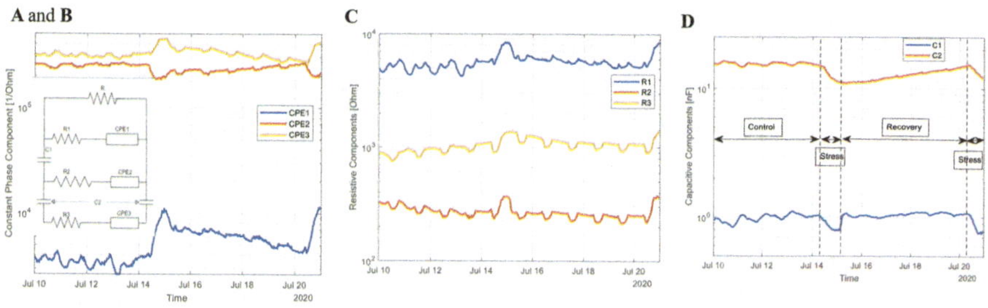

Fig. 22 Lumped element circuit parameters across time (**a**) The insert: the plan physically based lumped electrical model, (**b**) the constant phase elements (CPE1, 2 & 3), (**c**) resistive components, (**d**) capacitive elements vs. time, over 10 days

are shown in Fig. 22. In our case, a group of resistors, constant phase elements, and capacitors have been arranged based on the measurement setup and known plant physiology. Due to the plant stem physiology, the circuit assumes current flow across a collection of channels. Each channel is represented by a resistor and a CPE (constant phase element), while the different channels are capacitively coupled. Inspection of each ingredient in such a circuit across time will indicate their contribution to conduction across the specimen and the sensitivity of the system to change. This allows us to assess the significance of different components and estimate the system response to induced physiological stress. The presented parameters of the lumped element circuit are grouped by type in each graph, i.e., resistive, capacitive, and CPEs. Inspecting the different parameters, we may notice different behaviors. The different resistive components all show a change due to the stress.

In addition, across the presented control period, they show slight fluctuation. These changes may, in the future, be correlated to daily changes within the plant. Yet all resistive components behave similarly. Looking into the capacitive and CPE details, all parameters show an absolute value change due to the water stress introduced. Yet the direction of the change differs. This may indicate the actual physical mechanism the parameter represents within the plant. The most significant change we observe using this method seemingly occurs in the first CPE component (CPE1) (Fig. 23).

The daily fluctuations are enhanced, while the response to stress (observed as a peak both in the middle and at the end) increases by order of magnitude. Furthermore, it clearly can be seen that after the stress, the baseline of values is shifted during recovery. In modeling a system as in the plant stem, a constant phase element represents transport and diffusion within the biological specimen and can be expected to show higher sensitivity to a physiological change such as the water stress tested here. Comparing the change observed across the resistive elements, it can be assumed that they are more responsive to changes in ion concentration and therefore indicate more minor deviations. A better understanding of these changes requires a combined study with known plant physiology measurement methods. In addition, these qualitative changes shown mean the added value of the presented

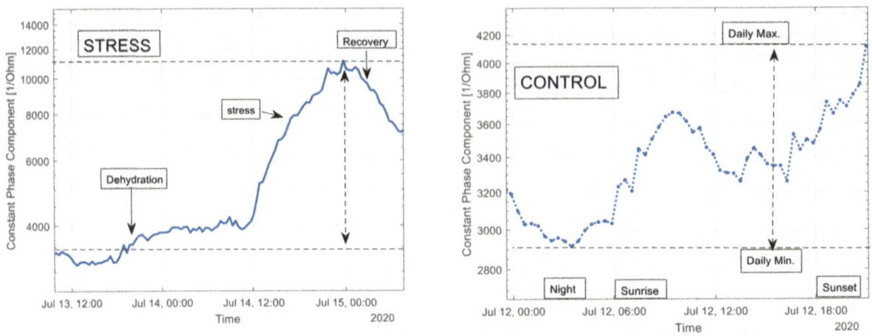

Fig. 23 A single element from the model as a function of time for normal conditions (e.g. daily watering) or under a draught induced stress. Showing the difference in response magnitude under the two conditions

continuous impedance measurement method, as it allows for data sensitivity to more than a single change across the plant. However, the lumped element approach with different parameters requires further study of the relations between the parameters and their physical meaning. The ability to detect variations across the different components is apparent.

References

Agarwal A, Lang J (2005) Foundations of analog and digital electronic circuits. Elsevier

Aloni R, Zimmermann MH (1983) The control of vessel size and density along the plant axis: a new hypothesis. Differentiation 24(1–3):203–208

Bar-On L, Peradotto S, Sanginario A, Ros PM, Shacham-Diamand Y, Demarchi D (2019a) In-vivo monitoring for electrical expression of plant living parameters by an impedance lab system. In: 2019 26th IEEE international conference on electronics, circuits, and systems (ICECS), pp 178–180

Bar-on L, Jog A, Shacham-Diamand Y (2019b) Four point probe electrical spectroscopy based system for plant monitoring. In: 2019 IEEE international symposium on circuits and systems (ISCAS), pp 1–5

Bar-On L, Garlando U, Sophocleous M, Jog A, Motto Ros P, Sade N, Avni A, Shacham-Diamand Y, Demarchi D (2021) Electrical modelling of in-vivo impedance spectroscopy of Nicotiana tabacum plants. Front Electronics 2. https://doi.org/10.3389/felec.2021.753145

Beck CB, Schmid R, Rothwell GW (1982) Stelar morphology and the primary vascular system of seed plants. Bot Rev 48(4):691–815

Brown CL, Volkov AG (2006) Plant electrophysiology. Springer, Berlin, Heidelberg

Cervantes-Godoy D et al (2020) Technical Report 4, 2014. UNICEF FAO. The State of Food Security and Nutrition in the World. Technical report

Garlando U, Bar-On L, Ros PM, Sanginario A, Peradotto S, Shacham-Diamand Y, Avni A, Martina M, Demarchi D (2020) Towards optimal green plant irrigation: watering and body electrical impedance. In: 2020 IEEE international symposium on circuits and systems (ISCAS), pp 1–5

Garlando U, Bar-On L, Ros PM, Sanginario A, Calvo S, Martina M, Avni A, Shacham-Diamand Y, Demarchi D (2021) Analysis of in vivo plant stem impedance variations in relation with external conditions daily cycle. In: 2021 IEEE international symposium on circuits and systems (ISCAS), pp 1–5

Garlando U, Calvo S, Barezzi M, Sanginario A, Ros PM, Demarchi D (2022) Ask the plants directly: understanding plant needs using electrical impedance measurements. Comput Electron Agric 193:106707

Georgia Tech Biological Science. Sugar transport in plants: phloem. Biology 1520

Hioki Ltd. Hioki Impedance Analyzer IM3570 (n.d.)

Irrometer (1978) Model 200SS WATERMARK Sensor WATERMARK Soil Moisture Sensor â MODEL 200SS

Jócsák I, Végvári G, Vozáry E (2019) Electrical impedance measurement on plants: a review with some insights to other fields. Theor Exp Plant Physiol 31(3):359–375

Keithley (2016) Low-level measurements handbook, pp vi, I–vi, 5

Lodi MB, Curreli N, Fanti A, Cuccu C, Pani D, Sanginario A, Spanu A et al (2020) A periodic transmission line model for body channel communication. IEEE Access 8:160099–160115

Luvisi A (2016) Electronic identification technology for agriculture, plant, and food. A review. Agron Sustain Dev 36:13

Luyet BJ (1932) Variation of the electric resistance of plant tissues for alternating currents of different frequencies during death. J Gen Physiol 15

Maxim Integrated (2011) MAX44009 industry s lowest-power ambient light sensor. p 1–20

Mogili UR, Deepak BBVL (2018) Review on application of drone systems in precision agriculture. Procedia Comput Sci 133:502–509

Repo T, Zhang G, Ryyppö A, Rikala R (2000) The electrical impedance spectroscopy of scots pine (Pinus sylvestris L.) shoots in relation to cold acclimation. J Exp Bot 51(353):2095–2107

Ros PM, Macrelli E, Sanginario A, Demarchi D (2017) Intra-plant communication system feasibility study. 62(11):2732

Ros PM, Macrelli E, Sanginario A, Shacham-Diamand Y, Demarchi D (2019) Electronic system for signal transmission inside green plant body. In: 2019 IEEE international symposium on circuits and systems (ISCAS), pp 1–5

Sack L, Tyree MT (2005) Leaf hydraulics and its implications in plant structure and function. In: Vascular transport in plants. Academic Press, pp 93–114

Sambuelli L, Socco LV, Godio A (2003) Ultrasonic, electric, and radar measurements for living trees assessment. Boll Geofis Teor Appl 44(3–4):253–279. Leonardo, (1)

Stamm AJ (1927) The electrical resistance of wood as a measure of its moisture content. Ind Eng Chem 19(9):1021–1025

Sze SM, Ng KK (2006) Physics of semiconductor devices. Wiley, New-Jersey

Taiz L, Zeiger E (2010) Plant physiology, vol 1. Sinauer Associates, Inc., Publishers, Sunderland

Texas Instruments. HDC2080 low power humidity and temperature digital sensor. 2018

Volkov AG (2000) Green plants: electrochemical interfaces. J Electroanal Chem 483(1):150–156

Volkov AG, Nyasani EK, Tuckett C, Blockmon AL, Reedus J, Volkova MI (2016) Bioelectrochemistry cyclic voltammetry of apple fruits: memristors in vivo. Bioelectrochemistry 112:9–15

Walter A, Finger R, Huber R, Buchmann N (2017) Smart farming is key to developing sustainable agriculture. Proc Natl Acad Sci 114(24):6148–6150

Yongzong L, Yongguang H, Asante EA, Kumi F, Li P-p (2016) Design of capacitance measurement module for determining the critical cold temperature of tea leaves. Sens Bio-Sens Res., Elsevier 11:26–32

Zhang J, Willison M (1991) Electrical impedance analysis in plant tissues: a double Shell model. J Exp Bot 42(11):1465–1475

Zhao DJ, Wang ZY, Li J, Wen X, Liu A, Huang L, Wang XD, Hou RF, Wang C (2013) Recording extracellular signals in plants: a modeling and experimental study. Math Comput Model 58(3–4):556–563

Zheng L, Wang Z, Sun H, Zhang M, Li M (2015) Real-time evaluation of corn leaf water content based on the electrical property of the leaf. Comput Electron Agric 112:102–109

Zimmermann MH, Jeje AA (1981) Vessel-length distribution in stems of some American woody plants. Can J Bot 59(10):1882–1892

Data Analytics in Agriculture

Ania Cravero Leal

Abstract Food security is a crucial global need, threatened by population growth, climate change, and decreasing arable land. Data-driven agriculture is the most promising approach to solving these current and future problems by improving crop yields, reducing costs, and ensuring sustainability. As the number of smart sensors and machines on farms increases and a greater variety of data is used, farms will become increasingly data-driven, enabling the development of smart farming. This is possible, thanks to new technologies that enable massive data storage, such as cloud computing and Hadoop, in addition to processing and analysis through Big Data and machine learning. In this chapter, we explain some practical examples of their use.

Keywords Data analytics · Agriculture · Big Data · Machine learning

1 Data Analytics in Agriculture

By 2050, the world is expected to face a substantial increase in the global demand for food, necessitating a significant boost in food production by as much as 25% to 70% (Hunter et al. 2017); for this reason, it is crucial to double food production per hectare by the time the world population stabilizes around 2100 (United Nations 2019). Food security is a fundamental global need, threatened by population increase, climate change, decreasing arable land, food waste, and living standards that focus on consumer preference for animal protein (White et al. 2021).

Increasing agriculture or food production rapidly is difficult (Ahmad and Huang 2021). For this, the agricultural sector needs to employ cutting-edge technologies such as cloud computing, Internet of Things (IoT), Big Data, and machine learning (ML) (Ahmad and Huang 2021; Gopal Maya 2020). Through these technologies,

A. C. Leal (✉)
Departamento de Ciencias de la Computación e Informática, Facultad de Ingeniería y Ciencias, Universidad de La Frontera, Temuco, Chile
e-mail: ania.cravero@ufrontera.cl

P. M. Priyadarshan et al. (eds.), *Digital Agriculture*,
https://doi.org/10.1007/978-3-031-43548-5_17

data-driven agriculture is the most promising approach to solving these current and future problems(Ahmad and Huang 2021), as it improves crop yields, reduces costs, and ensures sustainability (Torky and Hassanein 2020).

Digital agriculture, like agrotechnology and precision agriculture, is a new scientific discipline that promotes agricultural productivity while minimizing environmental impact through data analysis (Liakos et al. 2018). Data are extracted from farm operations using various sensors, satellite imagery, videos, and photographs. This is possible as data analysis enables more accurate decisions through better knowledge of crop dynamics, weather conditions, soil, and farm machinery use (Liakos et al. 2018).

As the number of smart sensors and machines on farms increases and a wider variety of data is used, farms will become increasingly data-driven, enabling the development of smart farming (Sundmaeker et al. 2017). The difference between precision farming and smart farming is that the former was developed for farm management, and the latter considers real-time situations triggered by an event (Wolfert et al. 2017). On the other hand, smart farming includes intelligent assistance in implementing, maintaining, and using information technology (IT), enabling farmers to react quickly to sudden changes, such as disease alerts or weather events (Nandyala and Kim 2016).

Li et al. (2020) explain that Agricultural Big Data belongs to a comprehensive, cutting-edge technology, as it contains specific concepts, technology, and measures, covering the whole range of agricultural activities, such as farming and planting. This technology allows the processing of a large amount of heterogeneous data such as lighting, temperature, the humidity of crop growth, and data on all aspects of the production process (Li et al. 2020). With the characteristics of informatization, intelligence, and precision, it can solve the problems encountered in traditional agriculture and provide new support for agricultural development. Agricultural Big Data can respond in the new era and promote the structural reform of the agricultural supply side (Li et al. 2020). However, the research on Agricultural Big Data is in the initial stage, so more researchers are needed to do more research and analysis (Li et al. 2020).

According to Gopal Maya (2020), due to the multimodal nature of data, it has several challenges, such as improving methods for data collection and selecting effective statistical and data analysis techniques to understand and support agricultural activities. To improve these aspects, the mechanism used in smart agriculture is ML, the scientific field that allows machines to learn without much programming. It has emerged along with Big Data technologies and high-performance computing to create new opportunities to facilitate, quantify, and understand data-intensive processes in agricultural operating environments (Gopal Maya 2020).

Experts indicate that agriculture can benefit from ML at all stages, such as spice management, field management, crop management, and livestock management (Gopal Maya 2020). ML is used in several agricultural applications, including yield prediction algorithms based on weather and historical yield data, image recognition algorithms to detect pests and diseases in plants, and robotics to harvest different types of specialty crops (Tibbetts 2018).

Agricultural Big Data is playing an important role by incorporating ML. Farmers are using data to calculate crop yield, fertilizer demand, cost savings, and even to identify optimization strategies for future crops (Gopal Maya 2020). For the case of crops, ML is being used for yield prediction, disease detection, weed detection, crop quality, and species recognition. In the case of livestock, it is being used for animal welfare and livestock production (Liakos et al. 2018). In this chapter, we explain some practical examples of their use.

2 Data and Storage in Agriculture

2.1 Data

In Agricultural Big Data and ML, structured, semistructured, and unstructured data are often used, which adds complexity to the analysis process, as their use poses a significant challenge (Saiz-Rubio and Rovira-Más 2020). Unstructured data come from archives, such as videos, satellite images, and surveys, which contain a large amount of information hidden from the data scientist and cannot be analyzed directly. On the other hand, semistructured data have been stored in spreadsheets and repositories containing both essential and unimportant data for the desired analysis. Therefore, it is also necessary to process these data to obtain essential structured data, allowing data scientists to perform the relevant analyses (Cravero et al. 2022a).

Unfortunately, processing unstructured data is not trivial, as it requires the use of specialized tools and the knowledge of subject matter experts. It also requires selecting the right types of repositories and databases for further processing and analysis (Šuman et al. 2020). Therefore, it is essential to identify available data, necessary processing, and potential studies based on the generated data, as ML requires test datasets of sufficient quality to achieve the expected learning (Bhatnagar 2018).

According to Nandi and Sharma (2020), the analytics that can be performed using ML can be descriptive, diagnostic, predictive, and prescriptive. Prescriptive analytics is the most complex, as it is responsible for finding a solution among several variants to optimize resources and increase operational efficiency: the more complex the studies to be performed, the more complex the data processing will be.

According to Firdaus and Hassan (2020), it is essential to know the data type before applying any algorithm. Therefore, data type plays a vital role in preprocessing and visualization. There are four main types of data: numeric, categorical, time series, and text. Numerical data are further classified into continuous and discrete. Categorical data types represent quality; concepts such as "good", "bad," and others define levels. These data must be processed to be described as numbers rather than text.

2.2 Agricultural Data

Data in Agriculture refer to variables or attributes that farmers need to carry out their business activities. The data can be specific agricultural records or parameters, such as crop varieties, yields, soil types in use, and acreage, and business-related information, such as products, suppliers, customers, and payments. They are classified into structured, semistructured, and unstructured data, depending on the storage format in which they are stored (Cravero et al. 2022a).

The data used in Agricultural Big Data come from sensors, IoT, satellites, cameras, global positioning system (GPS), databases, and data from farmers' expert knowledge. Figure 1 shows the main data used using a concept cloud. It can be seen the most used data are temperature, humidity, crop area, wind direction, and wind speed.

An example of the use of data is the work of Nóbrega et al. (2018), in which they used video data to learn about sheep behavior in a vineyard. To do so, they analyzed each video to obtain a set of rules processed by experts. Yang et al. (2017) used video data to monitor plant growth status. Rehman et al. (2020) used various types of data for crop analysis, allowing them to understand soil temperature, atmospheric temperature, and humidity, as well as data from sensors.

Another example is the work of Dutta et al. (2015), who used heterogeneous data from different sources to generate a knowledge system of agricultural processes in

Fig. 1 Cloud of data concepts used in Agricultural Big Data

conjunction with environmental processes. Data are obtained from a sensor network, large-scale simulated models, satellite imagery, meteorological data, and industry knowledge and experience to improve decision-making. The authors developed a Big Data system that incorporates unstructured, undocumented, and ad hoc knowledge into a structured rule base that allows for an improved decision support system.

On the other hand, Amani et al. (2020) also used satellite images to obtain information on terrain characteristics. The data are extracted directly from Google Earth Engine (GEE), as it improves the efficiency of data processing from the point of view of time and costs. In addition, GEE contains freely available remote sensing datasets and several classification algorithms, which can be accessed for various farmland applications.

Figure 2 shows the number of uses of different data sources for the generation or collection of Agricultural Big Data (y0, data type). The identified sources were categorized into six groups. These are sensors, cameras, databases, GPS, satellite, and people. Each of the groups is described below.

Satellites are an essential data source for obtaining data on sizeable agricultural land. An array of sensors attached to the satellite is used to capture the data, from which numerous products can be obtained, such as optical, synthetic aperture radar (SAR), or thermal images. Cravero et al. (2022a) identified the use of six different satellites: Google Earth (Amani et al. 2020), Sentinel-1 (Shelestov et al. 2020) and Sentinel-2 (Sitokonstantinou et al. 2020), Landsat 7 and Landsat 8 (Dutta et al. 2015), and MODIS (Dutta et al. 2015).

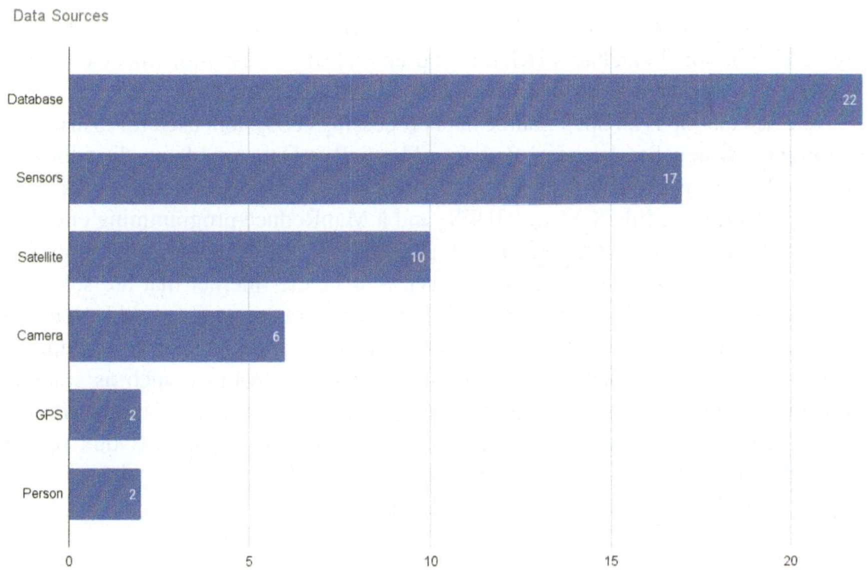

Fig. 2 Data sources used in Agricultural Big Data

The sensor group includes all IoT devices used statically in different locations to capture data. Many sensor types measure a single variable, such as temperature, radiation, and precipitation (Gnanasankaran and Ramaraj 2020). Similarly, devices that include several sensors, such as a weather stations or collars, are used in animals (Nóbrega et al. 2018). Using this data source usually requires dealing with IoT devices' deployment, connection, and maintenance. Some advantages of using sensors are that the data obtained are particular to the area or task in which they are used and that they will be captured in real time.

On the other hand, the temporal resolution of sensors tends to be low, from seconds to minutes, so large amounts of information are usually generated in a particular measurement period (Yang, J. et al., 2018). Wang and Mu explain that microsensors capable of capturing data on crop growth, land use, water use and characterization, and climatic variables, among other essential aspects, are being developed. The authors conclude that using these microsensors will enhance the development of artificial intelligence (Wang and Mu 2022).

Unlike sensors, databases allow easy and immediate access to a large amount of historical data, with accumulated records of up to 10 years. The vast majority of the identified databases are managed by public entities or government agencies, such as AWAP, CosmOZ, SILO, ASRIS, BOM, ISTAT, CNIR, IndiaStat, AAFC, ARPAS, ACIS, IMD, OGD, and KME (Cravero et al. 2022a).

2.3 Massive Storage

Two primary technologies have been employed in Big Data for massive storage. However, relational databases (RDBMS) prepared to process in-memory data and NoSQL databases that store unstructured data have also been used.

Apache Hadoop is an open-source data processing ecosystem used for distributed computing, which has been created to address Big Data problems. In addition, Hadoop has been expanded to use geospatial data. Hadoop generally contains a Hadoop Distributed File System (HDFS) and a MapReduce programming environment for data processing (Alkathiri et al. 2019).

Cloud computing provides various services over the Internet that are scalable. This technology allows resource sharing using the infrastructure owned by a cloud service provider. The provider's users or customers can access resources on demand by paying per use. It enables the abstraction of infrastructures, such as storage, network, and applications, through its three services: Platform as a Service (PaaS), Infrastructure as a Service (IaaS), and Software as a Service (SaaS) (Odun-Ayo et al. 2018). The fourth layer of services is Business Intelligence (BI), which contains applications to measure management indicators.

2.4 Massive Storage in Agriculture

Cravero et al. (2022a) analyzed 36 papers where Big Data and ML are applied for analysis in agriculture. Figure 3 shows the distribution of uses of the mentioned platforms, categorized into Hadoop, relational database, NoSQL database, and cloud.

The cloud category includes several cloud computing services, such as AWS (Amazon Web Services) or GEE (Google Earth Engine). A direct advantage of using these platforms is their large computational and storage capacity. They are suitable for working with Big Data and can be resized according to the user's needs. Another benefit is the free and direct access they provide to different data sources, such as satellite data captured by Landsat or Sentinel satellites.

Shelestov et al. (2020) used AWS's fast and easy access to Sentinel-1 and Sentinel-2 satellite imagery to work with datasets using up to 3 TB of memory space, eliminating the problems associated with downloading and storing data related to Big Data. Gumma et al. (2020) list the following reasons for using the GEE platform: easy access to Landsat satellite data, the powerful computational capability of the service, and the ability to perform parallel processing of the data, among others.

Wang et al. (2019) use MongoDB, a document-oriented database, as intermediate temporary storage for data collected by sensors, which are subsequently transferred to an implemented data warehouse. Sathiaraj et al. (2019) used the in-memory database REDIS, whose data model is key value, for the visualization system of

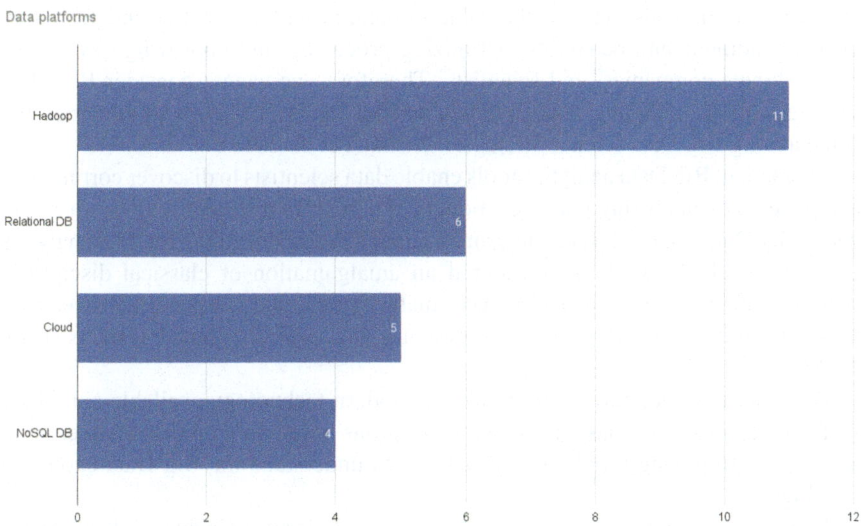

Fig. 3 Mass storage used in agriculture

the computational analyses performed, as it presents low latency when accessing the data.

Nóbrega et al. (2018) use PostgreSQL to store the data obtained by the collars placed on each sheep. They decided to use a relational database because their sensor collar network has several entities that can be efficiently designed in this database. Furthermore, they selected PostgreSQL among the available RDBMS options because it suits environments with system–critical data, security, and integrity mechanisms.

3 Analysis in Agriculture

3.1 Agricultural Big Data

Big Data is defined in four dimensions (4 Vs). The first V refers to the enormous volume of data being developed, stored, and processed. The second V refers to the high speed of data transmission in interactions and the rates at which data are generated, collected, and exchanged. The third V refers to the variety of data formats and structures (structured, semistructured, and unstructured) resulting from the heterogeneity of data sources (Sassi et al. 2019). The fourth V is veracity, which refers to the ability to validate the data quality used in the analyses.

Apart from the "4 Vs", another dimension of Big Data, its value, must also be considered. Value is obtained by analyzing data to extract hidden patterns, trends, and knowledge models through intelligent data analysis algorithms and techniques. Data science methods increase the value of data, providing a better understanding of its phenomena and behaviors, optimizing processes, and improving discoveries by machines, companies, and scientists. Therefore, we cannot consider Big Data science without including data analytics and ML as critical steps to numerate the value among Big Data science strategies (Elshawi et al. 2018).

In practice, Big Data analytics tools enable data scientists to discover correlations and patterns by analyzing massive amounts of data from different sources. In recent years, Big Data science has become an essential modern discipline for data analytics (Elshawi et al. 2018). It is considered an amalgamation of classical disciplines such as statistics, artificial intelligence, mathematics, and computer science, with its subdisciplines including database systems, ML, and distributed systems (Haig 2020).

Big Data in agriculture refers to all the modern technology available combined with data analysis as a basis for making decisions based only on data (Sarker et al. 2019). The following typology will help us to understand the Big Data evolution (see Fig. 4).

Precision agriculture collects real-time data on farm elements such as crops, air, and soil to protect the environment while ensuring profits and sustainability (Micheni et al. 2022). Incorporating ML techniques in farming has advanced aspects

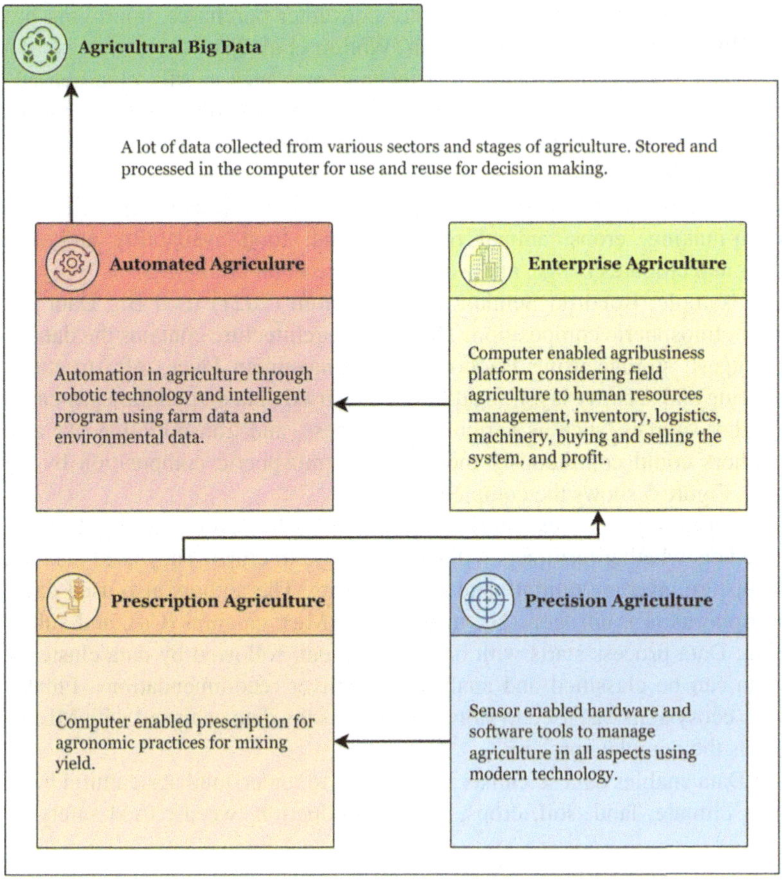

Fig. 4 Topologies in digital agriculture

such as crop and soil health, irrigation systems, crop disease identification, weed control, and recommended control measures. The adoption of a robotic farming system has a significant impact on crop production, efficiency, and sustainability. However, the success of precision farming is hampered by factors such as lack of training, low return on investment, high costs, and lack of Big Data analysis of precision farming.

Big Data has been used to improve various aspects of agriculture, such as knowledge about weather and climate change, land, animal research, crops, soil, weeds, food availability and security, biodiversity, farmer decision-making, insurance and farmer financing, and remote sensing (Kamilaris et al. 2017). It is also used to create platforms that enable supply chain actors to access high-quality products and processes; tools to improve yields and predict demand; and advice and guidance to farmers based on the responsiveness of their crops to fertilizers, leading to better fertilizer use. It has also led to the introduction of plant scanning equipment to track deliveries and enable retailers to monitor consumer purchases, improving product traceability throughout the supply chain (Wolfert et al. 2017).

Big Data has been used with other technologies such as ML, cloud-based platforms, image processing, modeling and simulation, statistical analysis, normalized difference vegetation index (NDVI), and geographic information systems (GIS) (Kamilaris et al. 2017).

There are Big Data solutions for different areas of agriculture, such as farmer decision-making, crops, animal research, land, food availability and security, weather and climate change, and weeds (Cravero et al. 2022b).

For example, Boudriki Semlali and El Amrani (2021) used Big Data tools to monitor atmospheric composition. The system architecture contains the data source layer, ingest, storage using Hadoop, data management layer, infrastructure, and monitoring and security layer. In addition, they used data on pollutant gas emissions from other sources, such as agriculture, business, and transportation. As a result, the authors could continuously monitor the atmospheric composition by remote sensing. Figure 5 shows the complete process.

Another example is Alex and Kanavalli (2019), who developed a Big Data system that predicts whether fertilizers will cause disease in crops. They used data such as soil moisture, average rainfall, and soil nutrients. The authors also used data such as phosphorus (P), nitrogen (N), magnesium (Mg), calcium (Ca), and sulfur (S). The Big Data process starts with data enrichment, followed by data clustering, so the data can be classified and analyzed to deliver recommendations. Finally, the Hadoop ecosystem was used to store and process the data analyzed with ML. Figure 6 depicts the complete process.

Big Data enables data scientists and farmers to understand agricultural behavior, such as climate, land, soil, crops, animal production, weeds, food safety, biodi-

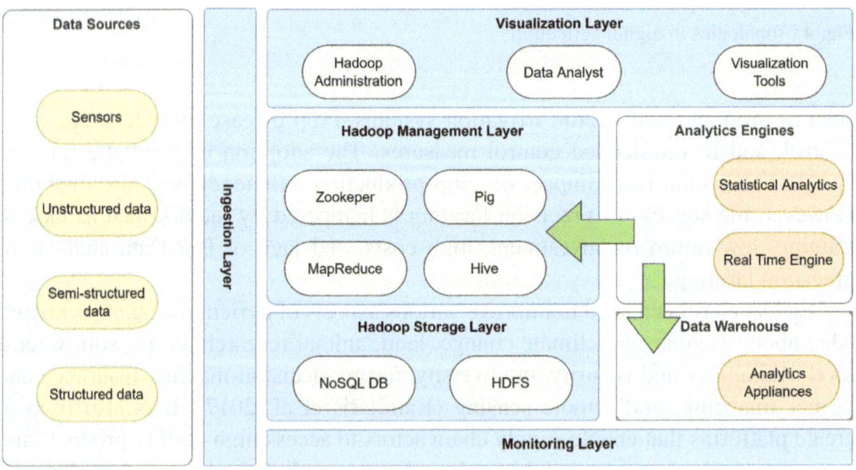

Fig. 5 Architecture of Big Data for atmospheric composition monitoring

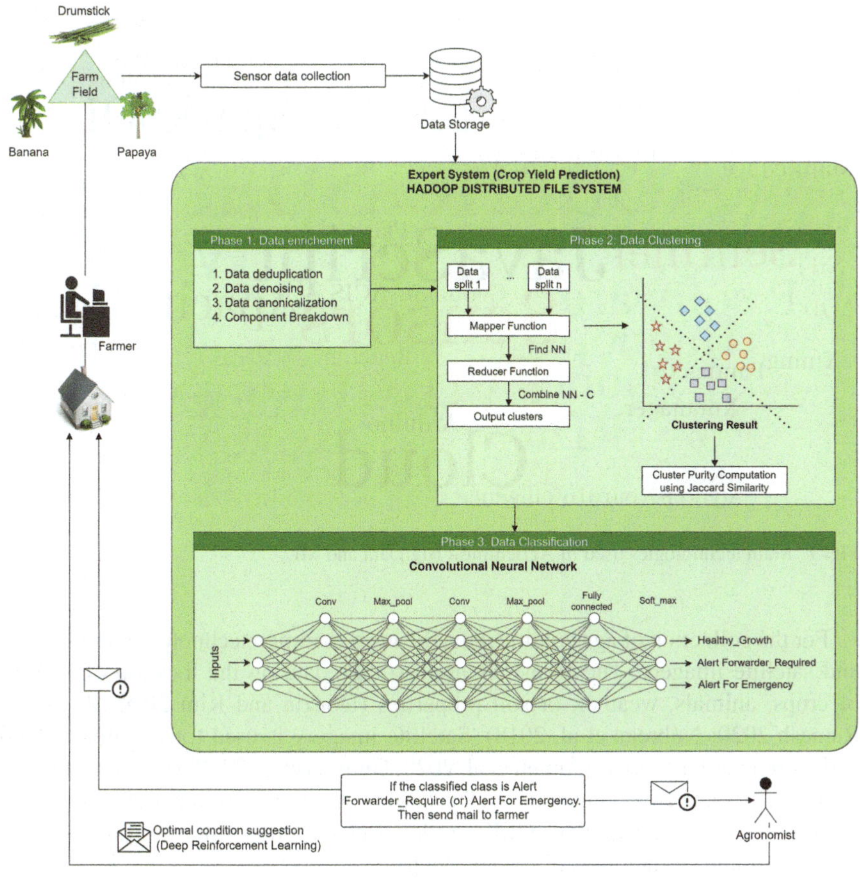

Fig. 6 Big Data architecture for fertilizer management and yield prediction

versity, remote sensing, farmer decision-making, insurance, financing, and climate change. It also enables the development of supply chain platforms, which allow players to access high-quality products, processes, and tools capable of improving yields, predicting demand, and targeting farmers based on crop needs, such as appropriate fertilizer use.

3.2 Agricultural Big Data Technologies

The technologies used for the implementation of Big Data and ML systems in agriculture are presented in Fig. 7. It can be seen that the most frequently identified technology was Apache Spark (Cravero et al. 2022b).

Fig. 7 Main technologies used in Agricultural Big Data and ML

For the collection of agricultural data, the most common technologies are sensors and satellite imagery. The former is used to take data on the location of interest of crops, animals, weather, or soil properties (Donzia and Kim 2020; Priya and Ramesh 2020; Nóbrega et al. 2018). Satellite imagery is used for monitoring land and crops over large areas (Amani et al. 2020; Gumma et al. 2020) (Sitokonstantinou et al. 2020). Data are obtained through satellites or services from external providers such as Google Earth Engine (GEE), a global positioning system (GPS), Sentinel satellites, Landsat satellites, or Google Maps.

In the implementation of Big Data systems, the most used file system is Hadoop Distributed File System (HDFS), because it allows to separate of datasets, storing them in a distributed way in several nodes of a cluster, and parallelizing operations on them (Sitokonstantinou et al. 2020). Most of the implemented clusters were configured with the various programs provided through the Apache Hadoop framework. Among these, the following stand out: Apache Hive and Apache Kafka. Apache Hive is used to configure data warehouses that streamline working with large datasets stored in distributed units (Wang et al. 2019) (Pandya et al. 2020). Apache Kafka is used to transmitting information or messages to different nodes of the designed Big Data architecture (Pandya et al. 2020; Donzia and Kim 2020).

The technology most often identified in the implementation of ML models is the Python programming language (Balducci et al. 2018; Gumma et al. 2020; Fenu and Malloci 2019). It highlights the latter especially together with libraries such as D3 for data visualization (Sathiaraj et al. 2019), Leaflet.js for displaying maps (Doshi et al. 2018), and React for building interactive user interfaces (Sathiaraj et al. 2019).

In (Wang et al. 2019), a Big Data system for agriculture was proposed and designed based on data collection, storage, analysis, and application. For collecting pear tree growth data (air temperature, soil moisture, light intensity, etc.), a high-precision wireless sensor network is used, sending collected data via TCP protocol to traditional databases (MySQL, MongoDB, etc.). These databases are used temporarily to store the data and serve as data sources for the overall Big Data system. For this purpose, data synchronization software such as NiFi, Sqoop, or Flume is used. Data sources are synchronized with the HDFS cluster responsible for storing all the data together. SparkSQL reads, filters, and stores data from the HDFS cluster to Apache Hive and Apache Hbase. The former is employed for data used for analysis, and the latter is utilized for data monitoring and visualization of data statistics. Apache Dubbo is used for running farmer management services in a distributed manner (Cravero et al. 2022a).

3.3 Machine Learning in Agriculture

ML is a highly interdisciplinary field based on different areas such as artificial intelligence, optimization theory, information theory, statistics, cognitive science, optimal control, and many other scientific, engineering, and mathematical disciplines (Cherkassky and Mulier 2007). ML has covered almost all science domains, impacting society significantly (Rudin and Wagstaff 2014). It has been used in various problems, including recommendation controllers, computer science and data mining, recognition systems, and autonomous control systems (Qiu et al. 2016). In general, ML is used to optimize the performance of a task through mining past examples or experiences, as it can generate efficient relationships concerning data inputs and reconstruct a knowledge schema.

ML has been used to solve different problems in agriculture, such as crop management, including yield prediction; disease detection, weed detection, crop quality, and species recognition; livestock management, including animal welfare and livestock production; water management; and soil management (Liakos et al. 2018; Benos et al. 2021; Bal and Kayaalp 2021).

An example of its use is in precise detection, as together with sensors, it allows accurate detection and identification of weeds without causing environmental problems or side effects. ML for weed detection has led to the development of tools and robots to destroy weeds, minimizing the need for herbicides (Liakos et al. 2018). In addition, accurate detection and classification of crop quality characteristics have increased the value of products and reduced waste.

The increased research interest in ML in agriculture is a consequence of several factors: the considerable advances in IT systems in agriculture; the vital need to increase the efficiency of farming practices while reducing the environmental burden; and the need for reliable measurements with the handling of large volumes of data (Benos et al. 2021; Bal and Kayaalp 2021).

ML is used in conjunction with Big Data, as it allows analyzing a volume of data that is generated after processing and filtering data coming from different heterogeneous sources. Agricultural Big Data has technologies that allow ML

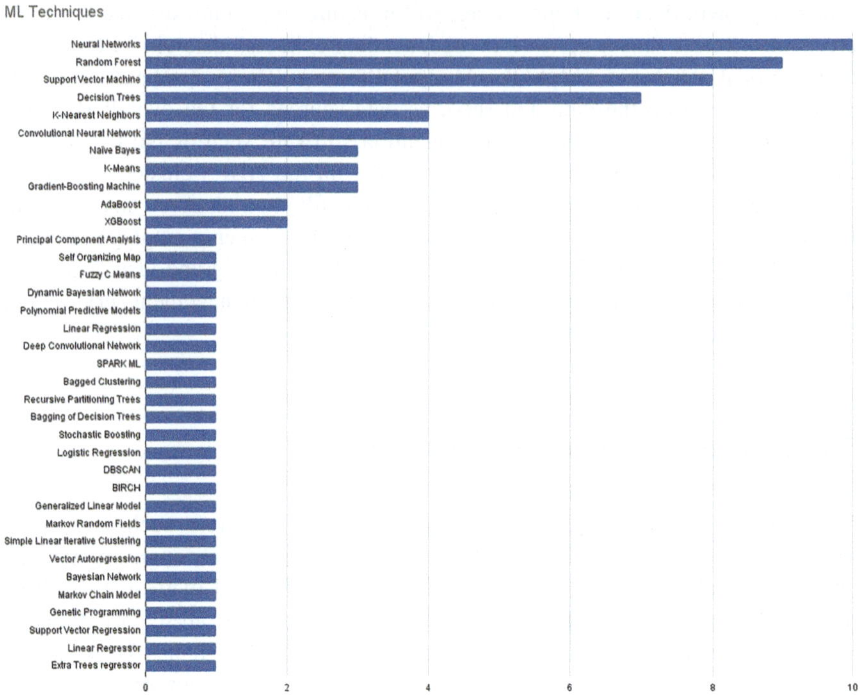

Fig. 8 ML techniques in Agricultural Big Data

algorithms to perform their work. According to Cravero et al. (2022b), the most commonly used ML techniques in Agricultural Big Data are Neural Networks (NNs), Random Forest (RF), Support Vector Machine (SVM), and Decision Tree (DT). Figure 8 shows a list of ML techniques and the number of times they have been used in Big Data in the last 5 years.

Some examples of their use are listed below.

3.3.1 Neural Networks

NNs are an excellent choice for working with large datasets because they have great flexibility to adapt to these, reducing the error produced by adjusting the weights and biases of each neuron based on the data it is trained with (Priya and Ramesh 2020). Saggi and Jain (2022) implemented the NN and compared its performance alongside other ML techniques. The NN was the best-performing technique, avoiding model overfitting and demonstrating excellent capabilities for estimating daily crop evapotranspiration.

Doshi et al. (2018) used NNs for automatic crop recommendation due to its built-in support for multilabel classification. In this task, the technique performed well with 91% classification accuracy. Shelestov et al. (2020) found that the most sensitive parameters for the classification accuracy of an NN are the number of hidden neurons and the regularization of alpha coefficients.

3.3.2 Random Forest

Some of the applications of RF are crop prediction, crop yield under adverse conditions, identification of climatic variables, and analysis of agriculture-related problems such as nitrogen emissions or drought prediction (Priya and Ramesh 2020). Furthermore, RF is ideal for working with massive datasets, as it needs less time for data preprocessing, is proficient in global time complexity, and works well with sparse datasets (Priya and Ramesh 2020).

Doshi et al. (2018) implement RF for crop recommendation due to its built-in support for multiple-label classification (MLC), highlighting that this technique is effective for handling missing values and is resistant to model overfitting. The latter feature is one of the reasons for its implementation in the classification of South Asian croplands (Gumma et al. 2020).

3.3.3 Support Vector Machine

SVM is suitable for handling small datasets that do not contain too many outliers, and its performance is increased when the dimensional space of the data is ample. However, the attributes are lower (Priya and Ramesh 2020).

In Nóbrega et al. (2018), different ML algorithms, including SVM, are compared to detect the conditions of an animal concerning posture data. Of the analyzed algorithms, SVM was the one that presented the worst performance; however, its results do not differ noticeably from the rest of the algorithms, and all of them had over 95% accuracy. A similar case was observed in Yang et al. (2018), where after comparing different ML techniques for predicting the growth state of a plant, it was observed that SVM had the lowest accuracy, although this was above 90%. In both cases, SVM was not the most suitable technique for the tasks performed, but it demonstrated a good level of accuracy.

Shelestov et al. (2020) found that the most sensitive parameters of SVM are gamma, C, and the type of kernel used. They performed measurements on the latter using Radial Basis Function (RBF) and sigmoid kernels and found RBF to be the most appropriate for crop classification tasks.

3.3.4 Decision Tree

DT is efficient in terms of computation and scalability. Moreover, its performance increases when the data are uncorrelated (Priya and Ramesh 2020). The efficiency of this technique is proven in Nóbrega et al. (2018), where they compared different

ML techniques for the classification of an animal's posture using data collected by an IoT collar. Of the compared techniques, the authors highlighted DT due to the low computational time required for model training and the easy subsequent interpretation of the model; they also presented one of the best values of accuracy and area under the curve (AUC) of the compared techniques.

On the other hand, Yang et al. (2018) investigated the prediction of the growth status of a plant using different ML techniques. The authors concluded that DT was the best algorithm compared due to the low time consumed and the high level of accuracy presented.

4 Challenges and Future Work

In general, data analysis for agriculture brings many benefits (Chergui et al. 2020) such as:

- Providing the farmer with helpful information and helping him to make decisions on how much, when, and where to apply nutrients, water, seeds, fertilizers, and other agricultural chemicals and inputs.
- Protecting the environment and helping to obtain healthy products, as it allows varying the number of inputs (irrigation, fertilizers, and pesticides) and even seeds used for crop production, and applying those inputs in exact amounts in each field.
- Data-driven management gives farmers access to sophisticated management solutions against climate change and other environmental challenges and natural phenomena. Thanks to these solutions, farmers can continuously monitor crop health and receive timely alerts about potential pests, disease problems, or climate change.
- From a marketing point of view, farmers can also benefit from advanced models that provide information about the market and which products could bring them the most profit.

According to Basnet and Bang (2018), there are still many challenges for data analysis in agriculture. The authors explain that improved technology will add more precision, accuracy, speed, and reliability to data analysis and reduce costs. On the other hand, it is crucial to achieve technology standardization to improve communication between agricultural equipment and research and open-source projects to improve the quality of technological solutions. The authors also explain that user-friendly technical solutions are required, as they must be adapted to local contexts and needs. It is also essential to improve the understanding of the use of Big Data by systematically promoting the concept, its practical use, the need for multidisciplinary work, and the value of its use, expanding education and awareness of the use in data analysis with Big Data.

Lassoued et al. (2021) analyzed the impact and potential of Big Data in agriculture. They identified several challenges related to data sources because not all value chain segments capture data in the same way. For example, they noted that there is no standard for data capture, which makes it difficult to harmonize and compile data from various sources. Another major obstacle identified is data governance. While most experts surveyed were willing to share their data under certain conditions, many expressed concerns about data privacy, security, cybercrime, and intellectual property protection.

On the other hand, Bhat and Huang (2021) examined the challenges of data collection and analysis. The combination of data from various sources raises concerns about the quality of information and its fusion. In addition, the volume of information collected causes safety and security concerns. The datasets collected are vast and complex, making it challenging to handle standard intelligent analysis procedures. These methods often do not work well when applied to agricultural data. The authors expect scalable and versatile methods to adapt to large amounts of information. Weersink et al. (2018) explained that data must be collected consistently and comply with protocols that allow them to be pooled on centralized servers. These servers must be protected from cyberattacks while masking the identity of the operations managers.

Regarding the use of ML in Big Data, a major challenge is to cope with a large volume of data. For example, the SVM algorithm has a training time complexity of O(n3) and a space complexity of O(n2), where n is the number of training samples. An increase in the value of n will drastically affect the time and memory required to train this algorithm and may even become computationally infeasible for large datasets.

A common assumption of ML is that algorithms can learn better with more data and provide more accurate results. However, massive datasets impose several challenges because traditional algorithms were not designed to meet such requirements (Cravero et al. 2022b).

In Agricultural Big Data, a combination of technologies is required, as data from experts, videos, and satellite images will be processed in batches. On the other hand, data from social networks and sensors will be processed by streaming. In the case of cloud-based technologies, there are several tools for ML use: Microsoft Azure Machine Learning, now part of Cortana Intelligence Suite; Google Cloud Machine Learning Platform; Amazon Machine Learning; and IBM Watson Analytics (Yang et al. 2017). Established vendors offer these services, which provide scalability and integration with other services and platforms.

Other challenges include understanding the statistical characteristics of the data before applying algorithms and the ability to work with larger datasets (Sukumar 2014). In addition, specific knowledge is required for certain problems in agriculture, such as increased production, quality improvement, and climate change, among others.

The future of Agricultural Big Data development and ML use is promising. This future will increase the effect of flexible Big Data architectures that consider various alternative ML techniques depending on the conditions of the data generated. This increase is possible thanks to developed and constantly evolving technologies. On the other hand, cloud computing will increase due to new professionals' training and network speed improvement. Cloud computing and other tools will include more alternative ML techniques, which will facilitate flexibility.

As for ML techniques, the use of DL and other techniques mentioned in this chapter, which were adapted to specific contexts due to problems with data volume, processing speed, variability, and veracity, will increase. However, these problems can be solved by classifying data storage through the Data Lake.

Future research should focus on implementing appropriate decision support systems for accurate crop decisions, natural resource management, and climate change mitigation.

References

Ahmad, Huang (2021) Big data and AI revolution in precision agriculture: survey and challenges. IEEE Access. Received 14 June 2021, accepted 30 July 2021, date of publication 3 August 2021, date of current version 12 August 2021. Digital Object Identifier. https://doi.org/10.1109/ACCESS.2021.3102227

Alex SA, Kanavalli A (2019) Intelligent computational techniques for crops yield prediction and fertilizer management over big data environment. Int J Innov Technol Exploring Eng (IJITEE) 8(12)

Alkathiri M, Jhummarwala A, Potdar MB (2019) Minería de datos geoespaciales multidimensionales en un entorno distribuido utilizando MapReduce. J Big Data 6:1–34

Amani M, Kakooei M, Moghimi A, Ghorbanian A, Ranjgar B, Mahdavi S et al (2020) Application of google earth engine cloud computing platform, sentinel imagery, and neural networks for crop mapping in Canada. Remote Sens 12(21):1–18. https://doi.org/10.3390/rs12213561

Bal, Kayaalp (2021) Int Adv Res Eng J 05(02):309–323

Balducci F, Impedovo D, Pirlo G (2018) Machine learning applications on agricultural datasets for smart farm enhancement. Mach Des 6(3):38. https://doi.org/10.3390/machines6030038

Basnet B, Bang J (2018) The state-of-the-art of knowledge-intensive agriculture: a review on applied sensing systems and data analytics. J Sens 2018

Benos L, Tagarakis AC, Dolias G, Berruto R, Kateris D, Bochtis D (2021) Machine learning in agriculture: a comprehensive updated review. Sensors 21(11):3758

Bhat SA, Huang NF (2021) Big data and AI revolution in precision agriculture: survey and challenges. IEEE Access 9:110209–110222

Bhatnagar R (2018) Machine learning and big data processing: a technological perspective and review. In: Proceedings of the international conference on advanced machine learning technologies and applications. Springer, Berlin/Heidelberg, pp 468–478

Boudriki Semlali BE, El Amrani C (2021) Big data and remote sensing: a new software of ingestion. Int J Electr Comput Eng 11:1521–1530

Chergui N, Kechadi MT, McDonnell M (2020) The impact of data analytics in digital agriculture: a review. In: 2020 international multi-conference on: "Organization of Knowledge and Advanced Technologies" (OCTA). IEEE, pp 1–13

Cherkassky V, Mulier FM (2007) Learning from data: concepts, theory, and methods. Wiley Intercience, John Wiley & Sons, IEEE

Cravero A, Pardo S, Galeas P, López Fenner J, Caniupán M (2022a) Data type and data sources for agricultural big data and machine learning. Sustainability 14(23):16131

Cravero A, Pardo S, Sepúlveda S, Muñoz L (2022b) Challenges to use machine learning in agricultural big data: a systematic literature review. Agronomy 12(3):748

Donzia SKY, Kim H-K (2020) Architecture design of a smart farm system based on big data appliance machine learning. In: 2020 20th international conference on computational science and its applications (ICCSA), pp 45–52. https://doi.org/10.1109/ICCSA50381.2020.00019

Doshi Z, Nadkarni S, Agrawal R, Shah N (2018) AgroConsultant: intelligent crop recommendation system using machine learning algorithms. In: 2018 fourth international conference on computing communication control and automation (ICCUBEA), pp 1–6. https://doi.org/10.1109/ICCUBEA.2018.8697349

Dutta R, Li C, Smith D, Das A, Aryal J (2015) Big data architecture for environmental analytics. Int Symp Environ Softw Syst:578–588

Elshawi R, Sakr S, Talia D, Trunfio P (2018) Big data systems meet machine learning challenges: towards big data science as a service. Big Data Res 14:1–11

Fenu G, Malloci FM (2019) An application of machine learning technique in forecasting crop disease. Paper presented at the PervasiveHealth: Pervasive Computing Technologies for Healthcare, 76–82. https://doi.org/10.1145/3372454.3372474. Retrieved from www.scopus.com

Firdaus H, Hassan SI (2020) Unsupervised learning on healthcare survey data with particle swarm optimization. In: Machine learning with health care perspective. Springer, Berlin/Heidelberg, pp 57–89

Gnanasankaran N, Ramaraj E (2020) The effective yield of paddy crop in Sivaganga District-an initiative for smart farming. Int J Sci Technol Res 9:2

Gopal Maya PS (2020) Big data challenges and opportunities in agriculture. Int J Agric Environ Inf Syst 11(1)

Gumma MK, Thenkabail PS, Teluguntla PG, Oliphant A, Xiong J, Giri C et al (2020) Agricultural cropland extent and areas of South Asia derived using landsat satellite 30-m time-series big-data using random forest machine learning algorithms on the google earth engine cloud. GIScience Remote Sens 57(3):302–322. https://doi.org/10.1080/15481603.2019.1690780

Haig BD (2020) Big data science: a philosophy of science perspective

Hunter MC, Smith RG, Schipanski ME, Atwood LW, Mortensen DA (2017) Agriculture in 2050: recalibrating targets for sustainable intensification. Bioscience 67(4):386–391

Kamilaris A, Kartakoullis A, Prenafeta-Boldú FX (2017) A review on the practice of big data analysis in agriculture. Comput Electron Agric 143:23–37

Lassoued R, Macall DM, Smyth SJ, Phillips PWB, Hesseln H (2021) Expert insights on the impacts of, and potential for, agricultural big data. Sustainability 13:2521. https://doi.org/10.3390/su13052521

Li H, Cheng Z, Wang H (2020) Research of agricultural big data. E3S Web Conf 214:01011. https://doi.org/10.1051/e3sconf/202021401011. EBLDM 2020

Liakos KG, Busato P, Moshou D, Pearson S, Bochtis D (2018) Machine learning in agriculture: a review. Sensors 18:2674. https://doi.org/10.3390/s18082674

Micheni E, Machii J, Murumba J (2022) Internet of things, big data analytics, and deep learning for sustainable precision agriculture. In: 2022 IST-Africa conference (IST-Africa). IEEE, pp 1–12

Nandi G, Sharma RK (2020) Data science fundamentals and practical approaches: understand why data science is the next. BPB Publications, Uttar Pradesh

Nandyala C, Kim HK (2016) Big and meta data management for U-agriculture mobile services. Int J Softw Eng Appl IJSEIA 10:257–270

Nóbrega L, Tavares A, Cardoso A, Gonçalves P (2018) Animal monitoring based on IoT technologies. In: 2018 IoT vertical and topical summit on agriculture - Tuscany (IOT Tuscany), pp 1–5. https://doi.org/10.1109/IOT-TUSCANY.2018.8373045

Odun-Ayo I, Ananya M, Agono F, Goddy-Worlu R (2018) Cloud computing architecture: Un análisis crítico. En Proceedings of the 2018 18th International Conference on Computational Science and Applications (ICCSA), Melbourne, Australia, 2–5 de julio de 2018

Pandya A, Odunsi O, Liu C, Cuzzocrea A, Wang J (2020) Adaptive and efficient streaming time series forecasting with lambda architecture and spark. In: Paper presented at the proceedings - 2020 IEEE international conference on big data, big data 2020, pp 5182–5190. https://doi.org/10.1109/BigData50022.2020.9377947. Retrieved from www.scopus.com

Priya R, Ramesh D (2020) ML based sustainable precision agriculture: a future generation perspective. Sustainable Comput Inf Syst 28:100439
Qiu J, Wu Q, Ding G, Xu Y, Feng S (2016) A survey of machine learning for big data processing. EURASIP J Adv Signal Process 2016:1–16
Rehman A, Liu J, Keqiu L, Mateen A, Yasin MQ (2020) Machine learning prediction analysis using IoT for smart farming. Int J Emerg Trends Eng Res 8(9):6482–6487. https://doi.org/10.30534/ijeter/2020/250892020
Rudin C, Wagstaff KL (2014) Machine learning for science and society. Mach Learn 95:1–9
Saggi MK, Jain S (2022) A survey towards decision support system on smart irrigation scheduling using machine learning approaches. Arch Comput Methods Eng 29(6):4455–4478
Saiz-Rubio V, Rovira-Más F (2020) From smart farming towards agriculture 5.0: a review on crop data management. Agronomy 10:207
Sarker MNI, Islam MS, Ali MA, Islam MS, Salam MA, Mahmud SH (2019) Promoting digital agriculture through big data for sustainable farm management. Int J Innov Appl Stud 25(4):1235–1240
Sassi I, Anter S, Bekkhoucha A (2019) An overview of big data and machine learning paradigms. In: Advanced intelligent systems for sustainable development (AI2SD'2018), vol 5: Advanced intelligent systems for computing sciences, pp 237–251
Sathiaraj D, Huang X, Chen J (2019) Predicting climate types for the continental United States using unsupervised clustering techniques. Environmetrics 30(4). https://doi.org/10.1002/env.2524
Shelestov A, Lavreniuk M, Vasiliev V, Shumilo L, Kolotii A, Yailymov B et al (2020) Cloud approach to automated crop classification using Sentinel-1 imagery. IEEE Trans Big Data 6(3):572–582. https://doi.org/10.1109/TBDATA.2019.2940237
Sitokonstantinou V, Drivas T, Koukos A, Papoutsis I, Kontoes C (2020) Scalable distributed random forest classification for paddy rice mapping Asian conference on remote sensing ACRS 2019. Paper presented at the 40th Asian Conference on Remote Sensing, ACRS 2019: Progress of Remote Sensing Technology for Smart Future, Retrieved from www.scopus.com
Sukumar SR (2014) Machine learning in the big data era: are we there yet? In: Proceedings of the ACM knowledge discovery and data mining: workshop on data science for social good, New York, NY, USA, 24–17 August 2014
Šuman S, Poščić P, Gligora Marković M (2020) Big data management challenges. Int J Adv Trends Comput Sci Eng 9:717–723
Sundmaeker H, Verdouw C, Wolfert S, Pérez Freire L (2017) Internet of food and farm. In: Vermesan O, Friess P (eds) Digitising the industry-internet of things connecting the physical, digital and virtual worlds. River Publishers, Delft
Tibbetts (2018) The frontiers of artificial intelligence. Bioscience 68(1):5–10. https://doi.org/10.1093/biosci/bix136
Torky M, Hassanein AE (2020) Integrating block chain and the Internet of Things in precision agriculture: analysis, opportunities, and challenges. Comput Electron Agric 178:105476
United Nations (2019) World population prospects. United Nations Department of Economic and Social Affairs, Population Division. Retrieved November 21, 2019 from https://population.un.org/wpp/
Wang Q, Mu Z (2022) Risk monitoring model of intelligent agriculture Internet of Things based on big data. Sustain Energy Technol Assess 53:102654
Wang X, Yang K, Liu T (2019) The implementation of a practical agricultural big data system. Paper presented at the 2019 IEEE 5th International Conference on Computer and Communications, ICCC 2019, pp 1955–1959. https://doi.org/10.1109/ICCC47050.2019.9064475. Retrieved from www.scopus.com
Weersink A, Fraser E, Pannell D, Duncan E, Rotz S (2018) Opportunities and challenges for big data in agricultural and environmental analysis. Annu Rev Resour Econ 10:19–37
White, Emma L.;·J. Alex Thomasson, Brent Auvermann, Newell R. Kitchen, Leland Sandy Pierson, Dana Porter, Craig Baillie, Hendrik Hamann, Gerrit Hoogenboom, Todd Janzen, Rajiv

Khosla, James Lowenberg-DeBoer, Matt McIntosh, Seth Murray, Dave Osborn, Ashoo Shetty, Craig Stevenson 2021

Wolfert S, Ge L, Verdouw C, Bogaardt MJ (2017) Big data in smart farming–a review. Agric Syst 153:69–80

Yang C, Huang Q, Li Z, Liu K, Hu F (2017) Big data and cloud computing: innovation opportunities and challenges. Int J Digit Earth 10:13–53

Yang J, Liu M, Lu J, Miao Y, Hossain MA, Alhamid MF (2018) Botanical internet of things: toward smart indoor farming by connecting people, plant, data and clouds. Mob Netw Appl 23:188–202

Part IV
Precision Agriculture Technologies

Sensing Systems for Precision Agriculture

Laura García, Sandra Sendra, and Jaime Lloret

Abstract Precision agriculture (PA) is becoming a staple part of the current primary sector. PA systems are comprised of different elements that allow monitoring variability in different agricultural parameters, transmitting data, and performing data storage and analysis. Sensing systems for PA focus primarily in acquiring all the information the PA system needs to gain knowledge of the state of the fields, the crops, the weather, or the available resources. All this information is key to providing farmers with predictions and suggestions to improve the quality and quantity of the produce as well as to aid in increasing the sustainability of agriculture. This chapter details the current state of sensing systems for PA from its evolution as a topic of research interest to the future trends to be expected in the next few years. Moreover, an overview of an architecture of a PA sensing system with complete functionalities is provided, as well as an overview of the sensing devices and technologies available for each PA domain.

Keywords Internet of Things (IoT) · Sensors · Precision agriculture (PA) · Monitoring · New technologies · Wireless sensor network (WSN)

1 Introduction

The introduction of technology in agriculture dates from the time humans began performing agricultural activities. This technology has evolved from learning planting schedules, learning when to irrigate, forms of irrigation, or how to fertilize, to the addition of chemicals to control pests, the evolution of tools and machinery, or the introduction of devices able to communicate with each other. This evolution was motivated by the will for increasing the yield, the quality, and the resilience of the produce so as to feed more people and trade for other goods with the excess produce.

L. García (✉) · S. Sendra · J. Lloret
Universitat Politècnica de València, Valencia, Spain
e-mail: laugarg2@teleco.upv.es; sansenco@upv.es; jlloret@dcom.upv.es

P. M. Priyadarshan et al. (eds.), *Digital Agriculture*,
https://doi.org/10.1007/978-3-031-43548-5_18

Therefore, agriculture nowadays is still on the path of evolving with the addition of Internet of Things (IoT) solutions and the incorporation of techniques such as artificial intelligence (AI) that analyze the information collected from different sensors deployed on the fields to automate management and perform predictions. This introduction of sensors and communication devices with the aim of managing spatial and temporal variability in agriculture has received the name of precision agriculture (PA) (Pierce and Nowak 1999). Thus, a successful implementation of a PA system is able to assess, understand, and manage variability in agriculture aiming for low inputs, high efficiency, and sustainability (Zhang et al. 2002).

The use of sensors is compulsory to deploy a smart agriculture system. Sensors are the devices that monitor variability for each factor or parameter of interest. Variability can be present in yield regarding historical and present distributions (Zhang et al. 2002). Field variability is related to the topography with factors such as elevation or slope. Soil variability depends on properties such as moisture content, water-holding capacity, conductivity, texture, or depth. Crop variability regards water stress, plant height, leaf-area index, nutrient deficits, or chlorophyll content. There is also variability in management such as crop rotations or the seeding process. Lastly, variability can be detected as well in other factors such as plant disease, pests, weeds, or the effects of adverse weather conditions. Each factor is monitored with the use of one sensor; therefore, sensors in PA may be part of multiparametric probes or devices comprised of an array of sensors that characterize one of the main areas of variability. The area of study or management zone (Zhang et al. 2002) is comprised of a part of the field that presents a homogeneous combination of crop-related factors. Thus, the design of the sensing deployment must consider the characteristics and number of management zones to determine not only the type of sensors that are needed but also the form in which the devices are connected to each other, and the information is transmitted.

Although sensors are the principal components of sensing systems for PA, the system would not be useful without the addition of devices to perform data storage, data analysis, and data transmission tasks. Most sensing devices usually include an SD module to store the data even if the system transmits information in real-time. This is done to avoid losing data in case of malfunction (Lloret et al. 2021). Furthermore, intermediate devices are needed to carry the data to the final destination. This way, a sink node or gateway is used to receive the data from the sensing devices and forward it to the database or even the final user, depending on the architecture design of the system. The selection of the communication technology must consider the distances between devices. Wired communications are usually avoided due to high costs and the interferences they cause to the machinery and the common activities in an agricultural field. Long-range wireless communications include technologies such as 3G, 4G, or LoRa. However, shorter-range wireless communication technologies encompass technologies such as Wi-Fi or ZigBee. Although many sensing systems proposed in literature choose to use only one wireless technologies, the creation of heterogeneous networks allows addressing specific needs by incorporating several technologies (Lloret et al. 2021).

This chapter provides a detailed view on the current state-of-the-art regarding sensing systems for precision agriculture. The rest of the chapter is organized as follows. Section 2 presents the evolution of sensing systems for PA in research. Section 3 presents an architecture for PA considering all domains of sensing systems. The sensing technologies employed in each dimension are discussed in Sect. 4. Section 5 presents the current challenges and future trends of sensing systems for PA. Lastly, the conclusion is provided in Sect. 6.

2 Evolution of Sensing Systems for Precision Agriculture

The concept of deploying electronic devices incorporating sensors to monitor all the aspects of agriculture that can be measured began becoming reality no more than 25 years ago. Since then, the interest in providing agriculture with smart capabilities has been growing. Figure 1 presents the historic of works in the literature on sensing systems for PA according to (WorldWideScience n.d.). As it can be seen, interest in this area experimented moderate growth up to 2011 before dropping for the next 3 years. Since 2015, the tendency has been positive with exponential growth in the last few years.

The most mentioned topics regarding sensing systems for PA in academic works according to (WorldWideScience n.d.) are presented in Fig. 2. Remote sensing is the most mentioned topic with sensors and unmanned vehicles being the most relevant subtopics. Control and report are the next most mentioned topics which are related to tasks such as monitoring. The term low-cost is another topic. The development of low-cost sensing systems is prevalent in smart agriculture literature, where most papers use low-cost sensors to obtain data on weather conditions, qualities of the soil, or the state of the plants (García et al. 2020a, b). Lastly, unmanned aerial

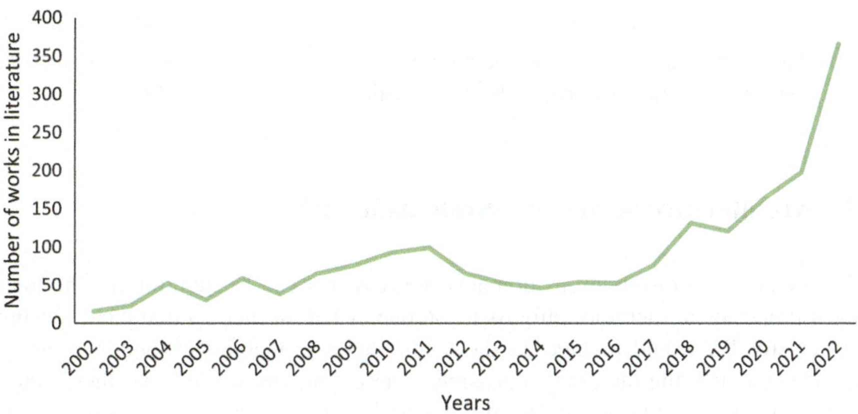

Fig. 1 Historic trend of interest in sensing systems for PA of academic literature

Fig. 2 Topics and subtopics of the academic works regarding sensing systems for PA

vehicles (UAVs) are a topic on their own as they are widely used for detection activities and to collect information so as to calculate vegetation indices.

3 Architecture of Sensing Systems for PA

PA is comprised of numerous domains or areas that can be monitored. Although most proposals in literature only focus on one or two of them, a complete sensing system for PA would include all the technologies available to monitor as many parameters affecting the crop as possible. These domains are soil, weather, water, plant, chemicals, and insects. The aim of classifying the parameters regarding agriculture that present special or temporal variability is to facilitate the understanding

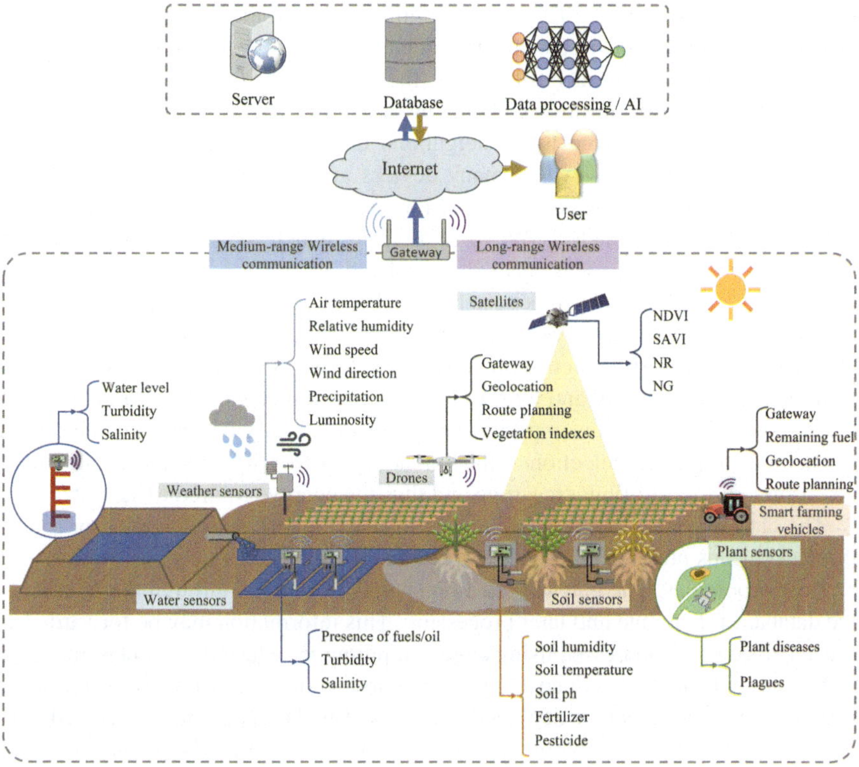

Fig. 3 Generic architecture of sensing systems for PA

of the main aspects of interest in sensing systems for PA. This classification is evolving as new subdomains are added as a result of the latest research on PA.

Figure 3 presents a generic architecture of sensing systems for PA that includes all domains. As it can be seen, there are specific nodes for monitoring the parameters of each domain in different areas. These nodes are comprised of a microcontroller that has the program to operate the device that may be part of an embedded system that includes the communication module, and the sensors that monitor each parameter. External storage like SD cards, modules such as a clock, or powering elements such as solar panels or batteries are elements that can be included as well. However, not only electronic devices with sensors can be used for sensing activities, but manned vehicles such as tractors, unmanned vehicles such as drones, and satellites are also used to gather data from the fields. Regardless of the platform where the sensor is installed, the device must consider the characteristics of the environment where it is going to be deployed so as to be placed inside a good encapsulation that protects it from physical harm.

These devices are then deployed on the fields. The type of plant and characteristics such as height and volume of foliage should be considered when designing the deployment (García et al. 2021) as the optimal height of the sensing device, or the placement of the antenna if it is not embedded, may vary due to foliage interferences in the wireless connection. Furthermore, the selection of the wireless technology used to forward the information gathered by the sensors also depends on the location of the device, the distance to its neighbors, the closest gateway, or the energy consumption. Thus, areas where many devices are deployed close to each other, for example, to monitor water quality (Lloret et al. 2021), may use medium-range wireless technologies such as Wi-Fi or ZigBee, which have coverages typically below 100 m. Other devices located in isolated areas, such as weather stations, may use long-range technologies such as LoRa or 3G/4G if there is an available infrastructure, which have coverages of kilometers. If energy consumption is the aspect of the network design with more priority, ZigBee and LoRa are wireless technologies, for medium and long-range coverage respectively, that are advertised as low power. However, it is important to consider the configuration as, in the case of LoRa for example, the selection of the Spreading Factor (SF) and bandwidth affects the airtime of transmitting a message and thus the energy required to send it.

The data is forwarded to the gateway either by direct communication or routed through various nodes. This is dictated by the topology of the network, such as star or mesh topologies. The gateway then sends the information through the Internet to the database for storage and later processing. This information may be forwarded as it is received or preprocessed using edge computing to reduce the amount of data to be transmitted, and thus reduce energy consumption, or to determine some real-time actions to be carried out by deployed actuators. The data is usually processed with the use of algorithms, in particular, AI algorithms are currently popular for uses such as crop selection aided by a decision support system, optimization or resources such as the reduction of water or pesticide usage which improves the sustainability of agriculture, detection of plant disease or pests, and management of crops and related risks (Bhat and Huang 2021). However, the use of AI for PA also presents some challenges such as the computing capabilities of the available infrastructure or the collection and management of Big Data. The results of processing the data can be accessed by the user through web services or by receiving alerts and notifications by SMS or instant messaging applications like Telegram (Marques Mostaço et al. 2018).

However, traditional farmers may still have challenges using IoT solutions for agriculture based on data analytics and machine learning (Akhter and Sofi 2021). Specifically, the results from a survey performed in India identified the following challenges. The key challenges regarding the commercial aspects of IoT are the costs, loss of manual employment, and the absence of immediate results after deploying an IoT system. The challenges regarding sectoral aspects are the size of the individual management zones, the interoperability of the available devices, and the supervisory challenges regarding data privacy and management. Regarding the technical aspects, the main challenges are the Internet connectivity in rural areas, data security, scalability and configuration, reliability, the selection of the technology to be used, and the optimization of scarce resources such as power usage. Lastly, the challenges regarding data analytics are data integration, knowledge mining, and visualization. It is, therefore, important to consider the real adoption possibilities of the proposed PA systems and promote training courses and the design of low-cost sensing systems to increase the chances of extending its use throughout the world and thus improve production, produce quality, and a sustainable use of resources.

4 Sensing Methods for PA Dimensions

PA typically involves the use of various sensing methods to gather data on crop growth, soil conditions, and weather patterns. Some common sensing methods used in precision agriculture include (Sishodia et al. 2020):

- Remote sensing: This method involves the use of aerial or satellite imagery to gather information on crop growth, soil moisture, and other factors.
- Ground-based sensors: These sensors are placed directly in the field and are used to measure factors such as soil moisture, temperature, and nutrient levels.
- Weather stations: These stations gather data on weather patterns such as temperature, precipitation, and wind speed, which can be used to optimize crop growth and management.
- Yield monitoring: This method involves the use of sensors on farm equipment to measure crop yield in real-time, allowing farmers to optimize planting and harvesting.
- Drones: Drones equipped with cameras and sensors can be used to gather data on crop growth and soil conditions, as well as to spray crops with pesticides and fertilizers.

All these methods have their own advantages and limitations. Therefore, the selection of the best method to use depends on the specific crop, the state of the field, and the goals of the farmer.

4.1 Soil Parameters

Measuring soil parameters is crucial in agriculture because it helps farmers understand the health and fertility of their soil. By analyzing soil samples, farmers can determine the pH level, moisture and temperature, among others, which are all important factors that affect plant growth and crop yields (Fazekaov 2012). Nutrient content is another important soil parameter to measure. Soil samples can be analyzed for the presence of essential nutrients like nitrogen, phosphorus, and potassium, as well as micronutrients like zinc and copper. Monitoring soil parameters is important for a variety of reasons. It can help farmers determine the best time to plant and harvest crops, as well as the best types of crops to grow in a specific area. It can also help to identify and address issues such as soil erosion and nutrient deficiencies, which can impact crop yields and overall soil health. Additionally, monitoring soil parameters can aid in conservation efforts by providing information on the impacts of land use changes, such as urbanization or deforestation, on soil conditions. Overall, monitoring soil parameters is essential for ensuring a sustainable agriculture and land management. Due to its importance, this subsection presents the main parameters to be monitored in soil with the aim of ensuring the correct development of crops.

4.1.1 pH Level in Soil

A soil's pH level, which measures its acidity or alkalinity, can greatly impact the availability of nutrients for plants. Measuring the pH level of soil is a simple process that can be done using a pH meter or a pH test kit (Sumner 1994).

- pH Meter: A pH meter is an electronic device that measures the acidity or basicity of a liquid solution, such as soil. To use a pH meter, simply insert the electrode into the soil sample and read the pH level displayed on the meter.
- pH Test Kit: pH test kits are available in the form of paper strips or liquid reagents. To use a paper strip, simply dip the strip into the soil sample and compare the color change to the color chart provided. For liquid reagents, mix a small amount of soil with the reagent and compare the color change to the color chart provided.

The soil pH can vary depending on the location in the soil. Therefore, it is recommended to take several readings from different parts of the soil and average the results to get a more accurate reading. It is also important to note that soil pH can be affected by various factors such as fertilizer application, liming and other management practices. Therefore, measuring soil pH regularly helps to ensure that the soil pH remains within the optimal range for the plants you want to grow.

In general, most plants prefer a soil pH between 6.0 and 7.0, which is slightly acidic to neutral. However, some plants have different pH requirements, so it is important to know what pH range is best for the plants you want to grow. If the pH level is too low or too high, it can be adjusted by adding lime to raise the pH or sulfur to lower the pH. However, it is important to note that changing the pH level of the soil can take several months to a year, so it is important to plan ahead and make adjustments well in advance of planting.

4.1.2 Soil Moisture

Measuring soil moisture involves determining the water content in soil (Sharma et al. 2018). There are various methods to measure soil moisture, including:

- Gravimetric method: This method involves taking a soil sample, drying it in an oven, and then weighing it to determine the moisture content.

- Time domain reflectometry (TDR): This method uses a device that sends an electrical pulse into the soil and measures the time it takes for the pulse to return. The time is used to calculate the soil moisture content.
- Neutron probe: This method involves using a neutron source to measure the amount of hydrogen (which is associated with water) in the soil.
- Capacitance method: This method uses a sensor that measures the ability of the soil to hold an electrical charge, which is related to the soil moisture content.
- Gypsum blocks: These are small blocks made of gypsum salt, which absorb moisture from the soil and change their weight in proportion to the amount of moisture present.
- Soil moisture sensors: These are electronic devices that measure the water content of the soil by measuring the electrical resistance or capacitance.

Regardless of the method used, it is important to take multiple measurements in different areas of the field or garden to get an accurate representation of the soil moisture content. It is also important to consider factors such as soil texture, plant growth stage, and weather conditions to get an accurate reading of the soil moisture. And regularly monitoring soil moisture will help farmers and gardeners to make informed decisions about irrigation and fertilization.

4.1.3 Soil Temperature

Soil temperature is an important factor in agricultural production, as it can influence the growth and productivity of crops and other plants. Measuring soil temperature is essential to ensure optimal growing conditions, and can be done with a variety of methods. The following are the most common methods used to measure soil temperature (Yin et al. 2021):

- Soil temperature is to use a thermometer specifically designed for soil use. This type of thermometer is usually inserted into the soil using a metal probe, and is able to measure temperature to a specific depth. The thermometer should be inserted at least 4–6 inches into the soil, and should be left for several minutes to allow for an accurate reading. The thermometer should also be placed in an area of the soil that is representative of the entire field, and should be periodically moved to different locations in the field to ensure an accurate overall reading.
- Soil temperature logger: This device is typically placed in the soil and records temperature readings over time. The logger is able to take readings at a variety of depths and can be used to track trends in soil temperature (Tongrod et al. 2009).
- Handheld infrared thermometer: This type of thermometer uses infrared light to measure the temperature of the soil (Martínez et al. 2016).
- Remote sensing technologies: It is possible to use satellites, aircraft, and drones for measuring soil temperature. These methods are advantageous because they allow for continuous monitoring of soil temperature over large areas in a short period of time. Remote sensing technologies are also more cost-effective

than manual or thermometer measurements, and can provide more detailed information about soil temperature (Bhat and Huang 2021).

As a summary, measuring soil temperature is an important part of understanding the effects of climate change on crop growth and assessing soil health. Accurate and reliable measurements of soil temperature can be obtained using a thermometer or temperature sensor, by hand, or with remote sensing technologies. By using these methods, agriculture professionals can gain a better understanding of soil temperature and how it affects crop growth and development.

4.2 Weather Monitoring

Weather monitoring in farming systems is a crucial aspect of modern agriculture. With the help of various weather monitoring technologies, farmers can predict and prepare for weather events that may affect crop growth and yields. This includes monitoring temperature, precipitation, wind, and solar radiation, among other factors. One of the most commonly used weather monitoring technologies in farming systems is the weather station. These stations are typically located on the crop. The data collected by these stations is used to predict weather patterns and to make informed decisions about crop management. Another important weather monitoring technology used in farming systems is remote sensing. This technology uses satellites, drones, and other aerial platforms to collect data on crop growth and weather patterns. This data can be used to create detailed maps of crop growth and to detect areas of stress or disease.

Remote sensing can also be used to monitor soil moisture, which is critical for crop growth and water management. Climate forecasting (Graham et al. 2011; Ukhurebor et al. 2022) is also an important aspect of weather monitoring in farming systems. With the help of computer models and other forecasting tools, farmers can predict weather patterns and plan accordingly. For example, if a farmer knows that a drought is likely to occur in the coming months, they can make plans to conserve water and to plant drought-resistant crops.

Smart irrigation systems are also becoming increasingly popular in farming systems. These systems use weather monitoring data to automatically adjust irrigation schedules and to conserve water. They can also detect areas of crop stress and adjust irrigation accordingly. This can help to save water, reduce costs, and improve crop yields. In addition to the technologies mentioned above, farmers can also use various mobile apps and online platforms to access weather data, as well as, the use of artificial Intelligence and big data to make informed decisions (El-magrous et al. 2019; Hachimi et al. 2022). These platforms can provide real-time weather data and alerts, as well as historical data and analysis tools.

This subsection shows the most commonly used technologies for measuring environmental parameters such as temperature, relative humidity, rainfall, wind characteristics, and solar radiation and luminosity.

4.2.1 Temperature

One of the most important parameters in precision agriculture is the measurement of ambient temperature. There are several types of temperature sensors available, including thermocouples, RTDs (resistance temperature detectors), and thermistors. Each type of sensor has its own strengths and weaknesses, and the best choice will depend on the specific application (Arman Kuzubasoglu and Kursun Bahadir 2020):

- Thermocouples are one of the most popular types of temperature sensors. They are relatively inexpensive and can measure temperatures over a wide range, from −200 °C to +1800 °C. They work by measuring the voltage difference between two different metals, and are most commonly used in industrial applications.
- RTDs, or resistance temperature detectors, are another popular type of temperature sensor. They are more accurate than thermocouples, and can measure temperatures over a narrower range, from −200 °C to +850 °C. They work by measuring the resistance of a specific metal, such as platinum, and are commonly used in laboratory and industrial applications.
- Thermistors are another type of temperature sensor. They are small, inexpensive, and can measure temperatures over a narrow range, from −50 °C to +150 °C. They work by measuring the resistance of a specific material, such as ceramics, and are commonly used in consumer and industrial applications.

In applications, such as agriculture, it may be sufficient to take readings every few hours.

4.2.2 Relative Humidity

Relative humidity (RH) is an important metric in agriculture as it can affect the growth and productivity of crops. There are several ways to measure relative humidity in an agricultural setting (Korotchenkov 2019).

- Hygrometer: One common method is through the use of a hygrometer, which is a device that measures the amount of water vapor present in the air. Hygrometers can be either analog or digital, and they typically have a range of 0–100% relative humidity.
- Psychrometer: Another way to measure relative humidity is through the use of a psychrometer, which is a combination of a thermometer and a hygrometer. A psychrometer, also known as a wet-and-dry-bulb thermometer, uses two thermometers, one with a wet wick wrapped around the bulb and the other with a dry bulb. As the air passes over the wet bulb, the evaporation of water cools the bulb and the temperature difference between the two bulbs is used to calculate the humidity.
- Electronic hygrometers: Electronic hygrometers use sensors to measure the humidity directly. One common type of sensor is the capacitive sensor, which

measures the electrical capacitance of a material. As the humidity changes, the capacitance of the sensor changes, allowing the humidity to be measured. Other types of electronic sensors include resistive sensors, and optical sensors.

The accuracy of these methods may vary depending on the specific device and how it is used, so it is important to calibrate and maintain the equipment properly. Determining relative humidity in agriculture is important since high humidity can lead to mold and mildew growth, which can damage crops and reduce yields. On the other hand, low humidity can cause stress on plants and can make them more susceptible to pests and diseases. By monitoring relative humidity levels, farmers can take steps to improve the growing conditions for their crops and increase yields.

4.2.3 Rainfall

Rainfall is essential for agriculture as it provides the necessary water for plants to grow. Adequate rainfall ensures that crops have enough water to support their growth and development, while insufficient rainfall can lead to crop failure and reduced yields. Additionally, rainfall helps to replenish soil moisture, which is important for maintaining soil fertility and supporting the growth of beneficial microorganisms. There are several methods for measuring rainfall in agriculture. The main ones are the following:

- Rain gauge: A simple device consisting of a funnel and a measuring cylinder that collects and measures the amount of rainfall in a specific location.
- Tipping bucket rain gauge: A type of rain gauge that uses a bucket to collect and measure rainfall. The bucket tips over when it reaches a certain capacity, and the number of times it tips is recorded to measure the amount of rainfall.
- Pluviometer: A device that uses a rotating drum or wheel to measure the amount of rainfall. The drum or wheel is turned by the falling rain, and the number of turns is recorded to measure the amount of rainfall.
- Radar: Weather radar uses radio waves to detect the presence of rain, snow, and other precipitation. It can provide a more comprehensive view of precipitation patterns in a given area.
- Satellite: Satellites can be used to measure rainfall by measuring the reflectivity of the Earth's surface. This can provide a broad view of precipitation patterns over a large area.

Recent researches are trying to mix new techniques such as fuzzy logic with classic measurements methods to obtain more accurate predictions (Janarthanan et al. 2021).

4.2.4 Wind Speed and Wind Direction

Wind direction and wind speed are important factors to consider in agriculture as they can have a significant impact on crop growth and yield. Strong winds can cause damage to crops, while weak winds can limit pollination and reduce crop yields. Therefore, measuring wind direction and wind speed in agriculture is crucial for farmers to make informed decisions about planting, irrigation, and pest control. Measuring wind direction and wind speed in agriculture can be done using various instruments (Oswald 2019):

Wind speed can be measured by:

- An anemometer is a device that measures wind speed by measuring the rotation of its cups or blades.
- A sonic anemometer is a device that uses sound waves to measure wind speed and direction. It works by emitting sound waves at a known frequency and measuring the time it takes for the sound waves to travel to a distant target and back. The wind speed and direction can then be calculated based on the changes in the frequency and arrival time of the sound waves. Sonic anemometers are commonly used in meteorology and wind energy applications.

Wind direction can be measured by using a:

- A wind vane measures wind direction by determining the direction in which the wind is blowing.
- Another tool to measure the wind is a windsock. It is not widely used in agriculture. Windsocks are most commonly found at airports to indicate the direction and intensity of wind to pilots. They are also used at chemical plants and oil rigs where they help to reduce and monitor the risk of air contamination. A windsock points in the direction the wind is blowing to, i.e., if the tapered end of the windsock is pointing to the north, this indicates a southerly wind. You can also estimate wind speed by looking at the angle of the windsock relative to the mounting pole as the wind blows through it. Each alternating orange and white stripe equal to 3 knots of wind speed, with a fully extended sock indicating wind speeds of 15 knots or greater (28 km/h; 17 mph).

To measure wind speed, it is commonly used the Beaufort scale. It is a widely used empirical measure that relates wind speed to observed conditions at sea or on land. The scale runs from 0 to 12, with 0 indicating calm conditions and 12 indicating the strongest winds, such as those found in a hurricane. The Beaufort scale provides a simple and intuitive way to describe wind conditions and is still in use today, although it has been updated over time to take into account advances in technology and understanding of wind phenomena. Using wind direction and wind speed information, farmers can make informed decisions about planting and harvesting. For example, wind direction and wind speed can affect the pollination of crops, so farmers can choose to plant crops that are more resistant to wind damage. They can also choose to plant crops that are more resistant to pests and diseases, as wind direction and wind speed can affect the spread of these problems. Wind direction and wind speed can also affect the irrigation of crops. For example, if the wind is blowing in a certain direction, farmers can adjust their irrigation schedules to ensure that the water is reaching the crops. This can help to reduce water waste and increase crop yields.

4.2.5 Luminosity or Solar Radiation

Luminosity is a measure of the amount of light energy received from the sun. It is an important aspect of weather monitoring in farming systems as it directly affects crop growth and yields (Babatunde 2012; Beyaz and Gül 2022). There are several ways to measure luminosity in weather monitoring systems:

- Pyranometer: One of the most common methods is the use of a pyranometer. A pyranometer is a device that measures the amount of solar radiation received on a flat surface. It typically consists of a sensor that detects the radiation and a data logger that records the data. The sensor is usually placed on a mast or tower at a height of around 2 m above the ground.
- Satellite data: Another method of measuring luminosity is through the use of satellite data. Satellites equipped with sensors can measure the amount of solar radiation received on a specific area. This data can then be used to create detailed maps of solar radiation over a specific region. This method is particularly useful for monitoring large areas of land and for predicting weather patterns.
- Light sensors: It is also possible to measure luminosity by using a light sensor. This sensor detects light energy and can be used to measure the amount of light received in a specific location. This method is commonly used in indoor farming systems to measure the amount of light received by plants.

In conclusion, weather monitoring in farming systems is crucial for modern agriculture. With the help of various technologies and platforms, farmers can predict and prepare for weather events, conserve water, improve crop yields, and respond to pests and diseases. This can help to increase productivity and profitability in the agriculture industry.

4.3 Plant Welfare and Its Physical Aspect

E-monitoring of plant diseases is a cutting-edge technology that enables the detection, monitoring, and management of plant diseases in real-time. This technology has seen an exponential growth over the last few years, owing to its ability to provide accurate, timely, and cost-effective solutions for plant disease management (Mohammad-Razdari et al. 2022). This subsection shows the most common used technologies to monitor different features of plants such as diseases, height, and foliage.

4.3.1 Diseases

Recent advances in the field of e-monitoring have led to the development of many innovative applications. The first of these is the use of remote sensing technology, which is capable of detecting symptoms of plant diseases in real-time. This technology utilizes satellites and remote sensing, aircraft, and drones (Wójtowicz et al. 2016) to capture images of plants and their environment which can be analyzed to identify plant disease symptoms. Additionally, this technology can be used to predict the spread and severity of diseases, thus allowing farmers to take preventive measures in advance. Another breakthrough in e-monitoring is the development of mobile applications. These applications use information gathered from satellite images and sensors to provide real-time monitoring of plant diseases. The data collected can be used to detect the presence of pests and diseases, as well as to monitor their spread. This technology has the potential to alert farmers to diseases in their crops before they can cause significant damage. In addition to remote sensing, mobile applications, and sensors, e-monitoring has also seen advances in the use of artificial intelligence (AI) (Misra et al. 2020). AI-based systems are capable of collecting, analyzing, and predicting the spread of diseases in real-time. This technology can help farmers make informed decisions about their crops and take the necessary steps to control the spread of diseases. Finally, the development of predictive analytics has also improved e-monitoring of plant diseases. Predictive analytics use data collected from sensors and satellite images to predict the spread of diseases and their severity. This technology can be used to provide farmers with advanced warning of disease outbreaks and help them to take the necessary measures to control them.

Image processing techniques are also widely used to detect problems in plants and they have become increasingly important in the detection of diseases in plants (Lloret et al. 2011). These techniques allow for the automated analysis of images of plants to detect diseases, saving time, and resources. The first step in the process is to collect the images of the plants. This is usually done through the use of cameras and other imaging equipment. The images must be of high quality and accurately represent the plants. The next step is to pre-process the images, which involves cleaning the images, correcting for noise, and making sure the images are properly formatted for further analysis. Once the images have been pre-processed, the next step is to extract features from the images. This can be done through a variety of methods, including edge detection, segmentation, and pattern recognition. By extracting features from the images, it is possible to create a model that can be used to identify the presence of a disease. The final step is to use the model to detect diseases in plants. This can be done through a variety of techniques, including supervised learning, unsupervised learning, and deep learning. These techniques can be used to classify the images into categories, such as healthy or diseased, and can also be used to detect signs of disease. Image processing techniques have become essential for the detection of diseases in plants. The process is relatively straightforward and can be used to automate the process of disease detection, saving time, and resources. By using these techniques, it is possible to create models that can be used to detect and classify diseases in plants quickly and accurately (Couliably et al. 2022; Jin et al. 2020). Image analysis techniques are particularly useful for diagnosing diseases which are difficult to see with the naked eye, such as leaf spot or bacterial blight. This technique is also useful for tracking the spread of disease over time, as images of plants can be taken at various intervals.

Interesting techniques based on taking pictures in different spectrum directly to the plants are the following ones (Sankaran et al. 2010; Fang and Ramasamy 2015):

- Fluorescence spectroscopy: Fluorescence spectroscopy refers to a type of spectroscopic method, where the fluorescence from the object of interest is measured after excitation with a beam of light (typically in the ultraviolet spectrum). Fluorescence spectroscopy can be used to analyze the biochemical composition of plants, detect early signs of disease, and measure the effects of treatments on the plant. Fluorescence spectroscopy can be used to detect the presence of certain chemicals, such as proteins, lipids, and carbohydrates, as well as to measure changes in the photosynthetic efficiency of plants. It can also be used to distinguish between healthy and diseased plants. Fluorescence spectroscopy is a quick, noninvasive technique that can provide insights into the biochemical and physiological changes caused by plant diseases.
- Visible and infrared spectroscopy: Visible and infrared spectroscopy can be used to detect the presence of pigments, proteins, and other components in plants. It can also be used to measure the amount of water present in plants, and to measure the changes in the concentration of water in plants due to disease. This can be used to monitor the progress of a disease. Infrared spectroscopy can also be used to measure the temperature of the plant material, which can be used to determine the degree of disease.
- Fluorescence imaging: Blue-green fluorescence is a phenomenon observed in some plants under certain conditions. It is caused by the presence of certain compounds in the plant, such as anthocyanins, carotenoids, and other pigments. Chlorophyll fluorescence refers to the light emitted from chlorophyll molecules when they are exposed to certain wavelengths of light. This fluorescence is used by plants to help with photosynthesis. Both blue-green fluorescence and chlorophyll fluorescence can be used to detect certain diseases in plants. For example, blue-green fluorescence can be used to detect disease caused by fungi, bacteria, and viruses, while chlorophyll fluorescence can be used to detect diseases caused by nutrient deficiencies or environmental stress. Both of these methods can help to provide an early warning of disease, allowing for more timely treatment and prevention.
- Hyperspectral imaging: Hyperspectral imaging is a technique used to detect and identify plant diseases, as it allows for the analysis of a large range of wavelengths that are emitted or reflected from the plant. In the hyperspectral imaging, the light reflected for each pixel is registered for a range of wavelengths in the electromagnetic spectrum. The resulting information is a set of pixel values (intensity of the reflectance) at each wavelength of the spectra in the form of an

image. Each spectral region provides unique information about the plant. This method allows for the detection of subtle changes in the spectral signature of the plant, which can indicate the presence of a disease. By analyzing the spectral signature of the plant, researchers can identify the presence of specific pigments, proteins, and other compounds. This can enable the detection of plant diseases at an early stage and allow for more effective management practices. Additionally, hyperspectral imaging can be used to identify the causes of disease, such as environmental conditions, pests, and nutrient deficiencies.
- Gas chromatography (GC): A completely different non-optical indirect method for plant disease detection involves the profiling of the volatile chemical signature of the infected plants. It can be used to identify different components within a sample and detect the presence of particular volatiles. GC is also used to differentiate between plant diseases and to provide a more accurate diagnosis. The pathogen infections of plants can result in the release of specific volatile organic compounds (VOCs) that are highly indicative of the type of stress experienced by plants, i.e., GC can be used to detect VOCs in the air surrounding a diseased plant, which can help to diagnose a particular disease. Additionally, GC can be used to detect and quantify metabolites in the plant tissue which can be used to assess the severity of the disease and the progress of the disease.

Finally, there are modern techniques based on laboratory assays which are extensively used to detect diseases in plants. These assays use a combination of chemical and biological tests to detect the presence of specific diseases. For example, PCR (polymerase chain reaction) can be used to detect the presence of a particular virus or bacteria (Tatineni et al. 2008). This technique is faster and more accurate than traditional methods of disease detection and can be used to quickly detect and identify a wide range of diseases. Finally, molecular detection is a modern technique that uses deoxyribonucleic acid (DNA) or ribonucleic acid (RNA) sequencing to identify the presence of a particular disease (MacKenzie et al. 1997). This technique is particularly useful for identifying new or emerging diseases, as it can quickly and accurately identify a wide range of pathogens. This technique is also useful for monitoring the spread of diseases over time, as the same sample can be used to compare the pathogen's DNA or RNA sequence at different points in time.

4.3.2 Height

Height is an important factor in precision agriculture since it can help to identify and measure the amount of crop residue, weeds, and other materials in the field. It can also be used to determine the exact amount of fertilizer and other inputs needed for optimal crop growth. Height can assist with mapping the field and creating an accurate record of the field's condition over time. In order to measure height, it is possible to use the most simple methods such as a tape measure up to the most modern ones based on AI and deep learning techniques. The following are the most modern methods used in PA:

- Machine vision: Machine-vision based height measurement in crops involves the use of stereo cameras and software to capture images of the crop and measure the height of the plants. The images are then analyzed to determine the height of each plant and generate a report. This can be used for tracking the growth of the crop, understanding the health of the crop, and predicting yields. The height measurements can also be used to inform decisions about irrigation, fertilization, and pest control (Kim et al. 2021).
- Light detection and ranging (LIDAR): LiDAR technology is a very accurate and reliable way of measuring the height of crops. It uses a laser beam to measure distances between the ground and the crop canopy, allowing for precise measurements without the need for manual labor or tedious surveying methods. In addition, LiDAR can provide detailed 3D models of the crop canopy, allowing for more accurate estimates of crop yield and crop health. LiDAR can also be used to detect changes in the terrain, allowing farmers to better manage their crop fields and maximize their yields (Canata et al. 2016).
- UAV-based Remote Sensing or drones: A drone can be used to measure the height of a plant. There are several methods; however, the most interesting one is the system based on a laser scanner. The system takes different series of data and processes and analyzes the data to determine the distance from the drone to the ground and to the top of the crops (Anthony et al. 2014).

In addition to the aforementioned advantages of measuring the height of a crop, quantifying the height of a plant can also help to detect diseases, pests, and other issues that may be affecting crop production, allowing for quick action to be taken to minimize any losses.

4.3.3 Foliage

Foliage monitoring in plants and trees has become increasingly important to ensure the health and growth of these essential elements of the environment. With the help of modern techniques, scientists and land managers can monitor foliage in real-time, collecting data that can be used to assess the health of both individual trees and entire forests. Foliage observation can also help to detect pests and diseases in plants and trees. The common parameters measured are changes in leaf color, shape, and size (Zhang et al. 2012). Considering the importance of the physical aspect of plants and trees in the detection of problems and diseases in a crop, this subsection describes some of the most modern methods to observe changes in foliage of plants and trees of different species. There are several modern techniques used to monitor foliage, including aerial photography and remote sensing, digital imaging, spectroscopy, and drone technology:

- Aerial photography and remote sensing are two of the most commonly used methods for foliage monitoring. By using aerial photography, images of a large area can be captured and then analyzed for changes in foliage (Megat Mohamed Nazir et al. 2021).
- Remote sensing takes this a step further by using satellite imagery to detect subtle changes in foliage. This allows for more accurate information on foliage health and changes in different areas.
- Digital imaging is another technique used for foliage monitoring. This involves capturing images of foliage with a digital camera and then analyzing the images for changes in color, size, and shape. Digital imaging is especially useful for determining the presence of pests and diseases in a particular area.
- Spectroscopy is a type of remote sensing which utilizes the electromagnetic spectrum to measure the spectral reflectance of foliage. This method allows for highly accurate measurements of foliage health, as it can measure subtle changes in color, texture, and structure. Spectroscopy is also useful for determining the presence of specific elements in the soil, such as nitrogen and phosphorus.
- Finally, drone technology is a relatively new method which has been used for foliage monitoring. By using a drone, images of a large area can be captured and then analyzed for changes in foliage. This method is especially useful for surveying large areas of land, as it can cover a wide area quickly and accurately.

In addition to sensors, advanced computer vision algorithms and convolutional neural networks (CNN) are used to analyze the data collected by the sensors. These algorithms can detect and identify individual tree species (Kumar et al. 2012), as well as measure the volume of branches (Zhang et al. 2020), size, shape, and health of each tree's foliage. This data can be used to assess the overall health of a forest and to identify potential problems that could be addressed.

Finally, it is important to highlight that Signal propagation of electromagnetic waves can also be used to determine the foliage loss in trees. Because plant matter has a degree of signal strength absorption, it is relatively easy to determine the amount of foliage lost in a set of trees. To do this we must determine the signal levels received by a receiver in the absence of vegetation and see how this value is modified due to the presence of a greater quantity of vegetation (Anzum et al. 2021; Lloret et al. 2009).

4.4 Water

Water is the most important resource in agriculture. However, water availability and managing the distribution and usage of water are not the only concern regarding water. Water in rivers, lakes, or groundwater may be polluted from fertilizers, pesticides, oils, and fuels, or different types of waste discarded into these masses of water. All this pollution makes water unsafe for human consumption, and irrigation with this type of water may result in unsafe and low-quality produce. Most PA systems focus on water management but do not include sensors to monitor water quality. One of the reasons is the need for chemical processes that cannot be done remotely. However, new sensors for water quality monitoring based on the study of physical properties have been developed in recent years (Rocher et al. 2021). Although they may have been proposed for other uses, these sensors can be added to PA sensing systems to add more functionalities and improve food security.

As water scarcity has been one of the main concerns in PA regarding the use of water, the use of wastewater for irrigation has been considered. Wastewater is collected by the sewage system and transported into a water treatment plant to clean it, so it can be released back into nature. Although the resulting water is not good for human consumption, some studies indicate that the use of treated wastewater would be safe for the irrigation of crops (Zhang and Shen 2019). However, although the water may have passed quality control before its release, monitoring water quality before irrigation adds another safety layer and helps in the detection of leakages of untreated or not enough treated water.

The following sensors can be included in a multiparametric sensing device to obtain real-time data on water quality:

- Turbidity: The use of water with high turbidity for irrigation may cause the reduction of hydraulic conductivity in the soil, which leads to increased surface pollution due to surface flow (Jeong et al. 2016). Germs and bacteria can also be attached to suspended solids and the reduction of turbidity levels can also reduce the germs in the water. Specifically, vegetables are susceptible to germ infection and thus some countries such as Spain or Greece have turbidity standards for irrigation water, 10 NTU (Nephelometric Turbidity Units) for vegetables, and 2 NTU for directly consumed crops respectively. Therefore, turbidity sensors should be considered in PA sensing systems to ensure the quality of the water remains within the standard.

  Turbidity sensors are based on optical parameters and can be based on nephelometry or optical-backscatter (OBS) (Rasmussen et al. 2009). They are available in a wide variety of price ranges, from low-cost sensors to high-end devices. The sensor presented by (Parra et al. 2018) has a price below 5 € and is comprised of color and infrared LEDs and their detectors, encapsulated in a waterproof case. The LEDs are placed on both sides of a translucent cylinder where the water flows through. Other light sources such as optical fiber have also been used in turbidimeters (Bin Omar and Bin MatJafri 2009).
- Ph: Changes in the pH of water can lead to less plant growth by affecting nutrient absorption, photosynthesis, or morphology (Zhao et al. 2013). It can also affect flower coloration and thus, its quality, when considering the sector of ornamental plants. Typical pH values considered in the recommendations of most countries range from 6.0 up to 9.5 (Jeong et al. 2016). Using water with extreme pH values such as 4.0 and 10.0 leads to the plant damages cited before (Zhao et al. 2013). However, pH is measured through chemical sensors that cannot be deployed for real-time monitoring and need frequent calibration. Some water quality proposals for real-time sensing use pH sensor probes (Das and Jain 2017). Others use

inkjet-printed pH sensors based on glass substrates (Qin et al. 2018). And others not intended for real-time monitoring even propose the use of smartphones for pH monitoring (Dutta et al. 2015).

- Salinity: Salinity is measured by the electric conductivity (EC); thus, it is often referred to as conductivity (Jeong et al. 2016). Therefore, both terms can be used interchangeably. Optimal salinity levels should remain below 700 μs/cm. However, standards recommend staying below 2000 μs/cm, with the stricter ones recommending 1000 μs/cm as the maximum. The principal effect of high salinity in water is reduced crop growth and permanent damage in extreme cases. Some conductivity sensors are based on measuring the voltage between electrodes which may be two (Shi et al. 2021), three, or even four (Ramos et al. 2008). Inductance-based sensors are another type of salinity sensor, which comprises two subtypes: transformer-based or Eddy current based (Harms and Kern 2021). The latter one is based on the use of coils where one is powered and the second one is induced. These sensors can be deployed in WSN for PA sensing for real-time monitoring at a low cost (Parra et al. 2014).
- Temperature: Water temperature is not often monitored in PA systems. However, it is a parameter to be considered if the selected crop is rice (Luo and Goudriaan 1999). Optimal rice leaf guttation is achieved with a water temperature of 30 °C, and higher water temperatures were able to increase the amount of dew by 4 times. Temperature sensors for water monitoring are widely available as they are common for smart solutions designed for other purposes such as aquaculture. Specifically, low-cost sensors such as the DS18B20 (Dallas Semiconductors n.d.) with waterproof encapsulations provide digital temperature readings at prices below 1€ per sensor.
- Oil/fuel: Oil or fuel spillages, from agricultural machinery for example, pollute the water making it inadequate for irrigation, as it would pollute the crops and the soil (Basterrechea et al. 2021). This also results in reduced crop production, decreased produce quality, fewer profits for farmers, and an increase in the microorganisms that attack the nitrogen-fixing bacteria. As a solution for monitoring oil spillage, specific sensors have been developed to be deployed in water channels. For example, reflection-absorption sensors based on different colored LEDs and light-dependent resistors (LDRs) can be used to determine the presence and amount of fuel in water.
- Bacteria: Bacteria is another important factor to consider, specifically in treated wastewater that is used for irrigation. For *Escherichia* coli, the WHO (World Health Organization) recommends maximum levels of 1000 cfu/100 mL (Jeong et al. 2016). However, other countries adopted more restrictive standards such as Spain, Portugal, or some states of the USA like California with maximum recommendations of 240 cfu/100 mL. There have also been studies of disease risk in farmers when manipulating the plants right after irrigation (Rhee et al. 2009). The recommendation when irrigating with wastewater with rather high bacteria levels is to wait for 24 h after irrigation to perform any farming activities (Kiziloglu et al. 2008). The detection of coliform bacteria can be performed with gas sensor arrays that measure the volatile materials they produce in their

growing period (McEntegart et al. 2000). Sensor nodes for bacteria monitoring in WSN have also been developed based on a chromogenic enzyme substrate assay method that detects color changes in the water (Kim and Myung 2015). These color changes in water are detected with the use of a camera and can be performed in real-time. However, the system includes a water sampling system consisting of a pump motor and bottles with the sampled water, the reagent, and the reaction chamber for the chemical reactions to be done automatically. This would need to be implemented on-site with an encapsulation that withstands weather conditions to be a viable option for real-time sensing systems for PA where all the sensors are deployed onsite.

- Organic matter: Organic matter is monitored by measuring the biochemical oxygen demand (BOD) (Jeong et al. 2016) which can be monitored by respirometers (Namour et al. 2010). Decomposing organic matter consumes the oxygen in the water and produces oxides in the soil that can affect the nutrient absorption of the plants. Different types of sensors can be used to monitor the amount of organic matter in the water. Photocatalytic sensors oxidize organic matter with the use of a light source (Namour et al. 2010). Chemiluminescence sensors measure the photons emanating after a chemical reaction. Optical sensors measure UV-visible absorbance. Furthermore, many commercial sensors are based on UV scanning paired with spectral deconvolution.
- Heavy metals: Excessive amounts of heavy metals in irrigation water can result in damaged crops (Jeong et al. 2016). Some of the heavy metals that can be found in irrigation water are zinc, arsenic, copper, lead, or aluminum, among others. These heavy metals cause leaf chlorosis, damage root growth, decrease the productivity of the crop, and can even be present in the produce to be consumed affecting human health. The FAO (Food and Agriculture Organization) has a set of recommendations for maximum concentrations of heavy metals such as 5.0 mg/L for aluminum, 0.10 mg/L for arsenic, or 0.20 mg/L for copper (Ayers and Westcot n.d.). Electrochemical sensors are one type of heavy metal sensor that can be used in sensing systems (García-Miranda Ferrari et al. 2020). These sensors are low-cost, portable, have a small size, and do not require complicated sample preparations. They are comprised of a transducer and a working electrode. Although they are based on chemical reactions, there are automatic heavy metal sensors based on anodic stripping voltammetry (ASV) that have been used for water quality sensing systems (Lin et al. 2020). Other sensors are implemented with semiconductors based on ion-sensitive field effect transistors (ISFETs) or AlGaN/Gan high electron mobility transistors (HEMTs) (Nigam et al. 2021).

Lastly, water quality monitoring systems for irrigation may include sensors that do not measure water quality but are used to monitor features related to water management such as water current in pipes or the water levels of water reservoirs for irrigation.

4.5 *Chemicals*

In this subsection, the technologies used for chemical use sensing in precision agriculture, specifically for fertilizer and chemical use, are presented.

4.5.1 Fertilizers

The use of fertilizers is paramount to addressing today's food needs. But the excess use of fertilizers can lead to environmental problems such as those experienced in the semiarid Mar Menor lagoon in Spain (Puertes et al. 2021). Therefore, the addition of sensors able to monitor fertilizer use in sensing systems for PA aids in making agriculture more sustainable. However, detecting fertilizer use may be difficult without the use of chemical sensors or larger-scale devices operated manually. For example, visible near-infrared (Vis-NIR) spectroscopy is used to detect nitrogen (N) and organic matter in organic fertilizer (Guindo et al. 2021). For that reason, it is important to develop sensors able to be deployed in WSN to enable real-time monitoring without the need for constant calibration, cleaning, or sample manipulation.

Nitrate is essential for chlorophyll production and plant growth. Direct nitrate measurement is primarily performed in laboratories. Electrical conductivity has been employed as an indirect form of nitrate monitoring. However, other techniques provide more direct nitrate measurements without the need for extended preparation, time, calibration, or maintenance (Sinfield et al. 2010). Regarding spectrometry techniques, near-infrared reflectance spectrometry (NIRS) has been utilized in the form of sensor probes with the use of optic fiber. Results show NIRS provides fast tests, good total nitrogen correlation, and portability. Other spectrometry techniques such as morphology-dependent stimulated Raman scattering (MDSRS) or copper/cadmium reduction (CCR) are used in laboratories and are not suitable for portable solutions. Regarding electrochemical techniques, nitrate ion-selective electrode (ISE) consists of two electrodes where one of which includes an ion-selective membrane. ISE sensors are portable and cost-effective while providing adequate accuracy for detecting low or high nitrate levels. However, more precise measurements performed with these techniques are not possible with on-the-go solutions. Nitrate ion-selective field-effect transistor (ISFET) has a short lifetime, needs cleaning between measurements, and requires the careful introduction of samples. Therefore, it is not adequate for WSN. Lastly, nitrate combination (CCR-ISE) is not suitable for on-the-go measurements. Finally, biosensors have a short lifetime but high precision and accuracy. As a result, they can be used for on-the-go measurements but not for WSN.

Regarding phosphate and potassium, spectrometry techniques require thorough sample preparation and automatic sample devices should be developed to adapt this technique to perform in-field testing as part of PA sensing systems (Sinfield et al. 2010). Specifically for Raman scattering. reflectance spectrometry is performed in laboratories and the available portable solutions require manual operation. Electrochemical techniques use materials with a short lifetime, ranging from 30 min to 1 month, which makes it not suitable for PA. Lastly, biosensors also have a short lifetime, which make them not suitable for WSN solutions.

Other techniques have been considered to detect the presence of a combination of fertilizers. Fiber optic color sensor probes have been used for monitoring NPK nutrients by classifying the readings into none, low, medium, and high levels (Ramane et al. 2015). It, however, does not differentiate between each nutrient. Colorimetry is used to detect NPK nutrients in soil samples with added reagents by using LEDs and a photodiode mounted on a robot that performs smart agricultural tasks (Amrutha et al. 2016). These tests can be completed in 40 min. A planar electromagnetic sensor was designed for nitrate and sulfate detection (Nor et al. 2013). Parallel, star, and delta sensor arrays were first simulated and then tested with different N, P, and K concentrations, determining the star configuration as the one with the best performance. Lastly, a carbon dioxide (CO_2) sensor that operates in the mid-infrared range based on NDIR (non-dispersive infrared) was designed for a fertilization system for greenhouses (Wang et al. 2016). The sensor was able to detect concentrations in the 30–5000 ppm range. These sensors were part of the sensor nodes that included condensation prevention based on waterproof membranes and wireless connectivity. CO_2 data was then analyzed using the Fuzzy-PID (Proportional, Integral, and Derivative) algorithm to determine optimal CO_2 values to regulate the greenhouse.

Monitoring gases in the air has been studied to determine if other less straightforward methods can detect fertilizers as well. Specifically, a MOS electronic nose has been used to measure volatile organic compounds (VOCs) emitted by vegetables with high accumulations of nitrates due to overfertilization with nitrogen chemicals (Tatli et al. 2021). This e-nose is comprised of a sensor array that measures alcohol, organic solvents, sulfur dioxide (SO_2), ammonia, sulfides, toluene, combustible gases, CO, CH_4, C_3H_8, and C_4H_{10}, among others. Furthermore, artificial neural networks (ANN), support vector machine (SVM), and linear and quadratic discriminant analysis (LDA-QDA) algorithms have been used to classify the data. Results showed different VOCs emissions for different treatments with urea-nitrogen fertilizer.

Remote sensing is one of the techniques that is being used to detect nutrient uptake in plants based on narrow-band hyperspectral images with a spatial resolution of 1 m (Gil-Pérez et al. 2010). However, instead of directly monitoring the chemicals in fertilizers, a relation between vegetation indices and foliar variables such as concentrations in chlorophyll a + b, iron (Fe), calcium (Ca), magnesium (Mg), nitrate (N), phosphate (P), or potassium (K) is observed. Therefore, the obtained results are not nutrient concentrations themselves.

Finally, the use of nanosensors and nanofertilizers has been considered as well as the research on nano-things keeps increasing (Rameshaiah et al. 2015). The use of nanomaterials in fertilizers can result in slow-release fertilizers that help in avoiding overfertilization. However, possible risk factors for human health with the contact or consumption of these nanomaterials should be considered.

4.5.2 Pesticides

More than 90% of the amount of pesticides used for treatment are said to miss the target pests, increasing in turn the contamination of the environment (Khairy et al. 2018). There are different types of pesticides according to their chemical composition.

The use of organophosphate pesticides is very extended for both pre-harvest and post-harvest treatments (Khairy et al. 2018). However, inadequate use of these pesticides can lead to health problems and environmental damage. High concentrations can also lead to human deaths and Parathion is one of the most toxic pesticides in this category. Although numerous countries including the US have banned its use, parathion is still purchased in some developing countries. Gas chromatography, flame photometry, or mass spectrometry are some of the techniques that have been employed thus far for parathion detection. However, the evolution of nanomaterials has helped in the development of new techniques with reduced costs and the possibility of being incorporated into sensing systems in the future. In particular, screen-printed electrodes modified with nickel oxide have been utilized to detect parathion (Khairy et al. 2018). These materials are being tested in laboratories but further development can lead to their integration into IoT systems (Banerjee 2022). Furthermore, portable solutions based on biosensors and equipped with a wireless communication module have also been considered for the detection of organophosphate pesticides (Kim et al. 2015). These types of devices could also be adapted for PA sensing systems.

Using the gases released by pesticides to detect their presence through the deployment of gas sensors has also been considered (Leccese et al. 2019). The electric nose, an array of different commercially available gas sensors (Marco et al. 2017), was developed as a first step in determining the possible uses of low-cost commercial solutions for this purpose within the framework of WSNs. This technology has been employed for the detection of pyrethroid pesticides in tea plantations (Tang et al. 2020). Particularly, neural networks were utilized to create models for bifenthrin, cyhalothrin, and fenpropathrin recognition reaching accuracies of 90% and above. Furthermore, their use for the detection of organic chloride pesticides (Ortiz et al. 2016) and organic phosphorus pesticides (Tan et al. 2010) has also been studied with accuracies exceeding 80%.

WSNs have also been used to monitor pesticide drift so as to avoid pesticides to reach areas further than the target crop (Santos and Cugnasca 2012). The factors determining pesticide drift are wind speed and direction, temperature, humidity, height of dusting equipment, and droplet characteristics such as weight and size. Therefore, environmental conditions are monitored by wireless sensors deployed on the field to generate new routes for dusting planes or equipment according to the obtained data.

4.6 Pests

Pest control is one of the main concerns in PA. Early detection helps in reducing crop damage and can lead to a reduction in the amount of pesticide needed to cover the target crop. The two main principal forms of sensing pests are acoustic and optical sensors (Azfar et al. 2018). These sensors can be easily installed in IoT to enable pest monitoring in real time.

Acoustic sensors detect noise from insects in the fields. These sensors are often deployed in above-ground devices for the detection of pests in crops such as sugarcane (Srivastava et al. 2013) or palm trees (Cardim Ferreira Lima et al. 2020) in the form of static sensor nodes or portable devices. Especially, larvae are located inside the produce of the woody parts of the plants. They are also employed to detect insects in the storage of grains (Fleurat-Lessard et al. 2006). However, acoustic sensors can also be used for the detection of underground pests with the use of underground sensor networks (Bayrakdar 2019). Underground sensors can communicate through wired or wireless connections. But wireless underground communications are still being developed and common wireless technologies such as Wi-Fi have their coverage greatly reduced in underground environments (García et al., 2020).

Systems based on optical sensors are mainly comprised of a camera and an embedded system that provides computing capabilities and wireless communication. However, while some solutions take pictures of the plants directly, others install cameras inside traps for better identification of the insect (Segalla et al. 2020; Wang et al. 2020). The collected pictures are then processed to determine the type of insect that is damaging the crops. Cameras are not limited to static devices deployed in the fields. Drones can use their cameras for pest detection applications as well (Refaai et al. 2022). Another type of optical sensor is the camera used in satellite imaging. Satellite images are widely used in agriculture for aspects such as the calculation of vegetation indexes. However, this data has also been used for the detection of desert locust plagues (Geng et al. 2018).

Artificial intelligence, specifically deep learning, is one of the most utilized techniques to perform identification tasks in recent years. This includes image processing for pest detection applications (TÜRKOĞLU and HANBAY 2019). However, some applications may not be able to send the images to a centralized server, leading to the need for local processing. So, the selection of the processing algorithms should consider the computing capabilities of the device. Therefore, it is important to test different algorithms to determine their accuracy for pest detection, which is a popular research line today (TÜRKOĞLU and HANBAY 2019).

Although the use of AI is very promising, good datasets are required to improve classifications and detect new types of species (Wang et al. 2021). The lack of data regarding some pests leads to the developed systems being useful for only some particular pests such as insects with beetle-like forms (TÜRKOĞLU and HANBAY 2019). Furthermore, laboratory images of pest specimens are not enough to develop systems with full functionalities in the field (Wang et al. 2021). More datasets of in-field images are necessary to improve pest detection for devices deployed in agricultural scenarios. An intermediate option is images taken from pests collected by a trap (Wang et al. 2020). This facilitates the identification of the insect as the trap may be manufactured with colors and materials that make insects more noticeable. However, it is limited to the insects attracted to that particular trap.

5 Future Directions and Challenges

The adoption of PA sensing systems is undergoing rapid growth which presents some challenges (Sinha and Dhanalakshmi 2021). *Standardization* is one of the main challenges. Many available IoT devices use different communication protocols or syntax that hinder interoperability among all devices. To achieve high interoperability, standardization agencies need to make a common effort to unify the criteria for IoT networks. *Securing* PA sensing systems presents several challenges as well. Apart from physically harming the device, PA sensing systems are susceptible to cyberattacks such as Denegation of Service attacks, man-in-the-middle attacks, jamming, or data theft. The low computational power of many IoT devices makes it difficult to implement complex encryption techniques (Quy et al. 2022), which increases their susceptibility to those attacks. Furthermore, failures in the operations of the devices can lead to data inaccuracy or corruption, which affects the performance of the system. The *reliability* of PA sensing networks can be affected by the characteristics of the environment these devices are deployed in. Extreme weather conditions, animals, and workers can cause deterioration in specific sensors and the overall sensing device, which can lead as well to faulty network connections. The *scalability* of the PA sensing solution depends on the specifications of the communication technology, which can limit the number of deployed devices that can establish communication with the gateway. There are, however, secondary factors such as cost that, although it does not affect the scalability of the system per se, may lead to farmers being unable to scale the system if the cost is so high it cannot be covered. Lastly, *powering* IoT devices presents a challenge as these devices may not have access to the power grid. Thus, powering must be done with batteries and the use of solar panels. However, even if the system is scalable, secondary factors such as cost may lead farmers to experience difficulties in scaling the system due to lack of economic resources.

On the other hand, the evolution of technology is leading to the following trends regarding the introduction of new techniques in PA sensing systems. *Artificial intelligence* is now a common theme in literature for sensing PA systems. It is now applied to every aspect of PA systems from predictions of irrigation needs or climate evolution to identification or classifications of plant diseases or pests. The rapid development of AI has made this technology affordable and highly available. Nonetheless, the use of AI in PA systems still requires qualified technicians to deploy and install these functionalities. *Blockchain* provides trustworthiness to information. It also allows for product tracking and traceability. For example, this can be applied to supply chains to track food origin or enable timely payments when combined with smart contracts (Sinha and Dhanalakshmi 2021). Furthermore, it can also be used to provide security in data transmission between wireless sensors and cloud servers (Qazi et al. 2022). The use of *robots* is still not very extended in agriculture. However, some advances have been done to automate some tasks such as sensing, harvesting, weeding, mowing, cutting, or applying herbicide (Gil et al. 2023). Some of the existing robots are being commercialized but their price is still an obstacle for many farmers. Another available option to use this technology is leasing, which is a revenue model that is currently expanding and allows farmers to use robots for agriculture at more affordable prices. Lastly, the use of *nanotechnology* in agriculture is being studied to improve soil health, plant resilience, production rates, crop yields, and resource efficiency, as well as to reduce pollution, or for the concept of *plants as sensors* (Zhang et al. 2021). However, nanotechnology presents some challenges such as the possible toxicity of the nanomaterials in plants and its effect on the environment.

6 Conclusion

Precision agriculture is now present in the primary sector of many countries. This leads to the need of expanding the knowledge of both farmers and researchers in this area. Through this chapter, the reader has been provided with a detailed introduction to sensing systems for PA. Firstly, the motivation for deploying sensing devices in fields and the evolution of PA as a topic of research interest have been presented. Furthermore, an architecture of a sensing system for PA with complete functionalities including all the domains currently available has been provided. The sensing devices and technologies that can be employed to acquire data for each of the PA domains have been discussed as well. Lastly, the current challenges and future trends of PA systems have been listed and commented on so the reader can obtain a clear view of what is expected for sensing systems for PA in the next years.

References

Akhter R, Sofi SA (2021) Precision agriculture using IoT data analytics and machine learning. J King Saud Univ Inf Sci. https://doi.org/10.1016/j.jksuci.2021.05.013

Amrutha A, Lekha R, Sreedevi A (2016) Automatic soil nutrient detection and fertilizer dispensary system. 2016 International Conference on Robotics: Current Trends and Future Challenges (RCTFC), pp 1–5

Anthony D, Elbaum S, Lorenz A, Detweiler C (2014) On crop height estimation with UAVs. 2014 IEEE/RSJ International Conference on Intelligent Robots and Systems, pp 4805–4812

Anzum R, Habaebi MH, Islam MR, Hakim GP (2021) Modeling and quantifying palm trees foliage loss using Lora radio links for smart agriculture applications. 2021 IEEE 7th International Conference on Smart Instrumentation, Measurement and Applications (ICSIMA), pp 105–110

Arman Kuzubasoglu B, Kursun Bahadir S (2020) Flexible temperature sensors: a review. Sensors Actuators A Phys 315:112282. https://doi.org/10.1016/j.sna.2020.112282

Ayers RS, Westcot DW (n.d.) Water quality for agriculture. FAO irrigation and drainage paper. Retrieved December 14, 2022, from https://www.fao.org/3/t0234e/t0234e00.htm

Azfar S, Nadeem A, Alkhodre AB, Ahsan K, Mehmood N, Alghmdi T, Alsaawy Y (2018) Monitoring, detection and control techniques of agriculture pests and diseases using wireless sensor network: a review. Int J Adv Comput Sci Appl 9(12). https://doi.org/10.14569/ijacsa.2018.091260

Babatunde EB (2012) Solar radiation. BoD – Books on Demand

Banerjee AN (2022) Green syntheses of graphene and its applications in internet of things (IoT) – a status review. Nanotechnology. https://doi.org/10.1088/1361-6528/ac6599

Basterrechea DA, Rocher J, Parra L, Lloret J (2021) Low-cost system based on optical sensor to monitor discharge of industrial oil in irrigation ditches. Sensors 21(16):5449. https://doi.org/10.3390/s21165449

Bayrakdar ME (2019) A smart insect Pest detection technique with qualified underground wireless sensor nodes for precision agriculture. IEEE Sensors J 19(22):10892–10897. https://doi.org/10.1109/jsen.2019.2931816

Beyaz A, Gül V (2022) Determination of low-cost arduino based light intensity sensors effectiveness for agricultural applications. Braz Arch Biol Technol:65. https://doi.org/10.1590/1678-4324-2022220172

Bhat SA, Huang N-F (2021) (PDF) big data and AI revolution in precision agriculture: survey and challenges. Research Gate 9:110209–110222. https://doi.org/10.1109/ACCESS.2021.3102227

Bin Omar A, Bin Mat Jafri M (2009) Turbidimeter design and analysis: A review on optical fiber sensors for the measurement of water turbidity. Sensors 9(10):8311–8335. https://doi.org/10.3390/s91008311

Canata TF, Molin JP, Colaço AF, Trevisan RG, Martello M, Fiorio PR (2016) Measuring height of sugarcane plants through LiDAR technology. 13th International Conference on Precision Agriculture, pp 1–13

Cardim Ferreira Lima M, Damascena de Almeida Leandro ME, Valero C, Pereira Coronel LC, Gonçalves Bazzo CO (2020) Automatic detection and monitoring of insect pests—a review. Agriculture 10(5):161. https://doi.org/10.3390/agriculture10050161

Couliably S, Kamsu-Foguem B, Kamissoko D, Traore D (2022) Deep learning for precision agriculture: a bibliometric analysis. Intelligent Systems with Applications, p 200102. https://doi.org/10.1016/j.iswa.2022.200102

Dallas Semiconductors (n.d.) DS18B20 datasheet. Retrieved December 20, 2022, from https://cdn.sparkfun.com/datasheets/Sensors/Temp/DS18B20.pdf

Das B, Jain P C (2017) Real-time water quality monitoring system using internet of things. 78–82

Dutta S, Sarma D, Nath P (2015) Ground and river water quality monitoring using a smartphone-based pH sensor. AIP Adv 5(5):057151. https://doi.org/10.1063/1.4921835

El-magrous A A, Sternhagen J D, Hatfield G, Qiao Q (2019) Internet of things based weather-soil sensor station for precision agriculture. 92–97

Fang Y, Ramasamy R (2015) Current and prospective methods for plant disease detection. Biosensors 5(3):537–561. https://doi.org/10.3390/bios5030537

Fazekaov D (2012) Evaluation of soil quality parameters development in terms of sustainable land use. Sustainable Development – Authoritative and Leading Edge Content for Environmental Management. https://doi.org/10.5772/48686

Fleurat-Lessard F, Tomasini B, Kostine L, Fuzeau B (2006) Acoustic detection and automatic identification of insect stages activity in grain bulks by noise spectra processing through classification algorithms. Conference Working on Stored Product Protection

García L, Jimenez JM, Parra L, Lloret J, Lloret J (2020a) An overview on IoUT and the performance of WiFi low-cost nodes for IoUT applications. An Overview on IoUT and the Performance of WiFi Low-Cost Nodes for IoUT Applications, pp 1–6

García L, Parra L, Jimenez JM, Lloret J, Lorenz P (2020b) IoT-based smart irrigation systems: an overview on the recent trends on sensors and IoT Systems for Irrigation in precision agriculture. Sensors 20(4):1042. https://doi.org/10.3390/s20041042

García L, Parra L, Jimenez JM, Parra M, Lloret J, Mauri PV, Lorenz P (2021) Deployment strategies of soil monitoring WSN for precision agriculture irrigation scheduling in rural areas. Sensors 21(5):1693. https://doi.org/10.3390/s21051693

García-Miranda Ferrari A, Carrington P, Rowley-Neale SJ, Banks CE (2020) Recent advances in portable heavy metal electrochemical sensing platforms. Environ Sci: Water Res Technol 6(10):2676–2690. https://doi.org/10.1039/d0ew00407c

Geng Y, Dong Y, Zhao L, Huang W, Ruan C, Zhang H, Zhang B (2018) Desert locust plague monitoring using time series satellite data. Int J Precis Agric Aviat 1(1):24–30. https://doi.org/10.33440/j.ijpaa.20200304.111

Gil G, Casagrande DE, Cortés LP, Verschae R (2023) Why the low adoption of robotics in the farms? Challenges for the establishment of commercial agricultural robots. Smart Agric Technol 3:100069. https://doi.org/10.1016/j.atech.2022.100069

Gil-Pérez B, Zarco-Tejada PJ, Correa-Guimaraes A, Relea-Gangas E, Navas-Gracia LM, Hernández-Navarro S, Sanz-Requena JF, Berjón A, Martín-Gil J (2010) Remote sensing detection of nutrient uptake in vineyards using narrow-band hyperspectral imagery. Vitis 49(4):167–173

Graham R, Yun W, Kim J, Kumar A, Jones D, Bettio L, Gagnon N, Kolli R, Smith D (2011) Long-range forecasting and the global framework for climate services. Clim Res 47(1):47–55. https://doi.org/10.3354/cr00963

Guindo ML, Kabir MH, Chen R, Liu F (2021) Particle swarm optimization and multiple stacked generalizations to detect nitrogen and organic-matter in organic-fertilizer using vis-NIR. Sensors 21(14):4882. https://doi.org/10.3390/s21144882

Hachimi CE, Belaqziz S, Khabba S, Sebbar B, Dhiba D, Chehbouni A (2022) Smart weather data management based on artificial intelligence and big data analytics for precision agriculture. Agriculture 13(1):95. https://doi.org/10.3390/agriculture13010095

Harms J, Kern TA (2021) Theory and modeling of Eddy current type inductive conductivity sensors. Engineering Proceedings 6(1):37

Huang R, Huang J, Zhang C, Ma H, Zhuo W, Chen Y, Zhu D, Wu Q, Mansaray LR (2020) Soil temperature estimation at different depths, using remotely-sensed data. J Integr Agric 19(1):277–290. https://doi.org/10.1016/s2095-3119(19)62657-2

Janarthanan R, Balamurali R, Annapoorani A, Vimala V (2021) Prediction of rainfall using fuzzy logic. Mater Today: Proc 37:959–963. https://doi.org/10.1016/j.matpr.2020.06.179

Jeong H, Kim H, Jang T (2016) Irrigation water quality standards for indirect wastewater reuse in agriculture: A contribution toward sustainable wastewater reuse in South Korea. Water 8(4):169. https://doi.org/10.3390/w8040169

Jin X-B, Yu X-H, Wang X-Y, Bai Y-T, Su T-L, Kong J-L (2020) Deep learning predictor for sustainable precision agriculture based on internet of things system. Sustainability 12(4):1433. https://doi.org/10.3390/su12041433

Khairy M, Ayoub HA, Banks CE (2018) Non-enzymatic electrochemical platform for parathion pesticide sensing based on nanometer-sized nickel oxide modified screen-printed electrodes. Food Chem 255:104–111. https://doi.org/10.1016/j.foodchem.2018.02.004

Kim K, Myung H (2015) Sensor node for remote monitoring of waterborne disease-causing bacteria. Sensors 15(5):10569–10579. https://doi.org/10.3390/s150510569

Kim BS, Kim GW, Heo NS, Kim MS, Yang KS, Lee SY, Park TJ (2015) Development of a portable biosensor system for pesticide detection on a metal chip surface integrated with wireless communication. Food Sci Biotechnol 24(2):743–750. https://doi.org/10.1007/s10068-015-0096-x

Kim W-S, Lee D-H, Kim Y-J, Kim T, Lee W-S, Choi C-H (2021) Stereo-vision-based crop height estimation for agricultural robots. Comput Electron Agric 181:105937. https://doi.org/10.1016/j.compag.2020.105937

Kiziloglu FM, Turan M, Sahin U, Kuslu Y, Dursun A (2008) Effects of untreated and treated wastewater irrigation on some chemical properties of cauliflower (Brassica olerecea L. var. botrytis) and red cabbage (Brassica olerecea L. var. rubra) grown on calcareous soil in Turkey. Agric Water Manag 95(6):716–724. https://doi.org/10.1016/j.agwat.2008.01.008

Korotchenkov GS (2019) Handbook of humidity measurement: methods, materials and technologies. In: Volume 2, electronic and electrical humidity sensors. CRC Press

Kumar N, Belhumeur PN, Biswas A, Jacobs DW, Kress WJ, Lopez IC, Soares JV (2012) Leafsnap: A computer vision system for automatic plant species identification. Computer Vision–ECCV 2012: 12th European Conference on Computer Vision, pp 502–516

Leccese F, Cagnetti M, Giarnetti S, Petritoli E, Orioni B, Luisetto I, Tuti S, Leccisi M, Pecora A, Maiolo L, Spagnolo GS, Đurović-Pejčev R, Đorđević T, Tomašević A, De Francesco E, Quadarella R, Bozzi L, Arenella V, Gabriele P, Formisano C (2019) Electronic nose for pesticides: the first study towards a smart analysis. Contemp Agric 68(1–2):17–22. https://doi.org/10.2478/contagri-2019-0004

Lin Y-P, Mukhtar H, Huang K-T, Petway JR, Lin C-M, Chou C-F, Liao S-W (2020) Real-time identification of irrigation water pollution sources and pathways with a wireless sensor network and Blockchain framework. Sensors 20(13):3634. https://doi.org/10.3390/s20133634

Lloret J, Garcia M, Bri D, Sendra S (2009) A wireless sensor network deployment for rural and forest fire detection and verification. Sensors 9(11):8722–8747. https://doi.org/10.3390/s91108722

Lloret J, Bosch I, Sendra S, Serrano A (2011) A wireless sensor network for vineyard monitoring that uses image processing. Sensors 11(6):6165–6196. https://doi.org/10.3390/s110606165

Lloret J, Garcia L, Jimenez JM, Sendra S, Lorenz P (2021) Cluster-based communication protocol and architecture for a wastewater purification system intended for irrigation. IEEE Access 9:142374–142389. https://doi.org/10.1109/access.2021.3119757

Luo W, Goudriaan J (1999) Effects of altering water temperature on leaf wetness in paddy rice crops. Agric For Meteorol 97(1):33–42. https://doi.org/10.1016/s0168-1923(99)00046-5

MacKenzie DJ, McLean MA, Mukerji S, Green M (1997) Improved RNA extraction from woody plants for the detection of viral pathogens by reverse transcription-polymerase chain reaction. Plant Dis 81(2):222–226. https://doi.org/10.1094/pdis.1997.81.2.222

Marco FL, Sabino C, Tuti GS, Luisetto I, Petritoli E, Pecora A, Maiolo L, Đurović-Pejčev R, Đorđević T, Tomašević A, Bursić V, Arenella V, Gabriele P, De Francesco E (2017) Electronic nose for pesticides detection: a first realization. 2017 IEEE International Workshop on Metrology for Aero Space (Metro AeroSpace), pp 403–407

Marques Mostaço G, Ramires Costa De Souza ´i, Barreto Campos L, Cugnasca CE (2018) Agronomo Bot: a smart answering Chatbot applied to agricultural sensor networks. Proceedings of the 14th International Conference on Precision Agriculture, pp 1–13

Martínez J, Egea G, Agüera J, Pérez-Ruiz M (2016) A cost-effective canopy temperature measurement system for precision agriculture: a case study on sugar beet. Precis Agric 18(1):95–110. https://doi.org/10.1007/s11119-016-9470-9

McEntegart CM, Penrose WR, Strathmann S, Stetter JR (2000) Detection and discrimination of coliform bacteria with gas sensor arrays. Sensors Actuators B Chem 70(1–3):170–176. https://doi.org/10.1016/s0925-4005(00)00561-x

Megat Mohamed Nazir MN, Terhem R, Norhisham AR, Mohd Razali S, Meder R (2021) Early monitoring of health status of plantation-grown Eucalyptus pellita at large spatial scale via visible Spectrum imaging of canopy foliage using unmanned aerial vehicles. Forests 12(10):1393. https://doi.org/10.3390/f12101393

Misra NN, Dixit Y, Al-Mallahi A, Bhullar MS, Upadhyay R, Martynenko A (2020) IoT, big data and artificial intelligence in agriculture and food industry. IEEE Internet Things J 9(9):1–1. https://doi.org/10.1109/jiot.2020.2998584

Mohammad-Razdari A, Rousseau D, Bakhshipour A, Taylor S, Poveda J, Kiani H (2022) Recent advances in E-monitoring of plant diseases. Biosens Bioelectron 201:113953. https://doi.org/10.1016/j.bios.2021.113953

Namour P, Jaffrezic-Renault N, Namour P (2010) Sensors for measuring biodegradable and total organic matter in water. TrAC Trends Anal Chem 29(8):848–857. https://doi.org/10.1016/j.trac.2010.04.013

Nigam A, Sharma N, Tripathy S, Kumar M (2021) Development of semiconductor based heavy metal ion sensors for water analysis: a review. Sensors Actuators A Phys 330:112879. https://doi.org/10.1016/j.sna.2021.112879

Nor ASM, Yunus MAM, Nawawi SW, Ibrahim S (2013) Low-cost sensor array design optimization based on planar electromagnetic sensor design for detecting nitrate and sulphate. 2013 Seventh International Conference on Sensing Technology (ICST), pp 693–698

Ortiz J, Gualdron O, Duran C (2016) Detection of pesticide in fruits using an electronic nose. Aust J Basic Appl Sci 10(10):107–113

Oswald E (2019, August 1) How is wind measured? Anemometers and wind vanes explained | weather station advisor. Www.weatherstationadvisor.com. https://www.weatherstationadvisor.com/how-is-wind-measured/

Parra L, Sendra S, Lloret J, Rodrigues JJPC (2014) Low cost wireless sensor network for salinity monitoring in mangrove forests. IEEE SENSORS 2014 Proceedings. https://doi.org/10.1109/icsens.2014.6984949

Parra L, Rocher J, Escrivá J, Lloret J (2018) Design and development of low cost smart turbidity sensor for water quality monitoring in fish farms. Aquac Eng 81:10–18. https://doi.org/10.1016/j.aquaeng.2018.01.004

Pierce FJ, Nowak P (1999) Aspects of precision agriculture. Adv Agron 67:1–85. https://doi.org/10.1016/s0065-2113(08)60513-1

Puertes C, Bautista I, Lidón A, Francés F (2021) Best management practices scenario analysis to reduce agricultural nitrogen loads and sediment yield to the semiarid mar Menor coastal lagoon (Spain). Agric Syst 188:103029. https://doi.org/10.1016/j.agsy.2020.103029

Qazi S, Khawaja BA, Farooq QU (2022) IoT-equipped and AI-enabled next generation smart agriculture: a critical review, current challenges and future trends. IEEE Access 10:21219–21235. https://doi.org/10.1109/access.2022.3152544

Qin Y, Alam AU, Pan S, Howlader MMR, Ghosh R, Hu N-X, Jin H, Dong S, Chen C-H, Deen MJ (2018) Integrated water quality monitoring system with pH, free chlorine, and temperature sensors. Sensors Actuators B Chem 255:781–790. https://doi.org/10.1016/j.snb.2017.07.188

Quy VK, Hau NV, Anh DV, Quy NM, Ban NT, Lanza S, Randazzo G, Muzirafuti A (2022) IoT-enabled smart agriculture: architecture, applications, and challenges. Appl Sci 12(7):3396. https://doi.org/10.3390/app12073396

Ramane DV, Patil SS, Shaligram AD (2015) Detection of NPK nutrients of soil using fiber optic sensor. Int J Res Adv Technol:66–70

Rameshaiah GN, Pallavi J, Shabnam S (2015) Nano fertilizers and nano sensors – an attempt for developing smart agriculture. Int J Eng Res General Sci 3(1):314–320

Ramos PM, Pereira JMD, Ramos HMG, Ribeiro AL (2008) A four-terminal water-quality-monitoring conductivity sensor. IEEE Trans Instrum Meas 57(3):577–583. https://doi.org/10.1109/tim.2007.911703

Rasmussen PP, Gray JR, Glysson GD, Ziegler AC (2009) Guidelines and procedures for computing time-series suspended-sediment concentrations and loads from in-stream turbidity-sensor and streamflow data. In US Geological Survey Techniques and Methods, p 53

Refaai M, Reda A, Dattu VS, Gireesh N, Dixit E, Sandeep CH, Christopher D (2022) Application of IoT-based drones in precision agriculture for Pest control. Adv Mater Sci Eng 2022:1–12. https://doi.org/10.1155/2022/1160258

Rhee H-P, Yoon C-G, Jung K-W, Son J-W (2009) Microbial risk assessment using E. coli in UV disinfected wastewater irrigation on Paddy. Environ Eng Res 14(2):120–125. https://doi.org/10.4491/eer.2009.14.2.120

Rocher J, Parra L, Jimenez JM, Lloret J, Basterrechea DA (2021) Development of a low-cost optical sensor to detect eutrophication in irrigation reservoirs. Sensors 21(22):7637. https://doi.org/10.3390/s21227637

Sankaran S, Mishra A, Ehsani R, Davis C (2010) A review of advanced techniques for detecting plant diseases. Comput Electron Agric 72(1):1–13. https://doi.org/10.1016/j.compag.2010.02.007

Santos IM, Cugnasca CE (2012) Pesticide drift control with wireless sensor networks. 11th International Conference on Precision Agriculture, p 2

Segalla A, Fiacco G, Tramarin L, Nardello M, Brunelli D (2020) Neural networks for pest detection in precision agriculture. 2020 IEEE International Workshop on Metrology for Agriculture and Forestry (MetroAgriFor), pp 7–12

Sharma PK, Kumar D, Srivastava HS, Patel P (2018) Assessment of different methods for soil moisture estimation: A review. J Remote Sens GIS 9(1):57–73

Shi X, Tan C, Dong F, dos Santos EN, da Silva MJ (2021) Conductance sensors for multiphase flow measurement: A review. IEEE Sensors J 21(11):12913–12925. https://doi.org/10.1109/jsen.2020.3042206

Sinfield JV, Fagerman D, Colic O (2010) Evaluation of sensing technologies for on-the-go detection of macro-nutrients in cultivated soils. Comput Electron Agric 70(1):1–18. https://doi.org/10.1016/j.compag.2009.09.017

Sinha BB, Dhanalakshmi R (2021) Recent advancements and challenges of internet of things in smart agriculture: a survey. Futur Gener Comput Syst 126:169–184. https://doi.org/10.1016/j.future.2021.08.006

Sishodia RP, Ray RL, Singh SK (2020) Applications of remote sensing in precision agriculture: a review. Remote Sens 12(19):3136. https://doi.org/10.3390/rs12193136

Srivastava N, Chopra G, Jain P, Khatter B (2013) Pest monitor and control system using wireless sensor network with special reference to acoustic device wireless sensor. International Conference on Electrical and Electronics Engineering

Sumner ME (1994) Measurement of soil pH: problems and solutions. Commun Soil Sci Plant Anal 25(7–8):859–879. https://doi.org/10.1080/00103629409369085

Tan SL, Teo HS, García-Guzmán J (2010, October 28) E-nose screening of pesticide residue on chilli and double-checked analysis through different data-recognition algorithms. 7th IEEE Electronics, Robotics and Automotive Mechanics Conference(CERMA)

Tang X, Xiao W, Shang T, Zhang S, Han X, Wang Y, Sun H (2020) An electronic nose technology to quantify pyrethroid pesticide contamination in tea. Chemosensors 8(2):30. https://doi.org/10.3390/chemosensors8020030

Tatineni S, Sagaram US, Gowda S, Robertson CJ, Dawson WO, Iwanami T, Wang N (2008) In planta distribution of "*Candidatus* Liberibacter asiaticus" as revealed by polymerase chain reaction (PCR) and real-time PCR. Phytopathology 98(5):592–599. https://doi.org/10.1094/phyto-98-5-0592

Tatli S, Mirzaee-Ghaleh E, Rabbani H, Karami H, Wilson AD (2021) Rapid detection of urea fertilizer effects on VOC emissions from cucumber fruits using a MOS E-nose sensor array. Agronomy 12(1):35. https://doi.org/10.3390/agronomy12010035

Tongrod N, Tuantranont A, Kerdcharoen T (2009) Adoption of precision agriculture in vineyard. 2009 6th International Conference on Electrical Engineering/Electronics, Computer, Telecommunications and Information Technology. https://doi.org/10.1109/ecticon.2009.5137152

TÜRKOĞLU M, HANBAY D (2019) Plant disease and pest detection using deep learning-based features. Turk J Elec Eng & Comp Sci 27(3):1636–1651. https://doi.org/10.3906/elk-1809-181

Ukhurebor KE, Adetunji CO, Olugbemi OT, Nwankwo W, Olayinka AS, Umezuruike C, Hefft DI (2022) Precision agriculture: weather forecasting for future farming. AI, Edge and IoT-Based Smart Agriculture, pp 101–121. https://doi.org/10.1016/b978-0-12-823694-9.00008-6

Wang J, Niu X, Zheng L, Zheng C, Wang Y (2016) Wireless mid-infrared spectroscopy sensor network for automatic carbon dioxide fertilization in a greenhouse environment. Sensors 16(11):1941. https://doi.org/10.3390/s16111941

Wang Q-J, Zhang S-Y, Dong S-F, Zhang G-C, Yang J, Li R, Wang H-Q (2020) Pest24: a large-scale very small object data set of agricultural pests for multi-target detection. Comput Electron Agric 175:105585. https://doi.org/10.1016/j.compag.2020.105585

Wang R, Liu L, Xie C, Yang P, Li R, Zhou M (2021) Agri Pest: a large-scale domain-specific benchmark dataset for practical agricultural Pest detection in the wild. Sensors 21(5):1601. https://doi.org/10.3390/s21051601

Wójtowicz M, Piekarczyk J, Wójtowicz A (2016) Application of remote sensing methods in agriculture. Commun Biometry Crop Sci 11(1):31–50. https://researchportal.amu.edu.pl/info/article/UAMe7eacbd80eee427abcb6f388b1eae8f6/

WorldWideScience (n.d.). Worldwidescience.org. https://worldwidescience.org/

Yin H, Cao Y, Marelli B, Zeng X, Mason AJ, Cao C (2021) Smart agriculture systems: soil sensors and plant wearables for smart and precision agriculture. Adv Mater 33(20):2170156. https://doi.org/10.1002/adma.202170156

Zhang Y, Shen Y (2019) Wastewater irrigation: past, present, and future. Wiley Interdiscip Rev Water 6(3):e1234

Zhang N, Wang M, Wang N (2002) Precision agriculture—a worldwide overview. Comput Electron Agric 36(2–3):113–132. https://doi.org/10.1016/s0168-1699(02)00096-0

Zhang X, Goldberg MD, Yu Y (2012) Prototype for monitoring and forecasting fall foliage coloration in real time from satellite data. Agric For Meteorol 158-159:21–29. https://doi.org/10.1016/j.agrformet.2012.01.013

Zhang X, Karkee M, Zhang Q, Whiting MD (2020) Computer vision-based tree trunk and branch identification and shaking points detection in dense-foliage canopy for automated harvesting of apples. J Field Robotic 38(3):476–493. https://doi.org/10.1002/rob.21998

Zhang P, Guo Z, Ullah S, Melagraki G, Afantitis A, Lynch I (2021) Nanotechnology and artificial intelligence to enable sustainable and precision agriculture. Nat Plant 7(7):864–876. https://doi.org/10.1038/s41477-021-00946-6

Zhao D, Hao Z, Wang J, Tao J (2013) Effects of pH in irrigation water on plant growth and flower quality in herbaceous peony (Paeonia lactiflora Pall.). Sci Hortic 154:45–53. https://doi.org/10.1016/j.scienta.2013.02.023

Robotics and Artificial Intelligence (AI) in Agriculture with Major Emphasis on Food Crops

Naman Gupta and P. K. Gupta

Abstract The current review deals with various aspects of robotics and artificial intelligence (AI) in agriculture with particular emphasis on food crops. The review starts with a brief account of what robotics and AI really mean, followed by a brief history of the development and use of robots and AI, in parallel with the development and growth of the concept of precision agriculture (particularly the development of unmanned ground vehicles [UGVs] including autonomous tractors). The major part of the review deals with modern technologies in agriculture with particular emphasis on development and use of robots for different agricultural operations including the following: planting (preparations of field and seeding), plant care and crop management (including weeding, thinning, pruning, plant protection measures), crop scouting (use of sensors and geo-mapping), picking and harvesting, and post-harvest technology (packaging, transport, and storage). This section is followed by a brief account of the following aspects: some case studies of using robotics and AI, cost effectiveness of using robotics and AI, and technologies for plant production under controlled conditions (greenhouse agriculture, in vitro culture, and gene banks). The next section deals with the future of robotics and AI in agriculture, which is also an important section that includes development and growth of technologies in agricultural operations in different parts of the world including Asia (Singapore, China, and India), the USA, Canada, Europe, and Australia. At the end, conclusions and prospects are provided.

Keywords Precision agriculture · Autonomous tractors · Robotics · Artificial intelligence

---

N. Gupta (✉)
Blue River Technology, Sunnyvale, CA, USA

P. K. Gupta
Department of Genetics and Plant Breeding, CCS University, Meerut, India

Murdoch's Centre for Crop & Food Innovation, Murdoch University, Murdoch, WA, USA

Borlaug Institute for South Asia (BISA), NASC Complex, DPS Marg, New Delhi, India

P. M. Priyadarshan et al. (eds.), *Digital Agriculture*,
https://doi.org/10.1007/978-3-031-43548-5_19

1 Introduction

Robotics and artificial intelligence (AI) have become an integrated important area of research in all walks of life and agriculture is no exception. Agriculture itself is associated with the problem of food security that has been an issue of worldwide interest, which in turn is associated with population growth (Verchot 2020; Torero 2020). The current world population of 7.6 billion is expected to reach 8.6 billion in 2030, 9.8 billion in 2050, and 11.2 billion in 2100 (United Nations 2022). In view of this, it is also estimated that food production will have to increase by ~68% by 2050 to feed the growing world population. This must be achieved despite the problem of climate change, urbanization, and land degradation/desertification, which will adversely affect food productivity and production (Fig. 1). The use of robots and AI in agriculture is one of the many viable approaches to deal with this problem of food security.

For achieving the food production level desired in 2050, one has to estimate the required annual global agriculture growth rate, which has been shown to be fairly volatile and has been decreasing ever since the period of the Green Revolution, when it used to be ~3%, to the current level of <1% (Fig. 2). During the last two decades also, it has been fairly volatile, ranging from 5.8% in 2005–2006 to 0.4% in 2009–2010 and as low as −0.2% in 2014–2015 (Deshpande 2017). Even in Southeast Asia (including India), the annual growth of food production and productivity of major crops has slowed down in recent decades, causing alarms. Major efforts like policy changes and biological improvements for increasing yield are, however, underway to deal with this problem.

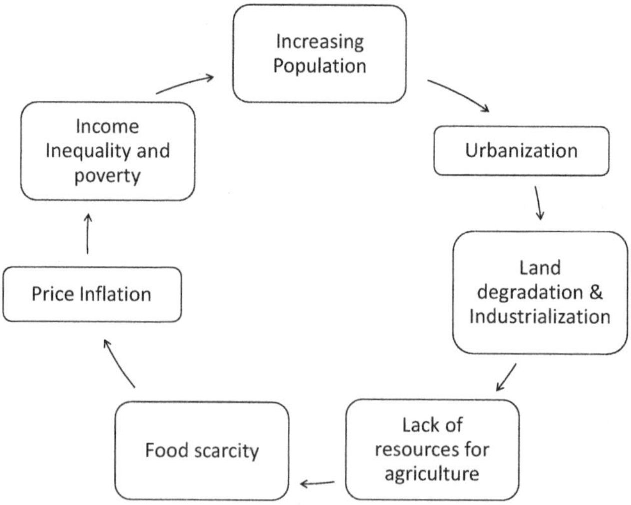

Fig. 1 Vicious circle of increasing population, urbanization, land degradation, food scarcity, etc

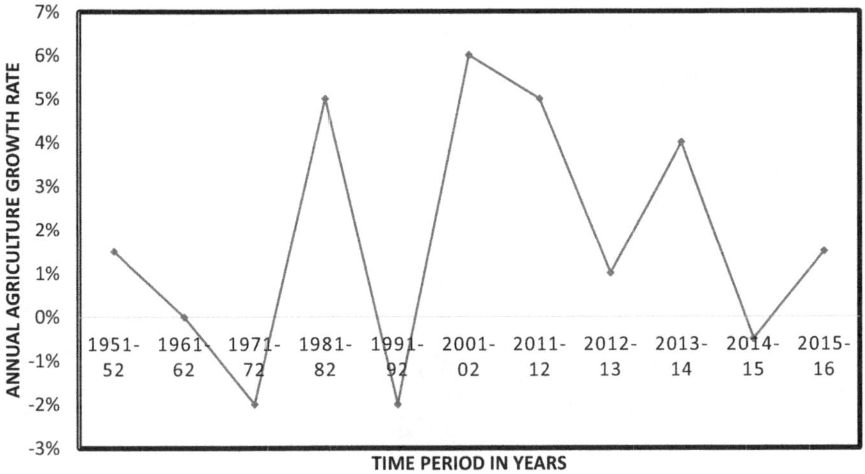

Fig. 2 Global agriculture growth rate during 1951–2015. (Agricultural Statistics at a Glance 2015; PRS Publication)

We also know that the Green Revolution helped us scale-up the food production with the help of improvement of the genetic architecture of major crops, particularly through development and introduction of semi-dwarf varieties with high photosynthetic activity and fertilizer-response, along with crop management practices, although concerns have, recently, been expressed regarding the damage done to the environment due to the Green Revolution.[1]

A digital revolution is taking place now, through AgTech (Agricultural Technology), which will engage a lot of robotics and AI with agriculture to solve the current bottlenecks in increasing the food production. Apart from other bottlenecks in agriculture, climate change has been both the cause and effect of problems in agriculture. It is debatable if all the effects and causes of climate change can be fixed with AgTech. However, the extensive use of herbicides and pesticides since the Green Revolution has caused significant land degradation affecting the yield and also other climate change issues because of their production and transport. AgTech can help reduce the use of these harmful chemicals by either using other technologies or targeted spraying. The food production system has also been shown to account for 26% of total global greenhouse gas emissions.[2] Other issues associated with climate change include increased temperatures around the globe, unpredictable rainfall, and other natural disasters, which have an impact on

[1] https://sciencing.com/harmful-effects-green-revolution-8587115.html

[2] https://news.un.org/en/story/2021/03/1086822

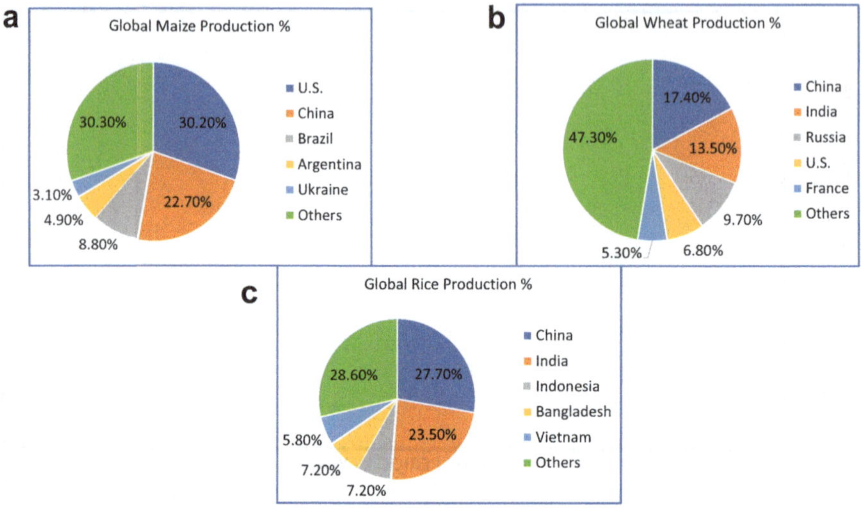

Fig. 3 The share of countries producing (**a**) maize, (**b**) wheat, and (**c**) rice at the global level. (Source: https://www.visualcapitalist.com/cp/mapped-food-production-around-the-world/)

agricultural production, and AgTech is evolving to better prepare farmers to face such adversaries.[3]

The major staple crops, which need attention toward achieving the solution to food security, include three major ones: wheat, maize, and rice. The shares of different countries involved in growing these crops are depicted in Fig. 3. Since these countries are also the ones with maximum demand, it is these countries that need to scale-up the production and productivity to deal with the problem of food security. Besides other measures including genetic improvement of crops, agronomic practices and the use of advanced technologies will have its own share. Among these advanced technologies, robotics and AI will have its own share, which will increase with time and will have their own set of limitations and challenges that might limit their usage across the globe. It is believed that robotics and AI can make all agricultural operations more efficient and at the same time reduce cost. Although the applications are not just limited to farms but also to greenhouses, vertical farming, and hydroponics, we will limit the scope of this chapter to farms only.

Some countries, especially the Western countries, have also witnessed a huge impact of labor shortage, although this is more prevalent in the West. In recent years, it is also becoming a common issue in most developing countries because of migration of workforce from countryside and villages to towns and cities, which happened because of the industrial revolution and globalization. Even with the use of machinery, this has been a challenge for many farmers because we still

[3] https://www.usgs.gov/faqs/what-are-long-term-effects-climate-change

need humans to monitor and manage the farms for using the farm equipment and machinery.

The influence of the industrial revolution on agriculture has also been witnessed. Firstly, it has reduced the productivity due to the shortage of labor in agriculture, and, secondly, it has increased the labor cost by a factor that is smaller than the increase in agriculture profits. The extent of manual labor needed depends on the crop, cattle, climate, farming method, topology, and unforeseen problems like pest attacks. Digital revolution aims to solve some of these problems in certain straightforward cases using robots and AI. Along with monitoring and supervision, the robots can also help with minimal tasks like controlling the weeds, harvesting, irrigation, and ploughing.

This chapter deals with use of robots and AI in agriculture and is divided into six main sections: Sect. 1, Introduction; Sect. 2, A primer explaining what are robotics and Al; Sect. 3, History of robotics and AI in agriculture; Sect. 4, Technologies for different agricultural operations; Sect. 5, Some case studies of using robots and AI in agriculture; Sect. 6, Cost effectiveness of using robotics and AI; Sect. 7, Robots and AI for plant production under artificial conditions; Sect. 8, The future of robotics and AI in agriculture; and Sect. 9, Summary and Conclusions. Several other related aspects including Internet of Things (IoT), unmanned aerial vehicles (UAVs), and vertical farming are discussed in other chapters of this volume, and therefore will not be covered in this chapter, except a brief account in Sect. 7.

2 What Are Robotics and AI? A Primer

The popularity of robotics and AI in the recent years has made many of us come across these technologies. However, the definitions regarding what are robotics and AI have been constantly evolving without a consensus. Robots were first used in science fiction plays, movies, and literature. All of them had human-like features; for example, they could either walk like humans, talk like them, or do tasks like humans that no other machines could do at that point of time. Since then, machines have become more sophisticated, and robots have also been developed for adaptation in various scenarios. The line between what we call a machine and a robot is very slim, and it is subjective to call something a robot or just a machine, but it is safe to say that all robots are machines but not all machines are robots. Just like machines and robots, robotics and AI have also been used interchangeably for some applications, which does spring up some confusion about how they are different from each other and what technology is at play in a certain technology. To prevent this confusion later in the chapter, this section will cover what the authors think should refer to as robotics and AI in this chapter.

2.1 Robotics

The word *robot* is derived from the Czech word *robota*, which means "forced labor." Since robots were built for repeated and dangerous tasks initially, this derivation of the word held some meaning in the past. As robots have evolved, the definition of robotics has become context dependent. Oxford English dictionary defines it as "a machine—especially one programmable by a computer—capable of carrying out a complex series of actions automatically." This definition is generic enough to encompass most of the robotics applications, but it is not specific enough, as this definition would allow a washing machine to be classified as a robot. A washing machine is not considered a robot because it is a pre-programmed machine that will do the same actions on a given input. It is not intelligent enough to select its own cycle based on the clothes in it.

Another definition classifies robots as those machines that can mimic human or an animal movement to complete a task like a human. "Shakey" was the first general-purpose mobile robot that was built in the late 1960s. It was able to do things with a purpose versus just a list of instructions. Space exploration vehicles or rovers, humanoids, and other legged robots are unanimously termed as robots as they can carry out tasks like humans can, and even go in places where humans cannot go. These robots were initially completely remotely controlled but gradually have been integrated with different levels of autonomy to perform tasks more efficiently with easier control and lesser human intervention. This definition of "completing task like a human" also does not include all types of robots because there exist autonomous machines that do tasks that no humans could ever do and sometimes, even do the same tasks better and more efficiently than humans.

There is no consensus on which machines qualify as robots but there is a general agreement among experts, and the public, that robots tend to possess some or all of the following abilities and functions: electronic programming, sense their environment, process data or physical perceptions electronically, operate autonomously to some degree, move around, operate physical parts by itself or physical processes, manipulate their environment, and exhibit intelligent decision-making. Furthermore, a robot might or might not be completely autonomous. It can have varied degrees of autonomy. Their usage can vary from being manually driven or tele-operated to being completely autonomous and needing no human intervention.

The most agreed difference between a machine and a robot is that a machine is an electro-mechanical appliance with a once programmed software on a low- to mid-level computer, while robots have sensors to perceive the environment and interact with the environment. With this definition, the robots can be broadly classified into the following two types, or a combination of the two: (i) manipulator arms and (ii) mobile robots (Trevelyan 1999). In turn, mobile robots are mainly classified into the following three types based on the environment they move in: (i) ground vehicles: for example, wheeled, tracked (like military tanks), or legged robots; (ii) unmanned aerial vehicles (UAVs): for example, drones and unmanned airplanes; and, (iii) underwater vehicles: for example, autonomous submarines.

In this chapter, we plan to restrict the scope to ground vehicles and manipulator arms attached to a moving base. In doing so, we will cover how these robots act intelligently and perform agricultural operations. There have been efforts to make robots of various levels of autonomy. For example, a leader–follower approach may be used where a leader tractor driven by a farmer is followed by an autonomous tractor. Swarm robots are another example where multiple robots talk to each other and complete a task collaboratively. There are also semi-autonomous robots, which do certain tasks on their own and ask for human intervention for more complex tasks.

2.2 *Artificial Intelligence*

The term "artificial intelligence" (AI) was coined in 1955 by John McCarthy, an American mathematician and computer scientist. Artificial intelligence refers to the ability of a computer-controlled machine or a computer to make decisions based on data. At this point, most of the AI applications include the use of machine learning (ML). Machine learning is an extension of a field in statistics called data science. Data science utilizes large-scale data and makes derived analysis from the data and generates or predicts the outputs for a given input by fitting a given non-linear mathematical function. The predictions are better with machine learning (ML) algorithms, which learn patterns from data and utilize that information to predict or generate an output that will help in decision-making. Learning patterns can be thought of as fitting an unknown mathematics function to the data, which specifically makes it different from the traditional data science methods. The complexity of learning these patterns depends on the quality and quantity of data, the number of data attributes associated with one output, and the complexity of dependence of these attributes on the output. The more complex the learning pattern is, the more computational resources are needed to learn accurately. Deep learning, which is a subset of machine learning, allows us to predict very complex patterns with a significant level of accuracy. However, these deep learning models act as a black box, meaning that they do not allow us to mathematically understand the function that converts the inputs to outputs.

Research in AI is constantly evolving and one of the focuses of the research community is to make better judgments using lesser data as that would need lesser resources and will lead to automation in agriculture (Jha et al. 2019). Recently, a new field of AI has sprung up, which is called "Generative AI," which explores generation of new unseen data based on old data. Another active field of research is "Explainable AI," which enables humans to understand the predictions made by AI and unbox the "black box."

AI, just like robotics, has different levels of intelligence. A simple AI algorithm optimizes a cost function against some constraints given by the user. For example,

AI can help find a path from point A to B using pre-defined distances and speed limits without any knowledge of real-time traffic. A more complex AI can learn from sensing the environment and actions of a human-driven car and then use that learning to drive a car by itself while making decisions about the fastest route and the fastest lane.

Robotics and AI are often used interchangeably and sometimes together because robotics often uses AI for autonomy and automation. AI has been used in all aspects of robotics including sensing (see the world using sensors), mapping (map the world around it), localizing (know where the robot is in the map), planning (make decisions to navigate the world), and control (send the right commands to motors to enable desired and accurate movement of the robot). AI can learn all these steps from sensing to control together or individually but robot will use the output of an AI algorithm to take an action and interact with humans or the environment. We will explore past, present, and future of all these types of agricultural robots and AI in the following sections in this chapter.

3 History of Robotics and AI in Agriculture

Industrial revolution, which itself started in mid-eighteenth century and early nineteenth century, led to occupational mobility, which led people to migrate from countryside or villages to towns and cities in search for better income. Standard of life also created an immense shortage of labor across the globe. Often, lack of skilled labor and high investment in agriculture by the governments are compared against the high cost of building and buying robotic technologies and good-quality camera and other sensor data for AI. This benefit–cost ratio changes with kind of agricultural operations, countries, crops, and even the state of the economy of the country.

In the early 1990s, there was a demand for improved efficiency, reduced costs, and automation in agriculture, and later robots and AI provided a solution for this demand. The world wars in the early twentieth century, which resulted in a shift of labor from agriculture to armed forces, also generated a demand for automation. A global increase in labor cost during the 1930s due to the so-called Dust Bowl involving poor farming practices and drought, thus devastating millions of acres of fertile land, also created a demand for automation for some of the trivial as well as major tasks in agriculture (Fountas et al. 2007, 2020). All these factors were responsible for regular and continuous research in the subject area of robotics and AI for agriculture. A timeline for major breakthroughs in the field of robotics and AI in agriculture during 1920–2020 is presented in Table 1.

Table 1 A timeline showing major breakthroughs in the field of robotics and AI in agriculture during 1920–2000s

Time period	Major breakthroughs that happened during that time period
1920–1945	Academic research in automation and electro-mechanical advancements in agriculture equipment
1941	Plowing of circular fields using a rope tied in the center as it wounds on itself[a]
1945	Teleoperation using radiofrequency of farm tractors
1959	Multichannel radio-controlled tractor employing hydraulic actuators
1964	Leader–follower operation of farm tractors (automatic follower farm tractor following leader farm tractor separated by a physical fixed distance)
1967	Localization using triangulation by detecting the exhaust pipe of a farm tractor using two fixed-in-place infrared sensors and then a control system to control the position of the tractor (other groups used radar instead of infrared sensors to do the same)
1970s–1990s	Development of control systems to control sprays for spraying herbicides and fertilizers (Harries and Ambler 1981)
1981	First use of optical sensors in agriculture to do non-contact sensing of the furrow wall
1992–2000	Integration of global positioning system (GPS) into agriculture vehicles for real-time position tracking and advanced control algorithms for precise navigation
1995–2000	Vision guidance systems was successfully implemented using optical sensors (Gerrish et al. 1997)
2000–2020	Industry and academic research working together on building various robotic platforms for agriculture

[a]https://livinghistoryfarm.org/farminginthe40s/water_09.html

3.1 *Robots for Autonomous Tractors*

During the 1950s, agriculture witnessed a rise in rudimentary robots. For instance, a tractor tied with a rope to the center of the field allowing the rope to wind itself as the robot went in circles was the simplest approach to automation. Thereafter, during the 1960s, tractors were controlled by radiofrequency remotes. Automation in tractors also resulted through research involving a leader–follower approach, which was a milestone in the history of using robots and AI in agriculture. However, the more difficult problem of making just a standalone tractor autonomous was solved by first solving the localization problem, which means that the robot needs to know where it is in the space in order to take an appropriate action. Earlier, triangulation was also a common approach for solving this problem because it is comparatively a simple mathematics problem. Triangulation is achieved by observing a common unknown point from two different known points. In the case of agriculture robotics, it was achieved by the use of infrared sensors placed at the two vertices of the field and measuring the distance of the tractor's exhaust pipe from these two sensors. Since, the sensor location and the distance between them was known, it was possible to find the tractor's location from either of the sensors. This was a step ahead in determining the location of the tractor and then precisely controlling the tractor based on its location. The late twentieth century also witnessed a boom in the use of computers and supercomputers, which were faster, involving use of complicated algorithms. These developments were also utilized in all aspects of robotics. It started with the invention of precise control of electro-mechanical systems, with usage of camera like optical sensors to detect furrow walls and algorithms that could use the data and run the appropriate calculations.

During the 1990s, Trimble released a Real-Time Kinematic (RTK) positioning device for precise control of the tractor. Since then, with increase in computer capabilities and semiconductor usage, sensors gradually became smaller in size with improved precision. Algorithms were also developed to utilize the sensor signals to correctly estimate the tractor's velocity and position in the field. Improvements in the processing speeds also led to online optimizations, which made these technologies safer and user-friendly. During the late 1990s, the research conducted on the first semi-autonomous tractor technology resulted in modern-day tractors using robotics and AI (Stentz et al. 2002).

3.2 Robots for Precision Agriculture

During the 1980s and 1990s, the concept of precision agriculture (PA) emerged as the science of improving crop yields through improved management decisions and use of high-technology sensor and analysis tools. This also involved analysis of the field used for a crop, including health of soil, so that optimum quantity of fertilizer could be used according to soil condition in different areas of the field and appropriate weeding operations may be planned. These operations paved the way for use of robots. Integrating the data from external sensors to the robot's computer allows the robot to know where an agricultural operation is needed. During early years of the present century, algorithms that allow precise control of these robots were also developed. However, the use of optimum fertilizers and weeding operations still continue to be an active area of research and development. During the last few years, computer vision (CV) involving use of imaging (through use of cameras) and machine learning (including deep learning) has made significant progress, and further steep rise in adoption of robotics and AI in agriculture is expected during the next few decades. More details about these developments will be covered in the following section.

4 Modern Technologies for Different Agricultural Operations

In recent years, automation in agriculture has become a necessary component of precision agriculture, involving operations that include seeding (planting), inspection, spraying, pruning, and harvesting (Bogue 2016). Automation generally involves use of robots for different agricultural operations to minimize physical work that needs to be undertaken by the farmers. Automation helps in increasing precision and efficiency, thereby reducing the costs. Therefore, the development of an efficient automation system involving use of robots and AI in agriculture is becoming an active area of current research in agriculture to achieve sustainability and food security (Mahmud et al. 2020).

The agricultural operation, which involves use of robotics, can be broadly classified into the following four parts: (i) planting, (ii) plant care, (iii) harvesting, and (iv) post-harvest storage and shipping. Each of these operations can be further classified into minor operations, as done in a recent article by Bera and Dutta (2021). Some of these operations are shown in Fig. 4. A brief account of each of these developments is provided in this section.

4.1 Robots for Planting

Planting involves three major operations, including preparation of beds, seed mapping, and seed placement. The use of robotics and AI in each of these operations will be briefly described.

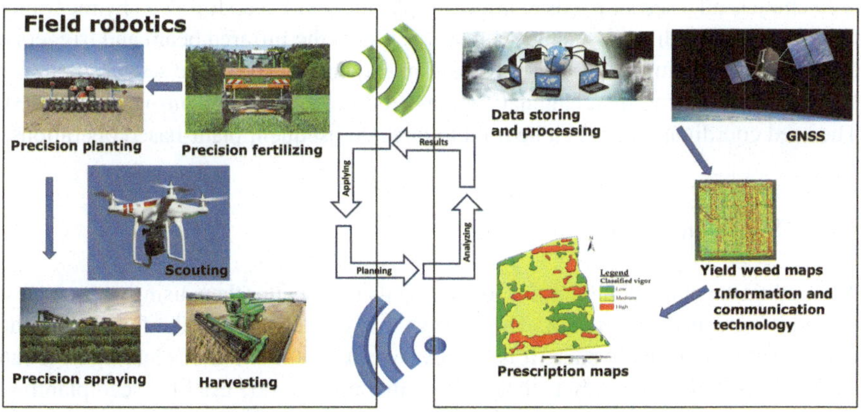

Fig. 4 Some agricultural operations involving use of robotics including ground robotics and drones (shown on the left) and artificial intelligence (AI) in planning (shown on the right)

4.1.1 Preparation of Seed Beds

Preparation of seed bed on a farm involves ploughing, which means inversion or mixing of topsoil to prepare a suitable seed bed. A small robot utilizing current technology does not have the energy density to sustain ploughing over a large area due to the high levels of energy needed to cut and invert the dense soil. As discussed before, autonomous tractors with various levels of autonomy have been used since the 1950s, but more recently, since 2010, companies like John Deere, Autonomous Tractor Corporation, Fendt, and Case IH have been moving toward mass production of their completely autonomous tractors. Current research has witnessed a rise in agriculture robots for multipurpose operations, including ploughing, seeding, and health monitoring (Chandana et al. 2020). These robots have the complete map of the farms using the satellite navigation systems like global navigation satellite system (GNSS), which they use to divide the farm into rows. The sensors on the robots like wheel encoders, inertial measurement unit (IMU), and RTK positioning provide the information about the robot's location on the farm and this precise location is used to position the robot in such a way that proper tillage is obtained. With advances in communication protocols, semiconductors, and software engineering, the control of the robots has become increasingly precise, fast, and efficient (Rus 2018).

4.1.2 Seed Mapping

Seed mapping is the concept of passively determining the geospatial position of each seed before it goes into the ground. The robot takes the input from a pre-existing model or from the user on how distant the seeds need to be placed. This information combined with an RTK GPS and infrared sensors mounted below the seed chute is all a seeder needs to know for placement of seed followed by recording the location of each seed. As the seed drops, it cuts the infrared beam and triggers a data logging system that records the position and orientation of the seeder. A simple kinematic model can then calculate actual seed position (Griepentrog et al. 2003). The seed coordinates can then be used to target subsequent plant-based operations.

4.1.3 Seed Placement

The traditional manual method for sowing seeds is to scatter them using a "broadcast spreader" attached to a tractor. This throws many seeds around the field while the tractor drives at a steady pace. It is not a very efficient method of planting as it can waste seeds. Therefore, robots that do seed mapping can be used for seed planting. Autonomous seeders have the mechanism to check the quality of the seed at a high speed using computer vision and weight sensing before placing them into the soil. The seeds are stored in huge bins on these seeders and a valve spits out one seed at a time. This seed goes on to the conveyer on which the quality is checked and is then put into the soil with high pressure so that it goes deep into the soil. The seed map is used to control the seed position. Most seeds are dropped at high densities within each row, while having relatively more space between the rows. First principles of agronomy dictate that each plant should have equal access to spatial resources of air, light, ground moisture, etc. Most crops use a hexagonal or triangular seeding pattern, since this is believed to be more efficient. All these models are put into the seeders and with the technologies discussed above, a seeder can precisely conduct the seed placement. Robots are also being used for seed placement in potted plants, and for picking and placing pots from one place to another, as done in case of Harvest Automation's HV-100.

4.2 Robots for Plant Care (Crop Management)

Crop growth and soil regulation are important to provide high yield, retain soil nutrients and microorganisms, and to deal with environmental uncertainties and natural disasters. Humidity, temperature, and nutrients of the soil also play a major role in crop yield and quality. It has been shown that global loss yield in rice, wheat, and maize due to weeds account for 46.2–61.5% of the potential yield losses. Similarly, losses due to pests account for 27.3–33.7% of actual yield losses (Oerke 2006). It has also been recognized that in the long term, soil treatment with chemicals (like fertilizers and herbicides) and operations including irrigation can cause substantial damage to the quality of top soil and may also contribute to climate change.

In view of the above, robot experts have also come up with appropriate solutions, which can limit soil degradation, herbicide resistance, and climate change by either using a better substitute to treat crops, weeds, and soil or limiting the use of chemicals. The plant care broadly includes the following four operations, each involving use of robots (Fountas et al. 2020): (i) weeding, thinning, and pruning; (ii) spraying: irrigation and fertilizers; (iii) disease and insect detection; and (iv) crop scouting (plant monitoring and phenotyping). Since extensive literature is currently available, only a brief summary of the most important literature, based on scientific and commercial resources, will be described here under four subsections.

4.2.1 Robots for Weeding, Thinning, and Pruning

Weeding is one of the most repetitive, tedious, and time-consuming activities within the crop production cycle, especially for developing countries and small farms. Weed management accounts for more than 40% of the labor effort (Fountas et al. 2020). For all the above-mentioned reasons, significant attention has been given to weeding robots both by private companies and academia (Ackerman 2015).

The removal of weeds has traditionally been done either by manual weeding or by spraying herbicides/weedicides over the entire farm with the help of sprayers.

The robotics community aims to improve the accuracy of weed detection (number of weeds found by the robot divided by the actual number of weeds in the farm) and hit rate (number of weeds killed given they were detected) and increase the speed at which the entire farm is weeded. As there are hundreds of crops and weed varieties, it is difficult to develop one robot that solves this problem for all crops around the world. Current solutions include selective spraying of herbicides (John Deere's See and Spray, QUT's Agbot II), burning the weeds with lasers (Verdant Robotics), use of lasers (Andreasen et al. 2022; Heisel 2001), and also mechanically uprooting/destroying the weeds (Naio Technologies' Dino, Garford's Robocrop, Bosch's BoniRob). All these robot cameras, lasers, radars, ultrasound sensors, infrared sensors, IMUs along with precision algorithm tools accurately detect the weeds above the ground in three dimensions (Table 2). These solutions not only reduce the time spent in weeding and cost associated with labor, but also reduce the amount of herbicides used, which saves cost for the farmers, prevents the weeds to become herbicide resistant, and protects the environment since minimum quantity of herbicides will get washed away into the oceans.

Thinning involves reducing the density of plants so that each plant has a better chance of growing. One of the popular thinning robots is LettuceBot, which was developed by Blue River Technology and also received an award in 2017 by the American Society of Agricultural and Biological Engineers (ASABE). It uses computer vision to detect lettuce plants as it drives over them and decides at that moment, which plants to keep and which to remove. Special robots like Ibex have also been designed to be used on hills (Cousins 2016).

4.2.2 Robots for Irrigation and Fertilizer Application

Irrigation automation is done using sprinklers placed across the field, which are controlled based on a timer. Weather-based irrigation controllers are also available, which allow farmers to achieve water-efficient irrigation scheduling. Research has also been conducted by academia to build robots that carry sprinklers so that they can move around and irrigate, but no commercial product is available, since it might not be commercially viable. However, a robot-assisted precision irrigation, if

Table 2 Robotics for weeding in different crops

Crop	Sensors	Weed detection	Weed control	Result
Maize	Cameras; optical/acoustic sensors	Yes	Chemical	No performance metrics
Rice	Laser range finders; IMU	Partly	Mechanical	Precision <62 nm
Potato	Webcam; gyroscope	No	Chemical	Accuracy: 98%; 89%
Tomato	Color camera; SensorWatch	Partly	Chemical	Poor
Sugar beet	Color camera	Yes	Mechanical	>90% weed removal

From Fountas et al. (2020)

available, can reduce wastage of water by targeting specific plants (Company Osiris Agriculture's Oscar). Ground robots can also autonomously navigate between rows of a crop and pour water directly at the base of each plant (Carpin et al. 2019).

Fertilizers help keep the soil fertile and nutritious for healthy crops. Precision farming tools and hand-held tools (for example, Trimble's GreenSeeker) have been used to measure the plant's needs. The health of soil is determined using the precision farming sensors and the health of the crop can be determined using NDVI (normalized difference vegetation index) value provided by tools like GreenSeeker. This information is enough to precisely determine the amount of fertilizer needed. Ground robots for fertilizer application have an advantage as they can access areas where other agriculture equipment cannot reach. For example, corn growers face a problem that the plants grow too quickly to make the operation of reliable application of fertilizer difficult. An important robot for fertilizer application is Rowbot (https://www.rowbot.com/) that aims to solve this problem as it easily drives between the rows of corn field and targets nitrogen fertilizer directly at the base of each plant.

4.2.3 Robots for Plant Protection (Pests and Diseases)

Robots and AI have also been used for integrated pest management (IPM) involving identification of acceptable levels of pests and diseases and preventive cultural practices. Different types of cameras (including high-resolution visible RGB [red-green-blue] to thermal, infrared, multispectral, hyperspectral, and ultraviolet [UV] cameras) and chemical and electrophysiological sensors are being used for collecting data on health of the plants. The sensory data and imaging information is usually coupled with a machine learning/AI engine that either flags anomalies in the datasets or detects specific patterns or objects. Anomalies can be a sign of a pest attack or a disease-stricken crop. These solutions can produce real-time risk alerts, outbreak projections, and predictions.

The data collected as above are also used to formulate a proper course of action and treatment to fix the problem. AI systems can calculate the most optimized course of action that provides the best control for the lowest price. The treatment prescribed based on data collected needs robots, which have become available. For instance, ultraviolet-C (UV-C) treatments, biopesticide sprayers, and robotic systems for dispersal of biological control agents are becoming commercially available.

4.2.4 Robots for Crop Scouting (Monitoring and Phenotyping)

Monitoring huge fields of crop is a big job. New sensor and geo-mapping technologies enable farmers to get a much higher level of data about their crops than they had in the past. Ground robots, sensors, and drones (Anderson 2017) provide a way to collect these data autonomously. Researchers have built ground robots, which can use cameras to record plant health indicators like the ripeness of the fruits, color of the leaves, and other characteristics, which allows the researchers to conduct phenotyping on those crops. There has been an extensive use of manipulator arms on ground robots for phenotyping even if it is made for a specific crop, since this allows the robot to interact and examine crop at different heights. A significant amount of AI and deep learning is used to detect fruits and color of leaves and estimate the size of fruit, plant height, plant width, plant volume, and surface area. Multiple cameras or complex computer vision algorithms are needed with highest accuracy to gain three-dimensional (3D) knowledge. Researchers and entrepreneurs have used a laser-based technology called light detection and ranging (LiDAR), which directly gives the 3D data about the surroundings. It is based on the principle of "time of flight" technology, which is also used in sound navigation and ranging (SONAR) and radars. For instance, the plant volume of perennial ryegrass was measured using a LiDAR sensor on DairyBioBot, which was correlated with the biomass (Xu and Li 2022). The canopy volume of almond trees was also measured using a LiDAR sensor on the Shrimp robot, which was shown to be correlated with the yield. The flower and fruit density of the almond tree was measured using RGB images obtained from camera.

4.3 Robots for Picking and Harvesting

Picking and harvesting include one of the most popular robotic applications in agriculture due to the accuracy and speed that robots can achieve to improve yield levels and reduce waste that is left in the field after the crop is harvested. Most of the robots that pick fruits also use phenotyping results either through the same robot or a different robot to gauge the level of ripeness of different fruits before picking. Fruit picking robots have been developed for tomatoes, berries (strawberries, blueberries, etc.), apples, capsicum (or bell peppers), cotton, oranges, radicchio, etc. (Daniels 2018; Foglia and Reina 2006). These robots, which use a suction gripper to pick fruits, can also be used for picking different types of ripe fruits, because ripe fruits are more easily removed from the stalk. For a number of vegetables and fruits, like those that are heavy or grow on the ground or below the ground, harvesters are used. These harvesters are more efficient compared to manipulator arms and can be used with multiple crops. However, it is not always possible to use harvesters, particularly for fruits that are delicate and need careful and soft handling to avoid damage to the fruit or the crop around it.

Combines are also used to harvest the produce in a variety of grain crops. These machines combine the following four separate harvesting operations in a single process: (i) reaping, (ii) threshing, (iii) gathering, and (iv) winnowing. Semi-autonomous combines (like John Deere's S700) have been in the market since 2017. This machine can auto-steer, calculate yield, and dispose off all the hay into a separate bin leaving no waste on the field. Since harvesting for grains is not a problem now, new products will generally aim mainly on fruits, which can also be harvested like grains, but none of these products are automated yet.

4.4 Robots for Post-Harvest Technology

Harvesting and post-harvesting are the most delicate operations. This is where a mistake by a robot can lead to direct losses (either by recalling a product due to contamination or throwing away), which could likely have been avoided if humans were handling. This is why farmers still do not trust robots that directly handle the harvest, and thus there are fewer robots in this space. However, robots that interact with humans during the post-harvesting stage are more readily acceptable. Automation of carts that carry plants and harvest is one such example. A Philadelphia (USA)-based startup, Burro.ai, has developed such robots that move around the farms, scouting fruit crops, carrying the trays of harvest placed by humans from farms to shipping area and carrying small pots of plants around the nurseries. They increase efficiency of farmers by automating the simplest but important tasks that farmers can reliably let a robot help with. A large number of robots are also available for handling, storing, and transporting fresh horticulture produce (Luna-Maldonado 2010; Luna-Maldonado et al. 2012).

In the past, conveyor belts along with cameras and infrared devices have been used either with or without human intervention to scan the quality of the harvested product and then trim, cut, or discard the item depending on the standards set by the producer. Because of varying dimensions and fragility (depending on the crop or over-ripeness/damage) of the harvest, a conveyor belt solution without the use of robot and humans will eventually fail to do the job always correctly. Robots are also available, and are being developed for post-harvesting operations, since this operation also sometimes causes significant losses in yield. Manipulator robots (arm robots) are programmed to take the information from cameras and handle the product in the most suitable way at a consistent speed thereby neither damaging it nor dropping it.

The handling of the product using mobile manipulator robots is needed in storage, packaging, and transporting. In storage, the robots need to take the harvest from outside to inside refrigerators. These huge refrigerators have a controlled or modified gas composition, which might not be safe for humans to work in without protective gear. IoT and AI can also be used in these applications, if we feed in the data about the type of produce, harvest date, quality of the harvest at the time of storage, and destination and put sensors inside the refrigerators, which can monitor the health of the produce. Such automation can possibly increase the shelf life of the produce.

In packaging, manipulation plays a very important role as a fixed quantity of produce is supposed to be packaged individually and robots are capable of picking and placing different types of produce efficiently and accurately. The data for such an operation can be collected and used to train machine learning models to facilitate this operation.

Transportation involves transporting from storage to the vehicles/trucks, when produce needs to be transported from one town to another. These products are packaged and contained in boxes or pallets. Autonomous forklifts have seen their applications in inventory management for industrial warehouses, but the applications could be extended to agricultural inventory too. For trucking, there are ongoing efforts, at the time of writing this chapter, to automate highway driving for trucks.

Most of the post-harvest robots need a structured and controlled environment as robots need to work 100% of the time with little to no errors. Here, the cost of waste is higher as compared to other operations due to the added cost of farming and supply chain. Grocery stores also employ robots to monitor and record the health of the produce kept, so as to always deliver fresh produce to customers.

5 Some Case Studies of Using Robots and AI in Agriculture

In crop production systems, one of the most significant issues concerns human labor-intensive operations. These operations mainly include field tasks, such as harvesting of delicate and sensitive fruits that are prone to damage during harvesting and transport. Another major labor-intensive operation is weed control between rows and within rows, which are difficult to be executed by traditional field machinery. This has received major attention of robot experts involved in agriculture, such that autonomous tractors and robotic platforms have become available, and their successful use has been demonstrated in several case studies, although some of these are still being developed and are the subject of intensive research. In this section we describe a few case studies demonstrating the effective use of robotics in agriculture. Such cases studies largely deal with weeding and harvesting; the less-studied case studies deal with disease detection and seeding robots.

5.1 Agri.Q for Hilly Areas in Italy

In Italy, substantial regions of agricultural production are in hilly or mountainous areas, particularly for orchards, olive groves, and vineyards. As a result, robotic platforms must overcome and deal with a variety of challenges, for example, to address traction concerns, wheel slippage, tight spaces between rows, instabilities associated with terrain irregularities and changing slopes, and poor GPS signal reception, dedicated awareness (location, terrain, environment) systems and architectures must be built. Agri.Q is an innovative unmanned ground vehicle (UGV) developed for precision agriculture applications in vineyards (Botta and Cavallone 2022; Botta et al. 2022). It is outfitted with various instruments and sensors to perform specific activities, such as field mapping and crop monitoring. It also allows soil, leaf, and grape sample collections.

Agri.Q is an articulated robot made of two skid-steering modules, each module having two locomotion units driving two tires. The robot's low weight of around 110 kg allows it to move easily over rough and soft terrains, minimizing soil degradation. It has been designed to fulfill mostly proximal soil and vine monitoring and sample collection, while a flying drone can use Agri.Q as a base whenever its remote-sensing capabilities are not required.

5.2 A Case Study for Harvesting Sweet Pepper

A case study was conducted by van Herck et al. (2020) involving improvement in use of robots for harvesting sweet pepper. They gave a parallel approach to "design" the crop and its environment to best fit the robot. A systematic methodology was presented to select and modify the crop "design" (crop and environment) to improve robotic harvesting of sweet pepper. A sequential field experiment was planned for three years, involving ten cultivars, two climate control conditions, two cultivation techniques, and two artificial illumination types. Results showed how the modification of the crop itself affects the crops characteristics influencing robotic harvesting by increased visibility and reachability. The systematic crop "design" approach also led to robot design recommendations. The presented "engineering" of the crop "design" framework highlights the importance of close synergy between crop and robot design achieved by strong collaboration between robotic and agronomy experts resulting in improved robotic harvesting performance.

5.3 A Study on Designing an Apple Orchard for Robot Harvesting

While designing a robot for an agricultural operation is a major activity, sometimes orchards are also designed to suit a robot to economize on harvesting operation. One such study was conducted by Bloch et al. (2018).

6 Cost Effectiveness of Using Robotics and AI

As emphasized throughout this review, agricultural machinery including robots is a device that significantly reduces the amount of human labor required for food and fiber production, particularly at a time when labor costs are increasing everywhere in the world, leading to increase in cost of growing crops for food, feed, and fiber. In agriculture industry, the robots provide numerous other advantages, including improved quality of fresh produce. Harvesting is one of the most common areas where robots are used in agriculture today. The discovery of robots for harvesting highlighted the significant benefit of robots in agricultural techniques for effective mechanized farming in the agricultural industry. Finally, agricultural robots are critical equipment for performing repetitive tasks faster, cheaper, and more accurately than humans in farm cultural practices, inspection, and harvesting, as well as post-harvest handling.

7 Robots and AI for Plant Production Under Artificial Conditions

Growing crops under artificial condition is becoming a norm rather than an exception since farmers can now grow plants under controlled climatic conditions and can optimize production. Vertical farming is another example of plant production under artificial conditions. The greenhouses are usually built in areas where the climatic conditions for the growth of plants in the field are not optimal. This requires some artificial setups to improve productivity. Automation in using greenhouse requires monitoring and controlling of the climatic parameters. Efforts are, therefore, being made to minimize the cost of maintaining greenhouse environments using new technologies. This also involves use of robots and AI in automated system to optimally monitor and control the environmental factors inside greenhouse by monitoring temperature, soil moisture, humidity, and pH through a cloud-connected mobile robot that can detect unhealthy plants using image processing and machine learning. The mobile robot navigates through a pre-defined map of greenhouse. Database server has created a facility to store gathered real-time data that are useful in planning agricultural operations. The necessary accurate data are represented by using proper application for analyzing.

8 Future of Robotics and AI in Agriculture

It has been repeatedly emphasized that the future of agriculture lies in robot farmers (Jenkins 2013; Harvey 2014). The is evident from the fact that global spending on smart agricultural technologies and systems, including AI and machine learning, is projected to triple in revenue by 2025, reaching $15.3 billion. The development of AI integrated with IoT-enabled Agricultural (IoTAg) monitoring is also projected to reach $4.5 billion by 2025. According to PwC also, the global robot market will grow from 4082.8 million USD in 2018 to 16,640.4 million by 2026, which translates into compound annual growth rate (CAGR) of 19.2% and 400% over a period of eight years (Report by Research Dive; https://www.pwc.in/assets/pdfs/grid/agriculture/redefining-agriculture-through-artificial-intelligence.pdf).

The future growth of AI and robots in different parts of the world may also differ as witnessed in the past. For instance, it has been shown that by 2026, the Asia-Pacific's (APAC's) agricultural robot market will have the highest CAGR in the world, which is currently estimated at 19.7% with revenue of 3798.3 million USD.

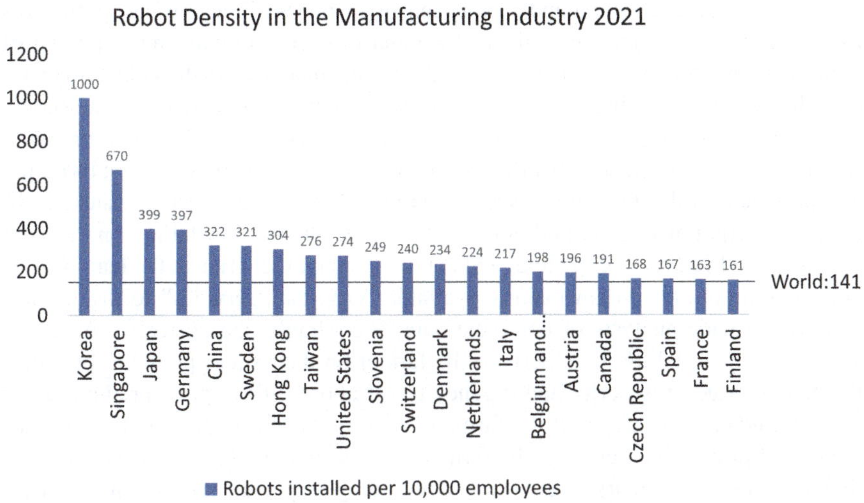

Fig. 5 Growth of robotics per 10,000 employees in different parts of the world. (https://ifr.org/ifr-press-releases/news/china-overtakes-usa-in-robot-density)

Given this promising data, countries in the APAC region are also taking steps and initiatives that aim to explore potential applications of robots in agriculture. The growth of robots per 10,000 employees in different parts of the world is shown in Fig. 5.

In this section we discuss the future of robotics in two parts, the first dealing with the expected future development of technologies globally and second dealing with the development of robotics and AI for agriculture in different parts of the world.

8.1 Future of Robotics and AI in Various Agricultural Operations

A number of reviews on use of robotics and AI have been written during the last five years (Kulothungan et al. 2018; Shieber 2018; Shamshiri et al. 2018; Mendes et al. 2019; Paquette 2019; Abdullayeva 2019; Jha et al. 2019; Fountas et al. 2020; Mahmud et al. 2020; Torero 2020; Chandana et al. 2020; Bera and Dutta 2021; Johnson 2021; Srivastava 2022; Papadopoulos 2022; Wakchaure et al. 2023). Based on these reviews and earlier studies, it is believed that the areas where robotics and AI will be used for agriculture in future include the following: (i) in autonomous disease detection and management; (ii) in use of large amounts of available data (digital solutions) using IoT (Internet of Things), including the following data: weather pattern, soil reports, rainfall, pest infestation, and imagery using drones and sensors; (iii) in gathering location data using proximity-sensing sensors installed in the field and remote sensing using Wi-Fi active hot spot towers for entire field coverage for in-depth field analysis, crop monitoring, scanning of fields, and remotely being able to control robots and machines in the field; (iv) in use of "Chatbot" for assisting farmers with answers and recommendations on specific problems; (v) in use of "Agri-E-calculator" for suitable crop selection along with resource estimation and affordability based on several dependency factors; its output will provide useful data on both estimation of fertilizers' cost/quantity, water, seeds, cultivation equipment cost, and labor day efforts/cost and distribution on calendar chart of crop life cycle, crop yield, and extrapolated market price at the harvest time and its profitability; (vi) in making available "crop care services" to the farmers that will provide analysis of complete data; accordingly, the corrective measures would be suggested through alerts to the farmer on their smart phone to prioritize the action based on severity and urgency to act upon; (vii) in price prediction and market guidance to safeguard the farmers from market fluctuation and to mitigate the risk of price loss; therefore, the farmers will be able to plan better for releasing their commodities to market; and (viii) in providing crop loan and insurance service, which will help the farmers by facilitating feasibility of getting crop loan, processing support, eligibility criteria, and loan limit as per the smart estimation made for the proposed crop; also, it would help to get the crop insured as a mitigation plan for crop failures due to any uncertainties or calamities.

AI will also make farmers more efficient and connected to the economy and government policies by providing some information services like interest rates for loans, price in the consumer market, etc. for which farmers had to earlier either guess based on experience, read/watch news, or go to the cities to inquire about. In the future, the number of robotics used in agricultural field is expected to increase considerably as autonomous robots (using solar energy power) are able to work for many hours. A photoelectric and a capacitive sensor were tested for localizing cutting along the row and proved to be suitable to be included in intra-row weeding machine.

8.2 *Future of AI and Robotics Around the World*

Although there are various common applications of AI and robotics that will be seen across the globe, operations and level of automation will differ in different countries. This aspect will be discussed for the following three groups of countries: (i) Singapore and China; (ii) India; and (iii) Australia, the USA, and Europe.

8.2.1 Activities in Singapore and China

Since Singapore is not one of the agricultural economies, a lot of companies that have sprung up in the area focus on vertical farming. Singaporean agriculture company, Singrow, has focused on using robotics and other related technologies to address issues of food security and agricultural sustainability. As reported by *Techwire Asia*, the company has already developed several technological models that integrate artificial intelligence (AI), including the following: (i) Automated Pollination: After a flower is automatically identified by a camera, agricultural robots will trigger a fan that toggles increased wind flow to encourage pollination effectivity; (ii) Strawberry Detection and Picking: Scanners are also becoming available, which identify and cross-check strawberries with an existing Singrow database; afterward, the agricultural robots can differentiate which strawberries are ready for harvesting and automatically pick them.

In addition to some private companies like Singrow, governments are also at the forefront of including robotics in the agricultural industry as evident from the Singaporean government's project "Agri-food Cluster Transformation (ACT) Fund." This initiative outright pledges an investment of 60 million Singapore dollars (SGD) for exploring ways in which the domestic agri-food sector can increase overall productivity; this is a matter in which robotics is of the utmost relevance.

Like Singapore, the Chinese Academy of Agricultural Sciences (CAAS) has also announced its commitment to exploring the deployment of technological solutions in China's agricultural industry. Among its intention to improve seed varieties, grain yield, and overall agricultural sustainability, the CAAS will also work on the automation of processes through robotics. By doing so, the government-affiliated organization hopes to innovate traditional farming operations to reduce costs and increase supply chain efficiency. The government policies in China have encouraged a lot of research and development in the field of AgTech.

8.2.2 Activities in India

During 2020–2021 and 2021–2022, the Indian government allocated funds amountiung to Indian rupee (INR) 1756.3 cores in 2020–2021 and INR 2422.7 in the fiscal year 2021–2022 for introducing new technologies including drones, AI, block chain, remote sensing, geographic information system (GIS), etc., in agriculture. The Indian government also allocated INR 7302.50 crores and INR 7908.18 crores in 2020–2021 and 2021–2022, respectively, to ICAR (Indian Agricultural Research Institute) for undertaking research and development in agriculture for developing new technologies, their demonstration at farmer's field, and capacity building of farmers for adoption of new technology. The government has recently also launched Digital Agriculture Mission (2021–2025).

Following are some other areas that have potential to improve agriculture in India with the integration of AI:

(i) Cognitive computing: Microsoft is currently working with 175 farmers in Andhra Pradesh to provide agricultural, land, and fertilizer advisory services. This initiative has already resulted in 30% higher average yield per hectare last year. The pilot project was completed using agricultural AI applications to communicate dates, soil preparation, fertilization based on soil tests, seed treatment, optimal spreading depth, and more. Mobile robots and field sensors also support digital agricultural robots; multidisciplinary cameras and laser scanners are used for facilities and areas of radiation that cannot be measured.

(ii) Proximity sensing, remote sensing, Internet of Things (IoT), and image-based precision farming are being used for intelligent data integration related to historical meteorology, soil reports, recent research, rainfall, and insect infections; also, drone imagery is being used for in-depth field analysis, crop monitoring, and field surveys.

(iii) The artificial use of image recognition using AI for plant identification, pest infestation, and disease diagnosis is also becoming prevalent. Using AI and machine learning-based surveillance systems to monitor every crop field's real-time video feed identifies animal or human breaches, and hence sending an alert immediately can become very useful to prevent crop damages.

(iv) Yield mapping to find patterns in large-scale datasets and understand the orthogonality of them in real-time, and optimizing irrigation systems to measure effectiveness of frequent crop irrigation are invaluable for crop planning.

(v) Due to shortage of labor, AI and machine learning-based smart tractors, agribots, and robotics are also considered to be viable options for many remote agricultural operations. These robots reduce operating costs.

(vi) Chatbots in local languages: Farmers will be able to ask queries about their farms and crops in their local languages, which would allow less educated farmers to operate these AI-enabled tools.

Although farming services and drones are being made available using AI, autonomous mobile robots have still a long way to go. Autonomous mobile robots need not only extensive and annotated data but also structured farming with clear boundaries for different crops and standard practices across the entire field. This is difficult to achieve if a lot of the tasks are done by humans in collaboration with machines and robots. Moreover, these robots are significantly more expensive because of the use of the highly precise sensors and actuators (motors). Country-wide issues like smooth Internet connectivity and literacy among farmers will also impede the use of such technology unless it is developed for the Indian market considering all the socio-economic factors.

8.2.3 Activities in Australia, the USA, and Europe

Countries like Australia, the USA, Germany, Spain, France, the UK, and Italy have a considerable share in the usage of robotic technologies in agriculture. Even though the scale of production is much lower in farms in Australia and Europe as compared to the USA, small autonomous field robots have been developed and used and the activity is increasing in these countries also. Although autonomous tractors and robots for picking of a variety of fruits are already popular, the adoption of robots and AI for all agricultural operations will increase at an exponential rate in all these developed countries.

In Australia, as an answer to labor shortage, robots are already being used for commercial farming in >405,000 hectares of Australian farmland, including farmlands in Queensland, New South Wales, and Western Australia. Queensland is the hub of these activities, where regional sales offices for many global robot retailers have started their activities. Companies like "Swarm Farm Robotics" (established in 2012) and LYRO Robotics are increasing their operations of manufacturing robots. Researchers at the Queensland University of Technology Centre for Robotics and the Australian Centre for Robotic Vision are also involved in building a Queensland ag robot knowledge-base. According to a 2020 report by McKinsey & Company, autonomous farm machinery in Australia could add up to $60 billion to global GDP by the end of the decade.

In the USA, dozens of companies are working toward improving the existing robots and developing new robots for future. Some of these companies include the following: Harvest Automation, Harvest Croo, Blue-White Robotics, and Tevel Aerobotics Technologies. Most of these companies are utilizing computer vision (CV) and Edge AI, the two subdisciplines of AI and deep learning (for image recognition), a subdiscipline of machine learning, for building up their capacity in developing robots for agriculture. The use of computer vision techniques in conjunction with image acquisition through remote cameras has already opened up a range of new applications in the agricultural sector, from saving production costs with intelligent automation to boosting productivity (https://viso.ai/applications/computer-vision-in-agriculture/).

In Europe there are several activities that will be used in future in agriculture. Their Horizon Europe Programme will boost innovations (including robotics and AI) in the agricultural sector. In addition, under Pillar II, Cluster 4, "Digital, Industry and Space" Programme, innovative technologies such as IoT, cloud and edge computing, AI, robotics, and block chain will be tested and validated for use in agriculture. Similarly, Cluster 6 will involve use of following advanced technologies in agri-food: drones, smart IoT, AI, upscaling real-time sensor data, 5G, and edge solutions for remote farming. Under Cluster 6 Programme, €9 billion will be invested in "food, bioeconomy, natural resources, agriculture, fisheries, aquaculture, and the environment," including the use of digital solutions for the agricultural sector.

9 Conclusion and Prospects

As described in this chapter, Robotics and AI are already being utilized on a fairly large scale for all operations in agricultural farmlands and horticultural orchards. Almost all operations, starting from preparation of fields and sowing to harvest and post-harvest operations, involve use of robots and AI. However, the use is not widespread, particularly in the developing countries, except in countries like Singapore and China (as discussed above), where private companies and the government, both are involved in promoting the use of robots and AI in agriculture.

The Government of India is also making massive investment in modernizing agriculture through use of robots and AI. In developed countries like Australia, the USA, and Europe also, several private companies are involved in developing robots using AI. These activities will certainly increase exponentially in future, thus bringing about an agricultural revolution to deal with the problem of food security, despite the anticipated problem of climate change and environment degradation.

References

Agricultural Statistics at a Glance (2015) Government of India Ministry of Agriculture & Farmers Welfare Department of Agriculture, Cooperation & Farmers Welfare Directorate of Economics and Statistics, pp 1–479

Abdullayeva A (2019) Impact of artificial intelligence on agricultural, healthcare and logistics industries. Ann SpiruHaret Univ Econ Ser 19:167–175

Ackerman E (2015) Bosch's Giant Robot can punch weeds to death. IEEE Spectrum. https://spectrum.ieee.org/ automaton/robotics/industrial-robots/bosch-deepfield-robotics-weedcontrol

Anderson C (2017) Agricultural Drones. https://www.technologyreview.com/technology/agricultural-drones/

Andreasen C, Scholle K, Saberi M (2022) Laser weeding with small autonomous vehicles: friends or foes? Front Agron 4:841086. https://doi.org/10.3389/fagro.2022.841086

Bera A, Dutta D (2021) Application of robotics in modern agriculture. Just. Agriculture 1:Article # 79

Bloch V, Degani A, Bechar A (2018) A methodology of orchard architecture design for an optimal harvesting robot. Biosyst Eng 166:126e137

Bogue R (2016) Robots poised to revolutionise agriculture. Indus Robot Int J 43(5):450–456. https://doi.org/10.1108/IR-05-2016-0142

Botta A, Cavallone P (2022) Robotics applied to precision agriculture: the sustainable Agri.q rover case study. In: Proceedings I4SDG workshop 2021, online, 25–26 November 2021; Mechanisms and machine science. Springer International Publishing, Cham, pp 41–50

Botta A, Cavallone P, Baglieri L, Colucci G, Tagliavini L, Quaglia GA (2022) Review of robots, perception, and tasks in precision agriculture. Appl Mech 2022(3):830–854

Carpin S, Goldberg K, Vougioukas S, Berenstein R, Viers J (2019) The use of intelligent/autonomous systems in crop irrigation. In: Robotics and automation for improving agriculture. https://doi.org/10.19103/AS.2019.0056.11

Chandana R et al (2020) A multipurpose agricultural robot for automatic ploughing, seeding and plant health monitoring. Int J Eng Res Techol (IJERT) IETE 8(11):57–60

Cousins D (2016) Self-driving Ibex robot sprayer helps farmers safely tackle hills. Farmers Weekly. Retrieved 2016-03-22. http://www.fwi.co.uk/machinery/self-driving-ibex-robot-sprayer-helps-farmer-safely-tackle-hills.htm. Cited by Wikipedia

Daniels J (2018) From strawberries to apples, a wave of agriculture robotics may ease the farm labor crunch. https://www.cnbc.com/2018/03/08/wave-of-agriculture-roboticsholds-potential-to-ease-farm-labor-crunch.html

Deshpande T (2017) State of agriculture in India. PRS Publication

Foglia MM, Reina G (2006) Agricultural robot for radicchio harvesting." (http://www.vagostudio.com/giulio/wp-content/uploads/2013/11/JFR06.pdf) (PDF). J Field Robot 23:363–377

Fountas S et al (2007) Decomposition of agricultural tasks into robotic behaviours. Agric Eng Int: the CIGR Ejournal. Manuscript PM 07 006 IX:1

Fountas S et al (2020) Agricultural robotics for field operations. Sensors 20:2672. https://doi.org/10.3390/s20092672

Gerrish JB, Fehr BW, Ee GRV, Welch DP (1997) Self-steering tractor guided by computer-vision. Appl Eng Agric 13:559–563

Griepentrog HW et al (2003) Seed mapping of sugar beet. Precis Agric 6:157–165

Harries GO, Ambler B (1981) Guidance system using opto-electronic techniques and a microprocessor based controller. J Agric Eng Res 26:67–75

Harvey F (2014) Robot farmers are the future of agriculture, says government. The Guardian. https://www.theguardian.com/environment/2014/jan/09/robots-farm-future

Heisel T (2001) Weeds in sugar beet rows - 1. Influence of neighbour plant on the beet yield - 2. Investigations of a CO_2 laser for in-row weed control. DIAS Report Plant Production (Denmark)

Jenkins D (2013) Agriculture shock: how robot farmers will take over our fields http://metro.co.uk/2013/09/23/robot-farming-4055079/

Jha K et al (2019) A comprehensive review on automation in agriculture using artificial intelligence. Artif Intell Agric 2:1–12

Johnson K (2021) These robots follow you to learn where to go http://www.wired.com/story/robots-follow-learn-where-go/

Kulothungan S, Kamalakannan K, Thirugnanam P (2018) Agriculture robot for irrigation and automation. Bull Pure Appl Sci 37 F:110–115

Luna-Maldonado AI (2010) Automation and robots for handling, storing and transporting fresh horticulture produce. Stewart Postharvest Rev 6(3):1–6. https://doi.org/10.2212/spr.2010.3.14

Luna-Maldonado AI, Vigneault C, Nakaji K (2012) Postharvest technologies of fresh horticulture produce. In: Luna-Maldonado AI (ed) Chapter 9: Horticulture, pp 161–172. https://doi.org/10.13140/RG.2.1.1351.4003

Mahmud MSA et al (2020) Robotics and automation in agriculture: present and future applications. Appl Model Simul 4:130–140

Mendes JM et al (2019) Nature inspired meta-heuristics and their applications in agriculture: a short review. In: EPIA conference on artificial intelligence. Springer, Cham, pp 167–179

Oerke EC (2006) Crop losses to pests. J Agric Sci 144:31–43

Papadopoulos L (2022) This new farming robot uses lasers to kill 200,000 weeds per hour. interesting engineering.com https://interestingengineering.com/innovation/farming-robot-lasers-200000-weeds-per-hour

Paquette D (2019) Farmworkers vs robots: will tomorrow fruit pickers be made of steel and tech? https://www.washingtonpost.com/news/national/wp/2019/02/17/feature/inside-the-rac e-to-replace-farmworkers-with-robots/

Rus D (2018) A decade of transformation in robotics. In: Towards a new enlightenment? A transcendent decade. BBVA, Madrid

Shieber J (2018) Your vegetables are going to be picked by robots sooner than you think. https://techcrunch.com/2018/08/08/your-vegetables-are-going-to-bepicked- by-robots-sooner-than-you-think/

Shamshiri R et al (2018) Research and development in agricultural robotics: a perspective of digital farming. Int J Agric Biol Eng 11:1–14

Srivastava S (2022) Food automation-how AI and robotics are transforming the future of the industry. https://appinventiv.com/blog/automation-in-food-industry/

Stentz A, Dima C, Wellington C et al (2002) A system for semi-autonomous tractor operations. Auton Robot 13:87–104. https://doi.org/10.1023/A:1015634322857

Trevelyan J (1999) Redefining robotics for the new millennium. Int J Robot Res 18(12):1211–1223. https://doi.org/10.1177/02783649922067816

Torero M (2020) Robotics and AI in food security and innovation: why they matter and how to harness their power. In: von Braun J et al (eds) Robotics, AI and humanity. Springer, pp 99–108

United Nations (2022) World population prospects 2022: summary and results. United Nations, New York. https://www.un.org/development/desa/pd/sites/www.un.org.development.desa.pd/files/wpp2022

van Herck L, Kurtser PL, Wittemans L, Yael Edan Y (2020) Crop design for improved robotic harvesting: a case study of sweet pepper harvesting. Biosyst Eng 192:294–308

Verchot M (2020) 5 Ways to improve global food security. Treehugger. from https://www.treehugger.com/ways-improve-global-food-security-4858809

Wakchaure M, Patle BK, Mahindrakar AK (2023) Application of AI techniques and robotics in agriculture: a review. Artif Intell Life Sci 3:100057

Xu R, Li C (2022) A review of high-throughput field phenotyping systems: focusing on ground robots. Plant Phenom 2022:9760269. PMID: 36059604; PMCID: PMC9394113. https://doi.org/10.34133/2022/9760269

Additional Literature Consulted for This Chapter

Anonymous Harvest Automation Inc. http://www.harvestautomation.com/

Anonymous How small robots may kill the tractor and make farming efficient. https://www.wired.co.uk/article/farming-robots-small-robot-company-tractors

Anonymous Autonomous seeder and planter projects. https://www.futurefarming.com/Machinery/Articles/2021/4/Autonomous-seeder-and-planter-projects-739206E//

Anonymous ecoRobotix plans spot-spraying robot for weed control. https://www.futurefarming.com/Tools-data/Articles/2018/5/ecoRobotix-plans-spot-spraying-robot-for-weed-control-283124E/

Anonymous How Blue River's technology could change agriculture forever. https://www.precisionag.com/service-providers/how-blue-rivers-see-spray-technology-could-change-agriculture-forever/

Anonymous Verdant Robotics launches multi-action agricultural robot for 'superhuman farming' https://roboticsandautomationnews.com/2022/02/23/verdant-robotics-launches-multi-action-agricultural-robot-for-superhuman-farming/49471/. Robotics & Automation News 23 February 2022

Anonympus Fighting world hunger: robotics aid in the study of corn and drought tolerance. https://phys.org/news/2017-03-world-hunger-robotics-aid-corn.html

Anonymous AgBot multi-function robot is powered by the sun. https://web.archive.org/web/20120331123358/http://goodcleantech.pcmag.com/future-tech/280063-agbot-multi-function-robot-is-powered-by-the-sun. pcmag.com. Archived from the original http://goodcleantech.pcmag.com/future-tech/280063-agbot-multi-function-robot-is-powered-by-the-sun

Anonymous (1995) IEEE international conference on robotics and automation, vol 3. Nagoya Institute of Electrical and Electronics Engineers, pp 2504–2509. https://doi.org/10.1109/ROBOT.1995.525635

Anonymous (2016a, 28 July) FarmBot DIY agriculture robot promises to usher in the future of farming. http://www.digitaltrends.com/cool-tech/farmbot-open-source-agriculture-robot/. digitaltrends.com

Anonymous (2016b) Agriculture robots at The University of Sydney. http://confluence.acfr.usyd.edu.au/display/AGPub/Our+Robots

Anonymous (2016c) Small Robot Company Tom, Dick, and Harry farm robots: the 200 Best Inventions of 2022. https://time.com/collection/best-inventions-2022/6225155/small-robot-company-tom-dick-and-harry/. Belton, Padraig. In the future, will farming be fully automated?. https://www.bbc.co.uk/news/business-38089984

Anonymous (2017) Automated agriculture: robots and the future of farming. http://www.theautomationengineer.com/markets-sectors/automated agriculture-robots-future-farming/

Anonymous (2018a) Global agriculture robots market size by type (driverless tractors,

Anonymous (2018b, Sept) Automated HarvestingMachine, Others), By Application (Field Farming, Dairy Management, Indoor Farming,Others) By Geography Scope and Forecast. https://www.verifiedmarketresearch.com/product/global-agriculture-robots-market-size-and-forecast-to-2025/. (Report). Verified MarketResearch. 3426

https://www.forbes.com/sites/louiscolumbus/2021/02/17/10-ways-ai-has-the-potential-to-improve-agriculture-in-2021/?sh=4ca5ba117f3b

https://www.wipro.com/holmes/towards-future-farming-how-artificial-intelligence-is-transforming-the-agriculture-industry/

https://www.v7labs.com/blog/ai-in-agriculture#h8

https://www.researchdive.com/52/agriculture-robot-market

Analysing Data from Open Sources to Manage Risks in Food Production

Nathaniel Narra, Reija Hietala, and Tarmo Lipping

Abstract Agri-food supply chains pose unique challenges. They are inherently complex due to natural processes, random processes, and unpredictability. In modern agriculture and food production, the supply chains necessary for operations are wider because of more components involved and longer because of more complex products and services that themselves depend on supply chains. From a primary producer's perspective, risk-sensitive components of the supply chain are not always fully identified. While the most important direct risks—such as energy, seeds, and fertilizers—are usually quite well understood, other deeper dependencies in the supply chain are not easily recognized. On the other hand, with data 'revolution' it has become easier to monitor processes. Advances in big data and analytics have made extracting information easier. Open data are also increasingly available and organized to the point where large amounts of data are shared and hosted through international collaboration. In this chapter, we seek to detail a producer's risks and map available sources of relevant data. Through relevant example cases we illustrate the utility of using big data to increase information content that can help alleviate risks. We foresee this strategy as a tool for producers to customize their own supply chains and gain a deeper understanding of their vulnerabilities.

Keywords Open data sources · Agricultural production · Risk management · Remote sensing · Crop selection · Pest monitoring

---

N. Narra · T. Lipping (✉)
Tampere University, Pori, Finland
e-mail: nathaniel.narra@tuni.fi; tarmo.lipping@tuni.fi

R. Hietala
Pyhäjärvi Institute, Eura, Finland
e-mail: reija.hietala@pji.fi

P. M. Priyadarshan et al. (eds.), *Digital Agriculture*,
https://doi.org/10.1007/978-3-031-43548-5_20

1 Introduction

Background
Improving production and staying competitive is often tackled via improving process efficiency, removing the presence of negative factors or mitigating unavoidable risk factors. Agricultural production, as the production of any commodity, is a part of a supply chain. With modern trends in agricultural practices, the reliance on an extensive supplier network has become increasingly important. Efficiency and productivity comes with various supplies and products that themselves are part of an extended supply chain. Combined with the market of global causes and effects and greater regulatory regimes, the task of a primary producer in staying relevant in the market and staying ahead of global trends and disruptions is ever more complex.

When considering preparedness, two kinds of changes in the operational environment can be considered. There are natural trends including, for example, changes in production volumes and market prices due to the yearly variability of weather and climate conditions. On top of these, there are disruptive events and crisis affecting the operational environment and rewriting the rules of operation. Of these, the recent global COVID crisis and war in Ukraine are prominent examples. While the natural trends can be anticipated and taken into account in forecasts, the occurrence and influence of disruptive events are virtually unpredictable. Also, the disruptive events usually possess the features of so-called wicked problems, first conceptualized by Rittel and Webber (1973). Wicked problems are unique so that the experiences gained from previous crisis are of limited use in dealing with them. There might be no true or false actions when addressing these problems and the consequences of the various actions might be difficult to anticipate (Schiefloe 2021). Wicked problems usually involve many dimensions such as social, (geo)political, behavioral, economical, etc.

Availability of data and the ability to interpret the data and extract useful information from them are crucial when addressing both kinds of changes—natural trends and disruptive events. Even if the wicked crises cannot be predicted or previous experiences are of limited value in addressing them, situation awareness coming from proper analysis and interpretation of data can give hints on the development of the crises, help to develop plausible scenarios and make forecasts as soon as the 'new normal' begins to emerge. The proper interpretation of data also enables us to better comprehend the changed rules and operation environment after a wicked crisis.

The Food and Agriculture Organization (FAO) of the United Nations has launched the Data Lab for Statistical Innovation[1] in 2019. They recognize that *in a globalized ecosystem, food value chains can be easily and quickly disrupted by sudden crises, putting food security and food safety at risk* and that *information needs in times of crisis are intrinsically different from business-as-usual needs*. While

[1] https://www.fao.org/datalab.

data play an important role in staying competitive and enhancing the sustainability of agricultural production also in business-as-usual operational environment, the significance of timeliness and the importance of having automatic data infrastructure in place become especially crucial in crises.

In light of the risks that production entails, a producer has to be aware of and monitor multiple risk factors. These factors can be either quantified or qualified through data of different forms such as textual information, performance categories, numeric data, or protocols. Open data collected by various organizations cover these types. Such heterogeneity in data characteristics and the necessary analyses requires involved handling. Although the interpretation of the data is mostly done by organizations such as ProAgria or the Natural Resources Institute in Finland or by technology companies, decision-making remains with producers, and therefore, general knowledge on the data sources and data infrastructure is highly recommended.

Potential for Data-Driven Analytics in Agri-Food Supply Chains

A lot of studies exist considering the Agri-Food Supply Chains (AFSCs) and their management (Zhong et al. 2085). Commonly, the main actors of the AFSC include the primary producer, food industry (food processor), retailer, and the consumer. Other actors may be marketers, distributors, caterers as well as various suppliers of energy, seeds, fertilizers, etc. (Dani and Deep 2010). Stone and Rahimifard (2018) consider in their review on the resilience of AFSCs the interrelationships between individual organizations and the supply chain as a whole in developing resilience. At both levels, similar phases are recognized; the actions of individual actors affect those of the supply chain and vice versa. The authors note that the actors of the AFSC both share general risks and face their own unique vulnerabilities.

The importance of data and data-driven decision support has also been discussed by several authors. As brought up by FAO, there are mainly two drivers when developing supply chains: competitiveness and sustainability. Banasik et al. (2018) review the use of Multi-Criteria Decision Making (MCDM) in designing Green Supply Chains (GSCs). In MCDM, trade-offs among environmental, social, economic, etc. factors are considered. The authors consider AFSCs in the broader context of GSCs. They note that using MCDM is a relatively new but emerging approach. Ivanov et al. (2019) take the idea a step further by considering data-driven digital twins of supply chains when simulating the effects of disruptions to supply chains. These kinds of models—digital twins or not—allow for the assessment of the effects of various factors on the Key Performance Indicators (KPIs) of supply chains through sensitivity analysis (van der Vorst and Beulens 2002). While digital twins may stay as the ultimate goal, it is clear that the nature of wicked problems is too complex and unpredictable to be fully comprehended by a digital twin.

The concept of Big Data has been around for some time by now and a multitude of studies consider its role in supply chain management (Talwar et al. 3509). Rejeb et al. (2021) provide a review on using Big Data specifically in AFSC management. While Big Data analytics has a lot of potential in mitigating risks, increasing efficiency and raising the level of sustainability, employing Big Data as it is commonly defined does not necessarily mean taking into account all the relevant aspects of supply chain development. In the studies reviewed in Rejeb et al. (2021), emphasis is on soil, water, crop, waste and traceability management, while social, financial, market-related, and regulatory issues are overlooked.

Study Goals
In this chapter we first consider a categorization of risks from the primary producer's point of view. We then make an attempt to identify sources of open data that can be used to mitigate these risks. Most of the identified risks and related data sources are common to the actors of the AFSC. Some case studies are described next on the usage of open data to provide decision support for the primary producers. Finally, recent initiatives to provide a comprehensive decision support platform for producers are considered.

2 Mapping Risks in Agricultural Production

In this section we first give a brief overview on the results of a set of interviews carried out in the Satakunta region, Finland, on the experiences of the actors of agri-food supply chain during the COVID crisis. The main focus here is on what data sources the actors were familiar with and what kind of data they would like to have for better decision-making. After that we provide a categorization of potential risks while suggesting corresponding data sources that can be used to mitigate these risks.

2.1 Usage of Open Data by Actors of the AFSC: Summary of Interviews

The interviews were carried out in Autumn 2021 as part of the project *Security of Supply and Sustainability of AFSCs During Recovery from Corona Pandemic in Satakunta* funded from EU Regional Funds. Altogether 10 farmers, 9 representatives of food industry and 2 representatives of retail were interviewed. While the overall scope of the interviews was wider, the following questions were asked regarding data usage:

- What data sources do the actors of the AFSC currently use?
- What data do the actors of the AFSC produce?
- What kind of data would the actors like to use if made available?

The main results of the interviews, concerning data sources, are presented in Fig. 1. Naturally, most of the farmers use weather data, either provided by the Finnish Meteorological Institute (FMI) or acquired using their own weather stations. Some weather data products such as cumulative temperature sum are also followed. In addition, farmers utilize data related to the diagnostics of their machinery (meaning

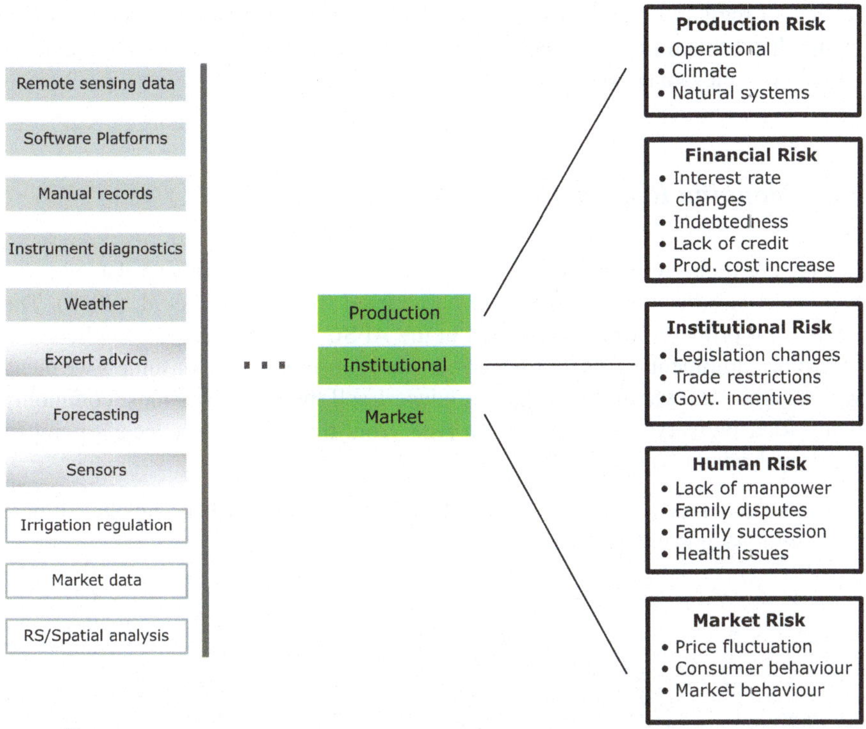

Fig. 1 Summary of the interviews regarding data usage by farmers. Grey boxes at the left side mark the data sources used by at least some of the farmers; data sources in shadowed boxes were used by some farmers while noted as desirable by others, while the data sources in white boxes were noted as not yet available but potentially useful

of error codes, for example), data provided at various open or proprietary portals (such as Farmit.net or ett.fi in Finland), remote sensing data, and data in their own manual records. Data sources such as expert advice, information on standards, market forecasts, or soil sensor data were used by some farmers but noted as desirable by others. Remote sensing data combined with appropriate analysis tools, information about regulations, and market data were noted as desirable if provided in a suitable form. The data produced by farmers included bookkeeping data about production, GPS data on the use of pesticides and fertilizers, or crop yield data. The data produced by farmers are mainly made available to the public stakeholders for statistics or to the food industry companies based on mutual agreements; there is yet no common marketplace for this kind of data.

In Fig. 1 the data sources are mapped to data categories according to the type of activities the data sources are related to. These data categories are in turn mapped to risk categories. For example, remote sensing or equipment diagnostics-related data are concerned with agricultural production and can be used to mitigate risks or support decisions on corresponding risks. On the other hand, data and information related to regulations fall in the category of institutional data and being able to comprehend this information enables us to anticipate the effects of the regulations and to take relevant actions.

2.2 *Mapping Risks to Data Sources*

In Fig. 2 a more comprehensive view of risk categories and related data sources is given. Classification of risks has been adopted from Rosales et al. (2015). The focus is on primary producers located in the AFSC between various suppliers (for obtaining equipment, seeds, fertilizers, etc.) and consumers. By consumers we mean various actors to whom the primary producers sell their products most commonly being food industry companies. Main risks from the primary producer's point of view can be categorized as follows:

- Production risks; i.e., risks related to the production such as labour, climate conditions, and environment

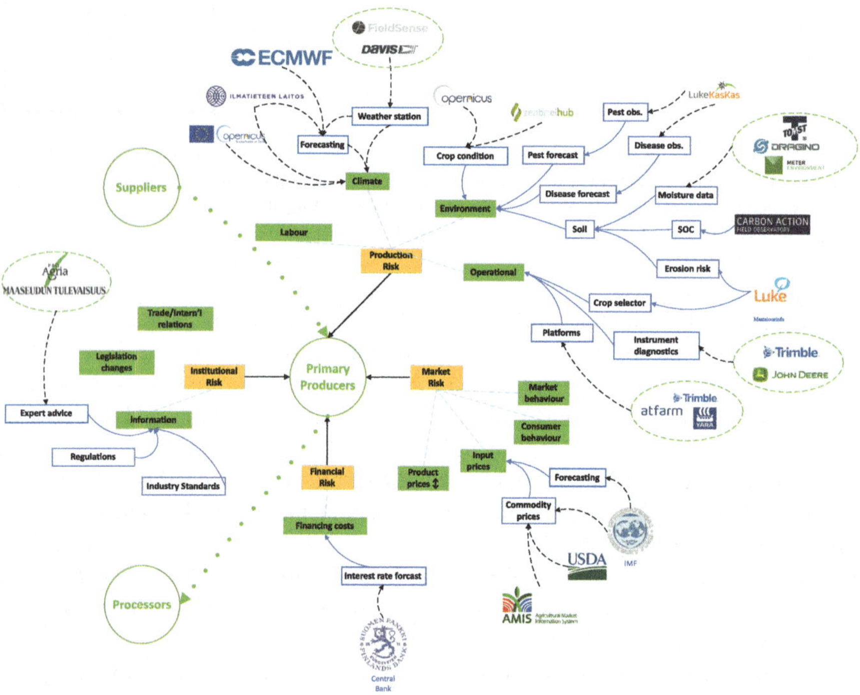

Fig. 2 A map of risk categories to relevant data sources from the primary producer's point of view

- Market risks; i.e., risks related to the prices of supplies and products, consumer behaviour, and market behaviour
- Financial risks such as interest rates, for example
- Institutional risks; i.e., risks related to regulations, changes in legislation, and changes in trade rules

For each risk (sub)category, a set of data sources has been identified that can inform the producers when dealing with corresponding risks. In the following we discuss each of these risk categories briefly.

Production Risks

All factors that affect the production processes are included here. While labour is often earmarked as a separate category, we chose to include it within the production category. This risk basically arises from the availability of appropriately skilled labour. This availability can be endemic to the local/national context or influenced externally through labour import restrictions and delays in the bureaucracy (including, for example, health-related restrictions as was the case with COVID). The other major components noted here are climate, bio-physical (environmental) factors, and operational issues.

Climate, separated here from the broader environmental category, specifically deals with the weather parameters. Data related to climate and weather are widely used among the producers, even meat producers. Short-term weather forecasting is widely available in most countries via the national meteorological institutes. Internationally, the European Centre for Medium-Range Weather Forecasts (ECMWF) provides potentially useful long-term (up to 4 months) forecasting. These long-term forecasts, although of limited reliability, can be useful for producers to plan ahead for the beginning of the growth season. Weather forecasting takes special importance when decisions on field preparation, sowing time, crop type, or irrigation are done. For example, decisions on irrigation can be done based on short-term forecasts, decisions on sowing time based on mid-term forecasts, and decisions on crop type and variety or purchase of irrigation equipment based on long-term forecasts. Weather conditions can have significant spatial variation caused by land forms or vicinity of large water bodies. When establishing one's own weather station, more accurate data can be obtained.

Another way of using the weather and climate data is to record and monitor year-to-year variability in the weather to acquire knowledge on how specific fields react to exceptionally dry/wet or warm/cold conditions, for example. Producers usually have some gut feeling about this based on their long-term experience and manual records. However, the proper visualization and systematic analysis of the data is needed to make evidence-based decisions. Combining this information with soil type, water table, and crop type/variety enables us to accumulate valuable information which can then be used for better decision-making and calibration of crop models.

Environmental risks are closely related to climate risks but consider a wider range of phenomena. Soil type with specific soil properties form a more-or-less permanent environment for crop growth, however, there are risks related to erosion or washing out of nutrients from the soil. Soil organic carbon has been found to significantly influence crop yield and, at the same time, mitigate the risks of climate change (Lal 2004). Information on how to sequestrate soil carbon is provided at https://carbonaction.org/en/front-page/ in the Finnish context. Soil moisture and soil temperature can also be considered as environmental variables as they depend on soil properties in addition to weather conditions. Soil sensors can be used to monitor these variables and to inform the farmer about the need for irrigation. Another environment-related issue is the occurrence of pests or crop diseases. As presented in Sect. 3.4, data sources estimating the probability of pests or diseases are available, however, there is clearly a need for better and more versatile data sources for that. Finally, remote sensing data can be used to monitor crop condition throughout the growth season and to inform farmers about unfavourable environmental changes.

In the subcategory of operational risks, the issues related to various actions by the farmer or the maintenance of equipment can be considered. These include the selection of crop type/variety, decisions regarding sowing time, selection of fertilizers and pesticides as well as the timing of their application, etc. The main data sources here include the platforms of the producers of these commodities. In Sect. 3.3 an example of a decision tool for farmers regarding crop selection is presented.

The overall production category is probably the widest and best researched among the risk categories. The factors within this category and the related data sources are somewhat specific to the type of production and the crop/variety and can have various subcategories when looked at in detail.

Market Risks

Market risks are mostly related to the volatility of the prices of various commodities (such as fuel, energy, equipment, machinery, fertilizers, etc.) as well as the prices of the produced crops. As a separate subcategory, market behaviour can be considered involving, in addition to prices, also the availability of the commodities and products. While apparently there is a clear connection between availability and price, geographical variation and temporal dynamics of the market as well as possible restrictions on trade due to political agendas or unexpected crises make the behaviour of the market difficult to predict even if changes in production volumes can be anticipated.

The market of agricultural products is also related to the consumer's behaviour which can change radically in modern globalized world. Currently, there are a lot of discussions on the price gap between meat and plant-based food products and to what extent the price can influence the sales and consumption habits of people (Garnett et al. 2021). While consumption is affected by price, there are also other factors that affect consumption such as trends and fashionability. These changes in consumption would affect prices if not globally, then at least within a certain economic region. Data sources that could potentially allow us to mitigate this kind of risks would include statistics on consumption or mining of social media and periodicals. While retailers collect data on consumer behaviour, these data are generally not available to producers. On the other hand, mining social media for detecting trends on consumption is difficult and the results unreliable.

Financial Risks

Financial risks are generally predominantly related to loans and the cost of loans. These are typically not extremely volatile if safety measures and risk mitigating packages are subscribed to with a financial institution. Financial risks are common to all businesses and financial institutions produce forecasts of interest rates regularly. At the moment of writing this chapter, inflation and interest rates are surging causing difficulties to many producers. It is difficult to say if any data source could have been useful in securing agricultural production against this situation.

Institutional Risks

The main institutional risks are posed by changes in the operation environment due to legislation and regulations. Also, political causes such as trade restrictions and sanctions can be considered as institutional risks as they are implemented via (often temporary) decrees or regulations. While normally changes in legislation are implemented after a careful evaluation of the consequences to the various groups involved and often there is a transition time to let citizens and businesses take actions, then in the case of crises the changes can be abrupt. Various compensation schemes can be designed, however, there tend to be always groups of people who suffer losses.

To anticipate the changes and prepare for institutional risks, it is useful to consider the causes that might drive these changes. A significant driver for changes in legislation is sustainability. Political parties have different attitudes on the urgency of these actions—whether there is an emergency and the actions should be taken as in the case of an acute crisis or there is time to proceed according to common procedures. However, it is clear that legislation will change to force environment-friendly ways of operation.

Another major driver for changes in regulations is technological development. Herrero et al. (2020) list numerous technologies that either have already or will in the near future change our operation environment. In their paper they consider technologies that accelerate the transition towards sustainable food system; however, technology changes the ways we operate also if sustainability is not the primary goal. For example, regulations related to the use of data may have primarily economic and social incentives.

When considering the sources of data to mitigate the risks due to changing regulations, the drafts of new laws and directives form, indeed, the most immediate source of information. It is clear that keeping track of these documents, making sense of their content and anticipating the consequences of their implementation cannot be expected from most of the agricultural producers. Therefore, disseminating this information by officials and explaining the opportunities and restrictions the new regulations will bring about from a farmer's point of view via seminars and workshops is important.

3 Case Studies

In this section several examples are presented on how various data sources can be used and visualized for decision support in agricultural production. The context of the cases is Finnish, i.e., some of the data sources are made available by Finnish organizations and concern operating in Finland. However, similar data sets are available in other countries.

3.1 Case 1: Using Weather Data for Mitigating Production Risks

As indicated in Fig. 1, weather data are the most common data source the farmers use. Weather is an important factor in agricultural productivity, especially in open-field cropping systems. While national (and international) meteorological institutes provide weather forecasts for at least 10 to 15 days ahead in various formats and visualizations, this is not the only way the weather data can be used (Lalić et al. 2018; Nobre et al. 2019). For example, visualizing yearly weather statistics (such as temperature and precipitation) in a way that makes it easy to grasp the characteristic features of each year's weather conditions and presenting this together with some target value (such as crop yield) enable to learn about the correlation between weather and crop performance for a specific field of particular soil type, water table, etc. Historical weather records can be obtained from an official data source such as Finnish Meteorological Institute (FMI) in Finland or by collecting data from one's own weather station. In Fig. 3 two ways of aggregating precipitation and temperature data are shown. For example, it can be seen that year 2020 has been exceptionally wet and cold at the beginning (heatsum starting to accumulate only at mid-May), while during the growth period, dry and wet seasons have alternated. The total heatsum can vary from year to year by as much as 20% (about 1500 in 2019 while over 1750 in 2018).

In addition to past weather data and short-term forecasts, long-term and seasonal forecasts can be useful for decision-making. For example, the European Centre for Medium-Range Weather Forecasts (ECMWF) offers various forecast products at different time scopes such as:[2]

- High-resolution short-term forecasts (up to 10 days)
- Ensemble forecasts (mid-term forecasts up to 15 days)
- Extended range forecasts (from 16 to 46 days)
- Long-range (seasonal) forecasts (monthly averages up to 7 or 12 months).

[2] https://www.ecmwf.int/.

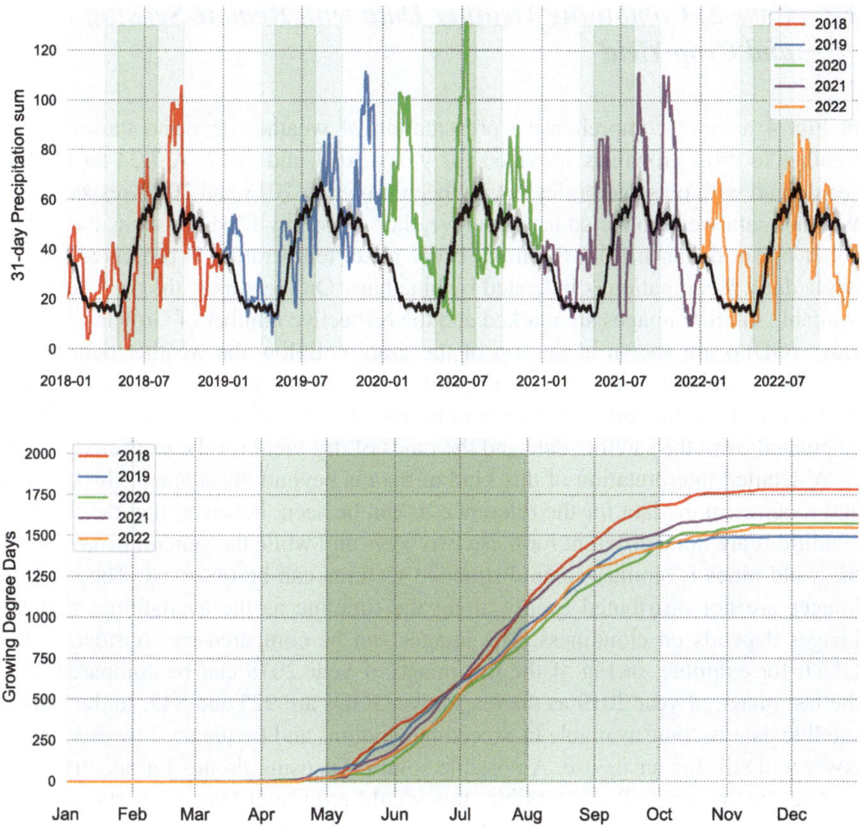

Fig. 3 Multiyear precipitation data (upper panel) and multiyear heatsum (lower panel) for Kokemäki station (N 61.25°, E 22.35°), Finland. In the upper panel, the black curve indicates long-term average precipitation over years 1981–2010, while the yearly precipitation (calculated within 31-day running window) is presented in different colours according to the legend. In the lower panel, heatsum curves in Growing Degree Days are presented

Naturally, the different forecast products have different level of confidence and thus require appropriate considerations when making agricultural decisions. These decisions range from scheduling field operations such as seeding to harvesting with various interventional operations in between (such as irrigation, fertilizer, and herbicide treatments). In 2011, Calanca et al. (2011) studied the efficacy of long-range forecasts as supplied by ECMWF in predicting soil water availability with encouraging results. Among topics for further investigation, they mentioned the need for promoting the dissemination of the utility among the agricultural community. However, in spite of significant advances in forecasting and software-based productivity tools, the prevalence of freely available agriculturally relevant long-range forecasts can be improved.

3.2 *Case 2: Combining Weather Data with Remote Sensing and Crop Yield*

In Fig. 4 a more comprehensive presentation of weather data and the effect of weather to crop growth is provided for years 2018 and 2019. A 7.24-ha field is considered with oats and barley having been grown in 2018 and 2019, respectively. Weather data were collected using the weather station by Davis Systems located at about 600 m distance from the field. Daily mean temperature is presented in red, while daily precipitation is indicated by blue bars. On the x-axis, the time points of available satellite images are marked and the respective number of Growing Degree Days (GDD) are shown at the top of the graphs. Below the weather data panel, the satellite images from the Sentinel Hub (bands of the L2A product) are shown. Still below are differential plots of Normalized Difference Moisture Index (NDMI) calculated from the satellite data and the maps of dry yield for the respective fields.

A detailed interpretation of this kind of data is beyond the scope of this chapter and requires more data for the reference. It can be seen, however, that the weather conditions are quite different for these two years and while the general behaviour of the yield maps is similar, some distinctive features can be observed. The satellite images are not distributed evenly along the timeline as the availability of these images depends on cloudiness. The images can be compared by considering the GDD; for example, in Fig. 4 the 6th image of year 2018 can be compared with the last image of year 2019 as the respective GDDs are 925 and 845. In the future, satellite data become available in exceeding amounts and frequency, however, cloud cover will still hinder its use. A possible solution is using drones for acquiring the remote sensing data. To be feasible, this would require at least semi-autonomous usage of drones (Kaivosoja 2022). It has been argued that by more efficient use of the vast amount of remote sensing data, available from different types of platforms

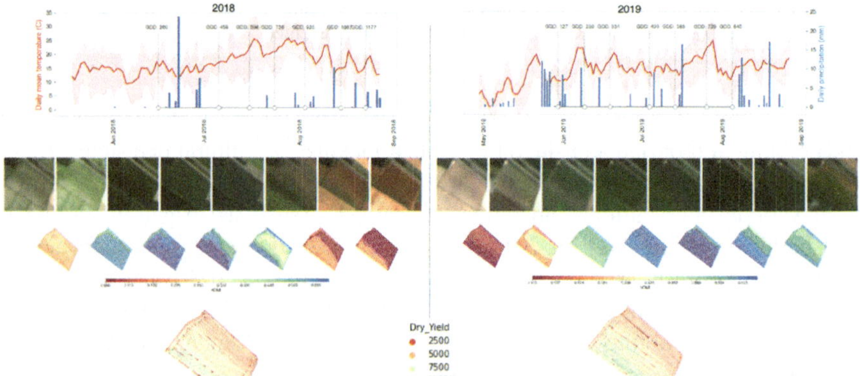

Fig. 4 Multiyear weather station data at about 600 m from the field. For more detailed explanation see the text in Sect. 3.1

(drones, airborne platforms, and spaceborne platforms) and of various modalities (multispectral, hyperpectral, LiDar, and radar), the resilience of agricultural food production systems can be significantly improved (Jung et al. 2021).

For even more comprehensive representation of the effect of weather conditions on crop growth and yield, data from other sources such as soil sensors or soil maps (based, for example, on electrical conductivity of the soil) can be incorporated. With an increasing amount of data, machine learning techniques can be used to learn these relationships and suggest measures for increasing the yield and optimizing the production costs (Nevavuori et al. 2022).

3.3 Case 3: Crop Selection Tool Based on Crop Variety Trials

An important decision for any farmer before a growing season is the choice of crop and variety. This choice can be affected by considerations such as soil type, field specific crop cycles, projected weather, and market outlook.

The Natural Resources Institute of Finland (Luke) maintains and updates an extensive database on the results of the nation-wide crop variety trials.[3] They also offer a web-based selector tool (Pesola 2021). The tool currently includes the following plant species: oats, spring wheat, barley, winter rye, winter wheat, spring canola, and winter canola. The tool works as a webform that presents to the user multiple options based on which the underlying database is queried and the results presented in the table format (Fig. 5, upper panels; available only in Finnish). For better usability, we developed a user interface on top of the database with sliders for various crop properties (such as expected yield, growing time, risk for lodging, etc.) and a radar chart to present the results (Fig. 5, lower panel). This crop selection tool is now being integrated into our peltodata.fi service for farmers.

While the selection of the crop variety depends on many other considerations such as previous experience of the farmer, existing marketing contracts, etc., this kind of tool may provide decision support and encourage producers to test new, potentially more productive varieties.

3.4 Case 4: Prediction and Monitoring for Pests, Diseases, and Weeds

The occurrence of pests, crop diseases, and weeds forms another important production risk. There are mainly two ways to provide data about these risks: either to predict the probability of occurrence of pests, diseases, or weeds based on past weather conditions (temperature, humidity, and snow conditions) or to

[3] https://px.luke.fi/PxWeb/pxweb/en/maatalous/maatalous__lajikekokeet/.

Fig. 5 Crop selector tool as offered at Luke's website (upper panels; see Sect. 3.3) and the user interface developed on top of the crop variety trials database

collect information about the observations and try to predict the spreading of the phenomenon. Naturally, the weather conditions are specific to the pest, weed, or disease. In Finland, Luke currently provides forecasts for the occurrence of three pests: *Psila rosae*, *Trioza apicalis*, and *Delia radicum* up to 5 days ahead. The information is provided in the form of a map[4] (see Fig. 6). The map cells are color-coded according to the predicted occurrence of the pest: green denotes areas of no forecasted occurrence, yellow denotes the areas where the occurrence probably starts, and red denotes areas of peak occurrence for the season. A platform for collecting observations has also been set up but is underused so far. One reason might be that producers are reluctant to provide this information as it needs to be geotagged and can thus be linked to the field and the farmer.

[4] https://maatalousinfo.luke.fi/fi/tuholaisennusteet.

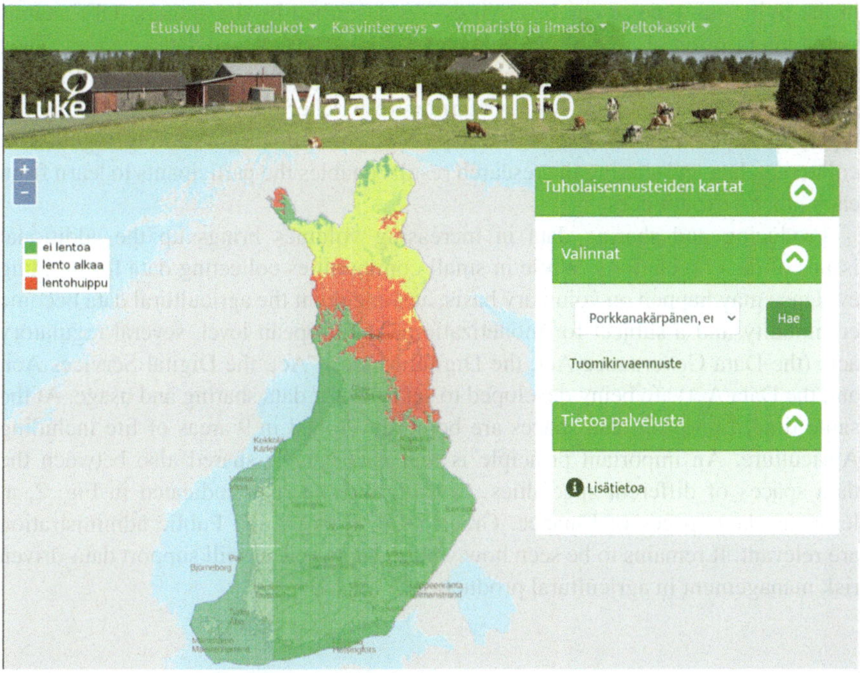

Fig. 6 Luke pest forecast map for *Psila rosae*

4 Discussion and Conclusions

As indicated in Fig. 2, the number and variety of open data sources useful to mitigate risk in food production is large. Clearly, the figure is not comprehensive and the individual data sources depend on the country and even region as some actors may collect data and offer services regionally. However, sometimes relevant data that is not offered by national actors may be available from institutions operating globally. While combining data related to the environment and production risks is quite common (although usually considering a certain subset of the production environment) incorporating market data, consumer behaviour, financial data, and the effect of regulations is more challenging.

Using data for decision support requires not only the data but the evidence as well. Evidence is built upon the data through research and data analytics. To a limited extent, research can be carried out using data collected from trial plots where the growth conditions are strictly controlled. However, the scale and efficiency of this kind of evidence building falls short when considering the requirements for competitive, resilient, and sustainable food production. To address these needs, Lacoste et al. (2021) have launched the initiative of On-Farm Experimentation (OFE) through restructuring farmer–researcher relationship. By the end of 2021 they had engaged already over 30,000 farms from more than 30 countries to participate in the living-lab type research activities. As described by the authors, OFE stands in the intersection of agricultural sciences, social sciences, and data sciences. They see that OFE has the power to transform global agriculture. In addition to more efficient evidence building, stronger farmer–researcher relationships and living-lab type experimentation have also the aspect of communality. Working together in collecting data and sharing the research results enables the participants to learn from each other's experiences.

Producing and sharing data in increasing volumes brings up the additional issues of data regulations. While in small communities collecting data for building evidence may happen on voluntary basis, at some point the agricultural data become commodity and a subject for monetization. At European level, several regulatory acts (the Data Governance Act, the Digital Markets Act, the Digital Services Act, and the Data Act) are being developed to set rules for data sharing and usage. At the same time, European data spaces are being developed in 9 areas of life including Agriculture. An important principle is that data can be shared also between the data spaces of different specialties. Considering the risks indicated in Fig. 2, at least the data spaces of Finance, Green Deal, Energy, and Public administration are relevant. It remains to be seen how well the data spaces will support data-driven risk management in agricultural production.

References

Banasik A, Bloemhof-Ruwaard JM, Kanellopoulos A, Claassen GD, van der Vorst JG (2018) Multi-criteria decision making approaches for green supply chains: a review. Flex Serv Manuf J **30**(3):366–396. https://doi.org/10.1007/s10696-016-9263-5

Calanca P, Bolius D, Weigel AP, Liniger MA (2011) Application of long-rangeweather forecasts to agricultural decision problems in Europe. J Agric Sci 149(1):15–22. https://doi.org/10.1017/S0021859610000729

Dani S, Deep A (2010) Fragile food supply chains: reacting to risks. Int J Logist Res Appl 13(5):395–410 (Publisher: Taylor & Francis eprint: https://doi.org/10.1080/13675567.2010.518564)

Garnett EE, Balmford A, Marteau TM, Pilling MA, Sandbrook C (2021) Price of change: does a small alteration to the price of meat and vegetarian options affect their sales? J Environ Psychol 75:101589. https://doi.org/10.1016/j.jenvp.2021.101589

Herrero M, Thornton PK, Mason-D'fCroz D, Palmer J, Benton TG, Bodirsky BL, Bogard JR, Hall A, Lee B, Nyborg K, Pradhan P, Bonnett GD, Bryan BA, Campbell BM, Christensen S, Clark M, Cook MT, de Boer IJM, Downs C, Dizyee K, Folberth C, Godde CM, Gerber JS, Grundy M, Havlik P, Jarvis A, King R, Loboguerrero AM, Lopes MA, McIntyre CL, Naylor R, Navarro J, Obersteiner M, Parodi A, Peoples MB, Pikaar I, Popp A, Rockström J, Robertson MJ, Smith P, Stehfest E, Swain SM, Valin H, van Wijk M, van Zanten HHE, Vermeulen S, Vervoort J, West PC (2020) Innovation can accelerate the transition towards a sustainable food system. Nat Food 1(5):266–272. https://doi.org/10.1038/s43016-020-0074-1

Ivanov D, Dolgui A, Das A, Sokolov B (2019) Digital supply chain twins: managing the ripple effect, resilience, and disruption risks by data-driven optimization, simulation, and visibility. In: International series in operations research and management science, vol 276. Springer, Cham, pp 309–332. https://doi.org/10.1007/978-3-030-14302-2-15

Jung J, Maeda M, Chang A, Bhandari M, Ashapure A, Landivar-Bowles J (2021) The potential of remote sensing and artificial intelligence as tools to improve the resilience of agriculture production systems. Curr Opin Biotechnol 70:15–22. https://doi.org/10.1016/j.copbio.2020.09.003

Kaivosoja J (2022) Future possibilities and challenges for UAV-based imaging development in smart farming. In: Lipping T, Linna P, Narra N (eds) New developments and environmental applications of drones. Springer, Cham, pp 109–119

Lacoste M, Cook S, McNee M, Gale D, Ingram J, Bellon-Maurel V, MacMillan T, Sylvester-Bradley R, Kindred D, Bramley R, Tremblay N, Longchamps L, Thompson L, Ruiz J, García FO, Maxwell B, Griffin T, Oberthür T, Huyghe C, Zhang W, McNamara J, Hall A (2021) On-farm experimentation to transform global agriculture. Nature Food 3(1):11–18. https://doi.org/10.1038/s43016-021-00424-4

Lal R (2004) Soil carbon sequestration impacts on global climate change and food security. Science 304(5677):1623–1627. https://doi.org/10.1126/science.1097396

Lalić B, Firanj Sremac A, Dekić L, Eitzinger J, Perišić D (2018) Seasonal forecasting of green water components and crop yields of winter wheat in Serbia and Austria. J Agric Sci 156(5):645–657. https://doi.org/10.1017/S0021859617000788

Nevavuori P, Narra N, Linna P, Lipping T (2022) Assessment of crop yield prediction capabilities of CNN using multisource data. In: Lipping T, Linna P, Narra N (eds) New developments and environmental applications of drones. Springer, Cham, pp 173–186

Nobre GG, Hunink JE, Baruth B, Aerts JCJH, Ward PJ (2019) Translating large-scale climate variability into crop production forecast in Europe. Sci Rep 9(1):1277. https://doi.org/10.1038/s41598-018-38091-4

Pesola P (2021) Aiheuttaako lajikevalinta harmaita hiuksia? luken uusi työkalu auttaa päätöksenteossa. Maaseudun Tulevaisuus

Rejeb A, Rejeb K, Zailani S (2021) Big data for sustainable agri-food supply chains: a review and future research perspectives. J Data Inf Manag 3:167–182. https://doi.org/10.1007/s42488-021-00045-3

Rittel HWJ, Webber MM (1973) Dilemmas in a general theory of planning. Policy Sci 4(2):155–169. https://doi.org/10.1007/BF01405730

Rosales F, Batalha M, Raimundo L, Mazuchelli J (2015) Agri-food supply chain: mapping particular risks. In: EurOMA 2015 - European operations management association. https://doi.org/10.13140/RG.2.2.16324.22402

Schiefloe PM (2021) The Corona crisis: a wicked problem. Scand J Public Health 49(1):5–8. https://doi.org/10.1177/1403494820970767

Stone J, Rahimifard S (2018) Resilience in agri-food supply chains: a critical analysis of the literature and synthesis of a novel framework. Supply Chain Manag 23(3):207–238. https://doi.org/10.1108/SCM-06-2017-0201

Talwar S, Kaur P, Fosso Wamba S, Dhir A (2021) Big data in operations and supply chain management: a systematic literature review and future research agenda. Int J Prod Res 59(11):3509–3534. https://doi.org/10.1080/00207543.2020.1868599

van der Vorst JG, Beulens AJ (2002) Identifying sources of uncertainty to generate supply chain redesign strategies. Int J Phys Distrib Logist Manag 32(6):409–430. https://doi.org/10.1108/09600030210437951

Zhong R, Xu X, Wang L (2017) Food supply chain management: systems, implementations, and future research. Ind Manag Data Syst 117(9):2085–2114. https://doi.org/10.1108/IMDS-09-2016-0391

Crop Modeling for Future Climate Change Adaptation

Andrés J. Cortés, Felipe López-Hernández, and Matthew W. Blair

Abstract Crop susceptible to drought and heat stress is increasing due to climate change. Consequently, new analytical strategies are urgently required to determine sources of adaptation, and pyramid them into new sustainable cultivars for food security. Here we offer an overview on how modeling analytical tools serve to predict crop adaptive responses to ongoing climate change. First, we will describe how climate data meet ecophysiology modeling in order to forecast in situ stresses. Second, we will encourage coupling these climate-based ecophysiological inferences with genomics, as proxy to model standing natural adaptation already contained within current crop landraces, and their wild relatives. Third, we will discuss genomic-enabled modeling alternatives to optimize the introgression of such adaptive genetic variation into elite customized cultivars. Finally, we will prospect alternative models that could boost de novo adaptive variation, such as in silico breeding models, speed breeding, and genome editing. Throughout this compilation of case studies and reflections, readers will be able to identify the need for more robust high-resolution ecological data, combined with explicit empirical summary statistics of the genomic diversity within crop genepools. Only then, ecophysiological-based models would meet genomic-enabled predictions of the adaptive potential in current crops, empowering sustainable food security in the face of climate change.

A. J. Cortés (✉)
Corporación Colombiana de Investigación Agropecuaria – AGROSAVIA, Centro de Investigación La Selva, Rionegro, Colombia

Facultad de Ciencias Agrarias – Departamento de Ciencias Forestales, Universidad Nacional de Colombia – Sede Medellín, Medellín, Colombia

Department of Plant Breeding, Swedish University of Agricultural Sciences, Alnarp, Sweden
e-mail: acortes@agrosavia.co

F. López-Hernández
Corporación Colombiana de Investigación Agropecuaria – AGROSAVIA, Centro de Investigación La Selva, Rionegro, Colombia

M. W. Blair
Department of Agricultural & Environmental Sciences, Tennessee State University, Nashville, TN, USA

P. M. Priyadarshan et al. (eds.), *Digital Agriculture*,
https://doi.org/10.1007/978-3-031-43548-5_21

Keywords Abiotic stress · Standing adaptation · De novo adaptation · Ecophysiology · Genomics · GWAS model · Genomic prediction model · Introgression breeding · In silico breeding · Speed breeding · Gene editing

1 Climate-Based Ecophysiological Models for Crop Stress

Climate-based ecophysiological modeling aims at clustering crop genotypes according to the stress gradient of the geographic regions where they come from. To accomplish so, high-resolution climate data are typically gathered from worldwide model repositories such as WorldClim (https://www.worldclim.org/) and Chelsa (https://chelsa-climate.org/) using the geo-referencing consigned in the passport data at the moment when the crop sample was collected. Not all crop samples are ideal to calibrate climate-based ecophysiological models. Researchers are encouraged to rely on germplasm accessions for which it is feasible to assume genotype–environment equilibrium, for instance in crop landraces, and their wild relatives (Hancock et al. 2011).

Once climate data have been compiled and depurated, the design phase of the underlying ecophysiological model would depend on the a priori hypothesis that the researcher has. For instance, if the a priori hypothesis is to study drought adaptation in crop genepools, the classical hydrological water balance (Calvo 1986; Thornthwaite and Mather 1955) could inspire an explicit potential evapotranspiration (PET) model that weights average temperature and radiation variables. Examples of this PET model include Thornthwaite's (Thornthwaite and Mather 1957), and Hamon's (1961). Estimated PET can then be contrasted against in situ accumulated precipitation (P) records to synthetize a drought index (DI) proxy (Cortés et al. 2013), as follows in Eq. 1 for the time period j at the i geo-referenced site where the genotype came from.

$$\mathrm{DI}_{i,j} = \frac{\mathrm{PET}_{i,j} - P_{i,j}}{\mathrm{PET}_{i,j}} \times 100 \tag{1}$$

This way, crop genotypes would experience in situ drought stress if PET > P, meaning DI > 0 up to a maximum of 100%, which is the extreme case when there is no precipitation in the study time window at that particular geo-referenced site. When PET = P, DI equals 0, which counter-intuitively must not be interpreted as perfect water balance because of the soil water-holding capacity, and the ultimate nature of PET and P as stochastic variables in the time series (Heyman and Sobel 1990). Therefore, researchers are encouraged to explore multiple time windows as part of the j component in a way to minimize time-wise stochasticity.

On the other hand, negative DI estimates occur when P > PET, and are suggestive of soil water saturation. Because of this, the same PET models could inspire targeting adaptation to other types of abiotic stresses that imply hypoxia at the root level, such as flooding (Perata et al. 2011). A flooding index (FI) may be re-customized as in Eq. 2.

$$\mathrm{FI}_{i,j} = \frac{P_{i,j} - \mathrm{PET}_{i,j}}{P_{i,j}} \times 100 \quad (2)$$

As before, crop genotypes would experience in situ flooding stress if $P > \mathrm{PET}$ or $\mathrm{FI} > 0$ up to a maximum of 100%. When $P = \mathrm{PET}$, FI equals 0, an ambiguous stage for flooding stress due to the stochasticity of the PET and P variables in the time series (Heyman and Sobel 1990). If $P < \mathrm{PET}$, flooding stress is rare ($\mathrm{FI} < 0$).

Interestingly, Thornthwaite's model (Thornthwaite and Mather 1957) includes a monthly heat component referred to as HIT or heat index from Thornthwaite (López-Hernández and Cortés 2019). It is defined as the average daily maximum temperature. This estimate could in turn be leveraged as a climate-based eco-physiological model for further abiotic stresses, such as heat stress. Specifically, heat stress may occur when Thornthwaite's HIT estimates exceed a predefined physiological temperature threshold (T_{H}) for the crop species. Alternatively, heat stress can be simplified as minimum night temperatures ($\min(T)$) that surpass the upper physiological temperature threshold (T_{H}) ideal for the species. After all, plants are particularly susceptible to warm night temperatures during the reproductive phase of the crop, when heat shock may lead to pollen unviability, flower abortion, and unsuccessful fruit set (Burbano-Erazo et al. 2021). Hence, a heat index (HI) proxy (López-Hernández and Cortés 2019) may be defined as follows in Eq. 3 for the night of the day j during the crop's reproductive phase at the i geo-referenced site where the genotype came from.

$$\mathrm{HI}_{i,j} = \frac{\min\left(T_{i,j}\right) - T_{\mathrm{H}}}{T_{\mathrm{H}}} \quad (3)$$

Once again, crop genotypes would experience in situ heat stress if $\min(T) > T_{\mathrm{H}}$, implying an $\mathrm{HI} > 0$. When $\min(T) = T_{\mathrm{H}}$, HI equals 0, which does not necessarily reflect absence of heat stress because T is a stochastic variable in the time series (Heyman and Sobel 1990). Finally, $\mathrm{HI} < 0$ occurs when $\min(T) < T_{\mathrm{H}}$, which may be indicative of lack of heat stress. For the heat stress case, researchers are not only invited to test various time windows across days during the reproductive phase as way to capture variability in the j component, but also to consider contrasting physiological temperature thresholds (T_{H}) according to the literature, and experimental trials for the crop species in concrete geographies.

Notice that the same logic would apply for climate-based ecophysiological models targeting cold stress. Broadly speaking, cold stress includes chilling (0–15 °C) and freezing (<0 °C) damages (Ding et al. 2019). The only difference is that instead of comparing the minimum temperature with an upper threshold of physiological viability, modeling cold stress requires relying on a lower temperature threshold of physiological activity for the plant species (T_L), which for many crops may be generalized to the 15 °C chilling threshold (Ding et al. 2019). At temperatures below T_L, plants cease vegetative growth, compromising yield potential. Therefore, T_L is often incorporated into the growing degree-days (GDD) estimations. Cold index (CI) could be generalized as in Eq. 4 during day *j* at *i* geo-referenced site where the genotype came from.

$$\mathrm{CI}_{i,j} = \frac{T_{\mathrm{L}} - \min\left(T_{i,j}\right)}{T_{\mathrm{L}}} \tag{4}$$

Crops would experience cold damage *s.l.* when $\mathrm{CI} > 0$ (or $\min(T) < T_L$), chilling damage when $0 < \mathrm{CI} < 1$ (or $0 < \min(T) < T_L$), and frost damage (FD) when $\mathrm{CI} \geq 1$ (or $\min(T) \leq 0$). Again, $\mathrm{CI} = 0$ or $\min(T) = T_L$ is a stochastic steady point. Finally, $\mathrm{CI} < 0$ or $\min(T) > T_L$ would suggest lack of cold stress at the particular location.

Alternatively, frost stress cold index can simply be modeled by considering a 0 °C threshold for frost damage (FD), as follows: $\mathrm{FD} = -\min(T_{i,j})$ for the genotype *i* during the day *j*. Nonetheless, the original definition at Eq. 4 is broader and therefore preferable.

Some of these climate-based ecophysiological indices have recurrently been used to model adaptation in crop species. For instance, they have served to assess adaptation in common bean (*Phaseolus vulgaris* L.) to drought (Blair et al. 2016; Cortés and Blair 2018; Cortés et al. 2012a, b) and heat stresses (López-Hernández and Cortés 2019). More recently, these models have been extended to study adaptation of tepary bean (*Phaseolus acutifolius* A. Gray) to drought stress (Buitrago-Bitar et al. 2021).

There are several benefits in using explicit climate-based ecophysiological models as compared to raw environmental variables when assessing in situ crop adaptation. For instance, climate-based ecophysiological indices capture more precisely physiological mechanism capable to confer abiotic stress tolerance. Second, ecophysiological indices are hypothesis-inspired from the very beginning of the research, while using raw environmental variables lacks a priori definition of target stresses. It may be insightful to consider the latter approach as a pilot exploratory phase, but in any case such blind inspection should redound in ad hoc hypothesis for more customized ecophysiological models (Fig. 1).

Despite the conceptual beauty of this climate-based ecophysiological-modeling approach, major improvements are required. First, standardized climate data are urgently needed to build more reliable climate-based ecophysiological models. However, it often lacks repeatability and systematicity (Waldvogel et al. 2020). Gathered climate data may further be coupled with remote sensing (Zellweger et al. 2019) and statistical downscaling (Zellweger et al. 2019) to achieve a better microhabitat resolution of the environmental heterogeneity (Ratcliffe et al. 2019).

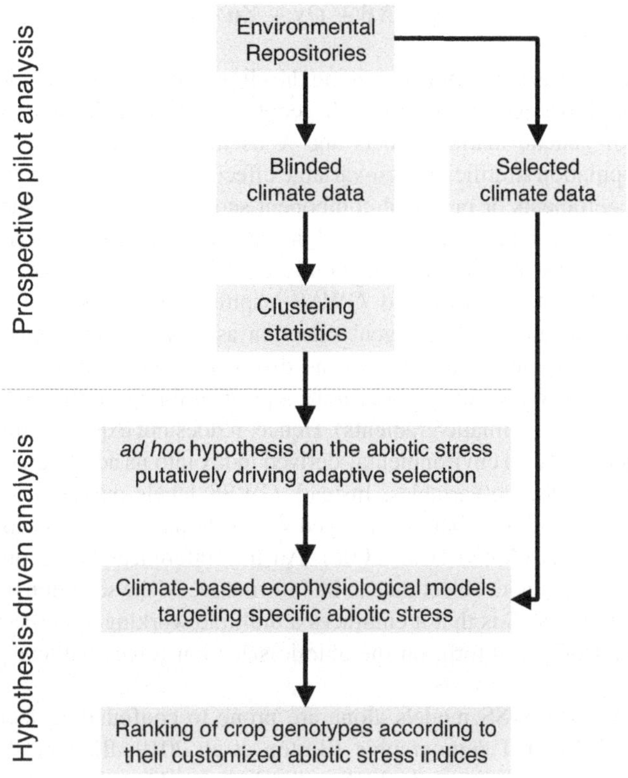

Fig. 1 Pipeline to leverage climate-based ecophysiological models in crops

2 Modeling Standing Genetic Adaption in Modern Crops

Modeling the genetic basis of environmental crop adaptation requires merging climate-based ecophysiological indices retrieved from the previous section with genomic screenings of the accessions putatively under adaptive divergent selection, which often correspond to crop landraces and their wild relatives. A type of linear mixed model (MLM) known as genome-wide association study (GWAS) inspired that purpose, by explicitly accounting for random effects such as population and kinship (Kruglyak 2008). To be precise, classical GWAS statistically converges phenotypes and genotypes. However, in order to distinguish it from models targeting the adaptive spectrum at the habitat–genotype interface, literature started coining the term genome–environment association (GEA) exclusively for MLM-based GWAS models that aimed retrieving the genomic architecture of adaptation (Cortés et al. 2022; Hancock et al. 2011). They are mechanistically identical, the only difference being the replacement of the phenotypic response variable by an environmental response variable, ideally a climate-based ecophysiological index. The underlying GWAS-inspired GEA model can be summarized as in Eq. 5.

$$\mathbf{y} = X\boldsymbol{\beta} + Q\mathbf{v} + Z\mathbf{u} + \mathbf{e} \quad (5)$$

In this model, **y** is the vector containing the climate-based ecophysiological index (or a weighted average of them) for all genotypes. Among the fixed effects, **β** is the vector of genetic marker effects and *X* its incidence matrix, while **v** is the vector of population stratification co-variable effects and *Q* its incidence matrix with admixture coefficients or principal component scores. Among the random effects, **u** is the vector of family effects and *Z* its incidence matrix containing kinship coefficient. By estimating **β**, the model ranks marker's adaptive value.

An alternative to MLM-based GWAS-inspired GEA models is to identify genome-wide signatures of divergent selection as a proxy for adaptation (Cortés et al. 2020a). This technique, known as divergence mapping via genome-wide selection scans (GWSS), does not require a priori training on the putative abiotic selective forces (i.e., climatic gradients). Hence, it does not explicitly model climate data, nor decompose an environmental-derived index into its additive genetic factors and its demographic co-variables. Instead, GWSS labels outlier markers from a genomic background distribution surveyed with summary statistics capable to spot selection footprints (Antao et al. 2008). Ad hoc inferences based on the outlier markers and genes under selection can be informative of the selection agents. Thus, the value of the GWSS is that it embraces a multiple working hypothesis paradigm (Chamberlin 1897) to inform on the abiotic selection forces without prior biased preferences.

Yet, GEA and GWSS models alone are prone to confounding factors (Maher 2008; Pennisi 2014) if demographic (Barton et al. 2019; Blair et al. 2012) and genomic features (Cortés et al. 2018), like linkage disequilibrium (Blair et al. 2018) and recombination rate (Galeano et al. 2012), are improperly accounted for (Ellegren and Wolf 2017; Huber et al. 2016; Lambert and Black 2012; Wolf and Ellegren 2017; Wray et al. 2013). It is then at the interface among GWAS, GEA, and GWSS models that a more cohesive reconstruction of the landscape (Barrett and Hoekstra 2011) of crop adaptation might arise.

3 Optimizing Standing Adaptation into Customized Crops

Modeling standing genetic adaption in modern crops using a combination of climate-based ecophysiological indices and genome-wide scans of selection footprints delivers candidate molecular markers for adaptation. These set of markers can then be utilized to guide breeding of customized crops for adaptive variation (Butcher and Southerton 2007; Sajad 2014; Stafford 2009). The strategy typically consists in pyramiding adaptive alleles from exotic donor accessions into the genomic backgrounds of elite cultivar. This introgression-breeding approach recombines allelic diversity conferring adaption with variants responsible for agronomic value (Burgarella et al. 2019; Kumar et al. 2020). The technique methodologically relies on backcrosses with elite parental accessions (Herzog and Frisch 2011). Such scheme has been implemented to introgress adaptation for heat and drought stress from exotic genepools of tepary bean (*P. acutifolius* A. Gray) and its wild relative *Phaseolus parvifolius* (Freytag), into commercially accepted common bean (*P. vulgaris* L.) cultivars (Burbano-Erazo et al. 2021; Mejía-Jiménez et al. 1994; Muñoz et al. 2003). Yet, a major difficulty of the introgression-breeding approach is to concurrently drag unrelated and unlinked alleles across the genome, a likely scenario for adaptation given its polygenic nature, in which many narrow regions across the genome confer modest adaptive effects (Barghi et al. 2020).

Genomic-enabled prediction models then appear as a feasible alternative to assist the withholding of exotic polygenic adaptation (Cortés et al. 2020b; Migicovsky and Myles 2017). Genomic-enabled models perform predictive breeding via genomic selection (Migicovsky and Myles 2017; Varshney 2021). Original genomic prediction BLUP models (Crossa et al. 2017; Desta and Ortiz 2014) relied on a molecular extension of the Fisherian infinitesimal polygenic model (Fisher 1930), as in Eq. 6. Yet, last-generation machine learning models offer new angles into predictive breeding (Crossa et al. 2019; Montesinos-López et al. 2021a, b; Tong and Nikoloski 2021).

$$y_i = g_i + e_i = \mu + \sum_{j=1}^{p} x_{i,j}\beta_j + e_i \tag{6}$$

Where y_i would be the predicted climate-based ecophysiological index for the accession j, g and e are genetic and environmental contributions, and $x_{i,j}$ is the genotype of the j marker weighted by its effect β_j and summed through all p markers.

Although genomic prediction models were originally designed to compute genomic estimated breeding values or polygenic risk scores, they could be redesigned to account for climate-driven adaptation, a genotype's genetic parameter worth defining as genomic estimated adaptive value, or GEAV (Arenas et al. 2021; Capblancq et al. 2020). Just as explained in the previous section when referring to the transition from GWAS to GEA, the training and validation of genomic prediction models should shift from forecasting phenotypes, into estimating climate-based indices and GEAVs. Genomic-assisted backcrossing can then arise as the ultimate introgression-breeding tactic (Fig. 2) during early phases of crop improvement for adaptation (Cortés and López-Hernández 2021).

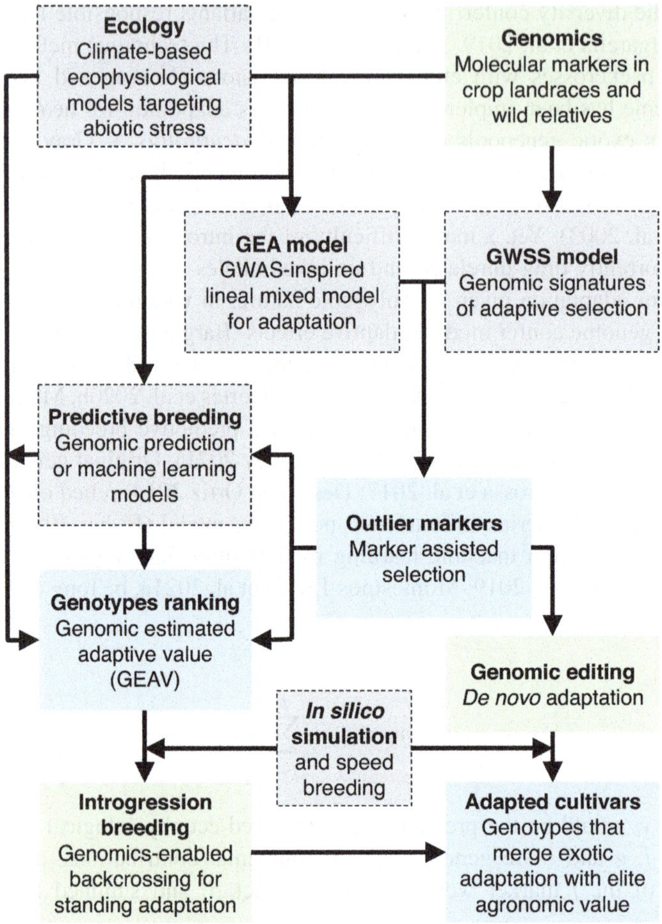

Fig. 2 Roadmap to harness standing genetic adaptive variation into adapted cultivars. *Gray* boxes with dashed edges mark crop-modeling approaches to predict and boost adaptation, as discussed in this chapter. *Blue* boxes pinpoint key outcomes of the crop-modeling approaches. *Green* boxes depict transgressive paradigms from the crop-breeding and genomics fields, which are adequate and promising to assist crop modeling for future climate change adaptation. Authors are reminded that an alternate fast track for introgression breeding is grafting

4 Predicting De Novo Adaptation for Future Crops

Even though heterogeneous environments may boost crops' standing adaptive diversity, in situ local adaptation is finite when it comes to genetic recombination. Therefore, alternative strategies are required to enable de novo adaptive variation within crop genepools to cope with climate change in a sustainable manner. For instance, modern gene-editing technology (Doudna and Charpentier 2014) is capable to mutate adaptive alleles, just as targeted mutagenesis did in the past (Holme et al. 2019), or silence those that prevent adaptation (Ahmar et al. 2021; Dort et al. 2020). When extrapolated from monogenic Mendelian traits to the quasi-infinitesimal paradigm of a typical polygenic adaptive trait, concurrent single-gene editions would lead to genome editing, an example of which is de novo domestication (Fernie and Yan 2019). Novel polygenic editions mediated by genome editing have already been implemented with success in complex genomic architectures of non-model orphan crop species (Lemmon et al. 2018).

There are other alternatives to boost de novo adaptation in crop species for which gene editing is unfeasible due to bottlenecks during tissue culture. One of them is speed breeding (Varshney et al. 2021b), in which the crop-breeding cycles are shortened by accelerating the reproductive phase throughout a combination of greenhouse treatments, hormonal management, and genetically induced precocity (Alves et al. 2020; Kumar et al. 2020; Watson et al. 2018). Speed breeding can even make classical introgression breeding backcrosses more efficient in time.

Another alternative to enable de novo adaptation in crop species is grafting, an ancient technique in which tissues of different plants, sometimes even from distinct species, are vascularly interconnected into a single viable individual (Goldschmidt 2014; Wang et al. 2017). Grafting can provide a fast track to combine locally adapted rootstocks with elite clonal scions (Migicovsky and Myles 2017; Warschefsky et al. 2016). A wide spectrum of crops and phenotypes have already been shaped by grafting, such as adaptation to soil toxicity (Fernández-Paz et al. 2021), soil borne fungal pathogens (Guevara-Escudero et al. 2021; Sánchez-González et al. 2019), and overall growth (Cañas-Gutiérrez et al. 2022) and yield traits (Reyes-Herrera et al. 2020) in croplands outside the center of origin. This way, grafting may simplify the utilization of exotic adaptive alleles into crop genepools, an alternative to classical introgression breeding.

Although we can agree on the utility of gene editing, speed breeding, and grafting as alternatives to enhance de novo adaptive variants during the design of future customized crops (Varshney et al. 2021a), they are methodological novelties that do not necessarily correspond to crop-modeling innovations. Nonetheless, in silico breeding models can help optimizing these crop improvement strategies. In silico breeding simulates and adjusts the selection intensity in a way that minimizes genetic erosion and the cycle length, without compromising overall genetic gains (Hoyos-Villegas et al. 2019). After all, simulation breeding guides scientists by modeling in silico crops with the target adaptations (Marshall-Colon et al. 2017).

5 Conclusions

Computational breeding for future climate change can benefit from standing genetic adaption already present in modern crops, mostly landraces and their wild relatives. For instance, climate-based ecophysiological models are informative of the natural adaptation to local environmental conditions. Meanwhile, these environmental indices can be inputted into MLM-type models that aim integrating landscape gradients with genomic diversity via GWAS-inspired GEA models at the habitat–genotype interface. Once a model for the genomic architecture of crop adaptation is construed, highly predictive molecular markers can be utilized to assist breeding by relying on last-generation genomic prediction and machine learning models. Genomic-enabled predictive models have the potential to accelerate the pyramiding of adaptive alleles via classical introgression pre-breeding approaches. For instance, backcrossing between exotic donors of adaptation and elite cultivars with agronomic value would be more efficient and precise if coupled with genomic estimated adaptive values. Finally, novel avenues to boost de novo adaptation for future customized crops are equally promising. Among these, in silico breeding models can optimize the introgression breeding, speed breeding, and genome editing pathways, which in turn could confer and sustain allelic novelty for adaptation in crop species.

6 Perspectives

Unlocking standing genetic adaptation in modern crops and boosting de novo adaptive variation for the future require transgressive phenotypes and trans-disciplinary innovations capable to unify the ecology, ecophysiology, agronomy, and genetic fields into the evolving research area of ecological genetics for crop adaptation. Transversal to these efforts are innovative modeling frameworks. For instance, crop modeling can assist prioritizing collection gaps targeting isolated pockets of cryptic adaptive diversity (Carver et al. 2021; Ramírez et al. 2010; Ramirez-Villegas et al. 2020). Crop modeling can also inspire the development of climate-based ecophysiological indices, and novel genetic mapping tools capable to label adaptive allelic variation. In turn, the results of these efforts serve to deploy exotic adaptive diversity into elite crop cultivars via introgressive breeding, which can be enabled and boosted by genomic prediction, machine learning, and in silico breeding models. In any case, the success of these modeling platforms relies on open-access multidimensional data (McCouch et al. 2016; Spindel and McCouch 2016), and efficient crop mobilization (McCouch et al. 2020). Therefore, open-access research networks are a prerequisite to vertically integrate the crop improvement pipeline for adaptation, which includes bridging ex situ conservation and parental screening (Blair et al. 2013) for adaptive value, with downstream seed deployment initiatives (Peláez et al. 2022). This way, ecophysiological-based models will be coupled with predictive breeding models of the adaptive potential, empowering sustainable food security under changing climate.

Acknowledgments The authors are indebted to I. Cerón-Souza, F.A. Monserrate, J. Mesa, P.H. Reyes-Herrera, and J.B. Valencia for insightful discussions on crop modeling for future climate change adaptation that respectively took place Colombia throughout the last decade. S. Madriñán is also thanked for encouraging the innovative work on adaptation modeling in landraces and wild relatives. Appreciations are also given to M.J. Torres for aid during the drafting of this chapter.

Finally, the editor is deeply acknowledged for foreseeing and encouraging submissions to this book on "Digital Agriculture: A Solution for Sustainable Food and Nutritional Security."

Disclaimer The authors declare no conflict of interest.

Author Contribution A.J.C. conceived this chapter with theoretical and practical insights from M.W.B. and F.L.-H. A first version of this chapter was drafted by A.J.C., later on edited by F.L.-H. and M.W.B. All authors made substantial contributions in preparing and editing this manuscript, and approved it for publication.

Funding The funding agencies Vetenskapsrådet (grants 2016-00418 and 2022-04411), Kungliga Vetenskapsakademien (grant BS20170036), and the British Council (grant 527023146) are acknowledged for sponsoring A.J.C. during the conception and drafting of this chapter. Similarly, the United States Department of Agriculture Evans Allen fund (TENX-07) is credited for supporting M.W.B., and AGROSAVIA's Department for Research Capacity Building is recognized for subsidizing the internship of F.L.-H. in 2018. A synergistic meeting among all authors was possible in Rionegro (Antioquia, Colombia) during the summer of 2019 thanks to the Fulbright U.S. Specialist Program granted to M.W.B. A prior version of this encounter happened between A.J.C. and M.W.B. in Nashville (TN, United States) during the spring of 2015, empowered through a scholarship awarded to A.J.C. by the Graduate Research School in Genomic Ecology (GENECO) from Lund University.

References

Ahmar S, Ballesta P, Ali M, Mora-Poblete F (2021) Achievements and challenges of genomics-assisted breeding in forest trees: from marker-assisted selection to genome editing. Int J Mol Sci 22:10583

Alves FC, Balmant KM, Resende MFR, Kirst M, los Campos G (2020) Accelerating forest tree breeding by integrating genomic selection and greenhouse phenotyping. Plant Genome 13:e20048

Antao T, Lopes A, Lopes RJ, Beja-Pereira A, Luikart G (2008) LOSITAN: a workbench to detect molecular adaptation based on a Fst-outlier method. BMC Bioinform 9:323

Arenas S, Cortés AJ, Mastretta-Yanes A, Jaramillo-Correa JP (2021) Evaluating the accuracy of genomic prediction for the management and conservation of relictual natural tree populations. Tree Genet Genomes 17:12

Barghi N, Hermisson J, Schlötterer C (2020) Polygenic adaptation: a unifying framework to understand positive selection. Nat Rev Genet 21:769–781

Barrett RDH, Hoekstra HE (2011) Molecular spandrels: tests of adaptation at the genetic level. Nat Rev Genet 12:767–780

Barton N, Hermisson J, Nordborg M (2019) Why structure matters. elife 8:e45380

Blair MW, Soler A, Cortés AJ (2012) Diversification and population structure in common beans (*Phaseolus vulgaris* L.). PLoS One 7:e49488

Blair MW, Cortés AJ, Penmetsa RV, Farmer A, Carrasquilla-Garcia N, Cook DR (2013) A high-throughput SNP marker system for parental polymorphism screening, and diversity analysis in common bean (*Phaseolus vulgaris* L.). Theor Appl Genet 126:535–548

Blair MW, Cortés AJ, This D (2016) Identification of an *ERECTA* gene and its drought adaptation associations with wild and cultivated common bean. Plant Sci 242:250–259

Blair MW, Cortes AJ, Farmer AD, Huang W, Ambachew D, Penmetsa RV, Carrasquilla-Garcia N, Assefa T, Cannon SB (2018) Uneven recombination rate and linkage disequilibrium across a reference SNP map for common bean (*Phaseolus vulgaris* L.). PLoS One 13:e0189597

Buitrago-Bitar MA, Cortés AJ, López-Hernández F, Londoño-Caicedo JM, Muñoz-Florez JE, Muñoz LC, Blair MW (2021) Allelic diversity at abiotic stress responsive genes in relationship to ecological drought indices for cultivated Tepary bean, *Phaseolus acutifolius* A. Gray, and its wild relatives. Genes 12:556

Burbano-Erazo E, León-Pacheco R, Cordero-Cordero C, López-Hernández F, Cortés A, Tofiño-Rivera A (2021) Multi-environment yield components in advanced common bean (*Phaseolus vulgaris* L.) × Tepary bean (*P. acutifolius* A. Gray) interspecific lines for heat and drought tolerance. Agronomy 11:1978

Burgarella C, Barnaud A, Kane NA, Jankowski F, Scarcelli N, Billot C, Vigouroux Y, Berthouly-Salazar C (2019) Adaptive introgression: an untapped evolutionary mechanism for crop adaptation. Front Plant Sci 10:4

Butcher P, Southerton S (2007) Marker-assisted selection in forestry species. In: Guimarães E, Ruane J, Scherf B, Sonnino A, Dargie J (eds) Marker-assisted selection – current status and future perspectives in crops, livestock, forestry and fish. FAO, Rome

Calvo JC (1986) An evaluation of Thornthwaite's water balance technique in predicting stream runoff in Costa Rica. Hydrol Sci 31:52–60

Cañas-Gutiérrez GP, Sepulveda-Ortega S, López-Hernández F, Navas-Arboleda AA, Cortés AJ (2022) Inheritance of yield components and morphological traits in Avocado cv. Hass from "Criollo" "Elite Trees" via Half-Sib seedling rootstocks. Front Plant Sci 13:843099

Capblancq T, Fitzpatrick MC, Bay RA, Exposito-Alonso M, Keller SR (2020) Genomic prediction of (Mal)Adaptation across current and future climatic landscapes. Annu Rev Ecol Evol Syst 51:245–269

Carver D, Sosa CC, Khoury CK, Achicanoy HA, Diaz MV, Sotelo S, Castañeda-Álvarez NP, Ramirez-Villegas J (2021) GapAnalysis: an R package to calculate conservation indicators using spatial information. Ecography 44:1000

Chamberlin TC (1897) The method of multiple working hypotheses. Science 148:754

Cortés AJ, Blair MW (2018) Genotyping by sequencing and genome – environment associations in wild common bean predict widespread divergent adaptation to drought. Front Plant Sci 9:128

Cortés AJ, López-Hernández F (2021) Harnessing crop wild diversity for climate change adaptation. Genes 12:783

Cortés AJ, Chavarro MC, Madriñán S, This D, Blair MW (2012a) Molecular ecology and selection in the drought-related *Asr* gene polymorphisms in wild and cultivated common bean (*Phaseolus vulgaris* L.). BMC Genet 13:58

Cortés AJ, This D, Chavarro C, Madriñán S, Blair MW (2012b) Nucleotide diversity patterns at the drought-related *DREB2* encoding genes in wild and cultivated common bean (*Phaseolus vulgaris* L.). Theor Appl Genet 125:1069–1085

Cortés AJ, Monserrate F, Ramírez-Villegas J, Madriñán S, Blair MW (2013) Drought tolerance in wild plant populations: the case of common beans (*Phaseolus vulgaris* L.). PLoS One 8:e62898

Cortés AJ, Skeen P, Blair MW, Chacón-Sánchez MI (2018) Does the genomic landscape of species divergence in *Phaseolus* beans coerce parallel signatures of adaptation and domestication? Front Plant Sci 9:1816

Cortés AJ, López-Hernández F, Osorio-Rodriguez D (2020a) Predicting thermal adaptation by looking into populations' genomic past. Front Genet 11:564515

Cortés AJ, Restrepo-Montoya M, Bedoya-Canas LE (2020b) Modern strategies to assess and breed forest tree adaptation to changing climate. Front Plant Sci 11:583323

Cortés AJ, López-Hernández F, Blair MW (2022) Genome–environment associations, an innovative tool for studying heritable evolutionary adaptation in orphan crops and wild relatives. Front Genet 13:910386

Crossa J, Perez-Rodriguez P, Cuevas J, Montesinos-Lopez O, Jarquin D, de Los CG, Burgueno J, Gonzalez-Camacho JM, Perez-Elizalde S, Beyene Y, Dreisigacker S, Singh R, Zhang X, Gowda M, Roorkiwal M, Rutkoski J, Varshney RK (2017) Genomic selection in plant breeding: methods, models, and perspectives. Trends Plant Sci 22:961–975

Crossa J, Martini JWR, Gianola D, Perez-Rodriguez P, Jarquin D, Juliana P, Montesinos-Lopez O, Cuevas J (2019) Deep Kernel and deep learning for genome-based prediction of single traits in multienvironment breeding trials. Front Genet 10:1168
Desta ZA, Ortiz R (2014) Genomic selection: genome-wide prediction in plant improvement. Trends Plant Sci 19:592–601
Ding Y, Shi Y, Yang S (2019) Advances and challenges in uncovering cold tolerance regulatory mechanisms in plants. New Phytol 222:1690–1704
Dort EN, Tanguay P, Hamelin RC (2020) CRISPR/Cas9 gene editing: an unexplored frontier for forest pathology. Front Plant Sci 11:1126
Doudna JA, Charpentier E (2014) Genome editing. The new frontier of genome engineering with CRISPR-Cas9. Science 346:1258096
Ellegren H, Wolf JBW (2017) Parallelism in genomic landscapes of differentiation, conserved genomic features and the role of linked selection. J Evol Biol 30:1516–1518
Fernández-Paz J, Cortés AJ, Hernández-Varela CA, Mejía-de-Tafur MS, Rodriguez-Medina C, Baligar VC (2021) Rootstock-mediated genetic variance in cadmium uptake by Juvenile Cacao (*Theobroma cacao* L.) genotypes, and its effect on growth and physiology. Front Plant Sci 12:777842
Fernie AR, Yan J (2019) De Novo domestication: an alternative route toward new crops for the future. Mol Plant 12:615–631
Fisher RA (1930) The Genetical theory of natural selection. Oxford University Press, Oxford
Galeano CH, Cortés AJ, Fernandez AC, Soler A, Franco-Herrera N, Makunde G, Vanderleyden J, Blair MW (2012) Gene-based single nucleotide polymorphism markers for genetic and association mapping in common bean. BMC Genet 13:48
Goldschmidt EE (2014) Plant grafting: new mechanisms, evolutionary implications. Front Plant Sci 5:727
Guevara-Escudero M, Osorio AN, Cortés AJ (2021) Integrative pre-breeding for biotic resistance in forest trees. Plan Theory 10:2022
Hamon WR (1961) Estimating Potential Evapotranspiration. Proc Am Soc Civ Eng 87:107–120
Hancock AM, Brachi B, Faure N, Horton MW, Jarymowycz LB, Sperone FG, Toomajian C, Roux F, Bergelson J (2011) Adaptation to climate across the *Arabidopsis thaliana* genome. Science 334:83–86
Herzog E, Frisch M (2011) Selection strategies for marker-assisted backcrossing with high-throughput marker systems. Theor Appl Genet 123:251–260
Heyman DP, Sobel MJ (1990) Stochastic models, 2nd edn. North-Holland, Amsterdam
Holme IB, Gregersen PL, Brinch-Pedersen H (2019) Induced genetic variation in crop plants by random or targeted mutagenesis: convergence and differences. Front Plant Sci 10:1468
Hoyos-Villegas V, Arief VN, Yang W-H, Sun M, DeLacy IH, Barrett BA, Jahufer Z, Basford KE (2019) QuLinePlus: extending plant breeding strategy and genetic model simulation to cross-pollinated populations—case studies in forage breeding. Heredity 122:684
Huber CD, DeGiorgio M, Hellmann I, Nielsen R (2016) Detecting recent selective sweeps while controlling for mutation rate and background selection. Mol Ecol 25:142–156
Kruglyak L (2008) The road to genome-wide association studies. Nat Rev Genet 9:314–318
Kumar S, Hilario E, Deng CH, Molloy C (2020) Turbocharging introgression breeding of perennial fruit crops: a case study on apple. Hortic Res 7:47
Lambert CG, Black LJ (2012) Learning from our GWAS mistakes: from experimental design to scientific method. Biostatistics 13:195–203
Lemmon ZH, Reem NT, Dalrymple J, Soyk S, Swartwood KE, Rodriguez-Leal D, Eck JV, Lippman ZB (2018) Rapid improvement of domestication traits in an orphan crop by genome editing. Nat Plants 4:766–770
López-Hernández F, Cortés AJ (2019) Last-generation genome–environment associations reveal the genetic basis of heat tolerance in common bean (*Phaseolus vulgaris* L.). Front Genet 10:22
Maher B (2008) Missing heritability. Nature 456:18
Marshall-Colon A, Long SP, Allen DK, Allen G, Beard DA, Benes B, von Caemmerer S, Christensen AJ, Cox DJ, Hart JC, Hirst PM, Kannan K, Katz DS, Lynch JP, Millar AJ,

Panneerselvam B, Price ND, Prusinkiewicz P, Raila D, Shekar RG, Shrivastava S, Shukla D, Srinivasan V, Stitt M, Turk MJ, Voit EO, Wang Y, Yin X, Zhu X-G (2017) Crops In Silico: generating virtual crops using an integrative and multi-scale modeling platform. Front Plant Sci 8:786

McCouch SR, Wright MH, Tung CW, Maron LG, McNally KL, Fitzgerald M, Singh N, DeClerck G, Agosto-Perez F, Korniliev P, Greenberg AJ, Naredo ME, Mercado SM, Harrington SE, Shi Y, Branchini DA, Kuser-Falcao PR, Leung H, Ebana K, Yano M, Eizenga G, McClung A, Mezey J (2016) Open access resources for genome-wide association mapping in rice. Nat Commun 7:10532

McCouch S, Navabi K, Abberton M, Anglin NL, Barbieri RL, Baum M, Bett K, Booker H, Brown GL, Bryan GJ, Cattivelli L, Charest D, Eversole K, Freitas M, Ghamkhar K, Grattapaglia D, Henry R, Inglis MCV, Islam T, King G, Kehel Z, Kersey PJ, Kresovich S, Marden E, Mayes S, Ndjiondjop MN, Nguyen HT, Paiva S, Papa R, Phillips PWB, Rasheed A, Richards C, Rouard M, Sampaio MJA, Scholz U, Shaw PD, Sherman B, Staton SE, Stein N, Svensson J, Tester M, Valls JFM, Varshney R, Visscher S, Wettberg E, Waugh R, Wenzl PWB, Rieseberg LH (2020) Mobilizing crop biodiversity. Mol Plant 13:1341–1344

Mejía-Jiménez A, Muñoz C, Jacobsen HJ, Roca WM, Singh SP (1994) Interspecific hybridization between commonand tepary beans: increased hybrid embryo growth, fertility, andefficiency of hybridization through recurrent and congruity backcrossing. Theor Appl Genet 88:324–331

Migicovsky Z, Myles S (2017) Exploiting wild relatives for genomics-assisted breeding of perennial crops. Front Plant Sci 8:460

Montesinos-López OA, Montesinos-López A, Pérez-Rodríguez P, Barrón-López JA, Martini JWR, Fajardo-Flores SB, Gaytan-Lugo LS, Santana-Mancilla PC, Crossa J (2021a) A review of deep learning applications for genomic selection. BMC Genomics 22:19

Montesinos-López OA, Montesinos-López A, Hernandez-Suarez CM, Barrón-López JA, Crossa J (2021b) Deep-learning power and perspectives for genomic selection. Plant Genome 14:e20122

Muñoz LC, Blair MW, Duque MC, Tohme J, Roca W (2003) Introgression in common bean x tepary bean interspecific congruity-backcross lines as measured by AFLP markers. Crop Sci 44:637–645

Peláez D, Aguilar PA, Mercado M, López-Hernández F, Guzmán M, Burbano-Erazo E, Denning-James K, Medina CI, Blair MW, De Vega JJ, Cortés AJ (2022) Genotype selection, and seed uniformity and multiplication to ensure common bean (*Phaseolus vulgaris* L.) var. Liborino. Agronomy 12:2285

Pennisi E (2014) Disputed islands. Science 345:611–613

Perata P, Armstrong W, Voesenek LACJ (2011) Plants and flooding stress. New Phytol 190:269–273

Ramírez J, Khoury C, Jarvis A, Debouck DG, Guarino L (2010) A gap analysis methodology for collecting crop genepools: a case study with *Phaseolus* beans. PLoS One 5:e13497

Ramirez-Villegas J, Khoury CK, Achicanoy HA, Mendez AC, Diaz MV, Sosa CC, Debouck DG, Kehel Z, Guarino L (2020) A gap analysis modelling framework to prioritize collecting for ex situ conservation of crop landraces. Divers Distrib 26:730–742

Ratcliffe B, Thistlethwaite F, El-Dien OG, Cappa EP, Porth I, Klápště J, Chen C, Wang T, Stoehr M, El-Kassaby YA (2019) Inter- and intra-generation genomic predictions for Douglas-fir growth in unobserved environments. bioRxiv

Reyes-Herrera PH, Muñoz-Baena L, Velásquez-Zapata V, Patiño L, Delgado-Paz OA, Díaz-Diez CA, Navas-Arboleda AA, Cortés AJ (2020) Inheritance of rootstock effects in Avocado (*Persea americana* Mill.) cv. Hass. Front Plant Sci 11:555071

Sajad Z (2014) Molecular Marker Assisted Approaches (MMAA) for enhancing low water stress tolerance in common bean: an update. Mol Plant Breed 5:1

Sánchez-González EI, Gutiérrez-Soto JG, Olivares-Sáenz E, Gutiérrez-Díez A, Barrientos-Priego AF, Ochoa-Ascencio S (2019) Screening progenies of Mexican Race Avocado genotypes for resistance to *Phytophthora cinnamomi* Rands. HortScience 54:809–813

Spindel JE, McCouch SR (2016) When more is better: how data sharing would accelerate genomic selection of crop plants. New Phytol 212:814–826

Stafford W (2009) Marker Assisted Selection (MAS): key issues for Africa. The African Centre for Biosafety, Melville, South Africa
Thornthwaite CW, Mather JR (1955) The water balance. Climatology 8:1–104
Thornthwaite CW, Mather JR (1957) Instructions and tables for computing potential evapotranspiration and the water balance. Climatology 10:185–311
Tong H, Nikoloski Z (2021) Machine learning approaches for crop improvement: leveraging phenotypic and genotypic big data. J Plant Physiol 257:153354
Varshney RK (2021) The Plant Genome special issue: advances in genomic selection and application of machine learning in genomic prediction for crop improvement. Plant Genome 14:e20178
Varshney RK, Barmukh R, Roorkiwal M, Qi Y, Kholova J, Tuberosa R, Reynolds MP, Tardieu F, Siddique KHM (2021a) Breeding custom-designed crops for improved drought adaptation. Adv Genet 2:e202100017
Varshney RK, Bohra A, Roorkiwal M, Barmukh R, Cowling WA, Chitikineni A, Lam H-M, Hickey LT, Croser JS, Bayer PE, Edwards D, Crossa J, Weckwerth W, Millar H, Kumar A, Bevan MW, Siddique KHM (2021b) Fast-forward breeding for a food-secure world. Trends Genet 37:1124–1136
Waldvogel AM, Schreiber D, Pfenninger M, Feldmeyer B (2020) Climate change genomics calls for standardised data reporting. Front Ecol Evol 8:242
Wang J, Jiang L, Wu R (2017) Plant grafting: how genetic exchange promotes vascular reconnection. New Phytol 214:56–65
Warschefsky EJ, Klein LL, Frank MH, Chitwood DH, Londo JP, von Wettberg EJB, Miller AJ (2016) Rootstocks: diversity, domestication, and impacts on shoot phenotypes. Trends Plant Sci 21:418–437
Watson A, Ghosh S, Williams MJ, Cuddy WS, Simmonds J, Rey MD, Asyraf Md Hatta M, Hinchliffe A, Steed A, Reynolds D, Adamski NM, Breakspear A, Korolev A, Rayner T, Dixon LE, Riaz A, Martin W, Ryan M, Edwards D, Batley J, Hickey LT (2018) Speed breeding is a powerful tool to accelerate crop research and breeding. Nat Plants 4:23–29
Wolf JB, Ellegren H (2017) Making sense of genomic islands of differentiation in light of speciation. Nat Rev Genet 18(2):87–100
Wray NR, Yang J, Hayes BJ, Price AL, Goddard ME, Visscher PM (2013) Pitfalls of predicting complex traits from SNPs. Nat Rev Genet 14:507–515
Zellweger F, De Frenne P, Lenoir J, Rocchini D, Coomes D (2019) Advances in microclimate ecology arising from remote sensing. Trends Ecol Evol 34:327–341

Index

P. M. Priyadarshan et al. (eds.), *Digital Agriculture*,
https://doi.org/10.1007/978-3-031-43548-5

The manufacturer's authorised representative in the EU is Springer Nature Customer Service Centre GmbH, Europaplatz 3, 69115 Heidelberg, Germany. If you have any concerns regarding our products, please contact ProductSafety@springernature.com

Printed and bound by CPI Group (UK) Ltd, Croydon, CR0 4YY
28/09/2026
02224065-0018